Hayden Lambson

Putting Sheep on the Mountain

Photograph by Marvin Clyncke.

Putting Sheep on the Mountain

The Foundation for North American Wild Sheep

Twenty-Five Years Dedicated to Wild Sheep 1974 to 1999

EDITED AND COMPILED BY
Dr. Roy A. Schultz, Daniel A. Pedrotti and Susan C. Reneau

FIRST EDITION
1999
The Foundation for North American Wild Sheep
720 Allen Avenue
Cody, Wyoming

The Foundation for North American Wild Sheep
Twenty-Five Years Dedicated to Wild Sheep
1974 to 1999

First Edition, First Printing

Library of Congress Catalog Card Number: 98-74607
ISBN Number: 1-888223-13-8
Published February 1999

Published in the United States of America
by the
Foundation for North American Wild Sheep
720 Allen Avenue
Cody, WY 82414

Phone: (307) 527-6261
FAX: (307) 527-7117
Web Site: www.fnaws.org or type "fnaws" at any search engine.
E-Mail: fnaws@fnaws.org

THIS 25TH ANNIVERSARY BOOK

is dedicated to
the entire membership
and supporters of the
Foundation for North American Wild Sheep
who have a passion for
wild sheep and the
magnificence of
the land they inhabit.

Photo courtesy of Bob Cassell.

Prologue

Putting Sheep on the Mountain

Dr. Roy A. Schultz, Daniel A. Pedrotti and Susan C. Reneau

Three members of FNAWS, Dr. Roy A. Schultz, Daniel A. Pedrotti and Susan C. Reneau, have spent more than a year working with hundreds of FNAWS members, leaders and supporters to compile this 25th anniversary book, ***Putting Sheep on the Mountain: The Foundation for North American Wild Sheep, Twenty-Five Years Dedicated to Wild Sheep, 1974 to 1999.*** Since each of us have a different perspective in relationship to this book, each of us has contributed our own thoughts for the prologue.

Dr. Schultz began the process of compiling this book more than five years ago, so he begins, followed by Daniel A. Pedrotti who continued the push to bring this book to reality, and followed by Susan C. Reneau who was asked by Roy and Dan to help pull this entire project together.

Dr. Roy A. Schultz

I wish to thank all the members of the Foundation for North American Wild Sheep for their help in completion of this book of the history of FNAWS.

Basil Bradbury, past editor of *Wild Sheep* and *Petersen's HUNTING*, as well as a world-class hunter, once said, "History will tell it like it is." With the membership's help, including help from past presidents and board members, I sincerely hope we've told it like it is and given a true picture of the beginning and growth of one of the greatest "grassroots" organizations in the history of wildlife conservation — The Foundation for North American Wild Sheep.

This book began as an idea from Jim Froelich, Lloyd Zeman and myself. This was an idea going back to the mid-1980s to record the history of FNAWS. After a couple of disappointing "false starts," the time, talent and financial support came together to go forward with the idea. The talents and experience of Susan C. Reneau, who has written or edited more than 40 books; the financial support and writing talent of Past President Daniel A. Pedrotti; the time of myself and the other co-editors to drive this project to completion and to record the efforts of hunters-conservationists, biologists and game management came together to "Put More Sheep on the Mountain."

I would also like to thank the present office staff of FNAWS for their help as well as FNAWS Board Member Duncan Gilchrist, Marge Hinz and Kristi Nielsen for their past help. Special thanks is extended to all past presidents and board members for their inputs.

But, most of all, I would like to thank Jan, my wife of 40 years, for her patience to put up with my passion for the wild sheep.

May the results of our efforts in some small way go down in history as making a difference in keeping pristine areas inhabited with wild sheep where modern-day men and women can go to once again to get in touch with their inner self to view and hunt the wild sheep of North America.

Daniel A. Pedrotti

The year 1974 was the turning point for the wild sheep of North America. A lot of conversation and concern existed among wildlife biologists about the decline of this magnificent species of North American big game but little funding was available to support their meager efforts to stabilize wild sheep populations. Several efforts to reintroduce these sheep to their ancestral habitat had been attempted but most met with failure or limited success.

A symposium on their plight was convened at the University of Montana in Missoula, Montana, in the summer of 1974. A total of 77 wild sheep biologists and conservationists from throughout North America attended this meeting, co-sponsored by the Boone and Crockett Club, National Audubon Society and the Wildlife Management Institute. The focus of the symposium was a discussion of the management biology of all North American wild sheep. Fortunately, a number of future founders of the Foundation for North American Wild Sheep were in attendance.

A keen interest and concern for this particular species, on the part of sheep hunters, dates back to 1947 when Grancel Fritz wrote an article titled, "The Grand Slam In Wild Sheep," and the formation of The Grand Slam Club in 1955 by the late Bob Housholder. Bob circulated an informal newsletter devoted to sheep hunting. One of his board members, Jim Froelich, called upon all interested sheep hunters to attend a weekend get together in the fall of 1974 at Mt. Horeb, Wisconsin. This is the gathering of hunters dedicated to provide the funding our state and provincial wildlife departments so desperately needed that ultimately led to the formation of the greatest wildlife conservation organization of the 20th century, the Foundation for North American Wild Sheep.

You might say that the rest is history except for one FNAWS member, Dr. Roy Schultz, who was determined to provide a written record of the many sacrifices and successes of not only the founders but the entire membership. Dr. Schultz had compiled a great deal of the early information from personal experience as a founding member and through contacts with many of the founders and past-presidents. His continued requests to the FNAWS Board to undertake this project caught my attention and my enthusiasm grew as numerous attempts to follow up on this idea faltered in the wake of other Board priorities.

Upon completion of my term as president, I made a pledge to myself and to FNAWS that this story would be pursued to completion into an extraordinary publication to be introduced at our 1999 convention. The year 1999 happens to be the 25th anniversary of the Mt. Horeb meeting that many consider to be the beginning of FNAWS. Our charter dates to 1977 but the idea of a continent-wide sheep organization had its inception in 1974.

The following pages recount the stories of these dedicated farsighted individuals and their successors who have continued this work. As you read through the text of the narrative history of the FNAWS, the section on wild sheep biology, the hunting stories and the outlook for the future, you can not help but be awed by the inspiration, unselfishness and hard work of a relatively small number of individuals who have achieved so much for a single species of wild game.

Had this group, composed primarily of hunters, not wished to put something back on the mountains in return for the many wonderful hours of quality experiences that had been theirs to enjoy because of the privilege to hunt our wild sheep, who knows what the condition of all North American wildlife and its habitat might look like today. We owe them and their supporters a debt of gratitude and an on-going obligation to carry on with the work that has led to the most successful recovery of a wildlife species in this century.

Inspired by the example of what has been accomplished by a few brave individuals, let us not be cowered by the immense challenges facing us today. We must look out to the years 2050 and even 2100 and do today whatever is needed to insure that our wild sheep populations will be as healthy and abundant for generations to come – hunters and nature lovers of the future – that do not yet exist in the minds of man.

Enjoy these stories and relax in the comfort of knowing that our heritage of wild sheep and our rights to hunt will become our "VISION" and will be preserved and continue into the future for endless decades. For as long as the Foundation for North American Wild Sheep continues to be led by members who are as dedicated as those who brought us the first 25 years, our vision of "ABUNDANT SHEEP ON THE MOUNTAIN" will be realized.

I wish to again thank Dr. Roy Schultz for his persistent demands upon the FNAWS Board to undertake this most important project, for his time and effort to record his experiences, his correspondence with other founders and especially for his continued involvement in the final production of this great narrative.

Susan C. Reneau

From the moment I first met the young members of the Foundation for North American Wild Sheep at a dinner with Boone and Crockett Club members in 1977, I was impressed with their passion and devout love for wild sheep. I sat beside a man who recounted his latest hunting adventure in the wilds of British Columbia, and for the life of me I can't remember his name. How I wish I could. He was probably Dr. Schultz or Lloyd Zeman or Fred McMurray or any number of early FNAWS members who joined with the Boone and Crockett Club for that formal dinner so many years ago. I was the young wife of Jack Reneau who worked for the B&C and had helped organize the dinner celebration at the conclusion of the 16th Awards Program of the Boone and Crockett Club in joint sponsorship with the National Rifle Association. Never in my wildest dreams did I imagine myself in 1997 and 1998 helping to compile and write the history of FNAWS with such outstanding American conservationists as Dr. Roy Schultz and Daniel A. Pedrotti and a whole host of other wildlife biologists and conservationists who are FNAWS.

Much history for me has transpired since that first meeting with FNAWS in 1977. For the past year I have emersed myself in the stories, photographs, challenges, failures, frustrations and tremendous triumphs of FNAWS members as they began their slow and steady climb up the mountain to strengthen and expand the native wild sheep herds of North America. Their story is one of perseverance and sacrifice, both personally and wholly as a group with one common goal.

Up until I began this project with Roy and Dan I felt as an outsider who stood at a distance away from FNAWS activity and marveled at their boundless energy and creative juices to raise necessary funds to help an animal that looks so strong, yet is so frail when placed in competition for habitat with humans who bulldoze their land and build more condominiums. The 1974 conference at the University of Montana was the seed of scientific concern that offered to dedicated wild sheep hunters a goal and set of projects that needed to be accomplished. Lack of money was the only thing standing in the way of these wildlife biologists from across North America from accomplishing the projects they proposed to help wild sheep. Less than six months after that conference, the core of FNAWS began thanks to hunters who loved the land so much they wanted to give back more than they had taken. These wild sheep hunters knew they had to do something, and something fast, if the species would survive to the next century.

Throughout my research I was amazed at the personal sacrifices these early members and present-day members made to help wild sheep thrive. Watching the auction as $405,000 was raised for an Alberta sheep permit in 1998 was a humbling experience for me. I am gratified that someone believed so much in FNAWS and wild sheep that he paid that much to help so many creatures.

This book began on the back of an envelope at a restaurant in Cody, Wyoming, in July 1997 when Dan Pedrotti, Dean Picl and many members of the Board of Directors talked with me about what they thought should be included in the pages of this book. I jotted down their thoughts on the back of an envelope as we ate dinner one evening and typed up the draft book outline for everyone to review and edit. Those same outstanding wildlife conservationists stepped up to the plate throughout the year to write chapters, proof chapters written by me, locate long-lost members and founders and mail to me never-before-seen photographs of early events that formed the foundation for a new and growing organization of people who were passionate about wild sheep. I scratched out a rough outline for the book that evening as they spoke to me and Dan and Roy finalized the book outline a few months later.

My first thanks goes to my co-editors and writers, Dr. Roy Schultz and Daniel A. Pedrotti. We were a team throughout this project and guided each other to people within FNAWS who needed to tell their stories. Both of these men have a deep sense of history and an appreciation for recording accurate records of FNAWS accomplishments in the permanent form of a book. My second thanks goes to the national staff in Cody, Wyoming, for their continual assistance. Thirdly, I thank Dean Picl, Dennis Campbell, Leland Speakes, Phyllis Carlson, John and Jane Babler, Lloyd Bare, Lowell Baier, Frank Bays and Ron McMillen for their comments and edits after reading the first few chapters of this book to make sure the book included all that it should. Fourthly, I thank Bob Model for allowing my husband and me to stay at his gorgeous Mooncrest Ranch near Cody, Wyoming, while I completed research and photograph selection with the staff at headquarters. Another round of applause needs to be extended to Dave Popelka, Ron and Becky McMillen and the entire staff at Sigler Printing and Publishing for their gracious sharing of their time and work space as I spread out all the chapters and photographs in their conference room for close to a week and met with Roy and Dan.

But, in reality, my greatest thanks goes out to YOU, the membership and supporters of FNAWS, who stopped what you were doing in your busy lives and wrote a chapter, answered my interview questions or shared your most treasured photographs with me. This book, we the editors hope, reflects you and what you care about most — wild sheep.

Thank you for sharing your passion and your lives with us. I've never become a life member of anything in my life, but I have pledged to become a life member of FNAWS after experiencing all that I have with you this past year. I don't think I will ever come close to your dedication, but I would like to try.

Dr. Roy Schultz, Susan C. Reneau and Daniel A. Pedrotti met in July 1998 at Sigler Printing and Publishing in Ames, Iowa, to organize and edit chapters, select photographs and discuss design ideas with Dave Popelka. Susan, standing, holds the final chapters in a black loose-leaf notebook as Roy and Dan discuss with her the order the chapters should appear in the 720-page book, Putting Sheep on the Mountain: The Foundation for North American Wild Sheep, Twenty-Five Years Dedicated to Wild Sheep.
Photograph by Dave Popelka.

Dr. Roy Schultz, Susan C. Reneau and Daniel A. Pedrotti, left to right, edited, wrote chapters and compiled the book, Putting Sheep on the Mountain.
Photograph by Dave Popelka.

Photograph by Marvin Clyncke.

Introduction

Putting Wild Sheep on the Mountain

LELAND SPEAKES, JR.

It was started by a small group of hunters and conservationists who were dedicated to the preservation and enhancement of the wild sheep populations on the North American continent.

These men envisioned an organization that would ensure that these wonderful animals were able to endure the stress of man's encroachment and that would put, above all things, these majestic wild sheep on the mountain. Through their effort there would always be sufficient resources to ensure that these sheep would be available for the enjoyment of many generations.

There weren't many of us around at that time and opportunities to hunt sheep were pretty well limited to a very few states and provinces that had huntable herds. It was always good to get together and talk about the fantastic experiences we had during those early days.

The Foundation for North American Wild Sheep quickly grew into a professional organization of conservationists, hunters and wildlife biologists.

The very first convention in Memphis, Tennessee, had a total of about 300 in attendance and a whopping $30,000 was raised to be spent on wild sheep. Today's conventions draw as many as 7,000 people who join together to raise in excess of $3.5 million for sheep-related projects.

I can't help but reflect upon those early years. New Mexico had an outbreak of scabies among their desert bighorn sheep, and the state game department had no money to battle this disease. One call was made to the Foundation's president who informed the people in New Mexico that we would help. He checked the FNAWS bank balance, which contained $10,000. Without hesitation, he instructed the treasurer to write a check to finance the treatment of this disease. Many of these animals were too far gone, but those animals that did survive grew into a huntable herd of sheep. Now that's dedication. Since that time the Foundation has continued to be just that responsive to the needs of mountain sheep throughout North America.

The mission and goal of FNAWS is to promote and enhance increasing populations of indigenous wild sheep on the North American continent, to safeguard against the decline or extinction of such species, and to fund programs for professional management of these populations, keeping all administrative costs to a minimum. FNAWS has always prided itself in being an almost totally volunteer organization. Our only paid personnel are those remarkably talented young ladies who serve on the headquarters staff. They have been well-trained to handle the workings of this prestigious national conservation and prohunting organization.

Our primary assignment in the arena of wildlife conservation is three-fold: (1) to attract conservation-minded hunters who share goals of preserving wildlife, (2) raising funds necessary to support the management of these species through transplants, habitat improvement and, most importantly, a scientific research to identify and guard against drastic die-offs as we have experienced in the past. On going work at the Cain Veterinary Center in Caldwell, Idaho, and at the Veterinary School of Medicine at Washington State University, together with other laboratories are currently making tremendous strides in wildlife disease studies.

In 1996, FNAWS recognized Hells Canyon as probably the most perfect sheep habitat in the United States. Hells Canyon is that gorge made by the Snake River along the state lines of Oregon, Idaho and Washington. It is 150 miles long and one and one-half miles deep. This is not a Grand Canyon type of area, but rather a very wide rolling landscape that is the perfect habitat for sheep that anybody could envision. History has shown this canyon at one time was the home of thousands of Rocky Mountain bighorn sheep. The encroachment of man and particularly his domestic sheep just about eliminated wild sheep in this canyon.

FNAWS, working with the U.S. Forest Service and the game departments of Idaho, Oregon and Washington have seen the withdrawal of almost all of the domestic sheep permits so that wild sheep may be placed here without the danger of contracting the fatal disease, Pasteurella haemolytica. Two transplants have already occurred, one from Cadomin, Alberta, in 1996 and the other in 1998 from Spences Bridge, British Columbia. Biologists have estimated this canyon is capable of supporting up to 10,000 wild Rocky Mountain bighorn sheep.

That we have huntable numbers of this species today must be attributed to the early conservation efforts of the Boone and Crockett Club, and more recently the work of the Foundation for North American Wild Sheep.

The "good-old-days" are right now. In the years from 1992 to 1994, five bighorn rams were taken in Montana with official scores in excess of 200 B&C points. This is a remarkable statistic when one considers the fact that in 1980 there were only 10 rams listed from Montana that exceeded a score of 190. Today more than 75 bighorn sheep are listed and all come from reintroduced populations. As native habitat for sheep continued to shrink during the late 1970s, wild sheep hunters actually increased. New herds were established in many traditional ranges where native populations had disappeared decades before. Existing herds became stronger and many more sheep hunting opportunities became available. In 1974, future FNAWS members met at Mt. Horeb, Wisconsin, to meet other sheep hunters for the express purpose of putting more wild sheep on the mountains of North America, and in 1977 that group officially became FNAWS. FNAWS is a 99 percent volunteer group of hunters and conservationists with headquarters in Cody, Wyoming. There are six full-time employees and an all-volunteer, 11-member Board of Directors that puts the largest percentage of gross revenues into wildlife projects of all the conservation groups in America.

FNAWS will celebrate its unofficial 25th anniversary of its first formation meeting in Mt. Horeb, Wisconsin, and its official 22nd anniversary at the annual convention in Reno, Nevada, in February 1999. During these years, more than $24 million has been funded directly into sheep projects. These funds were generated primarily at annual conventions, lotteries, fund-raising banquets by chapters and affiliates and through the sale of special sheep hunting permits given by the governors of the various states with huntable populations. These funds are controlled by the Board of Directors through approved Grant-in-Aid and chapter/

affiliate funding.

If it were not for this group, composed primarily of hunters who wanted to put something back on the mountain in return for the many hours of quality experiences that had been theirs to enjoy, who knows what the condition of all North American wildlife and their habitat might look like today. The recovery of wild sheep to their indigenous range, in abundance and healthy, is without a doubt the wildlife success story of the 20th century. The continued dedication of volunteer members of the Foundation for North American Wild Sheep will be a success story for the 21st century.

The accomplishments that FNAWS has made in protecting and increasing sheep populations has been a real revelation. It just makes me feel good to be a part of that. The restorations, the transplants, the enthusiasm of the people who love wild sheep — all make me proud. Sheep hunters are different. They're different because they always want to be on top of the mountain. For any lesser money, you never get to the top of the mountain like a sheep hunter.

You are welcomed and encouraged to join this great endeavor by becoming a member of FNAWS and a part of the "greatest show on earth." As a result of this effort by hunters and wildlife conservationists throughout North America, we have more big-game animals today than existed 100 years ago.

Leland Speakes, Jr.
President, Foundation for North American Wild Sheep, 1998-1999

This hardcover book, in regular and limited editions, is a permanent record of the Foundation for North American Wild Sheep's achievements and wonderful story from 1974 to 1999. We must know where we were to know where we are going. You will read about the men and women who, from the very beginning, made the Foundation for North American Wild Sheep as great as it is today. We have our founders and early members to thank for the strength and contributions made today.

The continued sale of the book underwrites a permanent wild sheep endowment fund established in 1998 as this book was written. FNAWS members and supporters who purchase this book help to continue to "put wild sheep on the mountain," and for that, we thank you.

I appreciate the hundreds of people who have contributed chapters, hunting stories and photographs to tell the story of the finest wildlife conservation organization I know. I am honored to be a part of this magnificent group that has contributed so much to magnificent wild sheep of North America.

Leland Speakes is president of the Foundation for North American Wild Sheep from spring 1998 to spring 1999. He joined FNAWS at the Memphis convention in 1978 as hospitality chairman, one year after this group of wild sheep enthusiasts officially registered their nonprofit organization in Iowa. Leland is owner of Leland Speakes Real Estate, Inc., of Cleveland, Mississippi. Leland has been a long-time supporter of FNAWS as well as his late wife Monnie who passed away in 1996.

Table of Contents

SECTION I

The History of Wild Sheep Conservation and the People Who Make It Happen

First Organizational Meeting

November 30 to December 1, 1974
Karakahl Inn
Mt. Horeb, Wisconsin
Organizer: Jim Froelich
Master of Ceremonies: Lloyd Zeman

Second Organizational Meeting

April 5 to 6, 1975
Playboy Club
Lake Geneva, Wisconsin
Organizer: Ron Kolpin
Master of Ceremonies: Lloyd Zeman

Third Organizational Meeting

April 9 to 11, 1976
Downtown Ramada Inn
Des Moines, Iowa
Organizer: Lloyd Zeman
Master of Ceremonies: Frank Bays

Fourth Organizational Meeting

March 24 to 27, 1977
Hyatt Regency Hotel
Dearborn, Michigan
Organizer: Frank Bays
Master of Ceremonies: Frank Bays

Chapter I

Birth of an Idea: 1974 to 1977

Dr. Roy A. Schultz and Susan C. Reneau

"Dedication, determination and sincere interest in North American wild sheep. That's the kind of sportsmen who make up the backbone of our club."

Lloyd Zeman wrote those words in 1976 to the members who would, on June 21, 1977, officially become the Foundation for North American Wild Sheep (FNAWS).

Prior to the official establishment of FNAWS, a small group of dedicated wild sheep hunters met informally in various Midwestern locations to share photographs and swap sheep hunting stories. Dr. Roy Schultz may have put it best: "Those of us who have had the privilege of collecting trophy rams decided we wanted to give something back in return." And, so they did.

Janet Froelich, known as Jan to friends and widow of one of the early founders, Jim, tells of her husband's efforts to locate avid sheep hunters for that first meeting: "He had nobody up north to talk to about the mystique of wild sheep hunting, and he was hungry to get together with other sheep hunters. I remember he wrote lots of letters to people whom he thought might come to that first meeting and he worked with Bob Housholder to locate them."

Jim Froelich wanted this gathering to be in a not-too-expensive place easily reached by most of the people. He had taken his Grand Slam prior to 1974 and wanted to share his rich experiences with others who understood his passion. He worked for a long time, according to Jan, digging out anybody who wanted to attend those first sheep hunting meetings. Locating kindred spirits became his passion.

Jim managed to take his Grand Slam in four years and used only one bullet for each sheep. He never thought he would be drawn for a desert sheep permit, but he was, and the year he took his desert sheep he took the largest ram in Arizona. All these adventures Jim had bottled up inside him, and he wanted to share with others who understood what it meant to stand at the top of a mountain.

Curtiss D. Judd wrote in his pocket-sized book, *I Call Myself a Sheep Hunter, Part II,* in 1997 that he first knew of this group through a letter received from Jim Froelich.

"At that time period, I was into sheep hunting. Jim had gotten a mailing list from Bob Housholder from the Grand Slam Club of Phoenix, Arizona. Members from Iowa, Minnesota, Wisconsin and Michigan were invited. I lived in Madison, Wisconsin, and the meeting was in Mt. Horeb, Wisconsin."

Curtiss was an ideal member, as were others, who were dedicated wild sheep hunters

Lloyd Zeman's Grand Slam as seen in 1998. Lloyd had completed his Grand Slam in 1970 a few months after Jim Froelich completed his.
Photograph courtesy of Don Schmidt.

Lloyd Zeman received recognition by FNAWS for his many years of outstanding service to the organization from the first organizational meeting in 1974 until he resigned as executive director in 1984. Lloyd Zeman was the driving force behind the forming of this close-knit group and became the group's spokesman from the start.
Photograph courtesy of FNAWS archives.

who wanted to share their unique experiences with others who also had the passion and to give back to the animals they so love. Curtiss had first hunted wild sheep in 1969 in the Brooks Range of Alaska and from 1971 to 1973 hunted in Alberta. He said Mt. Horeb was the ideal location for a meeting except that the days selected were "the eighth day of our nine-day deer season." Curtiss made the sacrifice anyway, and was present.

Letters written between Bob Housholder and Jim Froelich, and discovered by Executive Director Dennis Campbell of the Grand Slam Club (GSC) in his office archives, showed Jim's interest in forming a special group for wild sheep as early as 1970, which confirmed what early FNAWS members and founders said when interviewed for this book.

Jim Froelich wrote to Bob on June 9, 1970, "I am very much interested in a meeting of all interested `Grand Slammers' and your idea of meeting in San Antonio has merit."

Jim finished his Grand Slam in 1969 and by January 1970 was the "newest member" of Bob Housholder's Grand Slam Club. Jim worked fervently with Bob to design a shoulder patch to recognize the men and women who finished their Grand Slam, but only three members were interested, so the idea fell flat as reported in Bob's *Bulletin No. 9* dated September 1970 and resurrected again by Jim in January 1972 when he offered the patches at his own expense for each paid membership to GSC. Jim's patch idea became a tradition in FNAWS, and members today have him to thank for this idea that included embroidered portraits of all North American wild sheep species. The original sketches of bighorn, Dall's, desert and Stone's sheep were drawn by Jan Froelich. Since Jim had experience producing beautifully detailed shoulder patches for GSC, producing patches for FNAWS was a natural. Jan Froelich still has a few of the original wild sheep patches she saves in memory of Jim. Later in the history of FNAWS, around 1985, Gary Swanson painted a portrait of the Chadwick ram that became the basis for today's National FNAWS logo that is used on many FNAWS merchan-

Curtiss Judd, seen here among purple shooting star flowers, was one of the 12 men listed by Jim Froelich to attend the first organizational meeting in 1974 of the group that would become FNAWS in 1977. Curtiss authored several books about wild sheep hunting and continues his passion for sheep today. He is one of the most active participants in the construction of water projects for wild sheep in Arizona.
Photograph courtesy of Curtiss Judd.

Jim Froelich, grandfather to FNAWS, enjoys an early FNAWS convention. His passion for finding other like-minded hunters brought the first group together at Mt. Horeb, Wisconsin.
Photograph courtesy of Dr. Roy Schultz.

dise items. Read in the section "Presidents Speak" regarding the development of the logo.

First rewind to June 6, 1971, and the GSC files show Jim Froelich delicately asking Bob Housholder for the addresses of people in the Midwest who might enjoy getting together for a social event.

Jim wrote, "Bob, I know it is not the Club's policy to give out names and addresses of the entire list, but is it permissible to send me the address of Truman Wood of Wisconsin who is about number 102 on the list (right after my friend Lloyd Zeman)? I just want to contact him since he is another Wisconsin hunter with a 'Slam.'

The original motivating factors were friendship and shared experiences, but it quickly moved to an interest in conserving wild sheep for future generations, recalled Fred McMurray, another founder. "We thought we were a success from the very beginning because it just seemed to blossom. Every year it doubled and doubled and tripled and tripled."

Jim Froelich's underlying concern for ethical hunting practices and fair chase was also a motivating factor in forming this specialized group that glorified the sheep and not the humans who took them. In October 1971, Jim wrote to Bob Housholder expressing his disgust with GSC members who may have taken desert sheep illegally and wrote to Bob that no GSC member should be allowed to remain as a member if he were convicted of illegal hunting activity. Jim's attitude toward working for wild sheep and their habitat continued as the purpose of FNAWS when it became a formal organization.

Jim Froelich in 1967 with the Stone's ram he took near the Chischa River of British Columbia. Joseph James "Jim" Froelich died on April 15, 1996, at the Dr. Kat Neucomb Convalescent Center near his home in Wisconsin at age 75. He married Janet Estwing and they became parents of Jay, Joy and Julie. For more than 35 years Jim and Jan owned and operated the Wah-Wah-Taysee Resort in Sayner, Wisconsin. Jim was also a director for the Estwing Manufacturing Company, a company founded by Jan's father. Jim was active in Sayner civic affairs throughout his life and was a worldwide hunter in Africa, Europe, Canada, throughout the United States and Central and South America. From the time of their retirement, Jim and Jan enjoyed winters in Florida but always returned "home" to Wisconsin.

John Babler remembered that lots of Grand Slam Club members were invited to this first meeting, but only a handful of people showed up. Bob Housholder's list of hunters from the Grand Slam Club may have been the ones invited to the first meeting at Mt. Horeb by Jim Froelich. Midwestern Grand Slammers totaled 33 from Illinois, Iowa, Michigan, Minnesota, Nebraska and Wisconsin in May 1972.

The total number of people on Bob Housholder's master list as of May 1972 was 231 men and women. This master list was kept by Housholder and included people he reasonably believed had collected the Grand Slam. The list was actually published in May 1972 and was provided by Grand Slam Club Executive Director Dennis Campbell. He retrieved this information from the GSC archives. Even though Housholder had 231 peple on this master list, in reality only approximately 145 of them were active dues-paying members of GSC in May 1972. Dennis Campbell retrieved this information from issues of the GSC *Bulletin* published just prior to and just after May 1972. Of Midwesterners who were members of GSC, Michigan was the state with the most GSC members — a total of 14. The GSC list in 1972 included founders Lloyd Zeman of Iowa, Duane Smelser of Michigan, Dr. Roy Schultz of Iowa and Walter Wojciuk of Wisconsin who joined the ranks of GSC membership after Jim joined GSC.

Additional names and addresses were gathered by word of mouth and by talking to outfitters and taxidermists. Jan said Jim was focused on locating as many wild sheep hunters as he could find over a period of several years.

To this day, early members and past presidents of FNAWS express regret they didn't attend that first meeting, but for a variety of reasons, they couldn't. John Babler also recalled that additional people who attended that first meeting were wives and friends who didn't participate in the swapping of sheep stories and photographs.

"Jim Froelich had sent out a little letter to a number of Grand Slam Club members at that time, which was probably 1972 or 1973," John recalled. "We also had folks who responded to that letter at the Mt. Horeb meeting."

Jim was finally able to get a partial mailing list of the Grand Slam Club only once from GSC founder Bob Housholder. Bob was very protective of his mailing list and didn't want to share it with a new organization that might compete with his group. John couldn't recall the exact number of people who were sent letters, but he guessed it was several dozen Grand Slam Club members, especially those in the Midwest. Jan Froelich remembered Jim writing the notes and telling her he hoped somebody would respond because he had such a burning

Early members of the Foundation for North American Wild Sheep include, left to right: John Babler, Fred McMurray and Dr. Roy Schultz who are shown here at a 1988 convention function.
Photograph courtesy of FNAWS archives.

desire to talk to someone else who loved to hunt wild sheep as much as he did. Jim had such hope that enough people would respond to form a group that would help wild sheep in North America.

Unfortunately for FNAWS, Jim became a member of the GSC Board of Directors in July 1972, which eventually caused conflict of interest for him. Jim didn't sign the formal incorporation papers for FNAWS because he felt his directorship in GSC would complicate raising money and would involve Bob Housholder in those financial affairs. Basil Bradbury of California also joined the GSC Board of Directors at the same time and eventually joined the professional staff of FNAWS in the early 1980s as director of communications. Basil had taken two Grand Slams prior to Jim and the two were friends from the time they met at GSC.

For several years, Jim Froelich continued to write notes to anyone he thought would attend a wild sheep meeting. He made phone calls to sheep hunters to insure someone would show up at Karakahl Inn. For his years of effort to locate fellow wild sheep hunters and organize the first meeting, Jim is considered the "grandfather of FNAWS."

Without a doubt, Lloyd Zeman was the driving forces behind the forming of this close-knit group and became the group's spokesman from the start. Jim gravitated to Lloyd for his leadership abilities. Lloyd, a journalism graduate of the University of Wisconsin, was writing and publishing agricultural trade journals in Des Moines, Iowa. Lloyd owned an advertising agency called Creative Services in Iowa with partners and had an outgoing personality. Dr. Schultz had taken a Dall's, bighorn and Stone's sheep by 1970, and was on his way to taking a desert bighorn sheep to complete his Grand Slam when he picked up the *Des Moines Sunday Register* that featured Lloyd, who had completed his Grand Slam in 1970 a few months after Jim Froelich completed his. Doc Schultz completed his Grand Slam a few months after Lloyd.

Doc Schultz called Lloyd to meet him to "talk sheep." Lloyd's four sheep were mounted and displayed in his downtown office near the airport. The trophies scored high in the Boone and Crockett Club's records book. Doc Schultz and his wife Jan visited Barbara and Lloyd Zeman in their home outside west Des Moines.

Barbara Zeman, who is now Barbara Arkwright, remembered meeting Roy and Jan Schultz when they lived in Iowa, but not much detail about the visit. Barb and Lloyd were married from 1958 to 1977 at the same time she was producing their young family, so she was aware of Lloyd's dedication to FNAWS in its early years but didn't take part in it because of the children.

Barb credits Lloyd Zeman's parents, Ed and Frances, for instilling the Midwestern work ethic in Lloyd at an early age. His parents were dairy farmers from Wisconsin, and Lloyd was expected to do all the farm chores. He saved everything and had a passion for whatever he did. Lloyd was intense and driven, took everything personally and was a perfectionist. "In fact, when the kids and I (there were six children) went through Lloyd's papers after he died, we laughed because he had saved everything."

Much of those early FNAWS papers remain with Barb's grown children. Becoming a guide and outfitter was always Lloyd's dream and it became a reality after he left FNAWS and established Safari Outfitters of Cody. He found great comfort and satisfaction in FNAWS as he worked with guides and outfitters, wildlife biologists and fellow hunters who also adored wild sheep. From the beginning, FNAWS was his calling, in a way, Barb explained.

Lloyd graduated with a double major in journalism and agricultural journalism from the University of Wisconsin. He wrote for the *Rural Electric Missourian* in Columbia, Missouri, for the *Wisconsin Agriculturist* in Racine, Wisconsin, and for *Successful Farming Magazine* in Des Moines where he became the crops and soils editor. Barb said his love of writing, knowledge of farming, and passion for wild sheep combined to benefit the early sheep organization that soon became FNAWS.

These men wanted to find other hunters, so plans were made to gather sheep hunters together. They knew about Jim's efforts to pull together GSC members for a sheep meeting because they were also members of GSC. Both men became active members of the newly forming organization that began in 1974. Lloyd later moved to a house east of the airport where the Sheep Foundation office business was conducted with a typewriter and a copy machine. Later they relocated the early offices to John Babler's business in St. Paul.

The year 1974 was a critical turning point in wildlife management for wild sheep, according to early member Jerry Christian. "At this particular time, the wolf population was high in British Columbia where Stone's sheep were located. We had Board members like Dr. Jerry Waite and Doc Schultz who went on several bighorn hunts before even seeing a legal ram. The ram population was at such a low ebb that hunters would go hunting two and three times without getting a ram. This was in the late 1960s and early 1970s. As hunters, we noticed that the wild sheep population was suffering. Domestic sheep were up on open ranges. We would talk to the Wyoming and Montana people, and we could see that at the rate they were disappearing, we needed better management and better disease control. We decided we needed to do something NOW," Jerry emphasized.

This dedicated band of wild sheep hunters identified Lanny O. Wilson and Bill Montoya, wildlife biologists, as two of the early professionals who shared their concern for the health of all wild sheep herds. Biologists like Lanny and Bill agreed with the future FNAWS members to further study disease, habitat and sheep populations. Lanny later joined FNAWS as a member of the Board of Directors and was influential in helping wisely spend limited funds.

Early in the history of this newly forming wild sheep organization, professional wildlife biologists such as Lanny recognized that avid sheep hunters provided a wealth of information to them about these elusive and magnificent creatures. Jerry said he and other FNAWS members were asked to attend professional wildlife management meetings to provide information about the living habits and habitat of wild sheep.

"Sometimes these were not formal meetings. They would send us as representatives and they'd always ask for our opinion. We felt very privileged. They would say that what they were doing wasn't completely working but that we were doing it by the seat of our pants and were doing better than we should have. We liked giving our opinion and we didn't care who got the credit. That was always our motto. We didn't care who got the credit just so the sheep benefitted."

"These wild sheep needed protection from domestic sheep, and we needed to put money into research. (Dr. Schultz reported lung worms and mange mites were real problems on some ranges.) As it turned out later, we learned you could inject certain new medications into sheep to rid them of these diseases. We saw that wild sheep had disappeared from many historic ranges, such as from the Badlands in South Dakota. The Foundation put wild sheep back into the Badlands."

Early FNAWS members and wildlife biologists took the California bighorn, which is closely related genetically to the original Badlands bighorns, and transplanted them there. Members of this fledgling organization worked closely with wildlife biologists in different states to complete such projects.

Dr. Jerry Waite of Iowa, left, and Duane Smelser of Michigan both served the young wild sheep organization well before FNAWS was FNAWS. Duane attended the first organizational meeting at Mt. Horeb, Wisconsin, in 1974, and Dr. Waite attended the second organizational meeting that took place from April 5 to 6, 1975.

Photograph courtesy of FNAWS archives.

"We knew we had to start someplace, so we went to the biologists. We knew the states wouldn't allocate funds to help the wild sheep, so we knew the hunters would have to carry the whole load. We could see our monies multiply with federal money," Jerry explained.

In the summer of 1974 a symposium was attended by future FNAWS members and wild sheep biologists that was called to discuss what to do with depleted sheep habitat and declining wild sheep populations. Hunters and outfitters noticed the decline using their observations and practical hunting experience. This comprehensive gathering of 77 wild sheep biologists and North American conservationists was hosted from June 18 to 20, 1974, at The University of Montana in Missoula to discuss the management biology of all North American wild sheep. Future FNAWS supporters who presented papers and led discussions included Jim Ford of the Montana Department of Fish, Wildlife and Parks; John H. Batten of Racine, Wisconsin; Dr. Valerius Geist of Calgary, Alberta; and Lanny O. Wilson of the Bureau of Land Management in Boise, Idaho, to name a few. Both Lanny Wilson and Dr. Geist were later awarded and honored the "Outstanding Achievement Award" from FNAWS. Dr. Geist has contributed several chapters to this book.

The wild sheep workshop was co-sponsored by the Boone and Crockett Club, National Audubon Society and the Wildlife Management Institute. Lanny was chairman of the workshop group that reported on desert and Mexican bighorn sheep populations. Other biologists chaired workshops on the California bighorn sheep, Rocky Mountain bighorn sheep, and the Dall's and Stone's sheep. Dr. Geist offered theoretical considerations regarding the management of wild sheep. A complete list of wild sheep biologists and wildlife conservationists is included in the book from pages 297 to 302 and in the appendix of this book.

A far-reaching assemblage of biologists and avid sheep hunters from the United States,

Mexico and Canada shared their concern and knowledge with each other and produced a 300-page book titled, *The Wild Sheep in Modern North America,* published by the Boone and Crockett Club to permanently record their findings. Future FNAWS members took part in this workshop that recorded sheep populations as they were in 1974 and what needed to be done to improve their future populations and habitat. This was a significant starting point.

Money, or absence of it, to put workshop recommendations into action was the major obstacle facing these workshop attendees when it was discovered the extent of the problem and concern regarding the health or lack of health of wild sheep populations. A biologist from each state and province where wild sheep lived stood up at the meeting and recounted the problems, successes, population levels, professional efforts to help the sheep and recommendations for future help. They knew the problems and the challenges, but admitted they didn't have the money to put their recommendations into action.

James B. Trefethen, editor of the book that was created from reports given at this wild sheep workshop, wrote in the foreword: "Through the workshop and the publication of its deliberations, the sponsors are hopeful that the information and the suggestions contained herein will contribute to broader understanding and positive action to assure that the wild sheep of North America will persist to interest and benefit generations to come. With public support of the steps that are necessary, we have no doubt that this can be done.

"To the many scientists and managers who took part in the workshop, and to those representatives of conservation organizations who also participated, we express our gratitude and appreciation. This unique record would not have been possible without their dedicated interest and cooperation."

The book created from this wild sheep workshop, *The Wild Sheep in Modern North America,* is available through the Boone and Crockett Club at their national headquarters in Missoula, Montana.

This 1974 summer workshop was the serious backdrop of information for the meeting of wild sheep hunters in November 1974, although at the time of the Mt. Horeb gathering few attending knew about the professional gathering in Missoula only five months before. The passion for wild sheep felt by these Mt. Horeb hunters translated quickly into much-needed cash to support the professional recommendations of Lanny Wilson and Valerius Geist at the summer workshop. Even at this early meeting, these dedicated men knew they needed to figure out creative ways to raise money in order to put more sheep on the mountains.

The first meeting of enthusiastic wild sheep hunters took place from November 30 to December 1, 1974, after years of planning by Jim Froelich, Lloyd Zeman, Roy Schultz and other GSC members. Jim asked Lloyd Zeman to be master of ceremonies for the event because of his organizational abilities.

The first social gathering of 11 to 13 wild sheep hunters was at the Karakahl Inn in Mt. Horeb, Wisconsin, a resort near Madison, Wisconsin, that Jim Froelich thought would be convenient to people. His resort in northern Wisconsin, called Wah-Wah-Taysee, was too isolated for people to reach from across the Midwest. Sheep hunters who attended that first meeting in 1974 included John Babler, Jim Froelich, Curtiss Judd, Russ Kloosterboer, Dale Leonard, Jack and Joan Leeds, Fred McMurray, Duane Smelser, Walter Wojcuik and Lloyd Zeman, according to John Babler. Curtiss Judd listed the official founders attending Mt. Horeb as John Babler, Jim Froelich, Curtiss Judd, Russ Kloosterboer, Jack and Joan Leeds, Dale Leonard, Fred Murray (McMurray), Duane Smelzer (Smelser), Walt Wojciuk and Lloyd Zeman. Another list from Bob Housholder's newsletter, *Bulletin No. 22,* page 4, showed the sheep hunters who attended that first meeting included Dale Kreucher, rather than Dale Leonard,

Early donations of Estwing Manufacturing Co., Inc. tools and outdoor equipment by Jim and Jan Froelich added much-needed dollars to the young wild sheep organization's bank account. Here we see several types of Estwing tools that were donated to the young group from 1974 to the present day. Jim was director of the Estwing Manufacturing Company of Rockford, Illinois, and Jan Froelich is the daughter of the founder of the company, Ernest O. Estwing. Jan Froelich continues the tradition of donating Estwing tools to FNAWS for current national auctions, a tradition begun by Jim Froelich at Mt. Horeb.

Photographs of tools are courtesy of Estwing Manufacturing Co., Inc.

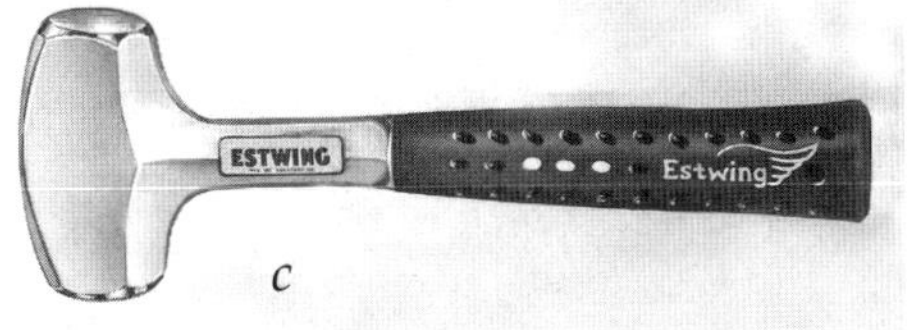

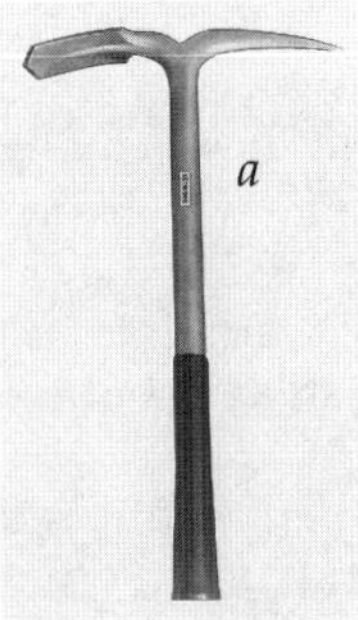

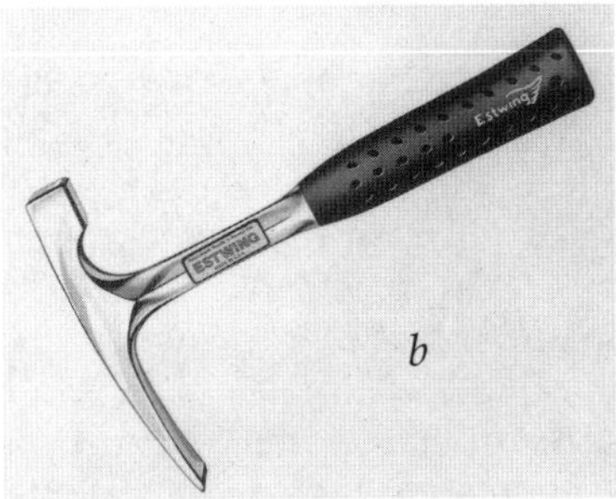

a: A hoe pick, catalog number EHP, is one of several types of geological tools donated to FNAWS to raise money at their annual auctions.

b: This nylon-vinyl grip tool to pick at rocks, catalog number E3-20PC, has been one of many tools donated to FNAWS by Estwing Manufacturing Co., Inc.

c: A 3-pound crack hammer, Estwing Manufacturing Co. catalog number B3-3LB.

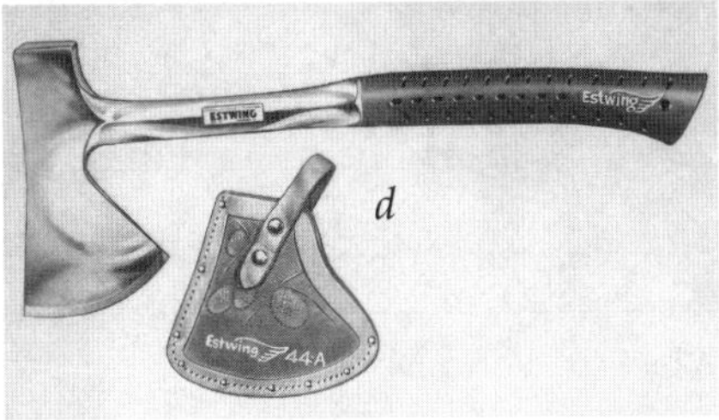

d: An Estwing camper's axe, catalog number E44A, with a sheath is one of dozens of tools donated to FNAWS.

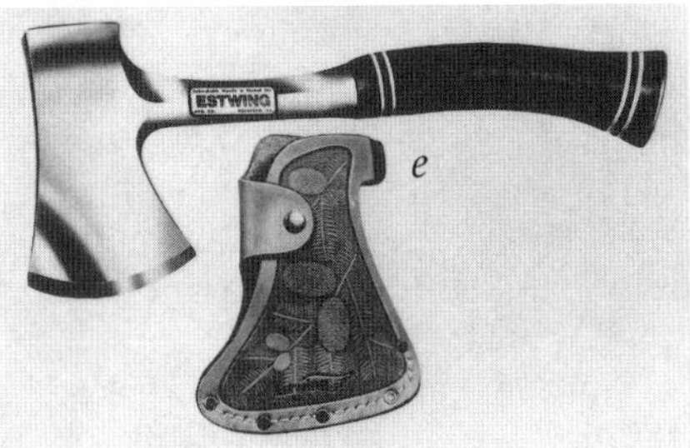

e: This leather grip sportsman's axe, manufactured by Estwing, was one of the first types of tools donated to early wild sheep hunter meetings.

f: A hoe pick manufactured by Estwing Manufacturing Co., Inc., catalog number EPS, is another donated item.

g: A camper's axe with a sheath, Estwing catalog number E45A, is one of dozens of tools donated to FNAWS to put more sheep on the mountain. This tradition of Estwing tool donations began with Jim Froelich.

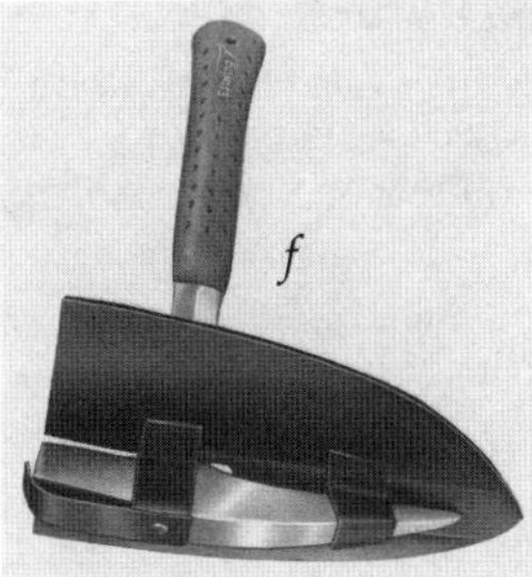

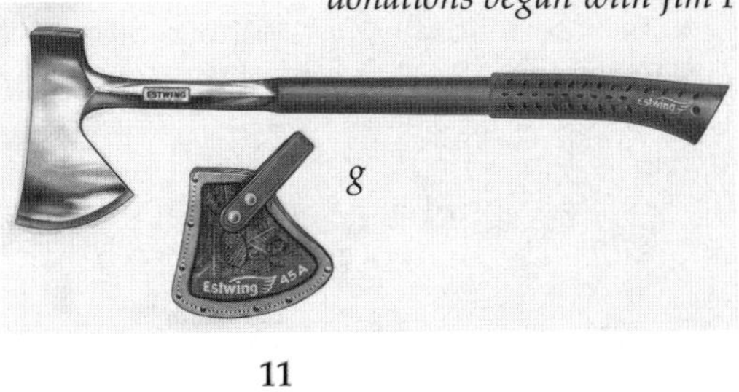

and Fred Wilson, as well as the other men already listed, but John Babler didn't remember them being at that first meeting. The list in the *Bulletin* was created by Jim Froelich.

Jan Froelich remembered 13 hunters at the first meeting based upon what Jim told her afterwards and because Jim loved the number 13, but Jim listed 12 hunters attending that first meeting. Fred McMurray remembered Jim Froelich tried to get a group of sheep hunters together for that first meeting.

"Jim called Duane Smelser and Duane had mentioned the meeting to me in 1973," Fred remembered. Fred was a good friend of Duane's. Duane explained that Jim was trying to pull a group of sheep hunters together to share stories, show movies and exchange pictures. "That's exactly what happened."

"The dates picked for this outing turned out to be the worst possible. It was not just a stormy weekend. It was a blizzard the morning my wife, Doris (Billie), and I were leaving home. I plowed my driveway, but by the time I got Billie in the car, my driveway was drifted over again so I re-plowed it (approximately 12 inches of light, soft snow), jumped in the car and started our trip from Detroit, Michigan, to Mt. Horeb, Wisconsin. There were times on the expressway that it was a complete whiteout. I had to guess where the road was, but we got to Mt. Horeb," Duane Smelser recalled.

"We didn't know each other. Jim had called Duane Smelser and Duane had mentioned the meeting to me. Duane was a friend of mine and was a member of Safari Club International at the time. Jim said he was just calling a bunch of sheep hunters together just to have a meeting and talk sheep. Everybody was asked to bring their pictures with them and if they had movies, bring those, too. That's exactly what happened. Everybody brought their pictures of sheep and movies and sat there all weekend and talked sheep and sheep hunting," Fred McMurray explained.

Lloyd Zeman wrote in 1976, "Our chapter began with a very informal gathering of 14 sheep hunters at Mt. Horeb, Wisconsin, during a stormy weekend in November 1974. Died-in-the-wool sheep hunters came from Iowa, Michigan, Wisconsin and Minnesota."

This hardy group of men and women who survived the blizzard and arrived in Mt. Horeb were in for a treat. The tie that bound these men together was sheep. "Most of us have had the thrill and the privilege of collecting one or more trophy rams — matching wits with the greatest trophy in North America, and probably the world — in the remote areas of the wild and steep, high country where these magnificent animals make their home."

"We talked sheep all weekend till the wee hours of the morning. I believe there were 12 or 13 of us. Most of us had our wives with us. We talked about getting together the next year and everybody would bring another hunter," Curtiss Judd said.

Jim wanted the wives to come to the Mt. Horeb meeting, and had asked Jan to organize special activities for them, but as the meeting neared, they didn't think any of the women would come, so Jan stayed home to take care of their children. She said this was really a party for guys who loved to sheep hunt. A few of the wives did come as it turned out, but Jan decided against the trip. She was just happy that Jim located fellow sheep hunters. Jan couldn't describe the joy her husband felt after the conclusion of this meeting. She described it as close to a religious experience for those who gathered. Ironically, Mt. Horeb is the name for a Biblical mountain where Moses received the Ten Commandments. They kept shaking each other's hands and talking. There was such a bond established that weekend that nobody wanted to leave.

Jim told her they talked until 1 a.m. each night and kept thinking they needed to sleep, but few did. And, besides, the snow was blowing so hard that no one could leave the hotel.

They all agreed to meet the next year so their friendship could continue.

"We ended up enjoying ourselves so much that weekend that we decided to get together next year. We talked about where to get together and we decided the Playboy Club in Wisconsin. We all decided we'd try to bring another sheep hunter with us," recalled Fred McMurray.

"The blizzard kept us confined to our hotel, which was all right as we had just come to swap sheep stories anyway," Duane recalled. The trips home for these men and women also included the same blizzard conditions that required many of them to seek shelter in area schools when the highways closed.

When Jim Froelich returned home he gushed to Jan about the length of time these avid hunters spent together that weekend.

He wrote a two-page report to Bob Housholder on December 3, 1974, only two days after he returned home.

Jim Froelich, center, was the first of the FNAWS founders who worked to gather names and addresses of Grand Slam wild sheep hunters as early as 1970. His wife Jan said that he worked for a long time to dig out anybody who wanted to attend those first sheep hunting meetings. Locating kindred spirits became his passion. For this effort, the Foundation for North American Wild Sheep named him "the grandfather of FNAWS" prior to his death at age 75 in 1996. With Jim in this 1990 photograph is Past President Arnie Johns, left, and Past President Harvey Kadlec, right.

Photograph courtesy of FNAWS archives.

"The 'get-together' of the members and associate members of the NASHA (North American Sheep Hunters Association) held this past weekend in Mt. Horeb, Wisconsin, was 100 percent successful in the opinion of the 12 persons attending. We would have liked to have more in attendance, but both the Wisconsin and Michigan deer season was on, plus the heavy snowstorm in the area could have been factors in not more showing up.

"Some of the members came from as far away as Mehtomedi, Minnesota, and Detroit, Michigan. Lloyd Zeman acted as master of ceremonies and besides yours truly, in attendance were: Duane Smelser, Fred McMurray, Curtiss Judd, Jack and Joan Leeds, Fred Wilson, Russ Kloosterboer, Walter Wojcuik, John Babler and Dale Kreucher.

"This was the most enthusiastic bunch of sheep hunters you can imagine. I had one hell of a time trying to get them out of my room (we missed lunch) to get some kind of meeting going. There never was a lag in the talk as everyone had four or five hunting trips to relate and, of course, the information on outfitters and territory was priceless.

"Lloyd and I finally got them pried out of my room and got some semblance of a meeting started. We started out with each one relating their sheep hunts and then got into movies and slides of the hunts. This continued until 1 a.m. when we were asked to adjourn by the (hotel) management. We broke up on Sunday but everyone again gathered in my room for a couple of hours of film on sheep hunting in Alaska. All in all, it was a great gathering and I am looking forward to the next one, tentatively scheduled for next March or April.

"Ron Kolpin (the gun case, etc. manufacturer) has volunteered to head up the next meeting because I will be in Florida most of the winter. I will furnish Ron the necessary names and addresses.

"Oh, yes, I forgot to mention that Lloyd Z. presented three copies (autographed) of *Grand Slam of North American Wild Sheep* as door prizes, and I presented an Estwing sportsman axe to the hunter traveling the farthest to attend the meeting.

"Bob, I'd certainly urge other hunters (sheep) in other areas to have a similar meeting. It was great. If you need any further information, let me know."

The December 1974 GSC *Bulletin No. 24* reported on this meeting, and Bob Housholder wrote in the newsletter to the rest of the GSC membership the following:

"I hope some of you others get the same idea. If you need names of members and associate members in your area, ask and I will be happy to provide them. Perhaps we can get several sub-club meetings going in the country and someday, somehow, we can hold a regular club meeting. That would be some undertaking but I am reasonably certain it would be a 'first' — a meeting for people whose main interest is in one animal — the mountain sheep of North America. If any of you have some comments to offer on any of the above, just fire away."

Jim Froelich continued to organize wild sheep hunter meetings, but, as history confirms, Bob Housholder really did not want sub-club meetings unless they were directly connected to GSC. He did not want any group raising money for wild sheep in a nonprofit setting unless he received a percentage of the money raised for GSC. Jim's leadership in the early years was the reason why this small group of sheep hunters stayed together. His passion to communicate with other sheep hunters motivated him, Jan explained. "Jim was always there in a pinch. He was the one who always stepped up to get the job done and always enjoyed working behind the scenes. He wasn't the speaker. He was the detail person."

Even at the first meeting, those who gathered on that stormy weekend in 1974 talked of serious concerns regarding the future of wild sheep in North America. They talked about doing something for sheep, putting them back on the mountain, so they would be there for our children. That was the beginning of this great organization. The goals of a not-yet-formed FNAWS were discussed. These passionate hunters of wild sheep knew they wanted an organization to promote the management of wild sheep native to North America and safeguard these species against extinction. These words became a part of the official FNAWS articles of incorporation several years later. Protection of wild sheep habitat, which, in turn, protected the habitat for all wildlife, became a theme for action by these men and women. Preserving the gene pool of wild sheep for future generations also was discussed, because once these animals were lost, the opportunities to discover their presently unknown values for the welfare of mankind would also be lost.

A tone of seriousness crept into the weekend as movies, photographs and hunting stories were swapped, and bonds of friendship were established between hunters who, up until this snowy weekend in Wisconsin, had not known each other. They were pulled together by the letters and phone calls of Jim Froelich and Lloyd Zeman. Jan said the experience for these men was such a strong force that movement toward a more serious organization to help wild sheep quickly became a priority for all of them. Their passion spread to others who also took up the drumbeat for wild sheep. They started thinking of ways to raise the necessary money.

A second informal meeting of this energetic band of men and women met from April 5 to 6, 1975, for the second gathering at the Playboy Club in Lake Geneva, Wisconsin. As had been agreed at the Mt. Horeb meeting, each one brought more sheep hunters and friends, so the second meeting grew from 12 to 42 hunters and their spouses. Ron Kolpin was the meeting organizer and Lloyd Zeman was the master of ceremonies.

Those attending the second meeting of the future FNAWS organization, based upon FNAWS headquarters records, included Mr. and Mrs. Ron Kolpin; Mr. and Mrs. Lloyd Bare; Roger Gipple; Mr. and Mrs. Merton Rowe; Mr. and Mrs. Dan Yaeger; Mr. and Mrs. Duane Smelser; Mr. and Mrs. Frank Bays; Mr. and Mrs. James Sinsky; Lloyd Zeman; Dr. Roy Schultz; Dr. Jerry Waite; William M. Criswell; Allen Hantelman; Pete Haupt; Dr. and Mrs. John Bullard; Allen Ripple; Bruce Schroeder; Arthur J. Pipp; Jim and Jan Froelich; Ned and Becky Plummer; Ron and Elaine Thomas; John A. Mueller; Linda R. Martin; John and Lois Suda; Roger, Gert and Jack Voelki; Rudy Jenny; Dave Bathke; and Fred and Betty McMurray.

This was the last time that a complete recollection of attendees could be made because the membership was growing by such leaps and bounds. Once again, the second gathering was "social in nature and a time to exchange hunting stories and pictures," Lloyd Zeman wrote.

A brief mention of the meeting was written by Bob Housholder in *Bulletin No. 26* in July 1975. The notice read:

"Ron Kolpin, Berlin, Wisconsin, M.C.'d a sheep meeting at the Playboy Club in Geneva, Wisconsin, with 42 in attendance. He showed my (Bob Housholder) film on the big desert ram. Roy Schultz, Avoca, Iowa, and Lloyd Zeman, Des Moines, donated five Grand Slam books as door prizes. Zeman and Board Member Jim Froelich of Sayner, Wisconsin, have volunteered to conduct the next meeting in April 1976."

Several sheep hunters who attended the Playboy Club meeting remembered it with fond memories.

"A bunch of us still don't know if we went to the Playboy Club for fun or if we went because of the wild sheep. Maybe a little bit of both," laughed Dr. Waite, who had missed the Mt. Horeb meeting but was present for the meeting at the Playboy Club.

"Our initial interest was to conserve sheep. We thought we were a success from the very beginning because it just seemed to blossom. Every year it doubled and doubled and tripled and tripled," said Fred McMurray.

"I drove to Des Moines and flew with Lloyd Zeman, stopping in Algona, Iowa, to pick up Dr. Jerry Waite. The meeting was once again filled with slides and stories of bighorn sheep hunting," said Doc Schultz.

The group continued to share hunting stories and photographs of past sheep hunts, but they wanted to create an organization to do something for wild sheep like Ducks Unlimited did for ducks. Fred McMurray wrote that by the end of the second meeting they were "all excited about getting something started."

The early founders were resolute in their purpose from the start. "We wanted to maintain the habitat, to transport animals to historic ranges and to study diseased animals," Dr. Waite said. He explained that Doc Schultz and he were both veterinarians, so they naturally gravitated to scientific methods of study. Other early members weren't doctors or scientists, but they knew involvement of wildlife biologists and state and provincial wildlife agencies were essential to their mission of properly conserving wild sheep and their habitat.

These new-found friends knew that to really help wild sheep they needed money and lots of it. Jim Froelich was one of the first to think of creative ways to raise money.

Jim was director of the Estwing Manufacturing Company of Rockford, Illinois. Jan Froelich is the daughter of the founder of the company, Ernest O. Estwing. Both Jim and Jan, with the help of Bob Youngren, Estwing executive vice president and director of sales and marketing, were instrumental in donating Estwing carpenter tools, hammers and hatchets manufactured by their company as auction items for the early meetings of this group as well as for current FNAWS auctions. Jim's report to Bob Housholder even noted that an Estwing axe was used as a door prize at the Mt. Horeb meeting. Dr. Schultz and Jerry Christian pointed to the donations of Estwing tools as one of the reasons why fund-raising efforts and early auctions were financially successful.

Bob Youngren provides more insight into the significance of the Estwing donations to the wild sheep hunters when he wrote about his friend on April 27, 1998. "I know that our former director, Jim Froelich, was very involved in the Foundation and he donated Estwing tools for FNAWS' early auctions and fundraisers. Unfortunately, our records of donations do not go back more than a few years, but we do know Jim was involved right from its earliest beginnings." Bob knew Jim for 27 years and knew he was dedicated to many conservation projects.

"Jim was greatly respected for his dedication to conservation causes and his respect for all living things. Jim was a wonderful man who did `make a difference' in the field, at work and as a director of our company," Bob wrote.

Bob worked with Jim to select tools for the FNAWS fundraisers and thinks Estwing tools in modern FNAWS auctions are the same as in the beginning of the organization. Axes and wood splitting tools have always been most popular. These are the tools of hunters, Jan explained. Jan continues to donate a variety of Estwing tools to FNAWS functions and its annual convention with the help of Bob in memory of Jim.

Jan Froelich remembered Jim as always trying to think of creative ways to raise money for wild sheep, including the creation of lapel pins made from colored enamel with the heads of the four North American species of sheep on them. Jan was the artist who drew the original sheep heads for these lapel pins that Jim sold at early meetings to benefit sheep projects. She remembers helping him sell the lapel pins. She drew a head portrait of each species and each species had a different colored background on its lapel pin. The bighorn sheep head was on a background of green for the Rocky Mountains, and the Dall's sheep was on a blue background. Later Jan designed a lapel pin in an oval shape that had the four heads of wild sheep species on it. The Stone's and desert sheep also had special colored backgrounds. Original lapel pins designed by Jim and Jan are collector items today.

"We were really committed to bringing good habitat back to wild sheep. We just wanted to return something to the mountains where the wild sheep were harvested. We wanted to do something so our sons and daughters could hunt them in the future," Dr. Waite recalled.

By now, the word of this fledgling group of dedicated wild sheep hunters was spreading, and from April 9 to 11, 1976, a third meeting took place that attracted 96 men and women, including guides and outfitters Earl and Marge Boose of British Columbia, Bill McKenzie of British Columbia and Bill Butler of Montana. Lloyd Zeman said this was the meeting where the Midwest Chapter of the North American Sheep Hunters Association of the Grand Slam Club was started, but within a year this group would soon become the Foundation for North American Wild Sheep. Significant to the people who planned this meeting was the fact that a fire destroyed the Des Moines Hilton where the meeting was to take place.

"The night before we got there they had a fire at our hotel, but with a little juggling around, Lloyd Zeman managed to find another hotel that was very nice," Fred McMurray reflected.

Lloyd Zeman, with Doc Schultz's help, made a quick switch to the Ramada Inn and even was able to organize the roasted pig dinner beside the swimming pool that had been planned for the Hilton. Lloyd was the primary organizer of the meeting, and Frank Bays was master of ceremonies, but Fred McMurray was the one who began the tradition of the hospitality room at this meeting.

"I started our hospitality room that I believed would keep our group together in the evening so we could talk more sheep!" Fred recalled. "I believe that was our first sheep judging contest and awards that received television coverage."

Lloyd Zeman took the podium at the Des Moines meeting, and asked the group assembled, "What should we do? What is our future?"

"Rather than just being a social group, we decided to start doing something for the animals we all cared so much about," Lloyd wrote.

Seven people were nominated from the floor to explore the future of this group and report back to everyone at the next meeting which was scheduled for Dearborn, Michigan, from March 24 to 27, 1977. The seven included Lloyd Zeman, Duane Smelser, Jim Froelich, John Babler, Dr. Roy Schultz, Dale Leonard and Frank Bays. They were charged with the task of creating a formal organization and structure. Jim Froelich was a national director of Bob Housholder's Grand Slam Club, so they decided to call themselves the Midwest Chapter of the Grand Slam Club.

"We had a very successful meeting and all enjoyed talking sheep all weekend," said Fred McMurray.

Prior to the fourth meeting in Dearborn, Lloyd Zeman and Dr. Roy Schultz contacted Bob Housholder. "We approached Mr. Robert Housholder (Bob) of the Grand Slam Club, asking if we could get a 501-(c)(3) tax status, so that we could raise funds to support a conservation effort for the Grand Slam Club as we were all members," Doc Schultz explained.

Bob Housholder said no. After repeated attempts of Lloyd Zeman and Doc Schultz talking to Bob Housholder on the telephone it was apparent Bob wanted to keep the Grand Slam Club for a profit center under his control. Jerry Christian said Bob had a "real strong grip" on the Grand Slam Club and didn't want to share profits with anyone. The new organization realized they needed to form a new organization to avoid any risk of raising money that Bob Housholder would want for the Grand Slam Club. Curtiss Judd agreed with Jerry that Bob Housholder didn't want to share profits with anyone, especially an organization registered as "nonprofit."

"Bob Housholder ran a profit organization and we wanted to be a non-profit to give back to sheep what they gave to us," said Fred McMurray.

John Babler agreed that connecting the sheep organization with the Grand Slam Club was a poor idea. Keeping control of money raised and of its purposes were two important reasons why John thought direct connections to the Grand Slam Club would not have worked. Jim Froelich was concerned that any connection to the Grand Slam Club would jeopardize the money raised by the sheep group, so Jan remembered Jim telling her that such a direct connection was probably not a good idea.

Next, the group turned to Safari Club International to see if there was any interest in a formal association. SCI founder C.J. McElroy sent Keith Bates of Chicago, editor of the SCI magazine, and Jack Leads of Minnesota, SCI Conservation Club president, to meet with Lloyd Zeman, John Babler, Dr. Schultz and Duane Smelser in Chicago.

"We discussed the possibility of our sheep group becoming a sheep chapter of Safari Club International. They took this suggestion back to their Board of Directors, who turned us down," said Duane.

The best the sheep hunters could muster, according to Dr. Schultz, was a suggestion that the group become a "sheep chapter" of SCI and money raised to help wild sheep would have to go through the main SCI organization.

"We did not have to make this decision as Safari Club International voted the idea down," said Dr. Schultz.

The stage was set for the official creation of the Foundation for North American Wild Sheep on June 21, 1977. A time to discuss the details of an emerging nonprofit organization took place in Des Moines, Iowa, remembered by many early FNAWS members including Doc Schultz who provided details about a clever fund-raising effort.

"One of the early fund-raising efforts was auctioning a "paddle wine list" requisition from the Pier Restaurant in Des Moines. The Board and friends of FNAWS were eating at a restaurant where the wine list was painted on a canoe paddle. Someone asked the waitress if any of the paddles ever disappeared. The waitress made the mistake of saying that many had tried to take a paddle, but no one had been successful. This was then a challenge. One Board member, Frank Bays, took up the challenge.

1998, past-president, Dr. Jerry Waite with the now famous paddle.

Photograph courtesy of Dr. Jerry Waite.

"He put the paddle in his pant leg, and walked out the door. At the meeting, the paddle was requisitioned by another member, and auctioned off for several hundred dollars. Strangely, the paddle showed up at the next convention and was re-auctioned. Somehow, it always seemed to disappear after the purchaser had paid, but would subsequently show up at the next convention (Memphis) to be re-auctioned. Several buyers of record included Frank Bays, Lloyd Zeman, Doc Schultz, John Babler, Doc Waite and others who contributed several thousand dollars over a few conventions, thinking they would end up with the paddle. It has only recently come to light that the paddle now decorates the wall of Dr. Jerry Waite's office in Fenton, Iowa. I have a picture for evidence. The wild sheep have benefitted from the 'paddle wine list' from the Pier Restaurant in Des Moines."

Prior to the fourth meeting of the fast-growing sheep group was a successful meeting of 10 men in Omaha, Nebraska, to continue organizational discussions and share sheep stories. This was the fall 1976 when the elected representatives met for a Board meeting in Omaha. In those days everyone paid his own way and continued to do so for at least 10 years.

At the close of the meeting, Lloyd Zeman commented to the group, "We don't have enough money for stamps." Each man present threw $100 on the hotel bed, and they had $1,000 for mailings. Dr. Waite remarked, with a note of awe, that this multi-million dollar organization now called FNAWS began with $100 from 10 men. A meeting in Omaha cemented friendships and secured seed money for the financial backing of this fledgling organization.

At the Omaha meeting, just prior to the Dearborn, Michigan, convention, Fred McMurray remembered a ride on a carousel. Actually, early members remembered the story because Fred was asleep.

"We were on a carousel and I had fallen asleep. My head was turned back and every time we came by this table of girls, they started pitching popcorn at my open mouth," Fred snickered.

Doc Schultz shed more light on this monumental event. "We stayed at the Omaha Red Lion Hotel. Fred McMurray went to sleep at the Carousel Restaurant waiting for dinner." The restaurant rotated in a circle at the top of the building, and the rest is history.

Mario Luis Cossio of Mexico was the invited guest of Frank Bays at the fourth organizational meeting of the fledgling wild sheep group from March 24 to 27, 1977, at the Hyatt Regency Hotel in Dearborn, Michigan. Mario reads the Boone and Crockett Club book, The Wild Sheep in Modern North America, *that was edited by James B. Trefethen.*

The half-mount Chadwick ram was the star attraction at the 1977 meeting in Dearborn, Michigan, shown here as it was seen on display in the Hyatt Regency Hotel. Frank Bays drove a friend's pickup truck to New York City to retrieve the ram from the Bronx Zoo where it was stored behind temporary and dusty walls.

William Harold Nesbitt, left, shakes hands with Lowell E. Baier when the National Collection of Heads and Horns exhibit opened at the National Rifle Association's firearms museum at 1600 Rhode Island Avenue in Washington, D.C. The collection, including the Chadwick ram, was on display at NRA from May 5, 1978, to late 1981 before they were relocated to the Buffalo Bill Historical Center in Cody, Wyoming.

Lowell E. Baier, FNAWS Charter Member 14, stands beside the half-mount Chadwick ram at the 1977 meeting of the organization that would become FNAWS on June 17, 1977. Lowell was the person who discovered the Chadwick ram and other trophies from the National Collection of Heads and Horns when he visited the Bronx Zoo on other business.
Photographs courtesy of the Boone and Crockett Club.

U.S. Rep. John D. Dingell (D-Michigan), right, talks with Sr. Mario Luis Cossio of Mexico at the March 1977 organization meeting of the wild sheep organization that would become FNAWS on June 17 of that year. Between the two men is the half-mount of the Chadwick ram.
Photographs courtesy of the Boone and Crockett Club.

Doc Schultz arranged for a memorable pheasant and quail hunt in Iowa for those who stayed over on Sunday and Monday following the meeting. His hometown of Avoca is less than an hour's drive from Omaha. Several men were pilots who owned airplanes, so they had more flexibility and stayed for the bird hunt, Dr. Waite recalled.

"This was a hunt we all enjoyed and won't forget. Thanks goes to Dr. Roy Schultz for our enjoyable hunt. In fact we saw so many pheasants that every time I shot a pheasant, Frank Bays would say I got him, and every time Frank would shoot, I said I got him. It seemed we always shot at the same pheasant at the same time. We sure had a good laugh over it. And, the next day when I took my first shot at a quail, my shotgun fell apart! I still get razzed about it. We sure had a fun time," Fred McMurray chuckled.

The fourth meeting of the sheep hunting group, calling itself the Midwest Chapter of the North American Sheep Hunters Association of the Grand Slam Club, took place from March 24 to 27, 1977, at the Hyatt Regency in Dearborn, Michigan, near Detroit. Frank Bays and Al Perry were organizers and Frank served as master of ceremonies to a group of men, women and children totaling between 255 and 300 people. Bob Housholder reported in his GSC Bulletin No. 35 in April 1977 that 265 men, women and children attended the function.

Bob wrote, "Sr. Cossio was present and had the authority to offer five sheep permits for Mexico. These, of course, have nothing to do with the Grand Slam Club." Bob encouraged his GSC members to join the newly organized group.

"We raised more monies by creating charter memberships for the first 100 members to donate $100 each for a charter membership. The first 100 would be all that would be considered as Charter Members. This charter membership list still exists in FNAWS headquarter's records," Dr. Schultz explained. The complete list of charter members is reproduced in the appendix of this book.

"I think Detroit (Dearborn) in 1977 was when we really started to become an organization. That was when we got our organization moving. This was when we realized we were attracting hunters," said Fred McMurray. The membership added Fred McMurray and Mike Valencia to the group of seven leaders who were appointed the year before to lead the young organization. Now the Board of Directors totaled nine. Lowell Baier was asked to serve as the FNAWS's first legal advisor and counsel at this point in time. Duane Smelser gave credit to Frank Bays for arranging a visit to the meeting by Senior Mario Luis Cossio, the past director general de la fauna silvestre of Mexico. Mario received the Midwest Sheep Hunters Achievement Award from the membership and brought with him four desert sheep permits to the first auction. The permits attracted sheep hunters to this first auction from as far away as 1,000 miles.

The exact wording of the Achievement Award read: "To Mario Luis Cossio for outstanding achievements in the field of sheep management, conservation and very much improved hunting conditions in Mexico. Dated 1977 by the Midwest Sheep Hunters."

"At the first convention in Detroit, we grossed $70,000. We thought we were on the top of the world!" Dr. Jerry Waite said. He also acknowledged that the desert sheep permits auctioned now sell for three or five times the amount collected in 1977, but money had a different value then. Frank Bays recalled that Lloyd Zeman was against auctioning off the desert bighorn permits, so people paid for chances to be drawn.

At the conclusion of the first auction, Waite, Zeman, Schultz and all the founders of this close-knit group of friends knew they were on to bigger and better things.

"We refined things and we made mistakes along the way, but as a good Scotsman like I am, you don't make the same mistakes twice," Dr. Waite said.

Highlights of the 1977 meeting also included a keynote speech by U.S. Senator John Dingell of Michigan and a trophy display that included the World's Record Stone's sheep taken in 1936 by L.S. Chadwick near the Muskwa River in British Columbia. This ram scores 196-6/8 Boone and Crockett points. Having the Chadwick ram on display was a great attraction to the people who attended.

"Frank Bays went to New York to pick up the Chadwick ram, the largest ram taken on the North American continent, which improved our ram display tremendously," Duane said. Frank said he stored the Chadwick ram in his house for several weeks and was always nervous that something would happen to it before taking to the wild sheep meeting. To pick up the ram, he drove a friend's truck to New York City because he didn't own a pickup truck at the time. He was only 23 years old in 1977 and thought this whole thing was quite an adventure. To this day, Frank Bays chuckles when he thinks of all he did for FNAWS and the crazy things the young members did to keep things going.

Dr. Waite shared a little-known tidbit of history regarding the Chadwick ram coming to the Dearborn meeting. Some of the sheep hunters discussed during their informal social meetings where the Chadwick ram was located or if it was still in existence. Early FNAWS member Lowell E. Baier (FNAWS Charter Member 14) visited the New York Zoological Society (Bronx Zoo) for unrelated business on January 20, 1977, and discovered that the National Collection, including the Chadwick ram, was hidden from public view.

"I found the Collection in a deplorable state, boarded up from public view, dusty, dirty, unhumidified, and many of the hides badly in need of repair," Lowell wrote in the 8th edition of the Boone and Crockett Club book, *Records of North American Big Game.* The once world-renowned National Collection, totaling 2,371 specimens in 1940, had dropped to 34 North American trophies when Lowell rediscovered them in early 1977.

The Chadwick ram remained in the collection, but 13 magnificent trophies, including the World's Record mule deer and Atlantic walrus, were stolen in 1974. The remaining big-game trophies were stored behind temporary walls. A complete description of how Lowell discovered the National Collection of Heads and Horns and mounted a year-long crusade to save the remaining trophies is detailed in *Records of North American Big Game,* on pages 72 to 96, which is still available through the Boone and Crockett Club in Missoula, Montana.

Those few sheep hunter members managed to convince the museum staff they would take good care of the Chadwick ram mount in time to bring the mount to the 1977 Dearborn,

William Harold Nesbitt of the National Rifle Association views the 34 remaining trophies from a once-glorious collection begun by William T. Hornaday in December 1906 as the National Collection of Heads and Horns in the New York Zoological Society (commonly called the Bronx Zoo). Here, Harold appreciates the collection at the NRA firearms museum where the collection stayed from May 5, 1978, to late 1981. Early FNAWS members, especially Lowell E. Baier of Potomac, Maryland, and Frank Bays of Whitefish, Montana, raised the money to preserve the remaining trophies that once numbered 2,371 in 1940.
Photographs courtesy of the Boone and Crockett Club.

Michigan, meeting. Dr. Waite remembered having to post a bond to remove the mount from the collection.

"Everyone had believed he was a shoulder mount, but he was a half mount," Dr. Waite said.

At that moment, Lowell Baier, alone and without help, realized something needed to be done immediately to save the Chadwick ram and the rest of the National Collection of Heads and Horns (NCHH). He orchestrated a one-man campaign for more than a year, trying to get the NCHH to be given to Safari Club International or some other wildlife or sporting group. The director of the Bronx Zoo would have none of it. He had already signed an agreement giving the NCHH to the American Museum of Natural History in New York City. Lowell threatened legal action and hired a New York attorney during the stand-off and I called the Boone and Crockett Club for help. They interceded through Samuel B. Webb who knew the director of the Bronx Zoo and the American Museum of Natural History, so he was able to "undo" the agreement provided the B&C would take the NCHH.

William Harold Nesbitt and Jack Reneau, who maintained the records-keeping program for the Boone and Crockett Club as part of their duties at the National Rifle Association (NRA), were contacted by Lowell. Lowell's knowledge and enthusiasm to save this unique collection sparked the aid of Harold who was NRA's director of hunter services at the time. Jack was the hunter information specialist for NRA's Hunter Information Services Department and Harold's assistant. Harold and Jack drove a U-Haul truck to New York during the winter 1978 and returned to Washington, D.C., with the remaining 34 North American big-game trophies from a collection that at one time numbered 4,000 in the early 1900s, according to the museum officials. When Harold and Jack arrived at the Bronx Zoo, they completed an inventory of the entire collection and discovered 238 remaining specimens. A total of 204 were from the other continents of the world. The international portion of the collection was deeded to SCI and hauled in two 18-wheeler trucks to a warehouse in Colorado where it was stored until SCI completed its wildlife museum. Today, the international specimens are showcased at SCI's international headquarters and museum in Tucson, Arizona. Lowell Baier headed the ad hoc committee to oversee this entire operation.

Museum guards at the Bronx Zoo told Harold and Jack that dozens of trophies had been thrown in the New York City dump as the museum staff made room for more exhibits and office space where the National Collection of Heads and Horns once was on glorious display. The 10,842 square foot Romanesque brick exhibit building — specifically constructed with private funds to house the entire National Collection at the New York Zoological Society in 1922 — was converted to office and storage space for the museum. Much of the private funds used to build the exhibit hall were collected from early Boone and Crockett Club members and future FNAWS members.

Lowell reported NRA generously renovated one section of their popular firearms museum at their national headquarters, which was then located at 1600 Rhode Island Avenue in Washington, D.C. The collection was on display at NRA from May 5, 1978, to late 1981, and then Boone and Crockett Club and FNAWS members made arrangements for the trophies to secure a permanent display area for the Chadwick ram and the National Collection.

Frank Bays of Michigan, one of the FNAWS founders and a strong supporter of Lowell's efforts to save this unique wildlife collection, organized an aggressive fund-raising program. Donors were offered a limited edition print of a handsome oil painting of the Chadwick ram that was donated to the Boone and Crockett Club by Michigan artist Harry Antis. Efforts by Frank Bays (FNAWS Charter Member 6) to raise money for the National Collection of Heads and Horns insured that the collection will be maintained and expanded for future generations.

A $3,000 donation by FNAWS in 1980 further helped preserve the National Collection and, eventually, the Buffalo Bill Historical Center in Cody, Wyoming, signed a long-term agreement with the Boone and Crockett Club to display the trophies at their facility in a large log hunting cabin within the Center, but it was the efforts of sheep hunters from an organization that would become FNAWS, especially Lowell E. Baier, who were responsible for locating the collection and the Chadwick ram in the first place.

"A donation by the Foundation would be uniquely appropriate, since it has chosen the Chadwick ram as its official emblem. Certainly the well-being of this truly outstanding trophy is of vital concern to the Foundation and its membership, and contributions to the NCHH Fund will certainly serve to help secure that future," wrote William Harold Nesbitt, director of the Hunter Services Division of the NRA, on March 7, 1980, to Lloyd Zeman.

The letter to Lloyd was in response to notification by Lloyd that FNAWS was giving $3,000 to the National Collection. The funds were used for the relocation of the collection from NRA to the Buffalo Bill Historical Center and for acquiring future outstanding trophies for the display.

At the successful conclusion of the 1977 meeting in March, the stage was set for this group to immediately form a separate group that would work to raise money for wild sheep. They called themselves the Foundation for North American Wild Sheep. They hired an attorney in Des Moines to draw up the articles of incorporation and to file legal papers with the Internal Revenue Service to qualify as a 501-(c)(3) corporation for the purpose of becoming a tax-exempt charitable, educational, scientific and environment protective organization. FNAWS didn't have enough money to pay for the lawyer, so Lloyd Zeman and Dr. Schultz dug in their pockets to pay the $678.30 legal fees to the Swift, Brown and Winick law firm.

Jim Froelich declined to sign the articles of incorporation because of his involvement with the Grand Slam Club. He feared his involvement with FNAWS would hinder its ability to raise money for wild sheep independently of the Grand Slam Club, according to John

Babler and Doc Schultz. The six who did sign the articles of incorporation, in order of their signatures on the legal papers filed in Polk County, Des Moines, Iowa, were Lloyd Zeman of Des Moines; Iowa; Frank Bays of Fenton, Michigan; Duane Smelser of Romeo, Michigan; Roy Schultz of Avoca, Iowa; John Babler of Mahtomedi, Minnesota; and Dale Leonard of Sedalia, Colorado. The first signature of Lloyd is dated June 17 and the final signature of Dale's dated July 7, but the effective date of the organization was June 17, 1977.

Lloyd Zeman invited Bob Housholder to the meeting, but Bob declined to attend, sighting surgery as the reason. A letter from Lloyd, dated March 29, 1977, to Bob in Phoenix, Arizona, informed Bob that the new group was no longer associated with the Grand Slam Club. Lloyd's letter clearly stated the reason why the separation was necessary.

"The purpose of the organization is to raise money to finance sheep management-related projects. The individuals feel that to accomplish their many objectives most successfully, it is necessary to have the benefits acquired by being a nonprofit organization."

Lloyd told Bob he would be active in FNAWS, and encouraged Bob to join FNAWS. He also told Bob that at FNAWS conventions he and other FNAWS members would always encourage its members to join GSC.

Bob told members of FNAWS that he had spent 20 years and a lot of hard work to organize and run GSC and didn't want to give up money to another group, especially one that was nonprofit.

The purposes of the Foundation for North American Wild Sheep remains the same as it was in 1977. The articles of incorporation read:

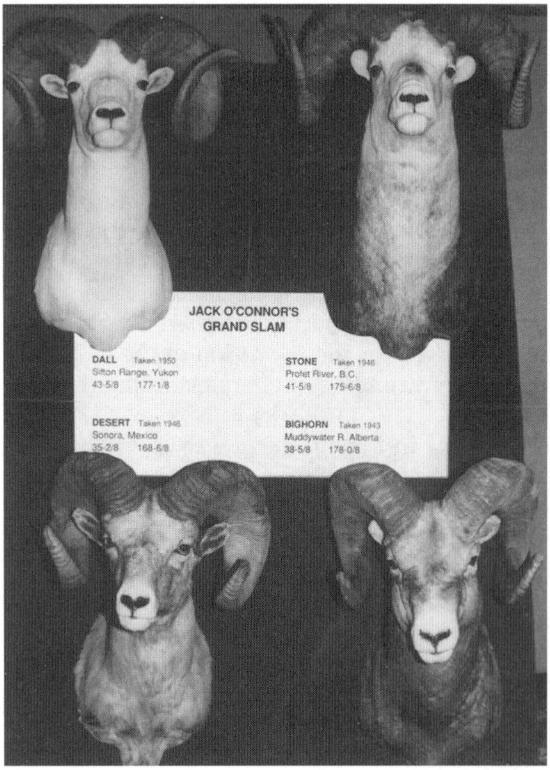

The Grand Slam of Jack O'Connor who has Grand Slam Club Number One. Top left is his Dall's sheep from 1950 taken in the Sifton Range of the Yukon scoring 177-1/8 points. Top right is his Stone's sheep taken in the Profet River area of British Columbia with a score of 175-6/8 points. In the lower left corner of the photo is his desert bighorn taken in 1946 in Sonora, Mexico, with a score of 168-6/8 points. In the lower right corner of the photo is Jack's bighorn from Muddywater River, Alberta, taken in 1943 with a score of 178 points.

Photograph courtesy of Hans W. von Barby.

"The purpose of the corporation shall be to promote the management of, and to safeguard against the extinction of, all species of wild sheep native to the continent of North America."

The purpose is steadfast and true today as it was in the beginning.

Dr. Roy A. Schultz is a swine practitioner and consultant from the Avoca Veterinary Clinic, Avoca, Iowa, where he has served for 38 years as senior partner with his son, Dr. Gary Schultz, and Dr. Jim Hoffmann. He is the author of, A Portrait of Sheep and Sheep Hunting, *and chapter author for Bob Housholder's book,* The Making of a Grand Slam. *Roy wrote a 10-year column titled, "Program for Health," in Hog Farm Management and currently writes a monthly column for the professional journal of the Latin American swine industry. Dr. Schultz received his D.V.M from Iowa State University and his M.S. from Iowa State University. Dr. Schultz graduated in 1993 as co-valedictorian from the Executive Veterinary Program at the University of Illinois. Dr. Schultz is a veterinary consultant for some of the largest swine operations in the United States, Poland, Brazil, Mexico, Venezuela and Japan. He is an adjunct professor in the Integrated Food Animal Medicine Program at the University of Illinois and collaborating professor in swine health management, swine economics and preventive medicine at Iowa State University. Dr. Schultz was awarded "Swine Practitioner of the Year" and "The Howard Dunne Award" for his outstanding service to the swine industry and the veterinary profession.*

Susan C. Reneau is the author of two books, Colorado's Biggest Bucks and Bulls *(1983, 1984, 1987, 1990) and* The Adventures of Moccasin Joe: True Life Story of Sgt. George S. Howard, *1872-1877 (1994, 1996). Susan is editor, production manager and marketing director for a variety of non-profit organizations, trade associations and national publishers, and has completed 38 books for them since 1983. She is a full-time outdoor writer and is producer for a popular syndicated morning drive-time show on politics and fast-breaking news events and a syndicated football show on Super News Talk 1290 KGVO Radio in Missoula, Montana, a CBS Radio affiliate. Susan is co-editor of 10 big-game hunting and wildlife conservation books published by the Boone and Crockett Club, including the all-time records book,* Records of North American Big Game, 10th Edition, *and* Records of North American Sheep, Rocky Mountain Goats and Pronghorn. *Susan holds a M.S. in business public relations management from American University in Washington, D.C., and a B.A. in education and speech communication from the University of Northern Colorado, Greeley, Colorado. She graduated with high honors from both institutions. She has owned Blue Mountain Publishing, Inc. since 1979. Susan received three awards from Associated Press (AP) regarding her live coverage of major news events in Montana from 1996 to 1998 and was a keynote speaker for Associated Press-Montana in late 1997 discussing the importance of freedom of speech and the First Amendment. She is the mother of three sons, John, Robert and Richard, a daughter-in-law, Elizabeth (married to John), and two black Labrador retrievers, Drake and Satie. She and her husband Jack have been married since April 6, 1974. She enjoys volunteering as the city-countywide chairman of Neighborhood Watch in Missoula and was co-chairman of the American Cancer Society's Relay for Life in Missoula in 1998 when more than $130,000 was raised for research and patient care from July 18 to 19. She is also president of her homeowners association and usher in her church, First Presbyterian Church of Missoula.*

1st Convention of Foundation for North American Big Game

February 2 to 4, 1978
Hyatt Regency
Memphis, Tennessee
Organizers: Bill and Wanda Walters
President: Lloyd Zeman, 1977 to 1980

2nd Convention of FNAWS

February 22 to 24, 1979
Thunderbird Motel (Now called Thurderbird Hotel and Convention Center)
Minneapolis, Minnesota
Organizers: John Babler and Jane Berken
President: Lloyd Zeman, 1977 to 1980

3rd Convention of FNAWS

February 28 to March 1, 1980
Galleria Plaza Hotel
Houston, Texas
Organizer: Dr. James "Red" Duke
President: Lloyd Zeman, 1977 to 1980

4th Convention of FNAWS

February 25 to 28, 1981
Hyatt Regency
Phoenix, Arizona
Organizers: Tom Saad and Dean Bowdoin
President: Dr. Roy Schultz, 1980 to 1981

5th Convention of FNAWS

March 3 to 6, 1982
Marriott Hotel
New Orleans, Louisiana
Organizers: Hobart Pardue and Dan Littleton
President: Jerry Christian, 1981 to 1983
Theme: The Hunter and His Partner

Chapter II

FNAWS Grows: Starting to Climb the Mountain 1977 to 1982

DR. ROY SCHULTZ, JOHN BABLER AND SUSAN C. RENEAU

An informal group of friends who loved to hunt wild sheep formed the beginnings of the Foundation for North American Wild Sheep, but this social basis for existence quickly evolved into a serious mission to raise extensive amounts of money to put more wild sheep on more mountains in more remote areas of North America.

Frustration experienced at the start of the 1977 meeting in Dearborn, Michigan, regarding the lack of association with the Grand Slam Club or Safari Club International, faded to lofty expectations and boundless enthusiasm for the new members of FNAWS. In mid-1977, as plans were made for the first convention using the legal nonprofit name of Foundation for North American Wild Sheep (FNAWS), the men listed as members of the Board of Directors for the Midwest Chapter of the North American Sheep Hunters Association of the Grand Slam Club became the first Board of Directors for FNAWS. Fred McMurray and Mike Valencia were added to the Board during the convention.

Minutes from the July 10, 1977, Board of Directors' meeting at the Holiday Inn Downtown in Denver, Colorado, reported that John Babler recommended money from the Dearborn meeting be put into a separate bank account in Minnesota with John in charge of the account as treasurer. Frank Bays (Charter Member 6) was assigned the task of compiling the income and expense records from the Dearborn convention to give to John Babler as soon as possible. Following the Board meeting, Dr. Schultz said they enjoyed dinner with Boone and Crockett Club members and award winners from their 16th Big Game Awards Program in the Denver Museum of Natural History. Co-author of this chapter, Susan Reneau, remembers having dinner with men and women from a new wild sheep organization at the Denver museum. William Harold Nesbitt and Jack Reneau, Susan's husband, were in charge of the awards program; Harold as the director of the North American Big Game Program of the National Rifle Association (NRA) and Jack as his assistant. Harold had attended the 1974 sheep conference at the University of Montana and knew several of the members of FNAWS.

At the July 1977 meeting, FNAWS donated the majority of funds, all but expenses, for the betterment of sheep and sheep management in Mexico with Frank Bays as coordinator of this project with Mexican officials. Dr. Schultz moved to have FNAWS reimburse Lloyd Zeman for all out-of-pocket expenses from his guiding the donated trip to Whitehorse. John Babler

Jane and John Babler were the first two staff of FNAWS prior to 1983. Jane had been hired by John in 1977 to work as a secretary/ receptionist for his new Signdesign business in St. Paul, Minnesota. Sue Burton was also hired by John to work at Signdesign and help FNAWS.
Photograph courtesy of John Babler.

seconded the motion and this was passed by the Board.

Jane, who at the time was Jane Berken, was hired by John Babler in June 1977 as a secretary / receptionist for John's new and growing Signdesign, Inc., a screen process printing business at 55 West Ivy Street, St. Paul, Minnesota at the same time FNAWS was new and growing. Jane immediately recognized and caught the enthusiasm of this small core of dedicated sheep hunters and was struck by John's energy.

"You couldn't help it! I was so attracted to his passion for the Sheep Foundation and wild sheep," Jane remembered.

Jane and John both remembered mailing "stacks and stacks of letters" to sheep hunters they thought would attend the convention that took place in Memphis, Tennessee, from February 2 to 4, 1978. More than 400 people attended. John, Jane and another Signdesign employee, Sue Burton, were the first office staff for FNAWS in its early years. Prior to office space in Signdesign, Lloyd Zeman operated from his home in Des Moines, Iowa, with only the bare necessities including a typewriter, telephone and copy machine, according to Doc Schultz. Tom Hinders of Ames, Iowa, a dear friend of Lloyd Zeman, said he helped Lloyd move from Des Moines and from St. Paul when the FNAWS headquarters relocated to Cody, Wyoming.

John's own business was booming, but he asked Jane and Sue to dedicate large portions of their work day to FNAWS business such as membership processing, meeting planning, typing letters, answering questions on the telephone from members and generally keeping the whole operation going. When asked who kept FNAWS going in that first year, the answer was clear.

"John did. John really did," Jane explained. "I was more the Girl Friday. Type this letter. Talk to this guy. Send him something, that kind of thing. But John really did it. John would ask us to do something."

Jerry Christian paid tribute to the Bablers. "These were two dedicated people. Jane was our secretary. We were working her about 40 hours a week plus she worked at her own business. We tried to pay her with what money we had. They donated space to us (FNAWS)."

John said the space for his business was 10,000 square feet and of that a room measuring

Here is the original 10,000-square-foot Signdesign building and office of the newly formed Foundation for North American Wild Sheep that was located at 55 W. Ivy in St. Paul. Prior to space at Signdesign, Lloyd Zeman conducted FNAWS business in his Des Moines, Iowa, home. This building served as the first FNAWS office from 1977 to 1982.
Photograph by John Babler.

Today, Signdesign, Inc. in St. Paul includes 65,000 square feet. John Babler established his business at the same time FNAWS was founded.
Photograph by John Babler.

18-by-18-feet was the primary FNAWS office for Lloyd Zeman, but Jane and Sue worked throughout the entire office space on behalf of FNAWS. If there weren't typing a letter or answering a phone, Jane and Sue were shipping lapel pins or membership patches.

The enthusiasm of Lloyd Zeman, Frank Bays, John Babler, Duane Smelser, Jim Froelich and Dale Leonard was contagious. Jerry Christian remembered with great fondness the early dedication of all those early founders.

"We all worked together, directors and nondirectors, who attended all meetings. Having the title as president or secretary was just that. We all did what we were best at doing. Duane Smelser and Frank Bays did fundraising. Lloyd Zeman worked with convention and membership. I worked with Paul Asper and Rich Roan on the auctions. John Babler worked on printing, conventions, and Jane kept us informed and kept communication going.

"Last, but far in front, was Roy Schultz. He kept us on course, united and moving us ahead on building a sound foundation on which the rest of the organizers would build. The Board and non-board people were all very dedicated, very bull-headed, extremely talented. When we reached a point of disagreement, Doc would either smooth things over or give us a plan we all could agree on."

Jane reflected, "You couldn't help but be caught up in the enthusiasm of the group of guys who started the organization. They were so full of energy and so excited to get this group of guys together who loved sheep hunting, and to think there were more people out there who wanted to be contacted and get this organization started. It started as a dream. You couldn't help but be caught up in that enthusiasm, just to see what could happen and to see where it would go."

"It was like starting off from scratch. I just got caught up in it. I mean I'm not a hunter. I've never hunted in my life, but I learned a lot. They were so enthusiastic. It was such a down-home kind of group with blue jeans, kick back, let's have a good time kind of a group. You could not help but have a good time with them," Jane reflected.

That down-home attitude remains today as FNAWS members enter the 21st century, but back then, John said every activity by his staff and him was on a volunteer basis. Except for direct costs like postage and printing of letterhead, time and energy were donated. Only later did Jane receive a salary from FNAWS of $100 per week as its executive secretary, starting on April 28, 1979, but not in the first two years of FNAWS operations. John paid her salary.

From the start, FNAWS members believed all money raised at conventions and other fundraisers should go directly to sheep projects, minus direct expenses. Doc Schultz made the motion, and John Babler seconded it, that all checks in excess of $500 require two Board members to sign the check with one Board member signing checks under $500. This business operation was approved by the Board of Directors. The summer 1977 Board meeting was also when Doc Schultz established the charter member category at $100 that included a special insignia or pin and a bumper sticker. John Babler seconded the motion and offered to investigate two types of FNAWS patches, one for the regular member paying $25 per year and one for the charter member paying $100. Only the first 100 people could become charter members. The complete list of the FNAWS charter members is in the appendix of this book.

A policy was established on April 29, 1979, to pay all Board members and officers for their expenses to and from Board meetings including transportation to and from the Board member's house, air fare, hotel, meals and miscellaneous expenses. Bill Walters (FNAWS Charter Member 5) made the motion and M. Francis Stubbs seconded it. To many past presidents and founders who were interviewed for this book, this policy didn't make sense. They wanted the money to go to sheep projects and not to Board members. This debate continues today.

"Mid to late 1977 was the real push for getting members. Three garbage bags full of letters were processed by the firefighters of Fire Station 22 in St. Paul, Minnesota, to help us recruit new members to FNAWS that summer and to get new members to Memphis for our first FNAWS convention," John remembered.

John estimated thousands of letters were mailed that were carried to the post office in St. Paul in 1977. Of the thousands mailed, John said 500 to 700 people responded and joined the new organization. Excitement mounted as each envelope was opened by Jane and Sue. Lloyd Zeman set a goal of 1,500 members by year's end, and Jane, Sue and John were delighted with the response to that summer mailing. John had suggested to Board members in 1977 that for financial reasons he make the first mailing by hand and provide the sorting and addressing. When money was available, John told the Board "some sort of mechanical mailing could be set up."

The money thrown down on the bed of the Omaha hotel paid for the postage to mail letters to future and potential FNAWS members. What excited John the most were the responses received after the mailing from people as faraway as Europe. The first international member of FNAWS in the summer 1977 was Jacques Henrijean of Belgium. Another early FNAWS member who responded to this summer letter was Watson T. Yoshimoto of Hawaii who became Charter Member 27. Both locations impressed Jane, Sue and John because the members lived in such faraway places. The original marketing letter couldn't be found in FNAWS archives. Much of the early correspondence has been lost. John and Jane Babler, though, remember it well.

"We actually made an attempt to start mailing to people all over the Midwest and the two coasts who were really interested in sheep hunting. We had no mailing lists. We'd talk to people who had hunted sheep, and they'd tell us the names of people in sheep camps. We'd talk to them, and they'd recommend other people. At first, it was just through association, Jerry Christian explained. From there, Jerry said, early members went to people like Keith Johnson, the outfitter from Alaska, who helped them locate the names of dedicated sheep hunters who cared about the habitat and wanted to do something positive.

Early members Dr. Roy Schultz and Fred McMurray, left to right, were involved with FNAWS early conventions. *Photograph courtesy of FNAWS archives.*

"We as hunters had taken from the mountain, so we wanted to give back. Our thing was we wanted to give something back. There was no organization at the time that would give back. They would pad their own pockets, but we wanted something that was a nonprofit organization," Jerry continued.

Keith Johnson told how he first got involved with FNAWS: "I met Duane Smelser at another show, and he advised me that I should be in this little clique of people. He tried to explain it to me. I didn't really understand what it was except that it was a group of sheep hunters who got together. I joined and gave him the money for the charter membership at a Safari Club International show. He was collecting money for that group."

Keith became FNAWS Charter Member 98 and assisted other early members to identify numerous wild sheep hunters he had guided over the years. Many founders and past presidents pointed to Keith as a good reason why FNAWS grew so quickly in the early years because of his willingness to contact potential members and donors. Keith later served on the FNAWS Board of Directors.

Efforts to locate other wild sheep hunters who wanted to join was by word of mouth. Early FNAWS members talked to their taxidermists and outfitters. They talked to other hunters who had a passion for wild sheep. The men and women who had attended early meetings were called and written to for suggestions of whom to contact. The reason why people joined FNAWS was similar.

"I was motivated to join because I was a sheep hunter," Craig Johnson explained. He first hunted California bighorn in 1973 and Dall's sheep in 1974 before FNAWS officially existed and was attracted to the men who founded FNAWS because of this underlying love and passion for wild sheep.

"The early years were all projects; transplants or habitat improvement after the Foundation formally began. I wasn't really into politics in the early 1980s," Craig said. Politics was not the focus during the early years, John Babler said. Locating kindred souls who loved to hunt and loved wild sheep of North America was the tie that bound these people who joined FNAWS on the eve of the Memphis convention.

Bill Walters, who became the Memphis convention chairman, provided more insight into why wild sheep hunters joined FNAWS: "I'm an avid sheep hunter. I just love the animal. I was impressed with what Ducks Unlimited did for ducks of North America, and I just wanted to do the same thing for the North American wild sheep. As a matter of fact, during some of our initial meetings we had extensive discussions about whether or not the objective of the Foundation would be sheep of the world rather than sheep of North America. After several meetings and discussions, we decided to restrict it to North American wild sheep because it was just too broad of a directive to go for the whole world and too many governments involved. It would have spread us out too much."

Lloyd Zeman was president of the Board and warned early members like Craig that if money wasn't raised, individual members would have to pull money from their pockets to keep everything going.

Bill Walters wrote, "We had some tough times early on and were on the verge of having to put up money from our own pockets to keep the Foundation running several times when money from other sources showed up in the nick of time."

John Babler served as treasurer and offered his office and warehouse space for FNAWS operations to save FNAWS money. Early members such as Frank Bays, Michael Valencia, Dr. Jerry Waite, Jerry Christian and Dr. Schultz did what was necessary to keep things going and encouraged a successful convention, Bill Walters explained. With regret, Bill admits some of these fine members haven't been active for many years, but he appreciates the guidance and inspiration they gave when FNAWS was in its infancy.

Dr. Waite knew the main reason why he joined and why he became involved from the start: "We wanted to maintain the habitat, to transport animals to historic ranges and to study diseased animals. Doc Schultz and I are both veterinarians, so we were interested in the biologists who specialized in wild sheep."

Bill Walters expressed why so many joined this new group: "I wanted to do something, particularly in range management. During the early years, Canada had the Olympics and they totally devastated a winter range for bighorn. Building the facilities for the winter Olympics, I wanted to be involved with that. The Canadian government seemed to want to propagate wolves. Wolves can have such a devastating effect on wild sheep. Mainly, habitat protection was what I was interested in. I felt like the sheep could take care of themselves if the habitat was taken care of. FNAWS could help by raising money and selling hunting permits. We put sheep on the mountain. We'd take that money and fight to save the habitat and to help the outfitters to perpetuate habitat in their areas. Americans have made a lot of mistakes. I was hoping the Canadians wouldn't make the same mistakes that Americans had made."

After the financial success of the Dearborn, Michigan, meeting, Dr. Waite knew this group of sportsmen were on to "bigger and better things." The Memphis convention, from February 2 to 4, 1978, was filled with optimism. Bill Walters was convention chairman and his wife, Wanda, was his assistant. John Babler remembered 400 to 500 people attended, which was a dramatic increase in attendance over the Dearborn meeting. Spirits were high as hunters and their spouses poured in from all over North America. The mailing Jane, John and Sue had organized in the summer paid off. The first official FNAWS convention took place with hardly a hitch at the Memphis Hyatt Regency from February 2 to 4, 1978. The cost per individual for the Memphis convention was $35 or $50 for a couple.

Leland Speakes, Jr., a true Southern gentleman from Mississippi and FNAWS president from 1998 to 1999, tells the story of how he and his wife, Monnie, as the hospitality suite organizers, provided liquid refreshment for the assembled masses in Memphis, a two- to three-hour drive from their home in the Mississippi Delta.

"Mississippi was a dry state. We were still under prohibition, believe it or not. But, at the same time, we had open liquor stores and state taxes. It was a black market. That's what it

Monnie Speakes and Leland Speakes, Jr., a true Southern couple from Mississippi, first became involved with FNAWS conventions and the hospitality suite when they transported "liquid refreshments" across Mississippi lines to the hotel suite in Memphis for the start of the 1978 convention. This photograph was taken of Leland and Monnie in 1996 only a few weeks before her death. From 1978 onward, Monnie was a regular volunteer at all FNAWS functions until her untimely death in 1996 at age 65 on July 8. Her gentle smile and warm laugh brought happiness to all who knew her. Monnie was born in Holly Springs, Mississippi, and graduated from Mississippi University for Women in Columbus. She married Leland on June 6, 1951, and became active in the First Presbyterian Church of Cleveland where they made their home from 1953 to the present. She was survived by Leland, two daughters, Margaret S. Tuttle and Lucy S. Capocaccia, one son, Leland Speakes III, and a granddaughter.
Photograph courtesy of FNAWS archives.

---

was. We had kind of a distributor and we went down to talk to him. We told him we needed to buy some booze for this party we were doing in Memphis — a few cases of this and a few cases of that. We wanted to take it to Memphis and what we didn't use we'd bring back and pay for it. The guy said, 'Great!' So, we had this old Buick station wagon. We loaded up the back of that thing with the booze and covered it up with a quilt, just to make it to Memphis about 125 miles away. We got there. We drove up to this hotel with all this whiskey in the back of the car. I pulled the bellman off to the side and told him, `This stuff needs to be sneaked into the hotel and taken up to my room.' The bellman said, `Yes, sir. I'd be glad to take care of it for you.'

"He got a four-wheel trailer out there and loaded it all up, covered it back up with the quilt and took it right up to our room. Every day we'd see how much we needed at the hospitality room. We'd carry them down there and open them up and have a little bar. Monnie and I were the bartenders. It was a combination of whiskey and everything. I had some help from a friend of mine, Charlie Gephart. We had a jar on the counter. People would come up and drop in $1 or sometimes they'd drop in $5 and sometimes they'd drop in nothing. It ended up we made $900 on that! We brought back what was left, turned it in to the distributor, paid for it and made a profit."

Jane Babler remembered the early hospitality suites and the effort put into them by such early FNAWS members as Leland. "Oh, those were the days. They still have hospitality rooms, but those hospitality rooms used to go ALL NIGHT LONG. They would supposedly close at like 1 a.m., but, you know at 3 or 4 in the morning you'd still see some hunters up there and some outfitters out there, still talking."

The hospitality rooms at FNAWS conventions have always been the place where friendships were cemented and stories were shared. Bill Walters and Leland were friends from a small group calling themselves the Delta Sheep Hunters made up of 12 to 20 avid sheep hunters. Once again, the tie that bound everyone together was sheep hunting. Bill's early involvement with FNAWS as convention chairman encouraged Leland's involvement. Being a part of the successful movement to protect and increase wild sheep populations was why Leland became a FNAWS member and assisted in making the Memphis convention such a fun success.

"It just makes me feel good to be a part of that. The restorations, the transplants, the enthusiasm of the people for the sheep, and plus, sheep hunters are different. They're different because they always want to be on top of the mountain. For any lesser monies, you never get to the top of the mountain," Leland said.

Lanny O. Wilson, a big-game wildlife biologist from the Bureau of Land Management and FNAWS Charter Member 87, received the first Outstanding Achievement Award in Memphis for his research on wild sheep. Lanny had been one of 77 biologists and wildlife conservationists who attended the sheep symposium at the University of Montana in the summer of 1974, and became one of the early leaders in FNAWS along with others who attended that same monumental program. A second award to famed outdoor writer and wild sheep hunter Jack O'Connor was scheduled to occur, too, but Jack died just prior to the start of the meeting while on a cruise back from Hawaii with his wife, Eleanor.

John Babler said life memberships were established in Memphis at a cost of $500. The winter 1979 FNAWS newsletter reported that a life member received a Fritze original silver-plated pin like those given the 100 charter members, except this pin said "Life Member" instead of "Charter Member" engraved on it. A gold-plated pin was made available to life members at a cost of $75.

The Chadwick ram was given to each life member and a red ribbon on their nametags at conventions identified them. Life memberships brought much needed money to help operate FNAWS. Additionally, FNAWS was classified as a nonprofit corporation with the Internal Revenue Service, which offered such business advantages as nonprofit bulk rate postage rates and encouraged people to make charitable donations since they received tax credits.

Bill Walters compared the Memphis convention with conventions of today: "The crowds in 1978 were much reduced (400 attended). We did have some exhibits and people from all over North America. We had a contingent from Mexico. We had several outfitters come from the Yukon, British Columbia and Alberta. I do remember that our total auction take was about $40,000, which included five very highly coveted Mexican desert sheep hunting permits that at the time on the open market would have gone for $5,000 a piece. We had a very good auctioneer from our area named Rodney Garrison who was equally as good as these guys still used today. We had very few sheep mounts on display, but I brought my bighorn that had won the gold medal in Detroit the year before and my Dall's sheep," Bill recalled.

"We had an outstanding desert sheep ram that we displayed that we called the Ed Stedman (FNAWS Charter Member 67) ram, which was at that time one of the largest sheep to be killed in Mexico. Ed kindly let us use that and he shipped it in for us. (The Stedman ram is ranked 15th in the world in the Boone and Crockett Club's records book, *Records of North American Sheep, Rocky Mountain Goats and Pronghorn*, with an official score of 187-6/8 points. The ram was taken in Baja, California, in 1976.)

"We also had for the banquet on Saturday night a very outstanding Dall's sheep from Alaska. I remember being very upset about that because I had made the arrangements through Bob Fannon, an outfitter in Alaska, to send it to us 10 days before the convention, and on the day before we started, it still wasn't there. I was responsible and I was trying to figure out what to do about this guy's ram. Delta Airlines said they knew nothing about it, and they couldn't do anything about the problem. I was really upset. I kept calling air freight at the Memphis Airport and they kept telling me they didn't have the sheep or anything for the Sheep Foundation. The banquet was about to start at 8 p.m. At 5 p.m. they called my room and said they had the sheep. They wanted to know what I wanted to do with it, so I told him I would pick it up. The sheep had been there the day after the outfitter sent it, but it was stuck back in a corner. It had been in the air cargo area for over a week.

"I had to guarantee that hotel 200 plates for the Saturday night banquet. We got down to a few hours before the dinner after selling advanced sales. We had 160 tickets sold, so we were 30 or 40 short of our guarantee. We had to pay for the 200 plates. We didn't know where to start. There was nobody to help. We just did the whole thing."

Money from the auction in Memphis provided an influx of much-needed cash that allowed the new organization to authorize Frank Bays to purchase a three-quarter ton, 4-wheel drive Ford pickup truck for the Mexican game department. Concern was expressed by several Board members regarding which truck had been purchased but the Board concluded that everything was done properly to insure the proper truck was delivered to Mexico. Keeping track of the incoming and outgoing money was a constant concern for founder and first treasurer John Babler throughout his involvement with the early years of FNAWS.

Other highlights of the Memphis convention included John Suda and Frank Bays treating a busload of members to hamburgers "one fine evening" at McDonalds and a special tour for the women to Graceland, home of Elvis Presley. Frank explained that the young members were still hungry after downing a banquet meal, so John and Frank took the matter into their own hands and directed the bus to pull into a McDonalds. More than 200 hamburgers were ordered and paid for in cash by John and Frank. The hamburgers were eaten back at the hotel hospital room, much to the merriment of the assembled group.

Bill Walters was in charge of the ram scoring clinic and continues today in that volunteer position along with other FNAWS members. He was also in charge of the ram contest in Memphis and for years to come.

The Board meeting at the convention included a discussion about future Board members. Names bantered around included Paul Asper of Pennsylvania, Bill Butler of Montana, Dr. James Duke of Houston, Butch White of Colorado, Lloyd Bare of Iowa, Fred McMurray of Michigan and Dr. Jerry Waite of Iowa.

"The main question is should we have Board members from widespread areas for a geographic membership draw or proven workers even if they come from an area where we already have Board members," minutes to the Board meeting recorded. The discussion was tabled, but this discourse continues today.

FNAWS was well on its way to becoming a financial powerhouse after the Memphis convention, and Jane, John and Sue continued running the day-to-day workings of the organization.

Bill Walters remembered being exhausted after the Memphis convention and evaluated the event: "We were not very well organized and did not really know what we were doing, but I was really surprised at the number and quality of people who showed up."

The Board of Directors met May 6, 1978, following the Memphis convention to analyze the successes and failures of their FNAWS first convention. Board members thought the registration table should be separate from the table selling raffle tickets and memberships, and the Board meeting room should not be in the hospitality suite. President Zeman recommended a lower pre-convention registration fee to encourage higher early registration. The Board agreed there was much to do to improve the convention, but this was a good start.

John Babler reported that cash on hand and in the checking account was $4,505.68. The savings account contained $11,397.85, and John said FNAWS spent $1,314 for two radios for the Mexican trucks. John Babler was given authority by the Board to spend $600 for a machine to address envelopes called "Master Addressers" rather than spending $1,000 for an automatic address machine.

Plans for the next FNAWS convention at the Thunderbird Motel (now called the Thunderbird Hotel and Convention Center), across from the Minneapolis International Airport in Bloomington, Minnesota, began immediately after the close of the Memphis meeting. At the August 1978 Board meeting, John Babler as treasurer reported they had $2,443.27 in their checking account and $5,574.64 in savings. That's not much money to plan a major convention, he noted. By December 1978, when the Board met at the Thunderbird Motel in Minneapolis, and after a marketing mailing had been organized by John, Jane and Sue, the bank-accounted assets totaled $8,746.37. To increase revenue to FNAWS, John Babler announced at this meeting the sale of belt buckles for $7 and pins for $11.75.

Obtaining the tax-exempt status with the IRS was a lively topic of discussion because such a status allows an organization to use lower postage rates to market their upcoming convention and fund-raising efforts. Paul Hietbrink of Swift, Brown & Winick was hired to work with the IRS on securing the tax-exempt status. In November 1978, the tax exemption was granted.

The first FNAWS newsletter was published by Lloyd with help from Jane on August 19, 1978, announcing details of the upcoming Minneapolis "shindig" to take place February 22 to 24, 1979. Jane remembered how the newsletter was produced in the early days. "Lloyd Zeman would write out the script that would describe what was going to be discussed in the newsletter, and then I would kind of fine-tune it, and type it again before it went to the printer. Lloyd would write the copy. He was the copy writer of all of that. He would write all of the auction lists, too. The little blurb for the auction list was written by Lloyd — what it was, who donated it. Then, we'd type all those auction lists, too, and then send those out to a printer. Then, all of that would come back to us and then we would type it again. It was a real team effort, and fun."

Significant in the financial development of FNAWS in Minneapolis was the first-time use of an exhibit hall was begun, thanks to the organizational skills of Jane Berken, who on October 5, 1979, became Mrs. John Babler. Cost per 10-foot booth space was $45, Lloyd reported, but if an exhibitor donated a hunt or other significant item for the auction, they received free exhibit space. This was proposed by Fred McMurray and agreed to by the Board of Directors. The exhibit hall also included a special section where guides and outfitters displayed their brochures and pictures.

"Notify John Babler if you want to take advantage of this," Lloyd Zeman advised in the FNAWS newsletter in 1979. "This year there's going to be a special outfitters section where they can display their brochures and meet with hunters. If John isn't there, his very capable secretary and assistant, Jane Berken, can undoubtedly help you!"

That "capable secretary and assistant" explained how the job was done and how they got people to sign up for that first exhibit hall: "We sent out letters saying that we were going to have this convention. I haven't a clue how many people received that letter. We contacted people and wrote letters to people whom we thought would want to exhibit like artists and outfitters we knew about. Anybody whom we thought had anything to do with sheep hunting or could sell something that would interest sheep hunters and would want to pay. I don't recall how many exhibitors there were in Minneapolis, but maybe 15 or 20."

Lloyd Zeman wrote in the winter 1979 issue of the FNAWS newsletter, "This year's event is going to be one heck of a do! We expect more than 500 people, maybe as many as 600 the largest gathering of sheep hunters, guides and outfitters and sheep enthusiasts ever to assemble."

The purpose was to get these men and women together who loved wild sheep and the habitat they lived in. "That's really what they wanted to do. That's all they wanted to do," Jane explained with fond memories.

To this day, Dr. James "Red" Duke remembers the success of Minneapolis. When he planned the Houston convention in 1980, Red called Jane to learn how she did it. So, how did the convention in Minneapolis come together? Jane tells it in her own words: "For Minneapolis, we contacted a convention bureau here in St. Paul. Actually she was in Minneapolis.

"I said, `Come over and tell us what we can do with these women.' I went to the convention bureau, and they put me in touch with a woman who did just that. She would plan activities for conventions in town. She kind of tossed some ideas out and then, I think we did say, `Okay, this is what we want to do. How much will it cost?' Then she would say, `We can do this, this and this and it will cost you that amount.' So we passed that cost on to the women. We'd say, `Ladies, this is what we can do. We can take you shopping. We can have lunch, and we can do this for $10, or something.'

"We would have buses that would pick up the women. Take them out to lunch. Let them go shopping. They thought that was GREAT. That's what we ended up with for the ladies. Mall of America didn't exist then. I think it was downtown Minneapolis. The ladies spread out. We had some drop-off points and then they'd be told there was this store, this store and this store in this area. Then we'd tell them to meet back at the bus at 11 a.m. and then we'd go and have lunch somewhere. I can't remember where." Minutes to a FNAWS Board meeting in 1978 prior to the Minneapolis convention indicated the women toured the Betty Crocker Kitchens and shopped at Crystal Court.

Jane continued with her memories: "Back then, the convention would start on Thursday and end on Saturday, so that must have been a Friday afternoon event for the ladies. The guys would be talking sheep hunting, and they'd be doing their thing, and we took the ladies and stuck them on the bus and took them shopping.

"We'd always planned a Friday night activity for the entire group. We hired buses and we went over to a place called Diamond Jim's Cafe. They had gals in cute little outfits who would swing from big swings from up in the sky. There was a stage show and we had a dinner. We sold tickets to that. We had a limited number of tickets.

"Diamond Jim's is still there today, but they're not operating as Diamond Jim's anymore.

They don't do the stage show. None of the fun stuff. We went and had dinner and they put on such a great show. We had such a great time. I remember calling this woman who had arranged it FRANTICALLY on Saturday afternoon and saying, 'Can we take more people?' Everyone wants to go."

Jane and John remembered about 200 people attending the convention in Minneapolis, but FNAWS records show 535 attending the various functions. Jane laughs, "The Bablers are very poor at dates and numbers."

Jane remembered the woman planning the dinner theater saying, 'Okay, we can take 30 more people, so maybe like 15 more couples.' It was just wonderful. I think everyone who wanted to go could go. I think we tried to accommodate everyone. We did not know if it

Harvey Kadlec at an early FNAWS auction.
Photograph courtesy of Tom Hinders.

would be a success. We thought these guys were going to get together, hang out, and we'd never get them away to do anything else. I couldn't have cared less, but I wanted these women to have a good time. You can just listen to sheep hunting stories so much and then you just need a little break.

"I figured they'd just have to grab the guys and say, `Look, honey, this is where we're going and we're all going.' That was always a really fun thing to do. That tradition stayed with the Sheep Foundation for a number of years," Jane concluded.

Because conventions in recent years have become so large, the tradition of shuttling people by bus to special events has ended. Jane explained that when you're feeding 3,000 people at a Saturday night banquet, you just can't shuttle them by bus.

"The numbers now are phenomenal. Changing convention sites, too, was something they did every year because we were small enough that people from Alberta and Saskatchewan and wherever else who would bring their wives wanted to do something in a different spot every year. That's why it moved every year. The numbers were so small that we could move it.

"Now we can't. We make up for it by having the ladies' luncheon, which has been really successful. As far as getting everyone together to go and do something fun, I guess one of the last places we did this was probably Phoenix in 1981, where we had busloads of people going out to the desert and having a barbecue. I'm thinking that was the last time this happened."

When Jane talks about "we" she means a group of volunteers who helped in Minneapolis and at other conventions. She recalls: "Sue, John and I would ask for volunteers from the Foundation. Women in the group helped. They were wonderful about helping out. Sandy Christian, wife of Jerry Christian; Sue Kadlec, Harvey's wife; Jan, Doc Schultz's wife; and Jan Froelich, Jim's wife would help out a lot. She's wonderful. A lot of that group were. We did a lot. We recruited a lot of women to work at the registration desk. We'd say, `Come and work at the registration desk just for an hour or so.'

"In Minneapolis, my mother, my sister-in-law and John's mother helped out. Gladys Noyes is my mother. John's sister is Linda Rulli. John's mother is Josephine Rulli. We had 17 exhibitors in Minneapolis! We had 500 for the banquet in Minneapolis! We also had Sue Burton and me, and a whole bunch of wives. We'd all run the registration desk, and sell memberships and raffle tickets.

"When you were at the convention in Reno in 1998, there was a Sheep Foundation booth and a registration booth. Back in the old days, all that was one thing. You were selling a day pass one minute, and a book the next. Oh, my goodness! And, we wore high heels because we wanted to look CUTE. So we'd be soaking our feet in cold water every single night sitting on the edge of the bathtubs in our rooms."

Minneapolis was also the first time the FNAWS membership was introduced to Rich Roan of the Bob, Chuck & Rich Roan, Inc. Auctioneers & Appraisers of Cogan Station, Pennsylvania. Rich and his family continued through 1998 as the auctioneers of the FNAWS annual auction. In comparison to the Minneapolis gross receipts, the 1998 FNAWS convention grossed more than $3.5 million! How things have grown.

Rich admits today that when FNAWS members encouraged him to continue after the Minneapolis convention, they promised his talent would allow FNAWS to raise millions of dollars for wild sheep. At the time, Rich didn't believe them and was unsure about continuing after Minneapolis.

"They asked me to go on to Houston, but I told them I didn't want to get too involved.

They wanted me to go. Mike Valencia, a really nice guy, was from California and a friend of Paul Asper. Mike was all excited because he kept saying we'd take in a million dollars if I'd stick with it. I thought he was talking through his ear, you know. But, he wasn't. Mike and Paul were the ones who convinced me to go to Houston."

Minneapolis was the first time the FNAWS membership was introduced to Rich Roan of the Bob, Chuck & Rich Roan, Inc. Auctioneers & Appraisers of Cogan Station, Pennsylvania. Here he turns to Harvey Kadlec as Jerry Christian, right, listens. *Photograph courtesy of Tom Hinders.*

Paul Asper of Pennsylvania (FNAWS Charter Member 95) and Mike Valencia of California (FNAWS Charter Member 7) recognized Rich's talent and wouldn't take no for an answer from him.

"Paul was from my part of Pennsylvania and was on the national FNAWS Board at that time. I told him I didn't want to get too involved because I had my own auction company and five children in the business. Of course, we knew about Ducks Unlimited back home.

"We already did Ducks Unlimited and other charity jobs as auctioneers as well as our own company business. We do household goods, antiques, estate sales and our own auction gallery. We do small and large auctions of all kinds. Here I am, still involved with FNAWS all these years later," Rich recalls with a smile.

"At the Minneapolis auction, they netted more than $42,000, which at the time was a lot of money. (The exact amount of profit from the convention was $42,985.48.)

"At that time, that was a lot of money for as few people as there were there. It seemed like there were less than 200 at the auction. The Thunderbird Inn or Motel was a pretty good-sized place; a nice place," Rich remembered.

"We went from there to Houston, and we maybe grossed $108,000. From there we went to Phoenix or New Orleans, I forget now. Then we went to San Diego, Hawaii, Reno and all that," Rich said. "I did so many charity auctions then and I do so many auctions now. I do a lot of charity auctions now for Ducks Unlimited, cancer, etc. I'm always happy to do them, but at that time I was doing these auctions alone. Then I went on to Houston, and we went downtown to the Galleria, I think. That's where we made $108,000 gross, I believe."

His 19-year history with FNAWS almost didn't happen for Rich because he had quite an adventure getting to Minneapolis. He relates the details of his trip to Minneapolis from Pennsylvania.

"Paul (Asper) was supposed to be there at the same time with me and, of course, he went to Africa to go hunting. I ended up going to the convention with his wife, Carol, and her girlfriend, Barb Whorl," Rich explained.

"What happened was we left Pittsburgh on the plane and, of course, the plane had trouble, so we ended up in Columbus, Ohio, where we stopped to change planes. We had to wait for a plane coming out of Florida. It took us nine hours to get there because of waiting and bad weather."

"We arrived in Minneapolis and there was rain and ice and snow. We got into a cab and arrived at the Thunderbird Motel in Minneapolis. We drove into the parking lot and had a little accident with another car because of the snow and the ice. We got into the motel and there were no rooms for us. Our reservations were all messed up. We hadn't had any food for 10 or 12 hours. It was really quite a mess, but we ended up quite successful with $42,000," Rich said.

Harvey Kadlec, FNAWS Charter Member 57 and FNAWS president in 1989-1990, worked for 15 years as the member behind the scenes of the auction organizing all the auction items, making sure items were ready for Rich to auction and printing a list of auction items for the bidders and auctioneers to follow. Harvey's involvement with the auction began in Minneapolis.

Harvey recalled his involvement: "I was asked to help out with the auction, and then it ended up I did the auction for the next 15 years. We went to Minneapolis, and John Babler was in charge of the convention there. He asked me if I would do the auction, so I did it. It just kept on going, and I kept on doing it. I'm the one who put everything together and cataloged it and gave it over to the auctioneer, Rich Roan. My son, Bradley, did it a couple of years and I helped him. He did the auction when I was president."

Harvey speaks with great appreciation for Rich Roan's abilities to get the crowd bidding. Rich's auction antics began in Minneapolis. "When permits went for a lot of money, Rich Roan used to jump on tables and everything else to get people to bid. Some people didn't like that because they thought it was unprofessional, but it gave color to the auction, and he brought in a lot of money. I just like the way he does the auction. He puts color into it. I've been to auctions, and they stand up there like a recording. Rich over the years knew the people. He knew who the bidders were. He could get money out of people when they were done, but they weren't done."

"He'd say, 'Come on. I'll throw in an extra $10 or $100 if you'll just bid once more.'

"Usually, he didn't have to because that guy would bid once more, but the guy who wanted it would bid again. Rich would just keep it going," Harvey remembered.

Rich Roan returns the compliment to Harvey for the work Harvey did for years as the "behind-the-scenes" man.

"The auction is a big job to get it prepared and to get it catalogued. Harvey Kadlec was one of the first and then his son followed through. That's important. Now the auction is more involved because it's bigger, but that early group of people were the ones who got it all going to make it so successful. What makes the auction such a success is all the people involved in the background. I'm not even sure who does all that today and gets that all arranged, but those are the people who are important. It's a lot of work," Rich said of Harvey and Brad Kadlec.

When Rich first told his neighbors he was an auctioneer for FNAWS, nobody knew what he was talking about. "Even now you have to explain to people what FNAWS is. You say this is the Foundation for North American Wild Sheep, and they question what I mean by wild sheep. Of course, they think of tame sheep. We have to explain to them about a bighorn and the conservation end of it, which is really good, and the hunting end of it. A lot of people do not know there are more wild sheep that die from natural causes than from being shot by hunters."

Auction aside, the Minneapolis convention chaired by John Babler with assistance from Jane and Sue was a success for this young group of wild sheep fanatics.

Highlights of the Minneapolis convention included the naming of Lowell Baier (FNAWS Charter Member 14), who had worked tirelessly to uncover and preserve the National Collection of Heads and Horns in 1977, as recipient of a special award for his efforts.

Jack O'Connor's son, Brad, accepted a special award for his father who died in 1978. Jack was recognized for his many accomplishments as a writer for the magazine *Outdoor Life* and author of numerous adventure books. Lloyd wrote in the winter 1979 FNAWS newsletter:

"Jack stirred Ovis pyrexia in most of us with his many inspiring articles in *Outdoor Life* about the wild sheep we all hold in such high esteem."

Dr. Valerius Geist, associate dean and professor of environmental science at the University of Calgary, was presented the Foundation Achievement Award for his contributions to wild sheep and their management. Dr. Geist was one of many wildlife biologists to attend the University of Montana conference in 1974, and continues to be distinguished for his knowledge of animal behavior and wild sheep of North America. His chapters grace this book.

Dr. Schultz was in charge of the program. He wrote, "In those days, we had hunters report and show slides of hunts they had purchased at the previous year's auction. This gave exposure and free advertising to the outfitters who donated hunts. This practice continued for several conventions under my guidance."

Some of the speakers at that convention included Garry Vince who was elected to the FNAWS Board of Directors and later became FNAWS president. Russ Reid, an outfitter from Wyoming and father-in-law to B. Joe Coy, told of sheep numbers and hunting conditions in Wyoming. B. Joe continues as an active outfitter participant and ring man during the auction.

Dr. Waite was crowned leg wrestling champion, and seven new national Board members were elected. The new Board members were Dr. Waite, Fred McMurray, Dr. James "Red" Duke, Jerry Christian, Paul Asper, Lanny O. Wilson and Francis Stubbs. Jane Berken was named as the FNAWS executive secretary. Notes from the Minneapolis convention Board meeting show that FNAWS helped the Nature Conservancy purchase the Dome Rock land in Colorado, wintering and lambing grounds for bighorn sheep.

Details about the accomplishments of the Iowa Chapter of FNAWS will be discussed in other sections of this book, but the Minneapolis convention was the official start of the Iowa FNAWS Chapter, thanks to a proposal by Dr. Waite of Fenton, Iowa. This was the first state chapter to be established. "It is thought that the local chapters can have a snowballing effect on national membership, increase participation at the local level, improve fraternalism by putting more fun back into smaller local meetings," Dr. Waite wrote in the Spring 1979 FNAWS newsletter. "Local chapters could get together, swap stories and show slides on a more personal basis. After all, this is how the Foundation got started."

FNAWS President Zeman and the rest of the national Board were in great support of the chapter concept. Lloyd wrote, "The chapters can be local, state or regional in nature, will need to follow bylaws set up by the National Foundation, be approved by the Board and make annual reports to the Board. Get started now. Be one of the first to organize!" Lloyd advised.

The Minneapolis convention ended with gross receipts exceeding expectations. John Babler reported to the Board on February 25, 1979, that gross receipts for the auction totaled

$48,200 and gross receipts from convention registrations, booth space rental and other convention sales totaled $59,000. "Guesstimate of net profit was $30,000," John concluded.

At the conclusion of the Minneapolis convention, plans began for the 1980 convention in Houston, Texas. Dr. James "Red" Duke, known to current FNAWS members as master of ceremonies for many Saturday night banquets and FNAWS president in 1986, was named as the Board member to head the Houston convention since that was his hometown. Dr. Duke presented a short movie and an oral presentation about Houston and possible hotel sites. He asked for volunteers to come to Houston and visit the hotel sites before the contract was signed. John Babler and Frank Bays stepped up to the plate.

Changes in the FNAWS Articles of Incorporation took place before the Houston convention on July 14 and 15, 1979, when the Board voted unanimously that no FNAWS president could succeed himself more than one consecutive term and that the president-elect would assume the office of presidency after his term ended. The office of first vice president became president-elect. Dr. Schultz was nominated by Lloyd Zeman to succeed him as president.

Prior to the Houston convention, as plans were forming, Board members decided to charge $90 per individual and $150 per couple for registration fees prior to the convention and $10 more per registration if individuals and couples sign up at the convention. That price included the following services: admission to all convention programs, movies, exhibits, displays, auctions, hospitality room parties and five planned meal functions and banquets.

A meeting in April 1979 included a report by Red Duke saying that Herb Klein's Boone and Crockett Club trophies might be on display at the convention.

Jane Berken has fond memories of her work with Red Duke as plans formed for the Houston convention set for February 28 to March 1, 1980, at the Galleria Plaza Hotel. Red selected as his convention theme, "The Hunter and his Role in America."

"Red Duke is a fun guy. Red Duke and I spent hours and hours and hours on the phone together planning the Houston convention. We put the whole thing together in Minneapolis and did all the planning and decided where it was going to be, and what were we going to do with 300 ladies in Minneapolis in February when it was 'freezing-our-butts-off' cold. So we kind of planned that whole thing, so then it was Red's turn to host the Houston convention, and of course, he just kept calling me all the time. We spent lots of hours on the phone.

Never had met the man in my life until one of the first Board meetings in Houston prior to the convention. And, you know, how can you NOT fall in love with him. He's just adorable," Jane recalled with a chuckle.

Jane gave him lots of information and remembered him wanting to know how the heck they pulled off Minneapolis. "How did we do it? What were the logistics? What did we do?" Red asked.

Red Duke has recollections about the Houston convention, too:

"We had a lot of funny stories about FNAWS coming down to Houston, like the time we hauled them around in a pickup truck. We were down where they wanted to eat Mexican food, and I knew the places in the business area would be full, so I took them out eight or nine miles where I lived to Ninfa's — a great restaurant chain — with everyone in the back of the pickup.

"We got out there and my beeper went off. I didn't know what it was, and I almost didn't answer it. But, then I did and this young man said, 'Dr. Duke, I'm glad you called.' He was excited. 'Sara has been in a wreck.' Sara was one of the daughters of a man that used to come to the FNAWS meetings at the time.

"I said, 'Where is she?'

"He told me the name of the street. We were at this shopping center that was adjacent to Memorial Hospital where she was supposed to be. She wasn't supposed to be off campus but they had gone down a street perpendicular to Echo Lane, which was what we were on. The three girls in the car were watching the football players work out. The girl driving ran the car into a tree and broke Sara's ankle.

"I told the guy to just sit tight and I'd walk over there, which was about 400 to 500 yards away. Most of the paramedics I had trained, so we put her in the car and sent them back.

We went back and did our rat killing and got Sara's leg operated on," Red explained.

Rich Roan, FNAWS auctioneer, has memories of the Houston convention and some of his secrets for getting the bidding going.

"I'm not sure I really have secrets. I guess I had a good teacher in my father who was in the auction business. I just picked up some of his techniques. I'm not sure what I do to get people to bid. I've been known to tell a bidder I'd throw in $10 if he would just keep bidding. I sometimes make it sound like I'm ready to close out the bidding. If somebody didn't jump in right away I'd do that. I created that but it's sort of a monster. It takes a little longer, no a lot longer, to sell an item, but I think I have them used to that. Sometimes I wish I wouldn't do it that way but it's something I do as a fundraiser for FNAWS.

"Every bidder is different. You have personal dealings with them. You know who they are. I'm terrible about remembering names. The techniques I use are to draw them out of it. I know they're not done. A couple of times I have sold short because a bidder held on too long, and then I sold it. At one of the Saturday banquet auctions a long time back I said going once, going twice, third and final call, and I sold it. The guy was very upset because he waited too long. But, I just said that's the way it is. I didn't intend to sell it short, but that was the way it was. I had to do something and I had to draw a line somewhere, so I did that.

"The only thing that gets more complicated is that I get older. It is a very physical job. What I do when I get there is get up every morning and go to the health club to work out. I get a massage. I rest. It's nice to have a cocktail now and then but I stay away from that.

"We originally started at 12 noon and then it went to 1 p.m. On Friday we started at 1 p.m. A traditional auction is where you auction things for one or two hours. The FNAWS auction goes on for five, six or seven hours each day! In the auction business like we do at our gallery back home we sell an average of an item a minute in the antique business. Here, you're doing 80 or 90 items."

Many of the early founders and past presidents interviewed for the book point to Rich Roan and his family for helping make the auction successful and interesting, especially during those early auctions.

Attendance records were broken in Houston with a count of 800, and founders were amazed at the enthusiastic embrace of FNAWS by wild sheep hunters, guides and outfitters, manufacturers, auction bidders and wildlife management professionals from across North America.

Awards were presented to the Wyoming Game and Fish Department and Director Earl Thomas for their outstanding support of FNAWS and continued wildlife research that benefitted wild sheep. The first Wyoming hunting permit was offered for auction in Houston. High bidder was Dr. Vogt for $23,000. Unlike other states and provinces, Wyoming gives FNAWS a hunting permit to auction each year with no strings attached. The tradition began in Houston and continues today. Over the years, the Wyoming permit has added millions of dollars to FNAWS so they can "put more sheep on the mountain."

Jim Froelich was named "Grandfather of FNAWS" as special recognition for all his efforts on behalf of the early organization. Jack Lee was honored for his dedicated service to hunters and to the guiding fraternity. Gary Swanson was selected as the artist for the Foundation's first Wild Sheep Conservation Stamp program that was established by Texan M.F. "Bubba" Wood. The 1980 summer issue of the FNAWS newsletter reported the stamp program was "a fantastic success." Lloyd Zeman said FNAWS' 22 dealers sold out of the limited edition prints in a matter of weeks. One dealer reported he sold his entire allotment within eight hours after they started accepting orders. Prints were listed as $100 each, but some sold for upwards of $300. The original bighorn sheep painting by Gary was auctioned at the Phoenix convention in February 1981.

Bob Kuhn was selected to do the second painting of a Dall's sheep. The Wild Sheep Conservation Stamps sold for $5 each from the 22 dealers or from the FNAWS office. The stamps were in sheets of 30 and each were numbered with their own three-digit block number. The book Dr. Schultz compiled and wrote, *Portrait of Sheep and Sheep Hunting*, was published by FNAWS and sold at the Houston convention. At the Houston Board of Directors' meeting on February 27, Fred McMurray commended Dr. Schultz for his "excellent work on the project," and the rest of the Board offered their congratulations.

FNAWS archives provided details about the first FNAWS book project.

The book project was first suggested by Dale Leonard of Littleton, Colorado. The book was edited and completed by Doc Schultz with the help of Marilyn Maltby, layout artist for John Babler of Signdesign, Inc.

The first printing of 3,000 was sold out in a year, and the second printing of 2,000 was sold out in two more years. The original cost was $6.78 per book and it sold for $20 each. The book has been out of print and stock for more than 10 years and commands $65 to $300 per copy in the secondary book market.

All of the profits from the book were donated to the Sheep Foundation for research, habitat and sheep transplanting projects.

The book was dedicated to those who have labored, and given of their time, talents and monies to perpetuate and expand the wild sheep populations.

Doc Schultz wrote in the introduction to the book, "Sheep are high, wild and free — often in pristine areas to which modern hunters can escape and once again get in touch with their inner selves."

On display were the late Herb Klein's Grand Slam, which included his 198-1/8 point bighorn taken near Saskatchewan Lake in Alberta in 1965; his 187-3/8-point desert sheep taken in Sonora, Mexico, in 1952; his 174-6/8-point Dall's sheep from the Sifton Mountain Range in Yukon Territory in 1950; and his 170-6/8-point Stone's sheep near Gataga River in British Columbia in 1963. Herb's grandson, Jimmy Deere, made the arrangements for the display with the Dallas Museum of Natural History that houses Herb's Grand Slam. The Harry L. Swank, Jr. Dall's ram, the World's Record with a score of 189-6/8-points, was also on display in Houston. The Swank ram was taken in 1961 in the Wrangell Mountains of Alaska.

"Honkytonk" invaded FNAWS in Houston. Lloyd wrote in the winter 1980 FNAWS newsletter days before the convention, "You'll need to grab a bite before boarding buses to go to Gilley's — the world's largest 'honkytonk.' Lloyd provided details of the event in a flier dated January 15, 1980, promoting registration by saying Gilley's was featured in the movie with John Travolta, "Urban Cowboy." The evening before the convention, the crowd witnessed Kathy Lee Crosby being instructed on how best to ride the bull for the movie, "Urban Cowboy." All the "fearless" FNAWS cowboys, outfitters and hunters rode the mechanical bull, much to the delight of all those who didn't have the nerve to mount the "bull." A bus to

Gilley's cost $5 before convention and $6 at convention, but food and drink was "Dutch treat," so everyone paid for themselves.

Doc Schultz bet Russ Reid he could ride the bull. Russ said he couldn't. Doc Duke said he was "crazy" as that was where he got a lot of patients. Doc Schultz went up to ride the bull and slipped the bull operator a $10 bill to insure the good doctor received an easy ride. Doc Schultz rode the bull but after he dismounted the operator swung the bull, hitting Doc in the butt and sending him sprawling across the mattresses that covered the bull's medal post. Russ Reid paid off his bet.

Friday night was spent attending the Houston Livestock Show and Rodeo in the nearby Houston Astrodome at a cost of $10 per person before convention and $12 at the door. An evening of wrestling also provided a night's worth of entertainment during the convention.

Winner of the men's leg wrestling and men's arm wrestling was Ray Woodward. Sue Kadlec was the ladies' leg wrestling champion. A Code of Ethics for FNAWS members and Board of Directors was adopted by those attending the Houston convention based upon the concerns expressed by John Babler at a special meeting of officers and Board of Directors at the Hilton Inn in Des Moines, Iowa, on Sunday, May 27, 1979, more than a year before the Houston convention.

The Code of Ethics is in effect today as it was when it was officially adopted in June of 1980:

"The Foundation shall adopt a Code of Ethics as the official Code of the Foundation, dated June 6, 1980. No officer or member of the Board of Directors of the Foundation shall exploit that position in an effort to obtain a hunt or other item not available to the general membership. Any 'gift of appreciation' over $100 in value to a director, officer or employee of the Foundation is the property of the Foundation and comes under control of the full Board of Directors.

"All members: I pledge myself to honorable and ethical participation in the conservation of wild animals and the sport of hunting. Emphasis will always be on sportsmanship and fair chase.

"Officers/Directors: (Pledge to be taken by incoming Directors): Recognizing the prestige, influence, and sensitivity of my position as an Officer and/or Director of the Foundation for North American Wild Sheep and of the trust and confidence placed in me, I subscribe to this 'Code of Ethics.'

"Directors will not consciously bring discredit to the image of the Foundation by any of their deeds. Directors shall not use the influence of their position to acquire for anyone, goods, services, favors, etc., on terms more favorable than those available to the entire membership of the Foundation in the ordinary course of business. Any violation of this code will be referred to the Ethics Committee of the Foundation for appropriate action."

Following the Houston convention, John Babler reported to the Board on February 27, 1980, that 761 people were regular members of FNAWS along with 23 life members. There was one unnamed corporate sustaining member. Dr. Schultz volunteered to approach other wild sheep organizations to become affiliates of FNAWS.

Life memberships remained at $500. Corporation supporting memberships were $500 per year. Family membership was $50 per year. Sustaining membership was $100 per year, and regular membership was $25 per year.

At the end of the 1980 convention, an out-of-print brochure listed numerous wildlife conservation projects as funded by FNAWS including money to the New Mexico Department of Fish and Game to capture and save scabies-infested sheep; money to the Yukon Wildlife

Branch to capture rams with diseased horns and to detect and analyze the problem; publication of a book on wild sheep of North America; funding to Dr. Thomas Bunch of Utah State University to study the causes and possible cures of frontal sinusitis in wild sheep; a study of desert sheep on Tiburon Island, Sonora, Mexico; money to Dr. Robert McArthur of the University of Calgary for heart rate stress and monitoring of wild sheep; funding for Dr. Jack Turner for water economy of desert bighorn sheep; a heredity study and transplant project; money to the Nature Conservancy to procure the Mueller Ranch in Colorado that is prime wintering range for the Pike's Peak bighorn herd; money to REAL Alaska Coalition for better land use and hunting as a game management tool on federal lands; and three 4-wheel drive, radio-equipped vehicles donated to the Mexican government for use in sheep management and protection.

Lloyd Zeman became the first executive director of FNAWS for one year with an option by the Board of Directors to renew his contract each year. Lloyd was hired by a unanimous vote by the Board of Directors. The Board voted to reimburse Lloyd $4,000 for January and February 1980. Dr. Schultz became the second FNAWS president. Jerry Christian became president-elect, and John Babler continued as treasurer.

A sense of great accomplishment flowed through the 1980 Board of Directors when it was reported by John Babler that the Houston convention netted a profit of $175,739! On June 6, 1980, the net worth, including inventory at cost, was $132,244. John estimated a budget of $60,000 to operate FNAWS for a year.

The seeds of financial planning were planted in 1980 when John Babler made a motion to set aside 20 percent of available Grant-in-Aid money each year for future habitat improvement projects not to exceed a fund of $100,000. This motion was seconded by Dr. Waite and approved by the entire Board. An additional $15,000 was set aside for future Wyoming projects and another $15,000 for British Columbia projects. Much of the summer 1980 Board meeting concerned ways to spend the money earned at the Houston convention for a variety of worthwhile wildlife conservation projects.

Dr. Duke pushed for an organized method of selecting wildlife management and research projects so money collected wasn't wasted.

"The only reason I agreed to be president was because for several years Dr. Schultz and I had reviewed all these applications for money. He and I are maybe pseudo-scientists, but no sheep biologists. It became evident that over and over again, this organization had been funding the same thing, which was stupid. I could see the idea that we would develop the Center for Sheep Studies. My thought was that we would use that center as a place to get the best brains together to ask the important questions and then take what money we had to fund appropriate studies with RFPs (request for proposals)."

Red Duke's and Roy Schultz's ideas for funding wildlife projects continue today as modern FNAWS members wrestle with the age-old question of how money collected from FNAWS's auctions and fundraisers should be spent. No Center for Sheep Studies was ever established, but FNAWS has funded research facilities.

Jane Berken reported total membership of FNAWS to be 971 in June of 1980, prior to the Phoenix convention. Jim Froelich moved, and the entire Board passed, a motion to mail the pre-convention marketing material to the entire mailing list to promote the upcoming convention.

Board members voted that only dues-paying FNAWS members receive all mailings, including the popular newsletters. The early newsletters written by Lloyd and typed by Jane Berken were an important source of communication for the fast-growing organization. Jane said the design was simple but the content of the newsletters was detailed. She laughed when asked about the numbering system for each issue of the early newsletters.

"I was always mystified at where Lloyd would come up with those numbers and letters to call each newsletter by volume and number, because when I would type them, he would write by hand, type a little, and cut and paste and lick and throw this thing on my typewriter back in those days.

"He'd say, 'Can you rough draft this before we send it to the printer?'

"So, I would retype the entire newsletter. And then, it would go to the printer from there. This always mystified me. I'd ask Lloyd, 'How did you decide that this was going to be Vol. I, Edition IV, or Vol. II, Edition III, when I don't remember Vol. I, First Edition?'"

Jane never did comprehend Lloyd's unique newsletter numbering system. As for the conventions prior to 1982, John, Jane and Sue were in charge of making sure everything ran smoothly in FNAWS. Jane explained:

"We did all the registrations through our office, and then, when we actually got there, we had name badges for people and packets for people with information. We were the registration desk!" Jane explained that she and the Signdesign staff continued working for FNAWS in every possible way through the Phoenix convention and the start of the New Orleans convention in 1982.

Lanny O. Wilson reported to FNAWS Board members that plugging of wild sheep horns was a good way to insure trophies were taken legally. Lloyd Zeman wrote in the summer 1980 FNAWS newsletter that nine western states (Arizona, Idaho, Montana, Nevada, New Mexico, North Dakota, Oregon, Utah and Wyoming) had permanent identification programs.

Plugging of trophy wild sheep horns continued as a discussion topic at the Phoenix convention and throughout the early 1980s since national Board members, under the direction of Lanny, supported the tagging program for all game and fish agencies throughout North America as an important method of stopping poaching. Lloyd encouraged FNAWS members to talk to their local state or province game and fish agency during hunting seasons if their home area didn't have such a program. Lloyd Zeman reported at the December 1980 Board meeting that Colorado decided to adopt the wild sheep plugging program.

The Board members also passed a resolution on December 13, 1980, offering a $1,000 reward to anyone giving FNAWS information leading to the arrest and conviction of someone illegally taking any of the four recognized North American wild sheep species.

Bubba Woods reported a minimum of $38,000 profit from the sale of the FNAWS Conservation Stamp and additional profits from the sale of the conservation prints. He expected a total profit of $52,000. Jerry Christian and the entire Board voted to commend Basil Bradbury for the publicity he gave to the FNAWS stamp and conservation print program. A few years later, Basil would become the director of communications for the group.

The first limited edition bronze was introduced to the FNAWS membership in 1980. Jim Stafford was the artist who planned to complete a Dall's, Stone's, bighorn and desert ram.

His first bronze of the Swank Dall's ram was on exhibit at the Houston convention as was the actual Swank mount. M.F. Wood was committee chairman for this fund-raising project. In addition to the bronze, a walnut base bronze desk set was commissioned with the FNAWS logo, the Chadwick ram, on the centerpiece of the base and a bronze of one of the North American wild sheep species on top. Paul Tadlock designed and produced the set. The set with two Sheaffer White Dot pens sold for $500.

For the first time, John Babler was given a receipt for $2,400 as rental of his business building for a year. John's business had given FNAWS an 18-by-8-foot private office for Lloyd

Zeman, use of his entire 10,000-square-foot building for mailings and other FNAWS activities, and use of two of his full-time staff for two years. Now FNAWS was financially able to pay John, although this amount was well below rental costs on the open market.

FNAWS reached out to Bob Housholder by inviting him to speak at the Phoenix convention for five or 10 minutes. Bob declined the invitation to speak, but attended earlier events instead. C.J. McElroy, founder of SCI, did speak at the convention, but declined to swap exhibit hall space as had been suggested by Francis Stubbs. SCI paid for a booth space at FNAWS instead. The FNAWS Board wanted to swap exhibit space at their convention with SCI so that FNAWS could exhibit at SCI, free of charge, but SCI declined. Harry Tennyson, CEO of Game Coin, also spoke at this convention.

Board members voted to review Lloyd Zeman's work and salary at the Phoenix convention, and John Babler recommended that FNAWS hire a professional accountant to handle the financial affairs. President Schultz appointed John, Bill Walters and Harvey Kadlec "to look into the accounting problem and make suggestions." The entire Board voted in favor of selecting an accounting firm in the near future to handle all bookkeeping. As for Rich Roan, the Board of Directors in June of 1980 voted to continue his pay at $300 per day plus expenses paid for himself, his wife and two children, all of whom helped with the auction. Jerry Christian and Harvey Kadlec made the motion to pay Rich. The unbelievable success of the Houston convention's auction confirmed the Board's confidence in Rich Roan and family to bring in the big bucks for wild sheep.

Plans immediately began in June of 1980 for the Phoenix convention that took place February 25 to 28, 1981, at the downtown Hyatt Regency. Tom Saad was the convention organizer, and Dean Bowdoin was his co-chairman. More than 1,000 men, women and children attended this four-day event and were treated to seeing the World's Record desert bighorn sheep taken by a nameless Mexican Indian and brought to the United States by Carl M. Scrivens. The ram scored 205-1/8 B&C points and was taken in Lower Baja California, Mexico, in 1940. The ram is now permanently on display in the B&C National Collection of Heads and Horns at the Buffalo Bill Historical Center in Cody, Wyoming.

A second outstanding desert bighorn ram, now ranked fifth in the world by the Boone and Crockett Club, was on display with the Scrivens ram that was taken by Javier Lopez del Bosque in Baja California, Mexico, in 1979 with a B&C score of 192-5/8 points. At the time, this ram was the largest taken by a hunter. Arthur R. Dubs, an avid auction bidder and supporter of FNAWS, successfully hunted an Arizona desert bighorn ram that surpassed the Bosque ram in 1988 with a score of 197-1/8 B&C points.

Rich Roan has fond memories of the Phoenix convention auction activities. "In Phoenix we ran into a guy in Colorado named Tom Brimhall who was one of the big spenders. I noticed him before the auction even began. He was wearing a tall Stetson cowboy hat. He was a pretty big-sized gentleman, tall, big blue snazzy sport coat on with a big buckle. He caught my eye.

"When you're up there selling, you kind of look around to see how you can improve the take. I just motioned back there to the gentleman in the back with the light-colored Stetson, and asked him if he wanted to bid on whatever I was selling. I don't remember what I was selling, but if you want to bid, just raise that big hat. It seemed like slow motion and everyone just turned around to look at him and that's when he raised his big hat. That year he paid $100,000 plus (FNAWS records show $153,000) for what he bought. He was a good bidder. He was good and active for a few years, but he's not in there anymore. I don't know what happened to him. Nice guy. Good bidder. I picked him out of the crowd. Being in this business, I looked back and I thought he'd be a guy who would spend some money. That's the job of a auctioneer. This guy manufactured homes in Colorado."

A question begs to be asked of Rich. How in heavens do you pick up on the people bidding when they barely raise their hands?

Rich's response is simple. "That's why we have ring men out there. You'll notice how I'll go out into the crowd after I've sold an item and I congratulate the underbidder and the bidder. Sometimes the ring man will say to me that someone doesn't want anyone to know he was bidding. That's sometimes not always true, but I always try to show my appreciation. I do as the auctioneer for FNAWS. FNAWS is thankful to the bidders. I am always appreciative of the underbidder and the bidder who brought this money into FNAWS.

"Sometimes I see a person rub their nose and scratch their head, and as a joke I ask, "Are you bidding?" They get all flustered. I do that at my own auctions, too. You just have your own things you pick up. They have auctioneer schools, of course, but I never went to one. I just did it on my own. I picked it up from my father, and my four boys picked it up from me. None of us went to school."

All of Rich's techniques for encouraging higher bidding worked. Gross receipts from the Phoenix auction alone totaled $430,000 as of March 1, 1980. Additional revenue came from arm and leg wrestling, totaling $500, and rifle raffles totaling $7,500. Thanks in large part to the unbelievable success of the Phoenix auction, Board members at their May 30, 1981, meeting at the Thunderbird Motel in Minneapolis committed $121,155 to disease and habitat research projects.

John H. Batten of Racine, Wisconsin, noted wildlife conservationist, author, editor and hunter, received recognition for his lifetime of dedication to wildlife conservation and the hunting community as the FNAWS Outstanding Achievement Award recipient. At this convention, a trade edition of his classic hardcover book, *The Best of Sheep Hunting*, was sold to FNAWS members by publisher James C. Rikhoff of Sporting Fraternity Limited and Amwell Press. Lloyd Zeman said the 1,000 copies of the collector's edition sold quickly, and additional copies were published with the FNAWS name on the title page. Copies of *The Best of Sheep Hunting* are still for sale through Sporting Fraternity Limited in Clinton, New Jersey. Special awards were presented to Basil Bradbury, editor of *Petersen's HUNTING*, for his promotions of FNAWS, to Bob Housholder for his 25 years of dedication to wild sheep hunting through his North American Sheep Hunters Association and the Grand Slam Club, and to Lloyd Zeman for his dedication to sheep hunters, outfitters, wild sheep and their welfare through his many actions. Lloyd also received an award from the Alaska Professional Hunters Association for his work to preserve hunting in Alaska. Dr. Tom Bunch and the Phoenix Zoo were recognized for their work with wildlife research and wild sheep.

Tom Saad made a name for himself at the Phoenix convention for organizing a very successful water hole project with Dean Bowdoin in the Plomosa Mountains of Arizona outside Phoenix where 45 FNAWS members built the water hole with 70 members of the Arizona Desert Bighorn Sheep Society (ADBSS). Helicopters were used to transport cement and other building material to the isolated project site. Tom was very active in ADBSS and served as its fourth president. Thanks to Tom's and Dean's leadership, and the efforts of dozens of FNAWS and ADBSS volunteers, the "Betty Lou" project was quickly completed during the convention. According to Curtiss Judd the project was called "Betty Lou" because this was a pet name for one of the radio-collared wild sheep in the area. Danny Nasca was a dedicated and hard-working member of ADBSS who provided the helicopter transportation for the project. Some years later he died while helping a film crew. His helicopter hit a bridge.

Some of the volunteers from FNAWS and ADBSS who built the "Betty Lou" water project in 1981.

Dr. Roy Schultz, left, and Dean Bowdoin stand beside the "Betty Lou" water project outside Phoenix. Dean and Tom Saad organized the project with volunteers from the newly organized FNAWS and the Arizona Desert Bighorn Sheep Society (ADBSS) in 1981.
Photographs courtesy of Dr. Roy Schultz.

"I can still remember the camaraderie around the campfire that night, and I'll never forget that Jim Froelich gave us all a hand axe that I still use to split firewood," Tom Saad told Doc Schultz in 1994.

Curtiss Judd also remembered the Phoenix water project as exceptional. He wrote in his 1997 book, *I Call Myself a Sheep Hunter, Part II*, "At least 30 of us newcomers went to a water hole project. The society (Arizona Desert Bighorn Sheep Society) has five or six projects a year. This was the first one for me. One minor problem for me was that I had had recent eye surgery. Our winter back home was mild since the surgery, but, being in Arizona was a little difficult. I walked considerably, so I was in good shape that way. My doctor said it was okay to go, but I was not to work. I took those instructions from my doctor very seriously. I watched while others worked. I always enjoyed watching others work." Curtiss recalls with fond memories his first water hole project, especially since he had official doctor's orders to "Watch, NOT work!"

Curtiss has been one of the most active of water hole workers over the years. As of September 1997 he had been on 30 of these projects. He remembered that "once I overdid it and screwed up my right arm." He winters in Arizona now and is often one of the first to arrive at the project. He often goes into the mountains where the project will take place a day before the rest of the group arrives.

Curtiss explained that whoever is in charge of a particular project gives him the order, "No Running the Air Hammer." Curtiss works a couple of hours before the big crew of men and women arrive ("Seriously, I do work," he says.) After everyone arrives, Curtiss finds a comfortable rock and watches everyone work! This tradition of attending water projects in Arizona began in Phoenix. Another Midwestern volunteer who also continues to help build water holes in Arizona is Richard Mitchell of Tipton, Iowa. Tom also led FNAWS members on a "Show Me Sheep!" overnight expedition prior to the convention at a game refuge with assistance from Dean Bowdoin and Danny Nasca. Ray Woodward once again became the men's leg wrestling champion, and Lori Johns received the same honors for women. Don Schmalz, who now serves on the FNAWS national Board of Directors and is an outfitter from Cody, Wyoming, was men's arm wrestling champion.

Jerry Christian was elected as FNAWS president for 1981-1982, and John Babler was selected as president-elect to serve as president during the year 1982-1983. John was also elected to continue as treasurer. Lloyd Bare was elected secretary and Francis Stubbs as vice president.

Paid advertising was accepted by a vote of the Board of Directors in Phoenix. A motion by Lanny Wilson and seconded by Dr. Red Duke was approved by the Board to publish four newsletters per year starting after the Phoenix convention. The Board also entertained a request from a man who wanted to purchase the FNAWS membership list. His request was denied by the Board, which established the tradition of not allowing such a transaction to occur.

The Board of Directors at the Phoenix convention voted to pay Lloyd Zeman $35,000 per year plus travel expenses not to exceed $7,000 and public relations expenses not to exceed $3,000. This salary and benefits package was arrived at after much discussion by board members. The Board also voted to provide for Lloyd paid hospital insurance and a recognized credit card system for records-keeping purposes. John Babler made the motion and Jerry Christian seconded it. The Board voted 8 to 1 to rehire Lloyd. Lloyd accepted the salary and benefits. Ironically, John Babler resigned from becoming FNAWS president the following year because of his concerns regarding Lloyd's business practices.

A memo to all FNAWS officers and directors, dated May 5, 1981, from Jerry Christian read, "A director may be removed from the Board for misfeasance or for any conduct which is contrary to the best interest of the Foundation by the affirmative majority vote of the full Board of Directors." This passage was an amendment to the FNAWS bylaws dated July 14 and 15, 1979. This same memo listed numerous amendments to the bylaws, but the above quote was noteworthy in light of what occurred in 1982.

Less than a month after the amendments to the bylaws memo from Jerry Christian were issued, Lloyd Zeman told the Board of Directors on May 31, 1981, that FNAWS needed a separate office and secretary, away from Signdesign, Inc. in Minneapolis. The Board voted to approve a relocation of the FNAWS office by October 15 at a cost of $6,000 per year or less with arrangements made by the president, treasurer, secretary and executive director.

Plans for the 1982 convention in New Orleans immediately began following the successful Phoenix convention, and minds were turned to planning a happy, successful and fun event in New Orleans. The New Orleans convention was set for March 3 to 7, at the Marriott Hotel in the French Quarter on Canal Street. "We walked out the back door and we were right there," Jane said.

Hobart Pardue (FNAWS Charter Member 26) was convention organizer with assistance from Dan Littleton. Carl Mahon of Utah received the FNAWS Outstanding Achievement Award for his overall concern and conservation work with the desert bighorn sheep populations. Special recognition was given to Bubba Woods for his "contribution to the financial success of the FNAWS Conservation Stamp Program." Lowell Baier and Tom Hinders were added to the FNAWS national Board of Directors, bringing the total to 17. On display was the Pancotto Grand Slam that was donated to FNAWS.

Jane Babler remembered how the early FNAWS office helped with the New Orleans planning. "I think in New Orleans there were so many people, and so much to do, we contacted the convention bureau and asked for some help — some bodies to help us. Mind you, all of the packets were stuffed in Minneapolis, well, St. Paul, in our office and transported down to New Orleans."

FNAWS membership stood at 1,905 at the start of the New Orleans meeting and 1,022 men, women and children attended the New Orleans convention, as reported by the minutes of the Board of Directors on March 2, 1982. A total of 13 Stafford bronzes were sold and only two life membership pins remained so Bubba Woods volunteered to investigate the creation of new life member pins.

Dr. Schultz became the building committee chairman after it was announced at the New Orleans Board meeting on May 2 and 3, 1982, that land had been donated to FNAWS for constructing its first national headquarters and education center on property owned by Mac Lewis, just east of Cody, Wyoming. FNAWS would eventually locate the land and a building across from the Buffalo Bill Historical Center in Cody, but New Orleans was the place where a relocation out West began for this young conservation group that loved wild sheep and their habitat.

Wednesday, March 3, in New Orleans was a Mardi Gras reception, with cocktails, hors d'oeuvres and a Dixieland band. Opening ceremonies on Thursday began with a breakfast and an evening floating party on an authentic stern-wheel steamboat called the "Natchez" for a two-hour cruise down the Mississippi River at a cost of $25 per person.

Memories of this unique Thursday evening event were recorded in the FNAWS newsletter written by Lloyd Zeman. A total of 532 FNAWS members, supporters and spouses assembled at the hotel at 5:15 p.m. on Thursday, March 4, and paraded to the boat dock about eight blocks away, complete with a Dixieland band and a police escort. The jazz band continued to play on the "Natchez" for dancing pleasure. Lloyd Zeman reported that all who enjoyed the cruise arrived back at the dock in time for a 9 p.m. dinner reservation!

The women toured Oak Alley Plantation, circa 1830, on Thursday prior to the steamboat cruise. A total of 168 ladies enjoyed Southern hospitality at the plantation.

The auction in New Orleans once again dominated the outstanding achievements of FNAWS with a record gross receipt of $478,000 and $22,000 for a silent auction item of an Alaskan safari. Top auction bidder was D.J. "Doc" Thurston, known as Doc Thurston, who was top bidder for the Wyoming permit at $43,000. A year later, he and Dan Pocapalia would earn top honors for their bidding abilities on the Wyoming permit, but in New Orleans, this was the all-time high in FNAWS history. Prior bids for the Wyoming permit in 1980 and 1981 had been $23,000 and $29,000, respectively.

Top men's leg wrestler was once again Ray Woodward, and Don Schmalz again became the top men's arm wrestling champion. Mayvon Platt won top honors as women's leg and arm wrestling champion. Financial operations and leadership by Lloyd Zeman concerned John Babler, so instead of becoming the fourth president of FNAWS after Jerry Christian, John stepped down from a position he really wanted and remained as treasurer after the New Orleans convention.

Jane Babler remembered John's decision to step down. "John and I spent a lot of sleepless nights discussing what we were going to do and what we could or couldn't put up with any longer. There were lots of tears shed and a lot of emotion. And, we talked about it a lot. In the end, the decision was his to step down. That decision was his, but it was based upon lots of conversations with me and what we saw as happening and where we thought the Founda-

Jerry Christian, left, Harvey Kadlec, center and Mahlon "Butch" White, right, with a painting that sold for $19,000 at the 1982 FNAWS convention in New Orleans.

There are some familiar faces in the crowd in this photograph taken in New Orleans in 1982. Do you recognize Sam Pancotto in the foreground or Bill Leech in the center with the plaid shirt? Sam and Bill were regulars during early FNAWS conventions.

British Columbia outfitter Red Sorensen, left, was in charge of the leg and arm wrestling championships for the early years. Here he is directing activities in 1982 in New Orleans.

Crowds of FNAWS members and supporters spilled into the 1982 FNAWS convention that was held at the Marriott Hotel.

Rich Roan, center, gives instructions to spotters at the 1982 convention.

Outfitter Gary Powell of British Columbia, who was later killed in a tragic airplane accident, took the microphone in 1982 as Auctioneer Rich Roan, right, talks with Doug Yajko.
Photographs by Dennis Campbell.

FNAWS President Jerry Christian, right, shakes the hand of Sam Pancotto for his support of FNAWS in 1982.

Ricardo Medem, left, an outfitter from Spain, donated a hunt at the 1982. Helmut Gruel, center, purchased the hunt. Auctioneer Rich Roan is to the right.

Outdoor Life's *Jim Carmichael, left, and Helmut Gruel examine the rifle Helmut just purchased at the 1982 auction in New Orleans.*

Lloyd Zeman, left, enjoys conversation with friends at the 1982 New Orleans convention.

A smiling Dr. Jerry Waite, center, congratulates Wyoming Outfitter Don Schmalz, right, as one of the 1982 arm wrestling champions.

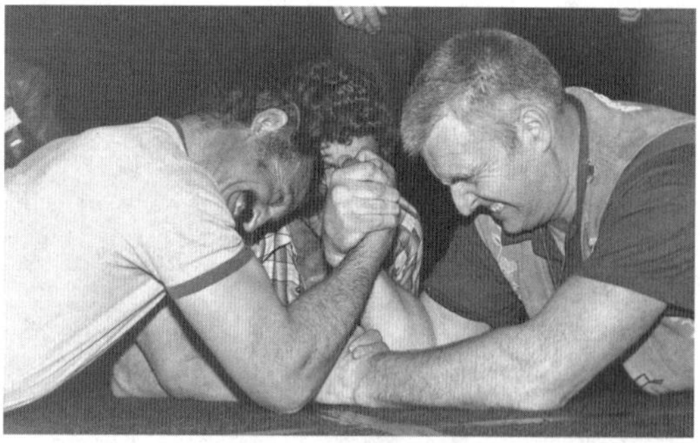

Arm wrestling at the 1982 New Orleans convention included a match between Bill Moynahan, left, and Ray Woodward with the late Jim Ford in the center as referee.
Photographs by Dennis Campbell.

Jim Carmichael speaks at the podium in 1982 as FNAWS President Jerry Christian, left, and Dr. James "Red" Duke listen.

Dr. Red Duke was Master of Ceremonies at the New Orleans banquet in 1982.

A young boy is congratulated by Dr. Jerry Waite, left, and Dr. Red Duke, center, at the 1982 New Orleans convention.

Early supporters of FNAWS, Sam Pancotto, left, and Jim Schmalz donated to the FNAWS Saturday night auction in 1982.

Bill Walters, left, and Dr. Red Duke at the 1982 FNAWS convention in New Orleans.
Photographs by Dennis Campbell.

Arnie Johns, left, and Earl Boose, center, received recognition from Jerry Christian during the ram awards ceremonies in 1982.

tion was going and this huge fear that the Foundation was going to be destroyed because of the stuff that was going on."

John never told the general membership during the New Orleans convention, and John and Jane continued to socialize with their friends, trying to act as though nothing had happened in their lives. Jerry Christian continued as FNAWS president until Duane Smelser became president in the spring of 1983. The Bablers didn't attend the following year in Reno but they returned to FNAWS as great supporters and to the friends they loved at the San Diego convention in 1984.

Jane reflected. "We never stopped being interested in what was happening to the Foundation, and we wanted to see that everything would turn and it would be good. There were friendships there that we still to this day only get a chance to renew at that convention. And, I probably wanted to go more than John did. It was much, much harder for John. With FNAWS leaving the Babler's office in Minneapolis and heading West, Jane and John could attend future conventions without the worry and stress that came with planning them.

"If I could put my finger on what FNAWS conventions mean to me — the friendships that we made and continue to have," Jane Babler said.

FNAWS was growing and changing. Finances were more complicated. And, as FNAWS entered the next year on their way to reaching a million dollars in auction revenue, as Paul Asper and Mike Valencia had predicted to Rich Roan in Minneapolis, new leaders of FNAWS looked with optimism to the future.

FNAWS was well on its way in the spring of 1982 to start climbing that mountain for wild sheep and to mature as it struggled to meet its goals and aspirations.

Jane and John Babler, center and right, received recognition from FNAWS Past President Garry Vince in 1995 for their steadfast service to FNAWS in its early years as recipients of the 8th annual Gordon Eastman Grass Roots Award. In 1988 Gordon Eastman established the "Grass Roots Award" to honor those who work so hard for FNAWS at the local level.
Photograph courtesy of FNAWS archives.

Early FNAWS member Charles Jagou was recognized in 1990 by FNAWS President Harvey Kadlec, right, who presented the Presidential Award to Charles' daughter, Nancy Jagou Holmes. Charles planned early FNAWS conventions and spent endless volunteer hours planning the Houston Safari Club events.
Photograph courtesy of FNAWS archives.

Dr. Roy A. Schultz's and Susan C. Reneau's biography are found in Chapter I of this book.

John Babler is co-owner of Signdesign, Inc., a screen process printing business in St. Paul, Minnesota, with his wife of 20 years, Jane. The business began in 1977 just as FNAWS was beginning and grew as FNAWS grew. Today Signdesign produces signage for retail companies worldwide and presently employs more than 100 workers. John worked as a full-time firefighter for the St. Paul Fire Department from 1963 to 1996 and retired as a deputy chief. John has remained an active FNAWS auction bidder participant, and both John and Jane attend all the annual FNAWS national conventions. Jane was hired by John in June of 1977 to work at Signdesign, Inc. as a secretary-receptionist. They were married on October 5, 1979.

6th Convention of the Foundation for North American Wild Sheep

February 9 to 12, 1983
MGM Grand Hotel
(now called Reno Hilton)
Reno, Nevada
Convention Chairmen: Steve DeRicco and John Turner
President: Jerry Christian, 1981 to 1983
Theme: What Does the Future Hold for Wild Sheep?

7th Convention of FNAWS

February 22 to 25, 1984
Town and Country Hotel
San Diego, California
Convention Chairman: John Turner
President: Duane Smelser, 1983 to 1984
Theme: The Challenge of the '80s; the Year of Protecting Wild Sheep and their Habitat.

8th Convention of FNAWS

February 20 to 23, 1985
MGM Grand Hotel
(now called Reno Hilton)
Reno, Nevada
Convention Chairman: Steve DeRicco
President: Tom Hinders, 1984 to 1985

9th Convention of FNAWS

February 26 to March 1, 1986
Honolulu, Hawaii
Convention Coordinator: John Turner
President: Tom Saad, 1985 to 1986
Theme: Challenging the '80s - The Year of the Outfitter

10th Convention of FNAWS

January 21 to 24, 1987
Opryland Hotel
Nashville, Tennessee
Convention Chairman: Arnie Johns
President: Dr. James "Red" Duke, 1986 to 1987
Theme: Hunting and Harmony: The Traditions Continue

Chapter III

FNAWS Moves: The Climb Continues 1982 to 1987

DR. ROY SCHULTZ AND SUSAN C. RENEAU

FNAWS was well on its way to reaching a goal of $1 million gross on auction sales when Board members met after the New Orleans convention to discuss the direction of this growing wild sheep organization.

Membership grew by leaps and bounds. In January 1980, total membership stood at 668. By the close of the New Orleans convention, membership had swollen to 2,567, with 530 of those memberships unpaid, but a notice to those members was sent to remind them to renew. By the close of the New Orleans convention, there were 102 life members, compared to 23 in early 1980.

Finances were equally impressive for this young group. Gross revenue of $40,000 at early FNAWS conventions were replaced with gross receipts of close to $500,000 in New Orleans. What a difference a few years make.

Dr. Waite may have said it best in 1997 when he wrote to Dan Pedrotti:

"We wanted to give something back to these great animals and for the rams we had harvested. The goals we set early on were far surpassed. We thought we were on top of the world when we grossed $70,000 in Detroit. As a result of our efforts, there are a great many more sheep on this continent now than when we started FNAWS."

Lloyd Bare (FNAWS Charter Member 11) served for many years as secretary to the Sheep Foundation and added to Dr. Waite's thoughts about the significance of FNAWS to him: "In those early days, we didn't always have much cash to help out projects, but I remember with pride that we did respond immediately to emergencies when other organizations responded with only pledges of help. His comments are typical of the early members of FNAWS who watched it grow and lived through its struggles.

Lloyd Bare was the one to step in as secretary of FNAWS in the early days when someone didn't take over the responsibilities after being elected when in Memphis. He was officially elected to the position in 1981 when he joined the Board and continued until 1982. "At the same time, I helped write and edit the newsletter, which has now evolved into a great magazine. I served as a "go-fer" and photographer at the conventions and, later, along with Dr. Jerry Waite, formed the Iowa FNAWS Chapter — the first local chapter of FNAWS."

Dr. Waite rose to the presidency in 1987 and presented the Outstanding Achievement

Award that year to the membership, not to one particular person. "We must never forget they are the backbone of our great organization."

Life-long friendships were the ties that continued to bind this close-knit group together. Lloyd Bare told Dr. Schultz as this book was being written: "The most gratifying aspect of being able to serve the Foundation was the fact that I was able to meet and form lasting friendships with fellow hunters, outfitters, guides and biologists from all over North America." He paid special tribute to John and Jane Babler who worked so hard to make it work in the beginning. He also recognized other dear friends "in no particular order," friends like Dr. Waite, Jerry and Sandy Christian, Dr. Roy Schultz, Bill Walters, Paul Asper, Michael Valencia, Dr. Red Duke, Lanny Wilson, Harvey and Sue Kadlec, Richard Mitchell, Paul Wiemer, Arnie and Lori Johns and Rich Roan, and a large cast of wildlife biologists who worked with early FNAWS members to complete early projects.

Questions arose as to Lloyd Zeman's employment at the regular Board meeting on March 2 and 3 during the New Orleans convention and at a regular Board meeting June 7 and 8, 1982, held at the Phoenix Hyatt Regency. A vote of confidence was given to Lloyd Zeman by a majority of Board members to retain his services for another year during the January 1982 Board meeting, but questions remained. Francis Stubbs, John Babler and Lloyd Bare voiced concerns, and Francis went as far to write a detailed letter outlining his objections to Lloyd Zeman and FNAWS operations. His letter was read to the Board of Directors on March 7, and Secretary Lloyd Bare responded to him in writing a few days later.

The letter to Francis clearly outlines the position of the Board at that time.

Lloyd Bare wrote, "I have been instructed by President Christian and the Board of Directors to respond to your letter of March 5, 1982, to Mr. Jerry Christian.

"Your first item concerned the legality of Mr. Tom Saad serving on the Board. The Board of Directors declares that there is nothing in the bylaws prohibiting an outfitter from serving on the Board. There is only a statement in the minutes that suggests perhaps an outfitter should not sit on the Board but since the Foundation has not adopted a formal policy manual, that statement is not binding.

"The second item concerns the dismissal of Mr. Lloyd Zeman as Executive Director and your request to bring the matter to the immediate attention of the Board for corrective action. President Christian declared that the matter was voted upon and settled on January 31, 1982, and that the motion by Duke (Dr. Red Duke) to look into the matter again at the New Orleans convention was out of order since no meeting was in session when he made the motion.

"The third item concerned the Foundation's tax-exempt status. At the Board meeting of March 7, 1982, a motion was passed to obtain the services of a tax attorney on a retainer basis to examine the Foundation's IRS Code 501-(c)(3) status and to make recommendations on how to best maintain that status.

"The fourth item concerned a director missing two consecutive Board meetings. President Christian declared that no Board member has missed two consecutive meetings without a valid reason and that the motion made at the Omaha meeting of December 10, 1977, concerning that policy is inconsistent with the bylaw. If you were referring to Mr. Smelser, he was excused by Dr. Schultz and/or Mr. Christian.

"The fifth and last item concerned the election of Board members at the convention on March 4, 1982. The Board of Directors on March 7, 1982, declared the election results invalid."

Jim Froelich reported that 178 men and women voted for members of the Board of Directors and five women wanted to vote in the election, but went on a tour to a plantation instead, so they voted on pieces of paper in front of a witness. The votes were then transferred in front of another witness, and three notarized proxies were received. Because of this action and the concerns expressed by Francis Stubbs, the Board voted to invalidate the election results, so all current Board members remained in place until the convention in Reno in 1983. Lowell Baier and Thomas M. Hinders, Tom to friends, joined the Board and the remaining Board members stayed on the board, including Michael Valencia and Francis Stubbs.

In June, Board members voted to remove Francis from the Board but confirmed confidence in John Babler and Lloyd Bare for "their valued service as both directors and officers of the Foundation." Francis Stubbs resigned from the Board at the New Orleans convention; John Babler resigned as treasurer in 1982; and Tom Saad became treasurer until 1983. Lloyd Bare remained as secretary until late 1982 when Dr. Schultz took over the duties of acting secretary at the January 1983 Board meeting.

Much time was spent in the summer 1982 discussing election procedures and the Foundation's bylaws. Board members concluded a Board of Directors' manual needed to be developed that included the articles of incorporation and bylaws, policies of the Board, operating procedures of FNAWS, standing committees and committee assignments, financial statements and a list of wildlife projects funded. The Board also agreed that from now on, Board elections would be conducted by mail, a method that continues today. Although the upset in New Orleans wrenched the bottom of the Foundation's soul, past presidents and founders who were interviewed for this book agreed that procedures were tightened and communication improved after 1982. From trauma came good.

In a letter to the general FNAWS membership in 1982, Jerry Christian said: "At a recent meeting of the directors attended by 13 of the 16 current Board members, as well as five of the original founders, various resolutions were adopted to assure that our future remains strong and confident. Not only were the wishes of the membership as expressed by their ballots, unanimously upheld, but also a detailed format was adopted to ensure the absolute integrity of all future elections.

"We offer to you the same determination, the same dedication, and the same sense of commitment that brought us so far from our birth at Mt. Horeb and will continue to sustain us in our efforts to make the Foundation for North American Wild Sheep the premier sportsmen's organization it was destined to be."

The letter was signed by Paul Asper, John F. Babler, Lowell E. Baier, Lloyd Bare, Jerry L. Christian, Jim Froelich, Thomas Hinders, Harvey Kadlec, Tom Saad, Dr. Roy Schultz, Duane Smelser, Michael Valencia and Dr. Jerry Waite.

To assist with financial matters, the Board hired the Phoenix accounting firm of Peat, Marwick and Mitchell to audit the Foundation's financial records from 1977 to the present. A final report was filed with their findings and conclusions in October 1982 that 100 percent of all monies donated to the Foundation went for sheep and related projects but that records keeping in all financial matters needed to improve. Discussions by Board members in meeting minutes indicate that sincere efforts were made by many Board members to improve records keeping in the years to come and efforts continue today. An example of this was the Board's vote after the New Orleans convention to not commit funds or spend money for wild sheep projects until checks and credit cards had cleared the bank after annual auctions.

Certainly, this upheaval distracted FNAWS members, but positive directions continued and FNAWS moved forward regardless of the conflict within the Board. Significant in the growth of FNAWS was the official vote by the Board of Directors in June 1982 to relocate national headquarters to Cody, Wyoming, thanks in large part to the help of Doc Schultz's friendship with Mac Lewis, a lifelong resident of Cody and great supporter of FNAWS. Mac

Home to FNAWS when they first moved to Cody, Wyoming, was the building seen here at 802 Canyon Avenue. The left portion of the building was recently added for current occupants, Coldwell Bankers Realtors.
Photograph by Sandra Nicola.

offered to donate to FNAWS an acre of land on the east side of Cody along highway U.S. 14, and across from the Cody airport.

"It is part of the property homesteaded by his dad. It is on a high piece of ground located between two small lakes, and we feel it is an ideal site to build a functional Foundation office and educational center," wrote Jerry Christian in the August 1982 newsletter.

Board members thought Cody was an ideal location for the FNAWS headquarters. "It is in the heart of some of the best Rocky Mountain bighorn country anywhere. This will put us close to where the action is, and make it possible to more easily monitor wild sheep problems or anything else that comes up," Jerry continued. Cody is 50 miles from the eastern entrance to Yellowstone National Park. Early FNAWS Board members thought a location in Cody provided visibility to the growing group as 1 million visitors to Yellowstone passed through Cody.

Long-time member of FNAWS and current national Board Member Don Schmalz, provides insight into one of the reasons why FNAWS relocated from Minneapolis to Cody.

"Lloyd and I were very good friends. In fact, back then, one of the reasons the Sheep Foundation moved to Cody, Wyoming, was because of a conversation I had with Lloyd about the seismic work in the springtime outside Cody. The sheep were running headlong around the base of Jim Mountain. I live at the base of Jim Mountain beside Shoshone National Forest (the largest national forest in the United States).

"When they (oil exploration company) set some of those charges off, it would look like a volcanic eruption. It rattled our windows and our dishes. It would rattle a long way across that valley. That was seismic activity caused by an independent oil company. It was something out of a Vietnam War movie. They set those charges up on sticks, and they'd rumble the ground. "They had these helicopters and they dropped these guys out with cables, running these cables up and down the hill. It was a bad time of year when the wild sheep were weak. It had been a hard winter, and the ewes were heavy with lambs. Running those blasts around the hill scared the sheep!"

"I called up Lloyd in Minneapolis. One of my first comments to Lloyd was, `What the hell is the Foundation for North American Wild Sheep doing in Minneapolis?!?!@##. There aren't any wild sheep in Minneapolis. You ought to be where you can see this stuff. I look out my window. How come the Sheep Foundation isn't here where everything is happening?'

"It wasn't but six or seven months later and the Sheep Foundation was out here in Cody because we were in sheep country. It was all that seismic activity that motivated me to pick up the phone and call Lloyd."

The building fund totaled $20,915 due to the sale of two original paintings donated by Peter Darro of Northbrook, Illinois, and a cash donation by Martin F. "Bubba" Wood of

Collector's Covey, Dallas, Texas. Sam Pancotto of Oak Brook, Illinois, offered to donate a building; Jim Schmalz of Bloomington, Indiana, pledged the brick and block; Duane Smelser offered to help with roofing; John Turner of Pacheco, California, offered the services of his architect, Bill Peterson; and Bill Feeley, artist and sculptor from Cody, offered to donate a lifesized sculpture of the Chadwick ram for the front of the building. Herman Bennett donated a kitchen.

While money was raised and plans were solidified for the construction of a national headquarters, the August 1982 newsletter reported the FNAWS address to be 1147 Sheridan Avenue, Suite 6, Cody, after September 1. Prior to that date the office remained at 295 South McKnight Road, St. Paul, Minnesota.

The MGM Grand Hotel in Reno in 1983 was renamed the Reno Hilton in the late 1980s but remains the Number One most popular place to hold a FNAWS convention to raise money to put wild sheep on the mountain.
Photograph by Dennis Campbell.

Amid all this activity in FNAWS, plans for the Reno convention were jelling. Lloyd Zeman wrote in the August 1982 FNAWS newsletter: "Mark February 9 to 12, 1983, on your calendars now. Those are the dates of our next annual convention that will be at the MGM Grand in Reno, Nevada."

The exhibit hall increased to 80 spaces at a cost of $300 for a 8-by-10-foot or 10-by-10-foot table. "But don't wait long because booth space has sold out well in advance at all past conventions," Lloyd advised.

Rich Roan recognized two outstanding bidders who began a history with FNAWS in the early 1980s. "Somewhere along the line, D.J. "Doc" Thurston of North Carolina and Dan Pocapalia of California came along. Of course, Doc Thurston has since passed away, but Dan still comes. Doc Thurston was a great supporter at the time. Back in 1984, Dan Pocapalia paid a tremendous amount of money for the Arizona hunting permit — $64,000. Dan also paid lots for the Wyoming permit — $50,000 — in 1983."

Dan Pocapalia reflected about his involvement with FNAWS in May 1998 and why he supports FNAWS so much each year:

"The first impression and excitement I had about bighorn sheep was when I saw my first Rocky Mountain bighorn while hunting elk with Jerry Abercrombie on the headwaters of the Buffalo River in Wyoming."

"Taxidermist Roy Ostermiller (Frontier Taxidermy of Cheyenne, Wyoming) gave me a rundown on how few of the sheep there were and how low on the priority of the Department of Fish and Game's budget bighorn sheep fell. Inasmuch as very few sheep permits were allowed to be sold, there was virtually no revenue for bighorn sheep. My interest was then primed to start helping to propagate, what I consider, the most magnificent animals in North America.

"Dr. Roy Schultz talked me into going to the FNAWS convention 18 years ago, when Governor Ed Herschler of Wyoming granted FNAWS a one-ram permit for auction at the convention. I ended up buying that permit for $50,000 and felt that I had made a major

Wyoming Governor Ed Herschler, center, began the tradition of donating a Wyoming hunting permit to FNAWS to raise money for wild sheep in 1980. Governor Herschler remained as a loyal and steadfast friend of FNAWS, and received many awards for his work for wildlife conservation. He is surrounded in this photograph by Tom and Joni Saad.
Photograph courtesy of Tom Saad.

Dennis and Nancy Campbell enjoy the 1983 auction in Reno.
Photograph by Dennis Campbell.

A couple of regulars at FNAWS conventions were Lad Shunneson, left, and Carlos Sanchez, shown here at the 1983 FNAWS national convention.
Photograph by Dennis Campbell.

In 1983, Dave Wiens, left, and his cousin Ed Wiens enjoy the auction.
Photograph by Dennis Campbell.

Red and Agnes Sorensen at the 1983 Virginia City party in Reno.
Photograph by Dennis Campbell.

The 1983 FNAWS auction was electrified by bidding from Dan Pocapalia and D.J. "Doc" Thurston, both shown here moments after the bidding stopped and the celebration began. Dan captured the high bid of $50,000 for the Wyoming permit. Left to right: Lloyd Zeman, Dan Pocapalia, Doc Thurston, Jerry Christian and Roy Ostermiller. Roy convinced Dan he should attend the 1983 FNAWS auction to help wild sheep. Dan and Mary Pocapalia have been attending FNAWS auctions ever since. *Photograph courtesy of Mary A. Pocapalia.*

---

contribution to bighorn sheep in a time of need. The rest is history!"

Dan's high bid electrified the convention, Rich Roan recalled. "What I'm saying is that the bidders are what make the auctions such a success. The first thing that really got everyone excited was the Wyoming permit. Dan Pocapalia was the first bidder who brought the BIG money. That was in 1983 when Dan paid $50,000 for the Wyoming permit. There were others before him, but not at that price. Doc Thurston was right in there, too, with his bids. Doc Thurston bought the Wyoming permit the next year for $52,000."

What goes through an auctioneer's mind when someone bids as high as $50,000 for one hunting permit?

"I thought everybody had fallen out of a tree. You're up there selling and you think what's wrong with these people that they'd pay so much for one hunting permit and maybe not even get an animal! You know what's nice about it is that even today you sell a permit for $405,000 to go hunt a sheep in Alberta, and you wonder what is going on. You have to be thankful for the people who have the money and who believe in the Foundation for North American Wild Sheep!"

A friendly bidding battle ensued at the 1983 convention between Doc Thurston and Dan Pocapalia. Lloyd Zeman explained in the May 1983 FNAWS newsletter following the convention. "Doc Thurston of North Carolina, who purchased Wyoming Governor Herschler's permit a year earlier for $43,000, started out the bidding at $25,000, and also had the second-to-last bid of $49,500."

"The best auctioneer in the world is no good without an underbidder, you know. The only difference between the underbidder and the bidder is that the underbidder doesn't pay the bill," Rich Roan explained.

FNAWS gained a great friend of the auction when Ron McMillen of Ames, Iowa, attended his first FNAWS convention in Reno to bid on a wild sheep hunt for himself. Ron was invited to Reno by his Ames, Iowa, friend Tom Hinders, and Tom soon recruited Ron to solicit hunts for future FNAWS auctions. Ron's knowledge and enjoyment of marketing and sales blended beautifully with Tom's knowledge of outfitter groups to visit. Tom was having health problems at the time and needed someone to visit these groups to secure donated hunts. Ron was single and available, so Tom offered to cover Ron's travel expenses if he would represent FNAWS at various Canadian and United States guide and outfitter groups. This sounded fun

A happy lineup in 1983 included, left to right, Dean Bowdoin, Tom Saad, Jerry Christian and Duane Smelser.
Photograph courtesy of Tom Hinders.

Gary Schoenborn, left, and his friend Dennis Campbell enjoy the 1983 auction action in Reno. Dennis was one of the official photographers of the 1982 and 1983 FNAWS conventions.
Photograph courtesy of Dennis Campbell.

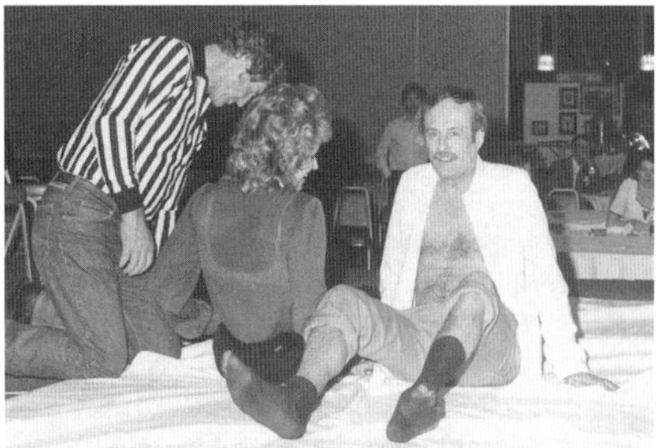

Jack Jonas didn't feel up to competing with the men in 1983 for the leg wrestling championship, so wrestling officials allowed him to compete with the women and lost.
Photograph by Dennis Campbell.

Arizona Senator Barry Goldwater, right, meets Mr. and Mrs. Sam Pancotto at the 1983 FNAWS banquet where Mr. Goldwater was keynote speaker.
Photograph by Dennis Campbell.

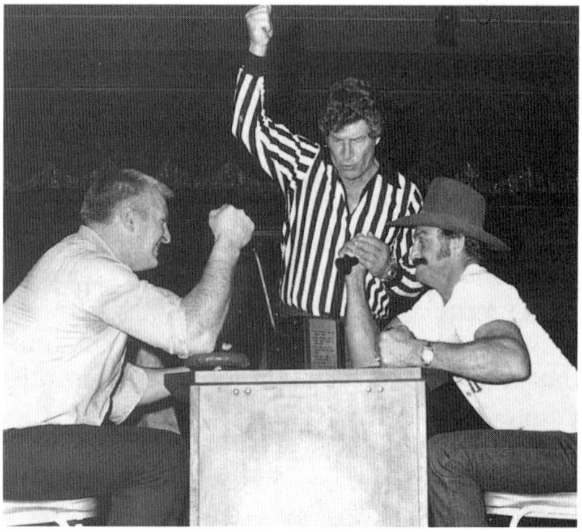

Between the 1982 and 1983 conventions, the arm wrestling championship changed dramatically when an arm wrestling machine was used for the event. Shown here in 1983 are Ray Woodward, left, and Bill Moynahan, right, wrestling with the late Jim Ford, center, as referee.
Photograph by Dennis Campbell.

to Ron, so he said yes.

Ron's first trip was to British Columbia. "I went to the Northern British Columbia Guides Association to make a pitch for outfitters to donate hunts for the Sheep Foundation auction. Tom told me to go up there and get all the donations I could. I remember Miles Bradford, Garry Vince, Dave Wiens and Gary Moore donating hunts. Those early outfitters in FNAWS who donated hunts encouraged others to step up and help."

From 1983 to 1988, Ron McMillen visited up to 16 guide and outfitter groups per year. Ron married Becky in 1988, so his extensive time on the road and up-in-the-air ended.

"I was just having a lot of fun. I would go from one to another outfitter meeting, from one state and province to another. By 1986 and 1987, I was going full swing," Ron explained.

Ron quit the active traveling after their daughter was born, but today continues to help the auction as the "auction liaison" to make sure the auction flows smoothly and on time. He ran for the national Board once, but he finished last, so he's never run again. He maintains his regular membership and has never become a life member. Ron attends all the Board of Director meetings (except two in the last 15 years) and enjoys the action of the annual auction. Later, Ron's company, Sigler Printing & Publishing, was hired by Basil Bradbury to publish the Wild Sheep magazine, a job his company still performs today.

Ron McMillen recognized the significance of such people as Doc Thurston and Dan Pocapalia. "Doc Thurston and Dan Pocapelia may be the reason why I got involved with FNAWS. When Doc and Dan started bidding those Wyoming hunts for $50,000 in 1983, people were standing on their tables, hooting and hollering. I was a younger man at the time, and was one of them. I was so excited and I wanted to be involved in the organization. I recognized the sale value of bringing those people in and we need to bring those people along. We need to recognize them, year after year."

The Reno convention broke all records with a gross take of more than $900,000. The Friday and Saturday afternoon auctions generated $597,375 alone. Attendance topped 1,500 men, women and children.

Outstanding Achievement Awards were presented to Senator Barry Goldwater and the Arizona Desert Bighorn Sheep Society, Inc. Convention organizer Steve DeRicco and assistant organizer John Turner were able to include Sam Pancotto's Grand Slam again after it made its entrance in New Orleans. Tom Hinders drove his truck and trailers and delivered the sheep to New Orleans and later to the Phoenix convention.

Governor Herschler congratulated FNAWS for its outstanding success in Reno. "I'm awfully glad to hear that this year's auction was such a big success and am pleased to know that my small efforts may have had some bearing on the outcome."

The second highest auction item was the special Utah desert sheep permit purchased by Bill Leech of Texas who paid a whopping $32,000. Lloyd explained that by auctioning this permit on the open market, rather than by sealed bid, FNAWS realized an extra $10,000 over what the permit brought for the three previous years.

Keith Connors of Finlay River Outfitters, Ltd. in Prince George, British Columbia, organized a backpacking contest and fully donated a trophy goat and caribou hunt as first prize. This began a tradition of backpacking contests for several years but it doesn't continue today due to insurance coverage at the hotel during the active competition. Winner was Ron Platt, an outfitter from Encampment, Wyoming, with a time of 1 minute, 15 seconds. He edged out 18-year-old Ed Evans by .06 seconds.

"This will be four straight years Keith Connors has donated hunts to FNAWS. And his backpacking contest added a new dimension and creative way to raise money while having fun at the FNAWS convention," Lloyd Zeman wrote in the *Wild Sheep* newsletter in May 1983.

Men's arm wrestling champion was Bill Moynihan of Alberta, who beat Rick Perrin of Alaska and long-time champ Ray Woodward of Alberta. Ed Wiens of British Columbia was men's leg wrestling champion. Women's arm wrestling champ was Debbie Overly of Alaska, and Mayvon Platt of Wyoming was again women's leg wrestling champion. The competition grossed $4,000.

The Nevada desert bighorn sheep tag sold by sealed bid to Robert Furguson of Texas with a donation of $21,000, and FNAWS gave the entire amount to the Nevada Department of Wildlife for use in sheep management. Arizona Game & Fish Department and Colorado Division of Wildlife considered issuing bighorn sheep permits to FNAWS for auction or raffle to raise money for wildlife projects benefitting wild sheep.

Significant highlights of the 1983 Reno convention included a keynote speech by U.S. Senator Barry Goldwater of Arizona, a special award to Doc Thurston for beginning his Grand Slam after age 70, and the President's Award to Buck and Raymond Allemand of Midwest, Wyoming, for their donation of land in Wapiti, Wyoming, for the new FNAWS national headquarters and education center.

The land donated by the Allemand brothers totaled 30.9 acres 18 miles west of Cody with an assessed value of $94,500. The property was in the heart of the Wapiti Valley only 34 miles from the eastern gate of Yellowstone National Park. Originally, FNAWS planned to build on the Mac Lewis property but the extension of the Cody airport runway expanded the clear air zone to include much of the Lewis property. Federal regulations prohibited the construction of a building in a clear air zone. "The property donated by Buck and Raymond comes as the answer to our needs," Lloyd Zeman wrote. In reality, FNAWS did purchase a building across the street from the Buffalo Bill Historical Center in Cody, but they neither built the headquarters nor used the land donated by the Allemand brothers. The FNAWS Board of Directors voted not to build a monument to themselves, but rather to spend the money for sheep research, transplants and habitat restoration — a tradition that continues today.

Election of new Board members was conducted by mail for the first time and the results were announced in Reno. A total of 867 FNAWS members voted using this new method. The Board was reduced to 11 members from 17 and new directors included Earl Boose, Arnie Johns, Keith Johnson and John Turner. Departing Board members included John Babler, Lloyd Bare, Lowell Baier, Jim Froelich, Paul Asper, Harvey Kadlec, Michael Valencia and Bill Walters. Duane Smelser became FNAWS president for 1983-84.

Shovels turned Cody dirt during groundbreaking ceremonies on July 16, 1983, for construction of the FNAWS national headquarters and educational center. Don Dexter, director of the Wyoming Game and Fish Department (WGFD) and Thermopolis WGFD Commissioner Alice Hayes joined FNAWS directors, officers and members on the land donated by the Allemand brothers. Building sketches were unveiled by Architect William F. Peterson whose services were fully funded by FNAWS director John Turner of Pacheco, California. The building totaled 14,800 square feet with a library, a wild sheep display area, research facility and observation deck facing Jim Mountain, one of the most critical winter range and lambing grounds in northwest Wyoming. Donations arrived from across North America as members and FNAWS donors stepped up to the plate to help construct the headquarters.

"For just $1,000 you can buy (donate) 20 square feet or 10 square feet for $500. All of your donation, and everyone who donates $500 or more will have their name inscribed on a special wall in the educational center. Those donating $2,500 or more will have their name inscribed as a special educational center supporting contributor," Lloyd Zeman said.

Donors giving $10,000 or more had a cornerstone of the building earmarked for him or her and was presented a wild sheep bronze. Iowa FNAWS Chapter donated the flagpole and Texas Bighorn Society donated hand-carved doors.

FNAWS was developing a name for itself as litigators and lobbyists for wild sheep and their habitat. Lloyd Zeman provided in the November 1983 *Wild Sheep* newsletter a clear summary of their accomplishments. FNAWS won an important case against the U.S. Forest Service and Curtis Tungsten, Inc. in the 9th District Court of Appeals requiring an Environmental Impact Statement (EIS) to be done regarding the opening of a road through desert bighorn sheep range and lambing area in California.

FNAWS played a large role in the attempt to open up 12 million acres of park land in Alaska by trying to get this land reclassified as a "preserve" so hunting could be allowed. A sheep management program in California was supported by FNAWS that allowed limited hunting of surplus sheep as a management tool. FNAWS temporarily stopped the drilling of a proposed well in a critical winter range area in the Shoshone National Forest by convincing the Bureau of Land Management (BLM) to do an EIS before making any decision on the well. This action was accomplished without going to court.

A lawsuit was filed in federal court against Shoshone National Forest to better evaluate the impact of repeated geophysical exploration and seismic testing on bighorn sheep habitat, so other forest users would receive equal consideration. Don Schmalz explained the case.

"You couldn't go to the U.S. Forest Service. They were the ones who gave the oil company the right to do the blasting. The Sheep Foundation was the only one that had something to do with the sheep and could have a group out there or someone big enough to do battle with them."

"A lawsuit was filed against the Forest Service and the Shoshone National Forest. Because of that lawsuit, Steve Mealey, who was the forest supervisor for the Shoshone National Forest, had to reverse the Forest Service decision regarding seismic activity."

Basil Bradbury was hired by the Board at their January 1984 meeting, just prior to the San Diego convention, to finish a sheep annual and "be in charge of all phases from content and ad sales, to printing." Basil also joined the FNAWS national Board to fill the unexpired term of John Turner. At the February 1984 Board meeting, Tom Hinders was elected president and Tom Saad was elected president-elect. Dr. Schultz became secretary, following the service of Jerry Christian as secretary. Dr. Waite was elected vice president and Arnie Johns became treasurer.

Tom Hinders pointed with pride at the work that Basil did to upgrade FNAWS publications. He wrote Dr. Schultz in 1994, "With the help of the Board of Directors, we started our magazine, known today as *Wild Sheep*, under the guiding hands of a world-renowned hunter and past editor of *Petersen's HUNTING*. This man was Basil Bradbury. Mr. Bradbury passed away in 1993 in Wyoming, a land he loved. To quote Brad, `See you farther down the stream.' Brad was a great man!"

Tom Hinders felt the move to Cody was also one of his accomplishments during his term. At the same time as the move, the Chadwick ram became the official, copyrighted symbol of FNAWS for use on any printed material or permanent building, sign or wall. Tom worked feverishly to raise money for FNAWS and is credited with starting an aggressive merchandising program during this time period using the Chadwick ram painting as a basis for embroidery on gift items, including shirts, aprons, towels and caps. Gary Swanson paintings of a male mallard duck taking flight, a lion and a bighorn ram's head were also featured embroidery on early FNAWS merchandise.

Gary Swanson's Chadwick ram painting, shown here, became the official logo of the Foundation for North American Wild Sheep. The painting was the basis for early designs on FNAWS embroidery that graced shirts, aprons, hats and other sale items prior to 1986 that Tom Hinders and Dr. Roy Schultz helped organize.
Photograph by Dr. Roy Schultz.

Charles T. Buck, president of Buck Knives, Inc., remembers during this time period an effort to raise money for FNAWS by selling limited edition Buck knife sets featuring the Grand Slam of Lloyd Zeman etched on the knives' handles. Charles, known as Chuck Buck within the outdoor sporting industry, said he and Tom Hinders worked for close to a year to design the etchings and coordinate the production of the knives that featured a sheep head on one side of a knife and a typical wild sheep habitat scene on the other side. A handsome walnut presentation box lined with royal blue velvet protected the four knives and a belt buckle with the four Zeman trophies etched on its surface.

"Tom Hinders was instrumental in lining up Lloyd Zeman's excellent Grand Slam pictures for the blades and belt buckle. We produced between 2,000 and 2,500 sets and immediately contributed 100 sets to FNAWS to raise much-needed funds for the sheep. The remaining sets sold for $500 and business was brisk. We at Buck are delighted to have been a part of the early days of a great organization," Chuck Buck recalled in September 1998.

Basil's creation of the Wild Sheep color magazine was also a source of pride for President-Elect Tom Saad, who served as convention chairman for the San Diego convention and rose to the presidency as *Wild Sheep* continued to evolve into today's publication.

Plans for the San Diego convention began even before the group left Reno. Duane Smelser presented a clear summary of the highlights of the San Diego convention February 22 to 25, 1984. The convention auction revenue grossed $882,000 and the convention netted $745,500 with gross revenue at $1.2 million. All figures were new records for FNAWS.

The Nevada permit was auctioned for the first time. Bill Leford of Texas was high bidder with a donation of $67,500. The first Arizona desert bighorn sheep permit was auctioned, and Dan Pocapalia was top bidder at $64,000. Daniel A. Pedrotti, Al Barone and Joe LoMonaco each successfully bid on Mexican permits, and Doc Thurston was successful bidder for permits in Wyoming and Utah. Jim Brockie was top bidder for the Nevada permit with a donation of $25,000. The Nevada permit sealed bid auction was won by Charles O. Hundley of Texas with a high bid of $25,002.

Duane Smelser told Dr. Schultz about some of the highlights for the San Diego convention: "The Town and Country Hotel in San Diego was the site of our 1984 convention. Our keynote speaker Saturday evening was an old friend, U.S. Secretary of the Interior James Watt. Working with our friends from Alaska, we put together 25 donated sheep hunts for our record-breaking convention. My wife, Doris (Billie) Smelser, hosted the first ladies' fashion show and auction. Her models included Jonie Saad, wife of Tom Saad; Jill Fanning, wife of Ken Fanning of Alaska; male model Skipper Bettis and others."

Grits Gresham, shooting editor for *Sports Afield* and Miller Lite celebrity, was keynote speaker at the Friday, February 24, luncheon. Long-time supporter of FNAWS and Wyoming

Wyoming Governor Ed Herschler, left, received a set of custom-made Buck knives from Past President Jerry Christian in 1985 for his donation of the first hunting permit in 1980 that put FNAWS on the map for fundraising that benefits wild sheep. The Buck knives had etched portraits of Lloyd Zeman's Grand Slam sheep on their handles. Tom Hinders worked with Charles "Chuck" T. Buck, president and chairman of the board for Buck Knives, Inc., of El Cajon, California, to reproduce these limited edition knives that sold to FNAWS members for $500 each. Buck Knives donated 100 knife sets to FNAWS, free of charge.
Photograph courtesy of Dr. Roy Schultz.

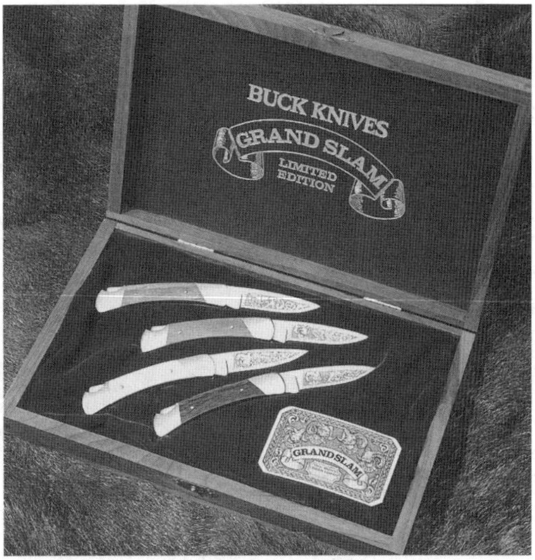

Four magnificently etched knives, produced by Buck Knives, Inc. and etched by Aurum, featured the Stone's, Dall's, bighorn and desert sheep taken by Lloyd Zeman when he collected his Grand Slam. Here is the publicity photograph produced in 1985 to sell the limited edition knife set that included a belt buckle featuring the full Zeman Grand Slam. A total of 2,500 sets were produced that sold for $500 each. The knives were presented in a handsome walnut case lined with royal blue velvet and the knife handles matched the colors of each sheep's fur.
Photograph courtesy of Buck Knives, Inc.

At left: Tom Hinders, left, and his 16-year-old son Matthew in July 1998, display the Buck knife set Tom purchased during an early FNAWS fundraiser. Tom worked for a year with Charles "Chuck" T. Buck, president of Buck Knives, Inc. of El Cajon, California, to create and produce 100 limited edition knife sets to be sold at FNAWS conventions to raise money to "put sheep on the mountain."
Photograph by Susan C. Reneau.

Gary Swanson, left, and Dan Pocapalia shared the hunting experience when Dan purchased an Arizona desert sheep permit in 1984 for $64,000. Dan was guided by Danny Nasca and Tom Saad. He tracked a specific ram nicknamed, "Balloon Horns," but never took the ram. Later, Tom Saad's son, Kyle, harvested "Balloon Horns" in the same hunting unit west of Kofa at age 13.
Photograph courtesy of Tom Saad.

Governor Ed Herschler was honored by the membership with a presentation by Duane of a set of Grand Slam bronzes created by Ben France, an artist from Cody. John Rinehart of the Wyoming Game and Fish was given a framed limited edition FNAWS bighorn sheep stamp and print set painted by Gary Swanson.

A desert bighorn sheep painting by Robert Bateman was announced in San Diego as the fourth in a series of limited edition prints and stamps sold through FNAWS to raise money for projects. The first print and stamp was of Gary Swanson's painting in 1980; the second print and stamp was of the Dall's sheep painting by Bob Kuhn in 1981; and the third in the series was a print and stamp of a Stone's sheep painting by Ken Carlson. The desert sheep painting completed the Grand Slam of prints and stamps for FNAWS.

Following the San Diego convention, on April 27, 1984, the Board of Directors voted no confidence for Lloyd Zeman after a secret ballot was conducted. Two members voted in favor of Lloyd and nine voted against him. One Board member abstained from voting. Severance pay for Lloyd Zeman included one month of full pay with benefits and six additional months at one-half pay with no benefits.

Board members present for the vote of no confidence were Tom Hinders, Tom Saad, Dr. Jerry Waite, Arnie Johns, Dr. Roy Schultz, Earl Boose, Steve DeRicco, Dr. James Duke, Keith Johnson, Duane Smelser, Marvin Wood and Jerry Christian. Visitors to the public meeting included John Babler, Pat Latham, Bill Ledford, Danny Nasca and Harvey Kadlec. The vote occurred in closed executive session without visitors.

Lloyd hand wrote his resignation on FNAWS letterhead stationery that read, "Board of Directors - I resign as executive director under and per the conditions suggested by the Board of Directors." A final settlement to Lloyd Zeman took place in April 1985 for $15,391.05. Former executive secretary Gretchen Stark settled for $4,000.

In July 1984 at the Board meeting, Keith Johnson moved to proceed with the planning for a permanent site and headquarters building in Cody, Wyoming, and the Board voted in agreement. Their goal was to complete relocation plans by the January 1985 meeting.

A new executive secretary was hired in September 1984 from a field of 46 applicants. The selection committee was commended for their excellent work. Five finalists were selected for final interviews, and the position was offered to Edward "Ted" Bryant for a one-year contract at a salary of $36,000, moving costs to a maximum of $3,000 and medical benefits as of the date of hire. Ted was present at the January 1985 Board meeting as the FNAWS administration director.

Questions arose throughout the development of FNAWS regarding use of the FNAWS mailing list and direct communication with FNAWS membership by Board members for business marketing purposes. At the September 1984 Board meeting, Arnie Johns made the motion, and Marvin Wood seconded it, that without Board authorization, no director or officer of the Foundation could communicate by letter, advertisement or any other means, any message of political or personal nature for his or her goals or on behalf of any other person he or she determines to need help of a political or personal nature. Authorization had to come from a quorum of the Board of Directors. The motion passed. Past presidents and founders point to the passage of this policy as helping to strengthen FNAWS. This policy is in effect today.

Dr. Red Duke and Doc Schultz pushed for a Center for Sheep Studies at the January 1985 Board meeting, and the Board directed Ted Bryant to work with Red on contacting universities across America that had wildlife management and biology degree programs. Red's plan was to establish a detailed network of communication with top-level wildlife researchers and biologists so money for wild sheep projects wouldn't be duplicated and repeated with the same individuals, government agencies and universities already receiving funding from FNAWS. "I thought I could finish this in two years during my president-elect and presidential years in FNAWS. That was my purpose." Red Duke reported in January 1986 that six universities had applied for the Sheep Research Institute and his committee, consisting of Dr. Schultz, Tom Saad, Dr. Fred Carsh of the University of Michigan, Dr. Oli Neilson of the University of Ontario, and Dr. R. Hensel of Cornell University, approved Texas A&M, Colorado State University and Idaho-Washington State. Unfortunately, Red's and Roy's idea of an institute was never funded, although today numerous respected research facilities are funded by FNAWS.

Reno was such a financial success in 1983 that the leadership voted to return to Reno in 1985. Reno remains the all-time favorite, and financially lucrative place for FNAWS to go — bar none.

The 1985 convention was held February 20 to 23 at the same white MGM Grand in 1983. Today, the hotel is called the Reno Hilton and remains a popular location for successful FNAWS annual conventions and auctions. Registration topped 2,000 and the convention grossed more than $1.2 million. Thanks to the friendly bidding duel between Doc Thurston and Dan Pocapalia, each man paid $40,200 for the Wyoming hunting permit for a gross profit of $80,400! Membership broke the 5,000 mark, and FNAWS founders were amazed. Never did any of them think that a small gathering of avid wild sheep hunters at Mt. Horeb, totaling 12, would spawn a membership of 5,000. Spirits ran high as the Board approved $600,000 in grant-in-aid funding of wild sheep and wildlife habitat projects following the 1985 convention.

Outstanding wildlife biologists and FNAWS volunteers were recognized for their service to wild sheep and their habitat. Duane Smelser received the Chadwick Ram Award. Outstanding Achievement Award was given to John Elliot, a wildlife biologist from British Columbia who worked closely with Garry Vince and early FNAWS members, outfitters and hunters to solve the problem of wolf overpopulation and habitat improvement. Outfitters and wild sheep hunters had recommended controlled burns for many years, and after exten-

Howard Pollock receives an award from FNAWS President Tom Hinders in 1985 for his service. The 8th convention of FNAWS was held in Reno from February 20 to 23, 1985, at the MGM Grand Hotel, which was eventually named the Reno Hilton.
Photograph courtesy of Dr. Roy Schultz.

Doc Thurston, center, was recognized at the 1985 FNAWS convention for his generous donations to help put wild sheep on the mountain. Here, Past President Jerry Christian, left, congratulates him as FNAWS President Tom Hinders smiles as he received the Gary Powell Memorial Award.
Photograph courtesy of Dr. Roy Schultz.

Past President Duane Smelser, left, stands arm-in-arm with FNAWS President Tom Hinders at the 1985 FNAWS convention when Duane was recognized for his service to the Foundation.
Photograph courtesy of Tom Hinders.

Jim Froelich, center, received special recognition in 1985 from FNAWS President Tom Hinders, right, for pulling together sheep hunters and organizing the first meeting at Mt. Horeb, Wisconsin. He was named "Grandfather of FNAWS." Beside them is Dr. James "Red" Duke who served as Master of Ceremonies.
Photograph courtesy of Dr. Roy Schultz.

Basil Bradbury, left, a dear friend of Tom Hinders, is congratulated by Tom for the publicity Basil gave to FNAWS in Petersen's HUNTING *magazine in its early years of existence.*
Photograph courtesy of Tom Hinders.

sive study, John agreed with Garry and others.

Jane Babler, who worked for years for no salary and later received only $100 per week from FNAWS as executive secretary, was presented the "Outstanding Supporting Member Award." Dr. Red Duke, who worked with Jane on the meticulous planning of the Houston convention, nominated her for the award at the January 1985 Board meeting just prior to the 1985 convention. The Board voted unanimously to present the award to her.

Jane wrote in the August 1985 *Wild Sheep* magazine: "I am never at a loss for words, but I sure was thunderstruck that evening. When Red called my name to come up on stage, I was totally stunned. Apparently many of my friends, to say nothing about my husband, were all aware that I was going to receive the award, but no one let the cat out of the bag, much to my chagrin."

Jane expressed thanks to all the friends she and John met in FNAWS and complimented the current leadership for a job well done. "It gives me great pleasure to see the Foundation as it is today — strong and ever-growing — and know that I was a small part of it."

Speaker of the House Gib Lewis of Texas and Tony Brummet were recognized as a national and international statesmen, respectively, in the field of wildlife conservation and management. Tom Hinders, Howard Pollock and Jim Froelich received special recognition for years of dedication to FNAWS and Doc Thurston received the Gary Powell Memorial Award for his boundless support of FNAWS at its annual auctions.

The 1985 FNAWS convention saw the formal adoption of a significant financial control that required all merchandise purchased at an auction to be paid for before the merchandise was released to the successful bidder. The successful bidder was required to sign an unconditional promise to pay in full for all goods or services on or before the close of the auction.

The agreement each successful bidder signed also stated that, "The Foundation for North American Wild Sheep offers no warranty, either expressed or implied, for goods or services offered at this auction and agree that any dispute arising in conjunction with this auction item will be settled directly between the donor and the bidder."

One of the reasons why FNAWS pulled itself together, professionally and financially, was the steadfast service by Dean Picl, as noted by Ron McMillen and others who were inter-

Outfitter Gary Moore was one of several early supporters of FNAWS who donated hunts. "Those early outfitters in FNAWS who donated hunts encouraged others to step up and help," recalled Ron McMillen. Gary is shown here in the early 1980s.
Photograph courtesy of Tom Hinders and Dr. Roy Schultz.

Johnny Drift, left, and Steve DeRico help with the 1985 FNAWS auction.
Photograph courtesy of Dr. Roy Schultz.

Lu Van Horn, left, is recognized by FNAWS President Tom Hinders in 1985.
Photograph courtesy of Dr. Roy Schultz.

John Turner moments after he was a successful bidder at the 1985 Nevada hunting permit that brought $42,000 to FNAWS.
Photograph courtesy of Tom Hinders.

viewed for this book. Dean is an attorney from California who began his service to FNAWS as its attorney in 1985. First mention of Dean at Board of Directors meetings was February 19, 1985. He has attended each Board of Directors' meeting and offers his legal advice on a variety of critical matters.

"Dean's commitment to the Foundation will never be recognized at the level it should. Not only has he volunteered his time as an attorney, which has saved us hundreds of thousands of dollars over the years, but he has saved us a lot of grief over the years," Ron McMillen explained. Ron said Dean's ability to write little passages and bylaw changes helped the Board of Directors express what they for a job wanted to say and do. "Plus, he's a marvelous speaker. You simply ask him to help, and he will help. He's a genius and a lot of fun."

Basil Bradbury, center, pours champagne at the Ladies' Luncheon in 1985.
Photograph courtesy of Dr. Roy Schultz.

Dean Picl continues today as the attorney, writer and friend to FNAWS and is often seen at conventions as master of ceremonies during regular convention meetings and banquets. Because Ron McMillen and Dean always attend each Board meeting, and have been since the mid-1980s, they have become the unofficial historians for FNAWS. "We've sat there for so many years, I'd like to think we have contributed. I feel that Dean and I are the only ones who have that historic perspective," Ron said.

Positive changes were happening in FNAWS in 1985 and founders and FNAWS members were optimistic. Tom Saad reported to the Board on February 24, 1985, that the Guides and Outfitters Advisory Council met for the first time at the Reno convention. Tom said, "The Council's input to the Board will be furnished for consideration prior to funding action." The consensus of the Board was that the Advisory Council could be most helpful during the Board's deliberations for funding of wildlife projects in various geographic locations, especially if the outfitters were asked their opinion from a laymen's standpoint as the appropriateness of the study.

Membership topped 5,000 for the first time after the 1985 Reno convention, and $600,000 was approved for many Grant-in-Aid projects. This was the convention where past presidents officially met for the first time and formed the Past Presidents Council, a social gathering for former FNAWS presidents and an opportunity for them to discuss FNAWS and its policies and activities with other former presidents. Past presidents could not re-join the Board of Directors, and the Board voted unanimously to not reimburse past presidents who decided to attend the Past President Council. This was also the first time the Idaho FNAWS Chapter was formed just as Idaho Fish and Game Department asked for emergency help from FNAWS because of a massive bighorn sheep die-off in Hells Canyon.

Another significant FNAWS leader made his entrance in 1985; Pat Latham of Texas offered to analyze the revenue and expenses at the Reno convention to provide guidance for future conventions. The Board unanimously authorized Pat on February 24, 1985, to analyze, investigate and study how administratively FNAWS should manage the investable funds of the Foundation and its several operating bank accounts. Lowell Baier made the motion to

Jerry Christian, left, and Tom Hinders pose by their one-shot Wyoming antelope in 1985.
Photograph courtesy of Tom Hinders.

The 1985 team for the Lander One-Shot Antelope Hunt in Wyoming who represented FNAWS included, left to right, Grits Gresham, Tom Hinders and Dean Buntrock. The FNAWS team were the winners because they each downed an antelope with a single shot! Grits was keynote speaker for FNAWS in 1984 and has remained a close friend of the organization ever since. The Lander One-Shot Antelope Hunt takes place each year. For more information about this unique competition, call them at 307-332-8190.
Photograph courtesy of Tom Hinders.

authorize Pat to perform these vital duties and Marvin Wood seconded the motion. An extensive financial report and analysis was presented by Pat at the May 1985 Board meeting. The Board agreed after Pat's presentation to let the FNAWS Budget and Finance Committee work to get a complete audit of accounts, payables and receivables and inventory as soon as possible. The Board also agreed with Pat that an annual audit of the books must be conducted annually and the results of the audit would be published in *Wild Sheep*. Nancy McDaniel of First Consolidated Service Corporation of Phoenix continued the bookkeeping that was audited by Peat, Marwick, Mitchell and Company each year.

A conference call on November 5 and 12 with the executive committee members President Saad, President-Elect Duke, Vice President Waite, Treasurer Latham, Secretary Baier and Past President Hinders centered on hiring Lee Hamilton to assist Basil Bradbury at a rate of $13,500 for six months until a location of the national headquarters could be determined. Prior to being offered the six-month contract, Lee was paid as an independent contractor at $26 per hour. On November 26, 1985, the same executive committee concluded the Foundation's financial deficit spending had to stop immediately. A payment authorization procedure for all future financial commitments was put in place. All FNAWS credit cards were cancelled immediately as well as telephones in Board of Directors' homes that were paid by FNAWS. No further cash advances would be given to Board members and all major expenditures would be deferred until the new budget was approved, especially including publication equipment. The leaders also recommended cancellation of the *Wild Sheep Annual* unless it was finished by the Hawaiian convention (9th convention).

The Center for Sheep Studies that Dr. Red Duke proposed was to be contained at the FNAWS national headquarters. At the September 1985 Board meeting, minutes stated: "The Board was of the uniform consensus that the location of a site for our research institute and our headquarters location should move in parallel rather than in sequence. Since we have $250,000 in our building fund, we have to be cognisant of its many contributors and their notions at the time of their donations as to where the headquarters would be eventually located." Lowell Baier as chairman of the Site Selection Committee was selected to visit potential cities in the West with fellow committee members Dr. Duke and Dr. Schultz at the January 1986 Board meeting. This year began with more than 5,722 paid regular members and 362 life members, according to Ted Bryant's administrative report. This was an all-time high for regular and life memberships for FNAWS.

Texas Gov. John Connolly, center, talks with Dwane Smelser, left, and Jerry Christian at a party in 1985.
Photograph courtesy of Dr. Roy Schultz.

Dr. Duke assumed the presidency in 1986 with Pat Latham as treasurer. "We met in Las Vegas. There was a lot of bickering going on, all kinds of craziness going on, lots of backbiting. It wasn't bothering me, but they were all fighting each other. The attorney general in Texas at the time had just busted a nonprofit foundation and its directors for not doing a good job. I took that clipping, the bylaws (of FNAWS) and the articles of incorporation, put on a suit and tie, and went down there and read them that stuff."

Dr. Duke told the Board he didn't need the job of president because he was already president of the Boone and Crockett Club that year, and he wasn't going to put up with any backbiting.

"I had no intention of putting up with a lot of bull _____," Dr. Duke explained, and said he asked the Board to vote again if they really wanted him as president because "I don't really need it. My ego doesn't need it, and I sure don't need to waste my time with it."

The Board elected Red Duke as president and before the meeting ended, the Board and Red Duke had cut 30 percent out of the Foundation operating expenses and 60 percent out of the directors' expenses. "Over the next year, we managed to do that," Dr. Duke said. He said in the middle of all this financial restructuring, the membership and Board decided to continue their search for a permanent headquarters.

In May 1986, the Board voted to send the committee members to the three final headquarters locations under consideration: Cody; Colorado Springs, Colorado; and Boise, Idaho. The total committee consisted of Dr. Duke, Lowell Baier, Dr. Schultz, Pat Latham, Marvin Wood and Arnie Johns. In August 1986 newly elected Board of Director member Don Schmalz of Cody was added to the committee.

John Turner's Nevada sheep hunt in 1985 is shown here. John purchased the hunt at the FNAWS convention for $42,000 and continued as a great supporter of FNAWS and a national FNAWS Board member.
Photograph by Tom Saad.

A debate about where to locate a national headquarters aside, the upcoming Hawaiian convention was set for February 26 to March 1, 1996, in Honolulu. The convention grossed $1,019,096 and 85 percent of the profits from this convention were given to a variety of Grant-in-Aid projects much to the delight of FNAWS membership. FNAWS accountant Nancy McDaniel reported in May 1986 that although the gross revenue was great, expenses topped $700,000, leaving a net income of $303,541 after the Board authorized purchase of a Ram Charger truck at a cost of $15,000.

Added to that, not as many FNAWS members attended the convention in Hawaii for a variety of reasons and those who did spent more time sightseeing and sunbathing than bidding at the auction.

Harvey Kadlec explained what happened in Hawaii. "Hawaii was a great place to go for the people but it was horrible for a convention because the members didn't want to be inside the hotel."

Ron McMillen had been traveling North America talking to guides and outfitters about donating wonderful hunts to be auctioned at the Hawaiian convention. When few people attended, Ron said his embarrassment mounted as the hunts he had received as donations from guides and outfitters didn't sell for as high as they should have.

"Before the Hawaiian convention in 1985, I believe I had gone out and gotten 21 sheep hunts. I had been told to go out and get as many hunts donated as possible. We were still in our infancy as far the auction was concerned. We didn't know how to do it or how to handle it. What happened was that not many people went to Hawaii. When we got there we didn't have enough buyers for the hunts. I had dragged a portion of the outfitters' livelihoods away from them, and then the hunts sold terribly. I had one outfitter whom I thought would start swinging at me. I thought he was going to start nailing me, but he didn't."

After the experience in Hawaii, Ron became more involved with the inner workings of the auction so duplicate species hunts would not be donated and auction items could be monitored. "Since Hawaii I have become the auction liaison. What I do is stand during the auction and monitor what happens. Before I started doing this, we were getting done extremely late and we were losing our audience. I make sure we stay on time. I also monitor and make sure someone is there to talk about an item. I just make sure the auction flows smoothly."

A "wish list" was established by Ron McMillen that meant FNAWS only sells two special hunts of the same species at Friday and Saturday auctions. "We won't take over so many caribou, so many moose, so many bear, so many Dall's sheep, so many Stone's sheep and so on. We don't want to offer more than we can get quality prices for. It's not fair to FNAWS, but

The 9th convention of FNAWS took place in Honolulu, Hawaii, February 26 to March 1, 1986. The Hawaiian convention attracted a friendly crowd of wild sheep supporters, shown here in front of the hotel at the start of all festivities. Hawaii was a great place to visit but not a great place to raise money. Reno remains the all-time-best place for FNAWS to have a successful auction.
Photograph by Tom Saad.

Three FNAWS supporters, all in a row with fur coats, include, left to right, Joni Saad, Fran Greaves and Phyllis Tucker during the Ladies' Luncheon and Auction in 1986.
Photograph courtesy of Tom Saad.

it is especially not fair to the outfitter. There are still people who have bad feelings from Hawaii and who have never donated since. This was an expensive lesson, but it was a positive lesson in the end," Ron concluded.

The Outstanding Achievement Award in Hawaii was given to Dick Weaver of California for his outstanding efforts to establish a sheep management program in and for California. Dr. Jack Frost was recognized as the first wild sheep hunter to collect a Grand Slam with bow and arrow. No international, national or statesman's awards were selected for the 1986 convention.

Significant to the evolution of FNAWS was the first time Montana Department of Fish, Wildlife and Parks donated a bighorn sheep hunting permit to the auction. Arthur Dubs of Oregon, a long-time FNAWS supporter over the years, successfully paid $79,000 and continued to be the successful bidder for the Montana permit through 1988. North Dakota joined FNAWS to donate a California bighorn sheep for the Hawaiian auction. Jim Rockstad bought it for $17,000. The Mexican permit sale price spiked to $25,000 thanks to the bidding success of Lee Hicks. Three Mexican permits had sold for $18,050 each in 1984.

Discussions by national Board members had centered on raising regular membership to $50 per year in light of concerns about finances, but the membership voted to keep the annual

Dr. Jack Frost of Anchorage, Alaska, beside his Grand Slam exhibit he took with bow and arrow, which was the first time such an accomplishment was achieved. His Grand Slam was on display during a FNAWS convention.
Photograph courtesy of Tom Saad.

membership at $25 at the Hawaiian convention. Anyone who paid $50 for annual membership received a refund. Family memberships that included parents and all dependent children remained at $75, and life memberships included the spouse and dependent children of the person paying for such membership.

The Hawaiian convention taught FNAWS members significant lessons, some positive and some negative, but the organization continued to grow and head toward the next annual convention set for January 21 to 24, 1987, at Opryland Hotel in Nashville, Tennessee.

The question of where the national headquarters would locate was finally settled in 1987 when the Board of Directors decided to move the offices to the current location at 720 Allen Avenue in Cody, buy the existing building, and not build a new headquarters and research center even though beautiful architectural drawings had been created.

President Duke appointed Lowell Baier as program chairman for the Nashville convention and asked Pat Latham, Tommy Caruthers, Charles Jagou and Daniel "Dan" A. Pedrotti to work with Lowell. Pat took care of general arrangements. Charles was in charge of all audio-visual equipment and operations. Tommy and Dan picked the programs and saw to it that all our speakers appeared on time.

Dan Pedrotti recalled his first involvement in FNAWS at the Nashville convention. "Pat Latham took care of general arrangements. Charles Jagou was in charge of all audio-visual equipment and operations. Tommy Caruthers and I picked the programs and saw to it that all our speakers appeared on time."

Dan continued. "In the process of meeting to do the preconvention work, it dawned on us that this was the heart of hound dog country and that we ought to have a surprise coon hunt for President Red Duke. We devised a plan whereby during opening ceremonies a pack of howling coon dogs would follow a well-scented trail down the middle isle, right up to Dr. Duke's feet.

"We got up very early and had the dogs in a janitor's closet. The trail was laid and, as we waited for the exact moment, the hotel convention manager noticed a rather strong odor emanating near the closet. As she approached the door to inspect, all hell broke loose as our hounds let out a series of ear-splitting yelps. Our prank was discovered and it took some tall talking by Tommy to keep the entire convention from being dislodged from the Opryland Hotel!" Dan Pedrotti concluded.

Tommy and Dan went on to work as Program Chairmen at every subsequent convention and when Dan was Convention Chairman in San Antonio, he asked Tommy to be in charge of the programs and food service. Pat Latham ran the rest of the operations. Unfortunately, Charles Jagou suffered a debilitating stroke after the Nashville convention and couldn't continue to work on their team.

President Duke pointed to his term of office during this period as "pulling FNAWS together financially and getting it more organized." He also recognized Treasurer Latham as a significant reason why 1987 was a positive turning point for the Sheep Foundation. "Pat and I really hit that ball hard and we also cut out the frills and got down to business," Dr. Duke explained. In the book chapter, "Presidents Speak," Dr. Duke presents a more poignant explanation of how he helped pull FNAWS together in 1986 and 1987.

Prior to the Nashville convention, the auditor's report was presented by Dave Lemoine of Peat, Marwick & Mitchell in May 1986. The audit ordered FNAWS to establish a better system of internal accounting control and to establish short- and long-term budget and fiscal planning. Treasurer Latham proposed that to save money and combine duties, administration and publication positions should be combined, thus eliminating one of the top two salaries in the Cody office. This proposal was rejected by the Board for a short time, but revisited later in 1987.

The 30.9 acres of donated land in Wapiti, Wyoming, was put up for sale at a list price of $10,000, and a complete inventory of art objects at FNAWS headquarters was evaluated to determine which pieces should be sold for the highest price. The property still remains under the ownership of FNAWS. President Duke wrote a letter to all FNAWS members asking for additional donations to cover operating expenses and Grant-in-Aid projects FNAWS had already committed to fund.

The Board voted at the May 1986 Board meeting to raise annual dues to $40 and to allocate $16.50 for the FNAWS magazine. Basil Bradbury was ordered to reduce the number of issues from four to three and to limit page count to 76. Basil's salary and salary for support staff and expenses to create the color magazine were paid from advertising revenue and not from the general operating budget. With advertising money going to fund the publishing staff, rent and utilities, Basil and two staff people were removed from the regular FNAWS payroll.

Ted Bryant was offered a second-year contract following the Hawaiian convention with no salary increase, remaining at the $41,000 level with authority to demand executive performance of the entire headquarters staff and keep within a set budget. Ted decided to resign effective October 31, 1986, and added that "the ladies in the office could run it quite well all by themselves."

A VISA and MasterCard credit card program, presented by Reed Taylor of Agriculture Insurance Administrators Company of Lewiston, Idaho, involving First Interstate Bank, was approved by the executive committee to help raise more funds for FNAWS with a small percentage of all purchases going to FNAWS. Any FNAWS members who used their credit card would contribute to FNAWS each time they bought something. This program began at the Nashville convention.

The sheep annual was cancelled and the Board authorized refunds to anyone who had prepaid for the book. Basil Bradbury was encouraged to complete the project on his own time and at his own expense, "hopefully for profit," Board notes reported. Following lengthy Board meetings regarding the economic status of FNAWS, the Nashville convention was a success.

Dr. Duke felt good about efforts in Texas to reintroduce wild sheep and to change a 1903 law that said if FNAWS got enough wild sheep and the wildlife biologists located another ram, then the governor could offer a permit for auction. The first Texas hunting permit was offered at the Nashville convention in 1987 because of Dr. Duke and many other FNAWS volunteers who worked with Texas government to change the 1903 law. "We kept trading antelope for sheep, so that's why we have 500 free-ranging sheep in Texas," Red Duke said.

Early FNAWS VISA card, this one with the National Bank of Commerce.

Doc Thurston successfully bid on the first Texas permit and paid $61,000 not to hunt anything, Red recalled with a smile. "The next year we got the permit again, and the guy never went hunting." The Montana permit that brought $79,000 from Arthur Dubs in 1986 jumped to $109,000 by Arthur in 1987, much to the delight of all attending Nashville. The Nevada sheep and antelope hunting permit went for $57,000 thanks to the bidding success of Mike Boyce. The spring issue of *Wild Sheep* in 1987 also thanked Jack Hayes for his purchase of the Arizona permit for $39,000; Robert Olander and Daniel Smith for their purchases of Mexican permits for $23,000 and $25,000, respectively; Bernard Quintiliani for his purchase of the Wyoming permit for $35,000; and, Jack Wanamaker for his purchase of the North Dakota permit for $21,000.

"Dubs' generous contributions to wild sheep management in the state of Montana has been one of great significance in their program's objectives," Basil Bradbury wrote in the magazine.

In 1987 Dr. Duke asked Ron McMillen to draw up a new donation form. "Here's what the deal was. Make it to where we make no deals. We have set cuts and people donate to that or we don't take it. Another thing, we're going to do this and we're not going to give anyone in the audience guarantees," Ron McMillen recalled what Dr. Duke told him.

Ever since Nashville, if something sells at the FNAWS auction, it sells as is, Ron explained. The one exception was in 1988 when the Idaho hunting permit was first auctioned. Dr. Bob DiGrazia had made a deal with Idaho Fish and Game Department regarding a minimum bid, but that deal was only good for 1988. After that, Idaho received whatever was bid at the auction with no guarantees.

The Nashville convention generated $162,050 for various Grant-in-Aid projects that year. Vice President Arnie Johns graciously took over all responsibility of the Nashville convention after John Turner resigned from the volunteer post.

Some of the hottest names in country music headlined events at the Nashville convention, including Lynn Anderson who sang, "I Never Promised You a Rose Garden," and her other hits at the Saturday luncheon. Tom T. Hall, nationally recognized songwriter and recording artist of more than 30 albums in 1987, headlined the Saturday night banquet. Gary Morris, Libby Hurley and Ed Bruce ("Momma, Don't Let Your Babies Grow Up to be Cowboys" and more) also performed at various FNAWS functions. The first annual photo contest began in Nashville with Jim Zumbo, current hunting editor for *Outdoor Life* and contributing editor in 1987, as the first judge. Two hundred exhibitors contributed to the excitement of the overall convention.

The ever-popular horse packing contest began in Nashville with a crowd of men, women and children cheering for their favorite team. Outfitter Stretch Main of Northwest Territory received first place with hunter Charlie Campbell of Mohn, Pennsylvania. The horse packing contest remains a popular social event at current FNAWS national conventions.

Once again, the arm and leg wrestling drew a large crowd. Women's leg wrestling champion was Bonnie Loche. The men's leg wrestling champion and the men's leg wrestling super champ was Shane Hansen. President Waite was runner-up as the men's leg wrestling champ! The women's arm wrestling champion was Joanie McKinnon. Irv Lozier was the lightweight men's arm wrestling champ and men's lightweight wrestling super champ, and Jim Rohrer was the men's heavyweight arm wrestling champ. Jim Harrower was the arm wrestling heavyweight super champ.

Regular membership per year in 1987 stood at $40 and sustaining membership per year was $200. Life memberships were $1,000 and annual family membership was $75 and corporate annual supporting membership was $1,000. Annual company membership stood at $200 and annual international membership was $100.

At the Nashville convention, FNAWS member Tommy L. Caruthers of Texas organized the first gathering of "Super Slammers" who were men and women who took 12 wild sheep species from seven different countries. Tommy presented each "Super Slammer" with a certificate attesting to their achievement and created a permanent plaque with each person's name engraved upon it to display at all future conventions.

Special thanks were given to FNAWS member Pete Sain, Jr., of Manchester, Tennessee, who contributed large amounts of time to the success of the Nashville convention by obtaining a variety of donations and manpower from businesses and organizations, and "many more coordinating efforts too numerous to mention."

Dr. Waite was elected president at the Nashville Board meeting, and Ron McMillen shared insight into how that happened.

"Red stood there and said, `Well, we're going to nominate Jerry Waite to the presidency.' Of course, he says this in his Texas drawl. `Jerry Waite has been the hardest working man for the Foundation. He is going to be president. All those in favor, say I.' Red said, `It is unanimous.' And he slammed the gavel down and never asked for a no vote because he said that was the way it was going to be. No one said anything. Of course, somebody could have said something, but they didn't. And, so Jerry Waite was president," Ron explained. Ron was on the sideline like he had been forever — at least since 1985. He said this was the only time it has ever been done this way and one of the rare times it was not done in closed session.

Basil Bradbury continued building *Wild Sheep* magazine but left in 1987 to pursue other interests. The editing and design of the magazine, plus the advertising sales efforts, were brought in-house for the spring 1988 issue of the magazine with the staff assistance of Sandra Nicola. Sandra continues today as the FNAWS national staff person responsible for advertising and publications. Ron's company, Sigler Printing & Publishing, in Ames, Iowa, continues as the printer and coordinator of *Wild Sheep* design and production. Ron's company has printed *Wild Sheep* since the Basil Bradbury era and explained how this relationship developed starting in 1985.

"Tom Hinders brought on board Basil Bradbury as editor and publisher of *Wild Sheep*. Basil took it from just a little newsletter to a nice publication. Basil brought the publication a long ways, but he had trouble with his printer at one time. Tom Hinders called me, and I gave Basil a price. It was a better price than he had been paying. He'd been having trouble getting it out on time, so we (Sigler Printing) turned it out."

"With Basil, there really was never a contract. It was a gentleman's agreement where we went from issue to issue. He grew the magazine. During Red Duke's presidency there was some controversy. Basil survived Red's presidency. I believe it was during Jerry Waite's year as president that Basil had built the magazine to the heights he could. The Board of Directors

The Spring 1988, issue of Wild Sheep *was the first issue where the advertising and content were handled by Sandra Nicola. It was also the first issue where the layout and printing were done by Sigler Printing & Publishing, Inc. This issue also started a tradition of using award winning photos from the annual conventions in the magazine. The photo used on the cover of this Spring issue was the 1st place photo in the 1987 FNAWS Photo Competition, it was taken by Ron Platt of Wyoming.*

felt it was time for a change and time to move on. They wanted to bring the magazine in-house," Ron explained.

Ron McMillen was grateful for what Basil Bradbury accomplished and how he got the magazine on track. The Board decided to take the magazine in- house with Sandra Nicola in charge of compiling most of the information submitted and attracting advertisers. The Board in 1987 asked for numerous printers to submit formal bids for the magazine and discovered Sigler Printing & Publishing was the least expensive for the product produced.

"Under John Harris' presidency, I got a formal contract. Some people questioned if I was in FNAWS for the magazine at that time, so they put it out for bid and found I was far and away the cheapest by a long shot. The Board gave me the contract. I've operated under that same contract probably for the last 10 years," Ron said. Ron explained that Sandra Nicola was hired by the Board as Basil was leaving FNAWS and has done a "super, super job" since the spring 1988 issue of *Wild Sheep*. Work on that issue by Sandra began in winter 1987.

Dr. Waite created the famous motto, "Putting Sheep on the Mountain," for FNAWS that endures today as the theme for everything this dynamic group does. This theme was used for the first time at the 1988 Reno convention as FNAWS entered its next 10 years of achievements and permeates the essence of what it means to be a member of the Foundation for North American Wild Sheep.

The first 10 years of FNAWS from 1977 to 1987 involved some trauma but, in the end, the organization thrived and evolved into one of North America's most successful wildlife conservation organizations — maybe THE most successful conservation organization ever created by a small group of men and women who knew they wanted to give back to wild sheep what the wild sheep had given them.

The next few years leading to the 21st century held promise, evolution and financial expansion for the Foundation for North American Wild Sheep and this progression of success continues as FNAWS enters the new millennium.

Dr. Roy Schultz's and Susan C. Reneau's biographies are found in Chapter I of this book.

11th Convention of FNAWS
February 24 to 28, 1988
Bally's Casino Resort (Reno Hilton)
Reno, Nevada
Convention Chairman: Arnie Johns
President: Dr. Jerry Waite, 1987 to 1988
Theme: Putting Sheep on the Mountain

12th Convention of FNAWS
February 15 to 18, 1989
Bally's Casino Resort (Reno Hilton)
Reno, Nevada
Convention Chairman: Tom Saad
President: Arnie Johns, 1988 to 1989
Theme: Putting Sheep on the Mountain II, An International Effort

13th Convention of FNAWS
February 28 to March 3, 1990
Bally's Casino Resort (Reno Hilton)
Reno, Nevada
Convention Chairman: John Harris
President: Harvey Kadlec, 1989 to 1990
Theme: Youth and Education for the Future of Wild Sheep

14th Convention of FNAWS
February 20 to 23, 1991
Bally's Casino Resort (Reno Hilton)
Reno, Nevada
Convention Chairman: Arnie Johns
President: John Harris, 1990 to 1991
Theme: Hunters Rights, Old Traditions, New Frontiers

15th Convention of FNAWS
February 26 to 29, 1992
Town and Country Hotel
San Diego, California
Convention Chairman: Arnie Johns
President: Howard W. Pollock, 1991 to 1992
Theme: The Partnership Continues

16th Convention of FNAWS
February 24 to 27, 1993
Reno Hilton
Reno, Nevada
Convention Chairman: Dan Artery
President: Dr. Robert DiGrazia, 1992 to 1993
Theme: FNAWS Membership . . . The Success of Our Past, the Promise of Our Future

17th Convention of FNAWS
February 16 to 19, 1994
San Antonio, Texas
Convention Chairman: Daniel A. Pedrotti
President: Craig Johnson, 1993 to 1994
Theme: Ethical Hunting; Our Heritage for Wildlife Conservation

18th Convention of FNAWS
March 8 to 11, 1995
Phoenix, Arizona
Convention Chairman: Pete Cimellaro
President: Garry Vince, 1994 to 1993
Theme: North America Celebrates the Conservation of Wild Sheep

19th Convention of FNAWS
February 21 to 24, 1996
Reno Hilton
Reno, Nevada
Convention Chairman: Leland Speakes, Jr.
President: Pete Cimellaro, 1995 to 1996
Theme: Wild Sheep . . . Our Vision - Their Future

20th Convention of FNAWS
February 5 to 8, 1997
Pennsylvania Convention Center
Philadelphia, Pennsylvania
Convention Chairman: Phyllis Carlson
President: Daniel A. Pedrotti, 1996 to 1997
Theme: 20 Years Putting Sheep on the Mountain, An Inspiration . . . A Triumph . . . Our Legacy . . . Our Challenge

21st Convention of FNAWS
February 18 to 21, 1998
Reno Hilton
Reno, Nevada
Convention Chairman: Royce Wood
President: Phyllis Carlson, 1997 to 1998
Theme: New Heights for Wild Sheep: Conquering Disease . . . Improving Habitat . . . Managing for the Future

22nd Convention of FNAWS
February 17 to 20, 1999
Reno Hilton
Reno, Nevada
Convention Chairman: Royce Wood
President: Leland Speakes, Jr., 1998 to 1999
Theme: New Beginnings - Wild Sheep in the 21st Century.

Chapter IV

FNAWS Matures: The Climb Goes On 1988 to 1998

Sandra L. Nicola and Susan C. Reneau

"Putting Sheep on the Mountain." His phrase and theme for the 1988 Reno convention was visioned in 1986 by then President-Elect Dr. Jerry Waite.

"If we really stop and think — this simple theme best describes FNAWS and our present and future goals. I hope we never forget the true meaning of the Foundation for North American Wild Sheep."

Dr. Waite told Dan Pedrotti how it was he thought of the saying that continues as the motto for FNAWS today. "During the start of my last three-year term on the FNAWS Board of Directors I realized I was in line to be one of its presidents. Every year picking a convention theme was a chore. It was hard to come up with a good idea for the Board to vote on."

"My phone was an open line, 24 hours a day, to my veterinary clients and FNAWS members, so I always had a pen and pad on my nightstand to jot down clients' calls as well as notes, etc., concerning FNAWS. One night I woke up and was writing parts of a FNAWS speech that I had to give at one of the Northern Guides and Outfitters meetings. One of the sentences contained this phrase and it suddenly stood out like a flashing neon sign. I underlined it and saved it for my term. Then I went back to sleep.

"It was the only idea brought up and was quickly voted on. Now it is a very popular phrase that is used many times. What else so aptly describes the goals of the Foundation for North American Wild Sheep," Dr. Waite concluded.

Dedicated to the memory of the late Jim Ford and Clark Engle, the 1988 Reno convention from February 24 to 28 was the largest convention to date, with more than 2,500 in attendance. The Foundation put sheep on the mountain by approving grants totaling $714,149. The state of Idaho joined the FNAWS permit auction arena, adding greatly to the funds available for our Grant-in-Aid program as well as boosting the Idaho Fish & Game trust account for bighorn sheep research and management.

Under the direction of FNAWS President Arnie Johns, the Foundation took a giant leap into the second 10 years. The international magazine of FNAWS, *Wild Sheep*, was brought in-house, giving the FNAWS membership, directors and staff an opportunity to be a part of the four-color membership publication. Sandra Nicola began her career at FNAWS and was asked to become the in-house contact for the magazine and advertising sales starting with

Scenes from the June 17, 1989, dedication of the FNAWS national headquarters in Cody, Wyoming.

The World Record Stone's sheep with a Boone and Crockett Club official score of 196-6/8 points is considered by many as the finest example of a wild sheep trophy ever taken. Each horn exceeds 50 inches. The Chadwick ram is the symbol of the Foundation for North American Wild Sheep and a painting of the ram by Gary Swanson is its logo. The trophy belongs to the Boone and Crockett Club's National Collection of Heads and Horns and is on display at the Buffalo Bill Historical Center (BBHC) in Cody, Wyoming. The national headquarters of FNAWS is located directly across the street from the BBHC. Early FNAWS member Lowell Baier of Potomac, Maryland, was responsible for recovering the Chadwick ram trophy from the Bronx Zoo where it was stored behind temporary walls in 1977.
Photograph courtesy of Boone and Crockett Club.

the spring 1988 issue, a position she continues today.

The FNAWS Visa Card Program was established with FNAWS receiving a royalty from card purchases. Tombstone Pizza Corporation stepped forward as the first FNAWS corporate sponsor of a convention program, relieving the Foundation of a convention expense and returning more money to sheep projects. Foundation headquarters were renovated to be a showpiece and information center for the promotion of our goals in the management of wild sheep.

In light of the severe drought and devastating fires of summer 1988 that swept the greater Yellowstone National Park area, the Foundation launched the project called Funds Available for Wildlife Now (FAWN). Our funding record had proven without a doubt that the Founda-

Displays at the national headquarters include a half-mount Grand Slam, donated through the Texas Bighorn Sheep Society (TBSS) by Pat Auld in memory of her late husband Dan Auld, Sr., of Kerrville, Texas. Pat continues as an active supporter of FNAWS and TBSS. Dan was an oil explorer, rancher and big-game hunter who was the recipient of numerous national and international awards for actively working on wildlife management projects throughout the world.
Photograph courtesy of FNAWS archives.

tion for North American Wild Sheep was the leading conservation organization in North America. Because of this commitment to the enhancement of habitat for wildlife, it was our obligation to launch a major fund-raising campaign with the funds dedicated to acquire winter range habitat, habitat restoration and re-vegetation (especially along river and creek banks where fire destroyed timber and grasses that erosion occurred), relocating and transplanting of wildlife and emergency feeding of wildlife. Project FAWN, under the direction of President Arnie Johns and Treasurer Martin Nielsen, worked in conjunction with the various government agencies in the fire recovery to fill the gaps from the lack of government funding for rehabilitation projects. Arnie met with officials from the U.S. Forest Service, U.S. Fish and Wildlife Service and the National Park Service on local and national levels to fund and monitor the FAWN projects.

Work was also done to introduce the Project FAWN program into public schools throughout North America and to provide information to the program through state and national organizations of the National Wildlife Federation. A 30-minute television show and a series of public service announcements were also produced to educate the general public about the work of Project FAWN.

On November 16, 1988, the Foundation for North American Wild Sheep, represented by FNAWS President Arnie Johns and the U.S. Forest Service, entered into a partnership agree-

Above the front desk at the national headquarters of FNAWS are a Dall's and a Stone's sheep. Also included at the national headquarters are four shoulder-mount Grand Slam trophies donated by Ingrid Nasca through the Arizona Desert Bighorn Sheep Society, in memory of her late husband Danny.
Photograph courtesy of FNAWS archives.

Sculptor Bill Feeley sits atop his magnificent Chadwick ram statue he created for FNAWS at their national headquarters in the summer of 1991. Each year Bill conducts an annual maintenance of the statue.
Photograph courtesy of FNAWS archives.

June 17, 1989, was the day FNAWS officially dedicated their permanent national headquarters at 720 Allen Avenue, Cody, Wyoming, shown here in the summer of 1998.
Photograph by Susan C. Reneau.

A sculpture garden to the side of the headquarters features the Grand Slam in bronze by sculptor and artist Mike Flanagan, cedar benches and flags from the United States, Canada and Mexico.
Photograph by Susan C. Reneau.

A visitor to the Cody, Wyoming, national FNAWS headquarters in summer 1998 walks to the front door past the magnificent steel sculpture, twice life-size, of the world-record Chadwick ram by Bill Feeley.
Photograph by Susan C. Reneau.

Teens who attended the NRA Whittington Center in 1991 thanks to scholarships by FNAWS included, left to right, top: Tonya Shelley, Scott Custer, Erick Sohr, Alan Bennett and Jeremy Roberts; left to right, bottom: Adam Cook, Andy Lewien, Nick Bednarczyk, Joseph Flipper and Loren Nielsen.
Photograph courtesy of FNAWS archives.

Bill Feeley in 1989, the sculptor who created the Chadwick ram that graces the front of the FNAWS national headquarters in Cody, Wyoming.
Photograph courtesy of FNAWS archives.

The National Rifle Association's Whittington Center near Raton, New Mexico, is one of the most comprehensive shooting facilities in the United States. Young people were sent to the Whittington Center Adventure Camp by FNAWS scholarships in 1990 and 1991. In those years, the camp offered a 14-day experience to young people ages 13 to 17 who focused attention on firearms and hunting activities. Here is one of the shooting ranges being used in 1990 by teens who were sponsored by FNAWS. For more information about current NRA Whittington Center programs, call (505) 445-3615.
Photograph courtesy of FNAWS archives.

ment by signing a Master Memorandum of Understanding (MOU) in Washington D.C. On the same day, an agreement was signed with Region II of the U.S. Forest Service. The key to the future of wild sheep continues to be dependent upon the conditions of their habitat, so the signing of the MOU established a new dimension for the Foundation to assist with the management of wild sheep found on national forest lands.

Foundation headquarters at 720 Allen Avenue, Cody, Wyoming, were officially dedicated on June 17, 1989, under the direction of President Harvey Kadlec. Many members from around the country were in attendance during the ceremonies to tour the sculpture garden, which today features the Grand Slam in bronze by sculptor and artist Mike Flanagan, and view the taxidermy displays.

Displays at the national headquarters include a half-mount Grand Slam, donated through the Texas Bighorn Sheep Society by Pat Auld in memory of her late husband Dan Auld, Sr., and a shoulder mount Grand Slam donated by Ingrid Nasca through the Arizona Desert Bighorn Sheep Society, in memory of her late husband Danny. They are very fine representations of the wild sheep of North America on display at headquarters. The magnificent steel sculpture, twice life-size, of the world-record Chadwick ram by Bill Feeley, received recognition from world-renowned artist Harry Jackson.

"The Chadwick ram that Bill Feeley has created is an authentic, time-dissolving masterpiece. This great work is of the same order as the sculptures of the Parthenon in the British Museum, the Lions of the ancient Assyrians, and finally the two Greek warriors discovered at Riacci, Italy. Bill and the artistic commitment expressed in this work are inspiring to me."

The majestic piece welcomes visitors to the international headquarters and is a fitting tribute to the Chadwick ram, the official logo of the Foundation for North American Wild Sheep.

Youth and education, a part of our heritage that cannot be taken for granted, was slowly vanishing as the times changed. The Foundation Board of Directors through its Grant-in-Aid funding sponsored 14 youth to attend the summer 1989 session of the National Rifle Association's Whittington Center. The education of the youth of today in hunting and conservation is very important, and FNAWS maintained a very high-profile interest and sponsorship of this program. The skills taught at the Whittington Center last a lifetime for young people who learned safe shooting techniques, hunting methods and team spirit. The adventure camp continues today and lasts 14 days in the summer with a strong

A pyramid of teachers in the summer of 1995 at the Safari Club International's American Wilderness Leadership School in Jackson, Wyoming. FNAWS sponsored teachers from across North America who attended the leadership school from 1989 to 1995. SCI founded the American Wilderness Leadership School in 1976 as a way to promote wildlife conservation as a way of life and to expose teachers to methods of natural resource management, including the value of hunting. The program continues today.
Photograph courtesy of FNAWS archives.

Early members of FNAWS enjoy a cocktail party at the 1988 FNAWS convention. From left to right, Sandy Christian, Sue Kadlec, Jane Babler, Jerry Christian, Harvey Kadlec, Leland Speakes, Jr., and John Babler.

Photographs courtesy of FNAWS archives.

Rich Roan, standing, congratulates successful bidder, Robert Robertson, in 1988 at the Saturday night banquet at Bally's Casino Resort (now called Reno Hilton).

Dean Picl, left, Gen. Chuck Yeager, center, and Basil Bradbury at a 1988 FNAWS reception.

Pat Latham, left, Lanny O. Wilson, center, and Jerry Christian at the 1988 convention.

emphasis upon hunting and the shooting sports.

Also in 1989 was the start of sponsorship of teachers who attended the Safari Club International American Wilderness Leadership School. Money from FNAWS paid for teachers to attend the summer school beneath the majestic Teton Mountains outside Jackson, Wyoming. This sponsorship continued through the mid-1990s. The curriculum at the school, still operating today, was designed to give teachers and students a well-rounded exposure to the considerations that affect natural resource management policies, specifically land use, wildlife management, community planning and resource development and utilization. Exposing teachers (many of whom were from large cities where outdoor recreation and appreciation was not common) to conservation techniques provided them an opportunity to share their new-found knowledge with students for years to come. The enthusiasm expressed by teachers who were sent to this summer school by FNAWS was truly gratifying as read in issues of Wild Sheep.

Wrote teacher Kathleen Kay Brunson: "What sort of school has no walls, a beautiful wilderness laboratory and a well-stocked nature library? The answer is the American Wilderness Leadership School at Jackson, Wyoming. All this intensive learning was made even more valuable by the ideas and activities that we teachers could take back to our classrooms and put to immediate use."

"It was refreshing to see first hand the worldwide efforts that are being made to conserve wildlife and also preserve sport hunting," wrote another teacher, Stephen L. Meyer.

The 1989 national convention took place in Reno from February 15 to 18, and featured a successful auction that included the purchase of a Colorado bighorn sheep and Rocky Mountain goat tags. This was the first time Colorado Division of Wildlife participated in the FNAWS auction by offering hunting tags. Robert Senter paid $43,000 for the Colorado bighorn permit and Simon Aragi paid $7,000 for the Colorado goat tag.

Ray Woodward, former owner of Arctic Red River Outfitters, submitted his caribou hunt donation to FNAWS in 1989 and proposed it be used for a youth-oriented project at the 1990 convention. The Board accepted Ray's donation and initiated the FNAWS Youth Essay Contest, with the theme, "What the Right to Hunt Means to Me." More than 90 essays were received for that inaugural contest, making the final essay selection challenging. Charles Darrell Hoovestol, a graduating senior at Bismarck High School, Bismarck, North Dakota, submitted the winning essay. The Youth Essay Contest continues today through the generosity of Kelly and Heather Hougen, the current owners of Arctic Red River Outfitters.

Another ominous cloud on the horizon in 1989 was the rapidly growing anti-hunting movement. "We need never be embarrassed about our very precious hunting heritage. It is conservation and game management in practice, as we scientifically control wildlife populations under regulated seasons and limits, furnish food for our table and funds for the management of both game and non-game species." wrote FNAWS President-Elect Howard W. Pollock in 1989, as he called for the membership to act to protect their rights as hunters. The Foundation became active in the Hunters Coalition, consisting of representatives from major conservation groups, the sporting industry and pro-hunting media to gather unity within the outdoor sports groups and to be ready to stampede the anti-hunting movement.

Foundation representatives were also in attendance during the first annual dinner fundraiser honoring National Hunting and Fishing Day, hosted by the Congressional Sportsmen's Caucus Foundation on October 9, 1990. The Sportsmen's Caucus Foundation was formed by 80 of our national lawmakers dedicated to protecting the rights of hunting, fishing and trapping.

"Youth and Education for the Future of Wild Sheep," was the theme as the Foundation

Rich and Ron Roan point to a bidder at the 1988 auction.

Photographs courtesy of FNAWS archives.

Early FNAWS member Guinn D. Crousen enjoys the 1988 FNAWS convention.

One of the joys in John Babler's life in recent years has been the purchase of the Parker/Pachmayr double-barreled, 12-gauge shotgun offered at the Saturday night banquet.

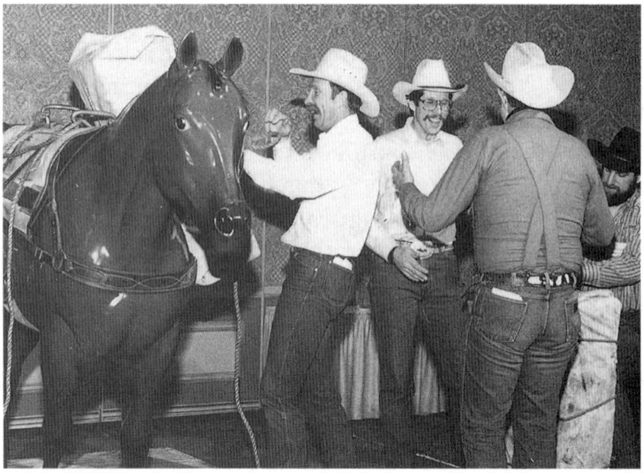

Horse packing contests continue as a popular competition at each annual FNAWS convention. At the 1988 diamond hitch competition, the team of Ross Peck and Joe Quarto took first place. Garry Merritt and Don Giottonini took first place in the Manty competition.

Tommy Caruthers checking the sound system at the 1988 convention.

returned to Reno for its 13th Annual Convention. This theme called out to our membership for continued action and set the course for the active role FNAWS plays in the fight to protect the future of hunting rights today.

As conservationists and hunters, a great deal had been accomplished by the Foundation by 1990 in "putting sheep on the mountain." The challenge as the Foundation entered the decade of the 1990s was membership growth, and the revenue required to fund the various research, transplant, habitat acquisition, water hole development and education projects. The membership growth goal was not to have a large membership, but to have an active membership made of members who participate and support the Foundation's work.

In a message to the membership in spring 1990, President John Harris wrote, "The only way that the Foundation can continue to build healthy sheep populations in North America is through continued member contributions. This support, whether it be time, energy or money is what the Foundation is all about, and allows us to grow."

Gordon Eastman, shown here with his beloved camera equipment, established a special award in 1988 to recognize FNAWS volunteers called the Gordon Eastman Grass Roots Award. Gordon died in 1996 but the award continues. A complete list of award winners is included in the appendix of this book.

He issued a challenge to each and every member to sign up a new member "who shares your beliefs and commitment to wild sheep and hunting." The Foundation established an "Outfitter Membership Drive," asking those member outfitters to sign up new FNAWS members. Who better to recruit new members than those in the field? The FNAWS member outfitters see first-hand the positive results the Foundation has in its conservation efforts. Through these membership drives, membership numbers of the Foundation have averaged 5,600, with a high exceeding 6,000. This reflection in member numbers shows a committed group of individuals willing to answer the Foundation's "call to action."

Historically, the volunteer spirit of the membership is the envy of the conservation industry. Drawing upon that membership pool, the Foundation developed the "Artist of the Year Program" in 1989 with Vivi Crandall as the first artist to be recognized. She donated an original drawing and painting titled, "The Buddy System." In 1990, Idaho Artist Hayden Lambson painted exclusively for FNAWS, "The Untouchables," a beautiful oil painting that was auctioned during the Saturday evening banquet with a limited edition print available for purchase directly from FNAWS headquarters. The auction proceeds and subsequent print sales added tremendous revenue to the Foundation's funding dollars. The program has continued with the support of the Foundation's member artists, and in 1995 was renamed to

Raffles at the national FNAWS conventions are a popular activity and a successful way to raise much-needed funds.
Photographs courtesy of FNAWS archives.

FNAWS supporters in 1990 listen to the items up for bid at the Friday afternoon auction.

Jim Reilly of James Reilly & Co., P.C., of Cody, Wyoming, official accountant for FNAWS, draws the winning raffle ticket at a FNAWS convention. Since the John Harris' and Harvey Kadlec's presidencies, beginning in 1990, raffle tickets are drawn in front of the entire FNAWS membership at the Saturday night banquet by someone independent of FNAWS to insure the raffle ticket drawings were handled fairly. Jim often draws the raffle tickets.

The ram measuring workshop offers hunters the opportunity to score their trophies in a relaxed atmosphere with official measurers of the Boone and Crockett Club who volunteer their time to help FNAWS members. Ram scoring workshops have been an important part of FNAWS conventions from the beginning. This is the 1992 scoring workshop.

Two steadfast FNAWS volunteers who help with the ram measuring clinic are, left to right, Bill Walters of Mississippi and James Hutcheson of Texas, shown here in 1992 at the national FNAWS convention.

In 1989 Vivi Crandall, shown standing beside her original artwork, donated this dramatic original drawing titled, "The Buddy System," for that year's auction and was named the first "Artist of the Year." For the first time, Vivi created a preliminary drawing of her idea for an original painting. This drawing was auctioned in conjunction with the original acrylic painting that shows two rams being stalked by a pack of wolves.
Photograph courtesy of FNAWS archives.

Flamboyant supporter of FNAWS for many years is artist Vivi Crandall, shown here at the 1991 Ladies Luncheon.
Photograph courtesy of FNAWS archives.

honor the late Jim Ford. The program is now called the "Jim Ford Memorial Art Competition." During the 1995 Phoenix convention, Jim's widow, Bonnie Ford, presented the winners with a special Gary Robert Swanson commemorative medal in her husband's honor.

Jim was a moving force in the Foundation, serving on the Board of Directors, conducting the ram scoring clinic, working the auction and the leg and arm wrestling competition, as well as the hospitality room. Jim was instrumental in getting legislation enacted enabling the state of Montana to donate a bighorn sheep permit to the Foundation. He was an avid sheep hunter, and hunted them all. Jim was an example for all sheep hunters. His efforts, along with others in Montana, made it possible for legislation to be passed that authorized the auction of the Montana bighorn sheep permit beginning at the 1986 FNAWS convention. The Montana bighorn permit continues today as one of the most popular and lucrative auction items, starting with Art Dubs who paid $79,000 in 1986. Today, the Montana permits regularly sell for in excess of $300,000.

The following member artists have been recognized by the Foundation as "Artist of the Year." The Foundation continues to sell the limited edition prints of the winning paintings from the yearly competition, which provides additional funds for the Foundation's projects.

The "Artist of the Year" winners include: Vivi Crandall in 1989 for her painting and drawing titled, "The Buddy System"; Hayden Lambson of Pocatello, Idaho, in 1990 for his painting, "The Untouchables"; George Lockwood of Santa Ynez, California, in 1991 for his painting, "Desert Sentinels"; Tom Mansanarez of Victor, Idaho, in 1992 for his painting, "Cliff Hanger"; Joseph Krausz of Canoga Park, California, in 1993 for his sculpture, "A Man's Dream";

Love was in the air in 1990 when Darla Shelton tied the knot with John Wayne Tanghe during the national convention since it was at this convention where the couple fell in love. John won the arm wrestling championships that year and married Darla in the hotel chapel. The bride wrote a letter to the editors of Wild Sheep *for the fall 1990 issue and enclosed a photograph of their wedding ceremony, which is shown here.*

Frank Pachmayr, shown seated in 1990 at the FNAWS national convention, donated seven of his fine quality firearms to FNAWS for auction items. In 1990 three of the seven shotguns were presented during the Saturday night banquet in the names of Gen. Harold Starke, Gen. Chuck Yeager and Gen. Joe Engle, shown in this picture surrounding Frank from left to right. To the left of Gen. Starke is Past President Arnie Johns. Past President Harvey Kadlec is behind Gen. Yeager and Past President Jerry Christian is to the far right. Pachmayr shotguns continue as popular auction items.
Photograph courtesy of FNAWS archives.

Trevor Swanson of Phoenix, Arizona, in 1994 for his painting, "A Time to Leave"; John Lewton of Whitehall, Montana, in 1995 for his sculpture, "The New King"; Martin Rathburn of Delta, Colorado, in 1996 for his painting, "Legends of Sacred Ground"; Kurt Jungblut of Golden, Colorado, in 1997 for his sculpture, "Desert Sentiment"; and Tom Mansanarez of Victor, Idaho, in 1998 for his painting, "Meeting the Challenge."

The FNAWS membership was honored with the presence and support of Life Member Frank A. Pachmayr during the 1990 Reno convention that took place February 28 to March 3. Frank bestowed seven of his fine quality firearms to FNAWS at that convention for use at auctions and three of the classic shotguns were presented in the names of Generals Chuck Yeager, Joe Engle and Harold Starke at that convention. Frank lived from 1906 to 1997 and was founder of Pachmayr Gun Works in 1929. He held 125 different patents on shooting accessories and his revolver grip adapters (U.S. Patent 1931) are some of the more commonly recognized shooting items along with his recoil pad (1932) and Signature grips (1965). Frank sold his company in 1981 but it maintained his famous name. He and his late wife, Nanita, founded the Pachmayr Foundation to continue their expanded charitable work with such groups as FNAWS, National Rifle Association, Ducks Unlimited, National Wildlife Federation and the Los Angeles Museum of Natural History.

Dan Pocapalia purchased the New Mexico hunting permit for $70,000 at the 1990 convention, the first convention offering the New Mexico permit. Frank Pachmayr, a legend in the

Sheep Camp Fundraiser

1990-91 FNAWS President, John Harris.

Tickets in your mailbox, available at the convention, or Call us!

Due to anti-hunting pressures in Mexico, the Foundation was unable to secure our desert sheep permits from the Consejo Nacional de la Fauna A.C., for the 1991 convention. Therefore, we have come up with an exciting hunting package sure to please.

FNAWS is proud to present a complete "Sheep Camp" experience. A chance in a lifetime for a first class Stone sheep hunt with expert outfitter, Garry Vince, of Muskwa Safari's LTD.. This hunt is for eighteen days on the Prophet and Muskwa rivers in Fort St. John, B.C.. Your hunt will begin on September 6, 1991 and will conclude on September 23, 1991. This is a beautiful time of year in Fort St. John, with autumn foliage at it's fullest, the scenery will be breathtaking.

Included in this hunt is your Stone sheep permit as well as a grizzly bear for a second animal. Other animals may be added for a mixed bag hunt at an additional cost. Hunters will be met in Fort St. John. To reach Ft. St. John you may drive or fly Time Air from Edmonton, Alberta, or Vancouver, B.C., and Air B.C. , Ft. St. John, is Mile 48 on the Alaskan Highway. Included in this excellent hunt are guides, riding equipment, saddles (roping free type) saddle bags, tents, stoves, gas lamps, food, (which is the very best possible with fresh produce flown in each trip). Pack and saddle horses are used on all trips, six to twenty four horses are used in each camp, depending on the length and size of the hunt. Garry has had a 100% success rate on his sheep permit quota, adding more excitement to this outstanding hunt, valued at $12,500.00.

To make your "Sheep Camp" complete, FNAWS has also included a complete personal outfitting package for the lucky winner of this great fund raiser. You will also win: a 280 Remington model 700 TM Mountain Rifle with bolt action and a 22" barrel, hunting coat, pants, boots, spotting scope, binoculars, mountain tent, sleeping bag, camp back mattress, backpack, knife, hat and gloves, making this complete "Sheep Camp" package valued at over $13,000.00.

The Foundation is very excited to be able to offer our membership such a great hunting opportunity, and with your participation we will be able to raise funds to help compensate for the loss of the desert sheep permits, to enable us to continue to keep "putting sheep on the mountain".

You may purchase one ticket for $25.00, or 5 tickets for $100.00. Tickets may be obtained by contacting FNAWS headquarters at (307)-527-6261. Tickets will also be for sale at the FNAWS fund raising area at the 1991 convention, with the lucky winner to be drawn during the 1991 FNAWS Saturday night Banquet, to be held on Feb. 23, 1991.

On behalf of the Foundation for North American Wild Sheep, I thank you in advance for your support in making this one of our greatest fund raising efforts ever.

I look forward to seeing all of you in Reno at the FNAWS 1991 convention, February 20 - 23 at Bally's Reno Hotel.

John W. Harris

John W. Harris
President

As membership to FNAWS grew, so did the lines at their national headquarters. Here we see registration lines at the 1991 FNAWS convention in Reno.

Photographs courtesy of FNAWS archives.

Dr. Robert Moses, Charter Member 32, of Worthington, Indiana, listens intently to auction action in 1991.

James J. McBride of Glen Ellen, California, left, talks to a "49er" at the country music night in 1991.

Richard Mitchel smiles for the camera as he purchases a shirt from national FNAWS staff person Kim Nieters in 1991 at the popular FNAWS merchandise booth.

Hans Hansen, center, won the Stone's sheep hunt in 1991 when it was auctioned at the Saturday banquet but donated it back, and FNAWS got another $18,000 from second successful bidder, Mel Kinsey (left)! It was actually a double fundraiser! With him in his celebration was Canadian outfitter Garry Vince, right, who provided the 18-day Stone's sheep hunt in Fort St. John, British Columbia.

Ron Pomeroy, left, starts the 1991 leg wrestling championship.
Photographs courtesy of FNAWS archives.

Norman C. Roettger, left, and Jack E. Beal enjoy opening ceremonies of FNAWS national convention in 1991.

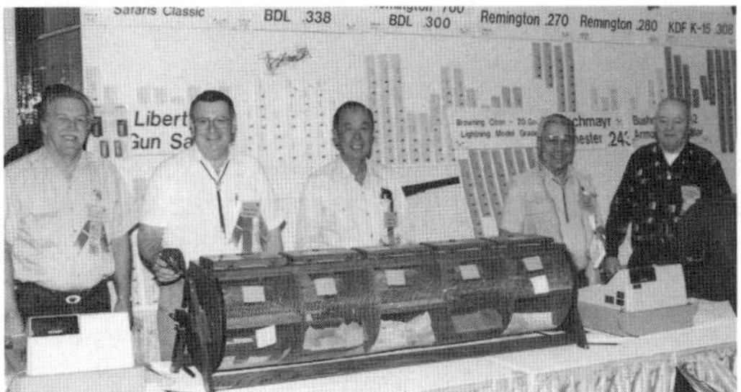

Volunteers who collect the raffle tickets each year help make the national FNAWS convention that much more successful. From left to right: Frank Blaha, Harold McEuen, Jimmy Low, Kern Chew and Earl Boose in 1991.

Auction spotter Richard Cox gives successful bidder Sherry Musgrove a hug for a generous donation to FNAWS at the 1991 Ladies' Luncheon.

Tom Pawlacyk sits in front of a painting he just purchased at the 1991 Friday morning auction and breakfast.

Men's tug-o-war in 1991.

Mark Twain (McAvoy Layne) spins a tale for a little one at country music night at the national FNAWS convention of 1991. McAvoy began his career as Mark Twain in 1975 when he found himself snowbound in a cabin for five days that contained Mark Twain novels. He memorized the stories and took on the voice of the real Mark Twain.

Dave Popelka, convention photographer, of Sigler Printing and Publishing with long-time FNAWS supporter and current FNAWS Board Member Virginia Frehner.

Photographs courtesy of FNAWS archives.

firearm and sport hunting world, bestowed seven of his fine quality firearms to the Foundation. Three of these classic shotguns were presented during the Saturday evening banquet in the names of General Chuck Yeager, General Joe Engle and General Harold Stark. The generous donation of firearms by the late Frank Pachmayr continues today through the Pachmayr Foundation, demonstrating the true dedication and commitment to the sporting fraternity by giving us the opportunity to help our unique industry and insuring this heritage will be available for future generations to enjoy.

The 1991 convention included the second auction of a hunting permit from the New Mexico Department of Game and Fish that was purchased by Pamela Atwood for $46,000. At this convention, the membership committed $10,000 to the National Bighorn Sheep Center in Dubois, Wyoming. More information about this outstanding center is included in the Rocky Mountain Bighorn Society chapter of this book.

This membership support was again demonstrated when, due to anti-hunting pressures in Mexico, the Foundation was unable to secure the desert sheep permits for auction and fundraiser during the 1991 convention that took place February 20 to 23 in Reno at the Reno Hilton.

John Harris, who had been elected president of FNAWS in 1990 through spring 1991, remembered vividly the elimination of the desert sheep hunts in Mexico. He recalled the details when he spoke to Susan Reneau at the 1998 convention as though it happened yesterday.

"In December I talked with Executive Director Karen Werbelow on the telephone while I was on vacation for Christmas. She indicated that the Mexicans had pulled the permits and were not going to have them at the convention. We had been out selling them and talking about them in sheep camps right before convention. This was a major fund-raising event for us. We tried to look at ideas to replace that money.

"I spent most of the Christmas holiday on the phone with Bob DiGrazia, Garry Vince of Muskwa Safari's Ltd. and Karen Werbelow discussing that issue. We came up with the idea of doing a sheep camp. It's tough to get a trip for that coming year. We wanted something that was significant enough to replace the desert sheep hunt, so I bought a Stone's sheep hunt for the Sheep Foundation auction from Garry Vince. We put that on the sheep camp.

"We got lots of criticism when we did it because they thought it would never make the money with the desert sheep camps we had. We made more money that year than we had the previous year with the desert sheep. Hans Hansen won the Stone's sheep hunt when it was auctioned at the Saturday banquet but donated it back, and FNAWS got another $18,000! It was actually a double fundraiser! We were able to put together something out of virtually nothing and out of disaster. I'll never forget that Christmas because we were really trying to figure out what we were going to do to replace that hunt," John Harris concluded.

A complete personal hunting package along with the Stone's sheep hunt was provided by Leupold & Stevens, Inc., W.C. Russell Moccasin Company, Johnson Camping, Frontier Gear of Alaska, Day One Camouflage, Bucks Bags, Americase, Rocky Mountain Sport Center and Cabela's.

We waited breathlessly to see the response from the members, and we found once again their dedicated support was tremendous. The Foundation netted $50,100, more money retained for "putting sheep on the mountain" than any one state sheep permit proceeds allotted the Foundation at that time. This revenue was achieved, in part, through the generous redonation by drawing winner Hans Hansen. This unselfish display of support allowed the Foundation to auction the "Sheep Camp" package for $18,000 to lucky bidder Melvin Kinsey. The success of the first "Sheep Camp" fundraiser broke new ground as a revenue basis, exhibiting immense patronage by the membership and has become an annual fundraiser for the Foundation.

As the membership stepped forward, answering our call for continued involvement, the Board of Directors was focused on the policies and distinct responsibilities of the governing Board, insuring the secure future of the Foundation. Amendments to the bylaws were proposed and passed, policies were enacted, and a committee formed that transformed the Board's administration of the Foundation.

President Harris pointed with pride that during his presidency through spring 1991 FNAWS strengthened its accounting system. He recalled, "I was the treasurer the second year in office when Harvey Kadlac was president. Harvey and I worked very hard to restructure the way we accounted for the funds that were made.

"We put in an accounting system and the use of cash registers. Harvey actually did that under his presidency the following year. He and I worked together on that, and we continued that process under my presidency. We wanted to be able to show where the money was coming from and show that we were responsible for it. It is virtually the system we have in place today. You buy raffle tickets and you'll see we have cash registers and an accounting system. We changed accounts at the same time, and the Board was split on that."

Recalled Harvey Kadlec, "I think my biggest accomplishment was the changing of the accountant and the accountability of FNAWS. I'm not going to say the negative, but there was a lot of it. We got a CPA on Board. We had more people running it. We had backups so you didn't have questions about who was doing what. When you're going through this amount of money, and only one person is handling this, how much was disappearing? We knew a lot of it disappeared, but we couldn't really prove it.

"In the next year, there was so much money, and we had about the same volume of sales. We used cash registers instead of a box. There was a paper that said how much was taken in and what it was taken in for. You have a trail of all the items you had. You know how much you had on the raffle. It's all rung up on a tape. It's all there. You don't have to pick it apart," Harvey said.

Ron McMillen commented on the accounting system changes and his involvement with FNAWS at that time. "There have been heartbreaks. Nancy McDaniels was a lot of fun and a dear person. They changed accountants to Jim Reilly (James Reilly & Co., P.C., of Cody, Wyoming), and now Jim has since become a dear friend. Nancy was out of Phoenix.

"Jim Reilly is the current accountant, and he took over when the Board decided to change accounting firms. They decided to bring the accounting in closer, so Jim Reilly took over. He's from Cody, and he's been at meetings. He sits on the sidelines with me, wears glasses, about 40, blondish hair. Nancy was out during John Harris' presidency. Jim Reilly has been the accountant for FNAWS ever since then. I know one of the areas that has always been of concern to the presidents has been the accounting system and how the money is spent," Ron McMillen explained.

John Harris told Susan Reneau one of the things that always kept cropping up in the early years were rumors of misspending of money raised from auctions and raffles. Members questioned what happened, so measures were created to tighten up the drawing system for raffles.

John gave more details. "That's why today you see the drawings for raffles occurring in front of people at the Saturday night banquet. The drawings had previously occurred behind closed doors. The problem with that is that it gives people the chance to speculate that something is wrong or something is rigged. We wanted to be open with it so when people spent their dollars they knew they had a chance to win something. They'd see it.

John also explained that the membership during his presidency came to the conclusion that Reno was the home base for FNAWS. "We were trying to get a system in place where we stayed in Reno two or three years, and then if we wanted to move, we'd move, but only move out for a year and come back to Reno. We always wanted to use Reno as our base. We always made money here because of the gambling and because of the other things the hotel could offer. We get cheaper rates on all the things that we do. We are a volunteer organization, and we do a lot of hands-on work, including unloading a truck and things like that. While unions are not bad, if you are a union state you can't do those things. The costs of putting on a convention are sky-high for us when we are working in that kind of environment."

In 1989, the bylaws were amended to read, "Past Presidents shall not be eligible for re-election to the Board of Directors." Those past president's will, as members of the Past President's Council for life, continue their work for the Foundation by participating in the conduct of the Board meetings and making recommendations, while having no vote on the final action taken by the current Board."

Subsequent amendments changed the title of "President-Elect" to "First Vice President," and the title of the office of "Vice-President" was changed to "Second Vice President." The purposes and objectives of the Foundation were expanded to include paragraphs embracing the education of mainstream America as to the lawful rights and privileges of recreational hunting. Wording also expressed the conservation benefits of such hunting to respond in a positive manner to those anti-hunting individuals or groups and to encourage all hunters to be responsible conservationists while hunting under fair chase and good sportsmanship conditions. The amendments written also advocated the funding of both federal and state wildlife departments through the purchase of hunting licenses, game stamps and the payment of excise taxes. The bylaws of the Foundation were updated, reprinted and distributed to the membership in June 1988, June 1992 and most recently in June 1997.

The Long Range Planning Committee, under the direction of Chair Tom Brown, developed a professional presentation package the "Case Statement," as well as initiated a charitable giving program. The early 1990s were busy and exciting times with growth and development for the Foundation.

A somber air reached the conservation community on August 10, 1991, when Lloyd Zeman tragically died in a horse accident on a trail headed toward the sheep mountains he dearly loved. Lloyd was born on May 28, 1933. Tributes were given by "Doc" Schultz, Dr. Jerry Waite and Duane Smelser.

"The world, and especially the sheep world, is much better for Lloyd's presence on this earth. His influence and work will live on. For all of us who are sheep hunters I say, thanks to Lloyd, for what he accomplished for us."

Barbara Zeman, who was married to Lloyd for 18 years and is mother of his six children, continues to speak with deep respect for Lloyd today for his passion and enthusiasm for the Foundation for North American Wild Sheep. She said Lloyd remained dedicated to the principles of FNAWS to the end, even when he was not executive director or president. Barbara said his horse stood beside his body for three days until he was found at the foot of a cliff, which indicates the loyalty this animal felt for Lloyd even though its foot was injured from the fall.

Duane Smelser wrote an eulogy to Lloyd Zeman and the entire eulogy reads:

"I met Lloyd in the year 1970, on a desert sheep hunt. As a matter of fact, there were other important hunters in camp, at the bay of De Los Angelos in Baja, Mexico, namely, Warren Parker, Jim Babala and Bell De Rosa. Most of us were members of the Grand Slam Club, run by Mr. Bob Housholder.

"Later, in one of Bob's newsletters, there was a note from James Froelich. He was asking if any members would like to meet and swap sheep hunting stories. So, one snow stormy weekend, we did. This first meeting was small, but the beginning of a great group. It is here I met Lloyd again. This group has since funded many millions of dollars for sheep conservation programs. We met four times in the next four years, twice in Wisconsin, once in Iowa, and once in Detroit, Michigan. After the Detroit meeting, the Foundation for North American Wild Sheep was formed, with Lloyd Zeman at the helm.

"Lloyd gave us his leadership for a total of eight years, including the four years preceding the forming of the organization, and three years as president. Under Lloyd's leadership we formed the Foundation — the Foundation for North American Wild Sheep. It is today as he created it. The club has continued to grow stronger over the past 15 years. It is hard to believe that it has only been a total of 19 years, since that first meeting on that snowy weekend. Our club is a monument to Lloyd Zeman.

"For all of us sheep hunters, I say thanks to Lloyd for what he accomplished for us and the many good memories."

The Minnesota-Wisconsin Chapter of FNAWS established the "Lloyd Zeman Water Hole Project," as a living memorial to his life and many years of contributions to the North Ameri-

Carl M. Scrivens of Jackson, Wyoming, third from left, listens as Robert "Bob" H. Hanson of Wapiti, Wyoming, second from left, recognizes FNAWS on June 27, 1992, for their work helping wild sheep populations. Dr. Robert "Bob" DiGrazia, FNAWS president at the time, holds a commemorative plate given to him by Bob Hanson in front of the Buffalo Bill Historical Center's exhibit of the National Collection of Heads and Horns that is owned by the Boone and Crockett Club. Carl Scrivens placed his Number One desert bighorn ram on display in the collection on that day. The ceremonies took place at the center which is in Cody, Wyoming. The desert sheep trophy, scoring 205-1/8 official B&C points, is seen behind Bob Hanson.
Photograph courtesy of FNAWS archives.

Groundbreaking ceremonies took place on September 16, 1992, for construction of the National Bighorn Sheep Interpretive Center (NBSIC) in Dubois, Wyoming. Wyoming Chapter of FNAWS President Jack Hildner and FNAWS Executive Director Karen Werbelow, shown shoveling dirt in the center of the photograph, participated in the ceremonies. National FNAWS and many FNAWS chapters and affiliates have contributed thousands of dollars to its construction and maintenance. The center sits at the base of Whiskey Mountain, home of North America's largest wintering herd of bighorn sheep and offers to visitors an opportunity to learn about all North American species of wild sheep. For information about exhibits and times of operation today, call the National Bighorn Sheep Interpretive Center at (307) 455-3429.
Photograph courtesy of NBSIC.

can wild sheep. That same month, avid hunter and president of the International Sheep Hunters Association James Schmalz died from coronary illness. He devoted considerable time to conservation causes, particularly the start-up of a fund to feed wildlife in Yellowstone National Park. Foundation headquarters today boasts a wonderful display of the world's sheep, donated from the estate of James Schmalz.

These mounts have been very useful in educating the public, giving them the opportunity to see and distinguish the difference between the native North American wild sheep and the exotic sheep of the world. These men, and many other departed members of the Foundation, have continued to "give something back" to the wild sheep that they were so dedicated to, through the FNAWS Memorial Fund.

The Wild Sheep Perpetual Memorial Trust Fund was established in June 1992 to accept contributions in memory of departed members. Contributions to this fund are put into a separate restricted account. The original gifts would not be touched, only the interest derived from this fund can be used for sheep-related projects. We have seen many members pass to the other side of the mountain, but their commitment to the Foundation remains through the generous contributions bestowed to the Foundation in their memory.

During the October 18 and 19, 1991, Board meeting, held at the Hanalei Hotel in San Diego, California, the FNAWS Board of Directors approved the reinstatement of the executive director position with the Foundation. The Foundation joined the ranks of other conservation organizations by expressing this vote of confidence in Business Administrator Karen Werbelow to act as the liaison to the Foundation's Board of Directors by becoming the Foundation's new executive director.

The Foundation returned for fun in the sun at the expanded Town and Country Hotel in San Diego, California, for the 15th annual convention in 1992 from February 26 to 29. This was a notable convention that allowed the Grant-in-Aid funding dollars to exceed $250,000. Of that spring 1992 funding, the Foundation earmarked money for the No-Wolf Option Committee, opposing the wolf reintroduction into Yellowstone National Park and central Idaho regions of the United States.

FNAWS committed $35,000 to the defeat of Proposition 200 in Arizona. This November 1992 ballot initiative would have stopped all hunting, fishing and trapping in Arizona with the potential domino effect reaching other states that would jeopardize our hunting heritage. The Arizonans for Wildlife Conservation was formed, with FNAWS Secretary Pete Cimellaro acting as chief spokesman, campaign manager and treasurer. With the support of the FNAWS membership, the conservation and sporting industry and Arizona sportsmen, this initiative was defeated 2 to 1 by the voters of Arizona after last-minute advertisements and publicity turned around public opinion. The generosity of the FNAWS membership in contributing directly to this effort, combined with the Foundation's funding, revealed an overall contribution exceeding $100,000 for this issues defeat. Craig Johnson

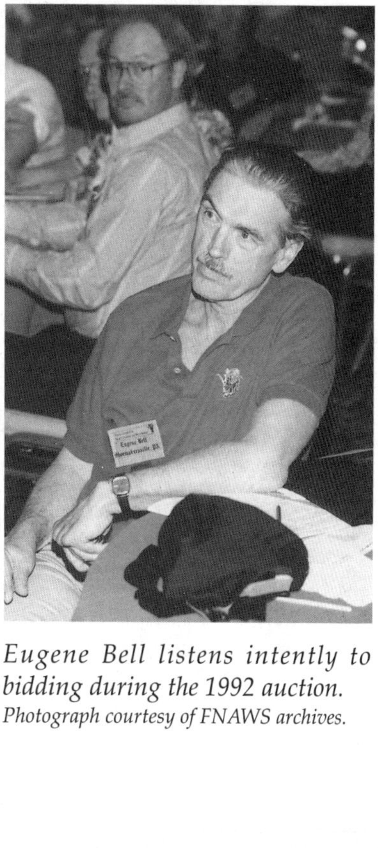

Eugene Bell listens intently to bidding during the 1992 auction.
Photograph courtesy of FNAWS archives.

Paul Inzauti and his son attend the men's luncheon in 1992.

Photographs courtesy of FNAWS archives.

Past President Arnie Johns as 1992 convention chairman helps with the sound system.

The 1992 auction "ringers" and women who hold auction items for bidders to view. Ringers and these volunteers were recognized by FNAWS auctioneer Rich Roan as why his job is so much easier each year.

Brad Kadlec, son of Past FNAWS President Harvey Kadlec, left, helped his father with the FNAWS auctions. With Brad at the 1992 Banquet is Tom Hoffman of Albany, New York.

Grits Gresham, outdoor writer and television personality, was a popular keynote speaker in 1992.

Connie Roberts and Janet Johnson, left and right, organize banquet seating at the 1992 FNAWS convention.

Joe Bishop does a two-step at the 15th Convention of FNAWS in February 1992 in San Diego.
Photographs courtesy of FNAWS archives.

Eastern Chapter leader Frank Blaha, left, appreciates the grandson of Harvey Kadlec who rises on Harvey's shoulders during a busy 1992 event.

The FNAWS national convention is a time for friends, good conversation and good food. Conventions have continued to grow in size since the first one in 1977.

A little one dances to the music in San Diego at the 1992 beach party.

"Fire and Ice" was the theme for the 1992 fashion show at the Town and Country Hotel in San Diego, California, for the 15th annual convention in 1992.

Lyle Foster of Mammoth Lakes, California, left, was one of the 1992 auction purchasers. He is congratulated by auction spotter Irv Lozier of Western Wyoming Trophy Hunts.
Photograph courtesy of FNAWS archives.

Douglas M. Johnson, center and seated, is congratulated by auction spotters at the Friday afternoon auction in 1992, for his purchase of the Arizona permit for $67,500. Doug has continued as a much-appreciated supporter of FNAWS.

Successful 1992 bidder Nicholas Coussoulis of San Bernadino, California, at the Friday afternoon auction.

A moment of congratulations are exchanged by 1992 auction participants, left to right, Spotter Carlos Sanchez, Bob Curtis and Daniel Boone.

Long-time FNAWS supporter Rocky Crate of Graham, Washington, makes a bid during the auction in 1992.

said FNAWS gave money to the Wildlife Legislative Fund of America (WLFA) to help defeat Proposition 200 in Arizona purchase television time and publish brochures and fliers. During the same spring 1992 meeting, the Board of Directors voted to restructure the membership dues of the Foundation, effective January 1, 1993. With the exception of junior memberships, this structure and dues schedule continues today. Regular membership is $45 per year; a 3-year renewal is $120; Family membership is $80 per year; Life member is $1,000; Senior life membership (age 58 to 64 years) is $750; Retired life membership (age 65 and over) is $500.

Oregon Department of Fish and Wildlife overwhelmingly approved FNAWS as the conservation organization to nationally sell a bighorn sheep tag for the first time. Sherwin Scott led the charge and was high bidder for the tag at $57,000.

FNAWS lost a life-long supporter on June 14, 1993, when Basil Bradbury died after a two-year battle with cancer. He was born on June 22, 1929. Basil served as editor of the national FNAWS magazine, *Wild Sheep*, and as director of communications before leaving FNAWS to establish another magazine in 1987. He worked for FNAWS as staff and under contract from 1984 to 1987. He established his reputation as a outdoor writer and journalist as editor of the popular international magazine, *Petersen's HUNTING*, prior to working for FNAWS. The 10th edition of the Boone and Crockett Club's all-time records book, *Records of North American Big Game*, was dedicated to him for his work as a wildlife photographer, wildlife conservationist and outdoor writer who "devoted most of his adult life to the out-of-doors."

A record $2.6 million was raised during the 1993 Reno convention that took place February 24 to 27. Sherwin Scott paid $205,000 for the Montana permit and Jerry Fletcher paid $303,000 for the Arizona permit. Lee Anderson paid $100,000 for the California permit. There was standing room only as auction permit bids reached all-time highs!

Rich Roan commented in general about the ability to raise money through art objects, a method of fund-raising financially successful in 1993 as has been the tradition in FNAWS from the start. "Art is a funny thing even in the antique world. You can take a piece of art that looks good to me but it looks ugly to you. It might bring more to you than it would to me, or whatever. That's the same with the art in FNAWS. You have to remember one thing. There are zillions of prints out there. There's not zillions of originals. It has to be somebody who is famous or a noted artist. Even though there are good artists who aren't famous, that's the way it is," Rich Roan explained.

"The Reigning King," an original oil painting by Gary Swanson, was the Saturday evening banquet auction item as Gary became the first artist selected in the Foundation's Living Legend of Wildlife Art program in 1992. What a fitting choice for the inauguration of this program. Gary's painting of the Chadwick ram became the logo for the Foundation.

The Living Legend program was developed with those renowned artists in mind, who helped the Foundation during its formative years. The appellation aptly describes the handful of outstanding wildlife artists whose donations to FNAWS so greatly enrich its coffers that it was able to undertake the myriad of projects to benefit wild sheep in North America. The art offered at auction under this program continues to bring phenomenal auction bids. The artists' commitment to the Foundation is extended through the Living Legends of Wildlife Art print series. Each of the featured paintings, offered at auction in the program, have been printed and are available for purchase through the Foundation headquarters.

Those artists receiving this recognition from the Foundation are: Gary Swanson of Phoenix, Arizona, in 1992-93 for his painting titled, "The Reigning King"; Fred Boyer of Anaconda, Montana, in 1993-94 for his bronze sculpture titled, "Look Who's Coming to Dinner"; Guy

Coheleach of Bernardsville, New Jersey, in 1994-95 for his painting titled, "Mutton Chase"; Douglas Van Howd of Auburn, California, in 1995-96 for his untitled marble sheep sculpture; Vivi Crandall of West Yellowstone and Casper, Wyoming, in 1996-97 for her painting titled, "Eye of the Storm"; Rick Taylor of Kaslo, British Columbia, in 1997-98 for his bronze sculpture titled, "Spirit of the Ram"; and Ken Carlson of Kerrville, Texas, in 1998-99 for a painting yet to be named as of summer 1998.

The core of our success at FNAWS has been the ability to raise funds through annual auctions and various fundraisers. Our gross revenue has continued to grow; however, so do the costs of the projects funded. The Board of Directors recognized the necessity to plan for the future, and in June 1993 established two funds to insure financial stability to continue "putting sheep on the mountain," while maintaining our hunting heritage.

Gary Robert Swanson, right, was named the first "Living Legend" artist by FNAWS in 1992 and stands beside his original painting, "Reigning King," that was purchased by Ron Simek, left, at the 1993 FNAWS auction.

Photograph courtesy of FNAWS archives.

The FNAWS Endowment Fund accepts contributions for investment, creating earnings that will be utilized to fund the general administrative expenses of FNAWS. All sales from this book go to this fund to permanently finance wildlife projects that "put wild sheep on the mountain." This fund, with the principal remaining the ownership of the contributor, will free up the fund-raising revenue for our project funding. The FNAWS Trust Fund was established with the existing life membership deposits on record, and continues to accept donations in any amount with the income from the Trust Fund dedicated to our Grant-in-Aid funding program.

The economic status of the nation was unstable as the Foundation headed to San Antonio, Texas, for the 17th annual convention and fundraiser from February 16 to 19, 1994. This uncertainty was not apparent as our enthusiastic membership and guests gathered in support of the Foundation raising over $2.9 million dollars! The sheep permit sales soared to new heights with Jerry Fletcher securing the Montana permit for $310,000, which remains the all-time high for the Montana permit through the 1998 convention. Dan Pocapalia purchased the Idaho permit for $50,000, and re-donated it back to the Foundation so the second auction brought $48,000. Stanley Boots paid $245,000 for the Arizona permit. The wild sheep fraternity demonstrated their desire to "give something back" in record-breaking fashion.

Rich Roan remembered the excitement generated at the 1994 auction when permits sold for such high prices. "The first big one (winning bid) after Wyoming was the Montana permit at $310,000 in 1994 thanks to Jerry Fletcher as the bidder. Stanley Boots was the underbidder. That was the one where he ignited the banquet. He jumped onto one of those big, round tables and was carrying on like a wild banshee. That was one where he didn't want anybody

Lee R. Anderson of Minneapolis, Minnesota, left, is a regular bidder at FNAWS auctions. He was successful bidder for the California permit in 1993 with a bid of $100,000 and in 1994 with a bid for the New Mexico permit of $58,000 and $110,000 for the Oregon permit. He was the successful 1995 Washington permit bidder, too. Lee is shown here in 1992.

Rich Roan leads the auction in 1992.
Photographs courtesy of FNAWS archives.

Auction purchaser Larry Hines at the Friday afternoon auction in 1992.

Ruby and Norman Etchison of Bakersfield, California, enjoy the Friday afternoon auction in 1992.

Ron and Vicki Pomeroy of Boulder, Colorado, at the Friday afternoon auction in 1992.

to know it. He had a ring man picked out. The ring man was taking care of everything. I think the ring man was B. Joe Coy." Rich always emphasizes that without the underbidder, a successful auction will not take place.

The Foundation stepped up its efforts in educating the public with the introduction of the "Animal Wheel." This colorful, hands-on wheel outlines animal facts and points of general interest concerning North American big-game animals while profiling the Foundation and its wild sheep conservation efforts. This educational tool was given to each convention attendee, and has been distributed throughout the country to schools, sports groups and meetings, hunter education instructors, and others dealing with youth or the public. The response to the "Animal Wheel" has been an overwhelming success.

The Animal Wheel is one of FNAWS many educational tools available from the headquarters office.

The San Antonio convention was our initial experience utilizing a separate convention center and, unfortunately, our expenses exceeded original budget estimates. Utilizing $200,000 of reserve money, the Foundation funded an exceptional $338,714 in projects. The Foundation developed a new awareness of the costs associated with a separate convention facility, and pledged to the membership that every effort would be made to keep future convention expenses to a minimum.

Craig Johnson, who served as FNAWS president from spring 1993 to spring 1994, recalled his memories about the San Antonio convention. "We had one of the biggest conventions ever in San Antonio. This was one of the most fun conventions ever. The Texas hospitality and the convention were great. The only negative was that the convention cost a little more. We had a decent year financially, but wasn't as good as it could have been. We had the exhibitor / donor party out at the Lone Star Brewery. We had a collection of trophies out there that was pretty marvelous. The Wednesday night party was around the river walk with a "Tex-Mex" theme. On one side was the Mexican-type foods, and the other side was the Texan-type food. We had a border in between and you had to get between the "border guards" to pay our dues. In order to cross the border you had to take a shot of tequilla and every time you went back and forth across the "border" and you didn't do it right, you would be "thrown into jail. It was a really fun party."

When Craig first became president he said the Board was in the process of redoing all the bylaws. "We'd go to meeting after meeting until 8 at night. We had a set of bylaws that included a president-elect but the president-elect had to be confirmed the year the president was elected. We called it president-elect, but technically a president-elect becomes president automatically. Our bylaws said it had to be confirmed.

"Unfortunately, Howard Pollock was president-elect one year, but the Board did not make him president. That's the only time that's happened. The Board decided to amend the bylaws to read that the president-elect won't become president and that anyone on the Board could become president. We made the position first vice-president. This was a mistake. We should have a president and president-elect."

Craig Johnson remains regretful today and explained that in 1994 the first vice president thought he would become president but because of a major conflict on the Board, Garry Vince was elected president. "The thing we learned from this is that a president-elect system is the best system. That's something that should be changed in the future. Right now it is first vice president and all people on the Board have the same chance of becoming president. I think the president-elect system would eliminate conflict, plus since we only have presidents for one year it would be a better grooming process for the president if he or she had a year to study. It makes a lot of sense," Craig Johnson concluded.

Craig Johnson shares a joke at the 1994 FNAWS convention.
Photograph courtesy of FNAWS archives.

Garry Vince provided insight into his FNAWS presidency in 1994. "I was appointed, not elected. I was very proud that I was the president from Canada. I was the first FNAWS president from Canada. When I first came on the Board, I was asked by one of the founders, Dr. Roy Schultz. I thought I was very much out of place to be a director of the Foundation for North American Wild Sheep, let alone becoming a president. For a mountain boy from the North, it was very much a privilege. Being an outfitter and a kid in Stone's sheep country, I thought I had a lot to offer from a different point of view on sheep and sheep management."

And Garry Vince did offer is expert knowledge. He explained to the Board that transplanting of wild sheep and sheep management wasn't all that needed to happen. Getting rid of wolves and other predators was also an important factor as well as burning habitat to open up the brush and other food sources. Garry's encouragement regarding the wolf-wild sheep issue helped direct funds to many projects. He was a big part of collecting funds from many sources. He felt his knowledge of the wild sheep and wolf populations of Canada was one of his greatest accomplishments during his presidency.

Grant-in-Aid funding after this successful convention totaled $343,214 and total GIA funding from the first convention forward to 1994 totaled in excess of $9 million! The wildlife management world certainly took notice of FNAWS with its uncanny ability to raise money to "put sheep on the mountain."

On December 7 and 8, 1994, representatives of the Foundation were in attendance as land managers, biologists and delegates from the tri-state area of Oregon, Washington and Idaho met to form a coalition with the goal of rebuilding the wild sheep herds of Hells Canyon, the finest bighorn sheep habitat in North America. Less than two years later, Hells Canyon experienced one of the most severe die-offs of their bighorn sheep population, as is detailed in a chapter later in this book by U.S. Forest Service Wildlife Biologist Tim Schommer.

President Pete Cimellaro pointed to the Hells Canyon Initiative and the securing of Mexican auction permits as the focus of his presidency. He was the Mexican liaison for six years and worked with Mexican officials and FNAWS leaders to establish a new Mexican program that worked for wild sheep and for people managing them. The successful sale of two Tiburon Island desert sheep tags at the 1998 auction was a source of great accomplishment for Pete.

As for Hells Canyon, he admits the project is still evolving but it began during his presidency.

"It is so large, involving so many states and people, that it's in the process of evolution. I can't say now where it is going or how it's going to get there, but certainly that is a project that FNAWS can be very proud of and certainly will be successful, in the long term." In the chapter, "Presidents' Speak," Pete details the direction he sees the Hells Canyon Initiative going for the next century.

With great pleasure, FNAWS members publicly displayed the prestigious sheep collection of Jack and Eleanor O'Connor during the 1995 Phoenix convention that took place March 8 to 11. This once-in-a lifetime opportunity was made possible through the efforts of the Idaho Chapter of FNAWS, with the generosity of the University of Idaho where the mounts had been stored for 20 years. The Buffalo Bill Historical Center, Cody, Wyoming, joined in the display with the inclusion of their two O'Connor sheep. Jack O'Connor, with his book, *Sheep and Sheep Hunting*, created a passion within many FNAWS members to pursue the wild sheep of North America. He is Grand Slam Club member Number One in their records keeping.

The 18th annual convention was a fun time with games and special events, including a private rodeo extravaganza. FNAWS and Federal Cartridge Corporation confirmed a joint commitment to youth education and hunters ethics with Federal's sponsorship of the program, "Welcome to Wild Sheep." President Ron Mason of Federal Cartridge Corporation attended the convention to present a $15,000 contribution to FNAWS for this 16-page supplement to Falcon, a children's nature magazine that is now published by the Rocky Mountain Elk Foundation (RMEF) and renamed WOW! At the time it was published by Falcon Press of Helena, Montana.

"Welcome to Wild Sheep" was distributed to the membership with the summer 1995 mailing of Wild Sheep. "Animal Wheel" and "Welcome to Wild Sheep" continue to educate the general public about the wild sheep of North America, hunters' rights and our conservation efforts. On March 8, 1995, during the convention, the RMEF-Canada and FNAWS signed a Memorandum of Understanding that was beneficial to sheep, elk and other wildlife by offering the 1995 Alberta sheep permit for auction. This was the featured permit on the Saturday evening banquet with exciting bidding. The winning bid of $225,000 was paid by Gary Hansen. Thanks in large part to the revenue from the auction, FNAWS was able to commit $308,794 to Grant-in-Aid wildlife projects.

The Foundation proved without a doubt at this Phoenix convention that FNAWS was truly the arena for marketing state and provincial sheep permits. It was time for the Foundation to move "from its own backyard" and reach the more populated areas with our conservation message. In that regard, Dr. Roy Schultz, Duncan Gilchrist and a whole host of past presidents, supporters and founders began compiling a book on the history of FNAWS and its accomplishments. The book was tentatively titled, *A Vision: Putting Sheep on the Mountain*, but has evolved to become this book, titled, *Putting Sheep on the Mountain: The Foundation for North American Wild Sheep, 25 Years Dedicated to Wild Sheep*, with chapter contributions by hundreds of FNAWS members and supporters, including "Doc" Schultz and Duncan.

In the fall 1995 the Foundation began promoting the celebration of our hunting heritage and the conservation of wild sheep in its 20th year that began with Daniel A. Pedrotti of Corpus Christi, Texas, being elected as its 17th president in Spring 1996. Plans began for the 20th FNAWS anniversary celebration and Philadelphia, Pennsylvania, was selected as the site.

FNAWS announced a 20-year commitment to the management and future of the wild sheep in Hells Canyon in fall 1995. This 20-year plan, titled, "Our Vision: Their Future,"

Tom Vaccaro, shown here in a spring 1995 photograph with an award from the Colorado Division of Wildlife for purchasing the 1995 mountain goat permit, died on October 21, 1995, when his airplane, a twin engine Cessna 421, crashed one mile from his Michigan home. Tom had purchased the mountain goat permit three times in four years. Tom was a great supporter of FNAWS who was an avid pilot, wildlife conservationist, sportsman and gentleman. To this day he is sincerely missed by all who knew him.

Photograph courtesy of FNAWS archives.

---

consisted of increasing the sheep population, a solid management plan for disease prevention and treatment, and continual monitoring of sheep and habitat within the 1 million acres within Hells Canyon that lie along the Snake River and many larger tributaries in the states of Idaho, Oregon and Washington.

In late December 1995, personnel from the Washington Department of Fish and Wildlife found a dead bighorn sheep while conducting census work. With no visible signs of a wound from a fall or gunshot, an immediate wave of concern swept through the neighboring states of Idaho and Oregon that an outbreak of the disease Pasteurella had once again found its way into the Hells Canyon habitat.

An outbreak in the winter of 1991 had almost completely eradicated the existing herds in the area. Test results taken at Washington State University confirmed the Hells Canyon herd was experiencing a Pasteurella related die-off, eventually affecting an estimated 200 sheep. The Foundation responded immediately to this emergency by allocating $25,000 to the trapping and transportation of 72 of the affected sheep to the Caldwell Wildlife Laboratory in Idaho for further research.

The Pasteurella outbreak brought together wildlife managers and biologists from Oregon, Idaho and Washington in the most comprehensive wildlife management efforts in the history of the agencies. The Hells Canyon fund-raising campaign was launched during the 1996 Reno convention from February 21 to 24 as members were asked to be a part of this long-range project.

On December 13, 1997, with a light covering of snow blanketing Hells Canyon, 39 healthy wild sheep were released, culminating many years of effort and planning. Thanks to the efforts of the Wild Sheep Society of British Columbia, the sheep were captured on December 11 at Spences Bridge, British Columbia, transported and subsequently released into Big Canyon in Idaho, Muir Creek in Oregon and near Asotin Creek, Washington. Many men and women volunteered to participate in this capture and transplant. They are all to be commended for their efforts of reestablishing the Hells Canyon sheep population. The following listing contains the names of the groups or organizations participating in this effort of put sheep on the mountain:

Wild Sheep Society of British Columbia
Washington Chapter of FNAWS
Washington Department of Fish and Wildlife
Washington State University
United States Forest Service, Enterprise, Oregon
Oregon Department of Fish and Wildlife
Idaho Chapter of FNAWS
Idaho Department of Fish and Game
British Columbia Ministry of Environment
King 5 TV, Seattle, Washington
British Columbia RCMP officers

FNAWS Outstanding Outfitter Award was first presented to Myles Bradford, Dalziel Hunting, Ltd., and to Bradford & Company Guide Service, Ltd., Dease Lake, British Columbia, Canada, during the 1996 convention. Chosen by his peers, Bradford has every reason to hold his head high for he symbolizes the very things for which the trophy stands — integrity, sportsmanship and conservation. The Foundation was honored to have Elizabeth "Betty" Golata in attendance for the inaugural presentation of "The Frank Golata Award." In her presentation she offered the following tribute:

"Through the efforts of FNAWS Past President Garry Vince, this outfitter recognition program was devised to honor the most outstanding outfitter in North America. This trophy, an 18-inch bronze statue depicting Frank Golata glassing the hills, was sculpted by Rick Taylor of Kaslo, British Columbia. So, also is the Foundation to be congratulated on sponsoring an award for which all outfitters in the future may strive — the "Oscar" of the sheep world. Such care and foresightedness should insure that we will always have sheep on the mountains of North America."

The 1996 Reno convention was attended by more than 7,000 people and raised $3.5 million dollars. The FNAWS Board of Directors approved Grant-in-Aid funding totaling $441,266.82. A Wild Sheep convention highlights souvenir edition was printed and offered to the membership for $5. The Nevada permit was not auctioned in 1996 because of new legislation passed in 1995 impacting the use of revenue generated from the sale of this permit. Significant developments in 1996 included the $15,000 contribution to the Idaho Sportsmen's Heritage Defense Fund to help battle anti-hunting initiatives on Idaho election ballots. For the first time, FNAWS membership exceeded 6,000.

During the August 9, 1996, Board meeting held in Denver, Colorado, the nominating committee met to select the candidates for the Board of Directors election for the 1997 year. A total of 13 members stepped forward and were nominated as candidates for this election to fill four vacancies. The nominating committee selected eight candidates, two for each open position on the Board based upon the a nominating policy established during the meeting.

In his fall 1996 letter to the membership, President Dan Pedrotti called on the membership to participate in the Board election process. "This is a historic Board election. For the past two years we have had harmony and progress due to the make-up of the Board. Now

Canadian outfitter and FNAWS Past President Garry Vince accepts the 1998 Frank Golata Award as the outstanding outfitter.
Photograph courtesy of FNAWS archives.

This photograph of world-famous outfitter Frank Golata was used by Canadian sculptor Rick Taylor as a guide to create the Frank Golata statue, titled "Men of Vision: The Frank Golata Award." An article penned by Frank's widow, Elizabeth, is included in this book.
Photograph courtesy of Elizabeth Golata.

Urban Rahoi of Ptarmigan Lake Lodge, Fairbanks, Alaska, left, received the Frank Golata Award in 1997. Presenting the award was 1996 Outstanding Outfitter Award winner, Myles Bradford of Dalziel Hunting, Ltd., center, and his wife Sherry.
Photograph courtesy of FNAWS archives.

---

we have the opportunity to bring fresh ideas, new faces and hopefully dedicated people who can work together solely for the purpose of putting sheep on the mountain. Remember these new Board members will set the policies and the direction of the Foundation as we move into the next century."

The Foundation for North American Wild Sheep joined the Bureau of Land Management in celebrating its golden anniversary by signing a Golden Anniversary Partnership Agreement. This agreement was signed by Karen Werbelow, FNAWS executive director, and BLM's Acting Director Mike Dombeck, at the Bureau's birthday celebration on July 18, 1996 in the historic Department of the Interior Building in Washington, D.C. This Golden Partnership spotlights the cooperative habitat improvement projects and underscores the various recreational opportunities available at 12 specific areas of public lands containing outstanding sheep habitat.

Preparations for the 1997 Philadelphia convention, held February 5 to 8, celebrating 20

The future of FNAWS is in the hands of young people like these two girls who share a laugh at the 1996 FNAWS convention.
Photographs courtesy of FNAWS archives.

Al Anton, Jackie Wells and Ron McMillen, left to right, at the 1996 convention.

Lari White performs at the 1996 FNAWS convention.

Bonnie Ford, wife of the late Jim Ford. She is a long-time supporter of FNAWS and a hunter.

Long-time auction spotter or ringer Jeff Reynolds convinces a bidder to bid higher at the 1996 Ladies' luncheon.

The merchandise shop, shown here in 1996, at the national convention offers FNAWS members and supporters clothing, gift items, hats and toys that sport the FNAWS logo and wild sheep embroidery. The first merchandising began with Jim Froelich in the mid-1970s who designed wild sheep patches based upon sketches created by his wife Janet. Sale items expanded through the years to its current level of activity.

Welcoming convention attendees at each FNAWS national convention is the FNAWS information booth, shown here in 1996 staffed by long-time FNAWS supporter Edwin Johnson, center, from the Royal Teton Ranch near Gardiner, Montana.
Photographs courtesy of FNAWS archives.

Teresa and Vern Bleich at the 1996 Saturday night banquet.

The volunteer team of Jimmy Low and Kern Chew, left to right, from California, are familiar faces to anyone who purchases raffle tickets at the FNAWS national conventions. These friendly fellows are always there with a kind word to all who support raffle drawings. They are shown at the 1996 convention.

years of wild sheep conservation, "An Inspiration...A Triumph...Our Legacy... Our Challenge," kept the focus of the Foundation Board during 1996. The 1997 Philadelphia convention was promoted as the convention offering something for everyone. The Saturday evening banquet was the stage for the premiere offering of the Alaska permit. The fireworks planned for the banquet conclusion were no comparison to the excitement generated by the energetic bidding for the permit with Jerry Fletcher winning the final bid at $200,000.

The 1996-97 fiscal year was filled with many successes and setbacks. The Philadelphia convention experienced a lower exhibitor and membership attendance, resulting in a break-even fund-raising event for the Foundation. The Foundation Board of Directors announced to the membership in the spring 1997 issue of *Wild Sheep* that they would not be funding any Grant-in-Aid requests for the year, but were still contributing to putting sheep on the mountain through the return of dollars generated by the sale of state and provincial auction tags.

After Philadelphia, the general consensus was that Reno was the best place for FNAWS to make money to put wild sheep on the mountain. Rich Roan explained why Reno has always been so ideally suited for FNAWS.

"Reno is under one roof. You can bring the family, and there's something for everybody to do. It has good facilities. The price is right. It is more central, and I think, overall, it's a good place. You hear some people say they don't want to go there every year, but you have to look at why they go and that's to raise money for FNAWS. The idea is to raise funds to put the sheep on the mountain. If you're going to go somewhere and loose money, you're not going to want to go there. We went to Nashville, and it wasn't the best. It was nice but it was expensive. We went to Hawaii and almost went broke. It was nice but it was expensive. Philadelphia was a bust because we came out with $2.3 million on the auction, but it was a bust from the other end (low attendance, high costs). It was something they tried and I guess they'll never do it again. The expenses were the facilities."

The FNAWS membership, donors, exhibitors and Board of Directors welcomed the scheduled return Reno, Nevada, at the Reno Hilton from February 18 to 21, 1998, to celebrate the 21st annual convention. A private performance by country western star, Tanya Tucker, was scheduled as the "Welcome Back to Reno" event. Enthusiasm for the convention was apparent throughout the year with increased donor support and early registrations received at headquarters. This excitement and devotion to the Foundation for North American Wild Sheep resulted in one of the most successful conventions held. FNAWS raised $3.5 million dollars at the 1998 Reno convention! This success was parlayed into $245,175 of Grant-in-Aid funding, approved for the 1998 funding year.

The Wild Sheep Journal, a special softcover book of the Foundation, was unveiled during the convention. Thanks to the generosity and special efforts of our advertisers, FNAWS chapters and affiliate organizations, major sponsors, contributors, writers and world-famous artists, this signature piece contains information about the four wild sheep species of North America, wonderful and exciting hunting stories, and shares the history, development and direction of FNAWS. *The Journal* was developed to be distributed to conservation, hunting and fishing organizations and gatherings to share and educate the public with our conservation message and goals.

On April 22, 1998, the Board of Directors accepted the resignation of Karen Werbelow as its executive director after 11 years of outstanding service to the Foundation. She departed with the Foundation's best wishes to pursue other goals. A dedicated administrative staff in place at headquarters in Cody, Wyoming, insures a smooth and progressive operation for the Foundation membership and Board of Directors. A decision by the Board of Directors was made not to continue the executive director's position.

With the competition for charitable aid and donations becoming more intense, the Foundation hired fund-raising and philanthropic management consultants, Robert F. Hartsook and Associates of Wichita, Kansas, to assist in realizing the benefits from a charitable giving campaign. Murray M. Blackwelder, senior consultant from Ames, Iowa, and Robert G. Swanson, vice president from Wichita, Kansas, were hired under contract in April 1998 to establish an endowment program for the continuation of FNAWS into the next century.

Rich Roan, FNAWS auctioneer from 1979 to 1998, has been replaced by Lex Madden, co-owner of Madden Brothers Auctions of Torrington, Wyoming. Lex and his brother, Shawn, received the nod from the FNAWS Board of Directors in the summer of 1998. The 1999 auction in Reno will be their first with FNAWS. On June 13, 1998, Lex Madden was named the World Champion Livestock Auctioneer by the Livestock Marketing Association of Kansas City, a competition that attracts outstanding auctioneers from throughout the United States and Canada. Lex has been an auctioneer since 1974. Lex and Shawn are also owners of Torrington Livestock Markets, Inc., since 1988; Frontier Realty of Wyoming since 1984; and Frontier Feeders, a commercial feedlot, since 1991. Lex and Shawn have owned their auctioneering business since 1979. At the 1999 FNAWS convention, Lex will bring Michael Schmitt as a ring man (auction spotter) to assist the FNAWS members who are volunteer ring men.

FNAWS members and its Board of Directors look enthusiastically toward the future millennium with many ideas and projects to review and implement, insuring the future and establishing the direction of the organization as we approach the new millennium.

Sandra L. Nicola has worked for FNAWS since 1987 when she began as a receptionist/public relations staff person. She was given the opportunity to develop the magazine, Wild Sheep, *when it was brought in-house in spring 1988. Sandra grew up in Casper, Wyoming, and was born in Montana where her family homesteaded in the late 1800s. Sandra is the mother of two sons, Nicholas and Christopher, and has been married to Jim since 1984. Sandra attended a "Becoming an Outdoors Woman" clinic in 1996 that renewed her appreciation of the great outdoors and the traditions that come with it. Sandra looks forward to sharing her love of the outdoors with her young sons as they grow.*

Susan C. Reneau's biography is found at the end of Chapter I.

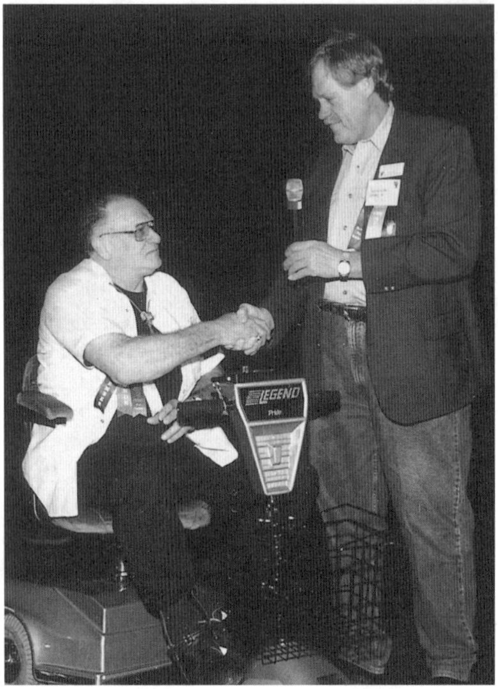

Dr. Jerry Waite, seated, accepts an award from National FNAWS Board Member Ted Schutte in 1998 for his creation of the motto, "Putting Sheep on the Mountain," that he coined when he became president in 1987-1988 and used this motto as the theme for the 11th convention from February 24 to 28, 1988 at Bally's Casino Resort (now called Reno Hilton).

The "Artist of the Year Program" was renamed in 1995 as the "Jim Ford Memorial Art Competition" to pay tribute to Jim Ford of Montana who worked tirelessly for wild sheep and FNAWS.
Photograph courtesy of FNAWS archives.

Devastating fires in Yellowstone National Park, shown here as a firefighter walked through a Yellowstone forest in 1988, prompted FNAWS to establish Funds Available for Wildlife Now (FAWN) that provided money to federal agencies involved with habitat restoration, watershed re-vegetation, relocation and transplantation of wildlife and emergency feed of wildlife when requested by agencies.
Photograph courtesy of FNAWS archives.

Chapter V

Staff of FNAWS

Susan C. Reneau

Without the women who run the office at the national headquarters, the projects, national conventions and all the other events FNAWS sponsors and coordinates would not be completed. Noteworthy is the fact that many of the staff have been worked for FNAWS for 11 years.

Volunteers are the backbone of FNAWS, but the lean, not-mean, working machine of FNAWS staff, totaling only six professionals, puts into reality the dreams, ideas and plans of the members. Unlike many non-profit organizations that hire hundreds and hundreds of people, FNAWS prides itself in the fact they accomplish great things without spending millions of dollars on salaries for a large staff. This means that by having a small, but effective staff, FNAWS is able to put more sheep on the mountain.

From a personal standpoint, I would not have been able to complete this book with Dan Pedrotti and Dr. Roy Schultz if it had not been for these women who wrote chapters, compiled data for the appendix, gave me leads to contacting former members and founders, sorted photographs, located photographs and did everything in their power to make this book a reality. And, through it all, the staff have been friendly, kind and enthusiastic. For this I am grateful. They work behind the scenes and are quick to give credit to a member for a job well done, but I can tell you they are the reason why so many projects and national conventions are successes. They care deeply for the work of FNAWS members and for the members themselves.

On April 23, 1998, it was announced by FNAWS that they had accepted the resignation of Karen Werbelow as its executive director. She left after serving FNAWS for 11 years to pursue other interests. She departed with the best wishes and hopes for every success from FNAWS. Karen was interviewed by me in late August after she had settled in Tracy's Landing, Maryland. She is engaged to Dennis Krus and will marry in October 1998. She is a wildlife political consultant based near Washington, D.C. Karen was born in Brooklyn, New York, and was raised on Staten Island, New York. She spent three years in San Antonio, Texas, substitute teaching in Bexar County before returning to New York and relocating to Cody, Wyoming, in 1980. She was first hired by FNAWS in 1987 and became office manager that October. A year later she became business administrator. Her final position was executive director, which she maintained for eight years. The historic Hells Canyon Initiative and meeting at the White House with Vice President Al Gore were projects she points to with accomplishment. Karen also secured the first grant for the Hells Canyon Initiative from the National Fish and Wildlife Foundation for $75,000. She is the mother of Christopher and Kourtney.

Prior to the staff in Cody, John and Jane Babler of St. Paul, Minnesota, coordinated the day to day activities of FNAWS in the early years, prior to 1982. Lloyd Zeman, as a founder and first president, worked with John and Jane in St. Paul until he moved to Cody to become the first executive director of FNAWS.

As the first three chapters of this book explain, there have been several executive directors since 1982 and several more support staff, but the fact remains that a small, energetic core of professional staff is the reason why work gets completed. The current staff biographies are listed here in alphabetical order since all are equally important to the workings of FNAWS. Each staff person added their personal touch to the paragraphs that presents a glimpse of the team that is the national staff of FNAWS as of September 1998.

Tami Clark

Tami Clark has worked for the Foundation for North American Wild Sheep since 1996. She is in charge of public relations and has most recently taken the responsibilities of exhibit coordinator. Tami is not only a Wyoming native but was raised in Cody. She and her husband, Jerry, have two children, Keenan and Erin. They are avid hunters and also enjoy camping, fishing, softball and many other outdoor activities. Tami's women's softball team took first place four years running. Tami has had a lot of experience working with the public. She worked for the airline industry in Cody for eight years prior to working for the Foundation. She especially enjoys working with and meeting new people.

Paula Karres

Paula Karres is a native of Los Angeles, California, and was raised believing hunting had no use in today's society. After moving to Wyoming, she was reeducated and is living proof that a little education goes a long way. Once a year, she and her horse make a pilgrimage to British Columbia to train with former captain of the Canadian Three-Day Equestrian Team, Robin Hahn. During the summer she competes at horse trials throughout the Rocky Mountain Region. Paula has been with FNAWS since 1993, working as a secretary to the executive director and as chapter affiliate coordinator. She is co-author of the chapter in this book on political activities with Former FNAWS President Phyllis Carson and compiled some of the information contained in the section of FNAWS chapters and affiliates. Paula lives in Cody with her husband David of 18 years, two children, Kelly and Amanda, six dogs and five horses.

Becky Layne

Becky Layne has worked for FNAWS since 1991. She is the Grant-in-Aid coordinator. She sees first-hand how FNAWS benefits wild sheep. She wrote a lengthy chapter of this book detailing some of the outstanding projects funded by FNAWS over the years and compiled the complete list of Grant-in-Aid projects from 1983 to the present for this book. Becky is an avid hunter and after attending the 1996 "Becoming an Outdoor Woman," she took an

Becky Layne, left, struggles to tie a black bandanna as Kim Nieters, Karen Werbelow and Paula Karres laugh at the 1996 Donor/Exhibitor Party.

Photographs courtesy of FNAWS archives.

FNAWS Grant-in-Aid Coordinator Becky Layne enjoys archery. Here she practices in the summer of 1996.

FNAWS President Pete Cimerello at the 1996 convention is surrounded by the national FNAWS staff who include, left to right: Becky Layne, Sandra Nicola, Kim Nieters, Karen Werbelow, Terry Ziehl and Paula Karres.

Taking a bow at the 1997 Philadelphia convention are, left to right: Sandra Nicola, Becky Layne, Tami Clark, Kim Nieters and Terry Ziehl.

interest in bowhunting. She and her husband both grew up in Wyoming and now reside in Cody. They enjoy camping and fishing with their two daughters, Ashley and Emily. Becky is also a Girl Scout leader for her daughter Ashley's troop. She feels it is extremely rewarding to encourage young minds to appreciate the benefits of conservation and the outdoors.

Sandra L. Nicola

Sandra L. Nicola has worked for FNAWS since 1987 when she began as a receptionist/ public relations staff person. She was given the opportunity to develop the magazine, Wild Sheep, when it was brought in-house in the spring of 1988. She is in charge of advertising and publications for FNAWS. Sandra grew up in Casper, Wyoming, and was born in Montana where her family homesteaded in the late 1800s. Sandra is the mother of two sons, Nicholas and Christopher, and has been married to Jim since 1984. Sandra attended a "Becoming an Outdoor Woman" clinic in 1996 that renewed her appreciation of the great outdoors and the traditions that come with it. She is the co-author of Chapter IV of this book and the person who located hundreds of photographs that appear in this book. Sandra looks forward to sharing her love of the outdoors with her young sons as they grow.

Kim Nieters

Kim Nieters was born and raised in Cody, Wyoming and has been working for FNAWS since 1988. Working in the membership and merchandise area of FNAWS is especially fun for her because she is able to be creative with styles and design productions for the many aspects of the Foundation. Raised in a hunting environment, Kim is proud to be working for an organization that she knows is "putting sheep on the mountain." Becoming acquainted with the membership and knowing many of them personally, is what makes her job memorable. Kim enjoys playing volleyball, softball with other staff member, Tami Clark and many other outdoor activities. Recently she began coaching AAU Volleyball for young teens during the summer. She and husband, Ron have a daughter, Shelbie, and enjoy camping together.

Terry Ziehl

Terry Ziehl has worked for FNAWS since 1987. She is currently the office manager and has worked with the exhibitors since her first convention in Reno in 1988. Working for FNAWS has been rewarding and challenging especially meeting people from all over the globe. Born and raised in Minnesota, Terry got married and found her future in Casper, Wyoming, with her husband Terry going to college. Her husband is currently employed with Marathon Oil. They were transferred from Casper to Cody in 1986. She and her husband have two children, Samantha and Kimberly. They all enjoy their time together camping, fishing, swimming and cross-country skiing. Terry also enjoys many hobbies, with stain glass being her passion. She had donated several pieces to past FNAWS auctions.

Sitting in the fundraiser car are the 1991 national staff of FNAWS, left to right: Sandra Nicola, Becky Layne, Marti Neilsen, Karen Werbelow, Kim Nieters and Terry Ziehl.

FNAWS President John Harris stands in the center of the 1991 FNAWS national staff who include, left to right: Sandra Nicola, Karen Werbelow, Terry Ziehl, Becky Layne, Kim Nieters and Marti Nielson.

The FNAWS staff in 1992 included, left to right: Marti Nielson, Becky Layne, President Howard W. Pollock, Karen Werbelow, Terry Ziehl, Kim Nieters and Sandra Nicola.

A "Welcome Back to Reno" night in 1991 found Terry Ziehl, Sandra Nicola, and their husbands Jim and Terry all dressed up.

Blowing kisses to the camera in 1992 are national FNAWS staff Marti Nielson, left, and Becky Layne.

Crazy for the Hulu Hoop is Kim Nieters in 1992.

The entire national FNAWS staff received the Presidential Award in 1995. They were, left to right: Exhibit Hall Coordinator Terry Ziehl, Grant-in-Aid Coordinator Becky Layne, Convention/Public Information Coordinator Paula Karres, Membership/Merchandise Coordinator Kim Nieters, Publications/ Advertising Coordinator Sandra Nicola and Executive Director Karen Werbelow. FNAWS President Garry Vince, right, presents the plaque to them.
Photographs courtesy of FNAWS archives.

Karen Werbelow and Dennis Krus at the 1998 FNAWS convention.

FNAWS Board Member Dan Artery shares a laugh with Karen Werbelow, left, and Convention Chairman Phyllis Carlson at the 1997 convention.

FNAWS President Craig Johnson in 1994 stood in the center of the national FNAWS staff who were, left to right: Paula Karres, Sandra Nicola, Karen Werbelow, Terry Ziehl, Kim Nieters and Becky Layne. Note that snappy hat.

The entire FNAWS staff at the 1998 convention sitting atop a Harley-Davidson motorcycle.

FNAWS Artist of the Year Prints

All prints are signed and numbered, and can be ordered by contacting

FNAWS Headquarters
720 Allen Ave., Cody, WY 82414
(307) 527-6261

"The Untouchables"
18" x 27" image size.
by Hayden Lambson
1990 Artist of the Year

"Desert Sentinels"
18" x 22" image size.
by George Lockwood
1991 Artist of the Year

"Cliff Hanger"
15" x 22" image size.
by Tom Mansanarez
1992 Artist of the Year

"A Time to Leave"
22" x 30" image size.
by Trevor Swanson
1994 Artist of the Year

All Artist of the Year Prints are $100.00 plus $10.00 shipping & handling each.

"Legends of Sacred Ground"
18" x 21" image size.
by Martin Rathburn
1996 Artist of the Year

"Meeting the Challenge"
18" x 27" image size.
by Tom Mansanarez
1998
Artist of the Year

FNAWS Living Legend Prints

All prints are signed and numbered, and can be ordered by contacting FNAWS Headquarters, 720 Allen Ave., Cody, WY 82414, (307) 527-6261

"Eye of the Storm"
24" x 32" image size.
by Vivi Crandall
1996-97 Living Legend Artist
$550.00 plus shipping & handling
(Fed-ex shipping only, due to the size of the shipping box.)

"The Reigning King"
27 3/4" x 20 1/2" image size.
by Gary Swanson
1993 Living Legend Artist
$250.00 plus $10.00 shipping & handling

Also Available
"Eye of the Storm"
Open Edition Posters
15" x 20" image size
$15.00 plus $10.00 shipping & handling

"Mutton Chase"
20" x 30" image size.
by Guy Coheleach
1994-95 Living Legend Artist
$250.00 plus $10.00 shipping & handling

Chapter VII

The FNAWS Youth Essay Contest Winners

Beginning in 1990, young people from across North America have submitted short essays to the Foundation for North American Wild Sheep. Winners of the annual essay contest receive an all-expense-paid hunting adventure from Arctic Red River Outfitters.

Ray Woodward began the tradition as the owner of Arctic Red River Outfitters in 1990 and when he sold his outfitting business to Kelly and Heather Hougen, they continued the hunting trip essay prize. Their continued support of the FNAWS Youth Essay Contest offers encouragement to the next generation of men and women who love to hunt and conserve wildlife.

The following essays and hunting adventures were submitted by the winners, starting with the first winner, Charles Darrell Hoovestol of Bismarck, North Dakota. The articles are included in the order of the year the essay contest was won.

To enter the Youth Essay Contest, write for current rules at the national headquarters of FNAWS, 720 Allen Avenue, Cody, WY 82414 or call them at (307) 527-6261.

Each year has a different essay theme but essays are to be 100 to 250 words in length. Male or female contestants must be between the ages of 16 and 19 years. The grand prize is a fully donated, 10-day caribou hunt in the Mckenzie Mountains of Northwest Territories with Arctic Red River Outfitters for the teenager. All accommodations and transportation to and from the young person's nearest airport is included along with hunting licenses and trophy fees.

The following is a list of Young Essay Contest winners:

1991 - Charles Darrell Hoovestol, Bismarck, North Dakota, "What the Right to Hunt Means to Me"

1992 - No contest held.

1993 - Cody Voermans, Whitefish, Montana, "Call for Education"

1994 - Marissa Johnson, Marysville, Utah, "Ethical Hunting: Our Responsibility to Protect Our Hunting Heritage"

1995 - Angie Wanamaker, Takysie Lake, British Columbia, "The Effects of Multiple Use Lands on the Future of Hunting"

1996 - Marcus D. Hite, Newark, Delaware, "What Do You Feel is the Single Greatest Danger to Wildlife Populations in North America?"

1997 - Justin Miller, Tracy, Minnesota, "Are the Actions of Animal Rights Activists Inhumane?"

1998 - Ryan Swanson, Fort Lupton, Colorado, "Should Wildlife Management be Controlled by the Ballot Box?"

1998 - Garrett B. McKitrick, Bowman, North Dakota, runner-up to Ryan Swanson

1991
What the Right to Hunt Means to Me

CHARLES DARRELL HOOVESTOL

The right to hunt above all is a privilege and must not be abused. The right to bear arms and harvest game is a responsibility that must be treated with a great deal of respect. Hunting has been an essential part of wildlife management and survival since the dawn of time. Althought hunting may not be essential to survival in this day and age, it is a critical part of wildlife management and is still as popular as it has ever been. Hunting for pleasure is a tradition that has been handed down from generation to generation. It is the supreme challenge, pitting the intellect of man against the heightened senses and speed of the wild animal. It is the oldest sport on earth and also the most natural; a direct interaction with nature. The average hunter is in the field to enjoy the outdoors as much as he or she is to harvest game. Is a Sunday afternoon better spent walking windbreaks in the fresh air, hoping to take home some dinner or sitting in front of the television watching the big game? To exclude the hunter from history is to deny mankind his roots; roots which still supply man with all the vital ingredients of life and evolution. Man is a predator by design. His jaws and teeth were not developed for living on roots and berries alone. He was obliged by nature to be a hunter and so as long as there is game to hunt, there will be hunters.

Editor's Note: Charles Darrell Hoovestol of Bismarck, North Dakota, was the first essay winner for a contest created in 1990 by Ray Woodward, former owner of Arctic Red River Outfitters. The current owners, Kelly and Heather Hougen, continue the essay contest to this day by donating a caribou hunt to the winner as the grand prize. Charles enjoyed a caribou hunt in the fall of 1991. He graduated from Bismarck High School in spring 1991. He was born on November 10, 1972, the first day of deer season so he wrote, "It would only be fitting that I am an avid hunter and have been since I was old enough to carry a gun." He enjoys hunting for deer, pheasants, grouse, rabbits, ducks, geese, coyotes and fox. He bowhunts and spends many of his summer days fishing. He is familiar with FNAWS because of their work in the Badlands and their contributions to the North Dakota Game and Fish Department.

1993
The Most Exciting Experience of My Life!

Cody Voermans

The 1993 Youth Essay Contest winner, Cody Voermans of Whitefish, Montana, center, is congratulated by Past President Phyllis Carlson, left, and FNAWS Publications and Advertising Coordinator Sandra Nicola.

I was fortunate and excited to win the contest this year. Little did I realize my prize caribou hunt with Kelly Hougen's Arctic Red River Outfitters would turn out to be the most exciting experience of my life.

My dad accompanied me on this trip, and our hunt began in the watershed of the Arctic Red River in the northwest corner of the Northwest Territories on September 8. Kelly hooked us up with guides that did everything in their power to ensure Dad and I had the best hunt possible.

On the first day of the hunt, my guide, Tavis, got me within range of two really good bulls I passed only because it was the first day. On the second day Tavis and our other guide, Alex, got us within 300 yards of two wolves. That, however, wasn't the end of the excitement. We saw northern lights the night of the third day. On our fourth day, our guides showed us more wolves and two different grizzlies.

The only problem with those first four days was that the caribou had moved out of the valleys we were hunting and could only be glassed in neighboring valleys too far to reach before dark. It was because of this that Kelly picked us up with the Super Cub and flew us to another camp the evening of the fourth day. He took us to a camp he called Beaver Lake where more caribou (many good bulls), had been seen in the days just prior to our arrival.

It was here on the fifth day that Tavis and I began what turned out to be an eight mile stalk, going after five mature bulls. We were lucky enough to catch up to them late that evening before they reached the next valley, and I took the best bull in the bunch. Even though he is a little smaller than the bull I passed on the first day, he is still close to the Boone and Crockett Club's all-time records book and the finest trophy I have ever taken.

Dad was more patient than I, and waited until the sixth day of our seven-day hunt to take a bull. His was beautiful bull with big palmated tops and long back points that would look fantastic on any wall. Our caribou were not the only trophies in base camp, however. Kelly's area, 1,000 square miles of the Northwest Territories, provided other hunters with caribou, sheep and moose that were also superb trophies.

Dad and I both agree the beauty of the Mackenzie Mountains, the trophy caribou, and the Arctic Red Rivers Outfitter's commitment to excellence made this trip the hunt of a lifetime.

I would like to thank FNAWS and Kelly and Heather Hougen for sponsoring the youth essay contest. I encourage other young people to participate in this contest in the future. The memories I have of writing the essay and hunting the Arctic have enrich my life, and will last forever.

Thank you!

The Role of Conservation in the Future of Hunting Call for Education

Cody Voermans

I believe that problems facing hunting today are a direct result of a lack of education. The only way to preserve the privilege of hunting for the future is for the conservationists to focus their efforts on educating everyone about the positive role hunting plays in nature.

It is increasingly apparent that many people feel hunting is wrong and should be stopped everywhere. These people rightly want nature to act independently without the interference of man, but I contend that they misunderstand how hunting helps maintain balance in nature. Many people, for example, don't understand that population explosions resulting from a lack of hunting and natural predation may cause rampant diseases such as rabies and tuberculosis in game populations. Furthermore, because many people don't live in rural areas, irreparable damage to available habitat and neighboring agricultural land resulting from overpopulation is not recognized. It is the conservationists' duty to vocalize the positive aspects of hunting to counteract the biased opinion many people have as a result of seeing only one side of the issue.

Being raised in a farming and hunting family, I have learned the interrelationships of the natural world, values concerning the conservation of habitat and wildlife, and the important role hunting plays in both. But not everyone has had the same opportunity as I to learn the importance of hunting in the working of nature. Consequently to conserve hunting in the future, hunters as conservationists must educate the public about how important preserving hunting is for the equilibrium of nature.

Editor's Note: This essay was submitted by 17-year-old Cody Voermans of Whitefish, Montana, for the 1993 essay contest. Cody was a junior in 1992 at Whitefish High School when he wrote this essay and participated in track and field, National Honor Society, 4-H and the Whitefish Junior Rifle Club. He was on the Junior National Running Target Development Team and continues to enjoy shooting a great deal. His fondest hobby is hunting. Cody from a hunting family and the sport has become a big part of his life.

1994
Ethical Hunting: Our Responsibility to Protect Our Hunting Heritage

MARISSA JOHNSON

It's been my privilege to grow up in a small, rural region where our hunting heritage played a vital role in its settlement and development. I vividly remember sitting on my father's lap as he retold stories of his ancestors hunting and exploring the wild, beautiful wilderness when it was pristine and untouched by man.

As civilization encroached and placed greater demands upon our natural resources, I realized the responsibility hunters established. For me and future children to continue to experience the enjoyment, beauty, thrill and artistic challenges of hunting, we must be ethically accountable for our actions.

Our fraternity must demonstrate to those in society who seek to criticize and destroy our "hunting heritage" that we are dedicated individuals representing all walks of life. We are devoted to and exemplify the following traits and principles that I believe promote ethical hunting.

- Love, appreciation and commitment to manage and preserve our natural resources.
- Respect property, landowners, and non-hunters' rights.
- Understand firearm safety.
- Development quality training programs for youth.
- Organize, raise funds, and promote new programs for the enhancement of wildlife.
- Enforce ethical hunting standards within our ranks.
- Know and obey hunting regulations.

Let's not procrastinate in protecting our hunting heritage. Let's race ahead as we chart courses to insure that "ethical hunting" cements the foundation for hunting in the future.

Editor's Note: Marissa Johnson of Marysville, Utah, was a 16 year old honor student at Piute High School when she wrote this winning essay. She was on volleyball and basketball teams in high school and enjoys backpacking, hunting, water skiing, 4-wheel driving and raising calves. She began backpacking and hunting with her father at age 10. Her father taught Marissa to love and appreciate the wilderness and the basic necessities of survival. Marissa's goal in 1993 was to graduate from college and her father collect his Grand Slam in wild sheep hunting of North America.

1995
Hunting with a Legend
A Successful Hunt for the Youth Essay Contest Winner

ANGIE WANAMAKER

Angie Wanamaker with her 1994 caribou.

My name is Angie Wanamaker. I am 16 years old, an average student and am now commuting an hour and a half to school in Burns Lake, British Columbia, Canada, each day as my old school only goes to grade 10. Being privileged to live in a beautiful rural area and being around loggers, hunters and guides, I have developed a strong interest in hunting, camping and horses. I am presently breaking my own colt.

My advice to anyone entering the essay contest is to use the wealth of information found in your community. Talk to and take a few notes from your local experts; the ranchers, loggers, trappers, guides, conservation officers, range management people and even politicians, the more outspoken the better. This is a trip of a lifetime and you will fall in love with the North. I know I have.

When the phone call came that I had won the 1994 Youth Essay Contest, I was surprised, happy and so excited that I could hardly talk. Then came the training to get in shape and all the target practice. I went through at least 250 rounds of .280 ammunition and several thousand .22 shells.

My guide on the hunt was the legendary Alex Van Bibber. At 80 years old, he is as tough as any 40-year-old mountain man going. We were moved by bush plane to Alex's "Hilton" on day one. We could not hunt but glassed and saw lots of caribou. What a great camp, complete with a large high wall tent, cots and a stove.

Day two was too warm for good hunting. Late in the afternoon we spotted a giant bull caribou with a pure white cape but couldn't get close enough.

We walked and walked for what seemed like forever, on day three, but saw no caribou. We stopped and glassed awhile and spotted a white timber wolf, took pictures of the beautiful scenery. We came upon some young rams that were as tame as "park" sheep. Finally we spotted a big bull, far in the distance. He was with a cow and headed towards our camp.

No success on day four, and we moved our camp to a little "village" of pup tents on day five. We joined the three other hunters that I came in with. We lounged and visited.

Day six — I got my caribou! We had only walked about an hour when we spotted seven bulls. We got within 100 yards. I aimed and fired. The bull ran and Alex told me I'd missed so I fired again and the bull went another 10 yards and fell. When the caping and skinning was done, it was obvious I had put both shots into his chest.

Alex cleaned the cape, turned the lips and ears and did a perfect job. Rick, another hunter, and his guide came in to camp with a beautiful caribou as well. The weather had changed to cold and caribou were everywhere.

My caribou is a very beautiful, double-shovelled bull that I will always be proud of. I cannot say enough good things about Kelly and his crew. They are the greatest. Thanks, guys. This is a dream fulfilled and is was all made possible by the generosity of FNAWS and Arctic Red River Outfitters. Thanks to all of you!

Editor's Note: Angie Wanamaker was fortunate to have had the opportunity to be guided on her winning hunt by Alex Van Bibber, a living legend in the North. Alex was born in 1916 "under a spruce tree on the Pelly River" in the Yukon Territory. He has guided hunters from all over the world since he got his first guides license in 1943 and ran his own outfit from 1948 until 1968. He guided many notable hunters such as Fred Bear, Johnny Caputo and guided Dr. Earl J. Thee to the Number One Boone and Crockett ram in 1948. Since 1969, Alex has guided for several outfitters and has been working for Kelly and Heather Hougen of Arctic Red River Outfitters for the past five years. Heather is Alex's granddaughter. Alex will be 80 years old this April and plans on celebrating his 80th fall in the mountains doing what he does best — keeping his hunters entertained and happy. He claims it will be his last year guiding, but nobody believes him. Angie rates this trip and hunt as the high light of her life so far and extends her appreciation to all those involved in making it such a memorable experience.

The Effects of Multiple Use Lands on the Future of Hunting

ANGIE WANAMAKER

Our future in hunting depends heavily on proper multiple use policies with all user groups banding together to ensure each other's survival.

Here in central British Columbia, we are no different from any other area. We are under constant pressure from well-financed international protest groups that would ban all hunting, logging and cattle grazing on government land. These protest groups would like everything left in the natural state without any harvest plan. Nature has a very cruel harvest plan: fires, starvation and disease. Forests and wildlife are like a crop of wheat. They need a well-managed harvest plan.

Old growth forests might look nice from a distance but take a closer look. They are rather sterile and provide very little feed for wildlife, no livestock grazing and very limited amounts of fur. Logging these forests with a proper management program of reforestation provides jobs, enhances vital winter habitat for moose, deer and other browsers, provides summer grazing for cattle and greatly increases the populations of fur bearers.

To increase wildlife populations and to ensure that hunters don't get slandered the way the trappers did by Green Peace, it is imperative that we all work together to ensure that proper multiple land use management objectives are met for the benefit of all resource users, animals and human alike.

1996
A Once-In-A-Lifetime Adventure

Marcus D. Hite

Earlier this year when I received a telephone call from someone who identified herself as Paula from FNAWS and who said that I had been chosen as the winner of the 1995 FNAWS Youth Essay Contest, I was in disbelief. In fact, I called the Foundation headquarters the next day to find out for sure that I was winner. I did win and the prize was an all-expense-paid trip to the Northwest Territories to hunt caribou with Kelly Hougen of Arctic Red River Outfitters.

As summer passed, I could not wait for the day when my father and I would finally be able to depart. When the time came to board the plane, I sensed this was going to be a trip I would remember for the rest of my life.

We departed from Harrisburg, Pennsylvania, flew to Atlanta and onto Salt Lake City and spent the first night of our trip in Edmonton, Alberta. We left early the next morning for the flight to Norman Wells in the Northwest Territories. We were met in Norman Wells by Monica Loomis who owns the Mackenzie Valley Hotel and is a very important part of Kelly's outfit. Monica went out of her way to make sure everything was okay for us.

The next morning we boarded a Twin Otter float plane in Norman Wells for the flight to Arctic Red River. When we arrived at Arctic Red we were greeted by a group of hunters who were just preparing to leave. Lying on the bank of the lake were the fruits of their hunts. I never saw so many trophies in one place at one time! Obviously the group had a very good hunt and seeing all these trophies only increased my excitement for what was to come.

After a hike to the base camp and meeting Kelly and his crew, we ate dinner and were quickly prepared for our back country flight to Kaleb camp, our home away from home for the next week or so. After a 20 minute flight in one of Kelly's Super Cub planes, we landed at camp and were met by our guide Alex Van Bibber. At first glance, Alex appeared to be about 60 years old, but when he told me he was 80, I was astounded!

We settled into the camp which, for a tent camp that was less than 100 miles from the Arctic Circle, was really very comfortable. For the remainder of our first day in camp we relaxed and glassed for wildlife every now and then. We awoke the next morning to a massive fog that completely "socked" us in. We walked for awhile, but it wasn't too long before Alex said it wasn't worth continuing any more that day.

Day three arrived and we were glassing early. Our first sighting that morning was a young bull moose that Alex claimed to be "crazy" after we watched it run straight up the side of the mountain. The only caribou we saw were about four to five miles away. Alex said it was too late and that maybe we would pursue them tomorrow.

Day four came and went like the three previous days. We saw a large group of caribou off in the distance and Alex said we might put a stalk on them the next day. On the walk home, we saw a bull caribou at a relatively close distance. It was small but, after the past three days and seeing nothing close, it was a very welcomed sight.

The next day Alex decided we should go after the caribou we saw the evening before. But to do so, he chose a different route that included crossing over the top of the mountain. After hiking the mountain, we could finally see the caribou but they were still several miles away. After walking more than five miles down the mountain through thick willows and grass and fording a river, we arrived. Alex studied the caribou and pointed out the largest bull in the group.

Taking my time, I fired and made good on a caribou I would later learn qualified for a Boone and Crockett Club award! A few minutes later, my father also scored on a caribou almost as large as mine. Little did I know, however, that our adventure was just beginning. After caping and quartering the meat, we loaded our packs and headed back to camp. We only walked for about an hour when Alex decided to call it a day and said that for the remainder of the evening we were going to si-wash it. For those of you who have never experienced it, be glad that you haven't because once you do, you will remember it for the rest of your life. Si-wash meant simply stopping where we were, building a small fire and sleeping under the stars. Our sleeping bags were back in camp!

It was one of the longest and coldest nights I think I have ever spent. I did not get much, if any, sleep that night! We started early the next day and began the trek home. We walked and walked and we walked some more. In fact, it took us almost 12 hours to get back to camp. And then just before we reached camp, we saw caribou everywhere, and I mean everywhere. We finally reached camp just before dusk and it was a great feeling. As it turned out, we had just experienced a 36-hour, 32-mile day!

The next day, Kelly flew out and returned us to the base camp where we spent the next two days relaxing and helping out around camp. While there I made friends with John and Chris from Battle Mountain, Nevada, who are two very loyal and dedicated members of FNAWS. We made plans to meet again in Philadelphia at the 1997 FNAWS convention.

At the end of our second day at the base camp, we were preparing for dinner when word came that the plane to return us to Norman Wells would be arriving in less than a half-hour. After some rather hectic packing an abbreviated farewells, we began the trip home.

As we flew over the Arctic Red River camp, I thought about how great Kelly and his entire crew had been to us. I think Alex said it best when he said that Kelly told his guides that "you have to be the hunter's friend; not just his guide." The statement typifies the attitude of Kelly and the entire Arctic Red staff.

For me, Alex Van Bibber became my very good friend throughout the hunt. His friendship, humor and knowledge of the outdoors all contributed to a once-in-a-lifetime adventure for me. He told me this was to be his last hunt and I can only hope that before next hunting season arrives somehow or someone will change his mind and Alex will again return to Arctic Red River, allowing someone else to benefit from his experiences as I certainly have. He is, to all those who know him, a legend among the guides throughout the Northwest Territories and the Yukon. I am glad I spent my hunt with him. Thank you Alex, for a great time.

Lastly, I would like to thank FNAWS for giving me a lifetime of memories and stories. Without you, this dream would never have been fulfilled.

I hope that all those who will read this account and who are eligible to enter the 1997 FNAWS Youth Essay Contest will do so. I hope the program will continue and there will be other young people who will be able to enjoy the outdoors as did with Kelly, Alex, and of course, my father. Thank you all for a great hunt.

Editor's Note: This winning essay was submitted by Marcus when he was a sophomore at the University of Delaware in Newark, Delaware. Marcus is majoring in public policy, with a minor in environmental sciences, and during the summer and winter breaks has been employed as a research intern for the Pennsylvania lieutenant governor. Hunting has been a tradition in his family for generations. He began hunting at age 12. He wrote with his essay the following letter:

"I think our founding fathers understood the importance of this tradition when they incorporated it in our Constitution by giving us the "right to bear arms" in the Bill of Rights, which they established as rights that no one should take away from us.

Today, this right is being threatened by the very same body who more than 200 years ago said these rights should never be rescinded. The debate has not only deterred now persons from taking up the sport of hunting, but in many families has stopped a tradition that has been carried on for generations.

If my generation does not act quickly, future generations may never have the opportunity to experience this great sport. The essay I wrote has one main goal: to show the people of my generation what is happening and what the end result may be if no action is taken. FNAWS is one of the groups that has taken the initiative to prevent this tragedy from happening. I can't begin to express my appreciation to FNAWS for the tremendous opportunity this has given me, and I am looking forward to hunting with Kelly Hougen later this year."

What Do You Feel is the Single Greatest Danger to Wildlife Populations in North America?

MARCUS D. HITE

In recent years, the public opinion of hunting has diminished to the point where fewer people from my generation are taking up the sport, while those who have enjoyed hunting in the past are finding it more politically correct to walk way from it altogether. As a result, one of the primary objectives of hunting, wildlife management, is being jeopardized. The fewer the hunters, the less game will be harvested. If the herds are not properly managed, including the harvesting from hunting, wildlife populations may soon exceed the levels for which particular species may coexist and survive. The end result may be an irreversible imbalance in our delicate wildlife ecosystem.

We need only to look at how wildlife management in this century has brought some species from near extinction, to where today these same species thrive. And ironically, the cost of this wildlife management has been funded not by taxes, but from license fees and voluntary contributions from sportsmen and the businesses that serve them. The success of organizations such as the Foundation for North America Wild Sheep in "putting sheep on the mountain" is a perfect example of hunters and non-hunters working together to protect and manage our wildlife.

Improving the public's perception of hunting and bringing a great awareness to the valuable role it serves in protecting and managing our wildlife populations should be the Number One priority for all sportsmen and groups like FNAWS as we prepare for the 21st century.

1997
A Memorable Hunt for the FNAWS Youth Essay Contest Winner

JUSTIN MILLER

On September 17, 1997, I departed from Sioux Falls, South Dakota, bound for Norman Wells, Northwest Territories, and the hunting trip of a lifetime.

For a high school senior from Tracy, Minnesota, who loves the outdoors, this was to be more than simply a hunting trip. It was going to be the adventure of a lifetime!

I arrived in Norman Wells on September 18, and was pleased to be greeted at the airport by the owner of the McKenzie Valley Hotel. A few other hunters bound for the same hunting camp and I were then taken to the hotel, where we were instructed to leave in locked storage everything except what we were taking into the mountains. Afterward, I met up with Darrell Smith of Kelly Hougan's Arctic Red River Outfitters. Darrell was to fly me into the base camp in the Mckenzie Mountains that evening.

We arrived at the base camp around 9:30 p.m. after a pleasant flight in a Piper Super Cub. I was very impressed with the camp because even though it's 100 miles from the nearest road, it has two runways, multiple buildings, three airplanes and three all-terrain vehicles (ATVs). The camp was clean and neat, and had storage sheds, a cook/dining shack, multiple bunkhouses, a small building for showering and washing clothes and even two meat storage shacks with battery-powered fences. After a wonderful moose supper, I spent the night in the guide's shack. The next day, I prepared my gear with my guide, Justin Patschke of British Columbia. As this was to be a backpack hunt, we needed to pack only the bare essentials, along with food and tent.

That evening, we were flown into a remote airstrip in a valley some 15 miles from base camp where we would spend the night. Canadian law does not allow hunting for 12 hours after riding in an airplane. This prevents the poor sportsman from simply landing near a herd of animals and shooting and ensures the animals being hunted have a fair chance of escape.

On Saturday, September 20, I began my caribou hunt. My guide and I hiked along a stream bed in a valley to a spot at the base of two mountains where we would be able to glass a large area of mountainside. This was important because above-normal temperatures had kept the caribou higher in the mountains rather than in the valleys where they are usually found during this time of the year.

Just as we started to glass the mountainsides, Justin excitedly told me he saw a small herd of about 20 caribou with a large bull about half mile up the stream. We didn't even have the tent set up yet, but this was a nice bull so we began a stalk. As we neared the herd, it became clear the dominant bull was a very nice specimen. We thought he would quite possibly make the Boone and Crockett Club's records book.

My guide and I decided to attempt to harvest this bull. We crossed a creek to get into position behind some brush. The bull wandered into view between some hedges, but before

I could shoot, a cow stepped in front of him, forcing me to hold fire. The herd proceeded to graze slowly up and out of the stream bed and up onto a hill, all the while partially obscured by brush and trees. The caribou were now another half mile away on a hill in the muskeg. We decided to wait until they had bedded down before we attempted another stalk. After what seemed like hours, the herd bedded down, and we began the arduous task of stalking animals in the rough terrain of the tundra. Once we were within about 130 yards of the herd, we belly crawled for 50 yards until we were behind a small hump in the terrain. Here I got in the prone position and simply waited for my bull to get up. After 15 minutes the caribou decided to graze again, and as my bull got up, I dropped him with a single shot from my .308 BLR.

The rest of the afternoon included a flurry of handshaking, picture-taking, field-scoring (just missed the book), and packing meat back to camp. Now that I had harvested my animal, the real work began. We spent the remaining days of the hunt trying to fill my wolf tag. We hunted up and down the valley for a few days, but I only managed to miss a 300 yard shot at a loping wolf.

I was not disappointed. The most pleasurable part of the hunt is simply being in the unspoiled wilderness. By the end of my hunt, I had seen hundreds of caribou (only one nicer than mine), seven wolves and even a yearling bull moose, which my guide and I sneaked up on and photographed. I also saw the sun in the middle of the sky at 3:30 p.m. and the most beautiful and clear sighting of aurora borealis imaginable.

Justin and I hiked out to a different airstrip on the last day of my hunt. It was a ten-mile journey and as we trodded along through the tundra and skirted the mountains, I couldn't help thinking about what an amazing place I was in. In today's high-tech world of computers, faxes and large cities, the place I hold dearest to my heart is where none of these things are known; where certain divinity ensures that the mountains and rivers remain pure and unscathed by man.

Someday I will return.

I would like to thank FNAWS, all of its members, and Kelly Hougan for providing me with this great opportunity. I also wish Kelly, the legendary Alex Van Bibber, and all of their crew (especially Justin) at Arctic Red River Outfitters—only the best of luck in the future. They are a great group of people, and they made my experience a truly memorable one.

Editor's Note: Justin Miller was encouraged to enter the essay contest by his social studies teacher and FNAWS member, John Coulter, and his agriculture instructor Paul Skoglund who attended the Safari Club International American Wilderness Leadership School on a FNAWS scholarship.

Justin was a junior at Tracy, Minnesota, Junior-Senior High School when he became the essay winner. He was active in Future Farmers of America (FFA), French Club, intramural basketball, peer counseling and peer mediating program. Justin has won state and regional trips in FFA for judging and public speaking. He attended the confirmation class at his local Catholic church, where he was a lector. In his spare time, Justin enjoys fishing of all kinds, hunting, tying of fishing flies, snowmobiling and being with his friends.

Justin wrote in a letter with his essay: "My future plans include college, marriage and, hopefully, pursuing a career as a writer. I would like to be an outdoors writer for a magazine."

Are the Actions of Animal Rights Activists Inhumane?

Justin Miller

There is a great argument in the United States today. The 10 percent of the United States population that hunts and traps are being called inhumane by the 10 percent that are so-called animal rights activists. The truth is, the activists themselves are the ones who are actually inhumane.

There have been many instances of activists being inhumane to both other people and the animals that they are helping. Activists have illegally broken into research laboratories and "freed" animals into the wild. These domesticated animals either came back to the lab or were immediately preyed upon by other animals. A hunter or wildlife manager would have known better. Also, they have discharged explosive devices in meat packing plants and in retail stores that deal in furs! Certainly this is an act of terrorism and an up-American act as well. In my own experience as a high school rodeo contestant, I have observed activists picketing our rodeos and harassing contestants. I know that the animals we are associated with are treated very humanely. The activists even treat other people inhumanely!

That is the difference between a normal person and an animal rights radical. They all seem to believe that animals deserve more rights than their human counterparts. Everyone is entitled to their own opinion, but I believe this is going too far. Even the Bible (Genesis 1:28) says may should make use of animals, without abusing them. I can only hope that 80 percent of the American public who are not active in the argument as of yet will not be swayed by the activists.

Breaking into labs, bombing trucks and stores and harassing kids are all illegal and immoral. The activists believe they are right, but I believe that they are the ones who are inhumane.

1998

Should Wildlife Management be Controlled by the Ballot Box?

Ryan Swanson

Anytime there is a vote over wildlife management, emotion over solid biological data will dictate the outcome of wildlife management issues. I've witnessed this firsthand here in Colorado. Bear hunting and leg-hold traps are two issues that have been directly affected by the ballot box.

I was born and raised into several generations of hunters. I started hunting with my parents and grandfather as soon as I was able to walk and "keep a little bit quiet," according to my Dad. Over the years I've learned the way nature takes care of things and how humans have changed the habitat for our use. Predator and consumer interrelationships have changed.

As the population of humans continues to grow and habitat diminishes, wildlife not just game animals need to be managed based on carefully collected data and informed decision making.

I remember when I was 5 years old, hunting bears with my Dad, baiting of bears in the spring was allowed and together we harvested several bears. It was exciting. However, due to a ballot initiative hunting bears with bait and dogs are no longer allowed. Ironically the hunters are now paying for damage done by bears through big-game license fee revenue. Likewise, I'm sure Colorado voters felt they were making the ethical choice in eliminating leg-hold traps. I wonder if they considered how their decision could potentially increase the number of predators and its effect on the ground-nesting birds.

Editor's Note: Ryan Swanson is currently attending Fort Lupton High School in Fort Lupton, Colorado. Unfortunately, Ryan could not take the donated hunt due to school enrollment conflicts, but his essay is recognized in Wild Sheep and in this book. He participates in football, wrestling and swimming. He is also involved in the National Honor Society and Future Business Leaders of America (FBLA). He is applying to various colleges to pursue a career in engineering and possibly playing football or wrestling at the collegiate level. Last year, he had a 30-1 record and was state runner up at 171 pounds. His family spends the Thanksgiving holiday hunting whitetail deer near Devils Tower in Wyoming. Although temperatures are bitter cold, they've become fond of a special canyon and lookout known as "The Rock." Last year he introduced his 11-year-old brother to this special place.

1998

Should Wildlife Management be Controlled by the Ballot Box?

GARRETT B. MCKITRICK

I think that controlling wildlife management through the ballot box wold be an enormous mistake. Putting the power to control the wildlife in the hands of people who do not have the background on the subject would be like the Heisman Trophy being voted on by every chess club in America. How are the citizens of major metropolitan areas such as New York and Los Angeles supposed to know what is needed for management on the prairies of the Midwest? The only evidence they ever see are reports on television and displays shown by extreme environmental activists. These are the people with the money and resources to get information to the general public, but they are also depicting an inaccurate analysis of the true situation. One-sided propaganda caused the voting public to lose sight of the complete pictures.

The only true way to judge the issue is to be properly informed on the situation. People who make it their life to know the proper wildlife management practices know that there are two sides to every issue. They are the people with knowledge to make rational decisions concerning proper management. Our U.S. Congress does not have the knowledge to decide on every bill brought before them. Instead, they send bills to specific committees that review the details and choose the appropriate course of action. In the same way, we should let the management issues be decided by wildlife experts. Let's leave the power to those who know.

Editor's Note: Garrett B. McKitrick of Bowman, North Dakota, was the first runner up in the 1998 Youth Essay Contest. Ryan Swanson could not take the hunting trip, so Garrett took the trip in September 1998. Garrett graduated from high school in May 1998. He lives near Bowman, North Dakota, that is located in the southwest corner of the state. The town is only about 1,500 people strong and his graduating class totaled 34 students. Garrett lives on a farm and ranch a few miles out of town where they grow spring wheat and raise cattle. In his spare time he enjoys riding motorcycles and snowmobiles (if it ever snows), hunting and shooting. He is one of five children with one older sister and three younger brothers. In fall 1998 he starts college in Fargo, North Dakota, where he intents to major in electronic engineering.

Faces from the Saturday afternoon auction during the 1998 convention in Reno.

Chapter VIII

Fund Raising for the Future of Wild Sheep

ROBERT E. DIGRAZIA, DDS

The Foundation for North American Wild Sheep's fund-raising program is the essential aspect that gives the Foundation its momentum. In looking through the history of the Foundation the sole issue that allows us to exist as an entity is the revenue we generate and how we generate it.

Members and supporters who donate to FNAWS and those who participate by buying donations number into the thousands. I would venture to state that each member to some level has participated in FNAWS fundraising. Countless outfitters and artists have donated to our auction programs. Thousands of members have bought hunts and artwork from our auction programs. In the auction program alone, FNAWS initially held auctions that grossed less than $50,000. At the 1998 convention alone, $3.5 million was raised. One bighorn sheep permit from Alberta brought $405,000 and the permit from Montana brought $300,000. Today FNAWS auctions generate revenue into the millions of dollars. FNAWS is famous for its ability to raise money to "put more sheep on the mountain."

In the annual fund-raising venue, FNAWS initially had a counter top that sold T-shirts. Now we have quality merchandise that members wear on a daily basis. The lottery section initially sold a few guns, but now markets sheep tags, cars, artwork and guns. The revenue generated has grown from a few thousand dollars to budget levels now well in excess of $500,000.

Wyoming was the only state permit sold at auction beginning in 1980. Today, various states, Canadian provinces and Mexico have some form of auction permit as is shown in the appendix to this book. The states offering auction permits throughout the history of FNAWS include Alaska, Arizona, California, Colorado, Idaho, Maine, Montana, New Mexico, Nevada, North Dakota, Oregon, Texas, Utah, Washington, and Wyoming. Alberta has offered permits since 1995, and the Yukon offered a permit in 1988. Mexico has donated auction permits, on and off, since 1984. These dedicated funds directly go to wild sheep management and habitat improvement. Millions of dollars have been generated throughout North America. The direct result of this revenue that was non-existent in 1980 has been the execution of projects that management agencies have had on the books for years but never had the money to complete them. For example, projects that were on the books for years but lacked funding to consummate are being done on a yearly basis now. The results are astronomical. The sheep populations have been growing. Herd sizes in some states and provinces have

Jim Reilly, FNAWS accountant, right, conducted the September 1, 1995, waterhole fundraiser drawing as FNAWS Executive Director Karen Werbelow watched.

All eyes turned to Bernie Fiedeldey of Cincinnati, Ohio, who was top bidder on the California desert bighorn permit with a total of $110,000 in 1995.

Lee R. Anderson of Minneapolis, Minnesota, gives a hoot as he becomes top bidder for the Arizona and Washington permits in 1995. He bid $233,000 for the Arizona permit and $67,500 for the Washington permit.

The Montana bighorn sheep permit went for $281,000 thanks to Jerry Fletcher of Phoenix, Arizona, in 1995.

Tom Vaccaro of Battle Creek, Michigan, captured the Colorado goat and Utah pronghorn permits in 1995.

Photographs courtesy of FNAWS archives until otherwise noted.

Mr. and Mrs. J. Curtis Newman were successful bidders on this wild sheep painting in 1995.

The Colorado bighorn sheep permit was purchased for $67,000 by Pocatello, Idaho, resident Carl G. Hoehner in 1995.

doubled or even tripled since the mid-70s.

Every member of FNAWS realizes that if this organization is going to succeed they must participate in fund raising to their level of comfort. Whether buying a state permit or a lottery ticket, we all participate to provide revenue to FNAWS. This revenue is the means by which we "put sheep on the mountain." It is the money raised that allows FNAWS to present its viewpoints in a public forum across North America and share our concerns about the environment as it directly relates to states and provinces that perpetuate wild sheep herds. FNAWS provides revenue for major research institutions to seek answers about diseases. It provides revenue for leaders to speak publicly on the right to hunt. In essence we have portrayed to the non-hunting public that there are more sheep on this continent, not in spite of hunting, but because of hunting. All of this was accomplished through FNAWS fund-raising activities.

Every donor participates in FNAWS fund-raising programs because of the same reason. Donors could sell their wares to others for their own personal gain, but choose not to do so. They do choose FNAWS because of our mission. Many times hunts sell below value, but donors donate repeatedly. On the contrary, FNAWS hopes donations sell above value. Purchasers participate for exactly the same reason. They know FNAWS is actively involved in protecting their privilege to hunt wild sheep on pristine North American habitat.

It is precisely the marriage between donors and purchasers that makes the fund-raising activities of the Foundation work. Whether it is a donor who donates an item that turns a bargain, or a purchaser who pays more than value, or any formula of revenue in between, the revenue generated by the Foundation is wisely used in its Grant-in-Aid program, political activity or simply the business of "putting sheep on the mountain" and protecting the right to hunt.

In our political and business atmosphere it now takes that third party to portray the concerns of the hunting industry. FNAWS is that third party that has taken hard and, at times, controversial stances on issues.

In examining this marriage further, the core issue on the workability of fundraising is

Mr. and Mrs. John R. Wilson of Cordova, Alaska, paid $14,000 for the Montana moose permit in 1995.

Michael Boyce, center with hat, grins for the camera moments after purchasing the Colorado mountain sheep license in 1992 for Dan Smith, Jr. when he bid $61,000.

Len Guldman of Aurora, Colorado, paid $8,250 for the 1995 Nevada special auction pronghorn tag.

Auction spotter Wes Hansen of California shakes the hand of top bidder Tom Pawlacyk of Neenah, Wisconsin, right, who spent $123,000 for the 1995 New Mexico desert bighorn sheep permit. Ronald Miller of Depere, Wisconsin, seated to the left, adds to the excitement.

Kathryn Irwin of New York was successful bidder on the North Dakota bighorn sheep tag with a tally of $31,000.

Steven C. Walker of Tiburon, California, was the winning bidder for the 1995 Oregon bighorn sheep permit in 1995 with a total of $73,000.

FNAWS volunteers. Donors do not have to donate. Buyers do not have to purchase. They do so because of their passion to hunt the wild sheep on this continent. FNAWS benefits, and the benefits are illustrative in our success. We have generated millions of dollars. There are more sheep. There is better habitat, and there is more opportunity to hunt. This has been accomplished through the generosity of donors and buyers.

History is now interpreting the success of the Foundation in remarkable means because of the volunteers and their generosity. There has been no other hunting organization in history that has done so much with so little resources. This is an epic story at which generations to come will marvel. Fundraising is the marriage of volunteers, donors and purchasers. It is a means to generate revenue that furthers our success.

There are donors who donate every year. There are donors who donate only occasionally. There are purchasers who purchase items yearly. There are purchasers who purchase only occasionally. There are members who buy thousands of dollars in lottery tickets, and there is the person who only can purchase one. The core essence of the Foundation is that each is as solely responsible for the success of this organization as the other. We are a body of people who come together to generate revenue through fundraising to protect and perpetuate the wild sheep on this continent. We argue for our right and privilege to hunt wild sheep. We do so that future generations can also do so. It all happens because of fundraising.

The privilege to hunt wild sheep has value and is worth perpetuating. Whatever the financial means of our membership, we place value and worth on generating funds to hunt these magnificent animals. The value comes from experiencing pristine country. The value comes from one's ability to deal with nature — steep mountains, cold winds and extreme fatigue. It is precisely the value that makes our level of fundraising remarkably easy.

In 1992 a U.S. senator from Utah spoke to me in the Halls of Congress: "The members of your organization are willing to walk steep mountains, in pouring rain, for the opportunity to take a mountain sheep?"

The answer I gave was, "Yes, indeed."

Further, no matter what the economic status of that individual may be, it is the only way

Novato, California, bidder Richard Garzoli took home the Utah desert bighorn sheep permit in 1995 with a tally of $41,000.

Part of the trick to raising funds for FNAWS is getting the word out to reporters, editors and broadcasters who tell the public about FNAWS projects and donations. In 1994, FNAWS Secretary Phyllis Carlson, left, and FNAWS Legal Counsel Dean Picl, right, spoke at the Outdoor Writers Association of America national convention in Maine and distributed information about FNAWS projects. FNAWS representatives attend OWAA national conventions each year.

During the 1995 Phoenix convention, FNAWS was presented a $15,000 contribution from Federal Cartridge for their support of a special 16-page supplement to Falcon magazine, which was published by Falcon Publishing of Helena. The children's magazine is now published by Rocky Mountain Elk Foundation and is called WOW!. Left to right are: FNAWS Publications/Marketing Coordinator Sandra Nicola, Federal Cartridge Manager of Conservation Activities Bill Stevens, FC President Ron Mason and FNAWS Treasurer Phyllis Carlson.

Stanley M. Boots, center, long-time Eastern Chapter supporter and national supporter of all FNAWS projects, purchased the sheep camp at the 1993 auction after Arthur Dubs drew the sheep camp ticket and donated it back to FNAWS for auction. The drawing and subsequent auction generated $107,271 for FNAWS.

Arthur Dubs in 1993 won the sheep camp but donated it back to FNAWS to auction. Stanley M. Boots purchased the sheep camp and went on a Dall's sheep, caribou and goat hunt with Stan Simpson of Ram Head Outfitters, Ltd.

Sherwin N. Scott of Phoenix, Arizona, is congratulated by Tom Pawlacyk of Meenah, Wisconsin, for taking the top bid for the Montana bighorn sheep license at $205,000 in 1993.

it can to be done. It is the physical nature of what we do in such pristine environment that bonds us all together. It bonds us to fundraising, furthering this pursuit for future generations. This congressman did scratch his head when he walked away.

There are many aspects to fundraising. It is not as simple as one thinks. Donations many times have to be solicited. Many board members over the years have spent countless days away from families and businesses. Board members have traveled 6,000 miles to obtain a donation for our annual convention. These board members have endured layovers, missed flights and even canceled flights. On average every board member has spent 30 days or more away from home fundraising for the Foundation.

On the other hand, the pressure is immense to donors of hunts. They feel they are under a microscope to perform and they are. They spend countless hours scouting for their clientele, but the pressure is up for a donated hunt.

Donors of artwork feel the same pressure. Their artwork is placed before the entire membership for critique. Purchasers sacrifice all privacy when purchasing donated items. Every one wants to know how they did or wants to come over and see their purchased artwork.

Working members at the lottery ticket boards over the years have seen people who have purchased many tickets win items, and have seen individuals who have purchased only one ticket win. They have seen some of the most superstitious activities. There are members who continually purchase the same tickets year after year. There are members who buy only incremental amounts of 13. There are members who will not fill out their tickets because they say it's bad luck.

The Board of Directors has seen outfitters donate to us even when the outfitter could not afford to do so. This is reflective of the commitment to excellence and FNAWS. Overall,

Wyoming was the state that started the famous tradition of donating wild sheep hunting permits to the annual FNAWS auction to raise millions for sheep and their habitat. Here in 1990 Wyoming Governor Mike Sullivan and his wife Jane pose during the state's centennial celebrations. Wyoming gives 100 percent of all money raised for the permits to FNAWS. The first Wyoming permit was donated in 1980 and those sales have generated more than $800,000 from 1980 to 1998.

Dan Pocapalia, winner of the 1989 Idaho bighorn sheep tag auctioned by the Idaho Chapter of FNAWS, receives his tag from Lynda Beach, IDFG license section supervisor. He paid $36,000 for the tag.

Shirts, hats, toys and other popular merchandise provides millions of dollars of revenue for FNAWS. Here FNAWS Past President John Harris and FNAWS Board Member Dan Artery model popular western-style shirts by Wrangler in 1990 sporting the FNAWS logo of the Chadwick ram.

Sale of gorgeous artwork and bronzes is a trademark of FNAWS fundraising efforts. Here, Mike Flanagan of Packstring Gallery and Foundry in Dayton, Wyoming, creates one of four North American wild sheep species in bronze that was sold in limited edition form to raise money for FNAWS. The Grand Slam bronze statues he created of bighorn, Stone's, Dall's and desert grace the sculpture garden at the FNAWS national headquarters in Cody. Ten life-size Grand Slam Signature Collection statues as well as two smaller versions of the same Grand Slam bronzes sold, beginning in 1989.

Fundraising means money to wild sheep and in 1988 this was no exception when FNAWS member Johnny Drift, right, presents President Dave Wiens of the Northern British Columbia Guides Association with a check for their projects.

Hossein Golabchi, left, purchased the New Mexico hunting permit in 1993 for $55,000.

FNAWS' fund-raising efforts have generated millions of dollars. You can see in other chapters of this publication how this revenue translates into field projects, political activities and research. But it is the camaraderie at conventions that brings this altogether. Have you ever seen so many people having fun for any purpose as to put sheep on the mountain and to go hunt these magnificent animals?

In the history with the Foundation, two episodes come to mind that reflect the passion of this organization. At the San Diego Convention in 1992, FNAWS was protested by the animal rights faction in that city. It did nothing but help us. When they showed up, the vitality of the auction jumped astronomically. There was an original piece of artwork auctioning at the time that sold for five times its value. A few years ago we used to auction a useless wolf hat. That hat brought more money than ever imagined. Members of states, chapters and friends would pool money to buy this hat.

We all feel proud about our concerns for the wild sheep of this continent. It certainly is reflective in our fund-raising ability. It is to be hoped it will continue.

Fundraising has taken some odd turns over the years. I can remember when I first became involved in FNAWS. Jerry Waite, Tom Hinders and other leaders would come to Idaho to solicit the Idaho Outfitters and Guides. They would be holed up in one end of a hotel, while other hunting organizations would be established elsewhere. These were all night affairs. Basically it was who could outlast the other. To the victor went the donated hunt. Times have changed and methods of soliciting donations have developed to a professional level, but this is how we started.

As I became involved in donation solicitation, I traveled many times to Fort St. John (always at Thanksgiving) for the Northern Guides and Outfitters Association. I was always impressed with the generosity of this group. Sandra Vince, knowing that we Americans were missing Thanksgiving, always invited us for a turkey dinner. I was impressed with this group's camaraderie. They fought their battles together, but when it came to celebrating, FNAWS and Safari Club International would sponsor a hospitality room. They would drink and sing

Earl Lunceford, Jr. of Coeur d'Alene, Idaho, was top bidder in 1993 for the Idaho bighorn sheep permit that sold for $31,000.

The late Frank Pachmayr of Los Angeles donated several special double-barreled shotguns to FNAWS as a unique Saturday night auction item.

FNAWS auctions wouldn't be the same without Ron Pomeroy, left, and Jim Travis, shown here in 1995.

Gene Majors of Asotin, Washington, gives a grin at the Saturday afternoon auction as a spotter, Wes Hansen, looks over his shoulder.

A big kiss on the cheek for donor Joe Klutsch at the Saturday night banquet in 1993.

Dr. Robert "Bob" DiGrazia stuck his neck out in 1993 by working to hire Patty Loveless who attracted record crowds to the FNAWS convention and set the tone for the entire auction because of the enthusiasm she generated. In the section, "Presidents Speak!," Dr. Bob tells the story of how he convinced the rest of the FNAWS board to hire Patty.

songs all night. Being a basic teetotaler, I would usually head to bed at about 2 a.m. In the morning I would go to clean up, and they were still partying. But in all, this group has always been there for FNAWS. We appreciated it very much, as we do with all donations.

The first auction I attended was in 1983. I remember Dan Pocapalia and "Doc" Jones Thurston got into a bidding war over the Wyoming permit. This also happened to be the first convention Dan attended. After a few beers (wine for Dan), they met in the men's room. Dan wanted to know what type of doctor could outbid him. He was surprised to discover Doc was not even a medical doctor, but a trucking and industrial magnate who founded Thurston Motor Lines and Thurston Aviation. Dan purchased the permit for $50,000 and Doc purchased it for $49,500. Two Wyoming permits were for sale that year. Many current members, including Ron McMillen, point to the 1983 Wyoming permit auction as a motivating factor for becoming more active in FNAWS. We were sad to loose such a supporter in Doc Thurston on November 24, 1993, but Dan continues that involvement. Dan and his wife, Mary Ann, annually attend our convention. Dan many times has bought state sheep licenses and donated back to FNAWS for our own coffers. In 1998 Dan purchased the Wyoming permit and gave it back to FNAWS, doubling our revenue. All he wanted out of it was a Wyoming cow elk permit.

Crazy things have happened at our auctions. In 1992, we were protested by People for the Ethical Treatment of Animals (PETA) in San Diego. It was at that time when bidding went out of control over a piece of art because people were mad at PETA. Virginia Frehner, a board member in 1998, purchased a piece of donated art from Joan Fleish for four times its value. Virginia is and has been a great supporter of FNAWS. When I was president in 1993, she honored me by purchasing a rattlesnake hunt in Texas. I hate snakes, so I gave it back to FNAWS. I believe that outing was sold three times, tripling the revenue for FNAWS. To this date, I do not know if anybody ever went on that snake hunt.

I have chaired the raffle fundraising at our annual convention from 1995 to 1998. I have been amazed over the consistency of various members to purchase the same tickets year after year. Butch Townsend from Wyoming always buys ticket Number 9. Larry Hunts always purchases the center and lower left corner tickets. He has won numerous times. Dave Colclough of California in 1998 purchased thousands of dollars of tickets. His sole reason was to support FNAWS. He brought his sons to convention with him. He wants them to enjoy hunting as he does. Dave won a calendar.

It has always amazed me over the years how members become involved in fundraising.

In 1990, the Idaho auction permit did not sell. The Idaho legislature, which is agriculturally dominated, placed a $30,000 minimum bid. They knew this would offend purchasers of the hunting community. It did. But when the auction was half over, Jerry Fletcher of Arizona came to us and stated that if we would reopen the Idaho permit, he would give the minimum bid. He got this permit for $30,000 and located a nice ram with Stan Potts. The rest is history. This started Jerry on his quest. Since this incident, Jerry has spent millions of dollars at FNAWS auction.

Fundraising is the essential and core reason why we have been able to place sheep on the mountain. By being members, purchasing art or outings and participating in raffles, we have done a great job.

We all should be proud of what we have accomplished.

Robert "Bob" E. DiGrazia, DDS, of Boise, Idaho, is a practicing dentist. Dr. Bob was born in Elko, Nevada, and attended grammar and high school in Wells, Nevada. He received a B.S. degree in mammalian biology from the University of San Francisco and a Doctorate in dental surgery (DDS) from the University of the Pacific School. He completed a general practice hospital residency with the U.S. Army at Reynolds Army Hospital in Fort Sill. Dr. Bob served in the U.S. Army as a general dental officer at the Dugway Proving Grounds, Utah, in 1975 but was promoted to Chief of Dental Staff in 1976. He was honorably discharged from the U.S. Army in 1977 and started his general dental practice the same year. For 1999 he will serve as Chief of Dental Staff for St. Lukes and St. Alphonsus Hospitals of Boise, a position he was also elected to in 1985. He first became active in FNAWS in 1980 and often refers to himself as "the second generation" of FNAWS members. He helped organize the Idaho Chapter of FNAWS in 1985 along with many others who also became active at the national level. His work with the Idaho legislature is legendary. In the chapter, "Presidents Speak," he tells the story of how the Idaho permit came to be. He has also been active in the Hells Canyon Initiative and most every aspect of FNAWS at the national and state levels. Dr. DiGrazia, known to friends as Dr. Bob, was president of FNAWS during the 1992-93 year. He was elected to the FNAWS Board of Directors in 1987. He is married to Mary Kay DiGrazia and has three children, Julia, Madeline and Joe. He has run nearly 25,000 miles and has been a runner in the Boston Marathon three times. His special interests include watching his family enjoy life and hunting.

Interview with Rich Roan Auctioneer to FNAWS, 1979-1998

Bob, Chuck & Rich Roan, Inc. Auctioneers & Appraisers Cogan Station, Pennsylvania

Susan C. Reneau

Just after the 1998 FNAWS national convention, I called Rich Roan to talk with him about his contributions to FNAWS since he began in 1979. My questions and his answers form the basis for this chapter. Some of his comments are included in the first four chapters of this book, but the editors of this book thought it was important for the readers to see the entire interview in one location. Rich can be reached by calling 1-800-955-ROAN.

What are some monumental auctions that immediately come to mind?

I remember the first auction we did $42,000 gross in Minneapolis. We went from there to Houston, and we maybe grossed $108,000. From there we went to Phoenix or New Orleans, I forget now. Then we went to San Diego, Hawaii, Reno and all that.

Were there particular conventions that stand out as especially successful financially?

I think they all were because number one you go back 20 years and $42,000 gross out of Minneapolis was a lot of money! They wanted me to go on to Houston and so on. Mike Valencia from California (Charter Member 7) and Paul Asper (Charter Member 95) were the two who got me involved with FNAWS. Paul was from my part of Pennsylvania and was on the national FNAWS Board at that time. I told him I didn't want to get too involved because I had my own auction company and five children in the business. Of course, we knew about Ducks Unlimited back home. We already did Ducks Unlimited and other charity jobs as auctioneers as well as our own company business. We do household goods, antiques, estate sales and our own auction gallery. We do small and large auctions of all kinds. Here I am still involved with FNAWS all these years later.

Who was the person who actually got you involved with the auctions of FNAWS?

Paul Asper asked me to come to Minneapolis for that first auction. That was the first one to talk about. He was supposed to be there at the same time with me, and of course he went to Africa to go hunting. I ended up going to the convention with his wife, Carol, and her girlfriend, Barb Whorl. Barb was a good friend of Carol's from up in the mountains. Paul is a life member of FNAWS. Paul and I are friends. We still are friends. He asked me to go do this job, and I said well I've got my own business to take care of but I'll do this as a favor to him. I didn't even know what FNAWS was at the time, you know. At the time very few

The line-up of Roans. Left to right we have son Randy, Rich Roan and son Ron. Rich and his family were the auctioneers for FNAWS from 1979 to 1998.
Photograph courtesy of FNAWS archives.

Harvey Kadlec, far right, reacts with a smile to comments made by Mahlon "Butch" White, center, as Rich Roan listens at the New Orleans convention in 1982.
Photograph by Dennis Campbell.

Moments after Doc Thurston successfully bid on the Wyoming permit for $42,000 in 1982 New Orleans, Doc and his friends and family lined up for this rare photograph. Left to right are Lloyd Bare, Bill Leech, Mr. and Mrs. Thurston and Jerry Christian, who was president of FNAWS at that time.
Photograph by Dennis Campbell.

An early supporter of FNAWS auctions, Helmut Gruel, center, holds a rifle he purchased during the 1982 auction. Jim Carmichael, left, and Rich Roan are there to congratulate him.
Photograph by Dennis Campbell.

people know about FNAWS. Even now you have to explain to people what FNAWS is. You say this is the Foundation for North American Wild Sheep, and they question what I mean by wild sheep. Of course, they think of tame sheep. We have to explain to them about a bighorn and the conservation end of it, which is really good, and the hunting end of it. A lot of people do not know there are more wild sheep that die from natural causes than from being shot by hunters.

That was in 1979 we were in Minneapolis. Well, what happened was we left Pittsburgh on the plane. We left Pittsburgh and, of course, the plane had trouble, so we ended up in Columbus, Ohio, where we stopped to change planes. We had to wait on a plane coming out of Florida. We went through all this. It took us nine hours to get there because of waiting on a plane and because of bad weather. We got into Minneapolis and there was rain and ice and snow. We got there and got into a cab and arrived at the Thunderbird Motel in Minneapolis. We drove into the parking lot and had a little accident with another car because of the snow and the ice. We get into the motel and there are no rooms for us. Our reservations were all messed up. We hadn't had any food to eat for 10 or 12 hours. It was really quite a mess, but we ended up quite successful with $42,000. At that time, that was a lot of money for as few people as there were there. It seemed like there were less than 200 at the auction. The Thunderbird Inn or Motel was a pretty good-sized place; a nice place.

Then they asked me to go onto Houston, but I told them I didn't want to get too involved. They really wanted me to go. Mike Valencia, a really nice guy, was from California and a friend of Paul Asper. Mike was all excited because he kept saying we'd take in a million dollars if I'd stick with it. I thought he was talking through his ear, you know. But, he wasn't. Mike and Paul were the ones who convinced me to go to Houston.

Would we say that you reluctantly said yes for the next year in Houston?

For some reason I just didn't want to get involved. I did so many charity auctions then and I do so many auctions now. I do a lot of charity auctions now for Ducks Unlimited, cancer, etc. I'm always happy to do them, but at that time I was doing these auctions alone. Then I went on to Houston, and we went downtown to the Galleria, I think. That's where we made $108,000 gross, I believe.

Going back to Minneapolis, were there any particular items for auction that stand out in your mind?

Well, not there. I really don't remember. Somewhere in Houston or Phoenix we ran into a guy in Colorado named Tom Brimhall who was one of the big spenders. I noticed him before the auction even began. He was wearing a tall Stetson. He was a pretty big-sized gentleman, tall, big, blue, snazzy sport coat on with a big buckle. I noticed him. He caught my eye.

When you're up there selling, you kind of look around to see how you can improve the take. I just motioned back there to the gentleman in the back with the light-colored Stetson, and asked him if he wanted to bid on whatever I was selling. I don't remember what I was selling, but if you want to bid, just raise that big hat. It seemed like slow motion and everyone just turned around to look at him and that's when he raised his big hat. That year in Phoenix he paid $100,000 plus (FNAWS records show $153,000) for what he bought. He was a good bidder. He was good and active for a few years, but he's not in there anymore. I don't know

Dean and June Picl, left, congratulate Doc Thurston for a successful bid at an early FNAWS convention.
Photograph courtesy of Roy Schultz.

Jerry Christian, left, with two of the great early FNAWS auction contributors; Doc Thurston, center, and Mahlon "Butch" White.
Photograph by Dennis Campbell.

Rich Roan, left, and FNAWS President Duane Smelser pause for a moment at the 1983 FNAWS banquet held in Reno with the theme, "What Does the Future Hold for Wild Sheep?"
Photograph by Dennis Campbell.

Dr. Roy Schultz, left, spent many years with Rich Roan, right, helping with the FNAWS national auction. Roy brought up items for Rich to auction and made sure everything was kept straight.
Photograph courtesy of Dr. Roy Schultz.

what happened to him. Nice guy. Good bidder. I picked him out of the crowd. Being in this business, I looked back and I thought he'd be a guy who would spend some money. That's the job of a auctioneer. This guy manufactured homes in Colorado.

Somewhere along the line, Doc Thurston of North Carolina came along and Dan Pocapalia came along. Of course, Doc Thurston has since passed away, but Dan still comes. Doc Thurston was a great supporter at the time. Back in 1984 Dan Pocapalia paid a tremendous amount of money for the Arizona hunting permit — $64,000. Dan also paid lots for the Wyoming permit — $50,000 — in 1983. This year (1998), Dan bought the Wyoming permit for $35,000 and put it back up for auction. He came up on the stage and said that the first $35,000 went directly to FNAWS, and then when we resold it the money went to the Wyoming Game and Fish Department. I think we got $30,000.

I can't remember when all these people I'm talking about attended those first auctions. What I'm saying is that the bidders are what make the auctions such a success. The best auctioneer in the world is no good without an underbidder, you know. The only difference between the underbidder and the bidder is that the underbidder doesn't pay the bill.

What are some of your techniques as an auctioneer to get people going? Please share some of your secrets.

I'm not sure I really have secrets. I guess I had a good teacher in my father who was in the auction business. I just picked up some of his techniques. I'm not sure what I do to get people to bid. I've been known to tell a bidder I'd throw in $10 if he would just keep bidding. I sometimes make it sound like I'm ready to close out the bidding. If somebody didn't jump in right away I'd do that. I created that but it's sort of a monster. It takes a little longer; no, a lot longer to sell an item, but I think I have them used to that. Sometimes I wish I wouldn't do it that way but it's something I do as a fundraiser for FNAWS.

The other thing that I think is interesting to note that the auction is a big job to get it prepared and to get it cataloged. Harvey Kadlac was one of the first and then his son followed through. That's important. The beginning Board of Directors was important because they got FNAWS where it is today. People like Doc Schultz, Doc Waite and Doc Duke got it where it is today. Now the auction is more involved because it's bigger, but that early group of people were the ones who got it all going to make it so successful. What makes the auction such a success are all the people involved in the background. I'm not even sure who does all that today and gets that all arranged, but those are the people who are important. It's a lot of work. I told them from the start that I wouldn't get involved in all that because that's all I'd be working on.

Going back to your recollections, what are some of your tricks that you use to get people interested in bidding who think they are done but they aren't?

Every bidder is different. You have personal dealings with them. You know who they are. I'm terrible about remembering names. The technique I use is to draw them out of it. I know they're not done. A couple of times I have sold short because a bidder held on too long, and then I sold it. At one of the Saturday banquet auctions a long time back I said going once, going twice, third and final call, and I sold it. The guy was very upset because he waited too long. But, I just said that's the way it is. I didn't intend to sell it short, but that was the way it was. I had to do something and I had to draw a line somewhere, so I did that.

The only thing that gets more complicated is that I get older. It is a very physical job. What I do when I get there is get up every morning and go to the health club to work out. I get a message. I rest. It's nice to have a cocktail now and then but I stay away from that.

We originally started at 12 noon and then it went to 1 p.m. On Friday we started at 1 p.m. This is a traditional auction where you're auctioning things for one or two hours. This goes on for five, six or seven hours each day. In the auction business like we do at our gallery back home we sell an average of an item a minute in the antique business. Here, you're doing 80 or 90 items.

We started in 1998 for the first time in the evening on Friday. It was good. This was the first time we did that. This is something they want to do again. They started at 5 o'clock on Friday and went on until 9:30 or 10 o'clock and we auctioned 90 items. That's a lot. This was the first time they had a Friday evening auction. It used to be 1 till 6. They eliminated the afternoon auction. They wanted to try this with the exhibitors. I wasn't even asked about the change. I'm not on the Board, and I don't want to be. I saw it advertised as such, and I thought maybe this was a good idea, AND IT WORKED!

I think they'd be foolish not to do the Friday night auction again. It worked just fine. The exhibitors were very pleased.

Reno really is one of the best place for the FNAWS convention as far as getting the money in to put sheep on the mountain.

Why do you think Reno is such a good place for FNAWS to have convention and auction?

Reno is under one roof. You can bring the family, and there's something for everybody to do. It has good facilities. The price is right up to this point. It is more central. I think overall it's a good place. You hear some people say they don't want to go there every year, but you have to look at why they go and that's to raise money for FNAWS. The idea is to raise funds to put the sheep on the mountain. If you're going to go somewhere and lose money, you're not going to want to go there. We went to Nashville, and it wasn't the best. It was nice but it was expensive. We went to Hawaii and almost went broke. It was very expensive. Philadelphia was a bust because we came out with $2.3 million on the auction, but it was a bust from the other end. It was something they tried and I guess they'll never do it again. The expenses were the facilities.

Please tell us some stories about items that really got people excited at auctions from the past?

The first thing we sent that really got everyone excited was the Wyoming permit. I think Dan Pocapalia was the first bidder who brought the BIG money, but I'm not sure. That was in 1983 when Dan paid $50,000 for the Wyoming permit. There were others before him, but not at that price. Doc Thurston was right in there, too, with his bids. Doc Thurston bought the Wyoming permit the next year for $52,000 in 1984.

What was going through your mind when the bidding got up that high?

I thought everybody had fallen out of a tree. You're up there selling and you think what's wrong with these people (that they'd pay so much for one hunting permit and maybe not even get an animal). You know what's nice about it after you think about it is that even today

Dan Pocapalia electrified the 1983 auction with his winning bid of $50,000 for the Wyoming permit. The moment of his success is captured on film. Roy Ostermiller gives Dan a hug as D.J. "Doc" Thurston, far right, reaches to shake Dan's hand.
Photograph courtesy of Mary A. Pocapalia.

Doc Thurston, second from right, and Dan Pocapalia, center, were two of the notables attending the 1983 FNAWS auction.
Photograph by Dennis Campbell.

(1998 convention in Reno) you sell a permit for $405,000 to go hunt a sheep in Alberta, and you wonder what is going on. You have to remember and be thankful for the people who have the money and who believe in the Foundation for North American Wild Sheep even though they want to have a record sheep. They have the right. They have the money to bid it. They know the money goes to put the sheep on the mountain. That money goes back to whomever donated the permit and to FNAWS. That's something that people should look at.

For instance, I heard that Paul Harvey said, "What is this? Somebody paid $405,000 to kill a sheep." The problem with that is that the news media doesn't come out and tell you what all that money goes for. And, Rocky Mountain Elk Foundation got the credit for the money. The news media isn't giving the right news. It's fine that the Rocky Mountain Elk Foundation got the sheep permit, but the Foundation for North American Wild Sheep, not Rich Roan, not you, not anybody else, raised the money. The credit goes to FNAWS for getting the money. The point was that it wasn't $405,000 to kill a sheep. It was to raise money to help wild sheep. They're probably anti-hunting. With the news media — I don't care if it is CBS, NBC, ABC or the newspapers — they will do what they see fit to say to fit whatever they want it to fit. That's why when I saw the NBC camera there (1998 convention auction) on the day of the auction, I told somebody that wasn't good. If we're going to be taking pictures of a guy who doesn't want anybody to know what he does when he's standing there, you don't know what they'll do with those pictures. That's why they asked the camera guy not to take pictures of the bidders.

Morgan Roan eats pepperoni pizza with a grin at the 1997 "Welcome Night."
Photograph courtesy of FNAWS archives.

Ron McMillen holds an enlarged photograph of Rich Roan at the 1991 convention in Reno with the theme, "Hunters Rights, Old Traditions, New Frontiers."
Photograph courtesy of FNAWS archives.

The famous "wolf hat" is auctioned in 1991 at the Saturday banquet February 23.
Photograph courtesy of FNAWS archives.

Randy Roan, right, was one of several sons who helped Rich over the years with the FNAWS convention. Rich is left and Jim Pirrung is center.
Photograph courtesy of FNAWS archives.

How in heavens do you pick up on the people bidding when they barely raise their hands?

That's why we have ring men out there. You'll notice how I'll go out into the crowd after I've sold an item and I congratulate the underbidder and the bidder. Sometimes the ring man will say to me that someone doesn't want anyone to know he was bidding. That's sometimes not always true, but I always try to show my appreciation. I do as the auctioneer for FNAWS. FNAWS is thankful to the bidders. I am always appreciative of the underbidder and the bidder who brought this money into FNAWS.

Sometimes I see a person rub their nose and scratch their head, and as a joke I ask, "Are you bidding?" Then they get all flustered. I do that at my own auctions, too. You just have your own things you pick up. They have auctioneer schools, of course, but I never went to one. I just did it on my own. I picked it up from my father, and my four boys picked it up from me. None of us went to school.

What are some common methods of communication by bidders at FNAWS?

I won't mention any names. A guy was from Texas and was standing at the bar in the casino talking to a couple of guys, maybe even my sons. I looked over at the guy and a lady were sitting there. The guy said, "Well, you owe me money." I looked at him and I thought I knew this guy from somewhere. Turns out he was one of the underbidders at the original Alberta permit auction. He was sort of pulling my leg because back when I first sold that in 1995 there was talk of this permit bringing $250,000, but the bidding died at $90,000 to $100,000. When it got to that price, I jumped off the stage.

I yelled to my son Randy that I was going down into the crowd. I'm yelling this over the microphone, "I'm not selling this permit for $110,000 or whatever it's selling for now!" I was sort of carrying on like I do. I went out through the crowd. This was the guy who was the bidder at the time. I worked with the crowd. I spent quite a lengthy time to sell it and we finally got $225,000.

In 1998 we got $405,000 for the Alberta permit, but that took a long time to sell, too. The first time when I got done, I didn't think I was going to make it. I was wringing wet from sweating. I'm a big guy. You'd think I'd lose 40 pounds, but I think I gain it when I do this! That's just the way it is. I know I was wringing wet, that's for sure.

I've got Randy, my oldest son, and Steve Leteer, who works part time for me, at the 1998 auction. I have also brought my son Ron with me and my other son Rick over the years. Rod, my other son, has also come at one time. I've had my whole family at the auction at one time or other.

How do people let you know they are bidding? Do they raise one finger?

Everybody has a different way. Some people wait until the very last minute to bid, which is tough on me. Physically, I'd sooner they'd do it a little faster. Like I said, when I'm selling at my everyday auction I go, "Hundred, hundred 'n a quarter, hundred 'n a half, one seventy-five, two hundred fifty, already done, SOLD." But here at FNAWS I put in, "Final call. Going once. Going twice. Final call. And, I sort of hesitate in between, and there's been comments that I shouldn't do that but I'm not trying to sell short. Number one, I'm selling a permit for maybe an outfitter who has donated an outstanding bighorn or Dall's sheep hunt. If I try to sell it short, that's not good for them. It's different than selling an antique or a piece of furniture. You do carry on a little more at the FNAWS auction. That's the reason for it. I've been coming for 19 years straight, since Minneapolis in 1979. I'm happy to do the job the way I'm doing it.

I think FNAWS recognizes what you mean to them.

I think they do. I'm not interested in bragging about Rich Roan, because, number one, let's go back to Dan Pedrotti. He's donated a lot to the Sheep Foundation and Doc Schultz, all of them. They've all donated their services. Jerry Waite. Harvey Kadlac. They've all donated their services. They've spent money. So, it's the same thing. The difference is I've never hunted sheep, and I'm never going to hunt sheep. I used to be a hunter in Pennsylvania, but I don't even hunt anymore. For fish, I go to the fish market to get my fish. I don't even fish anymore. Some people love that hunting, and I used to, and I think it's great if you want to hunt, but if you don't want to hunt, don't hunt.

Over the years, what has been the most popular auction item? Was it the Alberta tag? What was it?

It's been the Alberta tag now, but it used to be the Wyoming permit in the beginning. It's usually a 10 percent fee for selling at the FNAWS auction, so 90 percent goes back to where it was donated from. It's the best fundraiser, not only for wild sheep, but I think its the biggest fundraiser in the country, as far as I'm concerned. Nobody has beat the permit prices at FNAWS. I don't care if it's for elk or moose or whatever. They are foolish not to give the permits to the Sheep Foundation and the money goes back to them.

The Sheep Foundation should get a percentage of whatever is collected for a permit at the auction. If I find out that the Sheep Foundation doesn't get anything for a permit they auction, I get very upset if the Foundation doesn't get its fair share. Some of these things they sell they don't get anything for them. The Sheep Foundation should ALWAYS get a commission. They may not like me saying that, but I'm telling you right now, that's the only way I would take anything for the auction. I said something to somebody once and he said we'd lose the permit, but what are we going to lose? The members, the Board members, Rich Roan, all of us, we're all involved in raising money for FNAWS.

As for the Alberta permit that sold for $405,000 in 1998, was that the all-time high for any permit?

Yes. The first big one after Wyoming was the Montana permit at $310,000 in 1994 thanks to Jerry Fletcher as the bidder. Stanley Boots was the underbidder. That was the one where he ignited the banquet. He jumped onto one of those big round tables and was carrying on like a wild banshee. That was one where he didn't want anybody to know it. He had a ring man picked out. The ring man was taking care of everything. I think the ring man was B. Joe Coy. That's the other thing that needs to be stressed. Those ring men have worked all those years, and it's a lot of work standing out there all those hours.

The people who are your auction spotters (you call ring men), are they members of FNAWS? How do you train them?

B. Joe Coy, long-time FNAWS volunteer, takes a rest after a long day of auctions. He's been a ringman for more than 20 years. *Photograph courtesy of FNAWS archives.*

Spotter or Ring Man Jack Clark congratulates Don Mangels for becoming the winning bidder at the 1988 FNAWS auction in Reno. Thanking the bidders is an important tradition in FNAWS. *Photograph courtesy of FNAWS archives.*

They come into it on their own. Johnny Drift of British Columbia is someone I'm trying to get back into the Foundation. They've got them lined up and they know how to do it now. Jeff Reynolds does it. B. Joe Coy has been there for years. He's an outfitter from Cody. Carlos Sanchez is another one. I met him in Minneapolis, so he's been doing as long as me. There are others whose names I can't remember. The FNAWS staff in Cody probably have their names. I really rely on them. I can't do the auction without them. I don't know how many ring men I have on the floor, but Jeff Reynolds would know this. Those are the people I rely on to pick up those subtle hand movements and hat flicks by bidders.

For instance, a bidder might talk to a ring man before the auction to watch him when a certain item comes up for bid. With the Alberta permit in 1998 I think in the bottom line it came down to three bidders. They were competitive bidders, back and forth. I'm up there and I notice some things between a couple of them that seemed a little different. I found out later that two of the bidders knew each other and didn't want to bid against each other, so when they thought the other one was getting out of bid, they quit. But, then the third guy got in on it and he kept the bid going. So when he quit, the other guy jumped in. The bottomline was that they still bid against each other without realizing it at the end. The third person was the critical person in all this. Without him, we wouldn't have gotten $405,000. The third bidder ended up not getting the permit. It stalled out for a while at $100,000 because the other two guys knew each other.

How do you gauge when the bidding is over?

You have to gauge it. The other thing that goes along with this auction business is when people start telling rumors like what happened at the Saturday night auction (1998) for the Mexican permits. The big talk was that the permits would go higher than the Alberta permit. I've learned this a long time ago in FNAWS that if there is a lot of talk about something, then it's just a lot of talk. If a person is going to spend money, he's not going to walk around and tell everyone about it. I'm convinced of that, especially after the auction on Saturday night. They're talking about the Mexican permit bringing a half million dollars. When I heard this from different guys, I said naw. It's not true. It won't happen, I said.

I was surprised the second Mexican permit went for less than $200,000.

It's funny you'd say that because a lot of people said that was something but it only brought $200,000. We say that because we've got everyone spoiled. Only $200,000! That's a lot of money. The Mexicans walked away with almost a half million dollars. Nobody has to be told that's a lot of money. I'm surprised it even got to that level. Number one, that's why the Alberta tag brought more money because Jerry Fletcher was there and got a big sheep. Supposedly there is a better one there. That's what is going to happen in Mexico if things go the way they should. I don't understand the whole deal with the Mexican government.

How do you deal with auction items when things come up at the last minute? Is that difficult to do spontaneously?

They have all the cataloging done. I don't get involved with that. I won't get tied down to that. When you take the FNAWS fundraiser, that's very unique. You're in a territory where you know what things will bring because you know you have some big money people and you follow the people of FNAWS who believe in it. People say well it's easy for them to buy because those people have their money, but that's what we're out there for — to raise the funds! They could spend their money anywhere, but they decided to spend it on FNAWS.

The Saturday night thing in 1998 wasn't even planned. That skit just fell into place, and it wasn't even planned. Last year and this year we did, "Here, kitty, kitty, kitty." Over that you get $10,000 or $12,000 for a rifle

"Here, Kitty, Kitty." The impromptu skit with Rich Roan, left, and Dr. Red Duke at the 1996 brought down the house and was repeated again at the 1998 FNAWS convention. Red's tan hat in 1996 was auctioned.
Photograph courtesy of FNAWS archives.

that Dr. Duke donated. Over the fun and carrying on, you get that kind of money.

Red Duke's hat. That's another story. It's a real joy to be friends with him. There's nobody like him. I have a couple of letters that he wrote to me when he was president of FNAWS that hang in my office that are probably some of my most prized possessions. This was the only time Red Duke's hat was auctioned (1998 convention, Saturday night auction). We're going to auction it off again next year (1999). We brought $5,000 for it. Virginia Frehner bought that and then she gave it back. Now the hat sits in a bank vault somewhere for next year's auction. We did that years ago with a wolf hat out of Alaska. The hat got lost somewhere. We got as high as $10,000 for a baseball hat one year. The hat belonged to someone from Alaska, I think. I think it was bought by the Minnesota-Wisconsin Chapter. Phyllis Carlson might know.

The "kitty" thing came from one time I was doing a estate sale auction in Winston-Salem, North Carolina, at the Coliseum. We were sitting there visiting with some people.

The girlfriend of the head of the Coliseum said, "I can learn to be an auctioneer."

So I asked her, "How do you do that?"

She said, "You just say, here, kitty, you know, and you just go fast. And, each time you do it you go faster and just say hey, kitty. And say, kitty a quarter and kitty a quarter and a half, kitty and a quarter and a seventy-five and a kitty and a dollar."

I always brought that up to people kiddingly and would tell them that's how I learned to auction, and I can teach them. They'd say how and I'd say, "Hey, kitty, kitty." I really don't start with the words "kitty" but that's what I tell people. I pretend to teach people how to be an auctioneer.

Red Duke has been the m.c. two or three times for FNAWS. We kind of tag team together at those Saturday night functions. We're good friends. I met him at Paul Asper's house years ago when Red was on the national FNAWS Board of Directors. Paul lives just 20 miles away from where I live. This year we got $5,000 for his cowboy hat. I didn't come up with that idea to sell Red Duke's hat this year (1998). They asked me what I thought. I told them I thought it was a fine idea. I knew in advance that his hat would be auctioned. In fact I had some people from my town who asked to join. A lady came to me who had met Red Duke the night before. She said she'd bid on it. Red told her that he planned to buy it back, anyway. So I was up there selling the hat and she was bidding up to $1,500, and I told her to watch out. She didn't understand auctions and was having a good time. She could afford the hat, but she asked me why Red wasn't bidding, so I turned to Red and asked him why he wasn't bidding.

So Red says in his real slow Texan drawl, "I really thought it would go a lot cheaper than that." So he really didn't bid on it. The lady quit bidding and someone else jumped in. Red thought he was going to buy it back for 500 bucks! It surprised him that the hat brought a lot more. Red thought it was going to bring a lot less.

How about the art at the auction? How does that usually go?

Art is a funny thing even in the antique world. You can take a piece of art that looks good to me but it looks ugly to you. It might bring more to you than it would to me, or whatever. That's the same with the art in FNAWS. The art didn't bring a lot of money this year (1998). That will happen, but I have no idea why. Gary Swanson's paintings always do well but they went bust this year. I don't know why. I don't know what the problem was. Guy Coheleach's painting that was on the cover of the *Wild Sheep Journal* went quite low, and I don't know what the problem was. It's one of those things when you are auctioneering, you can almost

Rich Roan in 1984 auctions a Gary Swanson painting that is held by Jerry Christian.
Photograph courtesy of Dr. Roy Schultz.

tell. Even Vivi Crandall's painting didn't bring big money this time. You know, again the answer is to get a customer interested in the painting before the auction. I'm not saying they don't, but if you don't find a customer who is interested in the first place, it is tough to sell it. To have an auction, you have to at least have one person who is very interested. The bronze of the lamb went well. It went to $68,000 or $70,000. Rick Taylor is flying high right now, and he's good. Of course, if you say that to some people they get upset because he had his picture on the front of the auction catalog. Some people don't like that. Whoever is doing the promotions tries to do what is right, but then the next guy doesn't like it because he's not on there. That's human nature. This is what you run into. That's why I watch what I say.

What do I want to say? I guess what I really want to say is that I'm happy I can do what I have done with FNAWS. As long as I am able to do it, I hope I can do it. I have no contract, nor do I want one. If they want me, I'll be happy to be there. My children have been involved with it, and I hope my children will carry it on. If they don't I think they will lose, and the Foundation will lose, I think. That's entirely up to them. As far as they can keep the auction the way it's going, they may wish to change, but that's entirely up to the board.

You're right about art. We've had auctions in our local community in Missoula, Montana, for the local art museum, and prints have sold for more than original paintings, and I don't know why.

That's what happens with art. You have to remember something, and you probably do, but a lot of people don't think about it. There are zillions of prints out there. There's not zillions of originals. Again, it has to be somebody who is famous or a noted artist. Even though there are good artists who aren't famous, that's the way it is. Missoula is where Al Baron is located. He's the one who originally bought Vivi Crandall's paintings to start. He was there this year but he kept a low profile, and I don't know why. He's spent quite a bit of money over the years with the Foundation. You should make a note of that and check with him.

But, don't ignore the fact about who those people are because they like that recognition. Before you say you won't do it, I would ask them. They like people to know. For example, take the guy who bought the Montana permit. Now, don't print this. I'm just telling you this. I had always been told he didn't want recognition. When he bought the $300,000 permit this year, I went down to thank him for FNAWS. When I went to walk away, he said, "Wait a minute. You know I was the underbidder on about four or five different permits over a period of time." I told him I understood he had done this but that I wasn't supposed to let people know. He said that wasn't true, so that's why I went up on stage and told everyone that he was the underbidder up to a million dollars on four other permits. Everyone clapped, and that guy turned around on Saturday night and paid $190,000 for another permit. That's when I recognized him in front of the crowd. I had never done that before because I was told by one of the ring men or spotters not to say anything.

Getting a hug is Donald Snyder of Canton, Pennsylvania, who was top bidder at $36,000 for the 1995 Navajo Nations desert bighorn sheep permit.

Jim Frickert of Windermere, Florida, keeps track of auction items and successful bids in 1995.

Jerry Fletcher, left, sits with another top bidder, Stanley Boots in 1993. Jerry paid $303,000 for the Arizona desert bighorn sheep permit.

Stephen Walker of Tiburon, California, was top bidder in 1993 for the Mexican desert bighorn that brought $80,000 to "Put Sheep on the Mountain."

Lex Madden: New FNAWS Auctioneer

Interview with Susan C. Reneau on Tuesday, August 4, 1998

Lex Madden, 41, of Torrington, Wyoming, has been selected by the national FNAWS Board of Directors to replace Rich Roan as the official auctioneer of the Foundation for North American Wild Sheep beginning at the Reno convention in 1999. Lex and his brother Shawn, 44, are owners and operators of Madden Brothers Auctions of Torrington, a business they have owned since 1979.

Lex was named World Champion Livestock Auctioneer on June 13, 1998, by the Livestock Marketing Association during an international competition held in Sedalia, Missouri, and involving auctioneers from the United States and Canada. Prior to 1998, Lex had ranked second in the world in 1997 at the same competition and in the top 10 six out of seven times he has competed.

Shawn Madden has also been ranked one of the top in the world by the LMA. He has ranked in the top ten during competition eight out of 10 times that he has entered the event. He is not expected to attend the FNAWS convention in 1999.

Both brothers have years of experience as auctioneers. Lex began his career at age 17 so is approaching his quarter-century mark for auctioneer experience.

At the 1999 FNAWS auction Lex will bring Michael Schmitt as his company "ring man" and will rely on the volunteer ring men from FNAWS to spot bidders.

"I consider this a great honor for even being considered. I enjoy working with people and having fun with them during the auction. I hope to bring personality and charisma to the FNAWS auction," Lex said.

Lex and Shawn became famous to the world in 1981 when they sold all furnishings, art, building and property belonging to Dell Burke of Lusk, Wyoming, a well-known owner and operator of Wyoming's most famous brothel known as "Dell's Yellow Hotel." Lex said media from around the world descended upon Dell's Yellow Hotel for two days as the brothers sold everything belonging to Dell's estate.

"That was the auction that really put us on the map," Lex explained. He said that they hardly got paid anything for the effort and even paid for the advertising of the auction and the organization of the auction items prior to the sale. They were only paid 2-1/2 percent of the take which worked out to less than $7,000. Their expenses to organize the sale, advertise it and do the auction cost $6,300. Lex said they worked hard and the publicity was great.

Since 1981 their auctioneering business has really taken off and now they are often booked more than one year in advance to sell estates as well as livestock.

Lex and Shawn Madden also own Torrington Livestock Markets, Inc. (since 1988), Frontier Reality of Wyoming (since 1984) and Frontier Feeders (since 1991). Frontier Feeders is a commercial feed lot for cattle that feeds 6,000 to 7,000 animals at any one time. Most of their customers are from Wyoming but they also work with ranchers in Colorado and Montana.

For further information about Madden Brothers Auctions or the other businesses they own, call (307) 532-7079.

Doc Thurston

Doc Jones Thurston died November 24, 1993, at Presbyterian Hospital in North Palm Beach, Florida. Doc founded Thurston Motor Lines in 1932 and was its chief executive officer until its sale in 1979. He also was founder and chairman of Thurston Aviation.

Doc had been president of N.C. Motor Carriers, Inc. from 1941 to 1942 and served more than 50 years on its Board of Directors. He was Director Emeritus of it. He was also an officer and Board member of Motor Transportation Association of South Carolina, director of Virginia Highway Users Association and director of the American Trucking Association. He was chairman of the Common Carrier Conference from 1974 to 1975. Doc had been a member of the Board of Directors for Meredith College in Raleigh, North Carolina and a director of the University of North Carolina Charlotte Foundation. He was a Salvation Army Board member in Charlotte and Jacksonville, Florida. He chaired the fund-raising effort for the Thurston Arthritis Research Center and made a $3 million kickoff donation. He donated a laboratory at the Lineberger Cancer Research Center in Chapel Hill, North Carolina.

Survivors included his wife, Stella of North Palm Beach, Florida; sons, Doc Thurston III of Chapel Hill; William of Frankfort, Kentucky; daughter, Leigh Myers of Winston-Salem, North Carolina; and sisters, Ruth Stanford and Jessie Spencer of Raleigh.

In lieu of flowers, members of FNAWS were encouraged to donate to the various organizations Doc supported throughout his life or to a charity of their choice.

Dan Pocapalia wrote a fitting tribute to Doc in the spring 1994 issue of *Wild Sheep*. It read:

The first time I met Doc Thurston was also my introduction to the Foundation for North American Wild Sheep. I was talked into attending the convention by Roy Ostermiller of Frontier Taxidermists in Cheyenne, Wyoming. The governor of the state of Wyoming was donating one sheep permit to the Foundation, and it would be auctioned off to the highest bidder. Roy claimed that it would go for between $5,000 and $10,000. This was rather intriguing to me inasmuch as I had become an avid sheep hunting enthusiast and did not have the luxury of waiting for the state's drawing, which in all probability, would be an impossibility.

The bidding started out low enough and I got involved in the bidding. A character referred to as Doc was also bidding. I ended up with the Wyoming permit for $50,000. Doc dropped out at $49,000. At the conclusion of that purchase, I thought that pandemonium would take over the convention. It was extremely exciting. At this point, Roy Ostermiller introduced me to Doc.

He certainly portrayed the Southern gentleman and I was under the impression that he was a medical doctor because everyone was addressing him as Doc. We became good friends and were having dinner one evening when I asked him what segment of the medical profession he specialized in.

Doc started to laugh and said, "Hell, Dan, I'm just a lucky truck driver."

I later learned that he owned and operated a truck line. He really fooled me.

I would say that Doc was one of the Foundation's most generous donors and was liked by everyone.

Movers & Shakers are everywhere; donors, purchasers, bidders, volunteers, Board, staff and members.

Chapter IX

Movers and Shakers: Leadership in Action

RON MCMILLEN

It was back in 1983 when my involvement with the Foundation of North American Wild Sheep began. A good friend, Col. Don Mangels, retired, knowing I was interested in getting involved in big-game hunting, took me to a local watering hole in order to meet the current Foundation for North American Wild Sheep (FNAWS) President, Jerry Christian, along with the future president, Tom Hinders.

The two of them convinced me to attend the upcoming 1983 convention. Heading off to Reno that year, my only real intentions were getting involved in big-game hunting. Never would I have predicted FNAWS would evolve into the lifestyle and dedication that it has.

It all started when I decided to bid on a Stone's sheep hunt. I was told by many that Dave and Ed Wiens' Stone's sheep hunt was the best one available at the auction. I was ready to make that hunt mine.

However, I went up against Bill Leech of Texas. He was a true conservationist and had much deeper pockets than I ever dreamed of having. It was soon evident Bill was going to make this hunt his regardless. This put a light in my eye. I was truly impressed with Bill's commitment to such a worthy cause. It is this commitment that is demonstrated by so many individuals through their involvement in FNAWS that makes it such a successful organization for wild sheep in North America.

Perhaps, maybe Dan Pocapalia purchasing three state permits over the years and donating them back for re-auction would be an appropriate headline for this chapter: "Movers and Shakers."

Through the years I have been honored to be a part of FNAWS. I must credit the movers and shakers of this Foundation for they are the reason why I became involved to the extent I have. These people are more than just high-profile people. They include all the volunteers who donate their time and contributions to FNAWS. The volunteers who step forward to help at the convention, along with all the chapter and affiliate members, have helped to create the broad base that FNAWS is known for. More people than will be mentioned in this chapter help make the Foundation what it is. Please take reference to the appendix of the book that lists all donors and permit holders available through the National FNAWS headquarters. The listing was the most appropriate way to recognize all the contributors and volunteers. Unfortunately, all the donors to FNAWS can't be listed because computer records only include donors from 1993 to the present, but permit holders start with the first sale of a Wyo-

Dan Pocapalia, right, with his massive bighorn taken in 1990 using the New Mexico permit he purchased at the FNAWS auction for $70,000. The ram had a green score of 184 points.
Photograph courtesy of FNAWS archives.

ming hunting permit in 1980 to the present.

When thinking back to the start of FNAWS, Helmut and Gerta Gruel strike an important place with the Foundation. They were the first big contributors and buyers in the early years of FNAWS. Their style and charisma were truly fascinating but yet down to earth. The Gruels, much like the Thurstons (more to be mentioned on them later), were always dressed first class. Their gorgeous jewelry and classy tailor-made wardrobes were a bit intimidating for some of us. But, once again like the Thurstons, the Gruels made our early members want to come back. The new members learned it was so enjoyable to sit down and have breakfast or lunch with the Gruels because they were welcomed as true friends. In one form or another, I think their class and charisma has had a positive carryover in many of our lives.

Then, FNAWS reached a pinnacle with the involvement of two classic gentlemen in their own right: D.J. "Doc" Thurston and Dan Pocapalia. Doc Thurston, a true Southern gentlemen in every sense of the word, along with his always gracious wife, Stella, at his side, got his Grand Slam over the age of 70 with an artificial hip and all. He was very proud of that accomplishment as were the Foundation members. **If Yankee Stadium is the house that Ruth built, then I would list the Foundation auctions as the house that Thurston and Pocapalia built.** Back in the early 1980s, these two men went head to head to buy the governor's bighorn permit from Wyoming. That was 1983, and Doc Thurston with his charm and big smile, bought a Wyoming permit for $49,500. At this time, the life of wild sheep had just begun for Dan Pocapalia. Thanks to Roy Ostermiller, one of our great early members and owner of Frontier Taxidermy fame in Cheyenne, Wyoming, encouraged Dan to attend his first convention. Little did anyone know what an impact that introduction would have on wild sheep and the pro-hunting world in years to come.

If I can recall, Doc Thurston took the second Wyoming permit (there were two available at the convention) to $49,500. Dan was not to be denied when he reached the $50,000 barrier for the first time. This event created much entertainment for anyone who were present. The crowd was on chairs and tables hooting and hollering to watch these two white-haired gentle-

Two of the grandest supporters of FNAWS in its stunningly successful history are Earl and Marge Boose of British Columbia.

Dale and Greg Stires at the 1991 national FNAWS auction.

A "ringer" at the 1991 auction raises the hand of Pamela Atwood who successfully bid for the New Mexico permit at $46,000 to put sheep on the mountain.

A smiling C. Keith West of Rock Springs, Wyoming, at the 1996 FNAWS convention.

L to R: Lloyd Bare, Bill Leech, Mrs. & Mr. Doc Thurston and Jerry Christian. Doc had just purchased a Wyoming bighorn permit for $42,000 at the FNAWS convention in New Orleans, 1982. Bill Leech was the underbidder.
Photo by Dennis Campbell.

Nancy and Dennis Campbell enjoy a reception at the 1993 FNAWS convention. Dennis contributed dozens of photographs from early FNAWS functions and reviewed the first four chapters of this book. He is the executive director of the Grand Slam Club.

B. Joe Coy hugs Sue Kadlec at the 1991 FNAWS auction as another volunteer auction "ringer" or bid spotter, son Brad Kadlec, gives a hoot.

Artist Gary Swanson, center, was top bidder for this donated double-barreled, 12-gauge shotgun by the late Frank Pachmayr, left. The shotgun is held by one of the auction volunteers in 1993 and is one of ten shotguns donated by Frank to raise funds for FNAWS.

Auction action in 1988.

R.G. Harris and his family are successful bidders again at the 1989 FNAWS convention.

men battle to reach the first ever $50,000 plateau for a hunt. They created such electricity and made it so that all members bonded with such chemistry. All in attendance will never forget the first bidding wars that these two had with one another. It was truly one of our electrifying moments and helped set the stage for what followed.

One of Dan's favorite stories is never having asked Doc Thurston what kind of a doctor he was. Then, in a private moment in Dubois, Wyoming, Dan finally asked Doc the question. Doc responded by saying, "Heck Dan, I'm not a doctor. I own a trucking company." Dan said the two broke out laughing, and he is not sure if either one of them have ever quit. It was obvious they truly respected one another. These men eventually became dear friends, with each having utmost admiration for each other. Sometimes one of them would respectfully yield to the other in the middle of one of their bidding wars when they felt the price was high enough and it was "mission accomplished" for all concerned. They always put on such a great show.

Those members who haven't stopped to talk to Dan Pocapalia, will never know what they missed. His intellect, along with his charisma, character, and big heart, make him truly one of our most unique members. I hope he does not read this sentence, or Mary Ann will probably be all over me at the next convention giving me a hard time, but to coin the old phrase, "Behind every great man, there is a great woman."

These gentlemen commitments have encouraged many of us to not only volunteer our time but also to pay higher prices for all auction items. They made me realize the real bargain was to get top dollar for all items. This would help maintain the integrity of the donors and ensure everyone was working towards the cause we all love and believe in.

I remember one time trying to use these great men as mentors. I asked Doc Thurston to sum up the secret of success. With his great Southern drawl, he said, "Son, remember, you always need to plant the seed to reap the harvest." To this day, I still use that line many times and apply it to many things I do in life. It is a great little tip I definitely have taken advantage of.

The other thing I think the early men and women did (the Helmut Gruels, the Doc Thurstons, the Bill Leechs and the Dan Pocapalias) were to set the stage for the future movers and shakers to follow. I always think we will be indebted to these early members because of that. This is certainly not to slight people even before them (the Doc Schultzs, Daniel M. Galbreath, Butch White) who in the real early years were paying prices for items well beyond most people's comprehension. This included men like George Gamble who just quietly donated shares of stock. Some bidders were at the auctions with the attitude they were there to get a bargain based on price, instead of a bargain based on the fact the Foundation or any other great pro-hunting organization was going to use the revenue for such a worthy cause. Remember, this I did not begin my involvement with FNAWS until 1983.

We have another great mover and shaker who definitely must be mentioned. General Harold Starke was a true character in his own right. This man showed up from Canada (never quite figured out how he showed up) for many years and bought a lot of items. The ones he did not buy, he made sure went for a very good price. Starke added not only monetary value to the Foundation and our cause, but also brought a lot of color to our auctions. The world of conservation definitely lost a friend with the passing of General Stark. We will always see to it his dedication and commitment to the art auctions over the years will continue to enhance the population of wild sheep in North America for many generations to come.

Even though I already talked about Dan Pocapalia in the early years, his commitment has continued through the last convention when he purchased his third bighorn permit he fully donated back for re-auction. This most recent contribution was for the Wyoming permit in which 100 percent of the funds were earmarked for the Wyoming Fish and Game Department for their Wild Sheep Management Program. Then, when it went up again for auction, the Foundation was able to keep 100 percent of the proceeds. Another great example of this is in 1991 when Hans Hansen won a complete personal hunting package. Rather than accepting the hunt, he immediately donated it back for re-auction. For more on this story refer to Chapter IV of this book.

Arthur R. Dubs at the 1988 national FNAWS auction.

A similar gesture is that by Arthur R. Dubs, Art to friends. Who would not have been inspired along with the thrill of just being present when Dubs purchased the first auction item that went for over $100,000? It definitely has one of the highest ratings when it comes to the most enthusiasm I have ever seen at one of our auctions. This Montana bighorn sheep permit was sold at a dollar value most people thought would possibly stand forever. Little did we know that not only Dubs willingness to pay this kind of price along with his desire to pursue the largest North American Grand Slam ever taken would inspire others to create some of the greatest, most exciting and thrilling auctions in the pro-hunting/pro-conservation world. Art Dubs is the same strong supporter who purchased an art sculpture at one of our Saturday night banquets and immediately donated it back for re-auction because he did not believe the original price brought enough money to the Foundation.

Perhaps the memory that will be in my mind forever is when Dubs won Stan Simpson's Dall's sheep hunt and sheep camp fundraiser package at the Saturday night banquet. I got on the stage and announced we had come within $70,000 of having a $2 million auction at the convention. Moments later, Dubs donated this hunt back for auction. In true mover and shaker fashion, Dubs generously opened up the bidding on the hunt he had just won and donated back. Stanley Boots, one of our all-time great supporters, ultimately ended up purchasing the hunt to make sure we got our $2 million mark. This is truly one of FNAWS highlight moments.

Actions like these by such committed and generous men, can have immeasurable long-term benefits. Gestures of this nature not only add great entertainment value and obvious monetary benefits but encourage the rest of the members to put something back to the mountain in one form or another.

The same effects hold true when you see the Dietrich clan, Virginia Frehner, and perhaps our strong Texas supporters bidding well over value on certain items with each group having lots of fun, laughing and making excitement for all while at the same time making an obvious donation. I think I would be remiss if I did not mention that at the Saturday night banquets every year there are usually two large Texas tables of FNAWS supporters who always bid on a piece of artwork or a gun. They always make sure the auction items go for good prices and always make good contributions to our cause.

Even though, for the most part, I can not start naming names for there are so many dynamic individuals in our organization, I do need to make mention of one great example:

FNAWS Board Members Dr. Robert "Bob" DiGrazia and Phyllis Carlson, left and right, welcome new FNAWS member Jameson Parker, center, in 1992. Jameson is a Hollywood actor and director who strongly supports FNAWS and hunting. One of his best-known parts was as a detective on the popular television series, "Simon and Simon."
Photograph courtesy of FNAWS archives.

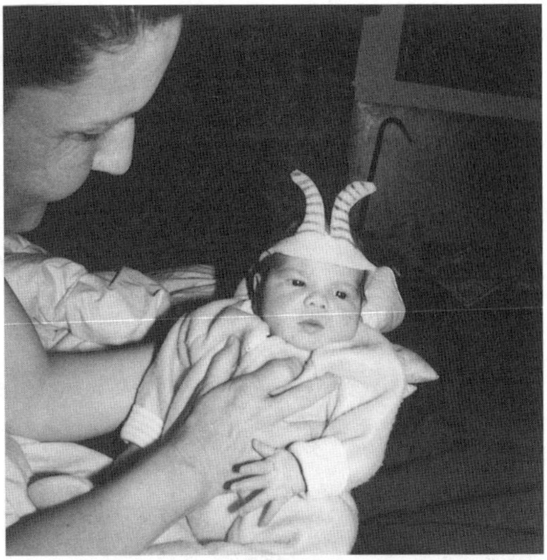

Allyssa Ann Madison, born April 2, 1991, is one of the future movers and shakers. The Don Madison family reported that her Boone and Crockett Club measurements were 8 pounds, 14 ounces. Here we see her as a newborn.
Photograph courtesy of FNAWS archives.

Bob Curtis, shown here in 1980 with a loyal and faithful friend leaning against him, has been a long supporter of FNAWS. At the 1991 FNAWS convention, Col. Donald K. Mangels, USAF Retired, interviewed Bob. His comments are included in the spring 1991 issue of the FNAWS national magazine, Wild Sheep.
Photograph courtesy of FNAWS archives.

Francis X. Bouchard, donor, bidder and purchaser at the FNAWS auction in 1996.

Judy Latham of Arlington, Texas, left, and Carolyn Pedrotti of Corpus Christi, Texas, next to Judy, with the rest of the crowd enjoy the San Diego beach party at the 1992 FNAWS convention. The gentleman to the far right is Paul D. Webster of Wayzata, Minnesota.

Mayvon Platt has been a long-time supporter of FNAWS. Here she enjoys the 1992 Ladies' Luncheon and Auction.

Long-time FNAWS supporter Joe Bishop of Arvada, Colorado, holds his newly acquired Pachmayr shotgun in 1992.

Tom McMillan of Reno, Nevada, is one of hundreds of regular auction bidders who donates funds to "put sheep on the mountain." Here he is a successful bidder in 1992.

Dietrich and Barger families, long time FNAWS supporters, are donors, sponsors, bidders and purchasers.

Daniel A. Pedrotti. Many of his donations are well-documented. However, his contributions behind the scenes that go without recognition are equaled by few if any. I have been fortunate to witness many of these with him. He never wanted to accept recognition for his generous gestures to the Foundation and sport. He is an extremely dedicated individual! As past president, he has put in countless hours and financial commitments that have made a huge impact. He has taken the initiative to the financing of this book by saying he would be responsible for the initial costs and dedicating all sales of this book to the newly established permanent endowment fund. When you buy a copy of this book you help put wild sheep on the mountain. By underwriting this book, he thought this would be a good way for every member to participate in kicking off the Foundation's endowment program. The revenue from each book goes into the fund. This makes it so his unselfish support creates a good avenue for all members to secure the future of FNAWS. He never wants any recognition for his gestures; therefore I'll just leave it at that.

While we are still on the subject of the Saturday night auctions, how can I not recognize someone like Joe Bishop who will walk up to me before an auction starts and ask me if I need any help in making sure an item goes for a high enough value? At the same time, he's prepared to write the check if he ends up with the item.

Movers and shakers come in forms other than state permit buyers and large contributions of cash. This is evident in the tremendous amount of volunteer members who work so hard on our Board of Directors and our chapters that are always so apparent at our annual conventions. There again, it would be fun to start listing names and everything they have done for the Foundation, but I think the best example lies with two Past Presidents Dr. Jerry Waite and Arnie Johns. These two past presidents both dedicated many years of their life volunteering countless hours to the Foundation when it was still in such a vital growth period. Both of these men traveled to numerous outfitter meetings in the United States, Canada and Mexico, attended years of Board meetings, ran all parts of the conventions, and worked on developing the headquarters and the staff. To use the phrase from the commercial they truly earned their Presidency the "old fashion way" would certainly hold true since these two men really earned their positions. The Foundation will forever be indebted to these movers and shakers who truly led by example. This type of contribution has established a

Mr. and Mrs. J. Curtis Newman were the successful bidders for the Wyoming permit in 1994 and 1995 by contributing $83,000 for those permits in two years.

Carol Bruckner, left, and Donna Howard enjoy the 1992 Ladies' Luncheon and Auction.

base of volunteers. No two men at the time had committed so much time and energy as these two men.

J. Curtis Newman is another strong supporter of FNAWS behind the scenes. He has supported the Foundation for so many years. In 1994 and 1995 he purchased the Wyoming permit and called his old friend (famous in his own right) Johnny Drift, to escort him on the hunt. This, at the age of 80, was his last bighorn sheep hunt. We are not sure to believe it, but we think he might take another one over the age of 90!

Red Duke as president has re-organized and brought true credibility to our auctions. He made it so much easier for the volunteers to solicit hunt donations and art items with his clear-cut approach. I had the pleasure of traveling with Red first to the Yukon Outfitters Association meeting in Whitehorse, Yukon, and then to be picked up in Dick Gunlogson's airplane as we flew to Anchorage for a meeting. Red has always told me I needed to lose weight. While leaving Whitehorse at temperatures of 35 degrees below zero, he told Dick he was flying in the back seat with me in case Dick had to put it down in a whiteout. He wanted to be as close to the BTU factory as he could get. I can't start telling Red Duke stories for that would take a book of its own. Red has served on the National FNAWS Board and as president as well as being the master of ceremonies for many Saturday night banquets. He entertains many of our members at the luncheons and annually donates to the auction and purchases auction items. He truly leads by example.

I would now like to shift to recognizing some of our famous outfitters. Every year the Sheep Foundation has a Golata Award that recognizes our Professional Hunter of the Year. This recognition has to do with the integrity, sportsmanship and support by individuals who are a part of the industry. Myles Bradford, Urban Rahoi and Garry Vince are the three men who have won this award. Bradford donated a 40-inch Stone's sheep that was considered big in the early days. As the list develops in the years to come, these will be our movers and shakers on the outfitters side of things.

It must be noted there are more movers and shakers who are not on the surface. I would

Robert "Bob" E. Petersen, founder and chairman of the Board of Petersen Publishing Company, Los Angeles, California.

Bob Housholder, founder and first executive director of the Grand Slam Club.

Jerry Fletcher, left, is presented the Outstanding Contributor Award by FNAWS President Craig Johnson.

Bernie Fiedeldey, left, paid $70,000 in 1994 for the Texas permit. He has also been successful bidder for the California permit in 1995, the Alberta permit in 1996 and the 1998 Arizona permit.

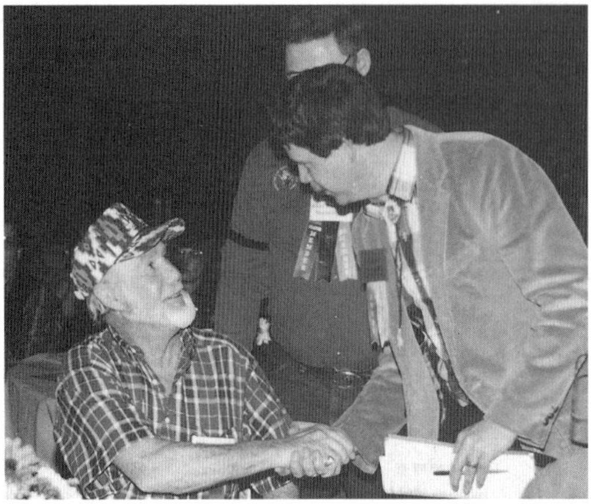

The Montana moose permit went to successful bidder Jack Wilkinson in 1994, shown seated as he is congratulated by the crowd.

J. Wesley Jones was top bidder in 1994 for the Utah bighorn permit when he paid $62,500 to "put wild sheep on the mountain."

Volunteers Ron McMillen, left, and Bob Carlson pause a moment for a photograph at the Friday afternoon auction in 1992.

Rich Roan tries to wiggle more bids from women attending the 1989 Ladies' Luncheon and Auction. Rich was auctioneer for FNAWS from 1979 to 1998.

Garth Frehner of Nevada, left, and Jerri Johnson of Utah enjoy the 1996 FNAWS convention.

Members of FNAWS were present in 1990 when Frank Pachmyer, center and seated, made the announcement that he would donate 10 12-gauge shotguns to FNAWS for future auctions. Frank is no longer with us but his ability to donate to FNAWS lives on. Shown standing from left to right are: Past President Arnie Johns, Gen. Harold Starke, Harvey Kadlec, Gen. Chuck Yeager, Gen. Joe Engle and Past President Jerry Christian.

---

hate to slight our outfitter members and guides because they are movers and shakers as well. These individuals are the ones who take our members hunting and keep our sport alive. This is certainly not to slight Frank Golata or his wife Mrs. Golata, one of the most classy women I have ever met. With his reputation as an outfitter, as a Board of Director for National FNAWS, and as an annual volunteer for the moose horn raffle, along with his lifelong companion, Mom Marge, at his side, I personally would not have objected to it being called the Earl Boose Award. But in our own hearts, we can recognize both these two as great outfitters.

Harvey Kadlec, Brad Kadlec, Doug Carlson and Hugo Kilian, along with all the people who helped is another group of volunteers who have over the years worked with all the auction items behind the scenes at the conventions. Many times they spent the entire day in the auction room assuring the auctions will run smoothly. Usually with their spouses by their side, these people have always been movers and shakers in my heart. Frank Pachmayr is another individual who needs to be mentioned. Frank made a full donation of two Parker side by sides and two Winchester Model 21s. All four were complete with Pachmayr upgrades, including many gold inlays. Until his passing in 1996, Frank made it a point to attend our convention and enjoyed watching the support our members showed while auctioning off one of his fully-donated shotguns. The donations still continue through his Foundation. Frank is truly a legend in the gun world. Many of our world known artists and members have truly enhanced and inspired all of us. Along with people like Al Barone who have helped these items always go for a good price. Gary Swanson and Vivi Crandall are two great examples. Gary has donated many original paintings for auction over the years along with the original oil bust for our Foundation's logo. He has painted, "The Reigning King," which was bought for more than $100,000 by Ron Simeck. Gary, our first Living Legend, not only has added many quality paintings to the auction over the years, but has generously purchased auction items well beyond value (like so many artist members have done). He has brought a true, colorful and charismatic atmosphere to our conventions. And then of course, there is Vivi Crandell, along with her husband/manager Gary, who in Philadelphia had her Living Leg-

Johnny Drift took the shirt off his back in 1988 to motivate Virginia Frehner to successfully bid on this gorgeous fur coat. His efforts worked!

Virgil Lair studies the auction items in 1992.

Al Barone at the 1990 FNAWS convention.

Past President Pete Cimerello models his wild sheep hat to the delight of the Saturday night banquet crowd in 1996.

Photographs courtesy of FNAWS archives.

Judy Nielsen of Las Vegas continues as an enthusiastic supporter of FNAWS and has devoted countless hours to convention organization. She has a distinct flair for organizing the Ladies' Luncheon and the exhibitor/donor parties. She is seen in this photograph at the 1990 FNAWS convention. She is pictured here with Debra Bradbury, wife of the late Basil Bradbury.

Pete Cimellaro, leaning forward, congratulates Wayne Weber at the 1992 Saturday afternoon auction.

Carl G. Hoehner of Idaho purchased the Nevada Nelson bighorn and sheep tag in 1992 for $65,000.

Lee Anderson and family after the 1994 Saturday afternoon auction. During the 1994 convention Lee purchased both the New Mexico and the Oregon permits. He was also a bidder on a number of other permits.

A party at the 1990 FNAWS convention found Sandy Christian, left, her husband and FNAWS Past President Jerry Christian enjoying conversation with Past President Arnie Johns.

end painting, "Eye of the Storm," break the record when it sold for $103,000 and was purchased by Ron Simeck. Vivi's contributions go well beyond her being a world-famous artist. The personal enthusiasm and color she brings to the convention along with her personal commitment and dedication to the FNAWS auctions and even more so to the ladies' luncheons is a commitment in which the Foundation will always be indebted.

Beth Artis, Gen. Chuck Yeager and Gen. Joe Engle enjoy the 1990 Saturday night banquet.

The Conservationist Award, previously the Outstanding Contributor Award, can be referred to in the appendix. Any individual having ever to receive this award truly has made a substantial commitment to FNAWS. Jerry Fletcher is the Number One winner of this award because he has won it more than anybody. Stanley Boots and Lee Anderson both have been nipping at Jerry's heels since the award's inception. I remember when I first walked over to the table and introduced myself and explained to Jerry a little about the North Dakota permit and how our auction works. Little did I know at the time that I was visiting with the person who would not only electrify and encourage so many of our other members to step forward but would end up investing more in our auctions over the past years than any other individual. These investments have truly paid high dividends in the enhancement of wild sheep and their cohabitants, conservation in general and the pro-hunter's rights. Always remember, we have been blessed with so many members who have approached these gentlemen-like bidding wars to get the top dollar for our common goals. I already referred to the enthusiasm when Doc Thurston and Dan Pocapalia bid $50,000 on a wild sheep permit. We thought that we had seen it all then. But, Art Dubs came back and bid $109,000 in 1987 for the Montana permit. Stanley Boots drove Jerry Fletcher up to $200,000. Once again we thought we would see this value stand forever. However, Sherwin Scott, Lee Anderson, Tom Pawlacyk and Jerry Fletcher bid it up to the $300,000 plateau. Then, of course, in 1998 when we thought we would not see a higher price, special friends of the Sheep Foundation from California, Texas and other states took Sherwin Scott to an excess of $400,000 ($405,000) for the Alberta permit. I kept saying, "I don't believe it!" I imagine the day we hit the $500,000 plateau a lot of tables will be collapsing for the all the people will be standing on them!

Whenever I question myself on the number of hours I volunteer to the Foundation, I think of someone like Lee Anderson, a quiet gentleman sitting on the side purchasing and bidding up items at our auction. I hope individuals like him understand the positive impact they have on volunteers like myself.

In conclusion, I can think of the Fiedeldeys and the Fletchers bidding Rick Taylor's (our 1998 Living Legend) bronze up to a new all-time high in 1998. The bronze was of a lamb wild sheep, representing the new beginning. I would like to think that this is the new beginning for the next 20 years of conventions where some of our old movers and shakers can lead the new generation of movers and shakers so that wild sheep, along with our other wildlife and

Tracking this Alberta ram with an 80-pound bow was one of Ron McMillen's best hunts ever. Ron, right, took this outstanding ram in Canmore, Alberta with outfitter, Randy Babala. The ram was harvested on the next to the last day of the 16-day hunt with guide Danny Leonard, left. The ram received the Archery Gold Ram Award in 1987.
Photograph courtesy of Ron McMillen.

pro-hunters, can benefit for generations to come. Remember, this chapter merely touches on a few of our movers and shakers. There are so many who have made great contributions in one form or another. Our members applaud you and recognize you for your contributions. Wild sheep and the sport of hunting will forever be indebted.

Ron McMillen was raised in Ames, Iowa. He has managed Sigler Printing and Publishing since 1972 and in 1979 became owner. His company started printing Wild Sheep *magazines for Basil Bradbury when he was publisher and editor. He prints the magazine to this day. In 1984, Ron started traveling and helping Tom Hinders solicit donations for the annual conventions. Having such a feel for the donations, he turned into the annual convention auction liaison to make sure the auctions ran as smoothly and accurately as possible. Currently, he still volunteers his time for that position. He married Becky McMillen in 1988. She had two little boys, Nate and Zach. A year later, they had a daughter Chelsea. He has backed off attending most of the outfitter meetings while raising his family. However, he must say he will always cherish warm receptions from this great group of people, and always remember fond memories along with having established life long friendships with many. He really considers himself lucky to have represented the Foundation for so many years. Ron, like so many other members, still needs his desert sheep to complete his Grand Slam. Being an eternal optimist, he knows it is just a matter of when, not if.*

Kim Mazza assisted with the writing of this chapter. She is an Illinois native, graduated from Iowa State University in May 1998 with a B.S. in finance. While earning her undergraduate degree she was a member of the ISU gymnastics team. She plans to attend graduate school in 1999 in order to work towards her Masters in Business Administration. Her future goals include continuing to work in an athletic setting.

Interview with Ron McMillen

President, Sigler Printing and Publishing, Inc.
Ames, Iowa

Susan C. Reneau

Just after the 1998 FNAWS national convention, I spent an hour with Ron McMillen, asking him about his involvement with FNAWS and what he thinks the future will be for this organization. My questions and his answers are included in this chapter along with photographs of the people who are dear to his heart.

What was your first involvement with FNAWS?

In 1982, I wanted to do some big-game hunting. I had gone small-game hunting in Iowa. I talked to Don Mangels of First National Bank, a retired Air Force colonel. He told me about Jerry Christian and Tom Hinders. Jerry Christian was actually president of FNAWS at the time and lived 10 miles from here (Ames, Iowa). Tom Hinders ended up president of FNAWS a few years later. He lived here in town. I originally met them at a bar one night, and they were visiting. They told me about the Reno FNAWS convention in 1983 that was just coming up in the next couple of months, so I decided to go to that convention. That was my very first involvement with FNAWS.

Don had done quite a bit of big-game hunting in his life. He had gotten to know Tom through the bank and Jerry through the Sheep Foundation. Just a few years ago, Don completed his Grand Slam for sheep. Yes, he was an active sheep hunter.

Did you go hunting with Jerry and Tom? How did you get involved with hunting sheep?

At that convention, I bought at auction. I first bid against Bill Leech of Texas, a true great gentleman, great hunter, a great supporter of the Sheep Foundation, and a great member of not only the sheep hunting fraternity but the hunting fraternity in general. I got a little crazy at that auction. I originally bid on Dave and Ed Wing's hunt who were partners at that time. Bill drove up the price with me, and I finally got out of that. I'm glad I got out of that bid because for 1983 it was quite substantial, especially for me. The final price was $13,000 for a Stone's sheep hunt in British Columbia. Bill beat me out for that hunt, but then I came back an hour or so later and bought another hunt for $10,000 from Vulker Gulnick that ended up quite a story in itself. I didn't harvest a Stone's sheep. It might be fun to describe my first sheep hunt.

Were you in the printing business when you came out to Reno in 1983?

Yes, I grew up in the printing business. I never had any intention to do printing for FNAWS when I came out to the convention in 1983. I just wanted to do some hunting. I got involved in more detail the following summer. What happened was Tom Hinders was getting more and more involved in the Sheep Foundation and was helping the president by doing quite a bit of traveling to outfitter meetings to solicit hunts for the auction at the annual convention. He had a health problem at the time. I was up in his office and I was single. He asked me to go to British Columbia for him to solicit funds. I said I didn't know anything about the Sheep Foundation and told him I wasn't qualified. He said he needed somebody there and we need somebody who's a salesman and level headed.

Tom said, "We'll give you an expense account and go."

So I said, "Sure, I'll go up. So I went up and had fun doing it."

Ron McMillen as Master of Ceremonies at the 1988 Ladies' Luncheon and Auction.
Photograph courtesy of FNAWS archives.

I went to the Northern British Columbia Guides Association to make a pitch for outfitters to donate hunts for the Sheep Foundation auction. Even to this day we have a lot of volunteers and directors who go to these outfitter meetings to represent the Sheep Foundation and to solicit hunts for the following convention.

Did you go there with a particular goal in mind or did Tom give you what he expected you to bring back?

This was a complete shot in the dark. He told me to go up there and get all the donations I can. I went up there and at the time there was a lot of momentum. Everything was going really strong. I remember Miles Bradford, Garry Vince, Dave Wing and Gary Moore all donating sheep hunts. I could go on and on. The British Columbia outfitters were very, very supportive. The early outfitters in FNAWS who donated hunts encouraged others to step up and help. Miles along with his wife Sheri were the first recipients of the Professional Hunter of the Year Award three years ago at the Sheep Foundation.

Did you serve on the FNAWS Board of Directors? What was your relationship with FNAWS?

I started going to more and more of the outfitter meetings. Because of that I would end up going to the FNAWS board of director meetings. I was a lot more focused on going to all the outfitter meetings at the time. It ended up that I was going to all of the outfitter meetings. It amounted to 12 meetings a year, then 14, then 16 a year. I was single and my business was doing okay. I was just having a lot of fun. I would go from one to another outfitter meeting, from one state and province to another. I began doing this in 1985 and 1986 and 1987 when I was in full swing.

Are you still at that rate of activity?

Oh, no. I got married in 1988. My wife Becky had a 1-year-old and a 4-year-old boy at the time, and a year later we had a baby girl, so that took most of my time. I usually go to Alaska every year. I like to still pick out some of the meetings. I have a hard time giving up all of them. I enjoy going to Alberta, northern British Columbia and Idaho. I've really enjoyed all of them. That's one thing that kept me tied in there was because I enjoyed the outfitters and enjoyed working with them. I was not bragging but I was pretty good about what I did as far as getting the donations for the auction and selling the outfitters on why they should donate hunts to FNAWS.

You asked me earlier if I had ever served on the board. I ran for the board. I don't remember which year, but it was probably two years before I got married. We always had a lot of laughs. I was extremely visible at these outfitter meetings, and I was helping with these auctions, and I finished dead last. I did all that work and never got elected to the board. To this day, when people tell me they are running for the Board I always tell them they don't have a thing over me. I ran and finished dead last. That was about 1986.

How many outfitter meetings were there that you were visiting in the mid-1980s?

Strictly a guess, I think about 14 a year. And, I went to the NRA shows, the National Rifle Association annual meetings. I didn't go to SHOT Shows. I went to all the FNAWS Board of Director meetings, though. Before the Hawaiian convention in 1985, I believe, I had gone out and gotten 21 sheep hunts. I had been told to go out and get as many hunts donated as possible. We were still in our infancy as far as the auction. We didn't know how to do it or how to handle it. What happened was that not many people went to Hawaii. When we got there we didn't have enough buyers for the hunts. I had dragged a portion of their (the outfitters) livelihoods away from them and then the hunts sold terribly. I had one outfitter who I thought would start swinging at me. I thought he was going to start nailing me, but he didn't.

After that I started getting involved in all the auctions, and I made what we call the "Hit List." What that means is that we don't want to sell over two Stone's sheep on Friday and Saturday. We won't take over so many caribou, so many moose, so many bear, so many Dall's sheep. We don't want to offer more than we can get quality prices for. It's not fair to FNAWS, but it is especially not fair to the outfitter. For longevity, of course, it really tears things down. We still haven't recovered from Hawaii. There are still people who have bad feelings from Hawaii and who have never donated since. This was an expensive lesson, but it was a lesson in the end.

Since Hawaii I have become the auction liaison. Isn't that something? What I do I stand there and anything that happens, I monitor it. A lot of things happen at the spur of the moment. A person might want to pull their hunt for a funny reason. A man walks up to me and says, "I'm not going to donate my hunt because one guy might buy it who is gay." I stand there and have to make an immediate decision about what to do about this. There is not time for a committee. I have to act on the spot. So I tell him, "You're opening yourself up for lawsuits, man. You're opening yourself for all kinds of things. You've just got to let things go," I said. So, the man listened to me and said okay.

There are other times when somebody says they want to donate this, and I say, "No we can't." Or another time I have someone come up to me who wants to talk about an item, so I ask, "Who are you?" We've had people come up to talk and all of a sudden we'll have the outfitter come up and say that person wasn't authorized to talk. I have to make decisions on the spot what's going to happen.

Before I started doing that, these people would come up to the auctioneer (Rich Roan). The auctioneer is the great barker in the world, the barker being the auctioneer, but he doesn't know the internal workings of the Sheep Foundation and the items he's selling, so he lets me handle all that. Before I started doing this, we were getting done extremely late and we were losing our audience. I make sure we stay on time. I also monitor and make sure someone is there to talk about an item. I just make sure the auction flows smoothly.

Are there some stories you can tell me that adds some life to the auction?

There are some behind things that I'd never tell in story form even without names because it would identify them too much. There is one state permit that was selling where I thought it would come to a fist fight. One man was simply not authorized to speak about the state permit, and he was going on about the state permit whether I liked it or not. I just kept the microphone away from him. It became a rather hairy situation, but in retrospect, it was quite humorous now, but it wasn't humorous then.

Wasn't Harvey Kadlec the auction organizer for a long time?

He and his son, Brad, and his wife did a tremendous amount of work behind the scenes. What they do, they go in a room and organize all the times in the order of everything. They're the reason that makes my job so easy is because I'm on the spot. They line everything up and make sure all the bronzes are there when they're to be sold. They make sure the actual coat or the actual art is there. If there is an add-on, they make sure the runners are getting taken care of. After Harvey and Brad, Doug Carlson took on this job. Doug resigned last year (1997), and Hugo Kilian from Cody, Wyoming, does it now. Hugo and his wife took it over last year.

This is quite a job. It really removes you from the convention. You really don't get to enjoy much of the convention at all. You find yourself in a room working all the time. They'll find me quite often. There are a lot of add-ons during the convention. People will come, and

Ron McMillen on stage in 1993 with his hand up joins the Roan family in the thick of things for the FNAWS auction. Leland Speakes leans forward toward Ron.
Photograph courtesy of FNAWS archives.

Ron McMillen, left, hams it up for the crowd in 1989 as Dr. Jerry Waite, center, and Randy Roan join in the fun. *Photograph courtesy of FNAWS archives.*

they'll just hand it to me, and I'll say yes or no. I look at the splits. The splits have to be aboveboard. Someone will hand it to me, "Here's a special. We're going to give you this, if it's sold here." And, I'll say, "Well, we can't sell it there."

Like Saturday night it takes a very special item to sell on Saturday night. Now I have had some items, such as a white sheep hunt, that was donated to us by Clayton Williams from Texas who ran for governor. He came up to me and told me this was a white sheep and not one of the traditional four North American species. It's a transplant species in Texas. He told me, "I guarantee you I'll get you a $20,000 donation for it." We had a lot of people, especially outfitters, who thought it was too controversial. You can't sell it. I told them, "It's $20,000 guaranteed in our pocket. We're selling it." This was a last-minute donation from Texas...Clayton Williams from Texas who ran against Anne Richards, and lost against her. He's not an outfitter. Clayton just owned a big ranch and had these white sheep roaming on his ranch. He made that donation and had a buyer in place.

Now we've had other buyers in place. It makes a tough call if they want special privileges. Before we used to cut deal after deal in our early stages. We'd buy hunts. We'd give different people different splits and all that. Then, of course, you'd catch yourself. One person would find out they didn't get as good a split as the other person. Red Duke, when he was president in 1986, he authorized for me to buy hunts when I was down in Tennessee. I told him there were some problems. People were finding out about these deals. I was in Alberta buying a bighorn sheep hunt, and I called him from Alberta.

Finally in Tennessee, Red Duke told me, "Here's what the deal is. Draw up a new donation form. Make it to where we make no deals. We have set cuts and people donate to that or we don't take it. Another thing we're going to do is we're not going to put anyone in the audience to where there are guarantees."

Ever since then, if something sells in our auction, it sells. The one exception to that rule since that date was the first year we sold the Idaho permit. Dr. Bob DiGrazia, who is one of the all-time great supporters of the Foundation, got the Idaho tag, but with a guaranteed minimum.

Dr. Bob came to me and said,"Ron, it will go over that, so it's not a problem."

I said, "But, it's our policy that we don't sell something with a guaranteed price to somebody where we're saying that if we don't get a certain price, we won't sell it."

So Dr. Bob felt bad about it but he was committed to it. He got up on stage that one time as we were going to sell this and said, "We're going to sell this, but it's against our policy. If it does not sell for $40,000, then it does not sell. We signed an agreement with Idaho, and we went against policy, and we wanted to make you aware of it because we want our members aboveboard on the whole thing, which we did." The members in the audience respected this.

The Idaho permit did go above price, so we didn't have any problems. Since then when Red Duke put that policy into place, we haven't had problems. Red didn't spend a lot of time at it, but he was very good and he accomplished great things in a short time he worked at this. He worked very efficiently.

Are you a regular member, a charter FNAWS member or a life member of FNAWS?

I'm a regular member of FNAWS and not a life member. Why I'm a regular member, I don't know for sure yet. Years ago, I saw lots of life memberships handed out free before Red Duke set all these things in place. They were handing them out because they wanted to recognized someone for doing something they considered great. Since then they haven't done that. I have always compared myself to a regular member, an outfitter and all those people who aren't the life members. I've always been hesitant to buy a life membership and it's ironic because I have urged so many to become a life member.

I've donated thousands of dollars over the years to the Foundation and I quite often go to convention and hand them a check, but I'm still a regular member. The board has done a great job of managing the Life Membership Drive program and I'm sure I will become a life member in the very near future.

Are you still doing the outreach you did before you had a young family since you're such a great sales person?

A lot of my time I've never counted. I've been to all but two FNAWS Board of Director meetings in the last 15 years. I go to all of the Board of the Director meetings, but I only go to the Alaska outfitters meeting now. As the kids get older, I might think of picking up one or two meetings.

Why have you gone to most of the Board of Director meetings? They are scantily attended. Why do you go?

In the beginning, I was getting so many hunts donated by outfitters that I felt obligated to them. I felt a moral obligation to make sure we did it right. So I started going and representing myself as the person who got those donations. And then a few presidents such as Jerry Waite asked me to attend the meetings. Some of those earlier presidents asked me to attend, so I started going to those director meetings. Tom Hinders said I should stay involved. I kept staying involved and I enjoyed it. I felt like

it was an obligation to the outfitters I was getting all these hunts from. Then what happened, Tom Hinders brought on board Basil Bradbury as editor and publisher of Wild Sheep. Basil took it from just a little newsletter to a nice publication. Basil had been with Petersen's HUNTING magazine. This had to have been 1983 or 1984. I can only guess. I can't remember. Tom Hinders worked hard with Jerry Christian and Duane Smelser during their presidencies as well. I don't think Tom was actually president at the time, but he was the one who brought Basil on board. Basil brought the publication a long ways. He had trouble with his printer at one time. Tom Hinders called me, and I gave him a price. It was a better price than he had been paying. He'd been having trouble getting it out on time, so we'd turn it out.

***When did you get the first contract to do* Wild Sheep?**

I'm thinking it was 1985ish. I'd have to look. With Basil, there really was never a contract. It was a gentleman's agreement where we went from issue to issue. He grew the magazine. During Red Duke's regime there was some controversy. Basil survived Red's presidency. In Tennessee, all the Board of Directors were sitting in the Board of Director's room and I was on the sideline like I had been forever. The only time it has ever been done this way and one of the rare times it has not been done in closed session was to select a president.

Red stood there and said, "Well, we're going to nominate Jerry Waite to the presidency." Of course, he says this in his Texas drawl. "Jerry Waite has been the hardest working man for the Foundation. He is going to be president. All those in favor, say aye."

Red said, "It is unanimous." And, he slammed the gavel down and never asked for nays because he said that was the way it was going to be. No one said anything. Of course, somebody could have said something, but they didn't. And so Jerry Waite was president.

I believe it was Jerry's year as president that Basil had built the magazine to the heights he could build it to. The Board of Directors felt it was time for a change, move on and bring it in-house and there quite a few changes internally with Lloyd Zeman and all that at the time.

Everyone I have been associated with in the Sheep Foundation has done a lot of nice positive things. That's what I tend to focus on. That's just my lifestyle. It's not like I'm trying to hide anything. It's just that I am very private when certain things happen. I think certain things are intended to be private, and I believe in privacy. I could be a great president because I wouldn't write a book when it was over.

I was very grateful for what Basil accomplished and how he got it on track. When he got that going, they put the magazine out for bid. They were going to take the magazine in-house with Sandra Nicola compiling most of the information and getting the advertising. So they took the magazine itself out for bid and I was one of six companies that bid it. I got the bid, and I got a contract under John Harris' presidency. Before then I was just printing it. Under John Harris' presidency I got a formal contract. Some people questioned if I was in FNAWS for the magazine at that time, so they put it out for bid and found out that I was far and away the cheapest by a long shot, so they gave me the contract. I've operated under that same contract probably for the last 10 years.

Sandra came on board during the time when Basil was doing the magazine, or was she hired to take over when Basil left?

She was on board during the phase out. Then she assumed the duties. We as a group were putting the whole thing together to kind of keep it all glued together. It ended up that she has been with FNAWS for over 10 years. She solicits advertising and compiles all the information that is sent to the office. She has done a super, super job for 12 years now.

Dan Pocapalia toasts FNAWS and his winning bid in 1983 with his daughters and wife. From left to right are daughter Diana, wife Mary Ann, Dan and daughter Ann. Dan's bid of $50,000 for the Wyoming permit inspired the young Ron McMillen and motivated him to become more active in FNAWS.
Photograph courtesy of Mary A. Pocapalia.

This will be the start of a new century, very soon. What do you see as the future of FNAWS? What would you like to see as the future of FNAWS?

There have been some major changes recently and I think the Foundation lost an opportunity three to five years ago, probably five years ago, to be a much bigger player than they are.

What happened five years ago?

I think that they had a few holes to fill in certain leadership areas and in some director areas. That's certainly not to take anything away from the directors we had. We've had phenomenal individuals, and I've made lifelong friendships with many of them. I think we probably needed a few more business people with vision. The way we've brought in our Board of Directors and voted on them we have left great individuals off our Board whom I think would have approached it a lot more businesslike. We kept it a little too much as a good old boy club. Therefore, we will have a place. We will always be a player. We'll always have a presence, a strong presence, but I think at one time we had a golden opportunity. Safari Club International was not doing well, and now they are absolutely thriving. A lot of it is because they brought on their board people we could have had on our board.

In long term, we will still have a solid presence for the enhancement of wild sheep and the other animals that surround themselves in the geographic area as wild sheep do. We are pro hunting. We have probably moved too far towards lobbying Washington and all that. That's not what we are all about. We need to get back to our roots and back to what we do best. I still fight for pro-hunting rights, but we can do that on a state basis. To me to put money in Washington is just a joke. The thing is what needs to get done with Washington doesn't get done with $50,000 or $100,000 contributions. They get done on $50 or $100 million contributions. We aren't even a grain of salt in Washington. We have to focus on where we can make the most impact. My advice to future FNAWS members is to put its money where it can do the most good. Absolutely.

I have spent so many years with it and have truly gained some wonderful friendships and have lost great friendships over the years. Clark Engle in Alaska and Jim Ford in Montana. They were both killed in plane wrecks. Doc Thurston in North Carolina and Dan Pocapelia may be the reason why I got involved with FNAWS. It has been a pleasure getting to know them. We could do a better job of recognizing people like this. That's why it's important to get the Chuck Yeager types in. When Doc Thurston and Dan Pocapelia started bidding those Wyoming hunts for $50,000, people were standing on their tables, hooting and hollering. I was a younger man at the time, and was one of them. I was so excited and I wanted to be involved in the organization. I recognized the sale value of that of bringing those people in and we need to bring those people along. We need to bring them in and recognize them, year after year.

How do you do that?

Doc Thurston and Dan Pocapelia are unbelievable contributors to FNAWS, beyond anything you'll probably ever write because psychologically they impacted so many people to get them more involved and to raise the bids. They also got more people to donate and get involved. Maybe FNAWS needs to recognize the big givers and focus more with them and target them more. We use the good-old-boy approach and we simply don't target ourselves to the big givers. We need to find out when their birthday is and send them a birthday card. Run it like a business.

Have you ever shared that simple idea of sending birthday cards to people?

No, I've shared bigger ideas than that. After they buy a hunt for $100,000, $200,000 or $300,000, every Board of Director and a good will committee should all write a personal letter to that individual thanking them and telling them how wonderful it was they were there. We should tell them we hope to see them next year. We should do a followup three months out before the convention and contact everyone of those individuals in person, making sure they are going to be at the convention, asking them if they need help with more convenient reservations.

I assume from these comments that this is currently not being done? Is that true?

Not to the extent we need to. It's a business approach. This isn't a criticism. I want this only as a suggestion. I don't even know if these suggestions belong in this history book. I've made so many dear friends. Dean Pic'l has turned into a dear, dear friend. His commitment to the Foundation will never be recognized at the level it should have been.

Why is that?

Not only has he volunteered his time as an attorney, which has saved us hundreds of thousands of dollars over the years, it has saved us a lot of grief over the years. It has let our meetings run a lot more smoothly. He's volunteered to write little passages, little bylaw changes. He would put in his own verbiage and through in his own personal vernacular that would tie in what the Board of Directors were trying to say. In a sense, he is a lot like myself by turning into a historian. Historians contribute if they contribute right, and Dean and I like to think we contribute right as historians at the Director's meetings. We help them not have to reinvent the wheel over things that have been year after year. We've sat there for so many years, I'd like to think we have contributed. I feel that Dean and I are the only ones who have that historian's perspective. I really hate to take that away from him. He's a marvelous speaker. You simply ask him to help, and he will help. He's a genius and a lot of fun. That's the reason why I go to the meetings because of people like that.

There have been heartbreaks. Nancy McDaniels was a lot of fun and a dear person. Then they changed accountants to Jim Reilly and now Jim has since become a dear friend. Nancy was out of Phoenix. Jim Reilly is the current accountant, and he took over when the Board decided to change accounting firms. They decided to bring the accounting in closer, so Jim Reilly took over. He's from Cody and he's been at meetings. He sits on the sidelines with me, wears glasses, about 40, blondish hair. She was out during John Harris' presidency. Jim Reilly has been ever since then.

I know one of the areas that has always been of concern to the presidents has been the accounting system and how the money is spent. Red Duke spent an awful lot of time talking to me about getting a scientific method in place to properly spend the money so that the sheep get the biggest benefit.

Do you agree with that and do you have anything to add to that kind of philosophy?

There is no question that it's the best philosophy, but there again, philosophically is how to approach that. When I was selling advertising to the Wild Sheep Journal just last year, I'd have certain state chapters say our bylaws say we only put money to putting sheep on the mountain. I finally got them to buy an ad. An artist noticed the ad in the Journal and donated a painting to them. They made more money from that painting that he's going to donate every year now to this chapter. They are going to sell this painting and they are making more money, thousands of dollars, every year because they placed this one ad. They more than paid for this one ad the first year, and they are going to continue to make money. They're putting more sheep on the mountain because they marketed it. When you say you are going to put sheep on the mountain, it doesn't mean you can't take some of that money and market it to build more money. This is just like building interest on your money by putting it on the stock market. You're not putting sheep on the mountain because you are growing your money. Philosophically, I don't fault somebody for that, but there again it goes back to dealing with business people in the right situations.

Hunters, politicians, conservationists, donors, purchasers, bidders, volunteers, Board and staff converge to Put More Sheep on the Mountain*!*

Chapter X

Working Together for Sheep, Sheep Projects and Grant-in-Aid

Grant-in-Aid to Put Sheep on the Mountain

DUNCAN GILCHRIST

I discovered many years ago that being involved with wild sheep can be one of the great joys of life. My first Dall's sheep was sighted in 1962 soon after moving to Alaska to work for the Bureau of Land Management (BLM). I saw it was seen from a Cessna 180 near Nabesna in the Wrangell Mountains. I have been obsessed with the animals ever since.

It is pure joy to hunt them, see them, photograph them, be in their home and most of all to help them. Yes, wild sheep seem to need help. They have all kinds of problems that we as humans can often rectify. Wild sheep die in vast numbers from diseases brought on by contact with domestic sheep, but populations can be brought back with reintroductions. They perish because of several habitat problems, but we can provide them with what they need to survive. Predators have become a major concern in many areas. The list goes on. I often tingle with the thought of having helped put or kept a sheep on the mountain.

Duncan Gilchrist is not a graduate biologist. He is a professional forester who has spent a lifetime working with wildlife as a guide, photographer, hunter, intense observer and outdoor communicator. Many of my friends are wildlife biologists including most of the top sheep names in the world. Instead of reading a popular magazine before bedtime, I usually study some wildlife research project.

Long before I became involved with FNAWS, I attended the Northern Wild Sheep and Goat Council (NWSGC), which meets once every two years. The first time I walked through its doors they didn't know if I should be allowed to stay. But stay I did and now Dunc is an integral part of the group. It is here where a person can really learn about wild sheep and their needs.

In an average year I spend some 140 days in the field with a gun, bow or camera in my hands. Much of this time is spent with wild sheep. My wife of 38 years, Pat, is totally devoted to what I do. Money seems to elude us, yet we live a quality of life that few enjoy. Pat's motto is have fun; Dunc's motto is have fun.

When I first became a director of FNAWS, and I knew absolutely no one on the Board;

they came from all walks of life, but most tended to know little about the fine details of sheep projects. Bill Foreyt, D.V.M., from Washington State University, who was the lead researcher in wild sheep diseases, had asked for Grant-in-Aid (GIA) funds for two projects. He was going to be turned down on his request because some of the members of the Board confused him with someone else who did not have Bill's fine reputation. I fought for Bill's two projects and finally got my way. Right then, at my first Board meeting, I found my niche with FNAWS which is to do my best to see that GIA funding is spent in the best possible way.

The political arena is not my forte. I can't even understand a balance sheet but, sure as heck, Dunc can see that money is spent wisely for wild sheep and that fellow sheep lovers learn about sheep projects. Each year I give many seminars around the country and Canada on sheep and sheep projects.

Ever been on a transplant? Most sheep captures require some volunteer labor. Usually all hands at the completion say they just enjoyed one of the best days of their lives. Wild sheep are caught in baited traps and driven into a net with a helicopter with baited drop nets or by net guns. Transplants certainly put sheep on the mountain and if a project is logical and cost-effective it is usually funded. Some are rejected for various reasons including too much monitoring costs or too high cost per released wild sheep.

Ever been on a desert sheep water project? There might be a team of more than 100 volunteers and wildlife management staff working together to provide water in some canyon where there is none. Some people work on building the water project itself, while others prepare food for the feast that is going to follow that night. Many volunteers have never hunted sheep a day in their lives yet they love to help this beast of the alpine.

Just because sheep drink from a water project doesn't automatically mean the project was a success. If not carefully designed, burros, elk or deer will benefit more than the sheep, leaving the wild sheep with diminished food supplies. A project is only a success if the sheep population increases. One biologist who designs water projects told me that after 17 years he could today build a better water project, and do it cheaper, yet he was unable to evaluate the project any better than he could 17 years ago.

Ever been on a sheep survey? Surveys are essential if there are to be proper levels of harvest. The survey is basic before proper management can be achieved. Some are done on the ground and some from the air. A Super Cub sometimes is used and sometimes a helicopter. I have been on more surveys than some biologists. I even have been handed the clipboard and told to go get the job done. When looking from the air for sheep, I become totally involved. This is life at its fullest.

FNAWS gets many requests each year for research projects. Many are not funded if the project is a repeat of past projects. At other times, research has the potential to put or keep more sheep on the mountain.

As the years pass, it seems like wild sheep, as well as all ungulates, have come under increasing pressure from predators. FNAWS tries to be as logical and as generous as possible when dealing with predator problems.

Each winter I wait with anticipation for the GIA request book to come from Cody. Countless hours are spent studying projects. FNAWS may have a million dollars in requested funding and a quarter of that to spend.

How to spend it most wisely? We have professionals who advise. We have chapters that advise on projects in their state. Because I know so many of the players I can call on the phone and ask questions. I can turn on my computer and ask questions via e-mail. Information may come flashing back at me via an e-mail attachment. Sometimes I get data from an interested biologist either for or against a project that amounts to hours of reading. I have one biologist friend in Montana who, in a single GIA season, sent me data on sagebrush that cost nearly $20 just for the postage to mail the information. I summarized the material and passed it onto the Board. Instead of trying to eradicate sagebrush we learned that sagebrush is important to maintain for the benefit of wild sheep.

I hope that FNAWS will choose to rely on my GIA input long after I cease being a director.

Duncan Gilchrist

Duncan Gilchrist was elected to the national FNAWS Board of Directors in 1992 and assumed office in 1993. He has served for two terms on the Board and was its secretary from spring 1995 to spring 1999. His term of office ended in spring 1999. He and his wife Pat live in Corvallis, Montana. Duncan, known to friends as Dunc, is a prolific writer and videographer of adventure and hunting books and videotapes on wild sheep and wildlife conservation. Some of his most recent books include Volume II: Montana — Land of Giants, Quest for Giant Bighorns, All About Bears, Hunt High, Successful Big Game Hunting *and* Quest for Dall Rams. *Some of his most recent videos include* The Hunters Spirit in Music and Song, Field Judging Southern African Plains Game, So You want to Experience Africa, Hunting Montana Rams, Evaluating the Bear, Dreams Fulfilled Aboard the Wilma Mae, Spot-Study-Shoot, *and* Adventures with Bighorn Rams. *To receive a complete listing of his videos and books, call Outdoor Expeditions and Books at (406) 961-4314. Prior to his writing and video career, Dunc was an Alaskan guide and outfitter from 1968 to 1985 and was certified in 18 of 26 hunting units. He has also been a forester and logger. Most years he spends about 150 days in the field with a gun, bow and camera.*

Working Together for Sheep: Grant-in-Aid

BECKY LAYNE

The backbone of the Foundation for North American Wild Sheep is its ability to raise millions of dollars to fund hundreds of worthwhile wildlife habitat and wild sheep projects across the United States, Canada and Mexico. Their ability to raise funds and put sheep back on the mountains of North America has been an astounding success story. Some of the projects and photographs to illustrate the work are outlined in this chapter. A complete listing of all funding GIA projects from 1983 to 1998 is found at the end of the chapter, organized by the year funded and by state, province or country where the money was sent.

PROJECT 84-118: Habitat Improvement for Mountain Sheep in Poudre Canyon, Colorado

This work was part of an ongoing cooperative project between the U.S. Department of Agriculture (USDA) Forest Service, Colorado Division of Wildlife (COW) and FNAWS. The project objectives were to improve forage quality and quantity; reduce tall vegetation to increase visibility; and to improve herd distribution.

On April 12, 1983, COW proposed three areas in the Cache la Poudre Canyon as preferred sites to use prescribed burning as a tool for habitat improvement to benefit mountain sheep. The use of prescribed fire to decrease amounts of dense cover on sheep ranges would also result in improved distribution of mountain sheep that avoid dense vegetation probably for predator evasion. By increasing forage areas and distribution, the occurrence of lungworm parasite may be reduced by drawing sheep off heavily contaminated areas where they presently concentrate. By reducing lungworm parasites, lamb survival may be increased from its present depressed level and increase overall health of the herd. Treatment was planned to occur during the period of January through March 1984. Three attempts were made on three separate days in February and March 1984, but they were unsuccessful due to sudden high winds. The project was put on hold until the next year.

Close-up of the area prior to burning. Note the number and size of juniper and ponderosa pine. Also the shrub density is high indicating that vegetative succession is advanced.

Beginning in February 1985, a two-man team burned

The burning prescription was designed for a fast hot fire to remove the overstory while protecting the organic layer of the soil.

A general view of the area the day after burning.

the lateral control lines. Snow drifts and rock outcrops were used for fire breaks on the upper slopes. On March 1,200 acres were burned as planned. An 11-man team ignited and controlled the fire. Information and photographs were contributed by Stephen R. Mighton, wildlife biologist.

PROJECT 84-145: Minam River Transplant, Oregon

A helicopter is used for the bighorn sheep release site along the Minam River in Oregon.

The release that took place on December 18, 1985, was in the Minam River Canyon on the Little Sheep Ridge.

Nine adult ewes and three rams were released with the aid of a helicopter on Little Sheep Ridge, Eagle Cap Wilderness, on December 18, 1985. Most of these bighorns remained near the release site during the winter but moved south and east of the range during the summer. One adult ewe moved to the Lostine Range in early January 1986, a distance of 12 miles, and has remained with that herd since. At least three lambs were born on the Minam Range. As of January 21, 1987, six ewes, three lambs and two 3/4-curl rams had been located. Eight of the animals were on Little Sheep Ridge near the release site and three ewes were seen on Goat Mountain, five air miles from the release site. Information and photographs were contributed by Victor L. Coggins, Wallowa District wildlife biologist, Oregon Department of Fish & Wildlife.

PROJECT 87-05: Paiute-Excelsior Waterhole, Nevada

Fiberglass storage tanks, overlain by a 16-by-32-foot water-collecting apron. The entire construction was done on September 19 and 20, 1987.

Around the guzzler a 240-foot steel fence was constructed.

The project consisted of burying two 1,500 gallon fiberglass tanks, construction of channel steel and corrugated metal collection surfaces, pipeline, float-controlled fiberglass watering trough, and pipe corral. It only takes a 1/2-inch rainfall to fill the tanks, which typically serve 20 sheep for five months without additional precipitation. The float-controlled water trough meters out water only as used. The pipe corral around the entire project is constructed to allow easy access by sheep, deer or other small game, while denying passage to larger animals such as cattle or horses. The project exemplifies cooperation between conservation groups with a common cause. Planning, design and supervision were performed by the Bureau of Land Management, Carson City District. Backhoe services were provided by the Nevada Department of Wildlife. Materials were purchased through a grant from the FNAWS and labor was supplied by Reno and Fallon Chapters of Nevada Bighorns Unlimited (NBU). Information and photographs were found in the October, November and December 1987 issues of the NBU newsletter.

PROJECT 87-15: Sharon Creek/Warner Rim Bighorn Sheep Restoration, Oregon

There are now more California bighorn sheep in more areas of Oregon than at anytime in the last 100 years. These animals of the high desert rimrock country had been extinct in the state for most of the 20th century until 20 bighorns were reintroduced at Hart Mountain in 1954. Since that original introduction, the Hart Mountain herd has served as a nursery stock, providing almost 200 animals for transplanting to other suitable habitat in the southeastern part of the state. By the end of 1986, there were more than 1,250 California bighorns in 11 separate locations. Department of Fish and Wildlife biologists returned to the site again on October 20, 1987, to trap additional bighorns and move them to new homes. Over the next three days, department workers and some volunteer helpers trapped and transplanted 43 bighorns to release areas in the Trout Creek Mountains and the Owyhee Canyon. The auction

Untangling a bighorn from the net after the capture.

Capturing a young ram with a net gun. The net is about to come down over the ram.

sale of a bighorn sheep tag last spring made this fall work possible. Information and photographs were provided by the Oregon Department of Fish & Wildlife Press Release

---

PROJECT 87-17: Foraging Ecology of Female Dall's Sheep in the Brooks Range, Alaska

Dall's sheep in the Brooks Range.

Monitoring of sheep activity, 24-hours-a-day, was continued using radio telemetry equipment and strip-chart recorders from an Alaska Department of Fish and Game field cabin and our tent camp.

Most wild sheep are primarily diurnal. Thus, extreme cold, darkness and limited quantities of low-quality forage during long winters above the Arctic Circle present a formidable challenge for sheep. Summer is particularly short at these high latitudes, providing little time for sheep to accumulate energy reserves for winter. This project studied the diet composition and selection, forage quality, nutrient intake and activity budgets of adult female Dall's sheep near the northern extreme of the range of wild sheep for two years and constructed a model of the energy relationship of these animals. Ewes primarily consumed forbs and grasses during the summer, and strongly selected forbs over other forages in accordance with the predictions of optimal foraging theory. Diets primarily consisted of grasses in early winter, shifted to sedges in February and back to grasses in early spring. Shrubs were consistently the least selected class of forage. In early summer, ewes foraged during all hours of the day when sunlight was present for 24 hours. Sheep restricted their foraging almost entirely to daylight hours near the equinoxes, and foraged during all available hours of light, as well as 2.8 hours in December and, when measured on a daily basis, was positively correlated with average wind-chill and day length. Duration of the short summer growing period was most important for weight gain, and presence of deep snow determined weight loss in winter. Information was provided by Michael Charles Hansen of Oregon and photographs are by Mike Hansen.

PROJECT 88-17: Thinhorn Sheep Habitat Enhancement, British Columbia

A fixed-wing aerial reconnaissance took place over four sites in the Todagin-Spatsizi Park to monitor snow and to delineate a fire break.

Poor weather made mopup unnecessary.

In April 1988, a fixed-wing aerial reconnaissance took place over four sites in the Todagin-Spatsizi Park area with the intent of monitoring the amount of snow and to delineate final fire break. It was apparent that the sites would not be ready for ignition until early June. An automatic weather station was established at Todagin Mountain. On June 6, 1988, ignition of the two Spatsizi Park sites commenced. Ignition strips were initially placed along the higher elevation portions of the slopes to create the proper drafting effect. Lower elevation zones were then ignited. The sites were visited the following day to determine the need for mopup. Due to the increasingly poor weather, mopup was not considered necessary and field crews then returned to Smithers. The two alternate sites at Todagin Lake were not ignited as good weather did not return to the area for several weeks after which the Ministry of Forests considered burning to be an unacceptable hazard. Information and photographs provided by Rick Marshall of the Wildlife Branch, British Columbia Ministry of Environment and Parks.

PROJECT 88-13: Timber Canyon Guzzler, Idaho

Leveling guzzler in its hole before filling in dirt.

Timber Canyon bighorn sheep guzzler work in progress.

The 1,000-gallon guzzler in the Timber Canyon was completed in August 1989. Due to the location of this guzzler, with very difficult access, more expensive and stronger materials were used in the construction. Previous to this year, corrugated roofing larger than 26-gauge could not be found. A source for 24-gauge roofing material, which is considerably stronger than the 26-gauge, was located. Instead of using roofing nails to pin the apron material to its frame, roofing screws, which will strengthen the apron and will not be as susceptible to the elements, were used. Also, the company making the guzzlers incorporated a "wonder" material called Kevlar into the fiberglass water guzzler. This material is virtually indestructible, thus increasing the longevity of the guzzler. Information and photographs were provided by District Forest Ranger Clarence M. Murdock, USDA Forest Service, Idaho.

PROJECT 89-02: Northern Sheep Habitat Enhancement, British Columbia

Under Ideal conditions the prescribed fire is often spectacular and cost effective.

Site Number 32 was burned on May 3, 1989. A total of approximately 1,500 hectares were burned.

Prescribed fire for wildlife is used to return vegetation communities to an earlier successional stage favored by the species involved. For grazers such as Stone's sheep the objective is to retard brush invasion of grassland-alpine winter ranges. For browsers such as moose and deer, the objective is to stimulate browse communities such as willows, aspen and birch. The desired result in all instances is to make increased quantities of higher quality forage available during the critical winter and early spring seasons, and to increase solar radiation and wind scour on south-facing slopes.

Seven sites were identified and approved by the Ministry of Forests in February 1989 as potential burn sites. Each area is characterized by having relatively steep south-facing slopes, primarily deciduous forest, shrub and grass communities on mid- and upper-elevation slopes, documented wildlife presence and definable fire guards. A prolonged and warm dry spell during the first three weeks of April 1989 resulted in several of the sites becoming free of snow by the end of April. Weather and fuel moisture codes were monitored and May 3 was recognized as the most appropriate time to commence the burn program.

The main ignition of the burn area, named site Number 32, commenced at 1 p.m. while winds were gusting from 15 to 25 kilometers per hour, the temperature was 20 degrees celsius and the humidity was relatively low and drought codes were high. All these indicators suggested that a fire would be volatile, spread quickly and most fuels would be consumed. By 1:15 p.m. the three ignition strips were completed and the ignition crew then waited for the fire to die down. It was noted the fire crossed the eastern fire guard and that mopup would be required. A four-man crew was airlifted into the site to ensure that fireguards remained established at both the eastern and western ends of the burn area. Information and photographs were provided by Rick Marshall of the Wildlife Branch, British Columbia Ministry of Environment and Parks.

PROJECT 89-49: Sawtooth Mountains Bighorn Sheep Guzzler, Wyoming

Obstacles were encountered during transportation of the guzzler to the Albany Peak bighorn sheep guzzler site. It was often transported over very steep and rough terrain.

North Sawtooth bighorn sheep guzzler site after guzzler installation is complete.

On August 24, 28 and 29, 1989, personnel from the Douglas Ranger District, Medicine Bow National Forest, Wyoming, installed two bighorn sheep guzzlers in the Sawtooth Mountains portion of the Laramie Mountain Range. This project was jointly funded by the FNAWS and the USDA Forest Service under the Master Memorandum of Understanding (MMOU) dated November 16, 1988.

Each guzzler was placed at the base of a rock outcrop so as to best facilitate water collection and escape habitat for the sheep. The guzzlers were placed approximately one mile from any permanent water source. The site is in a small, rocky valley at the top of a ridge that is used by bighorn sheep during the summer and fall. Transporting the guzzler to the site involved a 52-mile drive south from Douglas, Wyoming using four-wheel-drive (4WD) vehicles to get to within approximately one mile of the site. From the road to the site, the guzzler was carried by a four-person crew, up a 35 percent to 40 percent rocky timbered slope. This required 5-1/2 hours. Installation was completed by placing rocks under and around the guzzler. This area has no livestock use and therefore should be used exclusively by bighorn sheep and other wildlife.

The second location is at the southeast base of North Sawtooth Mountain. The site is located mid-slope on a rocky ridge leading up North Sawtooth Mountain. Transporting the guzzler to this site involved a 38-mile drive, and hand carrying the guzzler by a two-person crew. The topography in this area is gentler than the South Sawtooth area, but the site is still located approximately one mile from the nearest road. Once at the site, a shallow pit was dug to allow the guzzler to fit under a rock ledge and to allow better sheep access to the guzzler opening. Rocks were then piled around the base to support it. The transportation and installation of this guzzler required one day. This guzzler site is also inaccessible to livestock and therefore should be used exclusively by bighorn sheep and other wildlife. Information was provided by District Ranger Jack Cameron, USDA Forest Service, Wyoming. Photographs are by Tim Byer.

PROJECT 90-28: California Bighorn Sheep Lungworm Treatment, Oregon

California bighorn sheep at a worming block treatment site.

Placement of wormer blocks on site.

Many California bighorn sheep areas in southeast Oregon are experiencing poor lamb production and high lungworm loads. High lungworm numbers were confirmed through fecal analysis done by Washington State University. Biologists suspect the lungworm problem is the primary cause of low lamb production since the lambs are disappearing within the first month of life. The Department of Fish and Wildlife purchased a supply of fenbendazole wormer blocks and distributed them throughout the various California bighorn sheep ranges in southeast Oregon. Time-lapse photographic equipment has been placed at several of the treatment sites so it can record the extent of use by bighorn sheep. The real test of the success for this project will be determined in the next couple of years and will be measured by the degree of improvement, if any, in lamb recruitment within each herd. Information was provided by Assistant Staff Big Game Biologist Walt Van Dyke, Oregon Department of Fish and Wildlife, and photographs were provided by ODFW.

PROJECT 90-27: Pueblo Mountains-Alvord Peaks Radio Telemetry, Oregon

Side view of radio-collared California bighorn ram; capture crew preparing to remove the blindfold for release.

Radio-collared California bighorn ram upon release.

The project objective called for radio collaring 10 bighorn sheep within the roughly 600 square-mile range inhabited by 150 sheep to determine migration routes and herd ranges during a two-year sample period.

The capture technique utilized a Coda netgun fired from a Jet Ranger helicopter contracted by Idaho Helicopters of Boise. Sheep were handled by a set of muggers who untangled the nets and blindfolded, hobbled and attached the collars.

Ten sheep were captured in association with other project work on January 25 and 26 and February 8, 1991. Radio collars were attached to six adult ewes and four rams ranging in age from 1-1/2 to 7-1/2 years. Bighorn locations and movement patterns will be invaluable in the Bureau of Land Management (BLM) process of developing grazing allotment plans for the Pueblo Mountains. Fencing locations will be a major part of that process and a sound knowledge of bighorn use will help avoid impacts to the herds.

Financial support from FNAWS is greatly appreciated in this valuable cause to benefit bighorn sheep resources of southeastern Oregon. Information was provided by Assistant Staff Big Game Biologist Walt Van Dyke, Oregon Department of Fish and Wildlife. Photographs are courtesy of the ODFW.

PROJECT 91-09: Beckwith Plateau Bighorn Water Catchments, Utah

Two 1,150-gallon water catchments were constructed and are shown here. They were constructed on May 16, 1992.

Dramatic clouds and sandstone sculptures carved by wind were the backdrop for a successful water project constructed by volunteers from the Sportsmen for Quality Wildlife and staff from the Bureau of Land Management.

Two 1,150-gallon water catchments with exterior walk-in troughs were constructed specifically for Rocky Mountain bighorn sheep on May 16, 1992. The two units were flown into the remote locations of the Gray Canyon Wildland Management Area by helicopter on May 15. Eight volunteers from the Sportsmen for Quality Wildlife, a local sportsman group in the Price area, joined two people from the BLM to construct the units on the site.

The catchments were purchased with funds provided by the FNAWS. Flight time necessary to first select the water locations and to fly the materials and personnel was provided by the BLM. The sites for the units were selected jointly by the BLM and the Utah Division of Wildlife Resources. These sites were Last Chance Benches and Elliot Mesa. Information and photographs were provided by Utah BLM Wildlife Biologist David L. Mills.

PROJECT 92-55: Bighorn Sheep Range Informational Signs

In an attempt to better inform the public about the bighorn sheep program in Oregon as well as give credit to supporters of the program, the Oregon Department of Fish and Wildlife placed large signs at key locations at bighorn sheep transplant sites in three counties in the state. The text of the Winter Ridge informational sign reads: "California bighorn sheep were reintroduced to Winter Ridge in 1991 through a cooperative effort between the Foundation for North American Wild Sheep, Oregon Department of Fish and Wildlife, USDA Forest Service and private landowners."

Petroglyphs found in the area reveal the importance of bighorns to Native American culture. Bighorns were common in steep habitats around the Summer Lake Basin prior to Euroamerican settlement but disappeared during the early 1900s due to unregulated hunting and poor natural resistance to domestic livestock diseases.

This informational sign is one of several in three Oregon counties that tells residents and visitors of Oregon about the successful transplant program supported by FNAWS, Oregon Department of Fish and Wildlife, USDA Forest Service and private landowners. California bighorn sheep were transplanted into Oregon in 1991.

Look for these new residents in the steep rocky portions of winter ridge. Other areas around the Summer Lake basin also have had bighorns reintroduced. For information on these areas stop by the Summer Lake Wildlife Area Headquarters 1.5 miles south on Highway 31. Information provided by Klamath District Wildlife Biologist Ron Anglin, Oregon Department of Fish and Wildlife.

PROJECT 93-04: National Bighorn Sheep Interpretive Center, Wyoming

The Grand Slam, the wild sheep of North America in its final form.

"Sheep Mountain" scene depicting lambing areas.

The construction of the National Bighorn Sheep Interpretive Center (NBSIC) was completed June 28, 1993, with the official grand opening July 2, 1993. The total $25,000 Grant-in-Aid received from FNAWS was applied toward the construction and completion of the central exhibit, "Sheep Mountain." Total fabrication cost for this exhibit was $50,075. Since 1991, FNAWS has contributed $60,000 to the Sheep Center.

The central exhibit, "Sheep Mountain," will draw you into the world of the bighorn. It is 16-feet tall and features full-sized bighorns and many plants and other animals around them. It includes interpretive scenes about predator-prey relationships, lambing areas and dominance fights.

Just outside Dubois, Wyoming, stands Whiskey Mountain, home to the largest Rocky Mountain bighorn sheep herd in North America. A unique feature of Whiskey Mountain and the Dubois area is that it is possible to view the bighorn sheep up close. FNAWS has continued its support of the NBSIC by contributing $15,000 toward the construction of another exhibit entitled, "Managing the Whiskey Mountain Bighorn Sheep Herd." This exhibit is a unique, hands-on display that allows further education regarding bighorn sheep and wildlife management. The exhibit, designed with an interplay computer system, poses hypothetical questions to the visitors and lets them respond with their "management" decisions.

The crucial role of controlled, ethical hunting will be stressed as one of the most effective management tools in sustaining healthy, viable wildlife populations. Information and photographs are courtesy of the National Bighorn Interpretive Center in Dubois, Wyoming.

PROJECT 95-02: Wind River Canyon Reintroduction, Wyoming

Setting up the sheep trap on Whiskey Mountain while an eager crowd awaits.

The Burlington Northern Railroad employees loaded the Ewe Hauls onto a flatbed rail car for transportation to the mouth of Teeter Canyon.

It was January 7, 1995, when a train-full of wild critters and wildlife biologists prepared to leave a station just below Boysen Reservoir. Their destination was the northeast corner of the Wind River Indian Reservation. The critters were bighorn sheep. The biologists were from the Wyoming Game & Fish Department and the USDA Fish and Wildlife Service. The train was a Burlington Northern special dispatch.

The animals had been trapped south of Dubois that morning. They were being taken to the reservation for release into the Owl Creek Mountains. Wildlife enthusiasts hoped the train ride resulted in the return of a viable bighorn sheep population to native habitat that has gone unoccupied for more than 50 years.

"Burlington Northern provided a great assist in this cooperative operation," said Dick Baldes, a USDA Fish and Wildlife Service biologist. "Due to the track running on the west side of the Wind River, the train was able to get the sheep much closer to good winter range than trucks could have that time of the year."

Baldes, who advises the Shoshone and Arapahoe tribes on wildlife issues, says the release area, with its deep canyons and many cliffs, is excellent bighorn habitat, but he insists there is an additional benefit to the location.

"The extra beauty of this release site is that the canyon can be seen from the highway (U.S. 20)," he said. "We expect at least a few of the sheep to be visible almost year round."

Since bighorns aren't always visible, biologists will track their movements with radio collars donated by the FNAWS. Eight of the transplanted animals are wearing the high-tech devices. Another eight are sporting neck bands decorated by children from Fort Washakie and Thermopolis. A lot of people were involved in the bighorn sheep transplant.

"It's a great event when a large national corporation like Burlington Northern works with public and private wildlife organizations and with individuals," said Robb Hitchcock, a veteran conductor who coordinated the railroad company's involvement. "Burlington Northern was a proud partner in this historic event."

In the 43 years the Wyoming Game and Fish Department has trapped Whiskey Mountain bighorns, this was the first time the animals were shipped by rail. In the past, they've been transported by road to Nevada and Utah and to Wyoming destinations including the Black Hills, Shell Canyon and the Wyoming Range west of Big Piney.

Cody Beers wrote this information in the June 1995 Wyoming Wildlife magazine. Photographs were provided by Wyoming Project Leader Richard Baldes.

PROJECT 96-20: Midway Mountain Bighorn Sheep Water Enhancement, California

Hauling the 6,000-gallon fiberglass water storage tank up the wash to the site of drinker Number One.

Pouring the concrete steps in drinker Number Two.

During May 1996, Desert Wildlife Unlimited, Inc. (DWU) was notified they received a $16,000 grant from FNAWS. These funds were to be used to help construct two artificial wildlife drinkers in the Midway Mountains, Imperial County, California.

The two proposed drinkers were to be located in the Midway Mountains within the general location of the East Chocolate Range. The drinker design was developed by DWR and has been utilized in nearly every Western state, in Mexico and most recently, in Mongolia. The basic system that was installed in both locations included a 6,000-gallon fiberglass tank; a six-foot fiberglass step-down drinker; a concrete dam built at the designated water collection point; and a 6-inch ABS and galvanized steel piping to collect and transport the rain water runoff to the tank and drinker. Each water collection system was designed to store 7,100 gallons of water.

Midway Mountain Drinker One was installed August 1996, over a period of three weekends. DWU members from California and Arizona donated their time. An excess of 20 individuals worked on the project in temperatures approaching 114 degrees. All equipment and supplies were transported over a three-day period prior to the actual construction of the drinker. This involved the use of five 4WD pickups and a 4WD backhoe.

The first step was to excavate an area so that the drinker and storage tank could be set into the ground. The storage tank must be covered to protect it from degradation by ultraviolet light. The drinker and storage tank are connected by 2-inch flexible piping and are set into the ground at the same elevation.

While the drinker and storage tank were being backfilled, workers constructed a concrete dam approximately 6-feet long and 18-inch high at the water collection point. Galvanized steel pipe, measuring 80 feet, was installed between the collection dam and the drinker and was held in place by upright posts cemented into the ground. Six-inch ABS piping was used to connect the steel pipe to the drinker and the drinker to the storage tank.

Concrete steps were poured in the drinker to provide access for wildlife. Because this area contains wild burros, a burro exclusion fence will be constructed approximately one year after the mountain sheep begin to use this water source. The areas impacted by the construction were landscaped and native vegetation was planted. This drinker is fully operational and needs only rainfall to fill it.

Midway Mountain Drinker Number Two was completed during September 1996. Nine DWU members worked to complete this project over two weekends. The construction process was similar to that of Drinker Number One. These drinkers are 100 percent operational. Information and photographs were provided by Leon Lesicka of Desert Wildlife Unlimited, Inc., California.

PROJECT 94-55: Dean Bowdoin Tinaja Wildlife Water Development, Arizona

This mountain was the site of the Dean Bowdoin Water Development Project.

This project is located in the Saddle Mountains about 50 miles west of Phoenix. This project required the complete renovation of an existing but unreliable water development. The Saddle Mountain area has a thriving desert sheep population, but it has been declining for some time. This project was a major water development and the final cost was over $30,000. National FNAWS along with the Minnesota-Wisconsin and Eastern Chapters assisted in funding. The development site is a steep narrow gorge. There is natural water in the winter but it would normally dry out by mid-June. If there were no summer rains, there was no water at all for the sheep. A 12-foot high dam was constructed of concrete blocks and reinforced with rebar and the holes filled with concrete. A steel framework over the dam was constructed that was covered by 2-by-12-foot planks. This cover will reduce evaporation by a third. Upstream a gabon was constructed to block the entrance of contaminating rocks and debris into the pool itself. Materials were flown in by helicopter. Information and photograph was provided by FNAWS Director Duncan Gilchrist of Montana.

PROJECT 96-50: Petty Rock Bighorn Sheep Prescribed Burn, Montana

This project was initiated seven years ago. The concern was that forage areas had suffered a significant loss in productivity from over maturity and decadence of important browse plants and conifer encroachment into foraging areas due to 60 years exclusion of wildfire and planned fire ignition. Surveys of animal use indicated heavy overuse on portions of the suitable range and minimal use on a large percentage of the remaining available habitat. Treatments in 1992 resulted in the direct improvement of more than 1,000 acres of bighorn sheep range for more than 120 head of bighorns and numbers of elk, mule deer and whitetail deer. This treatment has improved forage production and reduced conifer encroachment that will benefit sheep and other wildlife species over the next 20 to 30 years. Benefits to these animals are expected to include increased reproductive rates, improved animal health, reduced parasitism, improved forage supplies and reduced predation. Information was provided by District Ranger Greg Munther, USDA Forest Service, Montana.

Because of the rugged terrain and access problems, a helicopter is the only feasible method for application of fire to planned burning units.

Becky Layne has been an employee of the Foundation for North American Wild Sheep since 1991. She enjoys her job as Grant-in-Aid coordinator because she sees first-hand how FNAWS benefits wild sheep. Becky is an avid hunter and after attending the 1996 "Becoming an Outdoor Woman" clinic, she took an interest in bow hunting. She and her husband both grew up in Wyoming and now reside in Cody. They enjoy camping and fishing with their two daughters, Ashley and Emily. Becky is also a Girl Scout leader for her daughter Ashley's troop. She feels it is extremely rewarding to encourage young minds to appreciate the benefits of conservation and outdoors.

All Funded Grant-in-Aid Projects of FNAWS 1983 to 1998

Becky Layne

The following section is a list of all funded Grant-in-Aid projects of the Foundation for North American Wild Sheep. This listing includes all projects receiving funds from FNAWS from 1983 to 1998, listed by the state, province or country where the money was sent. States are listed first in alphabetical order, followed by Canadian provinces and Mexico.

Projects receiving funds prior to 1983 do not exist in the National FNAWS computer records. The project number starts with the year of funding, so a project listed as 83-05 was funded in 1983, and so on. The numbers to the right of the year indicates its identification code. This listing includes the title of the project and the amount given.

ALASKA

Project 83-05 "Michigan Conservation Fund" $17,950
Project 83-14 "Wildlife Conservation Fund" $10,000
Project 83-25 "Citizens For Equal Rights" $10,000
Project 83-27 "Alaska Equal Hunt/Fish Rights" $5,000
Project 83-28 "Michigan United Conservation Fund" $16,500
Project 83-32 "Alaska Equal Hunt/Fish Rights" $3,254.22
Project 83-48 "Wildlife Legislative Fund of Alaska" $35,000
Project 83-50 "Benefit Dall's Sheep Hunting" $13,500
Project 83-75 "Conservation Fund" $40,000
Project 83-83 "Alaska Fish and Game Computer Loan" $7,000
Project 84-111 "Creating an Effective Statewide Alaska Hunter" $20,000
Project 85-09 "Winter Forage Ecology of Dall's Sheep in Brooks Range" $5,500
Project 85-13 "APHA Operations Funding" $5,000
Project 85-26 "Alaska Outdoor Council" $25,000
Project 87-17 "Winter Foraging Ecology of Dall's Sheep" $2,500
Project 87-43 "Reimbursement of Hunt Donation Monies" $6,250
Project 88-57 "Creating Effective Alaskan Hunters, Fishermen and Outdoor Recreationists" $20,000
Project 89-04 "Creating Effective Alaskan Hunters, Fishermen and Outdoor Recreationist" $15,000
Project 89-36 "Management of Dall's Sheep Full Curl vs. Permits" $5,000
Project 89-37 "Study of Wolf Predation on Dall's Sheep" $10,000
Project 89-38 "Study of Dall's Sheep Habitat and Population" $10,000
Project 89-48 "Dall's Ram Movements and Rutting Ecology" $3,600
Project 90-42 "Public Education of Guide Areas for Conservation" $35,000
Project 91-61 "APHA" $30,000
Project 92-48 "Establish New State Wide Comprehensive Guide Area Systems" $30,000
Project 93-EM06 "Legal Assistance in Alaska" $10,000
Project 94-EM01 "NW Arctic Borough Zoning Ordinance Issue" $15,000
Project 95-48 "Defense of State Guide Area System" $10,000

Project 96-17 "Sheep Predator Reduction in Tanana/Yukon Uplands" $15,000
Project 96-39 "Conservation of Alaska Game Resources Through Regulations" $10,000
Project 98-62 "Wildlife Management by the Ballot Box" $5,000
TOTAL FOR ALASKA *$446,054.22*

ARIZONA

Project 83-09 "Arizona Desert Bighorn Transplant" $24,000
Project 83-10 "Waterhole Project" $4,244
Project 83-21 "Free Release Transplant" $6,500
Project 83-33 "Arizona Waterhole" $12,806
Project 83-34 "Video Taping Waterhole Project" $5,756
Project 83-70 "Desert Sheep Transplant - Lake Meade" $10,000
Project 85-34 "White Mountain Apache Tribe/Technical Advice" $9,300
Project 85-60 "Mesa Desert Bighorn Sheep Reintroduction Program" $8,554
Project 85-85 "Arizona Sheep Society" $10,000
Project 89-27 "Ghost Flower Trick-tank" $4,000
Project 89-28 "Black Mountains Desert Bighorn Sheep Water Maintenance" $5,000
Project 89-29 "Muggins Mountains Tinaja Improvement Project" $3,000
Project 89-67 "South Needles Pothole" $9,569.44
Project 90-47 "Bunyon Peak Tank" $9,878.98
Project 90-53 "Arizona Arvapia Canyon Study" $19,844.38
Project 91-10 "Hunts Tank" $8,000
Project 92-64 "Chocolate Mountain Catchment" $8,000
Project 92-EM02 "Proposition 200 - Arizona" $25,000
Project 92-EM05 "Arizonians for Wildlife Conservation" $10,000
Project 93-23 "Jeff Sievers Tank" $8,000
Project 93-65 "Silver Bell Mountain Sheep Management Investigation" $10,000
Project 93-66 "Apex Slope Allotment" $15,000
Project 94-55 "Dean Bowdin Tank" $13,301.19
Project 94-64 "Silver Bell Mountain Sheep Management Investigation" $10,000
Project 96-02 "Silver Bell Mountain Sheep Management Investigation" $10,000
TOTAL FOR ARIZONA *$259,753.99*

CALIFORNIA

Project 83-36 "California Lamb Mortality" $6,000
Project 83-39 "California Waterhole" $9,450
Project 83-44 "California Bighorn Film" $20,000
Project 83-45 "California Lawsuit - N206" $3,000
Project 83-53 "Inbreeding" $8,500
Project 83-58 "Lamb Mortality" $2,500
Project 83-61 "Sheep Population Ecology" $9,500
Project 84-112 "Bighorn Capture and Reintroduction" $15,000
Project 84-152 "Population Ecology of Bighorn Sheep in Owens Valley" $9,500
Project 85-25 "Population Ecology of BH in Owens Valley" $5,000
Project 85-27 "Disease Investigation of Bighorn Sheep" $45,000
Project 88-28 "Characterizing Pasteurellas From Bighorn Pneumonias" $5,000

Project 88-36 "Population Ecology and Sexual Segregation of Mountain Sheep" $7,500
Project 88-58 "Ecology Investigations on Desert Bighorn Sheep in Baja, California" .. $42,500
Project 89-35 "Investigation of the Bighorn Sheep in Baja" $22,500
Project 89-66 "Big Game Guzzler" $6,500
Project 90-48 "California Water Project" $4,000
Project 90-EM05 "Anza-Borrego Desert State Park 100 Plus T-Shirts" $410
Project 90-EM09 "Yearly Census" $3,000
Project 91-11 "California Waterhole Project" $5,000
Project 91-39 "Population Ecology and Sexual Segregation in Old Dad Peak" $5,000
Project 92-46 "Little San Bernardino Mountain Bighorn Water Project" $10,000
Project 92-47 "Population Ecology and Sexual Segregation" $5,500
Project 92-65 "California Water Project; Kerr Guzzler" $5,000
Project 93-16 "Bighorn Sheep Interpretive Signs" $2,334
Project 93-22 "California Water Project" $4,000.02
Project 95-43 "Guadalupe Canyon Big Game Guzzler" $3,372.85
Project 95-EM01 "Replace Stolen Specially Modified Video Cameras" $7,967.23
Project 96-20 "Midway Mountain Bighorn Sheep Water Enhancement" $16,000
Project 96-34 "Cushenbury Herd Monitoring Project" $4,000
Project 96-51 "Californians for Balanced Wildlife Management" $20,000
Project 98-33 "San Gabriel Mountains Bighorn Sheep Habitat Improvement" $7,500
Project 98-40 "E. Chocolate Mountain Bighorn Sheep Water Enhancement" $10,000
TOTAL FOR CALIFORNIA $330,534.10

COLORADO

Project 83-51 "Stress in Bighorns" $10,000
Project 84-118 "Habitat Improvement" $3,000
Project 84-150 "Hunters Sports Fund" $1,000
Project 85-10 "Effects of Forage Availability on Bighorns" $3,000
Project 85-16 "Waterton Canyon Prescribed Burn" $5,500
Project 88-06 "Basalt State Wildlife Area Bighorn Sheep Guzzler" $2,595
Project 88-07 "Gore-Eagle's Nest Range-Bighorn Sheep Study" $2,400
Project 88-11 "Nutritional Ecology of Bighorn Sheep on Winter, Spring and Summer Ranges" $3,000
Project #89-43 "Develop Brochure on Wild Sheep on USFS System Land" $4,500
Project 89-52 "Tarryall Habitat Improvement Project" $4,000
Project 90-16 "Battlement Mesa Bighorn Sheep Habitat Improvement" $5,000
Project 91-21 "Evaluation of Desert Bighorn Habitat Model/Transplant Site" $4,500
Project 91-22 "Evaluation Effectiveness of FNAWS Funding Program" $8,700
Project 92-28 "Bighorn Sheep Habitat Improvement/Prescribed Burn" $3,000
Project 92-30 "Natural Arch Project" $2,000
Project 92-31 "Battlement Mesa Ridgetop Travel Lanes" $5,000
Project 92-32 "North Tarryall Peak Tree Overstory Removal" $1,000
Project 92-33 "Toner Creek Bighorn Habitat Improvement" $2,310
Project 92-35 "Battlement Mesa Bighorn Sheep Guzzlers" $4,000
Project 92-49 "Population Characteristics of St. Vrain Herd" $2,500
Project 92-EM04 "Colorado for Wildlife Conservation" $5,000
Project 93-01 "Ouray Bighorn Sheep Recovery Program" $5,000

Project 93-19 "Population Characteristics of St. Vrain Herd" $2,000
Project 93-29 "Ouray Bighorn Sheep Habitat Improvement" $4,000
Project 93-30 "Rampart Range Wildlife Habitat Improvement Project" $4,833.29
Project 93-32 "Battlement Mesa Ridgetop Travel Lanes Number Two" $2,500
Project 93-33 "Avalanche Creek Bighorn Habitat Improvement" $750
Project 94-47 "Battlement Mesa Ridgetop Prescribed Burn" $8,500
Project 94-48 "Almont Bighorn Sheep Prescribed Burns" $4,000
Project 94-49 "South Fork Prescribed Burn" $2,200
Project 94-EM11 "AG in the Classroom" $10,000
Project 95-12 "Deer Pen Gulch Bighorn Sheep Range Fertilization" $2,500
Project 95-15 "Pollock Bench Burn" $1,669.98
Project 95-20 "Almont Bighorn Sheep Prescribed Burn" $4,000
Project 95-22 "Amphitheater Interpretive Site" $2,500
Project 95-27 "Love Ranch Meadow" $7,500
Project 96-44 "COA Kids Kamp" $1,600
Project 96-56 "Love Ranch Meadow" $4,000
Project 98-04 "COA Kids Kamps" $1,600
Project 98-55 "Research on Long Term Effects of Prescribed Fire" $5,000
Project 98-64 "Tarryall Bighorn Sheep Study" $5,000
TOTAL FOR COLORADO $161,158.27

IOWA

Project 83-67 "Embryo Transplants in Wild Sheep" $8,750
TOTAL FOR IOWA $8,750.00

IDAHO

Project 83-63 "Translocation of California Bighorn" $7,000
Project 83-64 "Rocky Mountain Bighorn Translocation" $3,500
Project 84-115 "WILD Programs Unlimited Outdoor Education Symposium" $1,000
Project 84-137 "California Bighorn Sheep Relocation in Idaho" $22,000
Project 84-154 "Identification of Disease Affecting Bighorn Sheep" $2,500
Project 85-23 "Hells Canyon Disease Investigation" $8,950
Project 85-42 "Trapping and Transplanting of BH Sheep in Idaho" $15,807.43
Project 85-57 "Challis National Forest Burn" $1,700
Project 85-59 "Salmon National Forest Bighorn Sheep Enhancement" $9,200
Project 85-68 "Idaho Forest and Habitat Protection" $2,000
Project 85-75 "Biologists Workshop" $11,989
Project 87-27 "Panther/Clear Creek Bighorn Sheep Winter Range" $9,200
Project 87-31 "Goddard Canyon Wildlife Guzzlers" $1,090
Project 88-13 "Timber Canyon Guzzler" $1,390
Project 88-16 "Skull Canyon Wildlife Guzzler" $1,390
Project 88-44 "Bighorn Sheep Population Ecology and Migration Patterns" $10,000
Project 88-45 "WILD Programs Unlimited Pro-sportsman Seminar" $12,000
Project 88-53 "California Bighorn Sheep Capture and Transplanting" $7,000
Project 89-06 "Ramshorn Canyon Guzzler" $3,660
Project 89-07 "Burnett Canyon Winter Range Burn" $1,200

Project 89-13 "Assessing Rocky Mountain Bighorn Ewe-Lamb Herd Composition" $4,000
Project 89-17 "WILD Programs Unlimited, Pro-sportsman Symposium" $2,500
Project 89-EM03 "Assessing Rocky Mountain Bighorn Ewes" $4,000
Project 89-EM04 "Investigation of Mortality in Big Creek Herd" $4,000
Project 90-38 "Big Creek Wildlife Disease Study" $12,000
Project 90-54 "Saylor Creek Bombing Range" $1,000
Project 90-EM08 "Saddle Low Military Operating Area" $2,500
Project 91-17 "Trap and Transplant Bighorn Sheep" $5,400
Project 91-58 "Big Creek Wildlife Disease Study" $10,000
Project 91-EM05 "Bomb Range" $5,000
Project 92-38 "Lost River Mountain Range Guzzler Project" $10,381
Project 92-58 "Idaho Public Lands Protection Project" $5,000
Project 92-EM7 "Nils Nokkentved Written Report/Bomb Range" $1,000
Project 93-03 "Bighorn Sheep Transplant Vehicle and Trailer" $3,000
Project 93-13 "Bighorn Sheep Hunter Education" $450
Project 93-15 "California Bighorn Trap and Transplant" $30,000
Project 93-40 "Contribution of IOGA Industry to Idaho Economy" $10,000
Project 93-48 "Health Laboratory Research Sheep Facility" $4,000
Project 93-EM01 "Protection of California Bighorns-Big Springs Bomb Range" $28,268
Project 93-EM11 "An Idaho Hunters Rally" $2,000
Project 93-EM13 "Dr. DiGrazia's Presidential Gift Donation" $1,415
Project 94-31 "Contribution of Outfitting and Guiding to Idaho Economy" $10,000
Project 94-EM06 "Immunization of Sheep with a Multivalent Vaccine" $3,000
Project 95-54 "Bighorn Sheep Trapping and Transplant" $8,933.18
Project 95-60 "Foraging Behavior of Bighorn Sheep in Central Idaho" $3,500
Project 95-EM02 "Idaho Bear Initiative" $10,000
Project 96-EM05 "Idaho Bear Initiative" $5,000
Project 98-47 "Owyhee Canyonlands/The 45 Ranch" $10,000
Project 98-57 "Pasteurella Vaccine and Delivery Method Research" $3,700
TOTAL FOR IDAHO $332,623.61

MONTANA

Project 83-17 "Bighorn Social Behavior" $5,000
Project 83-59 "Bighorn Study - Petty Drainage" $470
Project 84-130 "Mineral Analysis/Bighorn" $2,500
Project 87-41 "Reestablishment of Rocky Mountain Bighorn Sheep in
Taylor-Hilgard Range" $9,000
Project 88-25 "Reestablishment of Rocky Mountain Bighorn Sheep in
Taylor-Hilgard Range" $8,000
Project 89-05 "Sheep Pass Guzzler" $3,000
Project 89-10 "Reestablishment of Rocky Mountain Bighorn Sheep in
Taylor-Hilgard Range" $7,700
Project 90-18 "Bighorn Sheep Habitat Improvement" $7,000
Project 90-33 "Cinnabar Mountain Bighorn Sheep Roadside Exhibit" $500
Project 90-35 "Wildhorse Island Bighorn Sheep Relocation" $2,000
Project 90-36 "Ten Lakes Bighorn Sheep Survey" $2,000
Project 91-12 "Petty Creek Bighorn Sheep Project" $9,000

Project 91-23 "Bighorn Sheep Range Water Developments" $4,500
Project 91-48 "Fence Removal - Buffalo Horn Pass" $728
Project 91-EM01 "Madden Grazing Lease" $1,500
Project 92-EM03 "Montana Habitat Survey" $5,000
Project 93-06 "Rocky Mountain Bighorn Sheep Population and Habitat Monitoring" . $2,000
Project 93-37 "Bighorn Habitat Availability Along the Rocky Mountain East Front" $1,135
Project 93-38 "Stillwater Bighorn Habitat Improvement Project" $3,000
Project 93-EM04 "Madden Lease Emergency Funding" $2,000
Project 93-EM05 "Emergency Mongolian Funding" $2,000
Project 94-24 "Mongolian Wild Sheep Investigations and Management" $3,000
Project 94-58 "Lawsuit - Sula Grazing Licenses" $4,000
Project 94-60 "Project Access" $9,810.22
Project 94-65 "Rock Creek Ranch Conservation Easement" $5,000
Project 94-EM15 "Relocation of Bighorns on Salisa Kootenai Reservation" $10,000
Project 95-16 "Munson Habitat Enhancement Project" $5,750
Project 95-25 "Movements and Habitat Use by Bighorns on the North Winter Range" $1,500
Project 95-56 "Mongolian Wild Sheep Investigations and Management" $2,000
Project 96-22 "Flathead Reservation Sheep Relocation Project" $5,000
Project 96-50 "Petty Rock Bighorn Sheep Prescribed Burn" $5,000
Project 96-55 "Community Based Inventory and Management of Argali" $5,000
Project 96-EM01 "Madden Lease Emergency Grant" $1,436.66
Project 98-17 "Health/Habitat Survey of Montana Spanish Peaks Bighorn Sheep" $1,875
Project 98-31 "Ford/Fairview Prescribed Burn" $5,000
Project 98-67 "Burgess Miller Ranch Habitat Restoration Project" $12,000
TOTAL FOR MONTANA $154,404.88

NORTH DAKOTA

Project 87-01 "Trapping and Transplanting California Bighorns in North Dakota" $6,000
Project 88-38 "Maintenance & Enhancement of California Bighorn in North Dakota" $10,000
Project 89-44 "Maintenance and Enhancement of California Bighorn in North Dakota" $7,500
Project 98-05 "Supplemental Bighorn Sheep Transplant" $8,000
TOTAL FOR NORTH DAKOTA $31,500

NEW MEXICO

Project 83-12 "New Mexico Desert Sheep Transplant" $4,000
Project 83-55 "Re-establishment of Bighorns" $2,000
Project 85-46 "Big Hatchet Mountains Water Catchments" $6,500
Project 85-48 "Peloncillo Mountains Pothole Developments" $2,000
Project 89-34 "Economic Study of New Mexico Hunting, Fishing and Outdoor Recreation" $10,000
Project 89-61 "Outfitters for Youth" $2,500
Project 90-51 "San Andres Bighorn Project" $6,000
Project 92-EM06 "Operation Game Thief" $500
Project 93-36 "Use of Salt to Alter Habitat Use in Desert Sheep" $3,282.56

Project 94-32 "Kiwanis Meadow Restoration" $3,000
Project 96-10 "Piedra Lisa Water Development" $988.76
Project 96-26 "Devil's Backbone Wildlife Waters" $9,000
Project 96-40 "Gap Spring Development" $2,500
Project 98-25 "1998 Desert Bighorn Council Meeting" $1,500
Project 98-27 "Pecos Wilderness Bighorn Capture / Transplant" $13,600
Project 98-35 "West Ladron Burn" $5,000
TOTAL FOR NEW MEXICO $72,371.32

NEVADA

Project 83-30 "Nevada Game and Fish" $21,000
Project 83-56 "Helicopter Replacement" $10,000
Project 84-153 "Hansen - Desert Bighorn" $1,000
Project 85-38 "BLM Desert Bighorn Water Developments" $12,000
Project 85-47 "Water Project Survey and Design" $3,700
Project 87-05 "Paiute-Excelsior Guzzlers" $4,000
Project 87-14 "Ely BLM District Desert Bighorn Sheep Water Projects" $9,600
Project 88-14 "Habitat Survey for White Pine Desert Sheep" $2,775.48
Project 89-08 "White Pine Desert Bighorn Sheep Water Development" $7,547.53
Project 89-63 "Nevada Waterholes" $5,000
Project 90-10 "Quinn Range Preparatory Work for Reintroduction of Desert Bighorn Sheep" $3,980
Project 90-57 "Bighorn Sheep Trapping and Transplant Program" $17,000
Project 90-EM02 "Muddy Mountains Slick Rock Project" $4,350
Project 91-01 "Sheep Pass Bighorn Water Development" $7,000
Project 91-59 "Frozen Toe Number One" $2,000
Project 93-47 "Spotted Range Water Developments" $9,000
Project 93-EM12 "Dr. DiGrazia's Presidential Gift Donation" $1,415
Project 94-EM02 "Spector Mountain Waterhole" $3,672.90
Project 94-EM10 "Muddy Mountains - 5 Rim Water Project" $3,000
Project 95-29 "Currant Mountain Desert Bighorn Sheep Study" $7,883.64
TOTAL FOR NEVADA $135,924.55

OREGON

Project 84-145 "Minam River Bighorn Restoration" $5,000
Project 87-15 "Sharon Creek / Warner Rim Bighorn Sheep Restoration" $5,000
Project 87-18 "Red Butte / Iron Mountain Bighorn Sheep Restoration" $5,000
Project 87-20 "Conservation of Genetic Resources in Bighorn Sheep" $5,000
Project 88-39 "Lower John Day / Van Horn Creek Bighorn Sheep Restoration" $9,000
Project 89-21 "Diablo Mountain Water Development" $5,000
Project 89-31 "Southeast Oregon Bighorn Sheep Restoration" $5,000
Project 90-27 "Pueblo Mountains Telemetry Project" $3,000
Project 90-28 "California Bighorn Sheep Lungworm Treatment" $2,910
Project 90-29 "Prescribe Burn-Silver Creek Winter Range" $2,550
Project 90-32 "Rocky Mountain Bighorn Sheep Reintroduction" $1,250
Project 91-52 "Lower Imnaha Forage Improvement" $9,000

Project 91-53 "Lone Mountain Release" $6,000
Project 92-52 "Orijana Canyon Guzzler" $2,500
Project 92-53 "Three Mile Rim" $6,000
Project 92-55 "Bighorn Sheep Range Informational Sign Program" $1,200
Project 93-08 "Winter Rim Water Development" $3,500
Project 93-57 "Orijana Rim Guzzlers" $3,200
Project 93-58 "Daughtery Rim Guzzlers" $5,200
Project 93-59 "Daughtery Rim Transplant" $3,400
Project 93-60 "Deadhorse Ridge Bighorn Transplant" $1,929
Project 93-61 "Imnaha Spring Development" $1,964
Project 93-64 "California Bighorn Sheep Lungworm Treatment" $1,600
Project 94-34 "California Bighorn Helicopter Surveys" $512
Project 94-35 "California Bighorn Sheep Lungworm Treatment" $1,615
Project 94-36 "South East Oregon Bighorn Sheep Restoration" $5,200
Project 95-30 "95/96 South East Bighorn Sheep Restoration" $72.55
Project 95-32 "California Bighorn Helicopter Surveys" $2,536
Project 95-33 "California Bighorn Sheep Lungworm Treatment" $1,581.25
Project 95-42 "Acquisition - Sargent Tract Addition" $5,000
Project 96-35 "Wildlife Educational Museum" $8,200
Project 96-36 "California Bighorn Helicopter Surveys" $2,232
Project 96-38 "1996-97 California Bighorn Sheep Restoration" $3,897.77
Project 96-EM03 "Oregon Initiative" $10,000
Project 97-EM01 "Genetic Discrimination of Domestic and Wild Exotic Sheep" $1,000
Project 98-19 "Southeast Bighorn Sheep Restoration" $10,000
Project 98-20 "Steens Mountain Bighorn Sheep Transplant" $3,500
Project 98-21 "Lower Deschutes Foraging Seeding" $2,000
Project 98-22 "Alder Creek Habitat Improvement Project" $2,000
Project 98-23 "Retrofit 2 Winter Rim Bighorn Sheep Guzzlers" $2,000
TOTAL FOR OREGON* *$155,549.57

SOUTH DAKOTA

Project 83-47 "Custer State Park" $2,500
Project 84-144 "Custer State Park" $4,000
Project 85-52 "Reproduction and Lamb Mortality in Bighorn Sheep" $2,000
Project 98-11 "Population Dynamics of Bighorn Sheep in the Black Hills" $12,000
TOTAL FOR SOUTH DAKOTA* *$20,500

TEXAS

Project 83-38 "Sheep Conservation" $20,000
Project 83-52 "Sinusitis - Desert Sheep" $2,500
Project 83-60 "Horn Abnormalities - Dall's Sheep" $3,290
Project 87-22 "Physiological Evaluation of Desert Bighorn Sheep" $6,500
Project 88-60 "High Lonesome Peak Water Project" $8,550
Project 93-17 "Development of Assisted Reproductive Techniques" $18,772.95
Project 94-EM07 "Apprentice Hunter Program" $3,000
Project 94-EM12 "Wildlife Expo" $800
TOTAL FOR TEXAS* *$63,412.95

UTAH

Project 83-18 "Waterhole - Utah" $1,000
Project 83-26 "Sheep Conservation" $22,500
Project 83-49 "Horn Aberrations in Dall's Sheep" $1,482
Project 83-65 "Locating and Data Gathering of Transplanted Population" $4,500
Project 83-73 "Desert Sheep Transplant" $32,000
Project 84-123 "Bunch Sheep Study" $9,000
Project 85-36 "Research of Biology and Habitat of Nelsons Bighorn" $10,000
Project 87-12 "Challis National Forest 1987 Projects" $3,310
Project 88-08 "Potash-Confluence Water Developments" $5,000
Project 88-09 "Red Canyon Water Developments" $5,000
Project 88-15 "Green River Corridor Prescribed Burn" $4,500
Project 89-03 "North Maxie Canyon Guzzlers" $5,589
Project 89-26 "Potash/San Juan Water Catchments" $12,000
Project 89-EM01 "Price River Resource Area" $5,000
Project 89-EM02 "Antibiotic Pellets" $92.42
Project 90-22 "Flaming Gorge Lakeside Burns" $3,500
Project 91-06 "Red Canyon Spring Development Maintenance" $2,000
Project 91-08 "King's Point Guzzler" $3,500
Project 91-09 "Beckwith Plateau Bighorn Water Catchment" $6,000
Project 92-07 "Desolation-Gray Bighorn Radio Telemetry Study" $1,800
Project 92-08 "Potash-Confluence Habitat Plan Bighorn Catchments" $5,000
Project 92-10 "Desert Bighorn Disease Study and Recovery Plan" $12,500
Project 92-37 "Sheep Creek Prescribed Burn" $3,000
Project 92-42 "Flaming Gorge NRA Wildlife Kiosks" $4,000
Project 93-20 "Hoop Lake Conservation Easement" $6,000
Project 93-21 "Rattlesnake Aum Purchase" $30,000
Project 94-37 "Desert Bighorn Sheep Disease Study and Recovery Plan" $5,000
Project 94-53 "Rattlesnake AUM Purchase" $26,224
Project 94-54 "Hoop Lake Conservation Easement" $6,000
Project 94-EM08 "Wolf Papers" $300
Project 95-35 "Floy Creek AUM Conversion" $10,000
Project 95-36 "Jenson AUM Conversion" $10,000
Project 95-37 "Sportsmen for Fish and Wildlife" $3,000
Project 96-31 "Jorgenson AUM Conversion" $7,000
Project 96-32 "Jensen AUM Conversion" $13,000
Project 96-33 "Sportsmen for Fish and Wildlife" $4,000
Project 97-44 "Jorgenson AUM Conversion" $15,000
Project 98-44 "Jorgenson AUM Conversion" $15,000
Project 98-45 "Sportsmen for Fish and Wildlife" $4,000
TOTAL FOR UTAH $316,797.42

WASHINGTON

Project 87-51 "Washington Landowner" $5,250
Project 90-17 "Pneumonia in Bighorn Sheep" $6,000
Project 90-44 "Mt. Hull California Bighorn Sheep Range Underburn" $1,750
Project 93-18 "Mt. Hull Radio-Telemetry" $2,250

Project 93-50 "Development. of Standardized Cytotoxicity Test for Pasteurella" $6,000
Project 93-51 "Pneumonia in Bighorn Sheep-Pasteurella Haemolytica" $6,000
Project 94-02 "Swakane Canyon Bighorn Sheep Research" $3,674.95
Project 94-40 "Identification of Bighorn Populations w/Genetic Resistance" $9,800
Project 94-41 "Susceptibility of Bighorns and Dall's Sheep to Pasteurella" $11,000
Project 95-05 "Little Vulcan Mountain" $1,443.50
Project 95-28 "Susceptibility of Bighorns and Stone's Sheep to Pasteurella" $5,200
Project 96-25 "Evaluation of Specific Strains of Pasteurella" $5,916.50
Project 98-29 "Controlling Noxious Weeds on Mt. Hull" $2,600
TOTAL FOR WASHINGTON $66,884.95

WYOMING

Project 83-01 "Documentation of Bighorn Pinkeye" $400
Project 83-02 "Treatment of Pinkeye in Bighorns" $2,500
Project 83-19 "Bighorn Sheep Study" $25,435
Project 83-57 "Trout Peak Bighorn Sheep Study" $26,278
Project 83-66 "Stress in Bighorns" $16,000
Project 83-72 "Educational Center Property Fund" $36,300
Project 83-80 "Shoshone Habitat Protection" $8,336
Project 84-128 "Water for Wildlife" $5,000
Project 84-132 "Heart Rate Response" $10,909
Project 84-135 "Park County Resource Council" $35,000
Project 84-141 "Trout Peak (2)" $20,000
Project 85-03 "Habitat Improvement and Ecology of Absaroka Bighorns" $10,000
Project 85-28 "Reintroduction of Bighorns in Wyoming Range" $3,850
Project 85-39 "Scabies Control in Bighorn Sheep" $10,205
Project 85-40 "Changes Associated with Stress in Bighorn Sheep" $10,000
Project 85-55 "Fish Creek Mountain Bighorn Habitat Enhancement" $3,500
Project 85-56 "Mt. Darby Bighorn Sheep Habitat Enhancement" $4,500
Project 87-08 "Water for Wildlife" $2,500
Project 87-09 "Population Decline of Introduced Bighorn Sheep Herd" $13,450
Project 87-13 "Absaroka Mountain Bighorn Sheep Study" $10,000
Project 87-44 "FNAWS Profile in Outdoor Magazine" $10,000
Project 87-54 "Wapiti Land" $100,000
Project 88-10 "Water For Wildlife" $2,000
Project 88-22 "S. Absaroka/Owl Creek Mountains Bighorn Sheep
Management Initiative" $6,000
Project 88-26 "Absaroka Mountains Bighorn Sheep Study" $10,000
Project 88-51 "Habitat & Population Regulation of Transplanted of Bighorn Sheep" ... $8,000
Project 88-61 "Desert Ram's Head Acquisition (Recaping)" $650
Project 89-20 "The Conservation of Genetic Resources in Bighorn Sheep" $5,000
Project 89-23 "Scabies Control In Bighorn Sheep" $5,530
Project 89-49 "Sawtooth Mountain Bighorn Water Development" $2,000
Project 90-11 "Cover Removal from Bighorn Sheep Watering Sites" $1,820
Project 90-EM03 "Cody Youth Center" $1,000
Project 91-15 "National Bighorn Sheep Interpretive Center" $10,000
Project 91-19 "Shell Canyon Bighorn Sheep Reintroduction" $6,000

Project 91-20 "Upper Southfork Shoshone River Land Exchange" $2,000
Project 91-41 "Genetic Diversity in Bighorn Sheep" $5,000
Project 91-57 "Restoration of Foraging Habitat in Bighorn Summer Range" $2,198.75
Project 91-60 "No-Wolf Option Committee" $3,000
Project 91-EM02 "Interpretive Sign: Annual Convention and Permanent Site" $2,000
Project 92-16 "No-Wolf Option Committee" $3,000
Project 92-29 "Controlled Burning For Bighorn Sheep" $2,144.09
Project 93-04 "National Bighorn Sheep Interpretive Center" $25,000
Project 93-11 "No Wolf Option Committee" $3,000
Project 93-26 "Shell Canyon Bighorn Sheep Reintroduction Project" $3,006.22
Project 94-23 "No Wolf Option Committee" $3,000
Project 94-25 "National Bighorn Sheep Interpretive Center" $15,000
Project 95-02 "Wind River Canyon Reintroduction" $3,600
Project 95-62 "National Bighorn Sheep Interpretive Center" $10,000
Project 96-14 "Operations Support of National Bighorn Sheep Interpretive Center" .. $15,000
Project 96-54 "South Fork Bighorn Sheep Habitat Improvement Burn" $2,700
Project 97-EM03 "Winter Feed Bighorn Sheep" $500
Project 98-01 "CMS Natural Habitat Area" $500
Project 98-13 "Biological Control of Dalmatian Toadflax" $2,500
Project 98-15 "Poison Creek Preservation Project" $10,000
Project 98-37 "Wyoming Hunting and Fishing Heritage Expo" $5,000
Project 98-42 "Operations Support of National Bighorn Sheep Interpretive Center" $5,000
Project 98-51 "Grey Hills Prescribed Burn" $1,800
Project 98-68 "Grizzly Encounter Education Program" $3,000
TOTAL FOR WYOMING $550,112.06

ALBERTA

Project 83-23 "Alberta Habitat Protection" $3,500
Project 83-76 "Alberta Sheep Alert" $5,019
Project 84-106 "Mount Allen Legal" $400
Project 84-107 "Mount Allen Sheep Monitoring" $11,000
Project 85-62 "Social Stimuli, Heart Rate Responses of Bighorns" $15,000
Project 93-35 "What Determines Number and Size of Trophy Bighorn Sheep" $500
Project 93-EM03 "Brochure Printing" $3,000
Project 94-52 "What Determines Number and Size of Trophy Rams" $1,500
Project 94-63 "Membership Directory for Alberta Outfitters Association" $5,000
Project 95-45 "The History and Management of California Bighorns" $7,000
Project 95-EM04 "POAA Funding" $15,000
Project 96-11 "Experimental Management Program to Increase Trophy Rams" $1,500
Project 98-03 "Experimental Management Program to Increase Trophy Rams" $5,000
TOTAL FOR ALBERTA $73,419

BRITISH COLUMBIA

Project 83-03 "Skeena River Project" $30,935
Project 83-04 "Toad River Burning Project" $8,000
Project 83-06 "Habitat Improvement" $20,000

Project 83-07 "Okanagon Game Farm Project"	$10,000
Project 83-08 "Bighorn Sheep Hunt Permit"	$3,700
Project 83-15 "Habitat Improvement"	$5,000
Project 83-37 "Lung Worm Research"	$4,000
Project 83-40 "Adult Sheep Mortality"	$12,300
Project 83-41 "Kechika Range Enhancement"	$11,070
Project 83-42 "Stone Battle Range"	$11,070
Project 83-43 "Four Mile Range"	$4,920
Project 83-77 "Bighorn Sheep Die-Off"	$15,782.50
Project 84-100 "Muskwa Range Enhancement"	$11,200
Project 84-101 "Kechika Range Enhancement"	$11,200
Project 84-102 "Muskwa Wolf Control"	$15,000
Project 84-103 "Kechika Range Wolf Control"	$15,000
Project 84-104 "Junction Transplant"	$3,000
Project 84-155 "Defense Against Greenpeace Anti Hunting Offensive"	$10,000
Project 85-02 "Comparative Study of Rocky Mountain Bighorns"	$18,877.50
Project 85-04 "Muskwa South Wolf Management"	$10,000
Project 85-05 "Muskwa North Wolf Management"	$3,000
Project 85-06 "Muskwa Mountain Sheep Range Enhancement	$10,000
Project 85-07 "Kechika Wolf Management"	$3,000
Project 85-08 "Kechika Mountain Sheep Range Enhancement"	$10,000
Project 85-22 "Junction Lamb Mortality Study"	$9,000
Project 85-31 "B.C. Wildlife Federation Reward Program"	$2,000
Project 85-32 "Vaseux Lake - California Bighorn Winter Range Acquisition"	$29,000
Project 85-43 "Taseko Lake Sheep Survey"	$3,000
Project 85-44 "Junction Poaching Problem"	$4,000
Project 85-49 "Thinhorn Sheep Range and Population Inventory"	$10,000
Project 87-46 "British Columbia Wolf Control Program"	$32,600
Project 88-17 "Thinhorn Sheep Habitat Enhancement"	$5,255.67
Project 88-20 "Churn Creek Sheep Survey"	$6,000
Project 88-56 "Muskwa Predator Control"	$22,500
Project 89-02 "Northern Sheep Habitat Enhancement"	$6,000
Project 89-58 "Bighorn Sheep Transplant to Nevada"	$2,500
Project 89-59 "Coyote Control"	$2,500
Project 89-60 "Bighorn Sheep Transplant To Oregon"	$2,500
Project 90-07 "Northern Sheep Habitat Enhancement"	$4,500
Project 90-46 "Akie-Pesika Wildlife Enhancement Project"	$3,000
Project 90-EM01 "Northern British Columbia Guides Association"	$10,000
Project 91-62 "Spring Range Burn and Habitat Improvement"	$20,000
Project 93-EM02 "Project WILD"	$24,000
Project 93-EM09 "Mission Ridge Sheep Transplant"	$4,485.40
Project 93-EM14 "Monument Installation"	$5,000
Project 93-EM15 "Churn Creek Emergency Funding"	$6,539.44
Project 94-EM14 "Herd Depredation Prevention Plan"	$10,000
Project 95-49 "Project WILD"	$7,100
Project 96-27 "Preserving the Opportunity"	$20,000
Project 96-28 "Muskwa-Kechika"	$20,000

Project 96-53 "North Eastern BC Collaring and Monitoring Project" $6,000.23
Project 97-EM05 "Ungulate Enhancement Program" $10,000
Project 98-02 "Wolf Radio Collaring Study in the Kechika" $15,000
Project 98-50 "Marble Range Sheep Migration Study" $5,500
Project 98-65 "Ensuring Effective Wildlife Management" $7,500
Project 98-66 "Spatsizi Park Sheep Habitat Burn" $6,000
TOTAL FOR BRITISH COLUMBIA $578,535.74

NORTHWEST TERRITORIES

Project 83-20 "Dall's Sheep - Northwest Territory" $45,000
Project 90-56 "Northwest Territories Outfitters Association" $10,000
TOTAL FOR NORTHWEST TERRITORIES $55,000.00

YUKON TERRITORY

Project 83-16 "Horn Aberrations - Dall's Sheep" $13,000
Project 83-24 "Kluane Lake Burning" $4,000
Project 83-74 "Trophy Identification" $650
Project 84-087 "Yukon Wolf Control" $15,000
Project 84-110 "Burning Dall's Sheep Range" $6,375
Project 85-65 "Finlayson Wolf Reduction Project" $20,000
Project 85-66 "Dall's Sheep - Wolf Relationship in Southern Yukon" $15,000
Project 85-67 "Dall's Sheep Winter Range Prescribed Burning" $7,500
Project 88-24 "Mitigating Impacts on Sheep Pop/Mt Mye, Curragh" $15,000
Project 89-15 "Mitigating Impacts of the Curragh Mining Development" $15,000
Project 90-08 "Relocation of Dall's Sheep to Caribou Mountain, Yukon" $2,096
Project 90-09 "Mitigating Impacts of Curragh Mining Development" $10,000
Project 94-13 "Yukon Outfitters Association. Legal Defense Fund" $10,000
Project 97-EM04 "Wolf Control" $6,000
TOTAL FOR YUKON TERRITORY $139,621

MEXICO

Project 84-156 "Mexican Population Study" $54,500
Project 85-35 "Ecology and Management of Desert BH in Sonora" $6,000
Project 87-48 "Mexico Population Study" $20,000
Project 88-59 "Desert Sheep Integral Program" $24,000
Project 89-69 "Desert Sheep Integral Program" $19,875
Project 90-58 "Desert Sheep Integral Program" $18,000
Project 91-EM03 "Census Survey" $54,375
Project 92-EM01 "Emergency Search for Helicopter" $5,000
Project 93-EM10 "Desert Bighorn Sheep Survey in Sonora, Mexico" $7,371.20
Project 95-EM08 "Bighorn Sheep Survey in Baja California" $68,527.90
Project 96-23 "Conservation Program for Desert Sheep in Mexico" $4,000
TOTAL FOR MEXICO $281,649.10

MULTI STATE/PROVINCE

(Projects involving more that one state or province)

Project 83-54 "Educational Film Strips" $6,250
Project 83-62 "Mineral Licks of North America" $4,000
Project 84-126 "Un-Endangered Species Filmstrip Program" $6,750
Project 84-143 "Wildlife Conservation Fund" $5,000
Project 85-11 "Combat Anti-Hunting Initiatives" $10,000
Project 85-17 "Un-endangered Species-Wildlife for Tomorrow" $15,600
Project 85-37 "Grass Roots Campaign of Wildlife Education" $5,000
Project 87-40 "Renew FNAWS Participation in WCFA/WLFA" $10,000
Project 87-53 "Sixth Biennial Northern Wild Sheep and Goat Council Symposium" $1,000
Project 88-04 "Renew FNAWS Participation of NAP/WCFA" $10,000
Project 89-22 "Renew FNAWS Participation In WCFA/WLFA National Association" $10,000
Project 89-25 "Desert Bighorn Sheep Management on Public Lands" $8,518.57
Project 89-62 "NRA Whittington School" $12,000
Project 90-05 "Seventh Biennial Northern Wild Sheep and Goat Council Symposium" $2,500
Project 90-19 "Renew FNAWS Participation in WCFA/WLFA" $10,000
Project 90-20 "Campaign for Reform of US Fish and Wildlife Service Division of Law" $24,000
Project 90-49 "NRA Whittington Center" $8,081.86
Project 90-55 "American Wilderness Leadership School" $1,781
Project 91-13 "American Wilderness Leadership School" $9,750
Project 91-37 "Renew FNAWS Participation in WCFA/WLFA" $25,000
Project 91-38 "Campaign to Reform U.S. Fish and Wildlife Service Policies and Practices" $22,500
Project 91-63 "NRA Whittington Center" $9,129
Project 92-09 "Eight Biennial Northern Wild Sheep and Goat Council Symposium" $2,000
Project 92-11 "Renew FNAWS Participation in WLFA" $25,000
Project 92-12 "FNAWS Participation in Campaign for Reform" $24,000
Project 92-62 "NRA Whittington Center" $2,880
Project 92-63 "American Wilderness Leadership School" $10,192
Project 93-24 "American Wilderness Leadership School" $11,146.80
Project 93-27 "Campaign for `Common Sense Amendments'" $35,000
Project 93-28 "Renew FNAWS Participation in WCFA/WLFA" $28,000
Project 94-03 "Ninth Biennial Northern Wild Sheep and Goat Council Symposium" $2,500
Project 94-04 "Renew FNAWS Participation in WCFA/WLFA" $25,000
Project 94-05 "Campaign for `Common Sense Amendments' to ESA" $35,000
Project 94-38 "America Outdoors - Strengthen Packer Programs" $10,000
Project 94-62 "American Wilderness Leadership School" $7,119.50
Project 94-EM13 "Voter Issues in Oregon and Arizona" $20,000
Project 94-EM16 "National 4-H Shooting Sports Program" $1,000
Project 95-06 "WCFA Sportsmen's Legal Defense Fund" $35,000

Project 95-07 "Renew FNAWS Participation in WCFA/WLFA"	$25,000
Project 95-52 "American Wilderness Leadership School"	$10467.80
Project 95-59 "Production of Educational Wildlife CD Rom"	$6,800
Project 95-EM06 "Wildlife Partners Network"	$10,000
Project 95-EM07 "Transportation of Diseased Sheep Herd"	$25,000
Project 96-01 "Pro-hunting Educational Programs in Schools"	$5,000
Project 96-07 "Tenth Biennial Northern Wild Sheep and Goat Symposium"	$2,500
Project 96-08 "American Wilderness Leadership School"	$5,653.87
Project 96-15 "Renew FNAWS Participation in WCFA/WLFA Program"	$30,000
Project 96-16 "Defeat Anti-hunting Ballot Issues/Support Pro-hunting"	$30,000
Project 96-46 "Support 40th Annual Meeting of Desert Bighorn Council"	$2,000
Project 96-47 "Wildlife Partners Network"	$12,500
Project 96-48 "Idaho Bighorn Sheep Trapping and Transplanting"	$1,066.82
Project 96-59 "Hells Canyon Initiative"	$100,000
Project 96-EM06 "National 4-H Shooting Sports Program"	$1,000
Project 97-00 "Hells Canyon"	$93,000
Project 97-54 "Wildlife Partners Network"	$10,000
Project 97-EM02 "Ballot Issues Coalition"	$5,000
Project 98-14 "Northern Wild Sheep and Goat Council 11th Symposium"	$2,500
Project 98-16 "Renew FNAWS Participation in WCFA/WLFA Program"	$5,000
Project 98-46 "National 4H Shooting Sports Program"	$1,000
Project 98-63 "American Wilderness Leadership School"	$2,000
Project 98-EM01 "Ballot Coalition"	$20,000
TOTAL FOR MULTI-STATE	***$896,187.22***

GRAND TOTAL FUNDED TO DATE $5,130,743.95

(EXCLUDING PERMIT SALES)

Whiskey Mountain

Bob Kiesling

In 1860 an estimated 175,000 bighorn sheep roamed the grasslands, foothills, and mountains of Wyoming. By 1960 Wyoming's bighorn population had declined more than 90 percent with fewer that 15,000 animals remaining. The litany of reasons for precipitous drops in population is long, well understood, and often publicized. But in wildlife management, as in so many other endeavors, the insights of hindsight are not terribly useful unless they become foresight.

A comparison of present day bighorn distribution and range with that of a century ago provides that graphic and mind-numbing insight that habitat loss correlates nearly identically with the decline in bighorn populations. Unfortunately, this insight seems merely a "statement of the obvious" and it is all to frequency employed bemoaning the mistakes of our past. Fortunately, however, bighorn numbers now are increasing modestly because the insight regarding the role of lost habitat is driving a few foresightful efforts to replace and restore traditional habitat.

There is perhaps no better example than the recent addition of over 25,000 acres of important habitat to the Whiskey Mountain Habitat Management Area in Fremont County, near Dubois, Wyoming. Thanks to the cooperative efforts of FNAWS, The Nature Conservancy, private landowners, the U.S. Forest Service, the Burlington Northern Foundation, the Bureau of Land Management and the Wyoming Game and Fish Department, it might now be possible to increase the Whiskey Mountain bighorn population by 200 to 300 sheep. With more than 25 percent expansion in herd size will come increased capacity to remove sheep from the Whiskey Mountain Area for release in other areas to supplement marginal herds or for reintroduction into historical habitat. Since 1949 more than 1,500 sheep have been removed from Whiskey Mountain for such purposes.

This remarkable capacity to increase herd size and, in time, to translocate more bighorns to other areas became possible because wildlife biologists of the Whiskey Mountain Bighorn Sheep Technical Committee recognized an opportunity to secure key habitat that could greatly leverage the existing 950 sheep carrying capacity of the Whiskey Mountain Habitat Management Area These biologists from the U.S . Forest Service, Wyoming Game and Fish Department, the Bureau of Land Management (BLM), and U.S. Fish and Wildlife Service work cooperatively to help establish management guidelines within the 34,000-acre Whiskey Mountain HMA.

When they became aware that an adjacent private ranch was for sale, the biologists understood immediately the benefit of securing the additional habitat for inclusion in the Whiskey Mountain HMA. Although the Red Creek Ranch consisted of only 3,200 acres of deeded land, it controlled an additional 22,000 acres of state and federal grazing leases and allotments. If the private land and public grazing leases could be secured by parties sympathetic to the idea of an expanded bighorn habitat area why not go for it?

Nowhere in sight were there any state or federal dollars available for such a project and the window of opportunity to purchase the ranch was quickly narrowing. Members of the Technical Committee pointed out the significance of the opportunity to the Big Sky Field Office of The Nature Conservancy and the Conservancy agreed to tackle the project.

Eighteen months, a quarter of a million dollars, and many, many hours of labor later the

Whiskey Mountain Habitat expansion dream became reality. How? The answer lies in one phrase: partners in conservation. The Nature Conservancy purchased the Red Creek Ranch with a cash loan from a private party sympathetic with the aims of the project. To help cover project costs The Nature Conservancy received a generous grant of $25,000 from the Burlington Northern Foundation. And in a highly creative and cooperative gesture, the Foundation for North American Wild Sheep pledged the gift of 36 acres of land to TNC, if the Conservancy succeeded in acquiring the Red Creek Ranch. Other funds for the acquisition will come from private gifts to the Conservancy and the sale of lands supplied by the BLM in exchange for TNC's Red Creek Ranch land.

In the wake of acquiring the Red Creek Ranch and securing transfer of the accompanying public land grazing leases, The Nature Conservancy is requesting that the domestic livestock grazing allotments be extinguished in order that the entire area be reserved for wildlife in general and bighorn sheep in particular.

Although it might take several more year for TNC and the BLM to complete their land, exchanges, the outcome will be the inclusion: of another 25,000 acres of prime habitat and critical winter range to the existing cooperatively-managed Whiskey Mountain Habit Area.

To be sure, this kind of opportunity to enhance the welfare of wildlife is not every where prevalent, but when opportunity knocks there is often no substitute for the mix of public and private partners, capital, talent, creativity and cooperation that constitute the hallmark of success at Whiskey Mountain.

Utah Capture Helps Sierra Diablo

FROM *WILD SHEEP*, AUGUST 1984

Texas FNAWS Chapter President Bill Leech was the high bidder for the Utah desert bighorn sheep permit for 1983, the first time such a permit was offered for auction at the FNAWS national convention. He paid $32,000 for the permit to hunt, but instead of hunting, Bill went to Salt Lake City and negotiated with the Utah Division of Wildlife Resources to capture a ram that was donated to the Texas Sierra Diablo Brood Facility. The state of Utah was so impressed with Bill's generosity that they decided two rams could be taken on the permit.

On January 26, 1984, Dr. Bob Glaze, Gene Holloway and Leech departed Fort Worth for Moab, Utah. After spending the night in the Canyon Lands National Forest, they were able to capture two young rams the next morning.

The two rams were transported back to Moab, loaded in a waiting airplane, and immediately flown to Van Horn, Texas. Dr. Red Duke, his camera crew and representatives from the Texas Parks and Wildlife Department were waiting. The young rams were released in the brood pasture before sunset.

Leech said, "We are very fortunate that the state of Utah agreed to give us two rams. There is no way that the Texas Chapter can express their appreciation to the state of Utah and its employees for the significant contribution they have made to our restoration program of the desert bighorn in the state of Texas."

Lone Mountain Bighorn Sheep Transplant

WALT VAN DYKE
FROM *WILD SHEEP*, FALL 1992

During the 1980s planning process for bighorn sheep in Oregon, the Lone Mountain area was selected as possessing native bighorn habitat characteristics. It was placed on the priority list for reintroduction of California bighorn sheep and went successfully through the Environmental Assessment process of the Bureau of Land Management.

In recent years, there have been periodic sightings of a band of bighorn rams on or near Lone Mountain. These probably come from a herd located about 15 miles north on the Square Mountain portion of South Catlow Rim. There are several square miles of potential lambing and rearing habitat available within an overall expected herd range of about 20 square miles. There are also connecting habitats to other occupied bighorn ranges and potential ranges.

A series of big game guzzlers was established on the dry sites of Lone Mountain in about 1965 to expand deer habitat utilization. Some use by bighorns was noted by the early 1980s around at least one guzzler. Perennial springs and seeps in the lambing area of rugged rock outcrops will make that habitat ideal, even on drought years.

Because of the excellent mixture of rock cliffs, outcrops and bunchgrass ridges intermingled with water sources, it is expected that the initial reintroduction of 15 bighorns will expand to 75 or 100 sheep when the habitat is fully occupied.

On January 24, 1992, during trapping efforts at Hart Mountain National Wildlife Refuge, 15 bighorns were captured for release on Lone Mountain. All 15 sheep were net-gunned from a Long Ranger 111 helicopter. One was from the North Mountain range and 14 from the South Mountain range of Hart Mountain.

The composition of the group was nine adult females, two adult males, one female lamb and three male lambs. Both adult rams were long yearlings (22 months old), while three of the nine adult females were yearlings. Radio collars were placed on four of the sheep for future monitoring of the transplant success. Transmitters were Telonics brand with mortality sensors. Three adult ewes and one ram received radio collars. All sheep were eartagged with orange Alflex cattle tags in the right ear and one inch round metal tags in the left ear. Blood samples were taken from all sheep to be used for a series of disease tests. Inoculations for lungworm, infection and stress were given at the base camp before sheep were loaded into the 4-wheel drive pickup transport vehicles with custom canopies. All sheep were weighed by a spring scale to provide a data base of size and physical condition from each of the herds that are trapped.

The Lone Mountain release site is about 40 air miles southeast of the capture area, and it took about two and one half hours to reach the site by the quickest route. The sheep were released from the transport vehicles at about 3 p.m. on January 24, 1992, near Scott's Cache Spring. All animals appeared to be in good condition and moved upslope into nearby rock areas.

Thanks to FNAWS For their continued support of Oregon's bighorn sheep program.

The Idaho Fish and Game Wildlife Health Laboratory Caldwell, Idaho

Karen Rudolph, David Hunter and Julia Mulholland

The Foundation for North American Wild Sheep (FNAWS), and most especially the Idaho Chapter of FNAWS, has worked long and hard with the Idaho Fish and Game Wildlife Health Laboratory in Caldwell, Idaho, since 1992. Prior to its current location and name, it was called the Wildlife Laboratory in Garden City.

The Wildlife Laboratory in Garden City, Idaho, was created in the 1970s to provide support services to wildlife biologists and enforcement officers. In 1989, Idaho Departments of Fish and Game and Agriculture created a joint state wildlife veterinary position. The facility was renamed the Wildlife Health Laboratory (WHL). Its mission was expanded to include disease investigations, collection baseline information on the health status of Idaho wildlife populations, and a provider of support services.

David L. Hunter, D.V.M., filled the first wildlife veterinary position in Idaho and has been director of the WHL ever since. During his tenure, he and his staff has served Idaho by providing veterinary services, including necropsies for wildlife, processing wildlife samples for disease screening, assisting in the monitoring and relocation of wildlife and conducting wildlife restraint and drug handling courses for Idaho Fish and Game and other agencies. The WHL staff provides recommendations for disease issues important to domestic livestock and wildlife such as pasteurellosis and brucellosis, serves as liaison with 18 other state and federal agencies to maintain healthy populations of wildlife in Idaho and other western states, represents Idaho in the Western Wildlife Health Cooperative, and directs and implements wildlife health and disease research.

In 1992, the WHL relocated to its current location on 26 acres leased from the University of Idaho in Caldwell. At first the WHL consisted of a building in need of renovation, lab space, a desk and a few pens for the animals. The WHL is now fully equipped with veterinary supplies and equipment for wildlife projects, research, rehabilitation and captures, multiple wildlife enclosures and a staff of three and a half full-time professional employees including Dr. Hunter (half of his salary is funded by Idaho Fish and Game); Julie Mulholland, wildlife health technologist; Karen Rudolph, wildlife laboratory biologist; a second wildlife laboratory biologist to be filled in 1998; and two part-time employees, Mary Ann Clemens, wildlife technician; and a part-time senior secretary to be filled in 1998.

The Idaho Chapter of FNAWS has been a strong financial supporter of the bighorn sheep research and management work carried out at the WHL. With the Idaho Chapter's annual donation of Idaho's bighorn sheep lottery tag monies, we have been able to employ our wildlife technician to maintain captive bighorn sheep for our research projects. The wildlife technician's duties include daily care, feeding and maintenance of the animals and enclosures. For bighorn sheep research projects, the technician coordinates and directs sample collections, treatments and health assessments of the animals with Dr. Hunter.

In addition to supporting our wildlife technician and per diem for captive bighorn sheep, the Idaho Chapter participates in an annual WHL-FNAWS clean up day, and invites us to participate in their program for the annual Idaho Chapter of FNAWS banquet. The Chapter members have volunteered to assist in bighorn sheep transplants and have worked diligently to coordinate support with the FNAWS Board of Directors when circumstances such as a major die-off in the state occurs. This especially came to light during the Hells Canyon die-off.

In 1998 we also received a generous donation from the Eastern Chapter of FNAWS totaling $5,500 for a clinical centrifuge and a sequencer. Bill Pastorek of the Wild Sheep Society of British Columbia has offered to lead a fund raiser involving all FNAWS chapters to benefit the WHL in February 1999.

In the same year the WHL relocated to Caldwell, the Idaho legislature mandated that $100,000 be transferred from Idaho Fish and Game to the University of Idaho's Caine Veterinary Teaching Center to carry out research pertinent to diseases affecting wildlife and domestic livestock. The WHL has collaborated with the University of Idaho on many research projects that have increased our knowledge and understanding of disease in bighorn sheep. Dr. Hunter oversees the research budget and disease programs at the University.

The first collaborative research project with the University of Idaho was undertaken in the early 1990s to address the hypothesis that bighorn sheep were not carriers of Pasteurella bacteria and that all Pasteurella species caused disease. Experiments carried out in the two labs and others demonstrated the hypothesis was incorrect. Bighorn sheep do carry Pasteurella bacteria, but the types of bacteria are generally different from those found in livestock. This research project was carried out with FNAWS Grant-in-Aid support.

Additional research with the University determined that in general, domestic sheep carry a majority of *Pasteurella haemolytica*, and lower numbers of *P. trehalosi*. Specifically, about 40 percent of Pasteurella strains carried by domestic sheep are called Biogroup 1. Bighorn sheep carry a majority of *P. trehalosi* strains of many types, while less than 1 percent carry *P. haemolytica* Biogroup 1 strains. In fact, we have found that Biogroup 1 strains cultured from bighorn sheep serve as a "red flag." They are associated with acute pneumonia with only a few exceptions.

From 1989 to 1991, the two laboratories also carried out a project to identify Pasteurella species associated with a die-off that was occurring in Big Creek, Idaho. A new technique was introduced in which DNA fingerprinting was used to track pathogenic Pasteurella strains. Today, DNA fingerprinting is used routinely during die-offs. This research project was supported by FNAWS Grant-in-Aid funding.

Continued research at WHL has led us down many avenues. We have tested an experimental Pasteurella vaccine in bighorn sheep. The vaccine did not protect the sheep from a pneumonia associated with a *P. haemolytica* Biogroup 1 strain. In 1998 we are testing the effectiveness of a different kind of vaccine that is made with live Pasteurella that cannot grow or divide. The irradiated vaccine is sprayed into the nose of the animal, rather than injected. Finding a vaccine that protects bighorn sheep from Pasteurella pneumonia would provide one solution for "keeping sheep on the mountain."

Additional research projects have been done to determine the importance of other disease agents including lungworm, viruses and other bacteria, and environmental conditions such as nutrition, stress and weather. The WHL has collaborated with other researchers at various universities such as those at Washington State University as well as state and federal agencies from Oregon, Washington, Colorado and British Columbia fish and wildlife departments, the Bureau of Land Management (BLM) and the U.S. Forest Service to carry out research that has served as a basis for making management decisions. An important outcome is the 1998 guidelines for domestic sheep and goat management in bighorn sheep habitats.

The draft guidelines from the Bureau of Land Management, dated April 2, 1998, desired

"Ichabod," a captive 3-year-old bighorn ram that lives at the Idaho Wildlife Health Laboratory in Caldwell, Idaho, butts a 50-gallon barrel as Mary Ann Clemens, wildlife technician, holds on for dear life and laughs. *Photographs courtesy of the Idaho Wildlife Health Laboratory.*

"Ickabod" prepares to head butt the drum again as Technician Mary Ann Clemens braces for impact.

"Ickabod" revs up for a big butt. Ickabod was born in captivity and lives at the research facility.

Karen Rudolph holds the head of captive bighorn ram, "Butthead," at the Idaho Wildlife Health Laboratory in preparation of collecting lab samples. After an injury and illness left him unable to live on his own in the wild, Butthead was transferred to the research facility. Butthead is tagged Number 18 and is 7-years-old.

A most undignified position for such a noble beast. "Butthead," the 7-year-old bighorn ram, is held by Wildlife Laboratory Biologist Karen Rudolph as Wildlife Technician Mary Ann Clemens administers a shot to its rump.
Photograph courtesy of Idaho Wildlife Health Laboratory.

A sign welcomes visitors and scientists to the Idaho Wildlife Health Laboratory in Caldwell. FNAWS has continued to be a major financial supporter of the laboratory as it studies wild sheep diseases and habitat for the purpose of "putting more wild sheep on the mountain."
Photograph by Karen Rudolph.

---

progressive native wild sheep management compatible with appropriate grazing on public lands by domestic sheep and goats. State and federal agencies recognized there are some disease agents that occur in both domestic sheep and goats and native wild sheep. There is evidence that if native wild and domestic sheep have close contact with each other, health problems and die-offs may occur. Some disease agents may be transmitted between both species. There is evidence indicating that some disease agents could be transmitted between domestic goats and native wild sheep.

The working draft guidelines also recognized in April 1998 native wild sheep die-offs that occur with no apparent relationship to contact with domestic sheep or goats; these two observations are both valid and not mutually exclusive. Bacterial pneumonia are not the only diseases of concern, although perhaps they are the most catastrophic, the draft guidelines state. The draft guidelines also indicate that risks of disease transmission are often unknown and they may be site specific. Reasonable efforts must be made by domestic sheep and goat permittees and wildlife and land management agencies to minimize the risk of disease transmission and to optimize preventive medical and management procedures, to ensure healthy populations of native wild sheep and domestic sheep and goats.

The draft guidelines for native sheep management includes 11 recommendations. The first recommendation says, "State wildlife and federal land management agencies, native wild sheep interest groups, and domestic sheep and goat industry cooperation and consultation are necessary to maintain and/or expand native wild sheep numbers. When an agency and industry agreement has been reached to maintain and/or expand native wild sheep numbers, the agencies and the domestic sheep industry will be held harmless in the event of disease impacting either native wild sheep or domestic sheep and goats. The draft guidelines further suggest that domestic sheep or goat grazing and trailing should be discouraged in the vicinity of native wild sheep ranges. Buffer strips surrounding native wild sheep habitat should be encouraged, except where topographic features or other barriers prevent physical contact between native wild sheep and domestic sheep and goats. Buffer strips could range up to 13.5 kilometers or nine miles, depending upon local conditions and management options. Cooperative efforts will be undertaken to quickly notify the permittee and appropriate agency to remove any stray domestic sheep or goats or wild sheep in areas that would allow contact between domestic sheep or goats and native wild sheep.

Significant to the draft guidelines is the statement that native wild sheep should only be reintroduced into areas where domestic sheep or goat grazing is not permitted and areas where wild sheep are reintroduced should be devoid of domesticated sheep or goats for at least two years before the reintroduction takes place.

The revised guidelines will be finalized after April 1998 and will be reviewed at least every five years by a work group comprised of representatives from the domestic sheep and goat industry, state wildlife agencies, BLM and native wild sheep organizations, including FNAWS. The process of cooperation and research is on-going.

In 1996 the FNAWS Grant-in-Aid program supported the establishment of a DNA molecular laboratory to carry out genetic research. State-of-the-art equipment for the DNA laboratory was purchased with monies from Shikar Safari International Club. FNAWS and the Rocky Mountain Elk Foundation provided money to carry out genetic research and DNA forensics work, respectively.

FNAWS has been an active supporter of the new DNA genetics research carried out at the WHL to identify a natural disease resistance gene named Nramp, which stands for Natural resistance associated macrophage protein. In earlier work done by Drs. Joe Templeton and Don Davis at Texas A&M, the Nramp gene was shown to protect 10 to 20 percent of elk and bison from bacterial diseases including brucellosis, tuberculosis and pasteurellosis. As a temporary employee with the Idaho Fish and Game in 1996-97, Karen Rudolph began a research project to find the Nramp gene in bighorn sheep.

There are two forms of the gene; one that protects against disease and one that does not. This is somewhat similar to the recent finding that some people seem to be resistant to AIDS while most others are susceptible. Karen determined that bighorn sheep do have a Nramp gene. She is currently working to find whether bighorn sheep have the protective form of the Nramp gene, and in how many sheep in a herd. To date she has collected DNA from several hundred bighorn sheep to finish the project.

In 1996, the FNAWS Board of Directors made a commitment to "put sheep back on the mountain" in the Hells Canyon National Recreation Area as a long-term project, now known as the Hells Canyon Initiative. A more detailed chapter in this book concerns the Hells Canyon Initiative. FNAWS has committed to raise $10 million in 10 years to reach its goal. Part of the plan included hiring a full-time wildlife biologist to assist in overseeing the lab research component of the Hells Canyon Initiative. The Board's plan came to fruition in April 1998 when Karen Rudolph was hired to fill a permanent position at Idaho Fish and Game, supported by FNAWS to the tune of $38,000 from 1998 to 2001.

A pivotal role in advancing bighorn sheep research and management has been filled by FNAWS. They have provided invaluable support through the years in the form of emergency money during catastrophic events such has major pneumonia die-offs. Most notable was the devastating bighorn sheep die-off in Hells Canyon in 1995 and 1996. Early in the die-off, a decision was made to remove 72 Rocky Mountain bighorn sheep to prevent the pneumonia from spreading to other sheep herds in the canyon.

The sheep were captured by helicopter-assisted net gunning and transported to the WHL.

An unprecedented disease investigation was carried out at the facility and in Hells Canyon, based on the cooperative efforts of several state and federal agencies, with emergency funding provided by FNAWS. Today the same agencies are actively involved in carrying out cutting-edge science as part of the Hells Canyon Initiative.

FNAWS has been a critical link in our commitment to establish and maintain healthy bighorn sheep populations for future generations to enjoy. Its members have provided tremendous support that encompasses financial assistance and volunteers from the organization in all projects that directly or indirectly involve bighorn sheep. To better address the future of bighorn sheep, we have maintained a long-term commitment of sharing with FNAWS all information gathered.

As we enter the new millennium, we look forward to continuing many more years of a productive working relationship with FNAWS. The WHL thanks the FNAWS Boards, past and present, and its membership for their unyielding support and commitment to bighorn sheep.

Dave Hunter, D.V.M., heads the Idaho Fish and Game Wildlife Health Laboratory in Caldwell, Idaho.

Karen Rudolph, Ph.D., received her doctorate in molecular biology studying drosophila (fruit flies) from Dartmouth College in 1991 and worked as a postdoctoral fellow for three and a half years at University of California at Los Angeles (UCLA) to study early development in fruit flies using genetics and molecular biology. Following her work at UCLA, Karen was hired by the University of Idaho at the Caine Center to study diseases in domestic and wild sheep for two years. She is currently the wildlife laboratory biologist for the Idaho Fish and Game Department. Her salary is supported by FNAWS. She uses her knowledge of molecular biology and research experience to understand the mechanisms of Pasteurellosis and other diseases in bighorn sheep to serve as a basis for making practical management decisions.

David Hunter, D.V.M., received his doctorate of veterinary medicine from Washington State University in 1976 and worked for a while in private practice. He also worked as a wildlife veterinarian for the California Department of Fish and Game and then became the first wildlife veterinarian for Idaho in 1989. His veterinary expertise has advanced the health and welfare of wildlife in Idaho and throughout the West.

Julie Mulholland received her B.S. from Colorado State University, Fort Collins, Colorado, in 1980. For seven years she conducted field work for the Peregrine Fund, New Jersey Department of Environmental Protection, Idaho Power Company and Idaho Department of Fish and Game. She was hired at the Wildlife Health Laboratory on a permanent basis in 1987. As the wildlife health technologist, Julie balances an uncommon mix of laboratory and field duties.

Hells Canyon Initiative
A Project for the New Millennium

Tim Schommer

Agencies and organizations interested in bighorn sheep in Hells Canyon conceived of the "Hells Canyon Initiative" in 1995 as a program to accelerate restoration of bighorn sheep. The initiative is a long-term, large score project to restore Rocky Mountain bighorn sheep to the tri-state region of Oregon, Washington and Idaho. This project is a partnership between the Foundation for North American Wild Sheep (FNAWS), U.S. Forest Service, Bureau of Land Management, the Idaho Department of Fish and Game, the Oregon Department of Fish and Wildlife and the Washington Department of Fish and Wildlife.

The project area encompasses more than 1.3 million acres of continuous, high-quality bighorn habitat. Most of the suitable habitat in the main project area is publicly owned and provides excellent year-round habitat. This represents some of the most extensive habitat in the lower 48 states, much of which is currently unoccupied by bighorn sheep. It is our vision to restore self-sustaining, healthy bighorn sheep populations to all suitable habitats in the Hells Canyon area. FNAWS has been an active participant in steering restoration activities, as well as undertaking a huge fund raising effort.

Reintroductions

Rocky Mountain bighorn sheep have inhabited Hells Canyon for thousands of years. They once thrived on the canyon's abundant grasses, mild winters and steep, open terrain. Bighorn sheep were almost eliminated from Hells Canyon and the adjacent Wallowa Mountains during the early 1900s. By 1940, bighorns in even the most remote areas were all gone. As elsewhere in the western United States, this was probably due to a combination of competition with livestock for forage, diseases introduced by domestic sheep and excessive hunting.

The first reintroductions of bighorn sheep to Hells Canyon and the surrounding area occurred in 1971 when the Oregon Department of Fish and Wildlife translocated 20 Rocky Mountain bighorn sheep from Jasper National Park, Alberta to the Hells Canyon Dam area, and 20 to the Lostine River in the Wallowa Mountains. Since 1971, about 368 Rocky Mountain bighorn sheep have been translocated into the Hells Canyon area. Bighorn sheep currently in Hells Canyon originated from 10 sources that include Waterton Lakes, Alberta; Jasper National Park, Alberta; Cardinal River, Alberta; Spences Bridge, British Columbia; Salmon River, Idaho; Wildhorse Island, Montana; Thompson Falls, Montana; Sun River, Montana; Tarryall, Colorado; and Whiskey Basin, Wyoming.

Herd History and Die-Offs

The Hells Canyon bighorn sheep population, or metapopulation, is composed of 16 ewe herds. Reintroductions have established 12 herds. Four herds were established by dispersal.

Disease has had a significant impact on Hells Canyon bighorn sheep herds. This has been primarily linked to transfer from domestic livestock. Seven population die-offs have been reported in the project area since reintroductions were initiated. Five die-offs have been linked to domestic sheep, one to a feral goat and one to drought and scabies.

Little information was collected during most early die-offs and most of it after the fact, when numerous dead bighorn sheep were observed by agency personnel. Pneumonia was the eventual cause of death in all cases where a cause of death could be identified. Typically, about 80 to 90 percent of the bighorns directly died from pneumonia. In addition, those ewes that did survive usually lost their lambs for the following three sessions. These factors have not allowed bighorns to recover.

The most recent Hells Canyon disease episode occurred in the winter of 1995-96. A population die-off believed to have started near 10-Mile Creek, Washington, and extended 40 miles to the Imnaha River, Oregon. About 235 bighorns died before it was over. Even though response to this die-off was immediate, and represented the most converted effort ever attempted to halt such a die-off, little could be done to stop it. This effort was launched by FNAWS, the U.S. Forest Service, and the states of Idaho, Oregon and Washington. Much valuable information was gained by the treatment and testing of the sick and dying animals, and yet in the end, the results were difficult for all of the parties to accept.

From this tragedy, a concept emerged. Agencies and organizations joined forces, and over the next year the concept became a plan. A plan to examine the loss, protect the vast habitat of Hells Canyon, restore bighorns to the area, and develop a consistent approach to research the problem of wild sheep disease. All of the parties involved in the management of wild sheep in Hells Canyon have come together to develop, guide and implement the plan. As of 1997, there were approximately 700 bighorn sheep in the 16 herds of the Hells Canyon project area.

Harvest of Hells Canyon Bighorns

Nearly 200 bighorns have been harvested in the project area since the first hunt in 1976. Harvest in all states is by controlled permit and limited to rams. Success rates are 80 to 90 percent. Idaho requires rams have a 3/4 curl or greater or be at least 4 years old; Washington and Oregon permit the taking of any ram. Each state has an auction and lottery tag.

Livestock Contact

Historically, contact with domestic sheep is believed to have contributed to bighorn sheep die-offs and limited the Hells Canyon sheep population. The number and extent of domestic sheep grazing in the project area has declined considerably in the last 25 years. All domestic sheep allotments in the heart of Hells Canyon have been removed and/or converted to cattle. Only one grazing allotment, Mudd Duck, remains in the project area, but outside of Hells Canyon. It has the potential to impact the Lostine bighorn herd. Plans are currently underway to solve this potential problem.

Habitat Improvement

At least 44 springs have been developed for wildlife use throughout the project area on state, federal and private lands. Several natural mineral licks occur in the study area. Salt blocks containing selenium and/or medicated (anthelmintic) blocks have been placed in Oregon and Washington. Hells Canyon is considered selenium poor for livestock, and bighorn sheep have selenium levels that are considered low for domestic sheep.

Four irrigation, cultivation and fertilization projects have been conducted in Idaho, Oregon and Washington. Nearly 70,000 acres in Oregon and Idaho have been treated with prescribed and natural burns since 1992. Extensive cooperative weed control efforts by all public agencies are ongoing in the tri-state area.

Three volunteers examine a bighorn transplant from British Columbia in 1998.
Photograph by Tim Schommer.

Aerial view of Hells Canyon.
Photograph by Tim Schommer.

Four volunteers involve themselves in a 1998 bighorn sheep transplant from British Columbia.
Photograph by Tim Schommer.

Billowing smoke from a controlled burn improves habitat for bighorn sheep.
Photograph by Vic Coggins.

A bighorn ram shows its dominance in the winter.
Photograph by Tim Schommer.

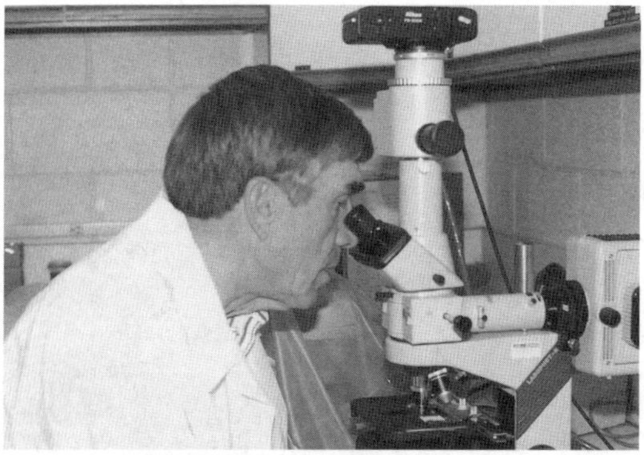

Pathologist Stuart Lincolning examines a Hells Canyon blood sample. Research continues on the Hells Canyon bighorn herds with funding from FNAWS.
Photograph by Duncan Gilchrist.

FNAWS members participated in a June 1997 Hells Canyon project. Seen leaning against a truck to the far left is FNAWS Past President Phyllis Carlson.
Photograph by Duncan Gilchrist.

Restoration Plan

To formalize the interagency working relationship at all levels, representatives from FNAWS and all five agencies, participated in several meetings from April 1996 to May 1997. Results of these meetings included development of a Memorandum of Agreement (MOA) and a Restoration Plan to provide a basis from which to coordinate long-term bighorn restoration in the Hells Canyon area.

The MOA was developed to provide a long-term commitment and continuity to the project by all agencies. The MOA establishes a 12-member, interagency committee whose goals is to restore bighorn sheep in Hells Canyon.

A restoration plan was completed in May 1997. It contains extensive information on the background and current condition of bighorn sheep, an overview of goals and objectives, and a summary of actions to be accomplished under the Hells Canyon Initiative. The plan identifies needs, funding requirements, and provides direction for prioritizing activities. The plan is divided into the following four parts.

The first is reintroductions of bighorn sheep will be conducted when animals from outside sources are available to fill unoccupied habitat and augment existing herds. Reintroductions are considered the most important factor in restoring bighorns to Hells Canyon. The generosity of other states and provinces will be essential for bighorn recovery. The plan identifies 18 new sites for potential releases and 10 sites for supplementation. Considering the difficulty of obtaining new stock, it will take many years to bring bighorns to all available habitats.

Secondly, bighorn sheep will not be released in areas with bighorn sheep that have survived a recent epidemic and are possibly carrying contagious diseases. In supplemental releases or releases adjacent to herds that have had a die-off, fall lamb:ewe ratios of the existing herd should be above 25:100 for at least two years. Where possible, a minimum of 20 bighorns are released per reintroduction with ratios of at least three ewes to rams. All bighorn sheep released in the project area are tested for presence of, or exposure to, pathogenic bacteria, viruses and parasites at capture.

Out of the box goes a bighorn ram that was released by Idaho Department of Fish and Game officials and FNAWS volunteers in the summer of 1995 on Jim Creek. Efforts to revive the Hells Canyon bighorn population continues by Idaho, Washington, Oregon and federal government officials with ongoing support from FNAWS and its members.
Photograph by Duncan Gilchrist.

Obtaining bighorn sheep for release has historically been a factor limited restoration. For the most part, bighorns have been obtained where available and relatively little attempt has been made to select specific source populations. Past releases have shown that releases of "non-migratory" bighorns, which are defined as bighorns that do not use elevationally distinct winter and summer ranges, have been more successful at establishing a self-sustaining population within 10 air miles of the release site than releases of migratory bighorns. Non-migratory bighorns may be better adapted to areas along the Snake River canyon, where winter and summer range habitats are at the same elevation. Migratory bighorns are better suited to the Wallowa Mountain area. Where possible, the source population is matched to the release site, and differences in movements and habitat utilization among different source populations will continue to be evaluated through monitoring.

In the past, Oregon and Washington have established two herds that are the Lostine and Hall Mountain herds, that have been used for release into new areas. Both herds are migratory. Numbers in Hall Mountain are currently poor and won't be available as a source for transplants for many years. Lostine herd is almost recovered from a die-off in 1986-87. Lostine sheep cannot be used as a source of transplants in Hells Canyon because of the closeness to their home area and their migratory nature. Until source populations are available within Hells Canyon, bighorn sheep will be relocated from outside Hells Canyon.

Population Monitoring and Research

Monitoring and research are designed to evaluate the success of the project, determine causes for success or failure, and guide future direction through adaptive management. Population monitoring is necessary to measure and evaluate the multiple interacting factors, including habitat, dispersal, predation and disease, that affect bighorn sheep population growth and productivity. Monitoring is reevaluated annually and adjusted as necessary based on the data collected.

Bighorns released in the project area are being radio-collared and regularly relocated. Goals are to document post-release movements and monitor the extent and causes of mortality. A monitoring plan has been developed for each release. This includes weekly relocations for the first month, and from two to four weeks for the life of the radio collar. If dead sheep are located, it is a priority to examine the carcass for the probable cause of mortality.

Blood or tissue samples are being collected from all bighorns released or handled in the project area. This information is being used in conjunction with movement data to determine the relative contribution of bighorns from different source populations and to assess the potential for genetically based disease resistance in bighorns.

Annual project-wide surveys are also being conducted. Some herds are being surveyed more intensively to monitor lamb survival. In order to provide accurate population estimates, radio-collared bighorns are being used to test an existing bighorn sheep sightability model.

Disease Treatment, Monitoring and Research

The plan brings together researches, veterinarians, universities, pharmaceutical companies and wildlife biologists to develop, implement and guide the project. Adequate funding has never been available to study wild sheep diseases. This project will offer funding at a new, consistent level.

Disease has historically reduced population growth rates by at least 40 percent in the project area. Disease monitoring, treatment and research are important components of successful restoration of bighorn sheep in Hells Canyon and elsewhere. All bighorns handled are being tested for disease according to a standard protocol. Blood samples are collected for bacterial, viral, chemistry, trace mineral and genetic testing. Viral and bacterial pharyngeal swabs, ear swabs, external parasites and fecal samples are collected. All bighorns that die will be retrieved where possible and necropsied using a standard protocol developed in this project.

An emergency disease response plan has been developed, and is available for implementation if a disease outbreak occurs. Its purpose is to rapidly respond to bighorn disease outbreaks to improve survival. Past experience with Hells Canyon die-offs has shown that sick bighorns given antibiotics and other drugs and released back on their range will have drastically higher survival rates over non-treated animals. A four-person response team has primary responsibility in Idaho, Oregon and Washington, but could be requested by other agencies elsewhere.

Investigation into identification of potentially disease resistant bighorn sheep is in progress at the Idaho Department of Fish and Game Health Laboratory at Caldwell. The procedure is focused on the Nramp gene (Natural Resistance Associated Macrophage Protein) that has been associated with genetic disease resistance in elk and bison. To date, domestic sheep DNA primers have been optimized for bighorn sheep and the Nramp gene has been recovered from bighorn sheep blood samples. Still remaining is the testing of these alleles for disease resistance. This will involve combining macrophages (cells that engulf other cells or organisms) with a virulent strain of Pasteurella in a test tube and evaluating whether survival varies among macrophages from different alleles of the Nramp gene. If so, the Nramp allele with high macrophages survival will be associated with resistance to Pasteurella. Blood samples collected in the wild can then be evaluated for this allele to assess resistance to Pasteurella in wild sheep.

The Hells Canyon Initiative has just completed a series of trials to test a new vaccine developed by the Colorado Division of Wildlife and Langford Laboratories. Low lamb survival following each die-off has been a significant problem for several years. The 1995-96 die-off provided a perfect opportunity to test the vaccines. The Pasteurella vaccine test was completed on 36 bighorn ewes in the wild and in captive bighorn sheep at the Idaho Wildlife Health Lab at Caldwell, Idaho. Unfortunately, the vaccine trials did not indicate any difference in survival of lambs from vaccinated and unvaccinated ewes. Neutralizing titers were found to be higher after vaccination, but apparently not high enough to increase lamb survival. The vaccine was also tested on bighorns at Washington State University. After vaccination, six healthy bighorns were exposed to a small domestic sheep herd. By day 60, all bighorns died. The vaccine failed to protect any of the bighorn sheep from virulent strains of Pasteurella received from domestic sheep.

Habitat Management

Land management agencies will continue to manage grazing allotments within the project area in a manner compatible with project goals. State agencies will capture bighorn sheep that have come into contact with domestic sheep and remove them from the wild. Education of private landowners grazing small domestic sheep flocks near bighorn habitat through dissemination of information and personal contact will continue to be a priority. Education efforts have focused on explaining the conflict between domestic and wild sheep, suggesting ways to reduce opportunity for disease transfer, and encouraging landowners to contact agencies when bighorn sheep come into contact with their livestock.

The Hells Canyon committee recently helped FNAWS develop and distributed a brochure on the incompatibility of domestic and bighorn sheep for distribution to the public. The brochure describes the problem of contact between the two species and urges livestock owners to contact state wildlife agencies if the two species are seen in proximity.

Several important tracts of bighorn sheep habitat have recently been acquired. In Oregon, purchase of the 80-acre Sargent inholding at the Wenaha Wildlife Area was finalized. This purchase was funded by a FNAWS grant along with money from other conservation groups, a private donor and receipts from a timber sale. In Idaho, 565 acres of habitat at Limestone Point was acquired with funds from the Idaho FNAWS auction tag, other conservation groups and private donors. both areas are important for bighorn sheep and were privately owned inholdings within public wildlife management areas. Support for land acquisition, such as these examples, will continue.

Habitat improvements are being designed to improve numbers or distribution of bighorn sheep. FNAWS is continuing to form funding partnerships with other agencies for prescribed burns, water developments, seedings, noxious weed treatments and fertilization.

Tim Schommer is currently the national bighorn sheep technical biologist for the U.S. Forest Service and forest wildlife biologist for the Wallowa-Whitman National Forest in Baker City, Oregon. He has worked as a wildlife biologist for five national forests in Oregon, Idaho and Colorado. He received a B.S. in wildlife biology from the University of Washington and a M.S. in wildlife habitat management from Washington State University.

Pasteurella Outbreak Hits Hells Canyon Sheep: FNAWS Plays Major Role in Rescue

FROM *WILD SHEEP* MAGAZINE, WINTER 1995-96 ISSUE

When Washington Department of Fish and Wildlife personnel found the first dead bighorn sheep while doing census work late in November (1995), an immediate wave of concern swept down on the neighboring states of Idaho and Oregon as well.

The animals didn't have any tell-tale signs of a gunshot wound or a nasty fall and wildlife biologists feared an outbreak of the disease Pasteurella had once again found its way into the Hells Canyon complex. In the winter of 1991, an outbreak had almost completely eradicated existing herds in the area.

Even one dead sheep is cause for alarm, but the additional 21 sheep found dead the next day sent fish and game personnel from all three states into a whirlwind of activity. Test results taken at Washington State University had confirmed what the biologists originally feared. The sheep had died as a result of breathing complications linked with Pasteurella.

Pasteurella is almost always fatal for bighorns, but biologists can save sheep if they are able to treat them on time. Representatives of all three states were consulted on a live capture operation. The Idaho Fish and Game has the Caldwell Wildlife Laboratory that could treat the captured animals and plans were immediately made to transport the sheep there. The Caldwell Wildlife Laboratory is funded by the Idaho lottery tag. The proceeds from this tag are TOTALLY dedicated to the facility. Since the legislature enacted this tag lottery in 1991, nearly $100,000, has been raised for this facility.

With the costs for trapping and transporting the animals expected to run into the tens of thousands of dollars, FNAWS generously stepped forward and immediately allocated $25,000 to the rescue and committed to funding all the operating costs.

Wildlife Helicopter Management was contracted as the company to haul sick bighorns to the processing center. They started capturing bighorns on December 1 and 2 (1995). The terrain is some of the most precipitous in the state and this capture effort was perhaps the most severe test of wildlife capture skills in North America. The right people were selected for the job and in two days they had captured 58 bighorns.

The Pasteurella outbreak brought together wildlife managers and biologists from Oregon, Idaho and Washington in the most comprehensive wildlife management effort in the history of the agencies. Each agency brought transport trucks and personnel to take care of captured animals and truck them to the holding facility in Caldwell, Idaho. Dr. Dave Hunter, Idaho veterinarian, volunteered to care for the bighorns and treat them with an antibiotic treatment program. Volunteers, including a local veterinarian from Lewiston, Idaho, joined in the effort to care for these sick bighorn and prevent the disease from spreading to neighboring areas. Television crews from five stations covered the bighorn capture and provided excellent media coverage. Local newspapers and radio stations covered the story from start to finish.

Before transporting them by truck from just outside Lewiston, biologists first administered antibiotics. The treatment has worked well so far and only time will tell exactly how successful the treatments were. The animals are expected to be quarantined for between one and two months and neither the public or the media will have access to the animals as there is tremendous stress involved with exposing the animals to people.

Fish and game officials are pleased that all the animals survived the transport efforts. The primary reason for bringing the animals out of the area was to prevent spreading the disease to nearby herds. All of the animals brought in were taken from the Washington side of the Snake River and they were located just north of the confluence with the Grande Ronde River. Hopefully, the joint effort will continue to be successful and no other herds in Hells Canyon will be affected. Follow up surveys in Washington, Oregon and Idaho will monitor the bighorns over the next few weeks.

"The Foundation for North American Wild Sheep is proud to have been a part of this rescue mission," said FNAWS Executive Director Karen Werbelow.

FNAWS is a leader in conservation, protecting sportsmen's rights and has spent millions of dollars for sheep habitat, research and restoration.

Chapter XI

Broadening the Base: Chapters and Affiliates

The Grand Slam Club: A History 1955 to the Present

DENNIS CAMPBELL

Many people in the hunting world have heard of the Grand Slam and the Grand Slam Club; however, there are many misconceptions floating around, especially concerning the Grand Slam Club. Most knowledgeable hunters can readily define what the Grand Slam is, but when it comes to defining the Grand Slam Club, far fewer hunters actually can do so. A popular misconception, amazingly, has been to believe that those who have taken all four sheep miraculously become members of some informal, almost mythical, fraternity. Such is not the case. There is, in fact, a Grand Slam Club! Let us begin where we should: at the beginning.

History

In 1947, the late Grancel Fitz wrote an article called "Grand Slam in Sheep" for an issue of *True* magazine. This designation was subsequently popularized by a few other outdoor writers, but no individual actually picked up on it until 1955.

Bob Housholder of Phoenix, Arizona, was the man who founded the Grand Slam Club. In 1955, he was guiding a Nebraska sheep hunter, Bernard Briggs, and realized that Briggs' desert ram taken on that hunt completed his fourth ram and his Grand Slam. Housholder became curious as to exactly how many others had accomplished this same feat. Being an outdoor writer, Housholder contacted several of his outdoor writer friends and put out the word he was looking for Grand Slam holders. Before long, he had a list of 20 names. One of those writer friends, the late Jack O'Connor, is listed as Member Number 1 of the Grand Slam Club. Even though O'Connor was not the first person to accomplish the feat, he was the first to document that he had done it legally.

The Club was officially founded in February 1956, and has been called in print "the most prestigious big-game hunting club in this country." The Grand Slam Club is also the forerunner of all the other sheep hunting organizations. As has been printed many times, FNAWS was birthed at Mt. Horeb, Wisconsin, in 1974. According to many of those who were present, FNAWS was conceived as the Midwest Chapter of the Grand Slam Club. That history, of course, you will find elsewhere in this book.

The Founder: Bob Housholder

People either liked Bob or had an extreme dislike for him. Yes, he was arrogant. Yes, he spoke his mind and didn't care whom he offended, but this man did a lot for the sheep hunting world. Probably his strongest suit was the fact that he was a communicator. Bob was a prolific letter writer, which is proven by the archives of the Grand Slam Club. There are hundreds of personal letters from some notable hunters, as they responded to Housholder's correspondence. Some of those notables include Jack O'Connor, Elgin Gates, Herb Klein and Warren Page. All in all, there are thousands of letters in the archives. Bob Housholder wrote and communicated with almost anyone who would respond, to feed his voracious appetite for wild sheep information.

Over the years between 1956 and 1967 the Grand Slam Club was very loosely knit, but Housholder's personal communication with the members amazingly, solidified the Club. In July 1967, Housholder published what he called *Bulletin No. 1* of the Grand Slam Club. Following are some interesting excerpts from that one-page Bulletin:

A young Bob Housholder prepares for a hunt.
Photograph courtesy of the Grand Slam Club.

Grand Slam Club Bulletin No. 1, July 1967

"It has taken a lot of doing, but I finally have assembled the names and addresses of the 140 names on the grand slam list. Enclosed is the latest list.

"For a start, I'd like to use this bulletin to pass on information on the sheep of N.A (North America). So if you people have some dope we all could use, send it in.

"I am in touch with 600 sheep hunters and about 50 sheep guides in N.A. and from them come an awful lot of dope."

Housholder ended that first Bulletin with these two lines: "Let's hope this first bulletin is the beginning of a strong organization in legislation and the preservation of the mountain sheep of North America."

"Shoot straight, keep your powder dry and keep a loose rein going over the pass."

Bob Housholder's Bulletin was definitely well-received. The sheep hunting world was quite small, but those within it welcomed the Bulletin. When Bob asked the members to send in information, they responded in a big way. The interest that was generated by that first appeal has never wavered. There was always plenty to talk about in Housholder's Bulletin, and the one-page newsletter quickly grew into a multipage affair. Sheep hunters around the world hungrily devoured every word that this interesting hunter/writer had to say about their beloved wild sheep. Of course, hearing about other sheep hunters and their exploits was a big part of the Bulletin.

Bob Housholder, besides his communication skills, was an interesting fellow in many other ways. He was a man of modest means in that he worked for the state of Arizona. Besides completing a Grand Slam, this man also took the N.A. 27, which is almost unbelievable considering his income. He was a gifted athlete in both high school and college. The awards heaped upon this man and the personal reports given indicate he was a spectacular

The late Jack O'Connor is Member 1 in the Grand Slam Club.
Photograph courtesy of Grand Slam Club.

The first formal meeting of the Grand Slam Club was held in San Antonio, Texas, on May 5, 1971. In attendance were: (back row, left to right) Bert Klineburger, George Landreth, Bob Speegle, Lynn Ross, Win Condict, Basil Bradbury; (center row, left to right) Ebb Warren, Steve Rose, Ken Vaughn, Keith Brown, Bill Bond, Joel Loveridge; (front row, left to right) Bob Housholder, Butch White, Estel Vaughn, Bill Lewis, I.D. Shapiro.
Photograph courtesy of Grand Slam Club.

athlete in the fast pitch softball arena. He was inducted into the Arizona All-Time Sports Hall of Fame. This award came because of his great achievements in track and field, fast pitch softball and, of course, hunting.

A Dark Day for Sheep Hunters

The Grand Slam Club did have a dark chapter in its history. In reality, it was not a dark chapter because of the Club itself, but because of some individuals within the Club's ranks. A few of the Club's members were found to have succumbed to the temptation of poaching desert sheep. Because of the misdeeds of these few, a question of integrity was cast upon sheep hunters in general. The Grand Slam Club was drawn into the investigation when a subpoena of Housholder's records was served in November 1970. Bob himself was totally and completely guiltless, and he cooperated fully with the investigation, which ultimately led to some having to suffer the consequences for their illegal activity.

This subpoena and investigation seemed to temporarily take some of the spark from Housholder's enthusiasm. In each of the Bulletins he published during that period of time, there were mentions of the problems, yet apparently Bob didn't know what to do about it. Then, in July 1972, some closure was brought to this dilemma for Housholder and the Grand Slam Club. It was in that month's *Bulletin No. 16* that this statement appeared at the end of the issue:

Upon receipt of this notice, each member who took his desert ram in Mexico without a legal permit issued from Mexico City, or who took his ram in California, is hereby asked to

Bob Housholder attended the 1981 FNAWS convention.
Photograph courtesy of Grand Slam Club.

resign. Those who have legal Mexican rams who wish to remain in the Club are now asked to submit evidence of legality of their desert rams. A Mexican ram appearing in B&C is prima facie evidence of legality. I realize many of you have already sent in copies of your legal Mexico permit, so this obviously applies only to those who have not.

The method of resignation is simple: do nothing. Do not write anything incriminating on paper. I shall assume by October 1, 1972, that those who have NOT contacted me no longer wish to remain members of this Club. This way, no one will know who resigned from the Club. Those with multiple Slams must also show legality of each desert ram, and if they cannot do so on all they claim, the number of Slams will be reduced accordingly and they will remain in membership with their legal, proven number of Slams.

Guess there's nothing more to say. Good hunting, always!" Bob Housholder

After this notice went out, eventually a new Grand Slam list was built all over again. Sadly, a lot of innocent people were dropped from the Club because they simply did not see this notice or inadvertently failed to respond. The fact that this notice appeared as just a two-paragraph ending to a *Bulletin* probably illustrates Housholder's distaste for this whole situation. Even though Housholder was usually a great communicator, in this particular case he must not have fully realized the extreme importance of getting the word out to the membership. In later years, some of those innocent people who were dropped from the Club's list of Grand Slammers were reinstated, but their prized early Grand Slam documentation number was lost forever to one who had been inserted in their place.

Housholder's Stroke

Bob Housholder's last *Bulletin No. 74* went out in July 1989. On September 16, 1989, GSC founder Bob Housholder suffered a serious stroke. His condition would not permit him to carry on his work with the organization. In reality, physically Bob came through the ordeal amazingly well. His long-term memory was even pretty good. However, his short-term memory never returned, and for all practical purposes Bob could not verbally communicate with those around him.

Bob's only immediate family member, his brother Bill Housholder, was left with the re-

Dennis Campbell with his Stone's ram taken in British Columbia in 1982.
Photograph courtesy of Grand Slam Club.

sponsibility of finding a new person to take over operation of the Club. Bill quickly discovered that Bob had a designated group of Directors. He also discovered there was even an Acting Deputy Director, Mahlon (Butch) White of Pueblo, Colorado, who was Number 71 on the Grand Slam list.

The following two letters went out to all the Grand Slam Club members:

November 16, 1989

Dear Grand Slam Club Member:

Both Bob Housholder's brother Bill and I are sad to report to you that Bob's health has deteriorated to the point that he will not be able to continue to handle the Club's business, nor write the newsletter.

Some years ago, Bob asked me to look after the Club's historic records if and when something should happen to him. I plan to work with Bill Housholder to identify and separate what we would determine to be the Club's archives. I plan to see that these are safely stored until such time as their final disposition is settled.

Apparently the Club's records, as to who has paid dues, who is delinquent, and for that matter, who is an active member and who is an associate member are, to say the least, confusing or incomplete. There appears to be only a small balance in the Club's checking account. Your name and address were obtained from the last mailing list we received from the printer.

I would appreciate any suggestions any of you might have as to the Club's future. I am not in a position to manage the Club, nor to write a newsletter. One should remember that to the best of my knowledge, the Club is not incorporated, does not enjoy nonprofit status, and has little or no financial records or cash resources.

Sincerely,
Butch White, Acting Deputy Director

To the Grand Slam Club:

My brother Bob Housholder suffered a stroke on September 16. I have placed him in the Veterans Hospital here in Phoenix after he spent a miserable five weeks in a local hospital.

The doctor's prognosis on Bob isn't very encouraging. There is no sign of any paralysis at this time, but his speech and memory have suffered.

I know the Grand Slam Club was extremely important to him, as he spent the greater majority of his waking hours working with the Club and its many members worldwide.

Best regards,
Bill Housholder

According to Bill Housholder, these letters generated quite a few responses. However, surprisingly only five or six people actually offered to take over the operation of the Club. Of that small number, only a couple seemed to be really serious about it. There was one letter that got Bill and Butch's attention. It came from an Associate Member of the Club, Dennis Campbell of Alabama. It read as follows:

November 30, 1989

Dear Bill, Butch and Bob:

Sorry to hear about Bob's health problems. I certainly hope that he will improve and once again return to publishing the unique Grand Slam Club newsletter. In the event that does not become possible, I submit the following:

Approximately two years ago I sent Bob a proposal concerning the possibility of continuing his work in the event of his inability to do so. Of course, I did not receive a reply, since Bob had made it clear that Butch was to be designated for that job. However, I am still interested, inasmuch as Butch is not able to continue.

I have enclosed a profile about myself, and also some examples of my past work. I do not write for a living, but simply as a hobby and the satisfaction that comes from it.

Some facts: I am a fanatic sheep hunter. I am 38 years old, and began hunting at age 7. Therefore, I am definitely not a newcomer. I have a 'down-home' style of writing, similar to the way Bob has always written.

I hope that the enclosed profile and examples are ample proof of my ability to organize a club such as the Grand Slam Club. I feel that the Club should continue in much the same direction that Bob has carried it. The Grand Slam Club is unique, and should remain so.

I have put a lot of thought into this idea. I realize that it will take a lot of dedication, with little monetary compensation, since the Club is not built upon a firm financial foundation. However, I have the means to make it go, and I feel that I am up to the task.

Sincerely,
Dennis Campbell

After this letter, a dialogue developed between Bill Housholder and Dennis Campbell. The two men came to know each other as well as was possible by phone. Bill and Butch finally made the decision to allow Campbell to come to Phoenix and have one final meeting together, with the full intention of turning everything over to him.

In February 1990, Campbell arrived in Phoenix, where he was met at the airport by Bill Housholder. One of the first items of business was to visit Bob Housholder in the nursing home. Even though Bob could not communicate very well, both Campbell and Bill agreed that Bob was pleased that the Grand Slam Club would continue.

Campbell left Phoenix with a document in hand, completely turning over to him the control of the Grand Slam Club. He and Bill had filled dozens of boxes with the archives Bob Housholder had developed over the years. These boxes were shipped back to Campbell's office in Alabama. Amazingly, there was not even enough money in the Club's checking account to pay for all the shipping.

A New Beginning

In March 1990, Grand Slam Club members received *Bulletin No. 75,* written by the new Executive Director Dennis Campbell. In that issue, Campbell began by introducing himself and trying to reassure the membership that things were not going to change that much.

Dennis Campbell published his first Grand Slam Club Bulletin No. 75 in March 1990. This continued the sequence by Bob Housholder's last newsletter, Bulletin No. 74, in July 1989.
Photograph courtesy of Grand Slam Club.

Following is the actual first two paragraphs of *Bulletin No. 75:*

Dear Member:
Hello! I am a sheep hunter (make that a fanatical sheep hunter), and my name is Dennis Campbell. But more about me later.
This is Grand Slam Club Newsletter No. 75, and by now you will have noticed that it has been quite a while since you received your last issue (July 1989). In October 1989, most of you received a notification from Bill Housholder, Bob's brother, and Butch White, deputy director, regarding the condition of this Club's founder, Bob Housholder. For those of you who did not receive that news, let's go over the details and get an update on Bob's progress thus far.

After that introduction, he went right on in to the same type of hunt reports, quips, quotes, observations and just general sheep stuff that had been so successful for Housholder. Campbell initially stuck with the old format of straight typewritten text on legal-size sheets of paper.

Campbell was immediately uncomfortable with the fact that the Grand Slam Club was not even incorporated. Not only that, but his desire was to see that the Club could be perpetuated in the event he was no longer able to continue as executive director. With these thoughts in mind, in November 1991 the Grand Slam Club was incorporated, and shortly thereafter was granted tax-exempt 501(c)(3) status as a conservation organization. (Actually, the tax-exempt status became retroactive to the November incorporation date.) Life Membership monies began to be invested to create an endowment account for the future.

When the Club was incorporated and bylaws were established, one of the main purposes was that the Club was to be the established documentation and records-keeping organization for legally taken Grand Slams of North American wild mountain sheep. In reality, the Club had fulfilled this purpose since 1972, when the original list was purged of any possible illegally taken Grand Slams. As of mid-1998, the Club was well on its way to having documented nearly 800 legally taken Grand Slams.

FNAWS and the Grand Slam Club

As stated above, Campbell was not happy with the Grand Slam Club being a privately held entity. Legally, he was responsible for paying taxes on any income generated by the

Gary Schoenborn (left) and Dennis Campbell (right) pose for the camera at the 1983 FNAWS convention in Reno. Dennis had just purchased a Dall's sheep hunting permit that was donated by Arctic Red River Outfitters as indicated by the cap Dennis holds. *Photograph courtesy by Grand Slam Club.*

Club. All organizations have to crawl before they can walk, and it took a lot of energy for Campbell to become accepted as the new guy by Club members. He also faced the task of rejuvenating the Club membership, as the roster of paying members had deteriorated dramatically in the last couple of years under Housholder's control. In all fairness, during that period of time Bob Housholder's health was failing rather severely.

After a little over a year at the helm, Campbell had things stabilized and began to dream. Yes, he knew the Club had to incorporate and apply for tax-exempt status, but there was more to Campbell's dream. He had been a member of FNAWS since 1981, when he had become infected with the dreaded sheep fever disease. This happened on a hunt in British Columbia that year. He had attended all the FNAWS conventions beginning with the 1982 New Orleans event. Campbell even helped out with the photography that year, and was one of the full-fledged volunteer photographers for FNAWS' head honcho Lloyd Zeman at the 1983 Reno convention. Being a loyal FNAWS member, he had the dream of somehow developing a closer relationship between the Grand Slam Club and FNAWS.

He actually acted upon this by contacting some of the FNAWS Board of Directors and the office in Cody. He was invited to the July 1991 Board meeting, held in Cody. At that meeting, Campbell seemed to be very-well received. He presented a proposal to the Board for a much closer working relationship. There was an active discussion, and a lot of questions were answered and ideas exchanged. After that meeting, FNAWS' attorney, Dean Picl, was instructed to work out an agreement for an alliance between FNAWS and the Grand Slam Club. Picl and Campbell communicated extensively about this, and the agreement was reached.

Dean presented the agreement to the Board of Directors via a conference call meeting in August 1991. There were some concerns raised as to the future direction that the Grand Slam Club would take. Dean asked Campbell to address these concerns, which he did with a letter dated August 19, 1991. At the next FNAWS Board of Directors meeting, the alliance between FNAWS and the Grand Slam Club was tabled. In reality, it did not come up again for some time.

Grand Slam Magazine

Campbell continued with the Grand Slam Club. He was keenly aware that the Grand

Slam Club Bulletin had been the cementing factor and real success vehicle of the Club. He pledged to the membership to never get away from the unique editorial style that Housholder had established for the publication. For the first three issues, everything was exactly the same as before. However, in Campbell's fourth attempt, *Bulletin No. 78*, Grand Slam Club members saw photographs and professionally typeset text for the first time.

Bob Housholder was still alive when that issue went out, and right there on the cover, as the very first photo ever, was Housholder himself. Even though Bob was not able to communicate well, his brother Bill reported that the organization's founder was most pleased with the addition of photos, and the fact that the Club was continuing.

With issue No. 85, published in July 1992, the publication finally got an actual name: *GRAND SLAM*. Then, in January 1994, the publication progressed to color covers and photographs, and a magazine format that has remained basically the same to this day. Today's *GRAND SLAM* magazine retains its down-home appeal to sheep hunters. The unique style that Housholder began has been maintained throughout all the changes. Because of its long-term appeal, Campbell knew there would be a mutiny of the membership should he ever stray from Housholder's original style.

In *GRAND SLAM*, Campbell often mentioned FNAWS and applauded the efforts of that fine organization. Occasionally he would even editorialize a little about how nice it would be for the Grand Slam Club to have a closer working relationship with FNAWS. At a 1993 FNAWS Board meeting, the subject of the Grand Slam Club came up again. Craig Johnson was president at the time, and was also a member of the Grand Slam Club. He knew Campbell because of that relationship, and also because they had communicated on many occasions. They even had become friends at one of the past FNAWS conventions. Craig was somewhat familiar with that initial proposal back in 1991, but could not remember the details. He called Campbell from a Board meeting, and had him fax all the letters and documents from the 1991 negotiations. Still, no alliance became a reality.

FNAWS Gets Interested Once Again

In mid-1994, certain members of the FNAWS Board of Directors again became interested in talking to the Grand Slam Club. Many Board members did not know the history of the Grand Slam Club or that an alliance had been proposed some years before. The decision was made to invite Campbell to the January 1995 Board meeting, to try to once and for all get all the questions answered, and clear up some misconceptions.

In fact, Campbell did attend that meeting, which became an extremely positive exchange. Board members were candid with their questions and concerns. Campbell was direct with the fact that the Grand Slam Club had desired an alliance with FNAWS for several years. An actual breakthrough came at that meeting. Dan Pedrotti was appointed to chair a committee to develop a "cooperative alliance between the Foundation for North American Wild Sheep and the Grand Slam Club. Negotiations began at that year's convention in Phoenix. Following is the important letter that came out of all those negotiations:

August 3, 1995

Dear Mr. Campbell:

I am pleased to enclose copies of the final draft of the agreement for the formation of a Cooperative Alliance between the Foundation for North American Wild Sheep and The Grand Slam Club. It seems that it has taken a lifetime to get this draft finalized.

I trust that it covers all your concerns and upon receipt of two copies signed by you, I will send it to our president, Mr. Pete Cimellaro, for execution and will return a fully executed copy for your files.

The entire FNAWS Board is looking forward to working more closely with The Grand Slam Club and the benefits that will inure to the wild sheep of North America.

Yours very truly,
Daniel A. Pedrotti, Chair
Grand Slam Committee, FNAWS

So, finally it was done! Two conservation organizations had agreed to work together for a common cause. The agreement was signed on August 14, 1995. The essence of the agreement is as follows:

(1) A cooperative alliance is hereby established between FNAWS and GSC, with the goal of fostering, promoting, and perpetuating their commonality of purposes of benefitting the wild sheep species of North America.
(2) GSC will accept the responsibility of acting as co-chairman of the FNAWS Ram Awards program each year at their annual convention.
(3) GSC will be allowed to have their annual meeting each year at FNAWS' annual convention, and the time of the meeting will be recognized in the convention's printed program.
(4) GSC will be allowed to solicit donations for the FNAWS annual convention auctions. The net proceeds from these auction items will be divided equally between the two organizations.
(5) FNAWS and GSC agree to cooperate in the funding of projects designed to promote their common purposes and refrain from the funding of projects which are injurious to their cooperative alliance.
(6) GSC pledges first right of refusal to FNAWS should GSC be dissolved. This would include assets, archives, documents, etc. This is done as FNAWS is the most logical successor, should GSC terminate its present function.
(7) This agreement is for an initial period of one year, but shall be subject to renewal for additional periods upon the mutual agreement of the parties.

This final agreement was almost identical to the one developed by Dean Picl after the FNAWS Board meeting in 1991. This is a testament to the commonality of the two organizations and the ideas of their leadership.

Success

The alliance between FNAWS and GSC worked perfectly. GSC was prominent in the Ram Awards program at the 1996 convention, and the auctions were filled with several different auction items that were secured through the Grand Slam Club. Then came the Grant-in-Aid funding meeting of 1996, and FNAWS and GSC worked cooperatively to fund thousands of dollars in worthwhile conservation projects.

The agreement was renewed for 1997, and again for 1998. The same successes of 1996 were repeated, but in an even greater way. The Foundation for North American Wild Sheep and the Grand Slam Club had accomplished what was thought to be impossible: a working cooperative alliance.

Dennis Campbell graduated from the Samford University School of Pharmacy in Birmingham, Alabama, in 1976, and became a licensed registered pharmacist in October 1976. He married his high school sweetheart, Nancy, in February 1968. He took over the reins of the Grand Slam Club in February 1990 with the title of executive director, a title he retains today. He and Nancy are parents to five children, Billy, Seth, Chris, Saul and Miriam. The Executive Director's position with GSC is totally voluntary on Campbell's part. His livelihood is gained from ownership of Campbell's Pharmacy, Campbell's Publishing and Campbell's Properties, all of Adamsville, Alabama.

The Fraternity of the Desert Bighorn: Early 1960s to Present

Edward Pribyl, Winona Denning and Virginia Frehner

The Fraternity of the Desert Bighorn (FDB), in cooperation with the Bureau of Land Management (BLM), Nevada Division of Wildlife (NDW) and the U.S. Fish and Wildlife Service (USFWS), is famous for its ability to finance and complete a variety of water projects.

The 1.5 million-acre Desert National Wildlife Range was established in 1936, primarily for the purpose of conserving the desert bighorn sheep. Shortly after its establishment, the U.S. Fish and Wildlife Service began developing water supplies for sheep. In the early 1970s the BLM stopped all projects on BLM lands until they could complete habitat management plans (HMPs). Since most of the projects were on BLM lands, the Fraternity went into hibernation for a few years for as long as it took to write the HMPs. After the HMPs were completed the Fraternity was reorganized, thanks in large part to Jerry B. Hughes and Stephen P. James and the Fraternity members who helped NDW, BLM and USFWS.

In the early 1960s Jerry B. Hughes and Stephen P. James met with agencies involved with land and game management. They appealed to have the desert bighorn sheep put on the big-game animal classification. They expressed their concerns for the lack of information available on the Nevada state animal. The information they received was that there would be needed changes in the state legislature and studies completed on both the animal and its habitat before any changes could be made.

Jerry and Steve were members of the Master Guides in Nevada. The length of time they spent in the field indicated to them there existed a "huntable" population of desert bighorn. There were no existing management programs and no accurate counts at that time. They found out what information they needed to bring about the desired results. Jerry and Stephen started to get the job done. They organized FDB. Steve was a plumbing inspector for the city of North Las Vegas at that time so he was familiar with political "red tape" and started working with the state legislature. Jerry, the talkative one, started recruiting friends and associates to help with the studies and desert bighorn sheep counts. Cooperating with USFWS and BLM, counts were made on horseback, foot and airplane. When money became available to rent air time and pilots, this small but determined group of men and women used helicopters for the counts.

Since its reorganization, the Fraternity has assisted the U.S. Forest Service with the construction of 17 new water developments on the Desert National Wildlife Range. They have also completed 13 major improvement projects at these and other developments on the range. To achieve these accomplishments, the Fraternity has utilized more than $38,500 of its budget and many thousands of dollars of financial aid from other organizations, including FNAWS, Safari Club International (SCI) and Nevada Bighorns Unlimited. Furthermore, the Fraternity has donated in excess of 5,000 hours of labor for this work.

"Without organizations such as the Fraternity and the others mentioned above, the future of the desert bighorn in Nevada would not be nearly as secure. It is the Fraternity's willingness to toil with their hand tools under extreme environmental conditions without compensation that truly makes these projects happen. Without this sort of dedication, which no amount of money can acquire, the objectives of the U.S. Fish and Wildlife Service could not be met," explained USFWS Wildlife Biologist Bruce Zeller of Las Vegas who is also active in the Fraternity.

Help expanded as new members joined and volunteered their time and services. We checked water holes, seeps, springs and vegetation. Herd composition was determined and a much larger population was found. The estimated number of bighorn sheep were between 500 and 600 animals. Not much was known about the sheep. As Dr. Charles Hansen said, "A sheep is a sheep is a sheep." That was all we really knew for sure. In cooperation with USFWS, BLM and USFS, members of the Fraternity of the Desert Bighorn learned more about this animal together. Within a matter of weeks after forming this group of volunteers, spring rehabilitation, trapping and water catchments were organized and completed. For those you who have been around for a while may have seen Curt Goudy's show, "Wide World of Sports." One of the features showed men and women trapping bighorns on Lake Mead with guest host Mark Harmon. Those workers were 99 percent members of the newly formed FDB.

Jerry B. Hughes, founder of the Fraternity of the Desert Bighorn, holds his last outstanding ram. Short in stature at 5 feet, 1 inch, Jerry was big in heart.
Photograph courtesy of Fraternity of the Desert Bighorn.

Some of our early members included Stephen P. James, Jr., as first president, Carol James, Connie Fuller as first secretary / treasurer, Jerry B. Hughes as second president and Nona Hughes as second secretary, Marvin Einerwald, Carl Ciliax, Manual and Jackie Hurtado, Sharon Reed, Hagan Thompson, Dennis "Bud" and Delores Walkington, Ken Kelly, Edward and Midge Pribyl, Ned Beardon and Royce "Woody" Wood.

A short section about Jerry Hughes is in order, since this chapter is dedicated to him. Jerry was an avid sportsman and hunter from a very young age. Born on a small ranch east of Mesquite, Nevada, he was the youngest son in a family of nine children. He rode horseback 14 miles to school. There were many days when a flock of ducks or geese would set down on the Virgin River and school would be forgotten. When he was in his early teens, Jerry helped his cousin, Milt Holt of Gunlock, Utah, guide deer and mountain lion hunts. This was the start of his career as a master guide in Nevada. Jerry guided many hunters from all walks of life. One of his favorites was Bruce Brady, editor-at-large for *Outdoor Life* magazine. Bruce took deer, mountain lion and sheep with Jerry. Bruce included stories of his hunts with Jerry in many of his articles. Bruce thought so highly of Jerry that the desert bighorn he collected on his hunt with Jerry was sculpted by Bruce in a bronze and a copy was donated to the Fraternity in appreciation to Jerry. In his book, *Bruce Brady's Game Trails,* included several of his hunting stories with Jerry. Bruce also donated two copies of his book for our annual auction.

Jerry Hughes was a master guide, father of five children and grandfather of nine. He loved the outdoors, hunting, fishing and photography. He had a phenomenal memory and was known as a storyteller deluxe. Bruce Brady said, "Jerry has the best game eye I've ever witnessed." Short in stature at 5 feet, 1 inch, Jerry was big in heart. He moved from Las Vegas to Mt. Wilson Ranch, located in midwestern Nevada, to be closer to wildlife. His sons Marl, Lomond "Lonnie" and Brian, and his daughters Tona Lee and Myrna were all enlisted at various times as camp cooks or as horse wranglers. From the time his children were born they spent as much time in the hills as they did at home. They were taught good camping, hunting and fishing practices and have retained a love of all outdoors.

Jerry was guiding deer hunters in an area called Table Top Mountain out of Tonopah, Nevada, when he suffered a fatal heart attack. We all miss him but remember the good times we've had. His tales were tall but he delighted us all with the story changing from telling to telling. Sometimes the same hunt i was on bore faint resemblance to what I remembered but it was entertaining if not strictly factual. Wherever he is now, if you find a circle of friends gathered, they are probably listening to one of Jerry's tales.

Since its founding, Fraternity members established five principles and followed them as they planned and completed projects. The principles are:

(1) Further the knowledge of desert bighorn sheep.

(2) Support the various governmental agencies directly responsible for the management of desert bighorn sheep.

(3) Become more aware of problems involved with desert bighorn sheep and advancing civilization.

(4) Educate and inform the public of the desert bighorns' habits and its welfare and utilize our united strength for their well-being.

(5) Encourage the utilization of the desert bighorn in every phase not determined to its welfare.

The Fraternity's objectives were being met, thanks to the volunteer workers, and the state legislature changed. The desert bighorn was classified as a big-game animal. Studies were completed. As our studies were completed and it was found there was a sufficient huntable population, next came the rules and regulations for the hunt. It was decided to issue tags on a drawing basis. Areas were designated according to the studies completed by the Fraternity and governmental agencies. Harvesting older rams that might be dying seemed the best idea. Average age for an older ram was from 7 to 14 years. Breeding age was mostly determined to be prior to age 7. This plan kept the population up and allowed hunters to harvest older animals that were in their downward cycle of life. Working together with governmental agencies, members of the Fraternity helped create a sensible criteria yet one easily met by hunters.

The plan that received acceptance was to harvest rams that were 7 years and older that had a minimum Boone and Crockett Club score of 144 points. A mandatory indoctrination course was required to teach hunters how to identify these animals in the field. Each hunter drawn for a permit was required to carry a spotting scope of at least 15x power and to use binoculars. This plan was put into effect and is now being used in most areas of Nevada. Some areas have been made "any ram" because of further knowledge about that particular population. After achieving such success through the five objectives, the Fraternity members became inactive for a short time. Then in 1972 it was revitalized with our objectives to develop water holes and reintroduce sheep into areas where there had been populations years before.

The Fraternity has grown to approximately 200 members from all over the world who are small in number but big in action. When it comes to getting work done there is nobody who can do it better than our group. When you read the list of past officers, board members and

Worker bees. That's what the members of the Fraternity of the Desert Bighorn do so well as they construct water projects to benefit the desert bighorn sheep of Nevada. Here worker bees buzz around a water project in rugged terrain.
Photographs courtesy of Fraternity of the Desert Bighorn.

"Sheep Camp" is a popular family event hosted by the members of the Fraternity of the Desert Bighorn that cements friendships and raises money for bighorn sheep. Here we see a typical Sheep Camp event, enjoyed good food, in 1997.
Photograph courtesy of Fraternity of the Desert Bighorn.

Two projects, completed at the same time but 35 miles apart, helped provide water to desert bighorn sheep. This is the "Jenny" project. Since its founding by Jerry B. Hughes and Stephen P. James and a core of Nevada volunteers in the late 1960s, the Fraternity of the Desert Bighorn has helped increase desert bighorn populations and improve habitat for all desert wildlife.
Photograph courtesy of Fraternity of the Desert Bighorn.

regular members, it is a who's who of desert bighorn supporters in Nevada. When the Fraternity of Desert Bighorn is mentioned, all our members stand tall and proud because they know that they are playing an important role in restoring this magnificent desert bighorn sheep to the mountain ranges of the great Southwest.

The who's who are not only the officers of the Fraternity but the people in southern Nevada who make it what it is today. Along with support from Nevada Bighorns Unlimited of Reno, Fallon and Elko, and Las Vegas chapters of SCI and other supporters, we continue to work to achieve our goals to establish desert bighorn sheep in all historic ranges and habitats.

The Fraternity members all have a bad case of "sheep fever." If you are a sheep person, you now know why we are so good at what we do. Let's talk about the real stars of desert bighorn sheep who we work so hard to help. Some of our favorites over the years have included "Longhorn," "Bags," "Snow Plow," "Leftie," "Rightie," "Chiphorn" and "Ugly." These great old rams roamed the canyons and peaks of the Nevada desert. Many a night around a campfire was spent telling stories about these great rams.

Our members have donated more than $1 million in labor, materials, equipment and funds for the more than 100 water projects completed since the organization began. All donations go directly to projects, as we have no salaried staff. Donations and memberships are put to good use for the desert bighorn sheep.

Trapping has become necessary for food and herd betterment. When herds become too large for the area they roam and feed becomes scarce, the chance of disease, especially soremouth, becomes high. Soremouth occurs when the feed is eaten so low that the sheep's mouth gets sore as it tries to graze. When this happens, infection begins and spreads. This can destroy a herd if not recognized in time. By trapping the animals from an over populated area this medical condition can be prevented. At the same time trapping augments our Nevada herds in other areas where bighorn populations are low or provides animals for other states.

Our water hole developments throughout the years have been accomplished with the cooperation of different groups including USFWS, BLM and USFS. Studies are made well in advance of the project and sites are chosen with care. Projects are carefully constructed to make them practical but also aesthetically pleasing. We are very careful not to impact any area to its detriment. The Fraternity members, in most cases, provides all materials and labor. Donations come from our individual members as well as such groups as FNAWS, SCI, Ducks Unlimited and many other groups that are eager to participate in a truly worthwhile endeavor. Our members spend thousands of hours of their time to help and put these sites in. Their completion is only the beginning as we maintain these developments and check them regularly. We work with the government agencies in locating new sources of water and keep current sources available to wildlife.

The Fraternity has received numerous national awards for its continual work on behalf of the desert bighorn sheep. The Nevada Wildlife Federation granted us the Conservation Award. The Presidents Award was given to us by the National Wildlife Federation. BLM presented their Conservationist and Management Award to us. USFWS gave us their Outstanding Contribution Award. We have also been given the Volunteer Action Award by northern Nevada. The Fraternity was given the Conservationist of the Year Award by The Wildlife Society's Nevada Chapter. They have also received the President's Volunteer Action Award and the Take Pride in America Award.

The amount of money raised and spent on behalf of desert bighorn by this group is astounding. From July 1994 to June 1998, the Fraternity has raised and spent in excess of $120,600 in equipment, labor and materials. For example, four 1,800-gallon polyethylene tanks costing $3,100 each were donated; a hypalon apron was donated that cost $3,954. Helicopters, all-terrain vehicles, airplanes and trailers have also been donated as well as more than 3,000 hours of skilled and unskilled labor during that same time period alone.

We have fundraisers twice a year. The annual "Sheep Camp" is lots of fun and is strictly on a donation basis. We host a barbecue-type cookout with all the food and drinks for two nights and days furnished to the participants. We have nature hikes, volleyball, horseshoe pitching, a live band for dancing. Many just enjoy conversation with friends beside the campfire. We also do a banquet at one of the hotels where we auction donated items. This is also an enjoyable event that keeps us going.

Over the years, Fraternity members have participated in water projects in the following areas:

Silver Peak Range
Arrow Canyon Range
Muddy Mountains
North McCullough Mountains
Highland Range reconstruction
Mormon Mountains reconstruction at Big Bertha, Rocky and Prospect
River Mountains cleanup
Last Chance Range
Government Springs reconstruction
North Hiko Range
South Hiko Range
Egan Range
Wire Grass Springs reconstruction
Renovated Hidden Forest cabin
Upper Deadman Spring reconstruction
Gravel Canyon Project repair
Bare Mountain
Specter Range
East Pahranagat
Highland Range pockets
Spotted Range
White Pine Peak
Devils #1, #2 and #3
Pahranagat Range
East Desert Range
Desert Range
Pintwater Range
Mud Spring enhancement
Las Vegas Range
Sheep Range, Joe May, Woody and Desperation Canyon

The organization meets the first Wednesday of each month and approximately six projects per year are completed that involve the welfare of desert bighorn sheep and that meet the principles of the group. To become a regular member, the cost is $15 per year. A family membership per year costs $25. An individual life membership is $250 and a corporate an-

Virginia Frehner and her hat smile for the camera at the 1996 national FNAWS convention. Virginia has been an active member of the Fraternity of the Desert Bighorn and coordinated the writing of this chapter.

nual membership is $500. To join the Fraternity of the Desert Bighorn, write or call them at P.O. Box 27494, Las Vegas, NV 89126-1494, (702) 385-2407 or (702) 873-5105.

We were asked to list all the past presidents and other members, but the Fraternity is about people — people who give up their weekends; who donate their equipment; and who contribute money. Most of all, these are people who love desert bighorn sheep.

Edward "Eddy" Pribyl has lived in Las Vegas, Nevada, since 1949. He was in the first graduating class at St. Joseph's Grade School and graduated from Las Vegas High School in 1955. He has been an auto mechanic since 1953 and a shop owner since 1985. He married his wife Midge in 1956 and has lived in the same house since then. Eddy is the father of four sons who are all active in wildlife projects and is grandfather to one girl and three boys. Eddy is current president of the Fraternity and has served before as president and member of the Board of Directors.

Winona Denning of Las Vegas was Jerry Hughes' first wife and mother of his five children. They married in 1955 when Winona was 16 and Jerry was 19. They started hunting together when Winona was 15, which she says holds special memories. She loves the outdoors and accompanied Jerry on many hunts. She was told on her first hunt that if she shot an animal that she had to care for it, so she has done so ever since. Winona was married to Jerry 17 years and remained good friends with him even after the divorce. She is now married to Gary Denning. She remains as active as possible even though she is recovering from a broken back injury. Gary and Winona love to travel but hiking and backpacking is not possible.

Virginia Frehner of Las Vegas is a newly elected member of the National FNAWS Board of Directors and is an active member of the Fraternity of the Desert Bighorn. She is a single mother of two adult daughters, Shannon and Kim, and grandmother to grandson Kyle. Virginia is an avid hunter and fisherman. She spends some summers packing her horses into the back country of Idaho with her grandson and girlfriends. She enjoys her Harley Davidson motorcycle. Virginia volunteers her time for FDB, SCI and FNAWS and has helped coordinate and instruct the first annual youth experience at Burns Hermitage Ranch in Lincoln County, Nevada. This experience allows young people to enjoy a weekend of instruction in archery, shooting, hiking, fishing. They also learn responsible uses of the outdoors. Virginia recognizes that youth are our future hunters and wildlife conservationists.

Society for the Conservation of Bighorn Sheep 1965 to Present

The Society for the Conservation of Bighorn Sheep based in La Canada, California, is dedicated to restore natural resources that support bighorn sheep. Through their work to construct water projects and to improve wild sheep habitat, young rams and ewes as seen here will continue their population expansion into the next millennium.
Photograph by Valerius Geist.

The Society for the Conservation of Bighorn Sheep (SCBS) has been registered as a California nonprofit corporation since 1965. We are the only organization in the state that is solely dedicated to restoring a natural resource that suffered a 98-percent population decline from the Gold Rush of the mid-1800s to 1877, when California's legislature declared bighorn sheep as a protected species.

Our mission is habitat improvement in the California deserts, on state and federal lands, military bases and mitigation areas. We build guzzlers, improve springs, clear tenajas, eliminate nonnative tamarisk and remove rubbish.

SCBS administers the Volunteer Desert Water and Wildlife Survey for the California Department of Fish and Game under a Memorandum of Understanding (MOU) signed in 1969. We raise funds, plan projects and work alongside DFG's habitat biologists and wildlife biologists on all major projects, such as construction, maintenance and repair of more than 70 big-game guzzlers that cost an average of $10,000 in materials. These guzzlers catch, hold and dispense up to 4,500 gallons of water each to resident insects, birds, reptiles and mammals.

SCBS has no paid staff. To perform our mission, we rely on an extensive volunteer network that ranges from Alaska to Maine and all points south. Our habitat improvement project mailing list for 1996 includes 1,300 volunteers and members. In the past 10 years, our volunteers have driven a minimum of 1 million miles and have logged at least 80,000 volunteer hours, all on weekends or spare time, for projects and wildlife census studies.

The Society has been reversing bighorn population declines by funding studies, improving habitat and reintroducing them to their historic ranges in accordance with DFG's Bighorn Management Plan of 1968. Their population has increased almost 80 percent since then.

The desert bighorn population had recovered sufficiently by 1987 that SCBS promoted legislation to permit limited hunting of them while leaving the Sierra and Peninsular subspecies on the state protected list.

Today, major obstacles to bighorn recovery include low birth rate, high infant mortality, increasing predation rate, domestic livestock diseases, feral burro competition in the same ecological niche, habitat loss, interstate highway system and federal laws.

To receive further information about the work of the Society for the Conservation of Bighorn Sheep, please leave a message at (213) 256-0463.

This article was submitted to the national headquarters of FNAWS by the Society for the Conservation of Bighorn Sheep.

The Arizona Desert Bighorn Sheep Society, Inc. 1967 to Present

STEPHEN WILLIAMS AND TRACEY MARTIN

Vision. In the business world, corporate leaders who have vision or the ability to anticipate that which will come to be, are the most successful. This analogy applies to wildlife conservation organizations as well. The Arizona Desert Bighorn Sheep Society, Inc. (ADBSS) was inspired by a vision of what was to be.

It all started on a desert bighorn sheep hunt in 1965. Bob Stonoff, a high school biology teacher, was the fortunate permit holder. While on his desert bighorn sheep hunt in the Buckskin Mountains, he watched a ram perched on a ledge above him, silhouetted against the horizon. This awe-inspiring spectacle prompted Bob to think about sharing his excitement and knowledge with other desert bighorn sheep hunters.

This knowledge was shared the next fall at an informational seminar on desert bighorn sheep. It was held at the auditorium of Jeff Sievers' Taxidermy Studio. All hunters who drew desert bighorn sheep tags were invited. The seminar was an overwhelming success. There was standing room only.

The hunter education seminar was held again the following year. That year, the discussion after the event turned to creating an organization devoted to the conservation and welfare of desert bighorn sheep. The core group of interested individuals had diverse backgrounds, but the strongest common thread among them was their desire to do something beneficial for desert bighorn sheep in Arizona. And so, the Arizona Desert Bighorn Sheep Society, Inc. was born.

The ADBSS's first year of existence was 1967. The officers for this year were President Bob Stonoff; Vice-President Jack Walters; Secretary-Treasurer Jim Pierce; Dean Bowdoin in charge of the ADBSS constitution; Ned Smith in charge of finances; John Houzenga in charge of promotions; John Russo in charge of resolutions; and Manuel Sinohui in charge of publicity.

ADBSS was incorporated on June 11, 1969, as a society "dedicated to the advancement and recognition of the desert bighorn sheep; to support the management, conservation, restoration and utilization of the desert bighorn sheep; and to oppose any action or legislation that would adversely effect the desert bighorn sheep or its habitat." ADBSS has grown from 62 charter members to approximately 1,400 sportsmen and sportswomen, wildlife conservationists, professional wildlife managers, educators, biologists, photographers and others who are interested in desert bighorn sheep.

Each October the Society conducts an annual sheep clinic training session for the lucky few hunters who drew sheep permits. The Society feels that the hunters who have attended the clinics have attained a statistically better record of harvesting higher trophy score rams and avoiding the mistakes of shooting at illegal immature rams than have the hunters who did not attend. The official publication, The Rams Horn, is the informational newsletter that is mailed to all members four times a year.

The goals of the Arizona Desert Bighorn Sheep Society are clear. They are:

(1) Development of water resources for desert bighorn sheep.

(2) Re-introduction of desert bighorn sheep into suitable historic ranges.

(3) Prevention of encroachment into vital sheep habitat.

(4) Promotion of research necessary to understand the needs of desert bighorn sheep.

(5) Control of feral burro populations, competition with domestic livestock and predators where necessary.

(6) Assistance of government agencies in sheep population surveys.

(7) Publicity to promote public awareness of desert bighorn sheep and their problems.

Securing a coveted desert bighorn sheep permit is a highlight of a hunter's life in Arizona.

The officers decided the priority for the new organization would be the construction of water holes in the rugged and arid mountains that comprise desert bighorn sheep habitat. An overture was made to Dan Schadle of the Development Branch of the Arizona Game and Fish Department to pool the ADBSS's labor and resources with the Department's engineering skills and equipment. A mutual understanding was reached. The two entities have worked together for 28 years. The record of achievement on behalf of desert bighorn sheep is incredible. To date, 149 cooperative water development projects have been completed. Water holes are designed to provide up to two years of reliable water during dry periods. Storage capacities vary from an average size of 12,500 gallons up to 100,000 gallons.

Through the years as the ADBSS became proficient in water hole development, it realized there were other aspects of desert bighorn sheep conservation needing attention. There was historic desert bighorn sheep habitat that was unoccupied and in need of transplants. Environmental impact statements needed to be reviewed and position statements prepared. Elected officials needed to know how the ADBSS stood on the various proposals that impacted desert bighorn sheep habitat. Educational outreach efforts were expanded beyond the Hunter Clinic to include presentations on desert bighorn sheep management to Kiwanis luncheons, Boy Scout troops, hunter education classes and National Hunting and Fishing Day observations.

In 1984, the ADBSS began fund-raising efforts to ensure desert bighorn sheep management remained at the forefront of the Arizona Game and Fish Department's planning and management efforts. With its characteristic enthusiasm. ADBSS held a series of 14 successful fund-raising banquets. Also, since 1984, ADBSS has conducted an annual raffle and auction for desert bighorn sheep permits, after obtaining special legislative and Arizona Game and Fish Commission authorization. To date, more than $2.6 million has been raised for desert bighorn sheep management in Arizona.

The ADBSS has a proud history of involvement with the Foundation for North American

Wild Sheep (FNAWS). Men and women of great dedication and energy have served both ADBSS and FNAWS through the years. FNAWS past presidents who have risen from the ranks of the ADBSS include Tom Saad, John Harris and Pete Cimellaro. The Foundation has recognized ADBSS members and Past Presidents Dean Bowdoin and Bill Hook as recipients of the Gordon Eastman Award for their hard work and dedication.

Past presidents of ADBSS include:

Bob Stonoff, 1966-1967; James Pierce, 1968-1969; John Houzenga, 1969-1970; Ed Shannahan, 1970-1971; Tom Saad, 1971-1972; Cecil LaVance, 1972-1973; Bob Gray, 1973-1974; Doug Reynolds, 1974-1975; Ned Smith, 1975-1976; John Krause, 1976-1977; Dean Bowdoin, 1977-1979; Jerry Petkovsek, 1978-1979; Tim Blank, 1979-1980; Tom Boggess, 1980-1981; Don Johnson, 1981-1982; Harry Hussey, 1982-1983; Bill Hook, 1983-1984; James Fiedler, Jr., 1984-1985; Joe Bill Pickrell, 1985-1986; Jim Marquardt, 1986-1987; Pete Cimellaro, 1987-1988; Dean Bowdoin, 1988-1989; Joe Machac, 1989-1990; Louis Coor, 1990-1991; John Harris, 1991-1992; John Vassel, 1992-1993; Richard L. Robles, 1993-1994; Matt Dominy, 1994-1995; Warren Leek, 1995-1996; Ken Meadors, 1996-1997; Gary Allen, 1997-1998.

Charter members of ADBSS, in alphabetical order, include the following:

George Atherhott, Jr., Ben Avery, Tom Barrett, Richard Beaudry, Lowell Berry, Dean Bowdoin, Eddie Brooks, Virgil Buerman, Lee Burcky, Robert Byrnes, H.W. Conrath, Vernon Currens, Wade Dominy, Jim Englemann, Richard Fits, William Foley, Drew Getgen, Allen Gilbert, M.D., Howard Gilmore, Fred Greenwald, Austin Haydis, Hy Hancock, Jr., Bob Hembrode, George Hill, P.D. Holt, John Houzenga, Howard Kincheloe, J.C. Kitts, Jr., Karl Koons, Terry Landrith, Claude Lard, Bernell Lawrence, J.E. Mason, Tommy Meredith, Killen Moore, William MacDonald, Robert McDonald, Danny Nasca, Robert Orth, Kenny Perkins, Hal Perry, S.D. "Pete" Peters, John Phelps, Jack Pierce, Henry Rodriguez, John Russo, Tom Saad, Frank Saller, Carl Scrivens, Edgar Shannahan, T.E. Shillinburg, Jeff Sievers, Ned Smith, Lester Stewart, Edwin Stockwell, Bob Stonoff, Jim Tankersley, James Watkins, George Welsh, Donal Williams, Don Wingfield, Leo Yates and Mark Zimmerman.

The volume of funds raised by ADBSS is outstanding. Starting with records from 1984 and continuing through 1999, the annual banquets and auctions provide the funds necessary to make the ADBSS projects possible. In 1984, the annual banquet and auction raised $32,234. Tom Bogges III was chairman of the event. Joe Bill Pickrell headed the 1985 event that raised $34,419. In 1986, Pete Cimellaro was chairperson for the banquet that raised $55,988. Paul Ruskaup was chairperson in 1987 and 1988 that raised $47,130 and $41,200, respectively. Once again, Joe Bill Pickrell headed the fund raiser in 1989 as he had in 1985. That year $50,394 was raised.

Jim Fiedler, Jr. was chairperson for the fund raiser in 1990 and $28,547 was raised. Richard Robles was chairperson in 1991 and 1992 when $36,000 and $41,013 was raised, respectively. A total of $50,265 was raised in 1993 when Don Yost was chairperson. Warren Leek headed the 1994 fund-raising effort when $35,267 was collected. Brian George was chairperson in 1995 when $43,188 was raised. Nancy Lewis chaired the 1996 fund raiser that collected $46,294. Again, Joe Bill Pickrell headed the fund-raising event in 1997, as he had done in 1985 and 1989, and $57,500 was raised.

As he had done in 1994, Warren Leek chaired the fund raiser in 1998 and the event raised $62,996. In 1999, James Roe is chairman for the fund raiser.

Chairmen who raised the most during their service as head of annual fund raisers included Joe Bill Pickrell ($142,313), Warren Leek ($98,263), Paul Ruskaup ($88,330) and Richard Robles ($77,013).

A typical water development work project.

Water development is the final project.

Sheep under the net during a capture operation at Lake Mead.

Captured sheep ready to move to a release site.

The Arizona Desert Bighorn Sheep Society, in association with the Arizona Game and Fish Department or the U.S. Fish and Wildlife Service, have developed an average of five water hole improvements per year from January 1969 through 1998. The complete list is too extensive, but the first four areas where projects took place were in the Palomas, Gila, Black and Upper Aquila mountains. The first projects were named Palomas Mountain Catchment, Dripping Springs, Wilson Ridge Spring and Thompson Catchment. The Lazarus Tank project in the Plomosa Mountains has been the focus of activity in February 1970, April 1973, March 1976, January 1980, January 1984, March 1989 and January 1991.

Twelve projects have each been worked on three times from 1971 to 1993. These projects are called "Recluse" tank in the Dome Rock Mountains in January 1971, March 1978 and March 1981. "Thanksgiving Day" tank in the Sauceda Mountains had work done on it in May 1971, March 1973 and March 1974. "Box Canyon" bucket tank in the Clanton Hills received volunteer labor in March 1972, January 1973 and March 1991. The "Betty Lou" cistern in the Copper Mountains worked occurred in January 1974, June 1981 and March 1982. The "Lost Drill Bit" tank in the Black Mountains was worked on in March 1974, May 1976 and May 1977.

Other projects worked on three times by ADBSS members and supporters include the "Socket" tank in the Tank Mountains (1976, 1979, 1993), "Van Deemen" tank in the Black Mountains (January and May 1977 and 1981), "Black" tank in the Crater Mountains (1977, 1979, 1991), "Little Bones" tank in Plomosa Mountains (1979, 1984 1991), "Dominy" tank in the Mohawk Mountains (1982, 1985, 1987), and "Aubrey Peak" tanks in the Rawhide Mountains (1985, 1986, 1989). "Triple Eye" tank in Eagletail Mountains received work four times on March 1979, April 1979, March 1980 and March 1987.

Many awards have been presented to ADBSS over the years in recognition of its outstanding efforts to conserve the desert bighorn sheep and its habitat. A sampling of awards include the Desert Bighorn Council "Ram Award," "Good Egg Award" from the Arizona Zoological Society for construction of a unique and beautiful desert bighorn sheep exhibit, The Wildlife Society's "Group Achievement Award," FNAWS National Conservation Award and the Arizona Game and Fish Department's "Commission Recognition Award."

To receive more information about membership in the Arizona Desert Bighorn Sheep Society, Inc. or to volunteer to help with water projects helping desert sheep, call or write ADBSS at 6557 E. Riverdale, Mesa, AZ 85215; (602) 832-8107.

Stephen Williams is an active member of the Arizona Desert Bighorn Sheep Society, Inc. and Tracey Martin is the ADBSS administrator based in Mesa, Arizona.

Rocky Mountain Bighorn Society 1975 to Present

Bob Tully, Hans W. von Barby, Bud Henderson,
Marv Clyncke and Dale Martin

There is a world of difference between ducks and bighorn sheep but, to a considerable extent, the origin of the Rocky Mountain Bighorn Society (RMBS) can be traced to waterfowl, specifically to Ducks Unlimited, according to RMBS founder Hans W. von Barby.

"During my tenure as Colorado state chairman for Ducks Unlimited (DU) in the early 1970s, I was impressed with the effectiveness of an organization that was totally dedicated to one cause. While some national conservation groups were trying to tackle every conceivable environmental problem, DU, with its single-minded purpose and a proven program, was able to do amazing things for waterfowl conservation," Hans wrote.

"It struck me that an organization dedicated solely to the conservation of the bighorn sheep could be similarly effective," Hans continued.

Research showed the idea was not a new one. Arizona, California and Nevada had established very effective organizations. Bob Housholder organized the Grand Slam Club in Phoenix many years before, but his efforts went in another direction, Hans explained. What was later to become the Foundation for North American Wild Sheep (FNAWS) was "in the embryo stage" at the time. Wayne Sandfort and Bob Tully, officials at the Denver headquarters of the Colorado Division of Wildlife (CDOW), encouraged Hans to organize a Colorado organization dedicated to bighorn sheep.

The RMBS was formed in Denver as a nonprofit corporation in April 1975 by a dedicated group of private individuals under the leadership of Hans of Evergreen, Colorado. The initial Board of Directors consisted of 14 individuals of diverse backgrounds and experiences, but with a common interest in the welfare of the bighorn sheep. The RMBS currently has 900 members in 42 states and a number of foreign countries.

The founding Board of Directors reflected an initial objective of bringing together a group of dedicated people whose interest in wild sheep conservation was based upon a wide variety of reasons. Harry Combs, a nationally known aviation executive and former member of the Colorado Wildlife Commission (CWC), envisioned our group as eventually raising sheep under controlled conditions for later introduction into the wilds, similar to trout farms. This was a commendable concept wildlife biologists unfortunately discounted as impractical and unworkable. Marty Stouffer, an Aspen resident and well-known wildlife photographer and wildlife film producer, provided early support, as did Clayton Freiheit, who was director of the Denver Zoo. Substantial guidance came from Colorado State University (CSU) research and teaching faculty thanks to Gene Decker, Chuck Hibler and many others. Over the years, the Society has made a concentrated effort to maintain this diversity within the board that represents the organization's leadership.

From the outset, the Society has enjoyed a close relationship with the state and federal land and wildlife agencies, whose efforts toward the management of the bighorn sheep the organization has actively encouraged and supported. A key objective has been the encouragement and promotion of the knowledge and enjoyment of the wonderful wildlife species

Colorado Governor Roy Romer, seated, signs the Bighorn Sheep Raffle and Auction Bill in 1988 that began a long and successful tradition of raising millions of dollars to put wild sheep on Colorado and western mountains. Members of the Rocky Mountain Bighorn Society join in the ceremony.
Photograph courtesy of Tom J. Brown.

The first U.S.D.A. Forest Service meeting involving many conservation groups in Washington, D.C. included, left to right, Rocky Mountain Elk Foundation President Gary Wolfe, Rocky Mountain Bighorn Society President Tom J. Brown and USDA-Forest Service Chief Dale Robertson.
Photograph courtesy by Tom J. Brown.

Hans W. von Barby with his beloved golden retriever, Fiddler. Hans was co-founder and first president of the Rocky Mountain Bighorn Society. He has served as president of RMBS many times.
Photograph courtesy of Hans W. von Barby.

The National Bighorn Sheep Interpretive Association operates the very-popular National Bighorn Sheep Interpretive Center in Dubois, Wyoming, that teaches the public about the western bighorn sheep populations. National FNAWS, FNAWS chapters and the Rocky Mountain Bighorn Society have contributed thousands of dollars to the center that opened July 2, 1993.
Photograph courtesy of National Bighorn Sheep Interpretive Association, Inc.

among the youth and the nonhunting public.

Our first fund-raising effort consisted of the sale of beautifully made bronze belt buckles by taxidermist-sculptor "par excellence" Jack Putnam of Franktown. The buckle depicted a bighorn ram's skull and horns, was made in a limited edition, signed by the artist and limited to 100. The buckles sold for a modest $15 each. Those who own one have a true collector's item of substantial sentiment and value. Jack later made a ram's head belt buckle designed in both bronze and silver that also became a part of Society history.

The first annual meeting was held May 1976 in Colorado Springs with the well-known outfitter and outdoor writer, Andy Russell of Waterton, Canada, as the keynote speaker. Subsequent annual meetings occur each year and have been most successful when located on the Front Range centered around Denver, Colorado Springs and Fort Collins where the concentration of membership is located. The group once ventured to Vail where there was an outstanding program but low attendance.

The most important activity of each year is the annual meeting and fundraiser held during the spring. This is the time many volunteers come together to rededicate themselves and their commitment to the sheep. Funds are raised that are needed to carry out the many tasks necessary to maintain and enhance our sheep populations, their management and their sometimes declining habitats. Whereas early fundraisers generated a few hundred dollars of income, recent results approach $50,000!

The one- or two-day annual fundraiser and banquet provides opportunities to have educational seminars and lectures by professional wildlife managers, noted authors and photographers, survival experts, conservationists and sheep hunters. Their contributions to the annual meeting assist the membership in having positive, lasting memories of their volunteer effort, hunting activities and fellowship.

We also recognize outstanding sheep trophies, sheep hunters, donors and individuals who have made special efforts during the past year or more. This is the time for one of our most rewarding annual educational efforts. Each year, school children are encouraged by teachers throughout Colorado to participate in a poster and essay contest about bighorn sheep. Winners are invited to a dinner and recognized by Society members and given ribbons for first, second and third place. The top winners receive a $75 U.S. Savings Bond. All the children are featured in the Society's quarterly newsletter, The Bighorn, after the award banquet. As many as 800 posters and essays have been submitted each year. We have conducted essay contests for school children as another way to instill interest and knowledge of bighorns for future generations.

Annual meetings for the Rocky Mountain Bighorn Society have bounced around Colorado throughout its history. The first organizational meeting was at the Denver Marriott on May 29, 1975. The first annual meeting took place May 1976 at Four Seasons Hotel in Colorado Springs; followed by May 21-22, 1977, in Denver; followed by May 23-24, 1980, in Colorado Springs; followed by May 21-22, 1981, in Fort Collins; followed by the 1982 meeting again in Denver.

The eighth convention was May 14, 1983, at Mt. Vernon Country Club; followed by the May 18, 1985, meeting at the Brown Palace in downtown Denver; the May 17, 1986, meeting at the Denver Ramada Inn beside I-70; followed by the May 16, 1987, meeting at the Lakewood Sheridan Hotel. The 1988 meeting took place on May 4 and 5 at the Denver Regency, which remained as the meeting place for RMBS through 1990. The Denver Marriott Southeast was the site on February 8, 1991; followed by the February 1992 meeting at the Denver Airport Holiday Inn, which remained the meeting place for the February 27, 1993, gathering.

A huge Rocky Mountain bighorn ram bolts from the hands of volunteers who help at an Estes Park County, Colorado, wild sheep trapping project that involved men and women from the Rocky Mountain Bighorn Society and staff from the Colorado Division of Wildlife in 1989.
Photograph courtesy of Tom J. Brown.

Jim Bowron, a young wildlife biologist in the making, holds a baby ram lamb in February 1998 during a Rocky Mountain Bighorn Society project. Involving young people in wild sheep projects is an important method of insuring enthusiasm for the future of wild populations.
Photograph by Mark Bowron.

Two past presidents of RMBS and avid wild sheep supporters, Hans W. von Barby and Dale Martin, explain to a visitor to their booth at the Denver Sports Show in Currigan Hall about the many projects of the Society in February 1997.
Photograph courtesy of Dennis R. Gardner.

Marvin and Judy Clyncke, left, enjoy conversation with Peter Kummerfeldt at a 1995 national FNAWS convention.
Photograph courtesy of FNAWS archives.

Colorado Governor Roy Romer, center and smiling, holds scissors to cut the ceremonial ribbon cutting at the dedication of a bighorn viewing station on Interstate 70 near Denver.
Photograph by Judy Clyncke.

The Georgetown County Wildlife Viewing Center near Denver, Colorado, is supported by the Rocky Mountain Bighorn Society. They supply volunteer interpreters who talk with visitors about the wild sheep populations and center. RMBS also contributes funds to the center.
Photograph by Bud Henderson.

A sheep viewing and visitor center in Empire County, Colorado, is partially funded by the Rocky Mountain Bighorn Society.
Photograph by Bud Henderson.

The 19th meeting was held at the Denver Regency on February 5, 1994.

For the 20th anniversary dinner and meeting on February 4, 1995, the Denver Ramada Inn was the host location, which was the location of the meeting the following year on February 10, 1996. The 1997 meeting was held April 19 at the Denver Renaissance Inn and the Holiday Inn near I-70 in Denver was host location for the April 25, 1998, meeting. Bob Tully said all of the meetings have been held in the Denver area except for two in Colorado Springs, one in Fort Collins and one on the west slope in Vail.

The Bighorn is the significant educational and informative publication the Society produces each quarter that totals 20 to 24 pages each issue. A total of 71 regular and several special issues have been mailed to members and made available to interested others since the group's founding in 1975. This publication is intended to inform the readers of upcoming activities where working participation is needed, where additional funds are required, what has been happening during the year, which on-the-ground projects have been completed, and who is doing what to assist the overall purpose of the Society to help sheep and their habitat. Successful sheep hunting stories, written by members, are included. Efforts were made by current Society members to locate all back issues of the newsletter for their historic archives. That effort continues today since the group has been around longer than FNAWS itself. Any current member of the Rocky Mountain Bighorn Society should contact them at P.O. Box 8320, Denver, CO 80201-8320 if they have past issues of newsletters or historic photographs to donate to RMBS for their history files.

Regular communication of an organization acts as a glue that binds everyone together and tells new members about the efforts of the organization. Rocky Mountain Bighorn Society has always communicated through the newsletter. It is mailed directly to each member. In the first few years the newsletter was an informal one that was produced on an unscheduled basis by Bev Andrews and Hans W. von Barby with help from other officers.

During 1977 the newsletter was officially named, *The Bighorn,* and was published through hand typing, photocopying and mailing to about 100 members. Dave Ward deserves credit for setting the standard of the newsletter that still exists today through its format and content. Today the newsletter is produced with the assistance of computers, fax machines, e-mail and photographic scanning. Today's newsletter is pulled together by a newsletter coordinator from people within the Society, CDOW, BLM, USFS and USFG. Articles are used from hunters, wildlife researchers, wildlife watchers and photographers. The editor finalizes layout and editing. Virginia Boyce of Lakewood, Colorado, as editor from 1987 to 1992, was recently awarded an Honorary Life Membership for her outstanding effort. She remains the editor today as of summer 1998.

RMBS has received many accolades for *The Bighorn* from members and other organizations. We are proud of its history and those who have labored to produce it for the last quarter century. We would like to recognize the following people who have been involved in its production. From 1975 to 1976, Hans W. von Barby and Bev Andrews were editors and production coordinators. From 1977 to 1978 Dave Ward took over the reins of newsletter production coordinator and editor, with assistance from Jim Huckabay in 1978. In 1979 and 1980, Clay Reeves; 1981, Joe Zufall and Hans W. von Barby; 1982 to 1984, Joan Bossert; 1984, Hans W. von Barby; 1985 to 1986, Steve Rehm; 1987 to 1992, Virginia Boyce with help from officers and board members; 1993, Ken Martinez and Bud Henderson; 1994 to 1997, Bud Henderson; 1998 to the present, Dennis Gardner.

In 1998 *The Bighorn* was mailed to more than 1,000 members, FNAWS chapters and friends of RMBS. Hundreds of copies of the newsletter are distributed at sports shows, hunter orientations and at the FNAWS national convention. Issues are produced quarterly, with specialized issues dedicated to hunting and habitat conservation.

The cover of each issue of the newsletter shows the RMBS logo, designed by one of our first board members, artist Veryl Goodnight, that features a ram, ewe and lamb. The logo exemplifies the Society's concern for the species; ewes in good condition for the production, lambs that survive to increase populations and rams for special enjoyment including trophy sport harvest and spectacular photographs. Veryl, one of the outstanding wildlife sculptors of our time and our first secretary, depicted in the logo the bighorn sheep family to demonstrate our Society's concern with all bighorn sheep, not just trophy rams.

The Society wants to recognize the early leaders for their hard work and dedication. Individuals who have served as presidents include: Dave Ward, Aurora; Dr. Joseph R. Zbylski, Denver; Jim Huckabay, Denver; Ron Powell, Golden; Tom Brown, Golden; Mike Baumann, Arvada; Marv Clyncke, Boulder; Glen Roberts, Buena Vista; Dale Martin, Aurora; Hans W. von Barby, Evergreen; Terry Sandmeier, Conifer; and Sam Knox, Highlands Ranch.

A significant effort in our educational program involved the publication from the book, *Guide to Sheep Hunting in Colorado*. The original version was prepared by Hans W. von Barby in 1983, with participation of other Society members. The combined efforts of Tom Nelson, Ron Powell and Dan Andrews produced an expanded and improved loose-leaf book version that is probably one of the best guides ever compiled on the subject of sheep hunting. It was first published in 1984, with subsequent annual updates. The Colorado Bowhunters Association actively participated in this project, and much information was provided by the Colorado Division of Wildlife and past successful wild sheep hunters. More than 1,300 copies were sold during the first year, and hunters still try to obtain copies. Unfortunately, the combination of cost, time to compile and write and other considerations brought a halt to this effort. It may be revived in the future, as many aspiring sheep hunters found the book to be a source of valuable information.

Since 1975 the RMBS has cosponsored the annual Colorado Sheep and Mountain Goat Hunter Orientation. This one-day seminar is conducted in partnership with CDOW, CBA, International Order of Rocky Mountain Goats (IORMG), and in some years, other organizations such as Safari Club International (SCI). This is a cooperative effort to offer licensed hunters an opportunity to expand their knowledge of the bighorn and Rocky Mountain goat biology, life history and habitat requirements, so hunters will, perhaps, have a better and more successful hunting experience. Knowledgeable volunteers cover such items as selection of clothing, trophy selection in the field, survival in the high country, care of the meat and cape in the field, hunting ethics, successful stalking techniques, evaluation of a legal ram, important statutes and regulations, and tips about hunting in the various management units. All licensed sheep and goat hunters are invited, and usually about 200 participate each year. Some folks have attended three to five times, indicating this is a very worthwhile project to continue.

An exciting and rewarding project that is nearly annual is a sheep safari and photo excursion to the Dubois-Whiskey Mountain, Wyoming, area. Seventeen trips have been taken since the first excursion in 1980. This four- or five-day trip during early December has provided great camaraderie, super photographs of sheep on winter range and expansion of the knowledge about bighorn sheep and their needs. Dan Andrews, whose organization of this trip with assistance originally from Jim Huckabay and later from Mike Baumann, has made this outing an annual highlight on the Society's calendar.

On one of those trips, Dubois city officials approached RMBS officials for input and sug-

Bob Tully, left, and Terry Sandmeier staff the Rocky Mountain Bighorn Society booth at a recent national FNAWS convention. Both men have dedicated themselves to wild sheep projects and RMBS.
Photograph courtesy of Dennis Gardner.

gestions regarding sheep. This discussion and other actions of RMBS and many other sheep organizations, including FNAWS, resulted in the July 2, 1993, opening of the National Bighorn Sheep Interpretive Center in Dubois, commonly called "The Sheep Center." The Center features a 16-foot tall exhibit of full-sized bighorn sheep mounts and native Wyoming plants and animals as the center of its dramatic educational presentation.

Just beyond Dubois stands Whiskey Mountain, home to the largest wintering Rocky Mountain bighorn sheep herd in North America. Guided tours from the Sheep Center to the Whiskey Mountain bighorn winter range take place in the fall and winter. The Sheep Center is dedicated to educating the public about the habitat and conservation needs of the Rocky Mountain bighorn and to encourage active stewardship of wildlife and wildlands. The center in Dubois is located on Highway 287/26, southeast of Yellowstone National Park and 90 miles from Jackson Hole. Summer hours extend from Memorial to Labor Day weekends from 9 a.m. to 8 p.m. Winter hours are limited, so call first. To receive more information about the National Bighorn Sheep Interpretive Center, call toll free (888) 209-2795 or (307) 455-3429. Fund raising by the Rocky Mountain Bighorn Society helped make this center possible in the first place, along with help from many other wildlife conservation organizations, including generous contributions by the Foundation for North American Wild Sheep and several of their chapters. RMBS members donate funds they raise at each year's auction to the Dubois center. Proceeds come from the sale of framed prints donated by Dan Anderson.

Interpretive and educational publications, signs and activities are noted throughout RMBS history. Pamphlets about sheep have been funded and written in cooperation with the U.S. Forest Service (USFS), *Full Curl* with the U.S. Forest Service), CDOW and CBA. Major signs were installed during 1993 at interpretive sites along the Arkansas River 15 miles above Canyon City, Colorado, in a cooperative effort with the Bureau of Land Management (BLM) and CDOW. In 1993 in partnership with USFS and CDOW, signs and spotting scopes were placed near a burn area frequented by bighorns in the Poudre River Canyon in Larimer County, Colorado, near Fort Collins.

With the USFS, we financially supported signs for a herd often observed near Ouray. Two large observation scopes costing $2,500 each were purchased and installed 40 miles above

Denver off U.S. Highway 40 in cooperation with the city of Empire. Former president Fred Ratliff donated the purchase funds. He and former president Mike Baumann and the Boy Scouts installed the scopes. We also provided framed photographs for their visitor center. Off Interstate 70 at the Georgetown exit, in cooperation with the city of Georgetown and CDOW, we supported construction of an observation building with signing and viewing scopes. RMBS volunteers assist each winter by being present during weekends to help visitors locate wild sheep on the mountainsides and talk about the life history, habitat requirements and conditions of the sheep. Members also annually clean the area and participate in the "Adopt-a-Highway" program.

More recently in 1994 and 1995, RMBS went in partnership with FNAWS, CDOW, CBA, IORMG, Boone and Crockett Club (B&C) and Colorado Foundation for Agriculture (CFOA) to publish the classroom publication, *Colorado Reader*, that reached more than 50,000 students with three issues featuring wild sheep and Rocky Mountain goats. Bob Tully, life member, advisor and past board member, wrote a majority of the material and a good deal of other information that will be used in future publications. Each eight-page issue was distributed to students through teachers, who were provided with a special teacher's manual.

How are these expensive projects funded? Private donations from individuals provide the bulk of our funding. If you read this chapter and book, then you have a special interest in bighorn sheep or other species of wild sheep. A person may purchase a resident hunting license for $150 and a non-resident license for $1,000, depending upon the state or province where you live. A person probably pays dues to sheep or other wildlife conservation organizations. A person pays taxes that are allocated to USFS, BLM, NPS and other land management agencies. A person may simply donate goods and/or funds for a project or activity because you care. A person might even buy a raffle ticket or purchase a hunting license through an auction. In Colorado and in most other states, the primary source of funds for sheep projects comes from budgets from state agencies that gain revenue from hunting license fees. A primary source of funds for the Society are membership fees, fundraisers of many types, individual donation of goods and cash for special projects obtained from corporate sponsors. Recent corporate donors to RMBS include Roche Chemical, American Web, Majestic Metals, State Farm Insurance, Cabella's and Phelps Tointon.

In recent years, Colorado's efforts on habitat improvement, research and education have been enhanced by funds from the auction and raffle of bighorn sheep and mountain goat licenses. Availability of these special funds originated with RMBS in 1985 when the Society developed a position statement "proposing the creation of a special sheep license to be auctioned for additional sheep work." In September of 1986, the Colorado Wildlife Commission "endorsed the effort of RMBS to seek legislation for a special bighorn sheep and mountain goat licenses." On April 20, 1988, Senate Bill 52 was signed by Governor Roy Romer that authorized the Wildlife Commission to issue one sheep and one goat hunting license for auction and one each for raffle. Under contract with the CDOW, RMBS joined with FNAWS annually from 1989 to 1999 in handling the auction, and has solely conducted the raffles since 1990. In 1989 the raffles were conducted through the cooperative effort of the Colorado Wildlife Federation, RMBS, IORM Goats and the Colorado Bowhunters Association.

A maximum of 10 percent of income from the auction and raffle are retained for expenses. Past RMBS President and Former National FNAWS Board Member Tom Brown was involved in securing successful legislation. The organization of Past Presidents Mike Baumann and current President Dale Martin, have handled most of the raffle work in recent years. During the past 10 years from 1989 to 1998, high bidders at the FNAWS national convention have

Encouraging young people to appreciate and learn about wild sheep is one of the major goals of the Rocky Mountain Bighorn Society. The annual poster and essay contests sponsored by RMBS involves school children throughout Colorado to participate in a poster and essay contest about bighorn sheep. Shown here on this page we see three recent poster and essay contest winners.

Brandon Thurston of Hochne School, Trinidad, Colorado.

Drew Widetich of Ranch View Middle School, Highlands Ranch, Colorado.

Mark George of Rampart High School, Colorado Springs, Colorado.

provided $470,000 for bighorn and nearly $80,000 for goat projects. Overall this amounts to nearly $750,000. Through cooperative efforts and other allocations and donations, this amount has been leveraged to more than $1.5 million. Proposed special projects are reviewed and recommended by an advisory committee that includes a representative from the Society.

Through the Society's efforts, habitat has been improved and preserved, needed research has been accomplished, and additional printed information has been made available to the public because of these special funds obtained from volunteer supporters. Some of the high bidders for the Colorado sheep tag took rams scoring more than 180 B&C points. The 1997 raffle winner, Tim Rushing of Colorado Springs, took a trophy that scored 191-6/8 B&C points. He received the FNAWS Gold Award for the largest ram taken in North America by a FNAWS member in 1997.

What has the RMBS done directly to help wild sheep populations? Members have not missed a year in providing baiting and trapping manpower to assist in capturing sheep for relocation, for disease treatment and for marking animals for studies. Funding has also been provided to purchase aircraft time, neck bands, radio collars, radio receivers, hay, apple pulp bait, trap nets and other equipment. As a result, numerous new herds have been established in unoccupied historic ranges and in desirable new habitats. Remnant populations have been increased in several locations.

For 10 years the Society and Past President Glen Roberts worked toward the purchase of the Love Ranch Meadows, a prime winter range in the Chalk Creek Canyon, northeast of Buena Vista, Colorado. Through cooperative efforts of the Colorado Wildlife Heritage Foundation, CDOW, FNAWS, Rocky Mountain Elk Foundation, this irrigated hay meadow surrounded by rocky cliffs and good timber became secure on September 1996. Habitat maintenance and an observation platform are current projects for this property.

The Society leases the 400-acre Horsethief property, another year-round and important winter range, from the State Board of Land Commissioners. The Society carries out land management activities to improve the property for sheep use. There are numerous areas throughout the state where the society has helped the CDOW, USFS, BLM and other landowners in planning and/or carrying out a variety of management practices to maintain or increase sheep population and survival. Waterholes have been constructed on the Grand Hogback near Rifle, Colorado, near Cross Mountain in Moffat County where both funds and manpower were furnished. Ranges have been fertilized near Ouray in western Colorado, in the Clear Creek area on the front range, to improve vegetative conditions in concentration areas. Significant funds as well as volunteers have been involved in numerous tree and shrub removal projects as well as small to large prescribed fire and burning activities. The Poudre River, Waterton Canyon, Ouray, Arkansas River, the Tarryalls of South Park and several locations in northeast Colorado, as well as Encampment Canyon in southern Wyoming, come to mind.

The 1988 publication, The Dilemma of the Rocky Mountain Bighorn Sheep, produced with the USFS, calls attention to habitat requirements and projects needed to have healthy populations.

In 1984 in concert with the Denver Water Department and the Martin-Marietta Company, prescribed burning was tested on small parcels prior to application on USFS lands farther up into Waterton Canyon. The RMBS, the Water Department, Jefferson County Schools and others scheduled an April 2, 1984, Waterton Canyon 10-K run of a thousand people. Funds for this habitat project were obtained through sales of a special art print by DWD's Bob Fletcher, T-shirts and items by RMBS. This run was also planned to call attention to the needs of bighorn sheep in that area and to celebrate the opening of the canyon to controlled walk-in public use.

RMBS has also provided direct funds for census and research. In 1984 and 1985, two students under the supervision of Colorado State University (CSU) were financially assisted to conduct census in the Sangre de Cristo Range. During 1986 video recording equipment was donated to the Department of Fisheries and Wildlife Biology at CSU for behavior studies. Rewards from $100 to $500 have been offered to assist in obtaining convictions in poaching cases.

Volunteer manpower has long been provided to the Colorado Division of Wildlife to assist in conducting ground counts on both summer and winter ranges. Volunteers have on occasion braved snowstorms and high altitudes counting sheep and mountain goats on and around Mt. Evans. We have made sheep counts on Pikes Peak near Colorado Springs, Poudre Canyon, Arkansas River Canyon, San Luis Valley, Georgetown, Ouray area and various other locations for more than 20 years.

Last but not least, we should mention numerous testimonials or position statements by the Society concerning hunting regulations of the Wildlife Commission, as well as similar modifying and support positions on land use through participation in local government, federal and state agencies and other planning activities, both in Colorado and other North American locations. RMBS supported the desert sheep introduction in western Colorado, opposed the front range Two Forks Dam project, opposed some of the sites proposed for the 1976 Winter Olympics in Colorado, supported a number of land acquisitions and leases to secure habitat, opposed the proposed site for the 1998 downhill competitions' Winter Olympics on Mt. Allen, Alberta, Canada, and provided a donation to the Sportsmans' Alliance in Maine for efforts in retaining moose hunting.

Whenever possible, the Society speaks out on behalf of sheep on a variety of issues. The RMBS is a strictly volunteer organization and will continue to put the Rocky Mountain bighorn as the focal point of its activities and funding. The Society will work with any state, federal and private organization to insure the future of bighorns in Colorado and North America. We are proud of our affiliation with FNAWS. The association has been very beneficial to both organizations and expanded cooperation and coordination are an objective for the future.

Past presidents, based upon the memory of many founders and a list published in the winter 1998 newsletter on page 13 include: Hans W. von Barby from 1975-76; Dave Ward, 1976-77; Dr. Joseph R. Zbylski, 1977-79; Jim Huckabay, 1979-82; Ron Powell, 1982-84; Hans W. von Barby, 1984-85; Sam Knox, 1985; Tom Brown, 1986-89; Marv Clyncke, 1989-90; Mike Baumann, 1990-93; Fred Ratliffe, 1993; Glen Roberts, 1994-95; Hans W. von Barby, 1995-96; Dale Martin, 1996-98; and Terry Sandmeier 1998-99.

At one time we envisioned the Society expanding into other Rocky Mountain states that contain bighorn populations. This explains the "Colorado Order" portion of our previous logo. Other states did not see the need for a similar effort and with the creation of the Foundation for North American Wild Sheep, the waterfront is adequately covered. We opted to drop "Colorado Order" from our logo.

The Rocky Mountain Bighorn Society is healthy, well and growing. Let us hope it will be active as long as there are sheep to roam on the mountains.

To join the RMBS or to make a tax-exempt contribution, write to the RMBS at P.O. Box 8320, Denver, CO 80201-8320, or call Dale Martin at (303) 693-4567.

———

Bob Tully was Chief of Wildlife Management for the Colorado Division of Wildlife and worked for them from 1956 to 1991. Bob graduated from Colorado State University (called Colorado A&M at the time) in Fort Collins with a degree in wildlife management. Bob has been an advisor for the Rocky Mountain Bighorn Society and a member since its beginning in 1975.

Hans W. von Barby was founder of the RMBS and was in the banking industry for more than 25 years. He is current Honorary German Consul in Colorado and Wyoming, a position he has held since 1975. Hans is president and principal in Westmark International, Inc., of Golden, that provides marketing and investment expertise to German American business clients. Since moving to Colorado in December 1969, he has been active in wildlife conservation activities. Hans has been president of the Colorado Wildlife Federation, state chairman of Ducks Unlimited, and founding president of RMBS.

Bud Henderson of Boulder, Colorado, grew up on a ranch in southwest Texas and then graduated from Texas Tech University with a math and physics degree. He worked for years in the aerospace industry as an engineer. For the last 25 years, Bud has lived in Colorado and operated his own computer graphics business, Bud Henderson Associates. He has been on the Board of Directors of RMBS for six years and was newsletter editor from 1993 to 1997.

Marvin Clyncke is owner of Clyncke & Sons Concrete, Inc. of Boulder, Colorado. Marvin is one of the original members of RMBS, a long-time volunteer and a past president.

Dale Martin was president of the Rocky Mountain Bighorn Society from 1996 to 1998. He has been active since 1993. He is a retired director of the Upjohn Company, a major pharmaceutical manufacturing company, where he worked for 31 years, from 1962 to 1994.

Iowa Chapter of FNAWS
First in the Nation as a Chapter

KATHY SCHUTTE

The Iowa Chapter of FNAWS sketch by Iowa Artist Brett Wingfield. This became the logo for the chapter. The original sketch was auctioned at a fundraiser and purchased by Charles Ruperto of Des Moines, Iowa. Charles is a charter member of FNAWS and an Iowa life member Number 2.

The Iowa Chapter of the Foundation for North American Wild Sheep (FNAWS) was the first chapter to be established and to form a partnership in conservation with National FNAWS. Started in 1979 by a group of dedicated Iowa hunters and wildlife conservationists, the Iowa Chapter has grown in size to a present-day membership of 190 members.

The original group establishing the chapter were Lloyd Zeman, Dr. Roy Schultz, Roger Hill, Paul Wiemer, Dr. Jerry Waite, Richard Mitchell, Roger Gipple, Lloyd Bare, Tom Hinders, Jerry Christian and others. Of this illustrious group, Dr. Schultz, Lloyd Zeman, Dr. Jerry Waite, Tom Hinders and Jerry Christian all became president of National FNAWS, as well as being active in the Iowa Chapter. From 1994 to the present, Ted Schutte, former Iowa Chapter president, has served on the National FNAWS Board and is currently first vice president of FNAWS.

The 1981 Iowa Chapter meeting was held at the Ames Izaak Walton League of America Chapter clubhouse on Sunday, October 19. Those attending that meeting included Dr. Roy Schultz, national president of FNAWS at the time. Others involved in that early Iowa meeting included Tom Hinders of Ames, Roger Hill, Paul Wiemer, Dr. Jerry Waite, Richard Mitchell, and Jerry Christian and others. A photograph taken at the time showed them holding examples of a Grand Slam of wild sheep.

From the beginning, the purposes and goals of the Iowa Chapter of FNAWS as well as National FNAWS were to promote the management of and safeguard against the extinction of all species of wild sheep native to the continent of North America. The Chapter members also worked to promote the protection and intensive management of the remaining wild sheep populations and their habitats in North America and the re-establishment of wild sheep populations in suitable historic habitat in North America.

The earliest chapter meetings were informal get-togethers where the members socialized, ate and talked about hunts, hunting and future dreams. Fund-raising consisted of merely taking in enough money to pay for the food that was consumed. Soon a raffle was started and then hunts and artwork were auctioned. Each year the fund-raising efforts of the Iowa Chapter have increased and by the 1998 winter banquet, nearly $100,000 was generated in one evening from auction items and raffles. This figure represents double the gross and net from any previous fundraiser.

Dr. Jerry Waite remembered having an Iowa Chapter meeting in Fenton, Iowa, in the early years of the chapter's history. He reported more than 350 people attended the gathering and more than 100 mounts were on display. Members and friends from all over came to visit and many brought their trophies for display. During the formative years, members from

The Iowa Chapter was started by these Iowa wild sheep enthusiasts who included, left to right, Dr. Roy Schultz, Roger Hill, Paul Wiemer, Dr. Jerry Waite, Richard Mitchell, Tom Hinders and Jerry Christian.
Photograph courtesy of Iowa Chapter of FNAWS.

A crowd of approximately 110 people attended a winter fundraiser near Spirit Lake, Iowa, to celebrate all the good works of the Iowa Chapter of FNAWS. The winter event remains a highlight of the Iowa Chapter since a majority of the money generated here funds many wild sheep projects.
Photograph courtesy of Iowa Chapter of FNAWS.

The summer Iowa Chapter meeting is more informal with an opportunity to exchange hunting stories. This 1994 event was held at Ted and Kathy Schutte's farm near Sibley, Iowa.
Photograph courtesy of Kathy Schutte.

Ted Schutte was instrumental in the resurgence of the Iowa Chapter in the early 1990s after a period of dormancy. Ted started the informative quarterly newsletter and was Iowa Chapter president from 1990 to 1994.
Photograph courtesy of Iowa Chapter of FNAWS.

Auctioneer for the Iowa Chapter in the summer of 1992, Wade Ellerbrock, is pictured relaxing while holding a wildflower bouquet from his wife, Donna, who won the lovely collection of flowers. The chapter raised $1,300.
Photograph courtesy of Iowa Chapter of FNAWS.

Auctioneer Jerry Langstraat takes a bid from spotter, Richard Mitchell at the February 1, 1992, Iowa Chapter auction. A total of 131 people attended the afternoon and evening events that grossed more than $30,000. Three past presidents of National FNAWS, Dr. Roy Schultz, Jerry Christian and Harvey Kadlec attended.
Photograph courtesy of Iowa Chapter of FNAWS.

different parts of the state hosted the chapter meeting to get more exposure for the chapter around the state. During the last eight or so years, the chapter's annual winter fundraiser has been held at the Iowa Great Lakes, near Spirit Lake, Iowa. This location has worked well for the Iowa Chapter and will probably remain as the site of many future winter fundraisers. An informal summer gathering is also held during the summer. Chapter members and other hunting enthusiasts come from every corner of Iowa and several surrounding states. An exceptional list of auction items including hunts, permits, fishing trips, artwork and sporting goods brings spirited bidding with the help of our dedicated auctioneers who donate their time and energy to wildlife conservation.

Efforts by FNAWS Board of Director member Ted Schutte revitalized the Iowa Chapter of FNAWS several years ago. Ted explained.

"I was not involved with FNAWS National from the start. I got active about six or seven years after it started. A friend of mine signed me up as a member as a gift. I went to one convention and became enthused with the organization, not necessarily wild sheep hunting.

"I saw that our state chapter in Iowa was very weak, so it would be sort of fun to bring it back. I took the lead and that's how I went up through the ranks. The Iowa Chapter of FNAWS had fallen down to a level that it was becoming inactive. The membership numbers were down to about eight due to a lack of communications.

"In a couple of years the membership grew to 100 and fundraiser and meetings were held regularly. I was appointed to the National FNAWS Board of Directors because of a resignation. I'm pretty outgoing and was known for my efforts in the resurgence of the Iowa Chapter. During my term as Iowa Chapter president, I wrote the Iowa Chapter newsletter four times a year. I would write about the various activities that were going on and hunts people had gone on using short liners like, 'John harvested a beautiful 39-inch ram with so and so,' or 'John Doe got so many animals in Africa I thought he must have stumbled into a zoo.' I

Tom Matuska (second from left) and Vicki Matuska (second from right) have been active in the Iowa Chapter through the 1990s. Tom has been President from 1995 to the present. Numerous fun gatherings have been hosted at their taxidermy studio near Spirit Lake, Iowa.
Photograph courtesy of Iowa Chapter of FNAWS.

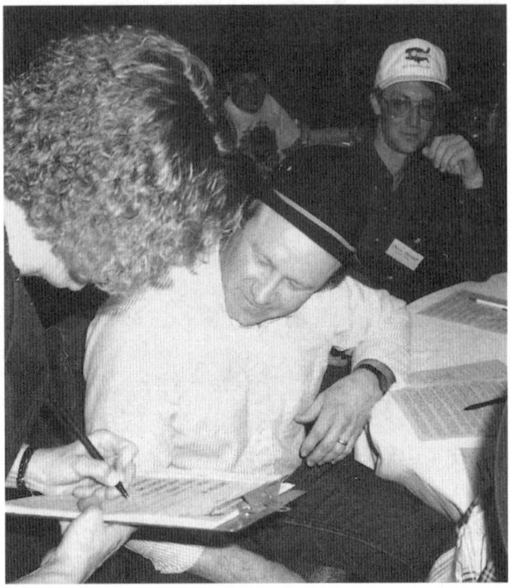

Iowa Chapter Secretary Kathy Schutte asks Ted Grenis sign as high bidder on the new John Deere riding mower he just bought at the January 30, 1993, fundraiser that grossed $30,016. The net profit of $9,214 went to the Grant-in-Aid program at the national level.
Photograph courtesy of Iowa Chapter of FNAWS.

joked around and then I'd get a little political at times. I had a lively newsletter," Ted concluded.

In the past several years, the Iowa Chapter has been able to deliver more than $75,000 to several Grant-in-Aid projects in the United States and Canada to support conservation efforts. The emphasis of these grants are wild sheep projects, hunter and outfitter rights, and support to National FNAWS. Since we are a wild sheep group, the Grant-in-Aid goes to states and provinces that have wild sheep. Dollars from Iowa Chapter members have paid for projects like transplants in Idaho, North Dakota, Oregon, Wyoming and Utah. Contributions to the Wildlife Legislative Fund of America have protected hunters' rights throughout the United States. Some Grant-in-Aid also has gone to outfitter groups in Canada to protect and help them in their business, knowing their challenges and the importance of the guiding industry to the Midwestern hunter. The Iowa Chapter funded the flagpole at the national headquarters in Cody as one of their first projects.

The chapter is governed by a seven-member Board of Directors. At present, the board consists of President Tom Matuska of Spirit Lake; Vice President Cy Elsinger of Rickardsville; Secretary-Treasurer Kathy Schutte of Sibley; and Directors Richard Meyer of Estherville; Vicki Matuska of Spirit Lake; Dennis Schmidt of Mason City; and Ted Grenis of Ely. This Board is also very instrumental in putting together the winter fundraiser and summer informal gathering. Many of the early officers became very active in National FNAWS when their efforts centered on the national scene.

During the mid-1980s a group in central Iowa kept the chapter moving. Some of the officers at that time were Richard Mitchell, Lloyd Bare, Ron McMillen, Don Mangels, Roger Hill, Dave Anderson and Larry Handsaker. Meetings at that time were held in Ames, Des Moines and Pleasantville, Iowa. The Iowa Chapter in the late 1980s became stagnant and membership dropped but was revitalized in 1990 when the fundraisers and meetings were moved to northwest Iowa. It was here, with the help of the older, avid supporters, that the Iowa Chapter started to grow again. With more communication and bigger and better fundraisers, membership has grown to its present high level.

Jeremy Houtsma of Sibley, Iowa, left, holds the .300 Winchester Model 70 he received as the lucky winner at the summer meeting of the Iowa Chapter in 1995. FNAWS Director Ted Schutte makes the presentation to Jeremy.
Photograph courtesy of Iowa Chapter of FNAWS.

Brett Wingfield of Spirit Lake, Iowa, was the artist who created the new logo for the Iowa Chapter in 1996. The original sketch was auctioned at a 1997 fundraiser and purchased by Charles Ruperto of Des Moines. Charles is an early member of FNAWS National and Iowa Chapter Life Member Number 2.

The Iowa Chapter mails to each member an information-filled newsletter four times a year. This quarterly newsletter was started in 1990 by then-President Ted Schutte and has continued by subsequent presidents. Members are anxious to read information on member hunts, accomplishments of fellow members and business of National FNAWS.

Iowa Chapter members support wildlife and also the outfitting business. They realize that time spent hunting is precious and by utilizing guides and outfitters, success is greater and worth the associated costs. These outfitters in turn have been generous in supporting the Iowa Chapter and its various causes.

The Midwestern attitude and culture have always been rural. Hunting and fishing are an integral part of family life. Many did so for subsistence in the early days and this attitude has carried to the present. Many people wonder, "How could FNAWS have been started by mostly flatlanders from the Midwest?" The answer has to be their love of nature, desire for adventure and wanting to put back on the mountain what they have taken.

Kathy Schutte joined the Board of Directors of the Iowa Chapter in 1990 and was Secretary-Treasurer from 1990 to 1994, while her husband, Ted, was president of the Chapter. From 1994 to 1996, Kathy served as vice president and since that time has returned to the job of secretary-treasurer, a position she continues to hold. Kathy and her family live on a farm near Sibley, Iowa, in northwestern Iowa.

Kathy Schutte
Photograph courtesy of Iowa Chapter of FNAWS.

Minnesota-Wisconsin Chapter of FNAWS 1979 to Present

MINNESOTA WISCONSIN

Foundation for North American **Wild Sheep**

JANET JOHNSON

The Minnesota-Wisconsin Chapter of the Foundation for North American Wild Sheep (FNAWS) was started in 1979 by a small group of 25 to 30 enthusiastic and dedicated sheep hunters and sheep enthusiasts.

We met at the Thunderbird Motel in Bloomington, Minnesota, for an informal get-together to talk sheep and hunting and decided to form a chapter of FNAWS. We picked our initial Board of Directors by asking for a volunteer from each table. One person said he had no choice as he was the only one at his table, and he is still a great supporter today!

Our first board consisted of: President Harvey Kadlec, Vice President Jerry Olsen, Secretary-Treasurer Geoff Meyer, David Loos and Roman Sinnen.

In 1980 the secretary-treasurer position was split, and Phyllis Carlson was asked to be the secretary and Geoff Meyer remained as treasurer. In 1981 the board had six members and our chapter received its non-profit status as a 501-(c)(3) corporation. In 1982 three more members were added to the board to make our current total of nine voting members.

The 1982 spring meeting had live auction receipts of $2,472. Total receipts came to $4,509.20 with expenses of $2,948.63 leaving a net profit of $1,560.57 for that meeting. The fall meeting also had a net profit of $885.48 so our grand total for that first year was $2,446.05. The Yukon requested funding for habitat improvement in the amount of $1,000. The chapter voted to grant them the money, and our funding program was off and running!

The auction that was held at the 1983 spring meeting featured a black bear hunt donated by Greg Johnson and Bill Klugow. At the 1984 spring fundraiser the preliminary auction list consisted of a duck hunt, a grouse and duck hunt, an archery deer hunt, a couple of framed prints and two guns. With a lot of hard work we have managed to expand our auction offerings significantly. In 1998 we had a live auction list of 63 items, including a Wyoming governor's bighorn sheep permit and silent auction offerings of 20 items on Friday night and 86 items on Saturday. With net proceeds of more than $84,700 from the 1998 spring fundraiser, our year was off to a good start. With proceeds from our annual 1998 fall meeting and fundraiser, our chapter funded about $100,000 in 1998. We expect to raise the same amount of money in 1999.

Since 1981 we have had two meetings each year; a fundraiser in the spring and our annual meeting in the summer or early fall. The spring fundraiser includes live and silent auctions, raffles, trophy displays, outfitter/donor booths and a banquet. The spring 1982 meeting at Bobbies in White Bear Lake, Minnesota, had 63 people in attendance and Harold McEuen brought cheese and crackers to snack on during the afternoon, thereby starting a tradition of potluck hors d'oeuvres. Soon we had contributions of sausage and jerky made from venison, elk, moose, etc., besides Wisconsin cheese, crackers, veggies — even meatballs. Not all of our meeting places are as accommodating as Bobbies, but a lot of them do let us bring in our own snacks for the afternoon as long as we still eat dinner since so many of our members came from Wisconsin so in 1983 it was decided at our annual meeting to change our name to the Minnesota-Wisconsin Chapter of FNAWS.

Our first meetings were well attended and have gotten bigger and better each year. It was getting hard to guess how many were going to attend so in the fall of 1985 we instituted our "Early Bird Drawing" for those who pre-registered by a certain date. This helped tremendously. In 1990 our spring fundraiser had grown to the point where we expanded to a two-day meeting. On Friday night we start with hors d'oeuvres, meet the outfitters and exhibitors and visit and share hunting stories with friends, old and new.

Saturday's events include a breakfast buffet, final set up, visiting, registration, visiting, videos, visiting, raffles, visiting, silent auctions, visiting, live auction and, of course, our banquet with plenty of visiting. The 1997 spring fundraiser had an attendance more than 450 people with a fantastic taxidermy display. Our fall meetings are usually much smaller with about 100 people in attendance who don't happen to be hunting at the time.

From those 25 or 30 initial members in 1979 we have grown to 335 members, including 94 life members in 1997. During 1998 we had a life member contest that brought in another 22 life members. Our monies and donations have grown along with our membership enabling our chapter to fund hundreds of thousands of dollars in sheep projects. Since we don't have any sheep in Minnesota or Wisconsin all our funding goes to other areas. From our first funded project of $1,000 in 1982 to the more than $135,000 funded in 1997 and $161,850 in funding approved for 1998, we have funded a grand total thus far of $779,740 — something of which we can all be proud.

The Minnesota-Wisconsin Chapter of FNAWS has funded many kinds of projects, including bighorn sheep studies, bighorn trapping and transplanting, habitat restoration and preservation, disease research and water holes. The first water hole project our chapter sponsored was in the Little Harquala Mountains of Arizona in January 1985. Not only does the chapter sponsor the water hole, but members are encouraged to go to the site and work on its construction. The Arizona Desert Bighorn Sheep Society members, along with the Arizona Game and Fish Department personnel, construct the water holes and they use all the volunteer labor they can get to help. We have funded the cost of many water holes over the years, but the most rewarding projects have been where we have gotten our youth involved.

In 1990 our chapter selected eight youth to participate in building a water hole. The chapter provided them their transportation, meals, lodging and chaperons. They provided labor and enthusiasm. The project was so well received by both the Arizona Desert Bighorn Sheep Society and our chapter participants that the chapter decided to sponsor 10 to 12 youth each year through an essay contest.

Lloyd Zeman died on August 10, 1991, after suffering injuries when a bank gave way, throwing him and his horse into a deep ravine while riding near Yellowstone National Park. The Minnesota-Wisconsin Chapter dedicated the Lloyd Zeman Memorial Water Hole Fund to him for his years of service to FNAWS and wild sheep as a living tribute to his dedication. The water hole fund was established in the fall of 1991 and continues today.

For the past several years, the ADBSS has donated a water hole project to the FNAWS national convention auction. Three of those water holes have been purchased by one of our chapter's life members, Oscar Carlson. He in turn has encouraged our chapter to sponsor youth to aid in the construction of these projects and has sponsored additional youth himself in order that more of our youth can participate. At the 1997 national convention of FNAWS, Oscar was honored with the prestigious Gordon Eastman Grass Roots Award for his dedication to perpetuate hunting and wildlife conservation through sponsorship of our young people.

Our method of communicating with our members has changed significantly over the years. At first a letter would go out to each member regarding the upcoming meeting and information on what had transpired since the last meeting. Then one of the directors, Steve Bruggeman, offered to get it printed on glossy paper at a discount from his company's printing department and we could include members' stories and photographs. Our first newsletters were four pages and have grown to the point where we are hard put to keep it to 16 pages. In 1988 Steve brought a copy of the Idaho Chapter newsletter to one of the board meetings that showed a logo. Our chapter's present logo, designed by Phyllis Carlson, was adopted shortly thereafter.

Over the years many of our chapter members have gone on to be national board members of FNAWS. Several have been elected president including Harvey Kadlec, Dr. Jerry Waite, Jerry Christian, Arnie Johns, Craig Johnson, Peter Cimellaro, Garry Vince, and most recently, Phyllis Carlson.

Dedicated leaders of the Minnesota-Wisconsin Chapter include the following officers and directors:

1979-1980

President Harvey Kadlec
Vice President Gerald Olsen
Secretary-Treasurer Geoff Meyer
Directors David Loos and Roman Sinnen

1980-1981

President Harvey Kadlec
Vice President Gerald Olsen
Treasurer Geoff Meyer
Secretary Phyllis Carlson*
Directors David Loos and Roman Sinnen

The Minnesota-Wisconsin Chapter has produced many national officers of FNAWS over the years. Shown in this 1988 are some of the national officers from the chapter who included, left to right, Future President Craig Johnson, Lyle Hanson, Past FNAWS President Jerry Waite and President-Elect Harvey Kadlec.

1981-1982

President Harvey Kadlec
Vice President Gerald Olsen
Treasurer Geoff Meyer
Secretary Phyllis Carlson*
Directors Wally Anderson, David Loos, Thomas Merkley

1982-1983

President David Loos
Vice President Wally Anderson
Treasurer Geoff Meyer
Secretary Phyllis Carlson*
Directors Russ Benoy, Craig Johnson, Harvey Kadlec, Dwight Lockerby, Tom Merkley and Jim Walden

1983-1984

President David Loos
Vice President Wally Anderson
Treasurer Jim Walden
Secretary Phyllis Carlson
Directors Jane Babler, Russ Benoy, Craig Johnson, Dwight Lockerby and Tom Merkley

1984-1985

President Wally Anderson
Vice President Craig Johnson
Treasurer Jim Walden
Secretary Jane Babler
Directors Russ Benoy, Steve Bruggeman, Phyllis Carlson, David Loos, Harold McEuen

1985-1986

President Wally Anderson
Vice President Craig Johnson
Treasurer Jim Walden
Secretary Phyllis Carlson
Directors Jane Babler, Russ Benoy, Steve Bruggeman, David Loos, Harold McEuen

1986-1987

President Craig Johnson
Vice President Steve Bruggeman
Treasurer Jim Walden
Secretary Phyllis Carlson
Directors Wally Anderson, Les Behrns, Russ Benoy, Brad Kadlec, Harold McEuen

1987-1988

President Steve Bruggeman
Vice President and Secretary Phyllis Carlson
Treasurer Jim Walden
Directors Wally Anderson, Les Behrns, Russ Benoy, Craig Johnson, Brad Kadlec and Harold McEuen

1988-1989

President Steve Bruggeman
Vice President and Treasurer Harold McEuen
Secretary Stewart Shaft
Directors Wally Anderson, Les Behrns, Phyllis Carlson, Jack Cook, Brad Kadlec and Pat O'Neill

1989-1990

President Steve Bruggeman
Vice President and Treasurer Harold McEuen
Secretary Stewart Shaft
Membership Director Janet Johnson
Directors Les Behrns, Phyllis Carlson, Jack Cook, Brad Kadlec and Pat O'Neill
1990-1991
President Brad Kadlec
Vice President and Treasurer Harold McEuen
Secretary Stewart Shaft
Membership Director Janet Johnson
Directors Les Behrns, Steve Bruggeman, Jack Cook, Neil McKenzie and Pat O'Neill

1991-1992
President Brad Kadlec
Vice President Marc Shaft
Treasurer Harold McEuen
Secretary Stewart Shaft
Membership Director Janet Johnson
Directors Les Behrns, Steve Bruggeman, Jack Cook and Neil McKenzie

1992-1993
President Marc Shaft
Vice President Brad Kadlec
Treasurer Harold McEuen
Secretary Stewart Shaft
Membership Director Janet Johnson*
Director Steve Bruggeman, Jack Cook, Tim McEuen, Neil McKenzie and Kathy Strong

1993-1994
President Marc Shaft
Vice President Brad Kadlec
Treasurer Tim McEuen
Secretary Stewart Shaft
Membership Director Janet Johnson*
Directors Steve Bruggeman, Jack Cook, Harold McEuen, Neil McKenzie and Kathy Strong

1994-1995
President Marc Shaft
Vice President Brad Kadlec
Treasurer Tim McEuen
Secretary Kathy Strong
Membership Director Janet Johnson*
Directors Steve Bruggeman, John Coulter, Jim Karau, Harold McEuen (resigned), Neil McKenzie, Bob Wall (completed Harold McEuen's term)

1995-1996
President Brad Kadlec
Vice President Curt Babler
Treasurer Tim McEuen
Secretary and Membership Director Bob Wall
Directors Steve Bruggeman, John Coulter, Janet Johnson, Jim Karau and Marc Shaft

1996-1997
President Janet Johnson
Vice President Curt Babler
Treasurer Tim McEuen
Secretary and Membership Director Bob Wall
Directors John Coulter, Jim Karau, Joe Lynch, Dan Scharmer and Marc Shaft

1997-1998

President Janet Johnson
Vice President Curt Babler
Treasurer Tim McEuen
Secretary and Membership Director Bob Wall
Directors Mark Almgren (resigned), John Coulter, Brian Iverson, Joe Lynch, Jerry Mariska (completed Mark Almgren's term) and Dan Scharmer
* Non-voting position.

Starting in 1979, the Minnesota-Wisconsin Chapter of FNAWS has hosted numerous successful fundraisers to benefit wild sheep and wildlife conservation. Theirs is one of the most dynamic and successful FNAWS chapters in North America. The first meeting occurred at the Thunderbird Motel in Bloomington, Minnesota, also the site of an early successful national convention of FNAWS. The following year, in 1980, the young group met at the Wilderness Hunting Preserve at Prior Lake, Minnesota, which became the site again in 1981. At the 1981 auction, two FNAWS prints and two guns were highlights of the donated items that helped raise money. A second 1981 meeting on September 26 took place at Sheraton Northwest in Minneapolis.

Two meetings in 1982 highlighted the fund-raising activities of the chapter. The first meeting was April 17 at Bobbies in the Park at White Bear Lake, Minnesota. The second meeting was October 9 at the same location when four additional people were added to the Board of Directors. Hudson House Inn in Hudson, Wisconsin, was the site for the two 1983 meetings that took place March 12 and August 27 and the two meetings in 1984 that were March 10 and August 25. Thanks to the leadership of Harvey Kadlec, the first water hole was funded in 1984 and work on it was done in January 1985.

March 9 was the date for the first meeting in 1985 at Hudson House Inn. The second meeting that year was at Trollhaugen Ski Resort in Dresser, Wisconsin. The March 22nd

In 1988 Janet Johnson, left, put together the reservation list for the Minnesota-Wisconsin Chapter along with Jackie Bruggeman, right, who helped collect dues.
Photograph courtesy of FNAWS archives.

The 14th annual Minnesota-Wisconsin Chapter spring banquet and fundraiser set new records for attendance, money raised and exhibit space sold and donor participation when it grossed more than $100,000. The event was held from March 26 to 27, 1993, at Hotel Sofitel in Bloomington, Minnesota.
Photograph courtesy of FNAWS archives.

Kimmel Rosenstein & Company bluegrass band entertained the Minnesota-Wisconsin Chapter of FNAWS at their 1997 fundraiser held at the Minneapolis Marriott SW in Minnetonka, Minnesota.

Photographs courtesy of the Minnesota-Wisconsin Chapter of FNAWS.

John Coulter, board member and live auction chairman for the Minnesota-Wisconsin Chapter of FNAWS, was their fall meeting auctioneer in 1995 at the Hotel Sofitel August 26 in Bloomington, Minnesota.

Neil McKenzie, a happy buyer at the live auction in March 1996, poses for the camera at the Minneapolis Marriott SW in Minnetonka, Minnesota.

A puppy and a smile tells the story for this young wild sheep supporter. Nichole Weber holds her new yellow Labrador retriever that her father bought for her at the live auction at the 1997 spring fundraiser held at the popular Minneapolis Marriott SW hotel.

Kathy Schutte of Iowa stands beside the Amish guilt she won in the Minnesota-Wisconsin Chapter raffle on August 26, 1995, at the Hotel Sofitel in Bloomington, Minnesota.

meeting in 1986 was also held at Hudson House Inn but the second meeting that year was held at Anderson House, Wabasha, Minnesota.

Hudson House Inn was the site for the first 1987 meeting on March 21 and the second meeting took place at Shoreview Plaza Hotel, Shoreview, Minnesota, on August 22. The year 1988 found the Minnesota-Wisconsin Chapter of FNAWS members enjoyed fellowship and fund-raising activities at Hudson House Inn on March 26 and the summer meeting at St. James Hotel, Red Wing, Minnesota, on August 20. The first meeting in 1988 attracted 176 people. The same sites in 1988 were the meeting sites in 1989 but the meetings took place March 11 and July 22. The 10th anniversary was celebrated at the Hudson House Inn on March 11.

The Thunderbird Motel once again hosted the chapter for its March 30 to 31, 1990, meeting. This was the first time a two-day event was organized and the national FNAWS Board of Directors attended to celebrate. This was also the first youth water hole that took place in February 1990. A second meeting in 1990 took place October 13 at Holiday Inn in Eau Claire, Wisconsin.

In 1991 the two-day fundraiser took place March 22 and 23 at Hotel Sofitel in Bloomington, Minnesota. The second meeting was August 10 at Hudson House Inn, Hudson, Wisconsin. Thunderbird Motel was the site for the March 20 to 21, 1992, fund-raising event and Hudson House was the summer site on August 15. Hotel Sofitel in Bloomington, Minnesota, was the site for the fundraiser March 26 to 27, 1993, and Hudson House Inn was again the summer site on August 14.

April 8 and 9, 1994, saw the Minnesota-Wisconsin Chapter of FNAWS going to the Minneapolis Marriott SW in Minnetonka, Minnesota, and the summer meeting to Hudson House Inn on August 27. Minneapolis Marriott SW was again the site the following year on April 7 and 8, 1995, for the big fund-raising event, and Hotel Sofitel hosted the summer gala on August 26. The same facilities were used in 1996 but the dates were March 29 and 30 for the Marriott event and August 24 for the Hotel Sofitel event.

The Minneapolis Marriott SW once again was the site for the 1997 and 1998 fund-raising banquet and events but the 1997 meeting took place March 21 and 22 and the 1998 event took place March 20 and 21. The chapter newsletter expanded to three issues in 1997. In 1997 the summer meeting took place August 23 at St. Croix Casino at Turtle Lake, Wisconsin, and the 1998 summer meeting took place at the Lac Courte Oreilles Convention Center in Hayward, Wisconsin, on August 15. We grossed more than $20,000 for this successful summer event. The 1999 fundraiser and banquet was once again at the Minneapolis Marriott SW from March 19 to 20.

Thanks to our dedicated volunteers and generous donors, we will keep growing and "putting more sheep on the mountain."

Janet Johnson was president of the Minnesota-Wisconsin Chapter of FNAWS from 1996 to 1998 and was in charge of membership for many years. She was the first woman president of the chapter who began her work as a leader in the Minnesota-Wisconsin Chapter shortly after she married Craig Johnson in 1981. She and Craig live in Hudson, Wisconsin. She was raised in the Twin Cities area of Minnesota and stayed in Minnesota until her move to Wisconsin in 1969. Craig and Jan are owners of Madsen-Johnson Corporation, a general construction company based in Hudson, Wisconsin.

Alaska Chapter of FNAWS 1980 to Present

Glenn Anderson and Warren E. Olson

The Articles of Incorporation for the Alaska Chapter of the Foundation for North American Wild Sheep (AKFNAWS) were approved by the state of Alaska in June 1980. The purpose for which the corporation is formed shall be to assist the national office, including the objectives of FNAWS. Those objectives are to promote the management of, and to safeguard against the extinction of, all species of wild sheep native to the contingent of North America, according to the principles, policies, standards and procedures as set forth by the Foundation from time to time. The three people who started the Alaska Chapter and were the first Board of Directors were Brent Jones, Warren E. Olson and Larry Smith. In 1991 the charter for the Alaska Chapter was revised and updated.

In 1986, "Sheep Days" was started. Sheep Days is the primary fund-raising event for a weekend in March where we have seminars to inform and educate the public and hunters on Dall's sheep or other subjects related to them. We sell merchandise and memberships in our chapter. On Saturday night we have a banquet, present awards and special recognition. There are several different types of auctions to raise money for our chapter. Various vendors display and sell their wares to the public of wild sheep. Sheep Days has continued annually through 1998.

Starting in 1989, the Alaska Chapter magazine, *The Ram,* was edited in-house by Chris Conway. Previously, *The Ram* had been sent out and commercially completed by Ray Orderica of North Light Communications.

In 1990, the Alaska Chapter of FNAWS had three goals.

(1) Educate the general public on how we assist in game management.
(2) Promote "Fair Chase" in our education courses and our hunts.
(3) Put more sheep on the mountains for ours and future lifetimes.

In 1990 the chapter and Alaska Bowhunters jointly funded an Alaska Department of Fish and Game (ADFG) aerial survey in Game Management Unit 14C. That same year during Sheep Days Alaska Chapter members gave away 300 National Rifle Association books, *The Hunter's Guide,* for children ages 6 to 16 years. They also sponsored a drawing in 1990 for a .22 rifle that was used to motivate youngsters to attend the show.

In 1991 the Alaska Chapter of FNAWS jointly funded the Alaska Fish and Wildlife Conservation Fund lawsuit on subsistence. Subsistence is defined as living off the land by hunting, fishing, growing and gathering food items for Native American use. The Sheep Days that year had booths and displays tailored toward teaching young people about nature and wildlife. A photo and taxidermy contest took place. The chapter donated $1,495 to ADFG for aerial survey of Game Management Unit 13D. That same year the Alaska Chapter erected a sheep information kiosk on the Seward Highway near Beluga Point. This is a popular area where the general public has an opportunity to view wild sheep and read about them. The project was coordinated with various state and federal agencies. Two other sheep information kiosks have been installed at Tern Lake in 1992 and in 1994 near Kenai Lake.

An aerial survey of Dall's sheep in Game Management Unit 14C took place in 1992 be-

The information booth of the Alaska Chapter of FNAWS is used by members to recruit new members and distribute details about upcoming chapter events. This booth is shown at Sheep Days in March 1997.

Photographs courtesy of Glenn Anderson.

The Alaska Chapter of FNAWS picnic in 1997 attracted Vice President Dan Crowson, facing and center, and Board Member Mark Chambers, beside Dan. Behind Mark is Chapter Past President Derek Tomlinson (president from 1996 to 1998).

Workers at the 1997 picnic cook hamburgers at the Alaska Chapter barbecue.

Jeff Baker, right, holds his winning sheep horn from the 1997 scoring contest. Congratulating him is Chapter Past President Derek Tomlinson, left, at the annual Alaska Chapter picnic.

An outstanding display of wild sheep taken by Alaska Chapter members entertained those who attended the 1997 Sheep Days. From left to right is Chapter Secretary Joe Breidinger, National FNAWS Board Member Matt Wolfe, Chapter Past President Glenn Anderson, Derek Tomlinson, Mark Chambers and Chapter Vice President Dan Crowson as he adjusts a full-body desert bighorn ram mount. The mount was created by Billy Dunbar.

cause of funding provided by the chapter. A sheep survey began on mineral licks in the Alaska Range, north of Tok on Sheep, Granite and Dry creeks by the chapter in 1992. The chapter funded a youth scholarship and airfare to the NRA Whittington Center. In 1992 and 1993 the Alaska Chapter funded the Lazy Mountain/Wolverine Mountain necropsy study. Another young person attended the NRA Whittington Center in 1993 because of the Alaska Chapter's donation.

Bylaws for the chapter changed to expand the Board of Directors from 10 members to 14 in 1993, and they funded Dall's sheep aerial surveys in Game Management Units 7, 14C and 15.

The year 1994 continued many accomplishments. A young person received a scholarship and airfare to attend the NRA Whittington Center and the first chapter picnic occurred, a tradition that continues today. Once again, the chapter funded Dall's sheep aerial surveys in Game Management Units 7, 14C and 15. Tony Russ became editor of The Ram in 1994. The Alaska Chapter made a donation to the Alaska Outdoor Council to help fight for hunting rights in the state.

Two women attended the weekend workshop, "Becoming an Outdoor Woman," thanks to scholarships paid by the Alaska Chapter in 1995. The scholarship was $200 each. Dall's sheep surveys in Game Management Units 7 and 15 took place with money from the chapter that year and the chapter along with Safari Club International (SCI) jointly held a raffle and the money raised was donated to the Rabbit Creek Rifle Range. This was also the year that the Alaska Chapter made a donation to the Constitutional Defense Fund regarding the McDowell II lawsuit. McDowell I and II were lawsuits that were saying that rural subsistence preference did not comply with the state of Alaska and the state constitution. With the passage of the Alaska National Interest Lands Conservation Act (ANILCA) in 1980, one section required subsistence preference to rural residents.

The Alaska Chapter of FNAWS in 1996 was the first conservation organization in the state to unveil their web wage and e-mail system on the internet. The web site for the Alaska Chapter of FNAWS is "http://www.alaska.net/~fnawsak" and their e-mail address is "fnawsak@alaska.net."

This was also the year that the membership category of life membership was established. As of July 1998, total number of life memberships was 13. The life members include Drew Browne (honorary life), Tom Carter, Murray Conklin II, James L. Davis, William "Billy" Dunbar, Terry Marcum, Bruce Moore, Charles Ricci (honorary life), Joe Romero (Aurora Wildlife), Duke Ruzika, Derek Tomlinson, Jerry Tyrrell and Michael D. Ward (Southwest Outfitters, Inc.).

In 1996 Cathie Harms of the Alaska Department of Fish and Game received a scholarship from the Alaska Chapter to attend the "Becoming an Outdoor Woman," along with a second woman, Tracey Conway, who also attended this weekend class. This was also the year that the chapter donated $5,000 to the Forty Mile Caribou Calf Protection Program.

The Alaska Chapter of FNAWS had as its objective in 1996 was to promote and enhance increasing populations of the Dall's sheep within Alaska; to inform and educate the people in Alaska concerning Dall's sheep, their habitat, health, population and general welfare; to promote the professional management of Dall's sheep and their habitat.

Efforts to work with the Alaska state legislature to approve the issuance of big game permits to qualified organizations for auction on behalf the Department of Game and Fish were a success in 1996. During the 1996 legislative session, Rep. Con Bunde (R-District 18) sponsored House Bill 59. The Alaska Chapter is allowed to deduct its administrative costs and not more than 10 percent of the net proceeds from the total sale price of the permit. The chapter was the first wildlife conservation organization in Alaska to receive a governor's permit to auction.

The Grand Slam series began in 1996. This series is a fund-raising activity where the chapter obtains a permit for one of the four North American wild sheep species that make up the Grand Slam to be raffled each year for four consecutive years. In 1996, a Stone's sheep and elk hunt combination hunt with Garry Vince as guide was raffled. In 1997, a Dall's sheep hunt was raffled for the 1998 hunting season, with Bill Stevenson as the guide. In 1996 at the Sheep Days banquet, we raffled the first Alaska Chapter life membership to Billy Dunbar and have continued this tradition each year.

In February 1997, the Alaska Chapter auctioned their first governor's permit for Dall's sheep at the FNAWS national convention in Philadelphia and the result was a $200,000 bid from Jerry Fletcher. The 1998 bid from Hossein Golabchi brought $50,000 and the caribou permit from Alaska brought $2,200 from William Cloyd. The Alaska Chapter began the gold, silver and bronze Ram Awards program in 1997, and the first awards were presented in 1998 to honor the largest Dall's sheep taken in Alaska by a chapter member. The first award winners included Bob Cassell, Harold Hollis and Tony Russ. In May the Alaska Chapter adopted the Twin Peaks Trail near Eklutna in the Chugach Mountains. Members in our chapter provide annual maintenance and work with the state of Alaska, Division of Parks and Outdoor Recreation.

The chapter auctioned another governor's permit in 1997, this time by telephone, for an Unimak Island brown bear permit. The high bid was $20,000. An administrative assistant and recording secretary, Kellie Mickelson, was hired in the summer of 1997. That same year, the Alaska Chapter contributed $3,000 to the organization, Scientific Management of Alaska's Resource Treasures (SMART). SMART is an organization that doesn't want our fish and game managed by the ballot box. The year of 1997 also included funding for an aerial wild sheep survey in the Mt. Harper and Tanana Hills areas of Alaska, and the funding of research by a graduate student who titled his project, "Nutritional Ecology of Dall's Sheep in the Chugach Mountains." Funding of this research continued in 1998.

Another aerial survey of wild sheep was funded in 1998 by the Alaska Chapter in Game Management Unit 13A. To raise money for various chapter projects, a raffle of a lightweight mountain rifle is conducted and has been done for several years. Raffle tickets continue to be sold in 1998. In 1998, the chapter members continued to contribute money to the Alaska Food Bank for the Hunters Against Hunger Program. The year 1998 also brought funding by the Alaska Chapter of another aerial survey of wild sheep in Game Management Unit 13A and the funding of "Youth Shooting Day" put on by the Alaska Gun Collections Association. In 1998, the Alaska Chapter of FNAWS and Safari Club International (SCI) jointly funded a rifle and shotgun raffle to help fund projects at the Rabbit Creek Rifle Range and youth hunter education programs.

Each year members, supporters and family from the Alaska Chapter of FNAWS enjoy a summer picnic in July. The picnic is scheduled to occur after the Alaska hunting permit winners have been announced. The party is an opportunity for permit holders to find a hunting partner, if they want one, and additional information about the area where they will hunt and where they have a permit. The picnic includes informational seminars on a variety of topics related to wild sheep hunting and the chapter sells merchandise and raffle tickets to raise money to "put sheep on the mountain." This is a social occasion for sheep hunters to get together with other hunters and swap stories and exchange information, plus its a great meal, too.

The Alaska Chapter of FNAWS honors the leaders who were elected as president of this dynamic organization from its beginning in 1980 until 1998. These past presidents include Brent Jones, Warren Olson, Roger Morris, Jim Morehouse, George Elledge, Matt Mathews, Chris Conway, Tony Russ, Derek Tomlinson and Glenn Anderson.

The Alaska Chapter of FNAWS and its members have distinguished itself for its arduous work on behalf of hunters and fishermen to preserve their rights to enjoy their sports and to have access to public lands and waters. Starting in 1982, Warren E. Olson and other members of the Alaska Chapter worked on the case of McDowell v. State of Alaska that sought to force state officials to abide by the Alaska Constitution, which says any citizen of the United States has equal access to public water and land for hunting and fishing under the 5th and 14th Amendments of the U.S. Constitution. This was won by McDowell and the plaintiffs, which fortified public trust law. Prior to that case, the state of Alaska showed preference towards local and rural Alaska residents to the exclusion of others. Although the plaintiffs won the case they lost politically and the state ignored the ruling.

Warren and 25 others decided to file a McDowell II case in June 1990 in federal court to strengthen the state ruling and received funding from the Alaska Chapter through the Alaska Fish and Wildlife Conservation Fund. After four and a half years of legal wrangling, Judge Russell Holland determined in 1995 that the Conservation Fund and the 25 others involved in the case were nine days premature in filing their case, so the most important portions of the case involving the 5th and 14th Amendments were not decided upon by him because of this technicality. The case was reestablished by Warren and 20 others in 1996 and in July 1998 was dismissed on technicalities in Federal District I Court. Efforts to resolve this issue continues in hopes that a favorable ruling will secure equal access and use to public waters and public lands (state or federal properties) by all U.S. citizens who want to enjoy the riches of Alaska. The Alaska Legislative Council, chaired by Rep. Ramona Barnes, continues to provide assistance to Warren and the Alaska Chapter.

The Alaska Chapter welcomes involvement in this issue and other chapter activities by contacting them at P.O. Box 240065, Anchorage, AK 99524. Their phone number is (907) 248-9010 and their e-mail is fanwsak@alaska.net. Contact them for more detailed information.

Glenn Anderson was elected president of the Alaska Chapter of FNAWS in April 1998 and will continue in that position through April 2000. He received a B.S. in forest management from Washington State University and has worked for the Bureau of Indian Affairs (BIA) since June 1967. He currently is stationed in Anchorage, Alaska, where he has works in the BIA Juneau Area Office since September 1980 as their fire management specialist. He is in charge of the forest protection program that includes fire management, trespass and forest insect and disease problems. He has been married to his wife Ann since 1965 and has one grown son, Dean. Much of this article was compiled with the able help of many members of the Alaska Chapter members, current Board of Directors and past officers.

Warren E. Olson has been a data technician for Anchorage Telephone Utility Company for 28 years. Because of his efforts to enforce the Alaska National Interest Lands Conservation Act for all hunters equitably, Warren has emersed himself in the state and federal judicial systems from Alaska to Washington, D.C. As an Alaska Chapter founder, Warren continues as one of its most enthusiastic supporters. Warren has contributed many articles on the cases in Alaska for Wild Sheep magazine. He and his wife, Jan, have been married since 1959 and are the parents of Gary and David who are also avid sheep hunters. He volunteers as an officer for SMART and finds time to hunt sheep near his home in Anchorage when fortunate to draw a permit.

The Texas Bighorn Society 1980 to Present

PAT HOLLOWAY AND DANIEL BOONE

The Texas Bighorn Society, an outgrowth of the Texas Chapter of the Foundation for North American Wild Sheep (FNAWS), provided the vital bridge from sure extinction to a present-day huntable population of more than 400 bighorn sheep in Texas. This small dedicated group, only 28 to start with, refused to accept failure of early transplant efforts and proved that restoration of this magnificent species to the mountains of west Texas was not only feasible but also economically sound.

Indigenous herds of desert bighorn sheep, Ovis canadensis texana, occupied most of the arid mountain ranges of the Trans-Pecos region of Texas. Pictographs, found in 1933 by Burch Carson in the Sierra Diablo Mountains, depicting two bighorns, document the existence of wild sheep in Texas to prehistoric times. It is estimated that in late 1800 more than 1,500 bighorn sheep inhabited this area.

Events associated with the settling of the Trans-Pecos area by Anglo-Americans adversely affected the desert sheep and although wild sheep hunting was banned in 1903, by 1943 as few as 150 wild sheep were left and the last of the species disappeared in the early 1960s.

A valiant effort on the part of the U.S. Fish and Wildlife Service, the Boone and Crockett Club, the Wildlife Management Institute and the Game and Fish Commissions of Arizona and Texas to reintroduce desert sheep to Texas began in 1953. The first transplant from Arizona to the Trans-Pecos area of Texas that occurred in 1957 was followed by a second transplant in 1959. These sheep flourished for a time but the project was plagued by losses from both disease and predators. Discouraged by these results, the Texas Legislature drastically reduced funding for the project and by the early 1980s the reintroduction of desert sheep in Texas was all but dead. All surviving desert sheep except for a few at the Chillicote Ranch were released into the Sierra Diablo Wildlife Management Area to fend for themselves.

Texas sheep hunters refused to accept the death of this effort and stepped forward to provide the courage and determination to see that the program was not only saved but also revitalized by the state legislature. Thus the Texas Chapter of FNAWS was formed in 1980, followed closely by the chartering of the Texas Bighorn Society. This is where our story begins.

The Texas Chapter of FNAWS, chartered in October 1980, held its organizational meeting in Fort Worth, Texas, on January 30, 1981. A total of 28 members attended and elected Paul Tadlock as the Texas Chapter's first president. Other officers elected were Joe Schaefer as treasurer, Bill Leech as secretary and Don Corley as vice-president. The 501-(c)(3) exemption was issued on February 20, 1981.

The Foundation for the Texas Bighorn Society (TBS) was formed in 1981 under the leadership of Dr. James H. "Red" Duke. Dr. Duke was chosen to become the first president of TBS and served in that capacity until he was succeeded by Daniel A. Pedrotti in 1987; Dr. Charles Drechsel in 1991; Pat Holloway in 1994; and current President Daniel Boone who took office in 1996.

Desert bighorn sheep have returned to Texas, thanks to the work of hundreds of volunteers from the Texas Bighorn Society.
Photographs courtesy of Texas Bighorn Society.

A group of volunteers from the Texas Bighorn Society complete a water guzzler project at the Black Gap Wildlife Management Area near the Rio Grande River. TBS was a group organized in 1981 with Dr. James "Red" Duke as its first president.

Photo 3:
A mule stands patiently as equipment for a guzzler project is packed on its back by two Texas Bighorn Society volunteers, Kennon Cantley, left, and Bill Scott. Two minutes after this photograph was taken, the mule ran off madder than, well, a mule.

Two young desert bighorn rams are reasons why Texas Bighorn Society works so hard and complete so many projects to expand their populations throughout the state.

Daniel Boone enjoys the action of the 1997 auction.
Photograph courtesy of FNAWS archives.

Presenting the 1990-1991 Texas Bighorn Society Presidential Award are from left to right, Society President Daniel A. Pedrotti, Carolyn Pedrotti, and award recipients Pat and Gene Holloway. The 1991 banquet and auction was held April 19 to 21 in Kerrville, Texas. Pat and Gene continue as strong supporters of TBSS.

The first task facing these organizations was to re-establish support for the desert sheep program in the Texas Legislature and the Texas Parks and Wildlife (TPW) Commission. Thus followed an intense lobbying effort. Bill Leech met with TPW Commissioner Perry Bass, Director Charles Travis and House Speaker Gib Lewis. He succeeded in getting the Texas bighorn sheep program refunded and revitalized. TPW personnel were directed to reassess past efforts and to develop a viable program for the successful reintroduction of desert sheep in the state.

The project was now underway. As Dr. Duke pointed out it had gone from "no enthusiasm" to "uncontrolled enthusiasm." It was decided that pen- raised sheep would survive in the wild and provide the best source on a continuing basis for transplant stock. Therefore, the plan designed by Jack Kilpatrick, TPW biologist, for four, l0-acre brood pastures was chosen as the first priority.

TBS volunteered to raise the funds, erect the pens in the Sierra Diablo Wildlife Management Area according to plans provided by TPW and then to turn the completed project over to the state for operation and maintenance. Bill Leech of Fort Worth, a director of both TBS and the Texas Chapter, took on the job of subcontractor. Dr. Duke became the solicitor of pipe and money. Gene Holloway and Ronnie Newman provided free trucking, and the project got underway.

In the meantime, the Texas Chapter, though heavily committed to the brood pasture facility, continued its work in other areas. In l979, Big Mama, a non-lactating ewe and her yearling lamb, were taken from the Sierra Diablos and transferred to the Glaze Clinic in Kerrville, Texas. This clinic was owned and operated by Dr. Bob Glaze, noted wildlife veterinarian. According to Ray Whatley of TPW, this ewe would have a lamb each year but her lambs had to be hand-fed with a bottle of goat milk in order to survive. The yearling lamb grew up to be named Billy and both were subsequently adopted by the Texas Chapter.

The brood facility was completed in June l983, less than nine months after construction began in October l982, at a cost of just over $200,000, donated by individuals, corporations and hunter-conservation organizations. Formal dedication ceremonies were held at the site on June l8 and the facility was donated to the state of Texas.

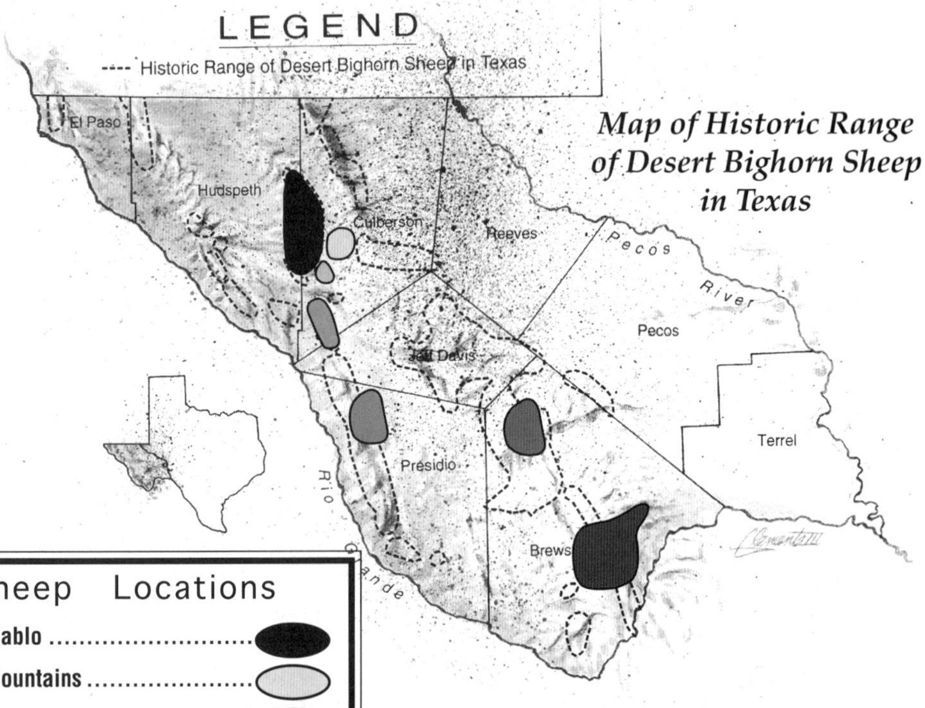

Map of Historic Range of Desert Bighorn Sheep in Texas

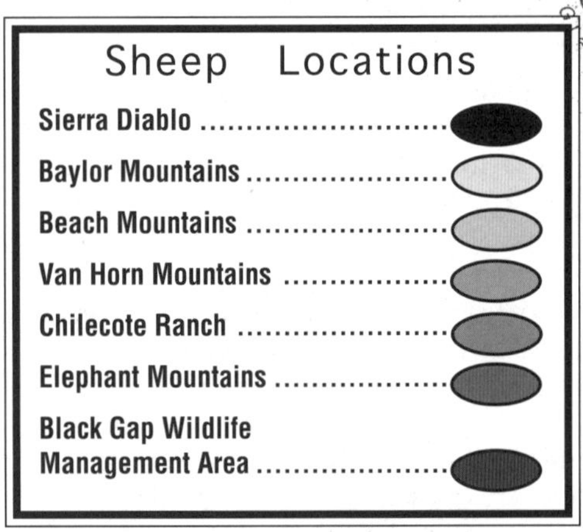

Stocking of these pens began in July 1983, when 15 sheep comprised of three ewe lambs, 11 ewes and one ram, provided by the state of Nevada, were transported by trucks donated by Gene and Pat Holloway and released at Sierra Diablo. In August 1983, seven ewes, one yearling ram and two ram lambs were obtained from Arizona. In September two free-ranging Sierra Diablo rams were trapped and placed in the pens. The state of Utah also supplied two rams in January 1984. To date approximately 175 lambs have been raised and released from these pens into the various desert bighorn habitats in Texas.

All of the excitement and success of the new restoration project caught the attention of C.G. Johnson who, in 1985, donated the 23,000-acre Elephant Mountain Ranch to the TPW to create a big-game management area primarily devoted to desert bighorn sheep. The first transplant of Sierra Diablo pen-raised sheep, totaling 20 bighorns, occurred at Elephant Mountain in February 1987. These new residents consisted of two adult rams, eight yearling rams and 10 ewes. Today this thriving herd contains 60 sheep and provided a 162-3/4 B&C-scored ram for Larry Holland, who won the 1996 Texas Grand Slam Hunt.

Texas issued its first sheep hunting tag in 1987. This was auctioned at the FNAWS convention in Nashville, Tennessee, and purchased by Doc Thurston for $60,000. The first legal Texas ram to be harvested since 1903 was taken in 1990 by David Abbey, who drew his permit through a statewide lottery. To date 11 ram permits have been issued by TPW and eight legal rams have been taken.

It became apparent that Texas needed only one sheep organization, so on July 27, 1985, the Texas Chapter of FNAWS merged with the Foundation for Texas Bighorn Sheep Society and became the Texas Bighorn Society. The combined membership grew rather slowly and leveled off to around 250, leading Dan Pedrotti to note, "We now had more wild sheep in Texas than members of TBS." This led to vigorous membership campaign, and today we have more than 400 paid and about 20 percent are life members.

The year 1998 marks the beginning of the 19th year of dedication to wild sheep in Texas. This epic example of successful wildlife conservation could not have occurred without the dedication and leadership of hunter-conservationists along with the complete cooperation of landowners; TPW personnel; major contributors; the sportsmen and wildlife departments of Arizona, Nevada and Utah; and FNAWS and other sheep and hunter-conservation groups throughout North America.

The story is not over. We will not rest until all of the mountain ranges in Texas that once were home to indigenous populations are restocked with viable herds. That will require more than 1,500 beautiful, healthy, free- ranging desert bighorn sheep.

Whether a person is a Texan, a hunter or just interested in being a part of this exciting conservation success, joining the Texas Bighorn Society provides members with opportunities to participate in the many enjoyable weekends spent working in the beautiful mountains of Texas. Members become partners in creating this legacy, "The Return of Wild Bighorns to Texas," for generations to come. As a member a person will be invited to all meetings, work projects and the annual banquet and fun weekend. Members receive reports on the progress of each of the TBS adopted herds through their quarterly bulletin.

To join TBA, write to Dan and Kathy Boone, P.O. Box 53145, Lubbock, TX 79453, or fax them at 806-745-1851 or call them at 806-745-3822. Renewal membership is $40 per year; new membership is $40 per year; family membership is $75; sponsor level is $200 per year; life membership is $500; and family life membership is $1,000.

Contributions to TBS helps keep wild sheep on the mountains and on Texas lands! Join today.

Pat Holloway is co-owner with her husband Gene of Holloway Company, Inc., Fort Worth, Texas, a company that manufactures steel water tanks and custom steel work. She and Gene founded the company in 1971. From the start of the Texas Bighorn Society, Pat and Gene have worked to help wild sheep in anyway possible. She and Gene first became involved with TBS when members asked them to haul cement pipe to make caves for wild sheep in 1981. Holloway Company provided the driver and the trucks and Pat and Gene followed behind the trucks to help. The caves were in the Sierra Diablo Mountains in west Texas. Pat is the mother of Chryste Denise and Lynton Vess and grandmother of Taylor, Nolan and Hunter. Daniel Boone says she has served TBS in just about every capacity, including a stint as president.

Daniel Boone is a direct descendant of explorer and adventurer Daniel Boone of the 1700s. Dan is president of the Three B Brush Manufacturing Company of Lubbock, Texas. Dan spends much of his free time working on projects for the Texas Bighorn Society. Dan and his wife Kathy live in Lubbock, Texas.

Texas Bighorn Society Projects and Activities

June 1983 - Sierra Diablo brood pens completed.

May 1983 - Cave shelters built for the brood pastures.

July 1983 - Transplanted 15 sheep from Nevada to the brood facility.

August 1983 - Transplanted 10 sheep donated by Arizona to the brood facility.

1983 - Made an embryo study by transplanting embryos from Big Mama into domestic sheep.

January 1984 - Bill Leech, TBS member, purchased a Utah sheep permit for $40,000 at the FNAWS national convention and, instead of harvesting a ram, donated it to the Texas restoration program, along with another ram donated by the governor of Utah.

1985 - Installed water guzzler and water lines for the sheep at Dr. Glaze's veterinary facility. C.G. Johnson, a TBS member, donated a 23,000-acre Elephant Mountain ranch to Texas, primarily for wild sheep.

1986 - 25 sheep from the Sierra Diablo brood facility are transplanted to Elephant Mountain Ranch.

1987 - Gene Holloway, a TBS member, donated and installed a spotting tower at Sierra Diablo.

- Hosted meeting of experts to discuss the upper respiratory problem at Sierra Diablo pens.

- Pat Latham, a TBS member, carved door and donated same in the name of TBS to FNAWS headquarters in Cody, Wyoming.

- Doc Thurston, a TBS member, purchased first sheep hunt in Texas since 1903 at the Nashville FNAWS convention for $60,000. This money funded the transfer of sheep captured in Nevada to the Van Horn Mountains and to the Baylor Mountains.

- 25 Nevada sheep translocated to the Van Horn Mountains on the Lado Ranch.

1988 - Watering hole in at Molcajete Springs.

- Elephant Mountain water guzzler installed on top of High Lonesome Peak on Lado Ranch in Van Horn paid for by Carlos Sanchez donation.

- Paul Webster donated $2,500 to TBS for the funding of a study of the desert bighorn.

- Paid $1,001.50 of water through pipe expense.

- Donated 20 solar-powered collars to Texas Parks and Wildlife.

1989 - Installed water storage tank on McAdoo Ranch.

March 1990 - Built horse and mule trail on Elephant Mountain.

- Funded a study by Dr. Swank for long-range planning for the desert bighorn.

- Co-hosted a roundtable in Van Horn.

December 1990 - First ram harvested out of Sierra Diablo Wildlife Management Area by David Abbey of Dallas.

- Installed a water guzzler on Beach Mountain.

- Assisted in the release of 27 sheep.

March 1992 - Installed two water guzzlers at Black Gap.

- Texana mascot refurbished. He's an original Texas bighorn sheep given to TBS by Col. Snow, one of Texas' first game wardens.

March 1993 - Installed two water guzzlers at Victoria Canyon. Principle contributor was TBS member David Terk at Victoria Canyon, Van Horn, Texas.

March 1994 - Installed two more water guzzlers north of Victoria Canyon with the sponsors Pat Auld, Joe Bishop, Marc Barrett and May Barrett.

October 1994 - 20 sheep transplanted from Nevada to the Black Gap Wildlife Management Area.

March 1995 - Installed two water guzzlers at Black Gap thanks to Alice and Gerald Sutherland as contributors.

March 1996 - Built two water sticks and one crash shack at Black Gap Wildlife Management Area.

March 1997 - Built Slick Rock guzzlers and rebuilt Molcajete Springs at Elephant Mountain and erected two water guzzlers on the Del Nortes. This was sponsored by Water for Wildlife Foundation of Lander, Wyoming, and TBS.

Nevada Bighorns Unlimited-Reno

Before the turn of the century, bighorn sheep were Nevada's most numerous big-game animal. John C. Fremont in the mid-1800s commented in his journal on his way through the Truckee River Canyon where Highway 80 is today that he saw bighorn silhouetted against the sky on every rock pinnacle.

Back then, Nevada was the only area in the nation to have three subspecies of bighorn sheep; the Rocky Mountain bighorn, the California bighorn and the desert bighorn. By the turn of the century desert bighorn populations had been drastically reduced while Rocky Mountain and California bighorns had become completely extinct within the state. This virtual extinction was caused by a combination of market hunting, loss of habitat and disease from the introduction of domestic sheep. Nevada Bighorns Unlimited-Reno, an affiliate organization of the Foundation for North American Wild Sheep, has as its mission since its founding in 1981 to promote and enhance increasing populations of wildlife in Nevada and to fund programs for professional management and habitat improvements. NBU was born out of the desire to put back what was lost. Since the bighorn release program's inception, bighorn sheep have been reintroduced into more than 40 mountain ranges in Nevada.

Leaders of NBU-Reno in 1998 include the following members of the Board of Directors: John Capurro, Mike Daniels, Rick Elmore, Larry Hicks, Gary Hull, Kevin Peterson, Mark Bohach, Grant Dalen, Ron Long, Kent Burroughs, Don Hanks, Larry Johnson, Larry Landa, Roy Walker, Jack Robb and Tony Zamboni. Founding members include Mitch Besso, Mike Ellena, Chris Ellena, Fred Church, Jim Burns, Lewi Chatelle and James Lathrop.

NBU-Reno is well known for transplanting big-game animals back into their original habitat, animals including not only bighorn sheep, but also elk and antelope. Today, the Nevada Division of Wildlife, funded by groups like NBU-Reno, has reintroduced these majestic animals back into their original habitat throughout the state.

The future of big game in Nevada is extremely promising due to a high percentage of public land and the fact Nevada has one of the most progressive division's of wildlife in the nation. NBU, along with Federal Pittman-Robinson matching funds, provides all funding for the Nevada Division of Wildlife's big game reintroduction program. A lengthy land management process that takes several years is necessary for a single reintroduction decision to be made. An average reintroduction consists of approximately 20 animals. Within five years, another augmentation of a similar amount to the same area from a different gene pool is then reintroduced. This process will produce a huntable population of viable sheep within a decade.

Bighorn sheep, when reintroduced into an ecological niche that was once their original habitat, usually experience a population explosion. For the first time in modern history, the state is opening up new management areas for hunting each year. At this time, bighorn sheep have been reintroduced into more than 50 mountain ranges in Nevada. In 1998, 44 management areas were open for hunting. Transplanted populations now account for more than 50 percent of the state's total quota.

NBU's actions prove that true sportsmen are consummate conservationists. In their efforts to create a more balanced and healthy wildlife population in Nevada and beyond, generations of hunters and non-hunters alike will benefit for years to come.

The founders of the Nevada Bighorns Unlimited-Reno include, left to right: Chris Ellena, Mitch Besso, Fred Church and Mike Ellena. Since its founding in 1981 members of NBU-Reno have worked to enhance and increase populations of wildlife in Nevada and to fund programs for professional management and habitat improvements.
Photograph courtesy of Nevada Bighorns Unlimited-Reno.

Volunteers from Nevada Bighorns Unlimited-Reno construct a guzzler on Table Mountain in Mineral County, Nevada.
Photograph courtesy of Nevada Bighorns Unlimited-Reno.

Two desert bighorn drinking from a guzzler that was constructed by the Nevada Bighorns Unlimited-Reno is motivation enough to continue work on water projects in the deserts of Nevada. This guzzler was constructed by NBU-Reno members in Mineral County in 1994.
Photograph courtesy of Nevada Bighorns Unlimited-Reno.

Greg "Lightening" William helps raise money for Nevada Bighorns Unlimited-Reno as their flamboyant auctioneer in 1996 at their annual fundraising banquet.
Photograph courtesy of Nevada Bighorns Unlimited-Reno.

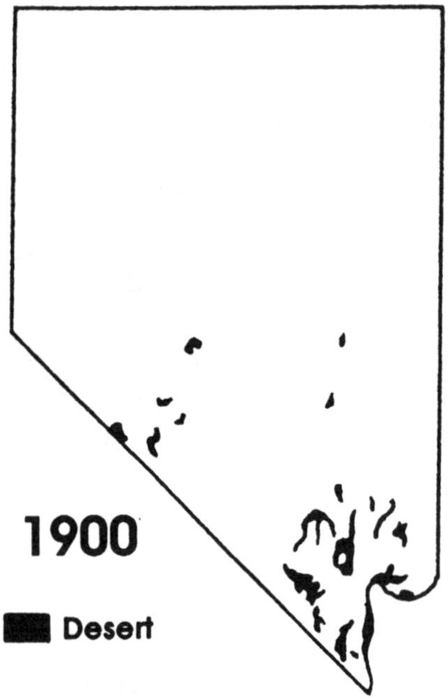

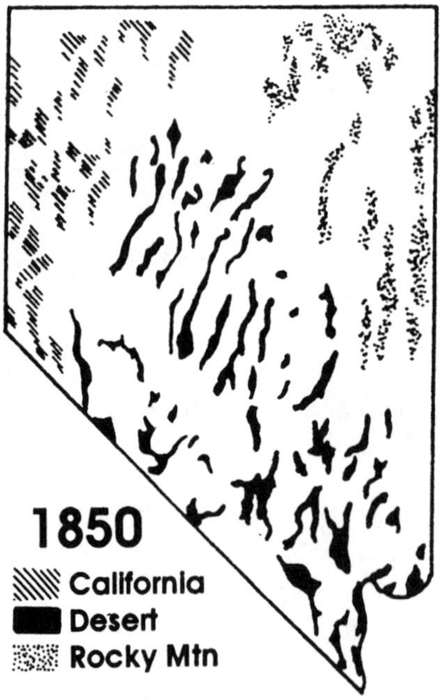

Maps of Nevada show the population decline and increase of desert bighorn sheep from 1850 to the present. Historic herds are being restored thanks to the volunteer efforts of Nevada Bighorns Unlimited-Reno.
Illustrations courtesy of Nevada Bighorns Unlimited-Reno.

NBU Supports Education

NBU faces the challenge of education head on determined to win. Popular sentiment over the last several decades has not supported the hunter. From prime-time media to our children's teachers, the true picture of hunters and their impact on the environment has been distorted and negative.

NBU fully believes that without educating the future generations by presenting facts and merits of hunting, sportsmanship, wildlife management and conservation, the results of other NBU endeavors will be a little or no benefit to the future of wildlife in Nevada.

To this end, there are several programs funded and supported by NBU that merit mention.

The Jim Lathrop Memorial Scholarship Fund was created by NBU in honor of the group's founder, the late Jim Lathrop. The fund represents a cooperative effort involving NBU, the Nevada Division of Wildlife and the University of Nevada. The scholarship was set up for post-graduate study in the fields of biology and wildlife management and has been extended to include funding of summer internships for selected individuals majoring in wildlife management. The objective of these studies will be the further understanding and development of big-game populations and habitat enhancement within our state.

"Project Wild" is an environmental and conservation education program emphasizing the benefits of enhancing wildlife populations. Its primary audience is teachers of students from kindergarten through high school. The goal of Project Wild is to assist learners of any age in developing awareness, knowledge, skills and commitment to result in informed decisions, responsible behavior and constructive actions concerning wildlife and the environment.

NBU's goal is to provide the teachers of our young people with the proper perspective on environmental, conservation and wildlife issues, and the role that we as sportsmen play. NBU sponsors the attendance of several teachers and conservationist to Project Wild's American Wilderness Leadership School.

The American Wilderness Leadership School has as its mission to offer a national and international outdoor instructional program that concentrates on natural resource management. The program provides the participant with a base of knowledge to better comprehend and evaluate current and future conservation issues.

Thanks to a generous donation by NBU, every 4th grader in Nevada, totaling 25,000 children, will receive five times per year the popular nature and conservation magazine, *Wild Outdoor World* (WOW), published by the Rocky Mountain Elk Foundation. This great magazine is designed to teach children about wildlife conservation and features educational articles, photographs and other features on a variety of wildlife topics.

"Operation Game Thief," similar to the "Secret Witness" program in other parts of the country, provides a way for concerned citizens to report wildlife violations to Nevada's game wardens. This program has become one of the most effective methods the Nevada Division of Wildlife has to combat poaching. Rewards are offered upon conviction of the violators and the caller can remain anonymous throughout the process. Reported violations range from minor infractions to major wildlife crimes. NBU helps fund this program through mass media, as well as posting rewards on special poaching cases. The toll-free number provided to report violators is 1-800-992-3030.

NBU Supports Habitat

Not much more than 100 years ago, Nevada's landscape was primarily that of grasslands and wooded mountains. This habitat supported a very large number of indigenous grazers, including elk, antelope and the three species of North American bighorn sheep.

With the spread of mining and ranching came the deforestation of the mountains and the destruction of the grasslands. Sagebrush then took over as the primary vegetation. Many of the native big-game animals become extinct in Nevada. Deer, never present in Nevada before, came to feed on sagebrush. The federal grazing laws of 1932 put a stop to uncontrolled livestock grazing and true wildlife and habitat management practices began to be implemented.

NBU supports Nevada's habitat with funding for a number of special restoration projects, as well las many volunteers donating hands-on labor for these projects. NBU's major habitat improvement programs take areas of poor quality and restore them to useable land that benefits all types of wildlife, including humans.

Reseeding projects, funded by NBU, helped Nevada land that has been devastated by range fires over many thousands of acres, including land used for big-game winter range and habitat. The reseeding significantly increases the animals' potential winter survival rates due to increased food. If these areas had not been reseeded with sagebrush and other natural grasses, a plant known as cheatgrass would have taken over almost immediately choking out all other forms of vegetation. Cheatgrass has no nutritional value. Deer have been known to starve with a belly-full of cheatgrass. Timely donations from NBU and other organizations have aided in purchasing seed and private helicopter services to assist in the reseeding efforts on critical range areas, saving the lives of hundreds, possibly thousands, of animals.

Water is often the limited factor in the expansion of wildlife populations. Nevada's climate ranges from arid in the South to semi-arid in the North, making access to a healthy water supply an even greater issue. NBU works with the U.S. Fish and Wildlife Service and the Fraternity for the Desert Bighorn's water projects in all facets of planning, design, funding and construction of water development projects. Water projects allow for the expansion of habitable ranges for big-game animals, including desert bighorn sheep, California bighorn sheep and antelope. Several of these water development systems, known as "guzzlers," have been completed with great success in Nevada over the past couple of years. As a consequence, big-game animals are not the only animals benefitting from these water developments. As anticipated, everything from coyotes to eagles have been sighted drinking from these guzzlers.

In the past several years, NBU has become more involved in legislative activities in order to further support the future of Nevada's wildlife. NBU helped organize a sportsman's conservationist group, known as the Coalition for Nevada's Wildlife (CNW), that provides a unified voice for sportsmen in the state's legislature. CNW represents all types of sportsmen, including big game, waterfowl and upland game hunters, fishermen, trappers, houndsmen, rod and gun clubs and general conservationists.

During the legislative sessions, the coalition distributes rapid dissemination of information to each coalition member group pertaining to relevant wildlife issues. Each group maintains its complete autonomy, but can join in with other groups on a statewide basis to provide real political clout. Through NBU's efforts a number of important victories have been won in the Nevada legislature. With continued support by groups like NBU and effective habitat management, the trend in Nevada today is a return to the grasslands of our past. This trend will assist in the state's augmentation efforts, significantly increasing the numbers of big-game animals across Nevada.

NBU Supports Reasearch

NBU-generated dollars assists in the funding of several important research programs conducted by several distinguished institutions such as the University of California at Davis, the Caine Veterinary, Teaching and Research Center in Idaho, and Washington State University (WSU).

The most recent research project is being conducted by WSU under the direction of Dr. William Foreyt, a veterinarian. The most important element of this research program has been the establishment of evidence outlining the devastating consequences upon wild sheep herds caused by bighorn sheep and domestic sheep. It is believed the main reason for bighorn sheep extinction throughout their original habitat is due to pneumonia contracted from a bacteria transmitted from domestic sheep herds.

NBU is a firm believer in the multiple use of public lands. They recognize the rights of domestic sheep operators assisted in their efforts to develop a laboratory test to determine the potency of the bacteria responsible for causing pneumonia in wild sheep. This test serves as the tool for discriminating between potentially dangerous and relatively harmless isolates bacteria. NBU-funded research is having an impact on policy- making decisions regarding the shared land use of bighorn and domestic sheep.

The research promises to contribute valuable information to facilitate management decisions regarding the transplantation of bighorn sheep populations. NBU expects management applications facilitated by this research tool will have a positive impact on the maintenance of healthy, flourishing wild sheep.

NBU Support Comes From . . .

Since its inception in 1981, NBU has held only one highly anticipated fund- raising event each year. In order to accommodate more than 1,500 enthusiastic NBU members, the annual fund-raising banquet is held at one of the finest and largest convention facilities in the Reno-Sparks area. The banquet includes an elaborate dinner, impressive displays of sporting and outdoor merchandise, fine wildlife artwork, outstanding taxidermy and an incredible display of some of Nevada's finest trophies.

The main event is the auction. Retailers, outfitters, manufacturers, taxidermists, artists, organizations, individuals and many others from around the world donate items for the auction. Everything from guided big-game hunts to river rafting trips to photographic safaris are auctioned off to the highest bidder. Silent auctions and raffles are also held for items such as fur coats, vacations and custom rifles. The evening's most anticipated and exciting auction items include the Nevada antelope tag and preeminently the Nevada desert bighorn sheep tag. Each year the auction price of these tags have increased dramatically with the proceeds from these two tags going directly to the Nevada Division of Wildlife. The remaining proceeds from the banquet help fund all the other NBU projects.

"NBU is one of the most outstanding citizen involvement organizations anywhere in the state and probably in the country. The work we've done here at the Nevada Division of Wildlife couldn't have been done without them. It is a partnership that makes the whole wildlife management system work," NDW Administrator William Molini said.

Awards are also given to honor the generous support of individuals and organizations for their contributions to NBU and Nevada's wildlife. The banquet's ticket price includes a tax-deductible membership in Nevada Bighorns Unlimited-Reno. The success of NBU's programs and projects involving big-game reintroductions, habitat improvement, education and wildlife research is only possible through the combined efforts and support of concerned community leaders, generous organizations and conscientious individuals. This is evidenced by the overwhelming success and participation each year at NBU's banquet.

Now more than ever, the involvement of each individual has a profound effect on the future of wildlife resources, not only throughout Nevada but throughout the world.

NBU is proud to have a membership that exceeds 2,500 sportsmen, sportswomen and wildlife conservationists from all over the United States. As the size of our membership increases, NBU is also able to wield more and more political clout in legislation that affects game management and the future of our sport.

Ticket information regarding their annual fund raising dinner is available at the NBU address. This premier hunting opportunity offers a hunter the chance to take a record-class ram.

For more information about Nevada Bighorns Unlimited and its many projects, please write or call, Nevada Bighorns Unlimited, P.O. Box 21393, Reno, NV 89515-1393, (702) 323-1177.

This article about Nevada Bighorns Unlimited-Reno was written by members of their Board of Directors. Kevin Peterson and Debra Deming coordinated the effort to submit this chapter.

Wyoming Chapter of FNAWS 1983 to Present

KEVIN HURLEY AND JIM COLLINS

As the second chapter of the Foundation for North American Wild Sheep, the Wyoming Chapter of FNAWS was founded in 1983 by five dedicated wild sheep hunters who were interested in the management and future of Wyoming's Rocky Mountain bighorn sheep population. Wyoming Chapter founders included Dave Steger, John Suda, Ron Ball, Alex Wolfer and Kent Stockton.

By June 17, 1983, Articles of Incorporation were filed with Wyoming's Secretary of State and the chapter was on its way to putting more sheep on Wyoming's mountains. Led by an all-volunteer Board of Directors and officers, with no paid staff, the Wyoming Chapter of FNAWS has a long record of actively working for Wyoming's bighorn sheep, other wildlife and conservation education. With the exception of financial assistance for out-of-state campaigns to battle anti-hunting initiatives, all money raised by the chapter is spent on Wyoming projects. Boasting the nation's smallest human population and the largest Rocky Mountain bighorn sheep population, the Wyoming Chapter is very active with almost 500 members, including more than 45 life members. In 1997 alone, regular and life memberships almost tripled. The current membership list includes bighorn sheep enthusiasts from 28 states, three Canadian provinces and international members from Zimbabwe, Africa and the Philippines.

In 1983, Dave Steger was elected president, and Ron Ball, vice president. Alex Wolfer was secretary/treasurer. Directors included John Suda and Terry Reach. No spring meeting was organized, but the winter meeting took place at the Stagecoach Motel in Dubois.

Dave, Ron, Alex, John and Terry continued as officers for the Wyoming Chapter in 1984. A summer meeting at Dave Steger's farm took place in Riverton and a winter meeting occurred at the Stagecoach Motel in Dubois.

In 1985, Wyoming Chapter re-elected Dave Steger and Ed Novotny became vice president. Larry Mathews was elected secretary/treasurer, and Gary Butler joined John Suda as a director. The spring meeting took place again at the Steger farm but the winter meeting transferred to the Rustic Pines, Dubois.

Dave Steger turned the leadership baton over to Ron Ball who became president in 1986, and Ed continued as vice president. Terry Reach was elected secretary/treasurer. Directors in 1986 were John Suda and Larry Mathews. The spring meeting was held at the VFW Club in Pinedale and the winter event occurred once again at the Rustic Pines.

Ed Novotny was elected president in 1987 and Ron Ball became his vice president. Secretary/treasurer for the year was Kathy Gay. Directors that year were Freddie Goetz and Dan Artery. The spring event took place at Marbleton Inn in Big Piney-Marbleton, and the winter meeting was again at Rustic Pines.

The same slate of officers and directors continued in 1988 but the spring meeting changed to the Holiday Inn of the Waters in Thermopolis. The winter meeting continued at Rustic Pines.

Dave Steger became president in 1989 and Ron continued as vice president. Kathy also continued as secretary/treasurer. Directors included Ed Novotny, Fritz Meyer and Kevin Hurley. Spring brought a meeting once again at the Thermopolis Holiday Inn of the Waters.

Winter meeting took place at the Rams Horn Inn in Dubois. The same slate of officers and directors in 1989 continued in 1990. Spring and winter meetings remained at the same places as in 1989.

The same slate of officers elected in 1991 continued in 1992, but members of the Board of Directors in 1992 included Dave Steger, Jack Hildner and Rod Bennett. The officers elected in 1991 and 1992 were Jack Hildner who became president; Jeff Reynolds as vice president; and Kathy Gay continued as secretary/treasurer. Rod Bennett, Dave Steger and Ron Ball were directors. Jeff Reynolds' ranch in Douglas was the site for the spring meeting and the Virginian Motel in Jackson was the site for the winter meeting.

Ron Ball was elected president of the Wyoming Chapter of FNAWS in 1993, and Jim Gay was elected vice president. After six years of service, Kathy Gay stepped down as secretary and turned those duties over to Mike Martin. Kathy remained as treasurer in 1993. Jim Collins, Jack Hildner and Freddie Goetz were elected as directors. The group returned to the Holiday Inn in Thermopolis in 1993 for the spring meeting and went to the Casper Hilton Inn for the winter meeting. The slate of officers and directors for 1993 remained in place for 1994, but Mike Pilch was added as a director. Rustic Pines in Dubois was the site for the spring meeting, and the Riverton Holiday Inn was where they enjoyed fellowship and fund-raising in the winter.

In 1995, Ron and Kathy continued as president and treasurer, respectively, but Jim Collins was elected as vice president and Jennifer Ball was elected secretary. Freddie, Jack and Mike were the 1995 directors. Meetings were held at the Thermopolis Holiday Inn in the spring and the Sheridan Holiday Inn in the winter.

After nine years of volunteer service to the Wyoming Chapter of FNAWS as its treasurer, Kathy Gay stepped down and turned responsibilities over to Jay Reddon in 1996. Ron remained as president, Jim remained as vice president and Jennifer remained as secretary. Spring meeting was again at the Thermopolis Holiday Inn and Little America in Cheyenne was the site for the winter meeting.

Jim Collins ascended to the presidency in 1997 and Cole Benton was elected vice president. Nancy Collins became secretary, and Jay continued as treasurer. Directors included Fritz Meyer, Mike Mohr, Randy Taylor and Carl Asbell. Once again, the Wyoming Chapter of FNAWS enjoyed fellowship in Thermopolis for the spring meeting at the Holiday Inn of the Waters. Winter meeting/fundraiser was held at the Inn of Rock Springs in Rock Springs.

The same slate of officers from 1997 continued in 1998. Directors included Fritz plus Dennis Claman, Jack Hildner and David Luzmoor. The same hotel in Thermopolis remained as the spring meeting site in 1998 and Casper Hilton hosted the winter meeting.

The Wyoming Chapter of FNAWS holds two general membership meetings each year during the first weekends of June and December. The June meeting is generally family-oriented, with fun pursuits like trap and rifle shooting, fly fishing demonstrations, whitewater rafting and a picnic-barbecue atmosphere. Out-of-state members have attended in recent years and would be most welcome anytime. The winter meeting is the principal fundraiser of the year, where artwork, hunts, firearms and numerous items are auctioned. A variety of exhibits and donor booths have proven popular with our members at the winter meeting. Technical presentations and seminars on Wyoming's bighorn sheep, outdoor survival, Sheepeater Indian history and funded projects have been offered.

Meeting locations rotate around the state to encourage participation and attendance by members and others interested in bighorn sheep. Members receive two semi-annual newsletters, a stick-on decal, periodic mailings, seminars and presentations on bighorn sheep, great

Whiskey Mountain lamb survival study, March 1998. Ron Ball, Wyoming Chapter legend, past president and co-founder of the chapter, holds a darted ewe as he waits for the vets to draw blood. The Wyoming Chapter of FNAWS has pledged $10,000 to create a war chest in matching funds of more than $30,000 for this study.
Photographs courtesy of Wyoming Chapter of FNAWS.

This is what the Wyoming Chapter of FNAWS is all about!
Photographs courtesy of Wyoming Chapter of FNAWS.

A young Rocky Mountain bighorn takes a walk.
Photograph courtesy of Wyoming Chapter of FNAWS.

A weekend of summer fun and excitement.
Photograph courtesy of Wind River Canyon Whitewater, Inc.

Rocky Mountain bighorns graze on traditional winter range on South Fork of the Shoshone River, site of the 1998 weed pulling weekend.
Photographs courtesy of Wyoming Chapter of FNAWS.

Sheep trapping on Whiskey Mountain. The Wyoming Chapter of FNAWS furnished the radio collars for these Rocky Mountain bighorn sheep that headed for the Bighorn Mountains in 1994.

A crowd gathered to watch the Rocky Mountain bighorn release in Wind River Canyon in 1995.

Ron Ball, Wyoming Chapter legend and one of the founders, sits beside a campfire at Lake Alice Camp.

Wyoming boasts the lowest human population and the highest Rocky Mountain bighorn population in the United States.

fundraisers and an opportunity to make a difference in Wyoming.

In recent years, the Wyoming Chapter has offered opportunities for its members to see some country, learn more about Wyoming's bighorn sheep, and become involved in hands-on projects. A BYOH (bring-your-own-horse) weekend in summer 1997 offered a tour of the North Fork of Owl Creek to members, a promising transplant site. In June 1998, the chapter sponsored a "Weed-Pulling Weekend," where sheep enthusiasts and others helped with physical control of dalmatian toadflax, a noxious weed invading crucial bighorn sheep winter range southwest of Cody. A ram aging and scoring clinic was offered by the chapter at the September 1998 Wyoming Hunting and Fishing Heritage Exposition in Casper.

In a history-making effort in the lower 48 states, 43 bighorns from Whiskey Mountain were transported by rail to the release site in Wind River Canyon in January 1995.
Photograph courtesy of Wyoming Chapter of FNAWS.

Since 1983, the Wyoming Chapter of FNAWS has funded more than $150,000 in projects to benefit Rocky Mountain bighorn sheep in the Cowboy State. Working with the Wyoming Game and Fish Department (WGFD), other agencies and interested individuals, Wyoming Chapter funding has been used for bighorn sheep transplants into suitable habitats throughout Wyoming. Since 1934, more than 1,500 Rocky Mountain bighorn sheep have been transplanted within Wyoming, with almost all of those originating from Whiskey Basin, the nation's largest concentration of wintering Rocky Mountain bighorns. The Whiskey Basin herd, near Dubois, has served as the primary source for Rocky Mountain bighorn sheep transplants around the western United States, with 399 sheep being provided to South Dakota, Idaho, Utah, New Mexico and Nevada since 1965. Wyoming Chapter funds have purchased drop-nets for trapping operations, secured apple pulp/bait to lure sheep to trap sites, modified WGFD's "ewe-haul" trailer, and provided radio collars and flight time to evaluate success and monitor movements of transplanted herds.

Wyoming Chapter funding has helped implement a variety of habitat improvement and land acquisition projects around the state. Prescribed burns along the upper South Fork of the Shoshone River, herbicide treatments of Whiskey Basin to control mat-forming cushion plants, forage evaluations in the Wyoming Range / Bighorn Mountains / Ferris Mountains, and noxious weed control on bighorn winter range near Cody have all been funded by the Wyoming Chapter of FNAWS. Water developments in the Ferris Mountains and on the Whiskey Basin Unit have enhanced bighorn habitat. To protect crucial bighorn sheep winter range along the Hoback River Canyon, the Wyoming Chapter and the Minnesota-Wisconsin Chapter jointly assisted the Jackson Hole Land Trust with acquisition of the Poison Creek property in 1997.

The Wyoming Chapter has funded research into causes of lamb mortality in the Ferris/ Seminoe Mountains and at Whiskey Basin; bighorn sheep interactions with mountain goats in the Clarks Fork Canyon; and University of Wyoming studies on bighorn sheep genetics throughout the state. The chapter has earmarked a $5,000 "Stop Poaching" reward for law

The annual shoot had a record turn out in 1997 and ferocious competition. Ernie Reynolds won the rifle shoot and Fritz Meyer won the trap shooting competition, and after a marathon "shoot oof" Fritz was declared the overall champion for 1997. Kevin Hurley was thanked for lining up the shooting club for the chapters use.
Photograph courtesy of Wyoming Chapter of FNAWS.

enforcement violations involving bighorn sheep, to further protect Wyoming's bighorns.

In recent years, the Wyoming Chapter has funded numerous educational efforts. Teacher tuition to wilderness leadership training, youth essay contests, "Becoming an Outdoors Woman" programs, and the Wyoming Hunting and Fishing Heritage Exposition have all been assisted by Wyoming Chapter funding. The chapter has been a consistent supporter of the various educational and outreach programs of the National Bighorn Sheep Interpretive Center, located in Dubois. Details about this outstanding center are detailed in the chapter about the Rocky Mountain Bighorn Society that also contributed to its construction along with other conservation groups. Unique viewing opportunities exist at Whiskey Basin, the South Fork of the Shoshone River and in Wind River Canyon, where bighorn sheep can be seen and photographed.

The Wyoming Chapter has stepped up on several occasions to defend sportsmen's rights and the hunting tradition, assisting with initiatives in Arizona, Idaho and Alaska. Thus far, Wyoming has escaped much of the anti-hunting debate raging around the western United States, but the Wyoming Chapter of FNAWS will be right there if challenges arise in Wyoming.

Founding President Dr. Dave Steger has continued to assist the chapter in their fund-raising efforts through the annual auction of one big-game license, made possible via Dave's ascent to and current position on the Wyoming Game and Fish Commission (WGFC). Other members of the WGFC have graciously aided the Wyoming Chapter with occasional donations of other big-game (i.e., pronghorn, deer, elk) licenses as well. Governor Jim Geringer, continuing in the tradition of past governors Ed Herschler and Mike Sullivan, has assisted greatly with periodic donations of statewide elk, moose and bighorn sheep licenses for auction.

In 1993, long-time Chapter President Ron Ball was presented the "Eastman's Outdoor World Grass Roots Award" at the FNAWS national convention in Reno, recognizing Ron's

Table I

Bighorn sheep population objectives and estimates for 16 herd units in Wyoming, plus Wind River Indian Reservation and Yellowstone National Park, winter 1997-1998.

Origin	Herd Unit	Objective	Estimate
Native	Targhee	125	100
Native	Jackson	500	530
Transplant	Darby Mountain	150	55
Native	Clarks Fork	500	470
Native	Trout Peak	750	580
Native	Wapiti Ridge	1,000	925
Native	Younts Peak	900	760
Native	Francs Peak	1,360	1,430
Transplant	Shell Canyon/Devils Canyon	100	50
Transplant	Douglas Creek	350	150
Transplant	Laramie Peak	500	250
Transplant	Encampment River	200	40
Native	Whiskey Mountain/WRIR	1,350	1,025
Transplant	Temple Peak	250	35
Transplant	Sweetwater Rocks*	400	0
Transplant	Ferris/Seminoe/Owl Hills	300	40
Native	Yellowstone National Park	—	285
	Statewide Totals	**8,735**	**6,725**

* *Scheduled transplant; no population established yet.*

contributions to wild sheep across North America. At the 20th FNAWS national convention in Philadelphia, WGFD biologist and active Wyoming Chapter member Kevin Hurley was presented the "State Statesman Award" for 1997, in recognition of his efforts on behalf of Wyoming's bighorn sheep. With true commitment, Kathy Gay spent nine years on the Wyoming Chapter board, holding dual offices of secretary/treasurer for many of those years.

While individual awards and recognition make a difference, the strength of the Wyoming Chapter lies in a core group of diehard members who are committed to putting more sheep on Wyoming's mountains. The success of the chapter also could not have been possible without the generosity of donor businesses, artists, outfitters, gunsmiths and members who provide items for our annual fund raisers.

Working cooperatively with local, state and federal agencies, Wyoming's school systems and the University of Wyoming; National FNAWS and other conservation organizations, the Wyoming Chapter has been a catalyst in ensuring the health, status and future of Rocky Mountain bighorn sheep in Wyoming. The future looks bright for Wyoming's more than 6,725 bighorn sheep, and the Wyoming Chapter of FNAWS is committed to ensuring that future.

As a nonprofit organization with 501(c)(3) charitable status, contributions to the Wyoming Chapter are tax-deductible. Annual membership is $15; a three-year membership is

$40; and a life membership is $300. Interested individuals who want to join the Wyoming Chapter of FNAWS are invited to send their dues to Nancy Collins, Chapter Secretary, P.O. Box 125, Chugwater, WY 82210. Won't you please consider joining us in our efforts on behalf of Wyoming's bighorn sheep, wildlife and sportsmen?

We thank you.

Kevin Hurley is a wildlife biologist with the Wyoming Game and Fish Department, stationed in Thermopolis. Kevin completed his master's degree thesis on the Trout Peak bighorn sheep herd, located between Cody and Yellowstone National Park. From 1982 through the summer of 1998, Kevin worked extensively with bighorn sheep in the Absaroka Mountains in northwestern Wyoming and has been involved with statewide sheep management. In 1997, Kevin received from FNAWS at their national convention the prestigious "State Statesman Award." Kevin has been an active member of the Wyoming Chapter of FNAWS since it formed in 1983 and became a life member in 1997. In addition to his duties with WGFD, Kevin volunteers as executive director of the Northern Wild Sheep and Goat Council, an international organization of wildlife managers, researchers and administrators with responsibilities throughout the western United States and Canada.

Jim Collins is a passionate sheep hunter and outfitter. He has been hunting since he was 19 years old. He owns and operates the Sand Creek Ranch in Alcova, Wyoming. He has owned the ranch for more than 20 years. Jim loves to hunt in North America and throughout the world and has friends throughout the world because of his hunting experiences. He was the 1997 and 1998 president of the Wyoming Chapter of FNAWS. Prior to that position, Jim served as vice president of the chapter and as a director.

Idaho Chapter of FNAWS 1985 to Present

Bill Louderback

Idaho is a state with a wealth of outdoor experiences for those who have ventured into its wild places. Large wilderness areas, crystal-clear lakes, fast-flowing rivers and streams, plus high mountains, semi-arid deserts and canyon lands are only part of its treasurers of the outdoors.

Idaho's history of dignitaries and famous outdoor writers visiting its wild places has left deep impressions on those of us who have made our home here. As hunters and outdoorsmen, we too have explored those wild places and others that forever leave lasting impressions.

Jack O'Connor, with his vivid descriptions of high country beauty, truly lived his glorious words. Ted Trueblood made us proud of our heritage through printed words and descriptions of his own experiences. Elmer Keith did it his way through guiding and showing us his large-caliber philosophy. Elgin Gates perhaps did it all by getting to those places and telling us about it in his published accounts.

All these men hunted other continents as well and spoke and wrote of the necessity for all of us who follow to be truly sportsmen, and most of all, to be guardians of the land that wild creatures inhabit.

Idaho received one of its most valuable gifts when a group of individuals with great vision and forethought explored the possibility of becoming a chapter of the Foundation for North American Wild Sheep (FNAWS). These people were already involved with the fight to preserve Idaho's wilds and wetlands as several belonged to a group called Sportsmen's Political Action Committee of Idaho.

They were very aware of the rapid development of cities, military bases, mining interests, livestock grazing, farms and ranches that were putting limits on big-game ranges. Dams flooded winter ranges and the Idaho Fish and Game Department (IFGD) had to seek alternatives to maintain stability for wildlife.

Idaho's wild sheep need vast acres of habitat for existence. Although they are federal ground, competition has become fierce to keep and maintain that ground. The U.S. Air Force continually keeps up the pressure for more land and airspace. The political climate and political decisions continually ignore the biological data in favor of continued economic development.

From this backdrop of information and history, the Idaho Chapter of FNAWS was founded in 1985. A total of 28 individuals stepped forward with $100 bills to do something for wild sheep. They were, and many of them continue to be, the cornerstone of the Idaho Chapter of FNAWS. The charter members of the Idaho Chapter of FNAWS include the following individuals:

1. Tom Hinders, Ames, Iowa
2. Bob DiGrazia, Boise, Idaho
3. Rick Dredge, Boise, Idaho
4. Ronald Morris, Boise, Idaho
5. Roger Michener, Boise, Idaho
6. Andy Thacker, Boise, Idaho
7. Burk Mantel, Boise, Idaho
8. Russ Biaggne, Boise, Idaho
9. James Kranz, Boise, Idaho
10. Wayne Schwabrow, Boise, Idaho
11. Mike Best, Eagle, Idaho
12. Jim Blaine, Rathdrum, Idaho

Bill Louderback with his Dall's sheep.

Founders of the Idaho Chapter of FNAWS throw in $100 bills to begin great traditions. Top row, 4th from left is FNAWS President Tom Hinders. Roger Michener is 6th from the left. Ron Morris is 7th from left. In the bottom row, first from the left is Wayne Schwabrow. Tom Schiermier is 3rd from the left.

Photographs courtesy of Idaho Chapter of FNAWS.

The first donated rifle is received by purchaser Roger Graham, center. Left of Roger is Wayne Schwabrow and to his right is FNAWS President Tom Hinders, who is Charter Member One for the Idaho Chapter.

Dr. Robert "Bob" DiGrazia, called Dr. Bob by many, talks to Idaho Chapter members regarding anti-hunting activity at the formation of the Sportsmen's Defense Fund in 1996. He stands beside his Dall's sheep.

Idaho Chapter President Bill Louderback presents Barbara Michener a 30.06 Ruger rifle at the 1995 banquet.

13. Mike Bland, Murtaugh, Idaho
14. Mike Brusco, Longview, Washington
15. Ed Carr, Twin Falls, Idaho
16. Harold Eshelman, Boise, Idaho
17. Ralph Garrison, Twin Falls, Idaho
18. Ron Gillette, Stanley, Idaho
19. Scott Hayes, Boise, Idaho
20. Ralph McClintock, Boise, Idaho
21. Tom Parker, Boise, Idaho
22. Tom Schiermeir, Twin Falls, Idaho
23. Ed Sweet, Meridian, Idaho
24. Jerry Wiesen, Lewiston, Idaho
25. Cheryl Hudgens, Caldwell, Idaho
26. Fred Wood, Burley, Idaho
27. Barry Wood, Shoshone, Idaho
28. Jerry Young, Twin Falls, Idaho

The first chapter president was Wayne Schwabrow. Wayne was excited about the prospect of forming a chapter of the national organization of FNAWS, and he was a friend of then-current National President Tom Hinders. Tom visited Boise and was first to produce a $100 bill. He was given the position of Charter Member Number One on the list of 28.

The strength of the national organization, plus the strength and leadership of the original charter members, and the mission of "putting more sheep on the mountain" was an obvious great match. Through the efforts of the original few, the mandatory paperwork was completed in the state of Idaho offices; and the articles of incorporation were signed on March 26, 1985, by Wayne Schwabrow, Roger Michener and Burk Mantel. The directors listed on the charter at this time were Ron K. Morris, Dennis Gratton, Russ Biaggne and Robert DiGrazia.

During this same period, Roger Michener organized the first and productive planning seminar and workshop with Idaho's Fish and Game Director Jerry Conley. The large group of original chapter members divided into groups and met in classrooms with Idaho's wildlife experts. Through a long afternoon of education, discussion and common-sense planning, the Idaho Chapter of FNAWS produced valuable input, data and concerns that eventually were incorporated by the IFGD into their five-year plan for Idaho's bighorn sheep.

The second annual Bighorn Sheep Seminar in 1986 was planned and coordinated by Burk Mantel. By this time, the chapter had made its first donation to the IFGD of $2,000. The informative sessions brought out some of the best professionals in the field of bighorn sheep management and research. Tom Schiermeir, taxidermist from Twin Falls, not only instructed the audience in taking care of trophies but donated yearly to the chapter auctions. The ever-present Stanley Potts lectured on qualities to look for in selecting a guide. He further demonstrated how to judge a trophy ram. Tom Thorn of Wyoming Game and Fish Department lectured on the problems bighorn sheep face with diseases from domestic sheep, goats and others.

Little did we know at the time, the Idaho Chapter, National FNAWS and other chapter affiliates from across the nation would become directly involved with the Hells Canyon die-off. Members from several chapters and game departments have spent countless hours of volunteer time, plus costly professional time, trying to conquer the disease that threatens wild sheep.

Through these efforts and those of Idaho's Dr. David L. Hunter of the Wildlife Health Laboratory in Caldwell, Idaho, and his associations with other leading research biologists, progress has been made. Another chapter in this book details the work of the Caldwell laboratory. The Idaho Chapter's lottery money is dedicated to wildlife research.

Also in 1986 Chapter President Dick Dredge and the Board of Directors voted unanimously to again pursue the availability of an auction sheep tag since earlier efforts had failed. This time a very determined group of directors — Dennis Gratton, Burk Mantel, Wayne Schwabrow and Dr. Bob DiGrazia — lobbied up and down the hallways of the capitol build-

Jim Thrash, center, was an active and dedicated member of the Idaho Chapter of FNAWS before his untimely death while fighting a raging forest fire in Colorado in 1994. He is holding his wife's antelope surrounded by son Nate, left, and Ginny, right, during a 1993 family hunt.
Photograph by Holly Thrash.

Jim Thrash of Boise, Idaho, took this outstanding Dall's sheep with Dick Gundlogsen in the Wrangell Mountains of Alaska in 1992. Jim died as a smokejumper fighting the infamous "Storm King" fire outside Glenwood Springs, Colorado, that took the lives of 14 firefighters in the summer of 1994. The Idaho Chapter of FNAWS established a memorial fund in his memory.
Photograph courtesy of Holly Thrash.

Bill Louderback recruits new members for the Idaho Chapter of FNAWS in their booth at the 1993 Idaho Outfitters and Guides Association convention in Boise in 1993. He rests his hand against a full mount Rocky Mountain bighorn ram.
Photograph courtesy of Idaho Chapter of FNAWS.

In 1985, Idaho Chapter President Wayne Schwabrow and Director Roger Michener award Craig Brown, center, a FNAWS rifle.
Photograph courtesy of Idaho Chapter of FNAWS.

ing while the state legislature was in session.

With the help of Representative Pam Bengson and Senator Ron Beitelspacher, success was finally achieved in the early summer of 1987. As Governor Cecil Andrus signed House Bill 130, Dr. DiGrazia declared this a "red letter day" for bighorn sheep of Idaho.

The chapter's first three years were quite remarkable. The first $2,000 donation was placed in a trust with the IFGD; and those funds were combined with Pittman-Robertson funds to generate approximately $10,000 for a sheep project in Idaho.

The Idaho Chapter board voted to dedicate more funds to Idaho Citizens Against Poaching and to the Idaho Sportsmen's Coalition. By mid-1987 the Idaho Chapter newsletter was recognized by its membership as a great communication tool for potential and successful sheep hunters. Joslyn-Morris Printing, a Boise firm, was donating paper and hours of time to produce a professional newsletter. Ron Morris, a charter member and board member, insisted on quality; and to this day, it has remained as he demanded it be.

Bill Louderback on a visit to Cody, Wyoming, and FNAWS national headquarters, nominated Dr. Bob DiGrazia for the national FNAWS Board of Directors. Dr. Bob, as he has become known throughout the sheep world, was elected on his first effort, receiving more votes than any other name on the ballot. Bob received word that his official start date was April 1, 1988.

The next seven years would bring dynamic changes and progress in the wild sheep world. Dr. Bob spent two years (1988-1989) as a FNAWS director. In 1990, he became first vice president; in 1991, he was president-elect, followed by first president-elect in 1992, and then national president of FNAWS in 1993. He also served as a voting member in 1994 in his capacity of past-president.

The 1988 Bighorn Sheep Seminar was held in Lewiston, Idaho, with Harold Eshelman serving the first of his two terms as chapter president. Harold felt more work could have been done to make the trip to Lewiston more successful. The good news was that prior to the Lewiston seminar, the first Idaho bighorn sheep auction tag had sold for $68,000 at the national FNAWS auction in February. Robert Senter from New Hampshire was the winning bidder. The complete list of Idaho auction tag buyers and their winning bids are included in the appendix of this book.

In 1989, Mike Best was elected president of the Idaho Chapter. Mike had just taken a new job and felt it was not in the best interests of the chapter that he try to do both. The chapter was grateful to Roger Warrick for stepping forward and offering his experience, faith and capable leadership. Roger served as master of ceremonies for many of the chapter functions.

Robert and Secretary-Treasurer George Law signed the amendment to the articles of incorporation defining the Idaho Chapter of FNAWS as an educational organization. Disease in bighorn sheep became a major issue for the chapter and state wildlife officials. Roger recognized that much of the die-off and disease was not understood by the general public and members. He convinced the chapter to allow time for the IFGD to scientifically identify the problem. The conclusion was that the hard winter had stressed the sheep in many areas and that they had died of natural causes.

Another major problem occurred at Pine Creek on the main Salmon River. A herd of domestic sheep was placed on property owned by the Girl Scouts of America. This area is a known habitat for a large population of wild sheep that moves miles up and down the Salmon River in Central Idaho.

By the fall of 1989, the domestic sheep were removed from the Girl Scout camp. However, the disease had begun to spread. Lloyd Oldenburg, biologist for the Idaho Department of Fish and Game said, " It will take years to see the end of this episode."

He was right. Immediately the following spring, the ratio on the main Salmon and along the Middle Fork was four lambs per 100 ewes. The following seven years saw dramatically low lamb-to-ewe ratios. Not only that — the U.S. Air Force decided that they needed to expand their training range in southern Idaho, which would cause an immediate effect on the unique and fragile California bighorns. The future for wild sheep in Idaho was questionable.

This was the beginning of an on-going dispute involving national FNAWS, the Idaho Chapter, and dozens of wildlife groups and individuals across the United States calling for the U.S. Air Force to back off. This negative cloud created by the U.S. Air Force and its supporters cost the taxpayers in this country millions upon millions of dollars. The dispute has cost Idaho Chapter members and other donating chapters large sums of money and countless hours of volunteer time. Public testimony throughout southern Idaho and the four major newspapers all overwhelming disapproved the expansion, along with biology and scientific evidence of the harm to wildlife caused by an expansion. Politics and politicians ignored the facts and promoted solutions that would be detrimental to the California bighorn population.

With the bad news came some good news as Dan Pocapalia from California purchased the 1989 auction permit for $36,000. Dan has some major business interests in Idaho and along with a fondness for the state, he continually supports the chapter's efforts. In 1990 the Idaho auction tag failed to sell because of a $30,000 minimum bid set by the Idaho legislature. FNAWS auction buyers actually "boo-ed" when the Idaho permit was offered and applauded when it did not sell. The bidding had stopped at $29,000. After some discussion, the Idaho tag was again offered, and Jerry Fletcher from Arizona purchased it for $30,000.

Because of the problems with the $30,000 minimum, Idaho Chapter President Scott McCracken, along with Bob DiGrazia, Harold Eshelman, Burk Mantel, Stan Potts, Mike Best and Wayne Schwabrow went back to the legislature. They successfully lobbied to get the $30,000 minimum removed. In the chapter, "Presidents Speak," Dr. Bob DiGrazia explains how that process took place. Through their diligence and forethought, they also earned approval for a lottery permit. All proceeds from the lottery permit go to the Wildlife Health Laboratory of Caldwell, Idaho, and to the efforts of Washington State University veterinarian, Dave Hunter. Not only did Representative Pam Bengson and Senator Ron Beitelspacher step forward again, but Senator Herb Carlson joined to assist the effort for the lottery permit.

As Wayne Schwabrow began his second term as Idaho Chapter president, some hard work was taking place to see that the 1991 special bighorn permit was offered. It ended up being sold by silent auction to Dave Colcough from California for $23,700. As usual, the overwhelming concern was the threat of the proposed realignment of Mountain Home Air Base and their range capacity. They continued to have little or no concern for the future of California bighorns in Owyhee County. Idaho Chapter members Dr. Bob DiGrazia and Herb Meyr were meeting weekly with wilderness groups and legal counsel and keeping the press and public informed. Since then, the public has been outraged at the persistence of the U.S. Air Force as they continue to threaten the Owyhee Canyon lands and California bighorns. Dr. Bob and the supporting groups have exposed document after document containing inaccurate information and incomplete environmental impact statements. This data has been published for all to see.

The Idaho Chapter is aware of the efforts of its members and the wildlife community. Their hours and hours of testimony have basically been ignored. The political agenda in Idaho, regardless of the biological data and input from top professionals on California big-

Chapter Members Roger Michener and Burk Mantel, left and right, organized the first and second fundraisers for the Idaho Chapter of FNAWS.
Photograph courtesy of Idaho Chapter of FNAWS.

Installation of a guzzler was a project for Idaho Chapter Director Elmer Huston and Idaho Department of Fish and Game personnel at the Cottonwood Wildlife Management Area.
Photograph courtesy of Idaho Chapter of FNAWS.

Participants of the 1996 Bighorn Sheep Day seminar receive valuable information about sheep permits and regulations and sheep habitat and mannerisms to prepare for the upcoming season.
Photograph courtesy of Idaho Chapter of FNAWS.

A skull and skeleton of an animal is all that remains after attempting a jump over a barbed wire fence, which motivates Idaho Chapter volunteers to replace the barbed wire with posts and rails at the Big Cottonwood Wildlife Management Area. Left to right are Elmer Huston, Harold Eshelman, Lloyd Oldenburg and Bob Ulshafer.
Photograph courtesy of Idaho Chapter of FNAWS.

Gerry Glazer and Elmer Huston install fence around a guzzler that allows sheep to go under the plain wire to reach the water. The lower wire is set 18 to 20 inches above ground.
Photograph courtesy of Idaho Chapter of FNAWS.

The Idaho Chapter of FNAWS watched in 1988 as their first sheep tag was auctioned at the national convention. Top bidder Robert "Bob" Senter, center, receives a pat on the back from Idaho Chapter members. He donated $68,000 to FNAWS to "put sheep on the mountain."
Photographs courtesy of FNAWS archives.

The Idaho Chapter held its annual banquet on February 1, 1992, at the Holiday Inn in Boise. Here Wayne Schwabrow, left, and Harold Eshelman congratulate each other for a job well done. A total of 180 men and women attended the event.

Participating in a fall 1995 field project were, left to right: Holly Thrash, Glen Gearhard, Dale Toweill, Tony Glaiser and Jerry Glaiser.

Idaho Governor Phillip Batt, left, drew the winning ticket in the annual Idaho state bighorn sheep tag lottery in the summer 1996. Idaho Chapter President Ralph McClintock, right, observed.

The Idaho Chapter refurbished Jack O'Connor's and Jess Taylor's ram collection. Bill Louderback, right, unpacks one of the Jack O'Connor trophies in 1995 with the help of Theresa Barner, left, and Harold Eshelman, center. The collection was exhibited at the Idaho Chapter banquet and at the FNAWS national convention in Phoenix.

horn sheep behavior, turns a deaf ear and ignores the facts. Their desire is to expand the base for economic reasons. As a result, the Idaho Chapter of FNAWS has little hope of ever increasing California bighorn permits to their potential.

In 1992 President Glenn Gearhard and his Board of Directors made several recommendations to the IFGD. They took measures to see that valuable capes and hides from deceased sheep at the Caldwell Clinic were properly cared for. With the idea that full-body mounts be created for educational and display purposes, they further requested the IFGD provide sheep horns from both California and Rocky Mountain species. Chapter members began their yearly tradition of spending a day at the Caldwell research facility, assisting Dr. Hunter and his staff with various cleaning projects and other chores to allow the professional staff more time to get to their mission.

In other business, the Idaho Chapter successfully operated its first lottery for a bighorn sheep permit, raising $19,760. Since then, under original Chairman Harold Eshelman and current Chairman Ralph McClintock, the annual lottery has raised $171,027 through 1997. Raffle winner for the first year in 1992 was Noel Poux of Montana. For the first three years, from 1992 to 1994, Harold Eshelman was raffle chairman. In 1993 Gerald Moody of Virginia was raffle winner and $11,780 was raised. A total of $16,150 was raised in 1994 and Richard Aznaran of Texas was the raffle winner. For the next three years, from 1995 to 1997, Ralph McClintock served as raffle chairman. In 1995 Joe Schram of Idaho was the raffle winner and $27,730 was raised. D. Kevin Moore of California was raffle winner in 1996 when $53,914 was raised, and $41,693 was raised in 1997 when Paul Black of Pennsylvania was the raffle winner.

The expanding chapter newsletter continued to keep members informed, and the winter of 1993 issue contained pictures of 16 successful sheep hunters. Harold Eshelman began his second year as president of the chapter and was involved in the Shoshone-Bannock Indian Nation on the Board of Directors. The annual work day for Idaho Chapter members continued under the able leadership of work projects coordination Elmer Huston. The IDFG gave the Idaho Chapter two sets of horns — one each of California bighorn and Rocky Mountain bighorn — on which to do full-body mounts. Board member and taxidermist Bob Ulshafer agreed to produce the mounts.

The bombing range again elevated concern as Governor Andrus was now more determined than ever to force the issue of the Big Springs Bombing Range. The chapter's Board of Directors and many members called all over the country to gain support from the Bureau of Land Management (BLM), Congress and other officials in putting an end to the threat to Owyhee County. Many times angry and frustrated outdoorsmen called the governor's office and anyone who would listen to stop the bombing range. National FNAWS, through Grant-in-Aid funding, had provided close to $40,000 to fight for California bighorns and to expose the politics. The issue was at an all-time high. During the winter of 1992-1993, the California bighorn herd lost approximately one fourth of its population. The U.S. Air Force, with its increased activity over the range, accepted no responsibility. Since the recent winter in Idaho had been severe, the U.S. Air Force pointed to that as a problem.

Then came another major threat: The wolf reintroduction program. At the FNAWS national convention in Reno, our Alberta and British Columbia friends could not believe the ridiculous effort by the United States government to reestablish wolves in Idaho. Of course, the Canadians jokingly said, "You can have all you want!" This project totally blind-sided the chapter and other Idaho sportsmen. Their input had no effect, and wolves were released near bighorn winter range. Of major concern to the public and hunters is that the federal government has the power to close large tracts of national forests and wilderness. If they feel the wolves are in danger, hunting could be banned in many areas.

The Idaho Chapter of FNAWS has evolved into a strong group of individuals who are active in wild sheep issues in Idaho. The chapter not only has dynamic, successful banquets and Bighorn Sheep Days but they also work the halls of the Idaho capitol building. They have joined forces with other Idaho wildlife groups to do whatever can be done to save winter habitat, summer ranges and wildlife. The chapter works hand-in-hand with IFGD managers and regional staff on projects that support the FNAWS mission to "put more sheep on the mountain."

The chapter promotes, and through volunteer efforts, raises thousands of dollars through the annual lottery and auction permits. It is truly amazing and very frustrating to watch and be part of the effort to preserve, protect and perpetuate bighorns for the state of Idaho and then have to deal with the organized efforts to stop our progress.

In 1994 Bill Louderback, president of the Idaho Chapter, got together with Lloyd Oldenburg from the IFGD and agreed to share a booth at the San Antonio national convention of FNAWS. It worked out well and the tradition has continued.

Idaho was proud that its own Harold Eshelman was presented the prestigious "Gordon Eastman Grass Root Award" for his work of many years on behalf of wild sheep. Another milestone was reached when the chapter banquet was moved to a larger facility in Boise to the Red Lion. A new state-of-the-art, custom-designed gooseneck trailer was unveiled to be used to transport wild sheep. The project was financed by national FNAWS and the Grand Slam Club. A major success was enjoyed by all Idaho Chapter members when we learned our lottery winner in 1994, Richard Aznaran, harvested a new Idaho record bighorn ram that scored 196 B&C points from Nez Perce County. His ram was officially entered into the 22nd Big Game Awards Period of the Boone and Crockett Club.

The Idaho Chapter and Idaho Outfitters and Guides Association (IOGA) suffered a great loss with the death of Jim Thrash in the Storm King fire outside Glenwood Springs, Colorado, that took the lives of 14 firefighters in the summer of 1994, including Jim. Jim was extremely active in Idaho wildlife politics and a staunch supporter of the Idaho Chapter. He was one of Idaho's premier bighorn sheep guides and a smokejumper based out of McCall, Idaho.

A major victory was accomplished for FNAWS and the California bighorns when the U.S. Air Force decided to do a supplemental draft environmental impact statement for the proposed Idaho Training Range (ITR).

"The Air Force realized finally that their negligence to include mitigation plans and alternative sites in their Draft EIS was in error," said Dr. Bob DiGrazia.

The spring 1995 banquet at the Red Lion in Boise was to date the best ever. Board member Teresa Barnes was chairman of the event and did an outstanding job. Dave Bull was the incoming president. The chapter had a most prestigious work project that captured our memories and created much interest within the community. We were offered the task of cleaning and refurbishing seven of Jack and Eleanor O'Connor's most famous trophy rams along with four of Jess Taylor's. We jumped at the opportunity; and as part of the agreement, the chapter was able to display the trophies at the 1995 chapter banquet.

Bob Ulshafer, board member and professional taxidermist, and taxidermist Charlie Paynter carefully instructed and supervised chapter members as they cleaned and prepared the trophies. As the Idaho Chapter banquet committee was preparing for the activities, a continual stream of visitors filed through the banquet area, interrupting preparations, to see the magnificent and legendary trophies. As far as we know, this was the first public viewing of these trophies. We then packed up the trophies and sent them to Phoenix for the FNAWS national convention. Chapter members set up the display in Phoenix and then returned the trophies to the University of Idaho following the convention.

Chapter President Bill Louderback, during his term of office, introduced the first annual Bighorn Sheep Day. Professionals from several outdoor fields were brought in to educate and entertain more than 200 participants. Duncan Gilchrist did a slide presentation; taxidermists Bob Ulshafer and Charlie Paynter demonstrated care of trophies; Scott Criddle discussed custom rifles; and REI's Eric Strickland demonstrated new lightweight camping and backpacking equipment.

The Hells Canyon situation was now becoming a major concern to the future of bighorn sheep. The three-state project involving Idaho, Washington and Oregon was beginning to unfold. From May 10 to 12, a diverse group of wildlife experts and FNAWS board members met to explore the deepest canyon in North America with its more than 100 miles of river and more than 1,000 square miles of sheep habitat. IFGD biologist Lloyd Oldenburg's dream of developing Hells Canyon for bighorn sheep was finally underway.

Then a major outbreak of the disease Pasteurella hit the Hells Canyon bighorns. This story made major headlines and television news across the country. FNAWS emergency funding mobilized game departments in Idaho, Oregon and Washington. The diseased sheep who were still alive were transported to the Caldwell wildlife laboratory for treatment. It was a massive effort with sad results for the 72 transported sheep. There was a positive input in the study of causes and effects of disease, stress and treatment. In all, 300 bighorn sheep died in the canyon. Ten have survived but will never be returned to the canyon because they carry the disease.

Then, another curse hit Idaho in the form of an anti-hunting group, the Idaho Coalition for Bears, funded largely by the Humane Society of the United States. This group filed petitions to stop spring bear hunting as well as hunting with bait and hounds. National FNAWS and the Idaho Chapter took the lead in combatting these organized forces. Bob DiGrazia and Herb Meyr met regularly with the Sportsmen's Heritage Defense Fund committee. Members of the Idaho Chapter of FNAWS met at the DiGrazia home and wrote checks to add to the fund. National FNAWS contributed $10,000; Idaho Chapter, $1,000; and SHDF eventually defeated the measure on November 6, 1996.

After the Pasteurella outbreak in Hells Canyon with its sad results, FNAWS President Pete Cimellaro and the national Board of Directors declared Hells Canyon as the future major project for FNAWS' mission. The 1996 Idaho Chapter President Ralph McClintock suggested the chapter be involved in the Hells Canyon Initiative. Bill Louderback, Dave Bull and Herb Meyr attended a two-day session at the Red Lion in Boise where the Hells Canyon Initiative was developed. Dr. Dave Hunter reported his findings on the disease outbreak and suggested appropriate ways to reintroduce bighorns to Hells Canyon. Idaho's Herb Meyr has remained on the Hells Canyon Committee.

The Idaho auction permit in 1996 at the FNAWS national convention sold for the first time for more than $100,000 — $101,000 to be exact! That amount, along with nearly $54,000 generated by the lottery permit, brought in more than $150,000 for Idaho's bighorn sheep.

Other highlights of 1996 included the U.S. Air Force returned with another proposal for an electronic and physical bombing range in Owyhee County. Interest in the plight of Idaho's bighorn sheep was a major issue; and in planning for the 1996 banquet and Bighorn Sheep Day, Chairman Bill Louderback brought in speakers and seminars to provide information about the current status of Hells Canyon, the Air Force proposal and other threats to Idaho's bighorn sheep population.

In 1997, Chapter President Herb Meyr developed a web site for the chapter, which is "http.//netnow.micron.net/~meyr/fnaws."

Herb launched a campaign to assist Alan Sands and the Idaho Nature Conservancy in purchasing the 45 Ranch on the South Fork of the Owyhee River. Idaho Chapter immediately contributed $5,000 to the effort. After years of not allowing hunting on the South Fork of the Owyhee, Chapter Secretary Bill Louderback and Lou Nelson of the IFGD met to discuss the possibilities. After several meetings, it was determined the population of California bighorns was stable, and four ram permits were allowed for each of the 1997 and 1998 hunting season.

Idaho Chapter members again assisted the Caldwell wildlife laboratory by painting the buffalo corrals in preparation for 40 buffalo that were being brought in for disease study. The chapter continued to support the Cottonwood Wildlife Management Area near Oakley, Idaho, by removing barbed-wire fence and replacing it with post and rail fence. A beautiful wood sign was purchased through cooperation with the IFGD. The sign identifies the area and displays the chapter logo, a bighorn ram.

President Herb Meyr continued to play an extremely active role in representing the chapter in all issues involving wildlife and bighorn sheep. In December 1997 eight chapter members participated in the capture and transplant of 39 bighorn sheep at Spences Bridge, British Columbia. Jerry Walters, Elmer Huston, Terry Henrickson, Russ Rhodig, Gene Swanson, Wendy Swanson, Jim Dufoe and Royce Tanner joined the monumental effort that involved several states and numerous volunteers. Idaho's Terry Henrickson's expertise was vital when a battery-powered net failed to operate properly. Of the 39 sheep transported to Hells Canyon, 10 were released in Washington; 16 in Idaho; and 13 in Oregon.

Attendance and participation at the 1997 banquet and Bighorn Sheep Day once again increased as it has each year. Paul Snider, life member from Lewiston, delivered a memorable dinner presentation, "Memories of a Sheep Hunter."

Idaho Chapter received another prestigious reward when Dave Bull, 1995 president, was elected to the national FNAWS Board of Directors. As a director, Dave makes it a point to communicate many times weekly with the Idaho Chapter president and board members to keep them informed and updated about the national board's activities.

Fresh from the successful Spences Bridge sheep capture and transplant, 1998 President Jerry Walters got off to a great start. The Bighorn Sheep Day seminars and evening banquet were the best ever. The chapter made room for 20 exhibitors to display their commercial interests. Attendance was outstanding in spite of competing athletic events taking place locally on the same day. Each attendee was given a copy of the newly produced Wild Sheep Journal and a foam-rubber cup holder displaying the Idaho Chapter logo. Duncan Gilchrist entertained the banquet attendees with his video, "Dreams of Ovis."

A special meeting was arranged with IFGD managers John Beecham and Steve Huffaker to go over the dedicated wild sheep money and to present to the department a list of 16 proposals for sheep management. These proposals were intended to show a common-sense approach to management of Idaho's wild sheep. The chapter's efforts are to increase numbers, harvest better trophies and offer more permits.

Chapter members continued the post and rail fence building project at the Cottonwood Wildlife Management Area near Oakley, Idaho.

The focus of the remainder of 1998 was to continue the search for bighorn sheep habitat and to continue our work with the IFGD. Other suitable bighorn sheep transplant sites are being explored by chapter directors in cooperation with IFGD. Past President Herb Meyr is meeting with community leaders in south Idaho. President Jerry Walters and Herb Meyr are currently seeking funding for a potential purchase. Much of the chapter's efforts will be directed to the Hells Canyon Initiative and to continue supportive measures of the South Fork 45 Ranch.

The Idaho Chapter of FNAWS is definitely a group of individuals and families who are willing to take on projects that benefit Idaho's bighorn sheep. This grassroots organization has removed miles of barbed-wire fence that is dangerous to wildlife and replaced it with post and rail fence that is not only attractive but wildlife friendly. Wild sheep, deer and elk can come and go freely through or over the fence while most domestic animals cannot pass through. This same group will walk the halls of the Idaho capitol building if needed or visit newspaper editors, IFGD officials or whatever it takes to educate people of the value of wild sheep. The chapter banquets and Bighorn Sheep Day in recent years have increased in numbers of participants and in dollars earned for wildlife projects. A sizeable and stable core of loyal individuals continue to support the efforts of the chapter. Life member numbers hovered around 40 for several years. In 1998, the number has grown to 55.

Idaho Chapter banquets don't reflect the fund-raising efforts of some of the larger chapters because the chapter's live auctions do not offer the premier hunts that bring in thousands of dollars. The chapter does make up for those funds by working diligently with the IFGD. In 1998 funds raised totaled $49,140. James Hascup of New Jersey was the raffle winner and Ralph McClintock was raffle chairman.

When banquet chairmen, Bill Louderback or Jerry Walters, call members for help, the response is always positive. George Law, FNAWS measurer, and Ralph Pehrson are always willing to spend a day scoring members' rams. IFGD personnel will willingly instruct a wildlife seminar whenever needed. Our banquets have been addressed by some of the best bighorn wildlife experts. Duncan Gilchrist, Jack Atcheson, Jr., and our own Paul Snider have entertained banquet attendees.

Whether it be slanted political agendas or the threat of being bombed, sonic-boomed into oblivion, poached by ignorant thieves or eaten by newly introduced wolves, Idaho's wild sheep are threatened continually by overwhelming odds. Idaho Chapter members are committed to keeping the pressure on to find a place for wild sheep. Roadblocks continue to show up whenever progress is eminent.

Recently, the chapter has honored those corporations that have given their support generously to the Idaho Chapter of FNAWS. Joslyn-Morris Printing, owned by Ron Morris, Larry Morris and Terry Flume, has for 14 years provided the printing needs of the chapter including the newsletter and fliers, stationary, logo, envelopes and more. Heating Supply Company, that is owned by Jim Hollister, has purchased and donated all of the chapter's special project hats for our banquets for the last couple of years. Their efforts turn into profits for Idaho's bighorn sheep.

FNAWS Grant-in-Aid Program has been very generous to the state of Idaho and its efforts to work with and for wild sheep. The relationship between the Idaho Chapter and National FNAWS has remained excellent from the very start. More than $1 million has been awarded for Idaho projects. The Hells Canyon Initiative is a fantastic undertaking by FNAWS. The three-state area has been designated by FNAWS to be the largest stronghold of bighorn sheep in the lower 48 states. It will require millions of Grant-in-Aid dollars. IFGD biologist Frances Cassiers has been given the task of keeping the project on track.

The Idaho Nature Conservancy, National FNAWS, the Idaho Chapter of FNAWS and IDFG have joined forces to support the Nature Conservancy effort to purchase the 45 Ranch, a stronghold of California bighorn sheep. This project is led by Alan Sands who coordinates the efforts to keep the ranch operational and gathers funds to retire the loan. Idaho and California bighorns, as well as all wildlife in the area, will benefit.

The Idaho Chapter is identified primarily, no doubt, as a hunting organization. The monumental efforts made by scores of members to preserve wildlife and propagate wild sheep in Idaho will probably never be adequately recognized.

In the eyes of the typical Idaho Chapter member, as in the eyes of most wildlife viewers, the sighting of a bighorn ram is a breath-taking event. The viewing of bighorn ewes with their lambs frolicking on steep hillsides is an event that happens very few times to most of us in our lifetime.

The Idaho Chapter membership, each and every one of us, wants to preserve this heritage for all to enjoy.

To join this dynamic group, write or call them at Idaho Chapter of FNAWS, 4984 Chinden Blvd., Boise, ID 83714, (208) 466-5132.

The Idaho Chapter is famous for its long list of work projects. From 1985 to 1998 the following projects have been completed:

* $2,000 to Idaho Department of Fish and Game, combined with Pittman-Robertson funds to generate $10,000.
* Donation to Idaho Citizens Against Poaching (CAP) Program
* Donation to Idaho Sportsmen's Coalition.
* Donation and work project with the creation of full-body mounts of a California bighorn ram and Rocky Mountain bighorn ram. Both rams are on display at the Idaho Fish and Game Department Headquarters in Boise, Idaho.
* Refurbished Dall's ram mount at Hillside Junior High School in Boise.
* Spring cleaning of the Idaho Fish and Game Wildlife Health Laboratory in Caldwell, Idaho.
* Removed barbed wire fence and replaced it with a post and rail fence at Big Cottonwood Creek Ranch Wildlife Management Area.
* Litigation to oppose proposed training range.
* Computer for Idaho peace officers.
* Donation to the Jim Thrash Memorial Fund.
* Donation to Citizens Against Poaching (CAP).
* Donation to the Phoenix national convention of FNAWS for the donors and outfitters dinner.
* Installed glass case to protect Dall's ram in Hillside Junior High School, Boise.
* Refurbished Jack O'Connor's and Jess Taylor's ram collection. Exhibited the collection at the Idaho Chapter banquet and at the FNAWS national convention in Phoenix.
* Donation to British Columbia outfitters and guides for wolf trapping.
* Worked to perform Little Jacks Creek lamb survival census with the Idaho Chapter and IDFG.
* Purchased and installed a guzzler and continued work on fence building at Big Cottonwood Creek Wildlife Management Area.
* Donated to create pedestal mount of California bighorn ram horns for educational purposes.
* Painted buffalo corrals at the Caldwell wildlife laboratory.
* Donated several times to the Nature Conservancy's 45 Ranch on the South Fork of the Owyhee to help retire the debt.

* Worked on the Spences Bridge sheep transplant.
* Donated a rifle and wolf hide to the FNAWS raffle at a Reno convention.
* Donated an art print to Deer Hunters of Idaho fundraiser.
* Donated life-size bighorn ram target to Traditional Bowhunters' shooting event.

Table 1

Presidents of the Idaho Chapter of FNAWS.

1985-1986	Wayne Schwabrow
1986-1987	Rick Dredge
1987-1888	Dr. Robert "Bob" DiGrazia
1988-1989	Harold Eshelman
1989-1990	Roger Warrick
1990-1991	J. Scott McCracken
1991-1992	Wayne Schwabrow
1992-1993	Glenn Gearhard II
1993-1994	Harold Eshelman
1984-1995	Bill Louderback
1995-1996	Dave Bull
1996-1997	Ralph McClintock
1997-1998	Herb Meyr
1998-1999	Jerry Walters

Bill Louderback is the current secretary of the Idaho Chapter of FNAWS and newsletter producer. He was chapter president in 1994-1995 and has been an Idaho Chapter and National FNAWS member since 1985. Bill has a B.A. in recreation and park administration from Chico State University, Chico, California. He was born and raised in Redding, California. He was an administrator for Job Corps for 30 years until his retirement in 1994. Bill has been on the Task Force for Personal Evaluation Performance in Washington, D.C. and continues as a consultant for numerous Job Corps sites throughout the United States. He has been married to his wife Barbara since 1963 and is the father of twin daughters, Sherri and Terri, and grandfather of two boys and two girls.

Eastern Chapter of FNAWS 1990 to Present

BARBARA BATES

The history of the Eastern Chapter of FNAWS began at the FNAWS 1990 national convention in Reno when a small group of FNAWS members from the northeastern states held an organizational meeting. The original officers and directors were appointed at this meeting to serve until an election could be held.

Early members included Reed Steinmetz of Reamstown, Pennsylvania; Frank Blaha, Jr. of Hopewell, Virginia; Ken Hess of Willow Street, Pennsylvania; Wes Blakeman of Shelton, Connecticut; and Stanley Boots of Baldwinsville, New York, who became the directors of the new organization. Dr. Bruce Thatcher of Williamsport, Pennsylvania; Simon Aragi of Brookfield, Connecticut; Darlene Stansfield of Carlisle, Pennsylvania; and Gerald Moody of Sutherland, Virginia, were added as directors a short time later.

C. Reed Steinmetz, one of the founding fathers, has been largely responsible for the success of the Eastern Chapter. He began his own business from a dream whereby his hard work and dedication made it grow into a very successful business. He had the same dream of starting the Eastern Chapter of FNAWS. He put that same hard work, determination and dedication into his efforts to put more sheep on the mountain and to preserve hunting rights.

"If it hadn't been for Reed, the Eastern Chapter of FNAWS wouldn't be anywhere it is today," Rod Krebs, 1998 Chapter Vice President, explained.

Reed has been a great inspiration to many. He puts his whole heart into the things he loves and believes in. It is very important for him to assure future generations to be able to be a part of the hunting world he loves so dearly. Reed unselfishly funded the Eastern Chapter for the first two years of its existence, showing the importance of the cause to other members.

In 1991, the Eastern Chapter of FNAWS held its first convention at the Eden Resort in Lancaster, Pennsylvania. Approximately 60 people attended the first fund raiser. The chapter grossed $65,000 that first year. Approximately 200 members were signed up at the first fund raiser. Following this first year, the Eastern Chapter moved to Host Resort and Convention Center in Lancaster from 1992 to 1995. In 1996 they have held their annual convention and auction at the Hershey, Pennsylvania, Hotel and Convention Center. In 1997 and 1998, their convention was held at the Host Resort in Lancaster.

During the first four years of the chapter's existence, the majority of the office work was done at Reed's home. Newsletter mailings, cancelled hunt notices, membership renewals and everything imaginable were done in his family dining room. Reed typed most of the articles for the newsletters. When there was too much for him to handle, a secretary was hired to come in on an "as needed" basis. Sandra, Reed's wife, has been an active part of the team since its beginning.

In 1996, the Board of Directors thought it was time to have the secretary come to work at a home office three days per week. Reed needed to be relieved of the entire workload. Each year the membership increased and the show got larger and larger. The larger the show, the more time and energy it took to make it successful.

Rod Krebs organized a successful Friday evening "Monte Carlo Night" for the first time in 1998 and will continue the new tradition.

Tables of Eastern Chapter members enjoy the first annual Monte Carlo Night.

Rich and Randy Roan, shown during the 1998 Eastern Chapter auction, have conducted the auction each year and are a large reason why the auctions have been such a success.

Life Member Cheryl Snyder, shown here in 1998, began a project in 1996 called "Adopt-a-Sheep Program." Her plan was to "adopt" a sheep for $100. All money from the "adoption" goes to protect and preserve wild sheep during stressful, emergency times.

The 1998 Board of Directors for the Eastern Chapter of FNAWS include, left to right: Wes Blakeman, Will Falcheck, Ken Hess, Mark Bansner, C. Reed Steinmetz, Frank Blaha, Rod Krebs, Dale Gaugler and President Stanley Boots. Since spring 1998, Richard Ziminski has been added to the Board.

Beginning in 1996, Ralph Gitz purchased a Washington bighorn skull from one of the Hells Canyon casualties that died from the Pasteurella outbreak along the border of Idaho, Oregon and Washington. He paid $10,000 for the skull and donated it back to the chapter to raise more money to find a cure for the disease. The chapter made a mold of the ram's skull and horns and created 10 reproductions of it to sell. Number One reproduction was made for Ralph and displayed for presentation at the 1998 convention and show. The other nine reproductions will be sold for the next nine years, beginning in 1996.

The exhibit hall at the Eastern Chapter's annual convention is a popular place to meet friends, book hunts and exchange stories. The Eastern Chapter attracted 60 exhibitors in 1998. Many of them return to the convention year after year.

Starting in the summer of 1996, the Eastern Chapter holds an annual golf tournament. The idea originated from Frank Blaha several years ago, even though Frank himself doesn't play golf.

The organizational meeting of the Eastern Chapter was held at the 1990 FNAWS national convention. The first group of Eastern Chapter members are shown.

The Eastern Chapter has a membership of more than 600 and 179 life members. John Rohrer takes care of the life memberships and sends each new life member a life membership card and a certificate suitable for framing.

Frank Blaha puts countless hours into securing donations from various outfitters for future auctions, and continues this task with his usual enthusiasm and energy. Frank dedicates his time to contact reputable outfitters and shows them that the chapter's goals benefit them as well as the chapter and wild sheep.

"You name it, Frank gets it done. He gets hunts from all over — Northwest Territories, Alaska, Mexico, everywhere," Rod Krebs said.

Wes Blakeman takes all the donation forms and creates an auction list for the Saturday afternoon live auction. Our 1998 auction had nearly 100 items including artwork, bronze sculptures, hunts and fishing trips. The 1998 convention generated gross receipts of $415,000. Approximately 300 attended our banquet and live auction. We had 60 exhibitors with many of them returning to the convention year after year. We offer seminars on topics of interest to hunters such as wildlife photography, wild sheep diseases and cooking with wild game.

Rod Krebs organized a successful Friday evening "Monte Carlo Night" for the first time in 1998 and will continue the new tradition. Our guests enjoyed the evening and are looking forward to this event for many years to come. Rod organized Black Jack, crap tables and roulette tables using "funny money." The "money" was issued to players who then cashed it in for raffle tickets. This was a great way to socialize and mingle. Everyone had something to say to each other.

Life Member Jack Svoboda donates his time and talents by becoming the disk jockey, which provides music and dancing on Saturday evening of the convention. In 1998, Jack did a marvelous job and saved the chapter the cost of hiring live entertainment. Director Dale Gaugler and his wife, Donna, organized a successful silent auction with 100 items, including jewelry, paintings, lamps and clothing.

Beginning in 1996, Ralph Gitz purchased a Washington bighorn skull from one of the Hells Canyon casualties that died from the Pasteurella outbreak along the border of Idaho, Oregon and Washington. He paid $10,000 for the skull and donated it back to the chapter to raise more money to find a cure for the disease. The chapter made a mold of the ram's skull and horns and created 10 reproductions of it to sell. Number One reproduction was made for Ralph and displayed for presentation at the 1998 convention and show. The other nine reproductions will be sold for the next nine years, beginning in 1996. Proceeds from the sales will be used to combat this almost always fatal disease of wild sheep. A full mount of the ram was created for display thanks to the efforts of Marty Hullhan who obtained a large enough cape. The ram rough scores 192 B&C points.

During the weeks that followed the 1998 convention, the Board received many calls and thank you letters from our guests and members. Most of the comments were expressing how the Eastern FNAWS stands apart from many other organizations because of a very personal and friendly atmosphere. From the time the chapter began, Reed sent a personal thank you letter to each and every person who donated to the fund raiser, no matter how large or small the item, hunt or service. Additionally, Sandra sends a thank you note to every buyer of every item sold at our live auction. The Eastern Chapter continues that tradition of thank you notes begun by Reed and Sandra Steinmetz. Showing and expressing appreciation to supporters is not only the right thing to do, but it motivates people to continue their support of the Eastern Chapter of FNAWS. People realize that their donations DO make a difference.

The Eastern Chapter holds an annual golf tournament. This is another opportunity for

First members of the Board of Directors for the Eastern Chapter of FNAWS, shown in 1990 when the chapter began include, left to right: Frank Blaha, Jr. of Hopewell, Virginia; C. Reed Steinmetz of Reamstown, Pennsylvania; Ken Hess of Willow Station, Pennsylvania; Wes Blakeman of Shelton, Connecticut; and Stanley Boots of Baldwinsville, New York.

Stanley Boots, right, and Donna Bessette at the 1988 national FNAWS convention. Stanley remains active at the national and chapter level of FNAWS.

Frank Blaha, Jr., left, and Johnny Drift work the Eastern Chapter's information booth at the Eastern Sports and Outdoor Show in Harrisburg, Pennsylvania, in 1992. The Eastern Chapter supplied the booth and wild sheep mounts. Approximately 80 new members of the chapter were signed up at this successful sportsmen's show.

Rick and Cathy Ship, left and right, receive special recognition from Eastern Chapter President C. Reed Steinmetz in 1994 at the chapter convention held at the Host Resort and Conference Center in Lancaster, Pennsylvania. The auction grossed more than $177,000 and the entire convention grossed $210,000, including the auction.

members to come together to benefit wild sheep and socialize with fellow hunters and wildlife conservationists. Rod Krebs heads the golf tournaments, and he does and outstanding job. The idea originated from Frank Blaha several years ago, even though Frank himself doesn't play golf. Reed also helped organize the event that began in July 1996 and continues each year during the summer. This is a day-long event that involves a lavish buffet brunch and evening meal of barbecue ribs, steak and chicken served at the Host Hotel and Convention Center in Lancaster. If someone is lucky and skillful enough to make a hole in one, they receive a donated sheep hunt. Ross Peck has donated a Stone's sheep hunt and Stan Simpson has donated a Dall's sheep hunt. So far, no one has made a hole in one, but Rod is confident this will happen soon! Players pledge money to the Eastern Chapter so cash is raised to put more wild sheep on the mountain.

Life Member Cheryl Snyder from the Eastern Chapter began a project in 1996 called "Adopt-a-Sheep Program." Her plan was to "adopt" a sheep for $100. All money from the "adoption" goes to protect and preserve wild sheep during stressful, emergency times. For the $100 adoption fee, a framed photograph of the adopted sheep is sent to the donor to proudly display among other hunting and conservation achievements.

She was inspired by a man named Albert Cooper. He was a master guide and outfitter in British Columbia who began his career as a horse wrangler in 1939 for Walter Nixon. When World War II called Walter way, Albert took over his guiding and outfitting service. In 1953, Walter went on his own as an independent outfitter. Since then, he has dedicated himself to the preservation and care of wild sheep. Out of his own pocket, he provided medication, water and food to wild sheep in and around his area during extremely hard times when their survival was dependent on assistance. This was long before National FNAWS was founded. Because of people like Albert Cooper, FNAWS is what it is today. Albert has been a hero to wild sheep for a long time but had never been recognized for his dedication until Cheryl began the "Adopt-a-Sheep" program. Cheryl thought we could do something special for sheep and something special for Albert.

Cheryl presented her creative idea to the Eastern Chapter FNAWS Board and offered to front all the expenses associated with the program, including printing costs for publicity. In less than six months, the program brought $5,300 for emergency situations that arise among wild sheep populations.

Dr. William Falcheck, Life Member of the Eastern Chapter, and Director in his second term, has taken on a special project within our organization. He is a veterinarian with knowledge of animal diseases who has given seminars on wild sheep diseases at our convention and is working closely with the Idaho Fish and Game Department's laboratory on wild sheep disease control. Dr. Falcheck organizes trips to Idaho, on his own time, to present the laboratory with much-needed funds to purchase equipment and work directly with department staff. The Eastern Chapter's Board of Directors have designed several auction items with the proceeds from their sale going towards this very worthwhile cause.

Below is a partial list of projects funded by the Eastern Chapter of FNAWS through 1998. All grant applications for funding are reviewed by the Board of Directors at their summer meeting. If you want information about this active FNAWS chapter, write to them at P.O. Box 907, Ephrata, PA 17517 or call at 717-733-7579.

The membership of the Eastern Chapter of FNAWS, the convention exhibitors and the chapter donors are the reasons why this chapter has raised in excess of $500,000 to "put wild sheep on the mountain" throughout North America since its beginning in 1990. Our gratitude goes out to all who have supported our efforts since we began.

Partial Listing of Wild Sheep Projects Funded by the Eastern Chapter of FNAWS:

$8,000 To Southern British Columbia Guides and Outfitters Association for the Elk Valley bighorn sheep transplant in the Quarry and Forsythe Creek drainages, upper Elk River, Wildlife Management Unit 4-23, Kootenay Region, British Columbia.

$5,000 To the California Department of Fish and Game to develop a demographic profile of mountain sheep in the Chocolate Mountains, Imperial County, California. The purpose of the study was to determine if there were mature desert rams in this range to justify additional hunting permits and to document these findings, which is required by existing California law. California doesn't provide funds for such studies.

$4,000 To the Utah Chapter of FNAWS to help purchase a domestic sheep grazing allotment in the Hole in the Rock and Hoop Lake areas. The U.S. Forest Service has committed to the Utah Chapter to do a controlled burn and merge two herds of about 50 bighorns into one that is separated from this domestic sheep grazing area.

$2,500 To Arizonians for Wildlife Conservation to help them in their efforts to defeat Proposition 200.

$5,000 To the Utah Chapter of FNAWS to help protect nearly 70,000 acres for the Rattlesnake Rocky Mountain bighorn sheep herd.

$12,000 To the Guides and Outfitters Association of British Columbia (GOABC) for its continued efforts to protect nonresident hunter rights. This grant was used to assure the nonresident hunter would receive a fair share of hunting permits in British Columbia.

$5,000 To install a bighorn bronze monument in Salma, British Columbia at their tourist information center to educate the public to the value of resident bighorn sheep and to curb the abuses of native hunting.

$12,050 To the British Columbia Ministry of Environment, Southern Interior Region, Kamloops, BC, to transplant California bighorn into an area near Seaton Lake, BC.

$1,500 To the Idaho Department of Fish and Game to purchase a computer for use in its covert program. This grant was shared, 50-50, with the Idaho Chapter of FNAWS.

$7,500 To the Southern British Columbia Guides and Outfitters Association for emergency funding to deal with its ongoing problems with hunting permit allocations in the Kootenay Region for nonresidents.

$10,000 The Eastern Chapter of FNAWS joined National FNAWS and the Minnesota-Wisconsin Chapter of FNAWS to fund the cost of constructing a water project in Arizona's Cabeza Prieta National Wildlife Refuge.

Frank Blaha, Jr., right, presents to Rocky Keen of Last Frontier Guiding and Outfitting, left, the Donor of the Year Award in 1995.

C. Reed Steinmetz, right, received the Service Award from the Eastern Chapter in 1995 for being the Eastern Chapter president from 1990 to 1994. With him in the celebration is Canadian Artist Rick Taylor, center, who sculpted the statue Reed holds and Frank Blaha, Jr., left.

Auctioneers Rich and Randy Roan were an important reason why the Eastern Chapter grossed close to $335,000 at their 1995 auction in the Host Resort and Conference Center in Lancaster, Pennsylvania. Nearly 500 people attended the events that netted $100,000 for wild sheep projects.

Auction spotter Booger Harris, left, congratulates Life Member Jim Nyce of Green Lane, Pennsylvania, on his purchase of the Ft. Belknap Community Council bighorn hunt at the 1996 chapter convention held at the Hershey Lodge and Convention Center in Hershey, Pennsylvania.

Dr. Bruce Thatcher of Williamsport, Pennsylvania, receives a handshake from Chapter Board Member Rod Krebs after Bruce purchased the Wyoming bighorn permit in 1996 at the chapter convention. Wyoming Governor James G. Geringer was thanked for making this permit auction possible.

Eastern Chapter President Stanley Boots, right, presents the Hunter of the Year Award to Life Member George Snyder of Chadds Ford, Pennsylvania, at the 1997 Eastern Chapter convention held at the Holiday Resort and Convention Center. Gross receipts totaled more than $350,000.

$9,000 — The Elkford Rod and Gun Club and the British Columbia Ministry of Environment to transplant Rocky Mountain bighorn sheep on the west side of the Elk Valley.

$1,000 — To Ron Fleming of Hazelton, BC, to send a teacher to the Safari Club International Granite Ranch School.

$3,000 — To the Yukon Outfitters Association to fund a sheep study that discharged government unfounded quota on trophy wild sheep hunting.

$500 — To the Wildlife Legislative Fund of America (WLFA) for their use in the protection of hunter rights.

$4,000 — To the British Columbia wildlife branch to assist in funding the Taseko sheep transplant in central British Columbia of California bighorn sheep.

$5,000 — To National FNAWS to help fund legal expenses related to the Idaho bombing range used by the U.S. Air Force near Hells Canyon where prime wild sheep habitat exists.

$7,500 — To purchase radio collars and fund sheep transplants in 1995.

$6,000 — To pay for a transplant and wild sheep study in 1995.

$10,000 — To the Alaska Professional Hunters Association to pay for legal fees to maintain the guide area system in 1995.

$5,000 — To the Utah Chapter of FNAWS to secure land from the U.S. Bureau of Land Management for wild sheep in 1995.

$2,000 — To the Northern British Columbia Guides Association for aircraft expenses during the trapper-assisted program for snaring wolves.

$8,000 To the Southern British Columbia Guides to study grizzly bear in 1995.

$5,000 To the National Bighorn Sheep Interpretive Association for an educational program in 1995.

$500 To the WLFA in 1995.

$4,050 To the Wyoming Game and Fish Department for a Fork Owl Creek bighorn sheep transplant in 1995.

$4,862 To fund emergency medical care for sick bighorn sheep taken from Hells Canyon in 1996.

$5,000 To the Alaska Chapter of FNAWS money for a caribou calf protection program in 1996.

$5,000 To the Utah Chapter of FNAWS for a Jensen Aum Acquisition Project in 1996.

$3,000 To the Fort Nelson Museum for the Alaska highway bronze in 1996.

$5,000 To the Fraternity of the Desert Bighorn for construction of an emergency water supply in 1996.

$500 To WLFA in 1996.

$10,000 To the Yukon Outfitters Association for legal fees related to land claims in 1996.

$5,000 To the What's Right Program for the research, management and education of wildlife in 1996.

$7,500 To the Wild Sheep Society of British Columbia for a sheep transplant and radio collars in 1996.

$12,000 To the Fort Belknap Community Council of Montana to fund a tribal biologist and wildlife technical position in 1996.

$1,000 To the Oregon Hunters Association to restore management of the state's wildlife in 1996.

$2,000 To the Kootenay Wildlife Heritage Fund for a wildlife feeding program in 1997.

$13,957 To fund helicopter deer surveys in Alberta in 1997.

$500 To the Native American Fish and Wildlife Society for a wildlife campaign in 1997.

$6,000 To SMART for the funding of legal fees in Alaska in 1997.

$6,000 To the Utah Chapter of FNAWS for the expansion and protection of existing bighorn sheep populations in 1997.

$1,000 To the Nature Conservancy of Idaho for a sheep population survey in 1997.

$2,000 To WLFA for management programs in 1997.

$8,495 To the Idaho Fish and Game Wildlife Health Laboratory for wild sheep disease research in 1998.

$48,375 To the state of Wyoming for wildlife projects throughout the state in 1998.

Barbara Bates, known as Barb to her friends, started to work for the Eastern Chapter of FNAWS on an "as needed basis" shortly after the chapter was organized in 1990. Mailings as well as convention registrations were done in the beginning from Reed and Sandra Steinmetz's home with the help of Barb's daughter, Kristi. Barb, Reed, Sandra and Kristi formed the "office crew" for this active FNAWS chapter. Kristi, Barb's husband, Jay, Sandra and Barb work at the annual convention these days and Barb remains as the secretary to the group. In addition to working for the Eastern Chapter of FNAWS, Barb coaches gymnastics at Berks Gymnastics Academy and loves to garden and train dogs in her home near the FNAWS office in Ephrata, Pennsylvania.

Utah Chapter of FNAWS 1991 to Present

Don Peay, Lee Howard and Fred L. Morris

The Utah Chapter of the Foundation for North American Wild Sheep (FNAWS) began with 100 members and a $12,000 fundraiser in 1991. The organization has continuously grown, finishing its 1998 fundraiser with a net profit of more than $200,000 and more than 500 members.

At the first banquet and fundraiser for the Utah Chapter, more than 220 people attended and donated nearly $21,000 for various auction items. There was a strong representation of officials from the Utah Division of Wildlife Resources, the U.S. Forest Service, Bureau of Land Management, Salt Lake City media, Rocky Mountain Elk Foundation (Kay Lewis and Bill Christensen) and national officers from FNAWS (Dr. Robert DiGrazia and Pete Cimellaro). More than 35 wild sheep trophies were on display and winners of the "Peoples' Choice" awards included Bruce Babcock for desert sheep, Dr. Larry Francis for Dall's sheep, Ken Jones for Rocky Mountain bighorn and Swen Mortenson for Stone's sheep. People from across the state attended including residents of Moab, Logan, Roosevelt and Cedar City. The first sportsmen allowed to hunt Rocky Mountain bighorns in Utah since the 1940s by the Utah Division of Wildlife Resources, Jay Anderson of Ogden and Shawn Labrum of Roosevelt, brought their trophies as did Jackie Ruder who took a permitted desert bighorn ram in 1991.

This dynamic organization of dedicated men, women and children with a passion for wild sheep has raised more than $1.6 million since its founding and 10 major habitat acquisition projects have been completed. This represents nearly $1 million in payments to ranchers. The focus of the habitat projects is to purchase federally owned grazing permits and convert them to "wildlife and watershed values." This action allows the Bureau of Land Management (BLM) and the U.S. Department of Agriculture Forest Service (USDA-FS) to more effectively manage watersheds, increase forage available for wildlife and eliminate disease transmission from domestic sheep to wild sheep.

"The Utah Chapter of FNAWS has pioneered and perfected the art of win-win resolution of public land conflicts. The privately funded, public agency-managed partnership has restored majestic bighorn sheep to millions of acres of Utah's landscape. It benefits ranchers, sportsmen and nonsportsmen wildlife enthusiasts," said Utah Governor Michael O. Leavitt in summer 1998.

On June 26, 1992, Don Peay and Lee Howard of the Utah Chapter and John Eyre, president of the Lyman Grazing Association, signed a conservation easement agreement effective for perpetuity that assured the success of the reintroduced bighorn sheep herd on the north slope of the Uintas. Under the easement agreement, the grazing association will never again graze domestic sheep, llamas or domestic goats on four sections of their private lands that border the U.S. Forest Service lands. The grazing association is composed of 14 ranchers who joined with the Utah Chapter in 1992 to sign this agreement. With 22 bighorn sheep reintroduced into the High Uintas area in January 1989, this agreement assured success of the reintroduction. Part of the 1992 agreement included construction of a mile-long fence on the easement property to help the ranchers protect the riparian areas from cattle grazing. Due to topography, the easement effectively provided the wild sheep with a five-mile buffer zone.

The three founders and continual leaders of the Utah Chapter of FNAWS include, left to right, Fred L. Morris, Lee Howard and Don Peay.
Photographs courtesy of Utah Chapter of FNAWS.

Lee Howard, right, discusses chapter activities with Collin Allan. The Utah Chapter of FNAWS began with 100 members and a $12,000 fundraiser in 1991. The organization has continuously grown and finished its 1998 fundraiser with a net profit of more than $200,000 and more than 500 members.

The first annual banquet and fundraiser in 1991 for the newly formed Utah Chapter of FNAWS attracted 220 men and women who donated close to $21,000 for various auction items. Here, Utah Chapter Vice President Fred Morris stands with Jackie Ruder and her 1991 Utah desert bighorn sheep skull, the trophy she took. There were more than 35 sheep trophies on display with peoples' choice awards presented to Bruce Babcock for desert sheep, Dr. Larry Francis for Dall's sheep, Ken Jones for Rocky Mountain bighorn and Swen Mortenson for Stone's sheep.

On June 26, 1992, Utah Chapter representatives Lee Howard and Don Peay, left and right of the sign, signed a conservation easement agreement with John Eyre, president of the Lyman Grazing Association, to perpetuate Rocky Mountain bighorn sheep in the High Uintas and adjoining private lands. The sign permanently records the event.

The fence construction and the signed conservation easement agreement was a tremendous example of how the U.S. Forest Service, domestic livestock interests and sportsmen can work together and produce a win-win solution. The first $12,000 payment to ranchers in 1992 took place as well as payments in 1993 and 1994 from proceeds from the Utah Chapter fundraisers.

Prior to the restoration program of bighorn to Utah's landscape, Rocky Mountain bighorns had disappeared and desert bighorn were isolated in two small bands in extreme southeastern Utah along the San Juan River and Colorado River. Beginning in 1980, the national organization of FNAWS began funding bighorn restoration accomplished via the professional wildlife biologists in the Utah Division of Wildlife Resources (UDWR).

Fred L. Morris bought the first Utah desert bighorn permit for $20,000. Those dollars were used to start taking sheep from the San Juan herds and re-establishing them to native ranges throughout Utah. The first national auction of a Utah permit took place at the 1983 national convention. Bill Leech was top bidder with a total payment of $32,000. The Utah permit has continued to be a popular auction item since that time. Through the 1998 convention, the Utah permit has generated $686,500 to benefit Utah wild sheep and wildlife habitat in general.

Fred L. Morris, Lee Howard and Don Peay are the founders of the Utah Chapter of FNAWS in 1991. Since then, the 500-plus members of the chapter, along with the Eastern, Iowa and Minnesota-Wisconsin Chapters and FNAWS National, have contributed more than $1.6 million to secure habitat on more than three million acres and provided management of bighorns. Rocky Mountain bighorn now number about 500 animals. The potential is to reach 2,000 animals.

Desert bighorn results are more dramatic. More than 2,500 animals currently live throughout Utah. Had the transplanting not begun in 1980, Utah desert sheep very well could be extinct. The San Juan herds suffered a catastrophic Pasteurella die-offs, beginning in 1985. By preserving habitat and providing the Utah Department of Wildlife biologists with operating capital to perform transplants, wild sheep populations have dramatically increased in Utah, thanks in large part to the dedicated band of volunteers from the Utah Chapter of FNAWS. Rocky Mountain bighorn populations in 1972 were nonexistent and in 1998 total 500. The goal of the Utah Chapter volunteers is to increase that population from 500 to 1,250 by 2005. As for desert bighorn sheep, the total population in 1972 was 500. By 1998 the desert bighorn population totaled approximately 3,000! The population target by 2005 is 5,000 desert bighorns. As for California bighorn sheep, the Utah Chapter volunteers hope to have 300 animals in the wild by the year 2005. In 1972 they were extinct and by 1998 their population topped 50 individuals.

The Utah Chapter of FNAWS has contributed $500,000 towards wildlife management programs, research and bighorn sheep transplants using helicopters. A permanent endowment fund, "The TRUST," includes $110,000 to establish this method of financing quality wildlife programs for future generations. A recent gift of $170,000 from the Jenson family of Salt Lake City has pushed the TRUST to nearly $300,000. Our goal is to raise $1 million by the year 2000. Using conservative hourly wage rates, the value of volunteer labor for fundraising, habitat acquisition negotiations and organizational development totals well in excess of $400,000. See Table 1 that lists the projects completed and money raised on behalf of wild sheep of Utah.

"Restoration of Rocky Mountain, desert and California bighorn sheep to Utah's landscape is a modern-day wildlife miracle. It is an example of the potential to solve problems

On June 27, 1992, 35 volunteers from the Utah Chapter of FNAWS built a mile-long fence on easement property to help ranchers protect the riparian areas from cattle grazing after the signing of a conservation easement agreement to perpetuate bighorn sheep in the High Uintas and adjoining private lands between the Utah Chapter and the Lyman Grazing Association.

Connie Brooks, right, who is a Utah Chapter and Utah Wildlife Board member, holds the winning ticket for the Utah Chapter fundraiser in 1997 with Utah Chapter President Lee Howard. The 1996-97 lucky winner of the desert bighorn sheep tag was Al Farace of Maryland who couldn't believe his name had been drawn. This desert sheep completes his Grand Slam. Money collected from this project went back to wild sheep projects in Utah.

Two long-time financial supporters of FNAWS at the national and Utah Chapter level are Ross Messerly and Kenneth Jones, shown here at the spring 1998 Utah Chapter banquet.

Five dedicated members of the Utah Chapter of FNAWS were recognized at their spring 1998 fundraiser as "Members of the Year." They are, left to right: Shawn Labrum, Lee Howard, Allen Yardley, Lane Parker and Swen Mortensen.

when private and public partnerships work together. The citizens involved in the leadership of the Utah Chapter of FNAWS and board members of the TRUST are very capable and highly respectable individuals in the Utah business and conservation community," explained Department of Natural Resources Executive Director Ted Stewart in the summer of 1998.

In a joint effort between the Utah Chapter of FNAWS, the Wild Sheep Society of British Columbia and the UDWR, California bighorns were transplanted to Antelope Island in 1997, making Utah a three-species bighorn state.

What makes the story of the Utah Chapter of FNAWS even more impressive is that it is done with all-volunteer labor. We have no paid staff. The millions of dollars generated has come from a core of 500 or so small business owners and individuals who believe in the mission of FNAWS and the Utah Chapter. We have begun the work and now reach out to others throughout North America to continue the effort. Actor Robert Redford, founder of the North Fork Canyon Land Trust, and basketball great Karl "The Mailman" Malone have joined us by contributing to wildlife habitat projects.

In 1998 the Utah Chapter of FNAWS, along with dozens of other wildlife conservation and sportsmen's groups, have joined forces to support a state constitutional amendment to protect Utah's wildlife heritage and professional wildlife management system. Calling themselves "Utahns for Wildlife Heritage and Conservation," this organization is a statewide, grassroots campaign coalition of wildlife conservationists, hunters, anglers, ranchers, farmers, business interests, trappers and Utah citizens.

The unified effort was prompted because of threats by the Washington, D.C.-based Humane Society of the United States, the Utah Predator Education Fund and the Utah Cougar Coalition to use the initiative process to manage Utah's wildlife by displacing biology with emotion and nonscientific rhetoric. The constitutional amendment Utahns for Wildlife Heritage and Conservation supports ensures that only trained professional biologists and wildlife managers will regulate wildlife with meaningful citizen participation through Utah's regional wildlife management public hearing process. The amendment supports the current scientifically based wildlife management system in Utah that has been in place for more than 100 years. Don Peay of Sportsmen for Fish and Wildlife is active in the Utah Chapter of FNAWS and finance chairman of this grassroots group.

Outfitters such as Elvin Hawkins, Pete Cimellaro, Larry Heathington and Roy Lerg have helped promote Utah and bring the "big money" hunters to Utah. The following supporters of the Utah Chapter of FNAWS are the outstanding contributors who annually donate to "put sheep on the mountain" in Utah for the Rocky Mountain bighorn, desert and California bighorn sheep:

Ken and Merlynn Jones, Glen Gold, Marc Bigham, Wayne Shelton, Ray and Susan Brown, Mark and Brenda Newman, Trent and Chris Wall, Ross and Todd Messerly, Steve and Judy Perry, Kevin and Diane Hatfield, Robert and Karen Jacobsen, Bob and Anita Kelez, Danny Bulter, Dennis and Romon Lower, Don and Mildred Mace, Dave Myrup, Russell and Cathy Phillips, Swen Mortenson, Allen Christensen, Don Todd, Brent Todd, Dave Wade, Randy and Connie Brooks, Bill Mamaless, Jay Ogen, Brett Caldwell, Randy Johnson, Bob Mastronardi, Ken and Shawn Labrum, Allen Yardley, John and Pamela Berman, Floyd Hatch, Ralph Kramer, Jeff Jensen, Dick Erickson, Chris Thurston, Stephen Walker, Carl Hoehner, Don Scheider, Rocky Crate, Rick Garzoli, Andy Anderson and Jim Tonkin.

In the past decade, a bright future for Utah bighorn exists thanks to the work of the Utah Chapter of FNAWS and many other dedicated wildlife conservation groups. The Utah bighorn populations have grown from three original herds in the southeastern corner of Utah in

U.S. Senator Robert F. Bennett (R-Utah) was keynote speaker at the 1998 Utah Chapter banquet. He stands with Utah Chapter President Lee Howard.

In front of a backdrop of outstanding wild sheep trophies at the Utah Chapter fundraiser in 1998 are, left to right: Utah Chapter President Lee Howard, FNAWS National Board Members Dave Bull and Duncan Gilchrist and FNAWS Chapter Coordinator Paula Karres.

1972 to 21 herds scattered throughout Utah, mainly concentrated in the southeastern corner of the state but now expanding to the other corners of the state. Populations include three of the four North American wild sheep species. The herds in 1972 were from North San Juan, South San Juan and Potash/Island in the Sky. After the Utah Chapter was established, the herds included the first three herds plus Professor Valley, Lockhart Basin, North San Rafael, South San Rafael, Dirty Devil, Little Rockies/Bullfrog, Escalante, Kaiparowitz, Paria River, Zion National Park, Elliot/Bighorn Mountain, Bear Top, Sheep Creek/Hoop Lake, Beaver Dam, Rocky Mountain High Basin, Pilot Mountain and Antelope Island.

Past relationships between public land management agencies, ranchers, wildlife conservation groups and the UDWR were confrontational and win-lose. The Utah Chapter as a private, non-profit foundation has created a win-win business arrangement with ranchers and public land agencies since its inception almost 10 years ago. With habitat secured, wild sheep populations have dramatically expanded by transplanting animals from Arizona, Wyoming, Colorado, Nevada and British Columbia. Utah is also working to transplant bighorn sheep from Montana and Idaho.

The Utah Chapter of FNAWS is a bottom line, results-oriented group of men and women who believe that preserving habitat and providing money to support wild sheep research and transplants through the Utah Department of Wildlife Resources will expand and enhance wild sheep populations for the future.

The future of Utah's wild sheep depends upon financial support of FNAWS members, the professional biologists of Utah's wildlife resources and public land managers. The future looks very bright! The volunteers who form the core of the Utah Chapter of FNAWS is a bottomline, results-oriented group of men, women and children who know they will succeed in "putting wild sheep on the mountain" in Utah.

To join this effective group, write or call them at 1174 South 300 West, Salt Lake City, UT 84101, (801) 294-6124.

Table 1

Chronological Order of Utah Chapter of FNAWS Habitat Projects.

LOCATION	*AMOUNT DONATED*
Hoop Lake, North Slope Uintas	$36,000
Rattlesnake, Western Book Cliffs	$180,000
Beaver Dam Mountains, St. George	$40,000
Paria River, Kabab	$2,500
San Rafael, Emery County	$125,000
Floy Creek, Western Book Cliffs	$125,000
San Rafael, Emery County	$185,000
Crescent Canyon, Western Book Cliffs	$95,000
Timpanogas, Utah County	$80,000
New Foundland Mountains, Box Elder County	$75,000
Bear Top Mountain, Flaming Gorge	$10,000
Total Project Funding:	$954,000
Funding for Management and Transplants:	$650,000
Total for Bighorn Restoration, 1991-1998:	**$1,604,000**

Bighorn Sheep Distribution in Utah

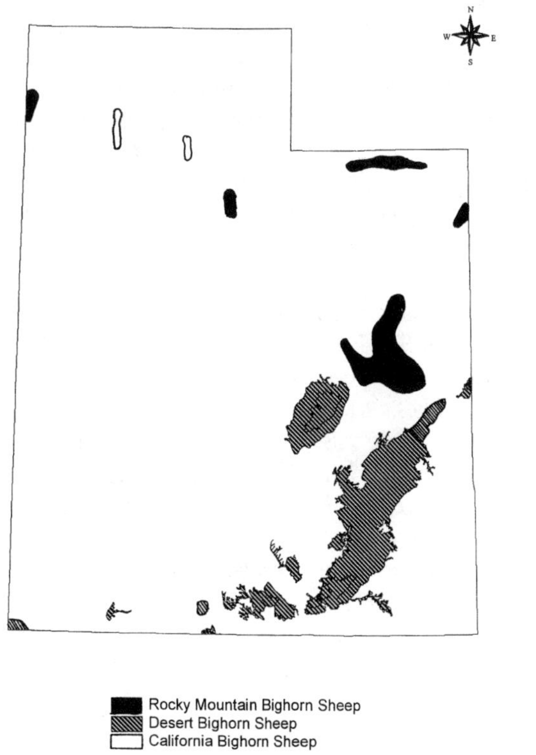

A map shows the current bighorn sheep distribution for the three species of wild sheep found today in Utah, thanks in large part to the efforts the Utah Chapter of FNAWS. Prior to the founding of the Utah Chapter, only a concentrated population of desert bighorn in the southeastern corner of the state existed. Rocky Mountain, desert and California bighorn sheep now live within the borders of Utah.

Table 2

BIGHORN SUBSPECIES	*DISTRICT POPULATIONS*		*# OF BIGHORNS 1972*	*# OF BIGHORNS 1998*	*GOAL 2005*
	1972	*1998*			
Rocky Mountain	0	5	Extinct	500	1,250
Desert	2	12	500	3,000	5,000
California	0	2	Extinct	50	300

Don Peay of Salt Lake City is executive director of Sportsmen for Fish and Wildlife. He earned a B.S. in chemical engineering from Brigham Young University and a MBA from BYU. He founded an environmental remediation consulting firm and sold it in 1995 to dedicate his energy full time as an outspoken advocate for wildlife habitat preservation. Don was awarded the 1996 Outstanding Achievement Award by National FNAWS. Don was appointed by Governor Leavitt to the School Trust Land Board that administers the management of more than 3.7 million acres of Utah land. Through Sportsmen for Fish and Wildlife activities, Don established a $2.5 million-a-year habitat preservation fund through the UDWR. Don and his wife Sue have three children, Brandon, Mckenzie and Kyle. They are involved in coaching Little League baseball, basketball and girls' soccer.

Lee Howard and his wife Carol own several small businesses in the Salt Lake City area. Lee holds a B.A. degree in history from Brigham Young University. Over the years, Lee has been involved in several habitat conservation groups, specializing in fund-raising efforts with groups such as the Utah Chapter of FNAWS, Rocky Mountain Elk Foundation and Ducks Unlimited. Lee has recently been appointed by Governor Leavitt to be part of the wildlife management decision-making board in Utah. He has served as president of the Utah Chapter of FNAWS since 1995. Lee and Carol are parents of four children, William, Becky, Rachel and Vaughn. The entire family is involved with wildlife conservation groups.

Fred L. Morris is a general partner of Pentalon Development Corporation, a real estate development firm based in Salt Lake City. He controls some $200 million worth of Utah real estate, including the 17-story Broadway Centre office building in downtown Salt Lake City and more than 5,000 apartment units. Fred has a M.S. in economic history from University of Utah and a B.A. from Brigham Young University. Fred was a U.S. Air Force fighter pilot in Vietnam and is an accomplished wildlife photographer. He has photographed wildlife in more than 110 countries from the North Pole to Antarctica. Fred and his wife Sue have three children, Zane, Amy and Lars, who are involved in their family business. Their son, Zane, a former Stealth fighter pilot with the U.S. Air Force, is also active in wildlife conservation projects. Fred was the first major donor to Utah's bighorn restoration program by donating $30,000 in 1980.

The Wild Sheep Society of British Columbia 1992 to Present

Bill Pastorek

As do many things in life, the Wild Sheep Society of British Columbia (WSSOBC) started from just an idea. The California bighorn sheep range in the Fraser River area of British Columbia was in trouble.

With the sheep population growing and the range deteriorating, it seemed a die-off was probable. Unless something was done, hunting opportunities for this species would be greatly reduced in the very near future. A small group of wild sheep enthusiasts decided to have a meeting to discuss if an organization could be set up to assist the Ministry of Environment in producing better habitat for the sheep. These far-sighted conservationists were Lex Ross, Kelly de Strake, Tim Cody, Clarence McIvor and Dave McBride. They met with Bob Hall from the local British Columbia government agency. With Bob's expertise, an application was drafted and sent to Victoria to form a society. An executive committee was formed and in March 1992 we finally received our certificate from Victoria. The Wild Sheep Society was born.

As is the case whenever a new organization starts with a limited amount of funding, membership was slow to increase. To attract new members, you have to have at the very least a newsletter to keep everyone informed. This requires funds—funds that become available from an increased membership. The growing pains were weathered and our membership slowly started to grow. It was quickly realized by the executive that to make the society successful, it would have to become a province-wide organization concerned with the welfare of all four species of British Columbia wild sheep. Advertising brochures were created and printed, and the society began to set up displays at gun shows to attract new members.

I well remember the day I was driving through the town of Lillooet on my way deer hunting. I stopped to fill up with gas, and when I went to pay, I saw this brochure lying on the counter. It was a membership form for the society. I remember reading the brochure while driving down the road and instantly knowing this was something of which I had to be a part. In hindsight, reading while driving wasn't all that clever, but I was reading about wild sheep. When I returned from my deer trip, I sent in my membership fees and sat back and waited. Before long a notice came that invited me to the annual general meeting in just a few months.

Along with meeting 64 other sheep enthusiasts at this, our first annual general meeting, the most important thing was the discussion as to whether we should become affiliated with the Foundation for North American Wild Sheep (FNAWS). Johnny Drift, then the first vice-president of FNAWS National, outlined the advantages of being associated with them. This option was extensively discussed and the yes vote resulted in our becoming one of the latest FNAWS affiliates.

Our first major project occurred in 1994. The Wild Sheep Society had been working with the Environment Ministry to coordinate and arrange funding for our first sheep transplant. A total of 23 California bighorn sheep were captured along the Fraser River near Big Bar and transplanted by truck and train to Seton Lake in the beautiful Bridge River Valley near Lillooet. The entire operation was filmed by British Columbia television (BCTV). We received exten-

Rose Pastorek, left, member of the Wild Sheep Society of British Columbia, helps a Utah wildlife official on a transplant of California bighorns heading south of the border in 1998.
Photographs by Bill Pastorek.

A helicopter flutters overhead while dangling a young California bighorn ram over Kamloops Lake during a successful transplant project supported by the Wild Sheep Society of British Columbia in 1994.

Dangerous collisions with wild sheep have been reduced in British Columbia thanks to dramatic signage along BC highways warning speeding motorists to slow down if they enter wild sheep habitat or see wild sheep along roadways.

Members of the Wild Sheep Society of British Columbia participate in their first bighorn transplant in 1992 along the Fraser River.
Photograph by Dave McBride.

sive coverage. A large portion of the funding for this transplant came from the Eastern Chapter of FNAWS. For this they should be thanked again. Wildlife Section Head Fred Harper of the Kamloops Southern Interior Region of the British Columbia Ministry of Environment, Lands and Parks threw his considerable support behind the project to make sure it would happen. After this successful transplant, there were many more projects to come.

Over the next few years our fund-raising efforts improved. Hand in hand with this went our ability to fund more projects. The Wild Sheep Society is now a rapidly growing province-wide organization of more than 500 members. We are dedicated to increasing the number of wild sheep on our mountains and in improving their habitat. Our membership provides both funding and manpower for the various projects that we do. Sheep transplants will always be the cornerstone of our efforts. They are a very effective way to move sheep onto once traditional habitat areas. By helicopter net gunning, and by baiting under a drop net, we can help to ensure the survival of the British Columbia wild sheep.

In many areas of the province, the wild sheep wintering range is adjacent to major highways. This results in many animal / vehicle collisions every year. By strategically placing our highway collision signs, we are reducing injury and fatalities to both sheep and people. We have also funded winter feeding and fencing projects to help the sheep during their time of greatest stress. In many cases, it always seems to come back to the wintering areas. This is where the sheep come into closest proximity to human populations. In the Okanagan Valley in central British Columbia there are California bighorns. There are also a great number of people. We have provided funding to Nature Trust to assist in the purchasing of critical wintering range.

The big question? Where is the future of WSSOBC? Our goal of putting more sheep on the mountains certainly has to remain at the forefront. If sheep populations do well, there will be more sheep tags for both resident and nonresident hunters. It is because of this goal that we now have a large number of international members. Sheep hunters all around the world are very glad to see we are watching out for their interests. As well, many of the guides and outfitters in the province are become aware we are both on the same side. Many are now joining as members since they see the good work we are doing. We look forward to helping the wild sheep of the province of British Columbia and ensuring their continued survival. As I am sure many of you are aware, hundreds and hundreds of sheep have been transplanted into the United States from British Columbia. Without our support, there would be very few, if any, California bighorns south of the border. Please keep this in mind when you are deciding which organizations to join and to donate your funds and time.

Past presidents of WSSOBC from 1992 to 1998 include Clarence McIvor, Lex Ross and Bill Pastorek. Vice presidents have included Lex Ross, Tim Cody, George Bikadi and Rick Green. Secretaries include Tim Cody, Kelly de Strake, Nate Ponte, Bill Pastorek and Kris Wrathall. Treasurers have been Kelly de Strake, Doug Parton and Kris Wrathall.

Other significant volunteers who have served as directors of WSSOBC include Bruce Ambler, Mitch Baker, Dale Copeland, Kelly de Strake, Pat Deuling, Johnny Drift, Greg Fournier, Les Husband, Andy Johnson, Glen Kuenzl, Ross LaFlamme, James Linton, Dave McBride, Andy Pezderic, Greg Sawchuk, Guy Scott, Mike Southin, John Todd, Pat Tomlinson and Dave White. More outstanding volunteer directors and officers will step forward in the years to come as the WSSOBC grows.

Major donors of WSSOBC, in alphabetical order by last name or business name, include Outfitter Bruce Ambler, Carver Stuart Antonenko, Arctic Red River Outfitters with Kelly and Heather Hougen, Ron Arnett, Bausch & Lomb Canada, Bearhill Taxidermy, British Columbia

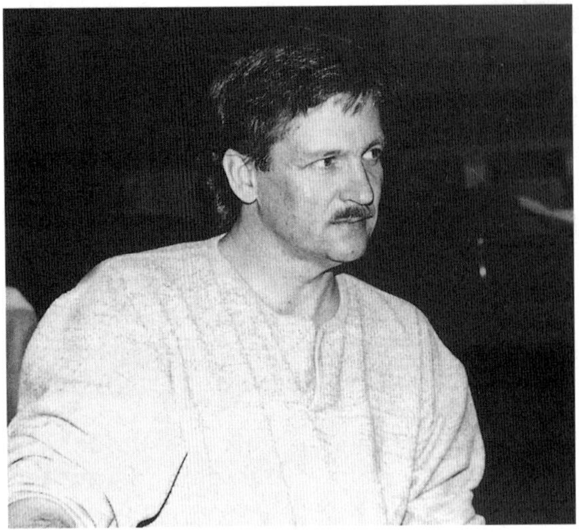

During the 1996 spring meeting of the Wild Sheep Society of British Columbia, Bill Pastorek was elected president. The banquet, dance and fundraiser was held at Cranbrook British Columbia on April 13.
Photograph by Duncan Gilchrist.

Wildlife Federation, British Columbia Wildlife Studios, Shane Black of Black's Furs, Biologist Tom Chapman, Etcher Jim Clayton, Mornay Duplessis of Faunafrica Safaris, Foundation for North American Wild Sheep, Glacier Air, Golden Rod & Gun Club, Ganders Taxidermy, Duncan Gilchrist, Frank and Arlene Gilbert of Fur & Feathers Taxidermy, Grand Slam Club and Sculptor Jack Gibson.

Other major donors include George Dennis Jewelers, Hugo Kilian of Kilian Galleries, Taxidermist Tony Judd, Artist Andrew Kiss, Artist Hayden Lambson, Sculptor Karl Lansing, Liard Air, Marc Laynes of Cascade Fishing Charters, Artist Tom Mansanarez, Ralf Martini Gunsmith, Tom McKenzie, Mark Meridith, Artist Steve Mills, Montana Chapter of FNAWS, Master Fisherman Peter Morrison, Barry Naimark, Only Africa Safaris, Artist Peter Popatoff, Rough Country Gun Safes, Greg Sawchuk, Artist Wilf Schlitt, Guy Scott of RiverJet Adventures, Mike Southin, Ron Strathern, Harvey Surina, Craig Swick of Paparazzi Furs, Sculptor Rick Taylor, Jim Turner of Thai Gems and Wilderness Wanderer Packs.

Any organization such as ours is only as good as its members. Many thanks have to go to those who have worked tirelessly to solicit donations for our fund-raising activities and who spend hours and hours of volunteer time on projects. Without them there would be no Wild Sheep Society of British Columbia.

Bill Pastorek lives near Vancouver, British Columbia. Bill owns Como Lake Garden Center. He says, "Our business months are March through July and December. This means that I can take most of the hunting season off. It's a tough life." He also owns a company called Grand Slam Marketing. With this company, Bill writes and markets books. One of his current projects is a book on British Columbia sheep titled, Dream Rams of B.C. Anyone taking a large ram in British Columbia who wants it recognized in his book should write to The Wild Sheep Society of British Columbia, Suite 539, 34A-2755 Lougheed Highway, Port Coquitlam, BC V3B 5Y9. This is the same address to receive membership and volunteer information about this worthwhile organization.

Montana Chapter of FNAWS 1992 to Present

Doug Dreeszen

In the early spring 1992, I received a postcard informing me of an organizational meeting for the purpose of organizing a state chapter affiliated with the Foundation for North American Wild Sheep (FNAWS). On May 9, 1992, men and women from all walks of life gathered at the Holiday Inn in Bozeman, Montana, where we started the Montana Chapter of FNAWS.

As a result of that May 9, 1992, meeting, a steering committee was established and the Montana Chapter of FNAWS was off and running.

This would not have been possible had it not been for several individuals, including Monte Schnur, Edwin Johnson and Mary Ellen Schnur who spearheaded the project and solicited a donation from the Montana Outfitters and Guides Association (MOGA), which was used for postcards and postage to send to FNAWS members and other wild sheep enthusiasts living in Montana.

With a large population of Rocky Mountain bighorn sheep in Montana, it would only seem natural Montana should become one of the chapters associated with National FNAWS based in Cody, Wyoming. Sheep hunting has always been a part of Montana's history and numerous records book trophy rams have been harvested, with many of them exceeding the minimum Boone and Crockett score of 180 points. Dozens of trophy bighorn are listed in the B&C records books. A total of 50 of these rams score in excess of 190 B&C points. Six currently score in excess of 200 points, according to the latest records book by the Boone and Crockett Club titled, Records of North American Sheep, Rocky Mountain Goats and Pronghorn that was published in 1996 and edited by Jack and Susan Reneau.

In the few years since its founding in 1992, the Montana Chapter has become one of the premier hunting and conservation organizations in the state. The highlight of each year's activities for the chapter is the annual banquet and fundraiser that is usually held in early February. The chapter's philosophy is that we do not want to have our fundraiser in the same city, so each year we move to another major Montana city. After our first fundraiser in Bozeman, the other banquets and auctions were in Butte in 1993, Billings in 1994, Bozeman in 1995, Butte in 1996, Bozeman in 1997, and Missoula in 1998. In 1999, the banquet is in Great Falls.

Each year the fundraiser is chaired by a local member and/or director from the host city who organizes the function. All members throughout the state participate and solicit auction items, including big-game hunts, fishing trips, artwork, jewelry, clothing, outdoor merchandise and a wide variety of other items. Our major funding for our projects is generated through the banquet and fundraiser each year.

Through the participation and cooperation of our membership, our first banquet and fundraiser was a huge success, showing us a bunch of sheep hunters can get a job done. We were fortunate to have one of the premier wildlife artists, Gary Robert Swanson, living on a ranch near Wilsall, Montana, who donated the original oil painting of a bust of a bighorn ram that has become known as, "The King." This oil painting and the limited edition prints created from that oil painting were also donated to our chapter. The painting and prints have generated many donations to the Montana Chapter, thanks to Gary. The painting has become

Dedication ceremonies took place at the Wayside Interpretive Bighorn Sheep Sign at Bighorn Canyon National Recreation Area. From left to right: Resource Management Chief Rick Lasko from the National Park Service; Tina Faxon and Jenyva Turner of Northwest Community College in Powell, Wyoming who designed the sign; and Montana Chapter President Doug Dreeszen.

Doug Dreeszen, left, and Montana Chapter member Scott DeMars of Billings work to sign up new members at the 1997 Montana FNAWS booth in Billings.
Photographs courtesy of Doug Dreeszen.

One of many fine bighorn sheep displays at the Montana Chapter fundraiser in 1996 in Bozeman.

Wild sheep participants enjoy the displays and exchange sheep stories at the 1996 Montana Chapter fundraiser.

Future sheep hunters, Cache Jensen, left, and Brandon DeMars, share the sunshine and a summer day with Leslie Jensen at a summer Montana Chapter picnic in Big Timber.

the adopted logo of the Montana Chapter of FNAWS. We cannot thank Gary enough for his generous donation, as well as taking the Montana Chapter under his wing.

The limited edition prints are used for a wide variety of fund-raising activities; however, the artist's proofs, numbered 1 to 50, were set aside to be given to members who signed up as a life member of the chapter. As of June 1998, Montana Chapter had 49 life members. The fee for life membership is $250; $15 for annual membership; and $40 for three years.

The Montana Chapter follows the bylaws of FNAWS by establishing a president, vice president, secretary, treasurer and five directors. Three board member positions are filled every summer by election. We hold board meetings bi-monthly, with a general membership meeting usually at our summer picnic. Our chapter's officers and directors have many contacts with the Montana Department of Fish, Wildlife and Parks, and keep an ongoing line of communication open. We also keep in close contact with National FNAWS in Cody.

With our membership currently at 150 members, we reach our membership through our quarterly newsletter, Wild Sheep. The newsletter gives information on upcoming chapter events, photographs of successful hunters, has articles of interest to the membership, and classified advertising for members. We expanded the advertising section in 1998 to help defray printing costs.

Over the last several years, we have tried different approaches in generating membership. Our most successful effort has become what is known as our "Trophy Mount Contest." This contest not only has registered new members, but the contest has generated a lot of interest from the sheep hunting fraternity. The rule is that hunters and taxidermists who enter the contest must be Montana Chapter members prior to February 1 of the year they enter; however, they are not required to be a Montana resident. Measurements and other information are given by the hunter or owner for each trophy. The winner receives $425 from the chapter to help defray the cost of the mount. The winner's name is drawn at the annual banquet. The 1998 contest had 12 entries. All successful hunters and entry photographs are published in the spring newsletter.

After each banquet, we critique it in order to improve future events. One of the highlights is the bighorn sheep display for the walk-in public and those attending the banquet. Outfitter booths, artists and various sporting merchandise vendors have booth space that is open to the public, too.

Our photo contest is similar to the National FNAWS photo contest and includes similar categories: hunter with trophy animal, wild sheep in the field, and hunter in the field. The quality and quantity of the photographs submitted has been impressive. We think this program emphasizes quality and ethical photographs.

In addition to the traditional live and silent auctions, we offer different styles of raffles and drawings and numerous rifles and sporting merchandise. Generally, we receive a rifle from National FNAWS headquarters that is used for our grand door prize. Each person purchasing a ticket to the banquet also buys a chance of winning the rifle. In conjunction with each fundraiser, we have a membership drive and give away a .22 rifle. We usually sign up from 25 to 35 new members at each fundraiser. The last several years we have raffled off a major item, including a 4-wheel drive all-terrain vehicle (ATV) and a custom-made Gentry big-game rifle.

One of the most popular raffles we conduct was the brainstorm of Director Bob Jensen of Bozeman. The raffle is known as "The Hat Raffle." We buy hats with our chapter logo imprinted on them that are numbered 1 to 50, which sell for $20 per chance. A number from 1 to 50 is drawn for the lucky winner who walks off with a very nice big-game rifle. Our chapter is fortunate to have corporate sponsors who purchase many of the rifles and shotguns and

The summer meeting near Big Timber brought the following Montana Chapter members to the table, from left to right: Timmer Reeves of Missoula; JoAnn McGovern of Billings; Harold Billings of Bozeman; Julaine Billings of Bozeman; Deb Solum of Rudyard; Jim Weatherly of Missoula; Larry McGovern of Billings; and Scott DeMars of Billings.

Future sheep hunter, Brandon DeMars of Billings, sports a new Montana Chapter cap and a catch of rainbow trout at the Big Timber meeting.
Photographs by Doug Dreeszen.

Bob Jensen, fundraiser chairman for the Montana Chapter of FNAWS, left, presents the winner Mark Justensen, center, with the grand door prize in Bozeman in 1996. Montana Chapter President Doug Dreeszen is to the right.

Steaming prime rib was savored by those attending a recent Montana Chapter banquet and fundraiser.

fully donate them for door prizes and raffle items.

Bob Jensen Electric of Bozeman is the traditional donor of the .22 rifle for the membership contest. Pat Connell of Connell Chevrolet-Buick-Toyota-Geo in Bozeman traditionally donates a rifle for the popular "Hat Raffle." In 1998, Jim Weatherly of WGM Group in Missoula donated a rifle with special FNAWS engraving on it for the grand door prize.

We are pleased to have David Gentry, custom gun maker, from Belgrade, Montana, on board as a supporter of our chapter. Mr. and Mrs. Gentry recently custom built a rifle that was one of the major raffle items during the 1998 fundraiser and donated the same talent and rifle for the 1999 fundraiser.

Revenues generated from our annual banquet and fundraiser are put back into a variety of projects and programs. We work with National FNAWS through their Grant-in-Aid program and sponsor or cosponsor some major conservation projects within Montana to "put more sheep on the mountain."

The challenge of "putting more sheep on the mountain" will continue. The Montana Chapter of FNAWS will be actively involved in pursuing several transplant sights in Montana as well as many other projects to benefit wild sheep.

The situation involving nonresident licenses, quotas, unlimited hunting districts and hunter opportunities will no doubt stir emotions and raise controversial discussions. The constant fight to maintain our hunting heritage and combat the anti-hunting movement are challenges that will continue into the next century. However, with the dedicated sportsmen and sportswomen within the Montana Chapter of FNAWS, the future looks promising.

In 1985, the Montana legislature established the Montana bighorn sheep auction license. The Montana Fish and Game Commission may issue one legal or larger bighorn sheep license each year through a competitive auction as defined in Montana State Statute 87-2-722.

National FNAWS, in appreciation of its continued efforts of supporting wild sheep management programs, has always been selected to auction the bighorn sheep license at its annual convention. Under the terms of the agreement, FNAWS will return 90 percent of the permit monies generated from the sale of the license to the Montana Department of Fish, Wildlife and Parks. The monies returned from the auction of the sheep license are required, by legislation, to provide money for the benefit of bighorn sheep in the state.

National FNAWS retains 10 percent of the money earned from the bighorn sheep permit for use in future Grant-in-Aid funding projects. To date, more than $2.1 million has been generated by the auction of the license, and of that, more than $1.9 million has been returned to the state for bighorn sheep management. Table 1 in this chapter shows the history of the Montana bighorn sheep auction license.

The Montana Chapter of FNAWS has been blessed with outstanding leaders since its founding in 1992. Officers and directors in 1998 included President Doug Dreeszen of Ballantine; Vice President Tim Magness of Anaconda; Secretary Deb Solum of Rudyard; Treasurer Rick Keller of West Yellowstone; and Directors Tom Grimes of Big Sky, Mike Murphy of Fairfield, Jim Weatherly of Missoula, Bob Jensen of Bozeman and Jim Hoschouer of Bozeman.

Past Presidents include Monte Schnur of Townsend and Larry McGovern of Billings. Past Directors include Lee Bridges of Eureka, Glen Erickson of Helena, Mary Ellen Schnur of Townsend, Tom Sather of Bozeman, Larry Altimus of Ennis, Harold Billings of Bozeman, Jack Atcheson, Jr., of Butte, Monte Schnur of Townsend, Larry McGovern of Billings and Doug Dreeszen of Ballantine.

To receive more information about the work and special events of the Montana Chapter of FNAWS, write to them at P.O. Box 3878, Bozeman, MT 59772.

Patty Dreeszen of Ballantine, left, holds her "Best of Show" photographic entry from the category of "Hunter with Trophy Animal." Patty received the award in Butte during the 1997 annual fundraiser. Doug Dreeszen congratulates her.
Photograph by Duncan Gilchrist.

Les Kish and Bob Jensen greet visitors to the Gallatin County Fair and tell them about the work of the Montana Chapter of FNAWS at their booth. The Chapter thanked Outfitter Lee Hart of Gallatin Gateway, Montana, for use of his booth backdrop.
Photograph by Harold Billings.

Left to right: Bob Fears, second from left, receives the rifle donated to the Yellowstone County Hunter Education Program by the Montana Chapter of FNAWS for the young hunter program. Also shown are, Larry McGovern of Billings, far left; Doug Dreeszen of Ballantine, third from left; and Pat Mullaney of Billings, fourth from left.
Photograph courtesy of Doug Dreeszen.

Artist Gary R. Swanson holds his original painting he created for the Montana Chapter of FNAWS that became the chapter's logo. With him is Steve Peterson who purchased the painting at the chapter's first annual and fundraiser. Gary donated his painting to the Montana Chapter of FNAWS.
Photograph courtesy of Doug Dreeszen.

Partial Listing of Wild Sheep Projects Funded by the Montana Chapter of FNAWS:

Donated funds to Montana Department of Fish, Wildlife and Parks, TIP-MONT Program, to assist with their reward program for information leading to the arrest and conviction of cases involving the illegal taking of bighorn sheep in the state. A reward of up to $1,000 is offered for this information. If you have information of any illegal hunting act, call 1-800-TIP-MONT (1-800-847-6668).

$1,000 Each year this amount is donated to the Wildlife Legislative Fund of America (WLFA).

$3,500 To study vegetation and habitat of bighorn sheep in Montana Hunting District 340, Highland Mountains. This funded project was with the Bureau of Land Management (BLM).

To fund study of parasite loads and their relationship to population size and herd health of bighorn sheep herd in the Highland Mountains by Montana State University graduate student Kerrie Hoar.

$2,000 To U.S. Forest Service, Lolo National Forest, Montana, for proposed controlled burn in Montana Hunting District 121 / Thompson Falls to increase feeding area away from highway to help decrease sheep mortality with vehicle and train traffic.

$5,000 To purchase emergency medication for bighorn sheep suffering from lung worm disease in Montana Hunting District 340 / Highland Mountains.

$2,919 To complete a helicopter bighorn sheep survey in Montana Hunting District 315 / Tendoy Mountains.

$5,000 To assist in the construction of a demonstration fencing project at the Pickering Ranch, Sanders County, Montana. This is in Montana Hunting District 124 / Paradise.

$1,500 To provide vegetation treatment in Montana Hunting District 270 / East Fork of Bitterroot Drainage.

$2,000 To provide vegetation treatment on the Lolo National Forest, Montana Hunting District 203 / Petty Creek.

$2,000 To fund interpretive wayside sign in Bighorn Canyon National Recreation Area explaining to the public about the local bighorn population, Montana Hunting District 503 / Pryor Mountains.

$1,448 To provide funds to the Montana Department of Fish, Wildlife and Parks, Enforcement Division, to purchase a sheep decoy to combat wild sheep poaching.

$1,000 To Washington State University to study Pasteurella.

Donated to Arizona to fight the anti-hunting and trapping Proposition 200.

Donated to Oregon to assist in the fight with anti-hunting movement.

Donated a rifle to the Yellowstone County Hunter Education Program for the young hunter program that offers an opportunity to youngsters who might not otherwise be able to get out in the field to hunt with an experienced hunter education instructor.

Donated funds and merchandise to "Montana Becoming an Outdoors Woman" program.

Donated funds to Make-A-Wish Foundation for a gravely ill youngster whose wish was to go hunting.

Co-sponsored Northern Wild Sheep and Goat Council meetings in Whitefish, Montana, April 26-30, 1998.

In 1997, Fred King, center, an official measurer for the Boone and Crockett Club and Montana Department of Fish, Wildlife and Parks (MDFWP) Region 3 manager in Bozeman, volunteered his time to measure rams at the annual banquet and fundraiser. He shows the finer points of measuring to FNAWS member Jim Hoschouer of Bozeman, left, and another MDFWP official. Fred has remained an active member of the Montana Chapter of FNAWS.
Photograph courtesy of Montana Chapter of FNAWS

At the 1994 annual meeting and banquet, Jack Atcheson spoke to the Montana Chapter of FNAWS about hunting opportunities at the Copper King in Butte, Montana, from March 11 and 12. More than 200 people attended the banquet and fundraiser. Gross receipts totaled more than $19,000.
Photograph by Duncan Gilchrist.

Montana Bighorn Sheep Auction License

In 1985, the Montana Legislature established what is known as the Montana Bighorn Sheep Auction License. The Montana Fish and Game Commission may issue one legal or larger bighorn sheep license each year through a competitive auction as defined in the Montana State Statute, 87-2-722.

The Foundation for North American Wild Sheep in appreciation of its continued efforts of supporting wild sheep management programs has been always selected to auction the bighorn sheep license at its annual convention.

Under the terms of the agreement, FNAWS will return 90 percent of the permit monies generated from the sale of the license to the Montana Department of Fish, Wildlife, and Parks. The monies returned from the auction of the sheep license are required by legislation to provide money for the benefit of bighorn sheep.

FNAWS will retain 10 percent for the use in future Grant-in-Aid funding projects. From 1985 to 1998, $2,138,000 in funds have been generated by the auction of the sheep license, while $1,923,000 has been returned to the state of Montana for bighorn sheep management.

Table 1

Year	*Hunter/Purchaser*	*Hunting District*	*Bid*	*Score*
1986	Arthur R. Dubs	H.D. 213	$79,000	196-6/8
1987	Arthur R. Dubs	H.D. 213	$109,000	197-5/8*
1988	Arthur R. Dubs	N/A	$93,000	No Harvest
1989	Guy Miller	H.D. 340	$74,000	196-2/8
1990	Greg Stires	H.D. 340	$61,000	Unknown
1991	Greg Stires	H.D. 340	$80,000	Unknown
1992	Glen Smith	H.D. 340	$88,000	194-3/8*
1993	Sherwin Scott	H.D. 124	$205,000	196-1/8 *(SCI score)*
1994	Jerry Fletcher	N/A	$310,000	No harvest
1995	Jerry Fletcher	H.D. 122	$281,000	196
1996	Robert Peterson	H.D. 216	$220,000	198
1997	Lee Anderson	N/A	$238,000	No harvest
1998	Anonymous	N/A	$300,000	N/A

* *Officially entered in Boone and Crockett Club records-keeping program. Other scores are based upon the Boone and Crockett Club method of measurement except where noted.*

Doug Dreeszen is one of the original organizers of the Montana Chapter of FNAWS. He has served as president for three years and was a director prior to his presidency. Doug is a Montana native who has hunted Big Sky country for close to 40 years. He first hunted sheep in 1972 in the Beartooth Mountains near Cooke City, Montana. He has been married to his wife and hunting partner, Patty, since 1970. Doug retired as a captain with the Billings Police Department in 1993 after 24 years of service. Doug and Patty live in the country near Ballantine, Montana, where they enjoy a quieter lifestyle and look forward to pursuing sheep whenever that opportunity arises.

New Mexico Chapter of FNAWS 1992 to Present

Kent Schauer

The New Mexico Chapter of the Foundation for North American Wild Sheep was formed in November 1992 by Wes Wilkening at the urging of Amy Fisher of the New Mexico Department of Game and Fish. Founding directors were Wes Wilkening, Lanny Romminger, Ed Benge, Kent Schauer, Mike Rowland and David Gilmore.

Current directors of the New Mexico Chapter of FNAWS through at least the year 2000 include: Kent Schauer, Lanny Romminger, Randy Martin, Ray Milligan and Jack Dilts.

We have been involved with five sheep transplants as volunteers or with funding of the transplants since our founding. We funded water transportation to the Pelincillos and have worked on two water catches on Ladrone Mountain. We have also funded the rebuilding of a dirt tank on Ladrone Mountain. We have worked on two water catches on Devil's Backbone in the Sacramento Mountains and cleared a spring in the Los Pinos Mountains. We have worked on a fence in the Manzano Mountains.

Our chapter and Wes worked closely with NMGF on several transplant and water projects. Wes Wilkening and Amy Fisher were instrumental in having desert sheep transplanted to Turner Armendariz Ranch in the Fra Cristobal Range. Our first fundraiser was in 1994 and was attended by FNAWS National Board of Director Pete Cimellaro who became president of FNAWS in 1995. Our second fundraiser was in 1996 and was quite successful, considering we had a membership of about 50.

A number of New Mexico Chapter members volunteered on the 1993 Pecos Wilderness transplant to Wheeler Peak. That transplant was so successful that in 1998 the wild sheep population is estimated to be approximately 100 animals. Two hunting permits will be issued in 1999 for rams in that in that herd, which is another strong indication of the success of that transplant.

In mid-1997 our chapter funded helicopter costs to locate and destroy some domestic sheep that were free-roaming on "M" Mountain near Socorro. At that time there were some desert bighorn sheep living on "M" Mountain. They had come from the Ladrone herd. In mid-1997 we also funded double fencing of domestic goats in the Manzano Mountains in preparation for an August 1998 transplant of Rocky Mountain bighorn sheep from the Pecos Wilderness. In October 1997 we paid for the purchase of 10 radio collars to be used in the Redrock transplant of desert sheep to the Pelincillo and Ladrone mountains.

In May 1998 we co-funded two graduate big-game biology students to follow sheep on the Pelincillo Mountains. They were able to document cougar or mountain lion kills resulting in two lions killed with tags. In June 1998 the New Mexico Chapter funded the rebuilding of a dirt tank on Ladrone Mountain. On August 1, 1998, a Rocky Mountain bighorn sheep relocation from the Pecos Wilderness to the Manzano Mountains took place. FNAWS National Board Member Duncan Gilchrist of Montana, FNAWS Grant-in-Aid Coordinator Becky Layne and Kent attended the project and assisted the officials from the New Mexico Department of Game and Fish. Duncan and Becky had to leave the trap site prior to any sheep captures but did help process and release the sheep. On Kent's fourth and final day in the mountains, he helped with the first capture of ewes and lambs. By the time this event took

Volunteers from the New Mexico Chapter of FNAWS helped with an August 3, 1993, Rocky Mountain bighorn sheep transplant that was so successful that in 1998 the wild sheep population was estimated to be approximately 100 animals.
Photographs by Wes Wilkening.

The successful 1993 transplant of Rocky Mountain bighorn sheep from the Pecos Wilderness to Wheeler Peak involved many members of the New Mexico Chapter of FNAWS, shown here blindfolding a sheep in preparation for the relocation.

Wes Wilkening, a founding president of the New Mexico Chapter of FNAWS in 1992, dressed in camouflage and helped with the 1993 transplant by netting an ewe.

Exhaustion overtook New Mexico Game and Fish officials in summer 1998 when Kent Schauer of the New Mexico Chapter of FNAWS participated in the August 1 Rocky Mountain bighorn sheep relocation from the Pecos Wilderness to the Manzano Mountains. National FNAWS Board Member Duncan Gilchrist of Montana and FNAWS Grant-in-Aid Coordinator Becky Layne of Cody, Wyoming, participated.
Photograph by Kent Schauer.

Eight members of the New Mexico Chapter of FNAWS worked on August 30, 1998, to complete a water catch on Ladrone Mountain. Left to right included: State Game Commissioner Steve Podilla, Kent Schauer, Jeff Blockwell, Dar Hourigan, Angel Montoya (knelling), Bob Jerrell, Chapter Director Randy Martin and Dave Heft. The volunteers coordinated with the Bureau of Land Management based out of Socorro. Dave Heft is with the BLM and was project organizer.
Photograph by Kent Schauer.

place, eight volunteers and state officials helped with the capture of 12 wild sheep under the drop net. This catch would have made great video footage, but nobody had time to take pictures until the sheep were all restrained. Kent was surprised that it was such an emotional experience. He looks forward to another wild sheep transplant, hopefully with a number of FNAWS volunteers.

In the opinion of Kent, the single biggest problem we face in New Mexico to put more wild sheep on mountains is control of mountain lions, especially predation of desert sheep. The New Mexico Department of Game and Fish has well documented that mountain lions aren't too hard on sheep as long as they have a good deer population as a buffer for a food source. New Mexico has a conservative estimate of enough mountain lions to consume 60,000 deer per year and the estimated deer population, statewide, is down to 250,000. Mountain lions continue to hammer the wild sheep populations. Almost every radio-collared wild sheep mortality in the past few years has been lion kills. Our desert sheep herd has gone from a statewide free roaming population of 360 head in 1995 to 235 in summer 1998. Unless these problems are properly managed, desert sheep will not survive in the state and will certainly not thrive in New Mexico. In 1998 the New Mexico Department of Game and Fish took a step in the right direction by extending the mountain lion season from four months to six months. Even that policy may be too little, too late.

In February 1997 while I was elk hunting on the Laguna Reservation, the remaining Board of Directors voted unanimously to disband the New Mexico Chapter. There were some serious differences with the New Mexico Department of Game and Fish. Upon hearing this, I volunteered to keep the chapter going. I do not have the time or energy to pour into the New Mexico Chapter as other dedicated FNAWS volunteers have for other state chapters and affiliate groups. These chapters could not be successful without those dedicated volunteers, but I am determined to keep the New Mexico Chapter of FNAWS going until another leader comes along.

The original directors have stayed on as members and one, Lanny Romminger, is a current director. We have increased our membership to about 70 individuals and have worked on several water projects. We have funded several outstanding projects from the money raised in 1996. I have two members who are always willing to help on projects. They are Dave Heft of the Bureau of Land Management and Director Randy Martin. All of our new directors are sheep hunters.

Before I step down, I want to get our raffle sheep tag through the New Mexico legislature and will hold my brother Ron to his promise of a significant monetary gift to our chapter. It was no coincidence that my brother, who was not a FNAWS member and took his only sheep in 1981, showed up in Reno in 1998 to purchase three sheep hunts at auction! He was high bidder for the New Mexico permit when he paid $77,500. We thank him for this outstanding contribution to wild sheep of New Mexico!

Kent Schauer is president of the New Mexico Chapter of FNAWS, living in Albuquerque, New Mexico. He is an optometrist who graduated from the College of Optometry at the University of Houston. He did most of his undergraduate work at the University of Texas. He has a passion for wild sheep and went on his 13th North American sheep hunt in Wyoming in 1998. Kent recognizes Brenda Schauer, his partner since 1989, for all the volunteer work she does on behalf of wild sheep and the New Mexico Chapter of FNAWS.

Washington Chapter of FNAWS 1992 to Present

O. Everett Rom

The following chapter history of the Washington Chapter of the Foundation for North American Wild Sheep (FNAWS) was written to celebrate the chapter's fifth anniversary. The information provided was taken from the official historical records, meeting minutes, newsletters and correspondence currently on file with the chapter historian, O. Everett Rom. Bighorn sheep data was provided by the Washington Department of Fish and Wildlife (WDOFW).

This chapter was written to provide organizational, charter, fund-raising activities, sheep transplanting, habitat improvement and other activities accomplished by the Washington Chapter. It is also written to provide credits to those individuals and donors who have steadfastly given of their time and talents to make the Washington Chapter of FNAWS a success. Our chapter motto is, "Bighorns for Future Generations," for the purpose of "putting sheep on the mountain."

In the Beginning: 1992 as the Pre-Charter Year

The Washington Chapter's organizational year started in 1992. Rocky Crate, a Puyallup, Washington, veterinarian and avid sheep enthusiast, dreamed of starting the Washington Chapter of FNAWS for several years. Rocky had many contacts within National FNAWS, various state organizations and other state FNAWS chapters because of his love of hunting and wildlife conservation. Rocky contacted the national headquarters of FNAWS. The staff provided him with a list of national members who lived in Washington.

Unbeknown to Rocky, another Washington resident, O. Everett Rom, was also working to form a Washington Chapter. Everett was compiling lists of individuals who might be interested in getting such an organization off the ground. Everett belonged to many wildlife organizations and helped with banquet fund-raising activities for many years. He was promoting and completing the banquet and fundraiser for the Seattle Chapter of Rocky Mountain Elk Foundation (RMEF) at the Washington State Convention and Trade Center. This huge undertaking grossed more than $100,000 on March 9, 1993.

Fate has a strange way of bringing people together. Rocky sent his letter of purpose and introduction to Washington state FNAWS members, and when Everett received the letter, he was elated. He immediately called Rocky and pledged his support to help establish the Washington Chapter of FNAWS. Everett pushed for an organizational meeting date. Rocky organized the first meeting on January 30, 1993, at Denny's restaurant near SeaTac airport in Seattle. Washington Chapter of FNAWS seeds had been sown and chapter organization was underway.

Year One: 1993, the Birth Year

On January 30, 1993, the first organizational meeting was held. Chapter member attendees included Rocky, Everett, Carl Phillips, Rusty Hunt, Terry Hunt, Kent Klineburger, Rollie Grant and Jim Young. These men donated $370 out-of-pocket to finance start-up operations. They discussed nonprofit organizations, distribution of elected officers and directors throughout the state, the time-frame for incorporation and the need for fund-raising activities. A slate

O. Everett Rom
Charter Member and First Secretary
Washington Chapter of FNAWS
Photograph by Olan Mills.

The first banquet committee of the Washington Chapter of FNAWS. Left to right, top row: Bob Manger, Blair Cooper, Terry Hunt, Carl Phillips, Dean Nelson. Left to right, bottom row: Jim Young, Rusty Hunt, Vic Schorr, Rocky Crate and Everett Rom.
Photograph by O. Everett Rom.

Rocky Crate, D.V.M.
Co-Founder and First Chapter President
Washington Chapter of FNAWS
Photograph by O. Everett Rom.

Terry Hunt
First Vice President
Washington Chapter of FNAWS
Photograph by O. Everett Rom.

James Young
First Treasurer
Washington Chapter of FNAWS
Photograph by O. Everett Rom.

Rolf Johnson, the sheep, moose and goat biologist for Washington Department of Fish and Game, displays a map showing the extent of bighorn populations in the state.
Photograph by O. Everett Rom.

Everett and Barbara Rom.

On December 16, 1994, under the direction of Bill Foreyt of the Washington State University School of Veterinary Medicine, a herd health day was held at Hall Mountain, 100 miles north of Spokane. These Rocky Mountain bighorns are used for transplanting stock into the Black Butte, Asotin Creek and Wenaha herds in southeastern Washington and to supplement herds in Oregon.
Photograph by O. Everett Rom.

Dan Yedinak, left, a Washington Chapter charter member, talks with a Washington Department of Fish and Game official about this transplant in 1994 that started a nucleus of sheep known as the Quilomene herd, which is as of March 1998 a population of 100 animals.
Photograph by O. Everett Rom.

of officers were elected and served from 1993 to March 1996: President Rocky Crate, Treasurer James Young, Secretary Everett Rom and Vice President Terry Hunt.

At the meeting it was decided that Everett Rom would recruit Board of Director members. Rocky worked on getting a charter granted from the state of Washington and asked Wenatchee attorney Kirk Bromley to help write the articles of incorporation for the new FNAWS chapter. Rocky made several trips to Olympia to succeed in getting the Washington Chapter incorporated. On March 9, 1993, the certificate of incorporation was signed by the Washington secretary of state and the Washington Chapter of FNAWS was a legal, nonprofit organization.

In the meantime, Everett was busy recruiting members to the Board of Directors and making plans for the first fund-raising banquet. Everett recruited Blair Cooper, Vic Shorr and Skip Clemens who were elected to the board at the next meeting. Rocky recruited Bob Manger and Dean Nelson to the board. The Washington Chapter was now gaining strength.

Everett, with the board's approval, contracted with the Aerospace Machinists Union for our first banquet facility at one-half price ($400). The first banquet date was set for February 5, 1994. Board of Directors met to plan this event and other chapter activities at Everett's home from 1993 to May 1995.

Everett Rom, Dean Nelson and Rocky Crate, along with other board members, solicited the WDOFW for a Washington governor's sheep tag. They were successful, and this tag is now sold at the national FNAWS convention every year, beginning in 1994. Tom Pawlacyk was successful bidder for the Washington permit that year with a total of $100,000. Lee Anderson was top bidder in 1995 with a total of $67,500. Stanley Boots was top bidder in 1996 with $55,000; Carl G. Hoehner in 1997 with $35,000; and Donald Schneider in 1998 with $27,500. Additionally, a Washington elk permit sold to Joe Balmelli at the national FNAWS auction in 1996 for $9,500 and for $8,500 to Virgil Lair in 1997. This represents a great success for the Washington Chapter's efforts to raise money to "put more sheep on the mountain."

Blair Cooper and Everett began the task of developing a chapter logo. Blair was associated with the printing industry and has for the first five years printed all the chapter's business cards, letterhead, newsletters, brochures and banquet programs. He has been a very valuable asset to the chapter in many ways. He and his wife Lisa have supported many wildlife organizations, including Ducks Unlimited (DU) and RMEF. Blair and Everett sorted through hundreds of pictures and magazines to come up with a magnificent bighorn ram used in our logo. They presented eight options to the Board of Directors. The board selected a green outline of the state on a white background with a full-sized bighorn ram in the center. The state background outline represented all of the people in the state and the ram depicted the conservation effort of the Washington Chapter. The white background represented the purity of the snow on Mt. Rainier and the color green represented the color of the Evergreen State — nickname for Washington.

Year Two: 1994

The Washington Chapter of FNAWS is very proud of its 1994 accomplishments. The year began with a successful first-year banquet that grossed more than $20,000 on February 5. Our first governor's tag was auctioned at the FNAWS national convention for $100,000. Phyllis Carlson, FNAWS national treasurer that year, flew to Seattle to attend our banquet. What a great lady she is. The chapter really enjoyed her visit. Rocky escorted her to and from the airport and the chapter gave her a huge bouquet of roses. Everett sold more than 45 new individual memberships to get the membership rolls started. The committee designed and

The animals in the 1994 transplant were crated in a record time of 2-1/2 hours and transported to the new herd site located 100 miles away by chapter members and WDOFW pickup trucks to a release staging area on the Columbia River. The crates were hand-loaded on a WDOFW small boat, transported up-river and released on the west bank of the river.
Photographs by O. Everett Rom.

A bighorn is removed from a helicopter and its gun net is unwrapped around it by Washington Chapter volunteers and Washington Department of Fish and Game officials.

Washington Chapter members pack water lines up 3,200 feet and two miles. Washington was not able to secure any sheep from Montana for our herd programs but the second water hole project was completed at the Quilomene herd's habitat area. The Washington Chapter now has two water hole areas located 15 miles apart that will allow the herd to expand.

A new tank with an existing spring is finished by, left to right, Matt Bell, Bret Forester and John Musser, the area biologist. The tank features a rodent escape ramp and an over-flow ramp to carry excess water 1,000 feet down the ravine to a smaller, secondary tank.

sold T-shirts and caps for fund-raising purposes, and the chapter members wrote thank-you notes and sent donation certificates to everyone who gave items or services to the fund-raising banquet.

On March 7, 1994, with only 12 hours notice from the WDOFW, the chapter funded $5,000 for our first sheep transplant. A helicopter team from Wildlife Management Capture, Inc., met with WDOFW game department personnel Rolf Johnson, Bill Essman, Roger Lloyd and John Musser at the Umtanum Canyon, east of Ellensburg, Washington, at 8 a.m. Washington Chapter members included Dan Yedinak, Bob Manger, Zack Wise, Terry Hunt, Rusty Hunt, Dean Nelson, Scott Hunt, Rocky and Everett. A total of 20 sheep were captured by helicopter gun nets (four young rams, 15 ewes and one yearling lamb). The sheep were vaccinated and throat and fecal samples were taken. The animals were crated in a record time of 2-1/2 hours and transported to the new herd site located 100 miles away by chapter members and WDOFW pickup trucks to a release staging area on the Columbia River. The crates were hand-loaded on a WDOFW small boat, transported up-river and released on the west bank of the river. Many hazardous trips up-river were taken to get the bighorns to the release site. The transplant was completed by 6 p.m. that day. This transplant started a nucleus of sheep known as the Quilomene herd, which is as of March 1998 a population of 115 animals. The herd has experienced some coyote and mountain lion predation but is growing and doing very well.

The Washington Chapter of FNAWS donated chapter caps to most of the 23 Washington chapters of the RMEF to help them with fund-raising activities and as partners in conservation.

Everett Rom solicited a $50 free booth for the Spokane Bighorn Show in 1995. Dean Nelson attended the Washington State Gambling Commission meetings in Olympia in order to understand state rules and regulations concerning gambling, raffles and ticket sales to the public. Dean shared his knowledge with Washington Chapter members and has been a very valuable asset in coordinating chapter activities within Washington state law.

Skip Clemens was elected as director of publicity and the newsletter. He and Blair Cooper set the wheels in motion for the chapter's first newsletter that was dated Spring 1994. Skip has worked very hard to format and publish our newsletter and the chapter is very thankful for his accomplishments. Skip has also donated the proceeds from his line of shirts and sweatshirts that are sold at our banquets. The chapter is very grateful for his generosity. Skip has provided a 3-by-7-foot Washington Chapter banner that is displayed at all major activities of the chapter.

Everett secured the banquet facilities at the Puyallup Elks' Lodge for the 1994 banquet where they have remained through 1998 at no charge for the room or the facility. The Washington Chapter members only paid for the food.

The chapter's first water hole project was completed October 22, 1994. Chapter members and WDOFW personnel who worked on the project included Randy Hein, Dan and Kelly Yedinak, Rolf Johnson, Bob Perleberg, John Musser, Larry Wadkins and John Morgan. The Collin Spring Project is at the Quilomene herd site. The project consisted of one 8-foot round tank, one 4-foot round tank and 3,200 feet of water line connecting the tanks. This was a backbreaking exercise because most of the supplies, tanks, lumber, steel posts, pipe, etc. were packed in by hand uphill two miles. The chapter appreciates the hard and valuable work accomplished by these men. More than 115 bighorns, 75 elk and many mule deer are now utilizing these tanks.

Everett built a display board showing the location and herd counts on a large Washington state map. This display will be utilized at various sportsmen's shows at the Washington

Chapter booth.

On December 16, 1994, under the direction of Bill Foreyt of the Washington State University School of Veterinary Medicine, a herd health day was held at Hall Mountain, 100 miles north of Spokane. A total of 21 veterinary students, Steve Zender of WDOFW and Everett Rom of the Washington Chapter participated in the capture, vaccination, fecal and blood sampling and mange evaluation. More than 45 sheep were at the capture site. The Hall Mountain herd is a non-hunted herd transplanted from Alberta, Canada, in 1972 and is the healthiest herd in the state. This herd is used for transplanting stock into the Black Butte, Asotin Creek and Wenaha herds in southeastern Washington and to supplement herds in Oregon. All of the sheep were released unharmed back into the Hall Mountain habitat area.

The Washington Chapter of FNAWS expresses its sincere appreciation for all the work, studies and data collected that was provided by Dr. Bill Foreyt, D.V.M., of WSU. Dr. Foreyt has studied bighorn sheep for more than 20 years and the research data provided to WDOFW and other organizations has been a valuable tool in sheep management. Dr. Foreyt has worked many long, hard hours in support of sheep programs.

Blair Cooper provided more than 175 plain white poster boards and 175 back stands. He also provided live and silent auction decals to Everett for assembly. Everett's wife Barbara and his daughters, Heather and Adele, spent more than 40 hours assembling the boards for the 1994 banquet. The poster boards are reusable year after year, which greatly lessens banquet set-up time and have at least a 20-year shelf life.

Year Three: 1995

On February 25, 1995, the second annual banquet of the Washington Chapter of FNAWS was held at the Puyallup, Washington Elks' Lodge. Gross receipts for the banquet were just over $32,000, which included private cash donations and the timeless rifle raffle ticket sales from the Spokane Bighorn Show. Net receipts were approximately $23,000 after expenses.

The 1995 membership rolls stood at 131 people, which included our donors and sponsors. The chapter that year contributed $500 to fight the "No Bear Hunting with Dogs" initiative in Oregon. National Board Member Duncan Gilchrist attended the February 5 banquet and provided a lecture and films on wild sheep. Washington was not able to secure any sheep from Montana for our herd programs but the second water hole project was completed at the Quilomene herd's habitat area. The Washington Chapter now has two water hole areas located 15 miles apart that will allow the herd to expand. Matt Bell and Bret Forester, chapter members, worked on the project that was funded by the chapter.

Everett worked with Lloyd Oldenberg of the Idaho Fish and Game Department to provide publicity for the Hells Canyon Initiative. This program's goals are to transplant and increase the Snake River herds to 10,000 bighorn sheep along the Washington, Oregon and Idaho lands of the Snake River. A detailed chapter about the Hells Canyon Initiative is included in another section of this book.

Jerry Tyrrell was welcomed as a new director in 1995. He had experience in fund-raising activities. Our new meeting place was at his business in Kent.

The Washington Chapter successfully lobbied the WDOFW for the issuance of one state raffle ticket per year to the public for a bighorn sheep hunting permit. The 1996 state legislature approved the issuance. The proceeds were earmarked for bighorn programs. Tickets sold for $10 each.

The chapter made donations to RMEF, Safari Club International (SCI), DU and other FNAWS chapters, the Northern British Columbia Guides and Outfitters' wolf project and other organizations to help them in conservation and fund-raising activities. The chapter purchased new window decals that would be issued to the members after future banquets.

Dean Nelson continued to work with the Washington Gambling Commission to get our gambling license for games of chances and raffles, etc. in support of annual fund-raising activities.

In the fall of 1995, a breakout of Pasteurella occurred in Washington's most valuable sheep herds along the Asotin drainage. The Wenaha, Black Butte, Asotin Creek and Joseph Creek bighorn herds and adjacent herds on the Oregon and Idaho sides of the Snake River were affected. Washington lost an estimated 150 to 200 Rocky Mountain bighorn sheep — a terrible blow to the state and its people. The Washington Chapter provided $2,500 of drugs and the vaccine, Ivomec, to Oregon, Idaho and Washington. The chapter and National FNAWS pumped $30,000 into a program to capture, vaccinate and re-release the diseased sheep into their habitat. Many sheep were captured and quarantined at Caldwell, Idaho. All of the sheep died even after these best efforts. The Washington Chapter of FNAWS thanks the three states involved for helping to try and save the bighorns.

Everett recruited Steve Kline of Spokane and presented his name to the board for approval, which was accepted. Steve had experience with fund-raising activities in support of RMEF, Boone and Crockett Club, National FNAWS, SCI and the Inland Wildlife Association. He became a director in 1996. Everett also recruited Matt Bell and introduced him at the September 1995 board meeting. Matt helped with the Quilomene water hole project. He became a director of the Washington Chapter in 1996.

Year Four: 1996

The Washington Chapter of FNAWS held its third banquet on February 17, 1996. A total of 138 members and donors raised just over $31,511 of gross receipts to help "bighorns for future generations."

On March 25 a new slate of officers was elected by the chapter board. Terry Hunt was elected president, Jerry Tyrrell and Blair Cooper were elected vice presidents. Rocky Crate was elected treasurer, and Dean Nelson was elected secretary. The committee agreed to set the price for life members at $300 each. The committee decided to have a timeless rifle raffle, statewide. The winning name would be drawn at the Spokane Bighorn Show in 1997.

The Washington Chapter funded a transplant of 24 Washington bighorns within the state boundaries. A total of 14 sheep were transplanted to a site behind Roza Dam. To Mount Clemons were transplanted 10 sheep. Chapter members helping with this effort were Blair Cooper, Lee Hughes, Jack Kominski and Rocky Crate assisting the WDOFW with the capture, vaccination, transplant and release of the sheep. The WDOFW originally planned to transplant at least 10 sheep to augment the few remnants of an old herd at Goldendale. The WDOFW cancelled these plans unexpectedly, which was a disappointment to the Washington Chapter.

Terry Hunt donated another rifle for our next banquet raffle. Terry has donated a rifle every year to help in our fund-raising activities. The Texas Bighorn Society offered a print donation to be auctioned at our next banquet.

The chapter also funded a sheep capture at Kamloops, British Columbia, on March 17, 1996. This transplant involved 31 California bighorns that were released into existing herds at Mt. Hull (3), Sinlahekin (4), Lincoln Cliffs (4), Quilomene (10) and Mt. Clemons (9). This was done to add new genetics to our existing herds. Rocky Crate, D.V.M., handled the veterinary work and members Larry Scharno, Glenn Scharno and Jerry Tyrrell worked three days to capture, vaccinate and transport the sheep back to Washington.

Shown is one of the 20 sheep that were captured by helicopter gun nets (four young rams, 15 ewes and one yearling lamb) on March 7, 1994. This was the Washington Chapters first sheep transplant.
Photography by O. Everett Rom.

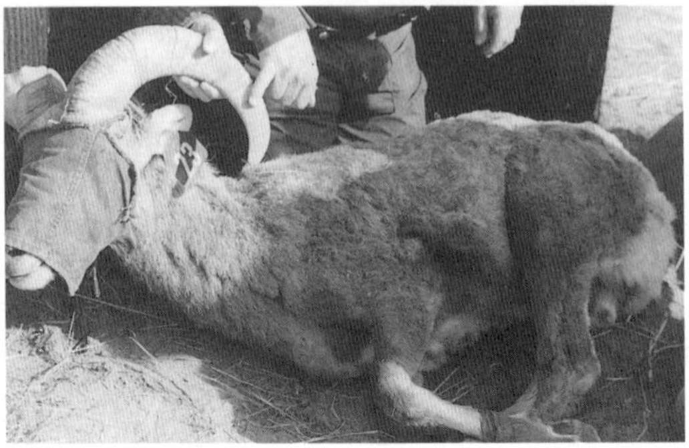

On March 7, 1994, with only 12 hours notice from the WDOFW, the chapter funded $5,000 for our first sheep transplant. A helicopter team from Wildlife Management Capture, Inc., met with WDOFW game department and FNAWS volunteers to help with the work.
Photograph by O. Everett Rom.

The WDOFW donated two sets of horns to the Washington Chapter that were auctioned off at our banquet. WDOFW also donated one set of horns to the Eastern Chapter of FNAWS for auction. The Eastern Chapter set went for $9,000 at auction. Our thanks to WDOFW.

A Spences Bridge, British Columbia, sheep capture and transplant had been planned for the spring of 1996. The Washington Chapter was terribly disappointed when the operation had to be called off because of the 1995 Pasteurella die-off in Washington, Oregon and Idaho. These valuable sheep were needed for the Hells Canyon Initiative. New sheep could not be transplanted into a diseased area so the operation was cancelled for 1996.

Steve Kline built a new booth that was used for a display at sportsmen's shows and other public events. The committee decided to change the logo that no longer includes its uniqueness of the silhouette of Washington state. It is now a copy of the National FNAWS logo using the round circle with a ram's head in the center.

Year Five: 1997

The Washington Chapter's fifth banquet was held on February 7 with 184 attendees. The banquet grossed $33,774. Directors and members noted there were many new faces in the crowd and it appeared to be a young group of people in attendance. This was viewed as a positive sign for future organization activities. The Washington Chapter had booths at both the Puyallup Sportsman's Show and Spokane Sportsman's Show. The chapter sent a letter to WDOFW reporting our concerns about cougar predation. Washington has a no dogs/bait law and our cougar population is viewed by many to be out of control. The chapter is concerned about the possible loss of our sheep population to predation.

The chapter approved appropriating $6,150 to fund a water tank project in the Chief Joseph Wildlife Area. The Bureau of Land Management has already purchased two 2,500-gallon tanks and funds are needed to complete the project by year end. The chapter funded $1,000 for Dr. Bill Foryet of WSU for bighorn research projects. The chapter also funded the cost of a mold and cast reproduction of a bighorn sheep bust. The 197-plus ram is a new

Idaho state record. The Washington Chapter will buy 20 busts and donate one to National FNAWS for the next convention.

During the week of December 7, chapters members Jim Hawk and Everett participated in the capture and transplant of 40 Rocky Mountain bighorns at Spences Bridge, British Columbia. This transplant had been delayed for two years because of the Pasteurella die-off in 1995 in Washington. After quarantine the sheep were transplanted by horse trailer to release sites in three states.

Idaho received 17 sheep, Oregon received 13 sheep and Washington received 10 sheep. Washington's sheep were released into the Asotin drainage adjacent to Hells Canyon. Washington's sheep were radio collared. Since the release, the WDOFW found one of the young rams dead from a cougar kill. One of the Oregon sheep died from stress and trauma and one of the Idaho sheep died from injuries suffered at the capture net. The WDOFW transplanted eight California bighorns from the Umtanum unit to Dead Canyon and 15 sheep from the Quilomene herd to the Tieton unit during the year.

Spences Bridge Sheep Report

What enthusiasm! During the week of December 7, 1997, about 60 men and women dedicated to the principles of FNAWS congregated at Spences Bridge, British Columbia, to capture and transplant Rocky Mountain bighorn sheep from British Columbia to Idaho, Oregon and Washington in support of the tri-state cooperative effort.

Spences Bridge is located approximately 200 kilometers north of the United States border on Canadian Highway 1 along the Thompson Fork of the Fraser River. The original 50 sheep were transplanted by railroad cars from Banff National Park in 1932. The herd has increased and decreased several times. The herd count estimates have been as low as 75 and are currently as high as 350 to 400 animals, according to the British Columbia Ministry of Environment (BCME) big-game officials who participated in this transplant. There are between 50 to 100 sheep that winter on the city borders from year to year. BCME game officials indicated 60 sheep are year-round residents. These sheep feed on the local citizens' flower gardens and shrubbery. They also create traffic hazards on the very busy Canadian Highway 1. One of the local area business owners told me that more than 30 sheep are killed each year by automobiles right in front of her restaurant.

On Tuesday, December 9 at 7 a.m. about 70 people met for capture instructions and assignments. The first net drop was scheduled for that morning. The sheep didn't cooperate! Only eight sheep were under the trap net and three of them were older rams that would not be transplanted. The team leader decided not to drop the net until more sheep were captured. The remainder of the day was spent working the sheep that had spread out into smaller groups throughout the town into one large group. The volunteers herded the sheep towards the trap net where the sheep would bed down for the night. This effort was successful and everyone looked forward to a good trap the next day.

Wednesday morning dawned with great enthusiasm. The morning's strategy meeting was held at 7 a.m. at the Quarter Circle J Motel that was further away from the trap site and would create less disturbance to the sheep. As the dawn broke over the beautiful British Columbia mountains, the sheep began to move to the trap net. More than 20 sheep were under the net and all systems were go. When the net drop mechanism was triggered, the net wouldn't drop. Several attempts were made with no results. Net failure was due to the minus 6 degrees Celsius (26 degrees Fahrenheit) cold temperature that resulted in insufficient battery current to release the huge drop net. The team rescheduled activities for the next day.

Sheep were vaccinated and throat and fecal samples were taken.
Photograph by O. Everett Rom.

On Thursday morning, December 11, the net was successfully dropped and 23 sheep were trapped (five rams and 17 ewes). On Thursday afternoon, the net was successfully dropped again and 17 more sheep were trapped (five rams and 12 ewes and lambs). The sheep were vaccinated and blood, fecal and throat samples were taken on all of the sheep.

- The Washington Chapter of FNAWS acknowledges the following men and women who participated in the project.
- British Columbia Royal Canadian Mounted Police - traffic and demonstration control.
- Wild Sheep Society of British Columbia - President Bill Pastorek and members Pat Deuling, Greg Sawchuck, Art Black, Steve Ruthven and Kelly Destrake.
- Washington Chapter members Everett Rom and Jim Hawk.
- Washington Department of Fish and Game leader Mark Grabski and biologists Brett Lyndaker and Joey McCanna.
- Washington State University sheep veterinarian, Dr. Bill Foryet and third-year veterinarian students Susan Allen and Andy Allen.
- King 5 Television, Seattle, film crew John Stofflet and Keven Strehle.
- U.S. Forest Service, Enterprise, Oregon, employees Kevin Martin, Cheri Miller and Tim Schommer.
- Oregon Department of Fish and Wildlife big-game personnel Vic Coggins, Don Whittaker, Craig Foster, Jim Lemos and Jim Torland.
- Idaho Chapter of FNAWS members Jerry Walters, Terry Henrickson, Elmer Huston, Wendy Swanson, Eugene Swanson, Jim Dufoe, Royce Tanner and Russ Rhodig.
- Idaho Department of Fish and Game employees Frances Casirrer, Dave Hunter, D.V.M., Jay Crensaw and John Beechum.
- British Columbia Ministry of Environment big-game personnel Kurt Kier, Fred Harper, Mike Peters, Helen Schwantze, Steve Willett, Bob Workmen and Don Eden.

The Rocky Mountain bighorn sheep were transported after a 24-hour quarantine by game department horse trailers to the release sites in three states. This capture and transplant effort is the result of many years of planning by FNAWS and the tri-state game departments. Known as the Hells Canyon Initiative, this cooperative project released bighorns on the Idaho and Oregon sides of Hells Canyon and in Washington on the Asotin Creek drainage adjacent to Hells Canyon. Idaho received 16 sheep; Washington received 10 sheep; and Oregon received 13 sheep. One sheep died due to injuries suffered under the drop net.

The citizens of Spences Bridge are warm and friendly and were very good hosts for this event. This area has excellent steelhead fishing, river rafting adventures and pockets of bighorn sheep populations that can be seen from the highway, year round.

Table 1

Washington Governor's Bighorn Sheep Tag Auctioned at the Annual National Convention of FNAWS.

YEAR	*SUCCESSFUL BID*	*SUCCESSFUL BIDDER*
1994	$100,000	Tom Pawlacyk
1995	$67,500	Lee Anderson
1996	$55,000	Stanley Boots
1997	$35,000	Carl G. Hoehner
1998	$27,500	Donald Schneider

Table 2

Money Raised at Washington Chapter Banquets and Auctions.

DATE	*TOTAL GROSS*	*ATTENDEES*
1994	$20,904	100
1995	$29,904	131
1996	$31,511*	138
1997	$40,844*	213
1998	$33,774*	184

* *Total gross figures for 1996 to 1998 include proceeds from a timeless, statewide rifle raffle and all yearly cash donations by friends and members of the Washington Chapter.*

Table 3

Washington Sheep Herd Populations from Washington Department of Fish and Wildlife Trend Report from 1997.

CALIFORNIA BIGHORN HERD	*POPULATION*
Tucannon	51
Vulcan Mountain	65
Mt. Hull	59
Sinlahekin (Okonogan)	35
Swakane	12
Quilomene	115
Umtanum (Selah Butte)	112
Cleman Mountain	103
Lincoln Cliffs	112
Colockum	0
Dead Canyon	6
Tieton	12
Cottonwood Creek	21

ROCKY MOUNTAIN BIGHORN HERD	*POPULATION*
Hall Mountain	35
Asotin Creek	23
Joseph Creek	40
Wenaha	55

Table 4

Officers of Washington Chapter of FNAWS, 1993-1999.

President

Rocky Crate, 1993-1996
Terry Hunt, 1996-1997
Jerry Tyrrell, 1998-1999

Vice President

Terry Hunt, 1993-1996
Blair Cooper, 1996-1997
Jerry Tyrrell, 1996-1997
Steve Kline, 1998-1999
Dean Nelson, 1998-1999

Treasurer

Jim Young, 1993-1996
Rocky Crate, 1996-1999

Secretary

Everett Rom, 1993-1996
Dean Nelson, 1996-1997
Rusty Hunt, 1998-1999

Directors, Past and Present, in Alphabetical Order

Matt Bell
Skip Clemons
Blair Cooper
Rocky Crate
Mary Hunt
Rusty Hunt
Terry Hunt
Steve Kline
Bob Manger
Dean Nelson
Carl Phillips
Everett Rom
Vic Schorr
Steve Stewart
Jerry Tyrrell
Zack Wise
Jim Young

The Washington Chapter of FNAWS certainly works hard to "put more sheep on the mountain." To join the Washington Chapter, write to them at 18334 SE 284th Street, Covington, WA 98042-5300, or call (253) 631-5321.

Since the inception of the Washington Chapter of FNAWS, O. Everett Rom has been one of its primary supporters. Everett was a charter member and founder of the chapter. He has solicited a large percentage of the chapter's donations, has worked on a year-round basis to sell memberships, promote events and has had hands-on capture and release experience with more than 100 bighorn sheep. Everett was appointed chapter historian in 1996 and has written articles for newspapers and for the FNAWS national magazine, Wild Sheep. His wife Barbara has spent hours typing meeting minutes and letters of correspondence, and his daughters, Heather and Adele, have helped with chapter display work. Everett is a life member of the Foundation for North American Wild Sheep and Rocky Mountain Elk Foundation. He is affiliated with Safari Club International, National Rifle Association, Mule Deer Foundation, Boone and Crockett Club, Ducks Unlimited and the Washington State Big Game Council. He completed a master's hunter course through the WDOFW. He has spent 10 years raising money for RMEF and five years raising funds for the Washington Chapter of FNAWS. Everett retired from Boeing Company. He and Barbara live in Covington, Washington.

International Sheep Hunters Association

JOSEPH W. QUARTO

The International Sheep Hunters Association (ISHA) is not an affiliate or chapter of FNAWS but continues as a strong supporter of the work FNAWS does for wild sheep and wildlife conservation. Their accomplishments and objectives compliment the purposes of FNAWS, so a short article about this fine organization is included in this section on FNAWS chapters and affiliates. ISHA meets each year at the annual FNAWS conference.

In 1974 a meeting of the greatest sheep hunters of all time was held and together they formed the International Sheep Hunters Association (ISHA) to honor those adventurous sportspersons that went beyond the North American continent to collect and study the rest of the world's wild sheep in their natural habitats.

Present were Elgin Gates, George Landreth, Chris Klineburger, Gene Rubeck, Joe Quarto, Matt Zonailo, Jacques Henrijean, Jack Greer, Ozzie Davis, John Batten, W. T. Yoshimoto, Butch White, Frank Hibbens, Jim Wilkinson, Bob Speegle, Hal Hallick and Vin Somavia. The following purposes and objectives were drawn up:

(1) To join together persons interested in international wild sheep hunting and wild sheep conservation.

(2) To gather, record and exchange information and ideas among members about wild sheep of the world.

(3) To inform and educate the people of the world about international wild sheep, their general welfare and habitat.

(4) To maintain lists and records of members activities in collecting sheep from around the world.

(5) To maintain scientific and biological data on wild sheep of the world.

(6) To endeavor to conserve, develop and obtain more hunting areas around the world and make them available to members and future generations.

(7) To maintain, record and oversee Super Slammers Club.

(8) To promote professional management and enhance wild sheep populations.

(9) To encourage wild sheep conservation by giving of our time, money and knowledge to safeguard against the decline or extinction of any species of wild sheep of the world.

(10) To put in practice the wild sheep "Ten Commandments."

This same group established and put into being one of the greatest challenges of the hunting world; completing the super slam by taking 12 different species of wild sheep from around the world. ISHA kept moving forward collecting sheep information, while passing on and exchanging information to its members. ISHA's current ultimate achievement is the completion of the book, The Encyclopedia of Wild Sheep of the World, which is moving along fine and will be the Bible of wild sheep. Even though ISHA does not solicit for new members, it's membership just keeps on growing.

To receive more information about ISHA, contact them at P.O. Box 2769, Riverside, CA 92516 or call them at (909) 369-1235. Their fax number is (909) 782-8616.

SECTION II

The Wild Sheep of North America

John Herbst is a long time FNAWS volunteer, donor and seminar giver during the annual conventions. Pictured in upper right corner is John Herbst giving a seminar during the 1997 convention.
Photograph by John W. Herbst.

Chapter I

Wild Sheep: Creatures of the Ice Age

Valerius Geist, Ph.D.

Wild sheep are creatures of the Ice Age and very successful ones at that. They spread across the mountain ranges of cold-temperate Eurasia and crossed the immense distances of Siberia into North America. This huge sweep of geography was aptly named by the late James L. Clark, *The Great Arc of the Wild Sheep*. Archeological records indicate sheep were abundant even when heavily used by ancient man, including in pre-Colombian North America. They also provided us with one of our most important domestic animals, as did the sister group to sheep, the true goats. We have been fellow travelers for quite some time.

The fate of sheep is intimately tied to the advances and retreats of huge continental glaciations. During the last two million years, the northern hemisphere was periodically overrun by such. These momentous climatic events have a periodicity of — roughly — a 100,000 years. Of this about a quarter are warm interglacial, such as we are experiencing currently, and three quarters is a time of huge glaciers creeping over the continents. Sheep followed these glaciers. Even during our warm interglacial periods, sheep are still glacier followers, for where glaciers exist, sheep are likely to reach abundance in numbers as well as large body and horn size. Sheep had to be very adaptable to survive successive glaciations and melt-offs for climates and landscapes differ greatly between glacial and interglacial times.

The huge ice masses are not necessarily hostile to life, quite the contrary. Glaciers grind away the underlying bedrock turning it into highly fertile rock dust. Melt waters sweep the tiny rock particle out beyond the glacier, where it muddies the runoff as silt. When waters recede, the silt is left high and dry. Desiccated by the sun, the fine rock dust is picked up by the wind and blown far and wide, and deposited as fertile "loess." So much fine rock powder is in the air during glaciations that wind currents carry the fine dust all the way to the poles. Thus glacial times can be recognized in ice cores drilled from Greenland glaciers by the rock dust in the annual layers of snow that were compressed into ice over tens of thousands of years.

The glacial dust spells fertility, but so does the water that regularly flows from the glacier, and so does the sunshine and blue skies over the lee sides the melting sides, of glaciers. Moreover, the earth is tilted during glacials so as to absorb more sunlight in summer. All that generates productive habitats wherever the glacier's fertility and water may reach. Great steppes form on the nutrient-rich, young soils where the rock dust settled, allowing large herbivores and carnivores to flourish. Sheep were among those that flourished, and still flourish where glaciers are large and active. Here they seek out the rich, seasonal forages that allow them to grow fat, despite the high costs of raising young, restoring winter-depleted bodies and growing thick hair coats and large, ornate horns. Modern humans also benefit from the glacial dust laid down by wind and water so long ago, for these deposits are the bread baskets of today's agriculture. By contrast, the old soils of Africa are leached of minerals and infertile. Africa's flora and fauna are adapted to live with very little fertility; African antelope carry little fat. Glaciers restore soil fertility.

Because sheep were such ardent glacier followers, we are faced with a rather poor fossil record for sheep and their relatives. Glaciers bulldoze rocks. Moving rocks crush any bones deposited. So do rock slides, avalanches and gushing melt waters in the mountains that thoroughly rearrange the rock rubble they meet. Whatever may be left of the bones after carnivores are through with them, there is little chance of them escaping being ground to pieces by moving rock, ice and water.

Consequently, in deciphering the origin of wild sheep we have to turn to a variety of sources, and cherish every fossil we manage to lay our hands on. Good science is meticulous detective work, and suggests that sheep have been around for quite a while. The most ancient of sheep appear to be the urials of southwestern Asia. They probably gave off a very early radiation of sheep, a radiation that reached North America probably half a million years ago. These are the sheep of the subgenus *Pachyceros*, of which the most extreme members are the American bighorn sheep *(Ovis canadensis)*, and the most primitive is the snow sheep race *(O. nivicola)* residing in Kamchatka. These sheep are very different from the Old World sheep such as the urials, mouflons and argalis, suggesting an early split from the stem group, a split that may date back to the very beginning of the major glaciations about two million years ago.

The bighorns, Dall's and snow sheep differ from the Old World sheep by adopting an ibex-like predator escape strategy, namely seeking refuge in steep cliffs, with a secondary option of moving deep and unpredictably into forest cover. As a consequence our sheep have a rather stocky ibex-like body build. That is, they are superb rock-hoppers with massive haunches, powerful backs and shoulders, and relatively short legs. Sprinting to cliffs also limits their body size to one just adequate to produce a single nine-pound young. Moreover, the pregnancy period of our sheep is a full month longer than that of Old World sheep of comparable size. Unlike Old World sheep, our sheep do not twin. The long gestation period and the low birth rate suggests that the ancestors of American wild sheep faced very long, cold winters with little feed. They had to minimize reproduction and compensate by living longer, and by closely conserving their offspring. Consequently, bighorns, Dall's and snow sheep virtually do not twin, live significantly longer than Old World sheep, live at higher latitudes in colder climates, and conserve the young they raise to maturity by closely leading them from seasonal home range to seasonal home range. This avoids the costly loss of life inherent in pioneering for living space; as is practiced, for instance, by yearling moose and whitetail deer.

From the outset, Old World sheep had a different way of escaping predators. Unlike North American sheep they specialized not in rock-hopping, but in outrunning predators. The urials, mouflons and argalis left rock-hopping to ibex and wild goats with which they usually lived in close proximity. Goats and sheep co-evolved as competitors in the Old World, which made them take on different anti-predator strategies to minimize competition. This effectively made goats residents of the mountains and cliffs, and sheep of the rolling foothills and plateaus. Old World sheep evolved into long-legged runners with large chests and surprisingly light skeletons. In this type of running, the runner places its long legs precisely about minor obstacles while maintaining a high velocity of travel. To be able to do that, long-legged runners run with elevated heads in order to see in good time what's ahead. Short-legged runners, such as saiga antelopes or reindeer, run with the head lowered and follow trails or run over ground with few obstructions. Their short legs allow running with a minimum of body lift, which conserves energy and stamina while maintaining high speed.

However, running is a very demanding way to avoid predators. It requires a young that is large at birth, quick on its legs and quick to grow up. That necessitates a mother able to shunt a larger amount of resources from her own maintenance requirements and body growth towards the growth of the young and the synthesis of rich milk. Rock-hoppers, however, can depend on the protection of cliffs in the early days of their young, and thus do not require as highly developed young at birth or as rich a milk. Consequently, we expect in Old World sheep relatively larger young at birth and richer milk than we expect in goats and ibex. A comparison of milk solids indicates, as expected, that sheep have richer milk than goats. In males, nutrients and energy in excess of maintenance requirements can go into the growth of luxury (bragging) organs, such as horns, bibs and neck manes. Therefore, we expect that Old World sheep will have relatively heavier and larger horns than goats. That is also found.

The geographic distribution of Old World sheep and goats indicates that the rock-hopper security adaptation is the superior one, allowing goats to spread farther than sheep. This is true, even though Old World sheep are the superior dispersers, having been seen hundreds of miles from their normal ranges following droughts. Moreover, while one finds every race of Old World sheep accompanied by a distinct race of goats or ibex, one finds wild goats widely distributed without sheep. Wild goats and ibex occupy a greater geographic area than do Asiatic sheep. The rock-hopping North American sheep and their relatives have also spread further and colonized more land than did Old World sheep. Old World and New World sheep, superficial similarities to the contrary, are thus quite different animals.

Body size in sheep is a function of genetics and the quality, abundance and seasonal availability of the food on their ranges. Why cannot bighorns grow just as large as the largest of argalis, the Altai argali? A bit of "bio-engineering" explains why. Adult body size appears to be a function of the "safe size "of a newborn. Both species often face in spring rather cold temperatures, and both bear lambs about nine pounds in weight to countermand the death on newborns from hypothermia. However, the argali ewe will bear two such lambs and the bighorn ewe only one. For two lambs to be produced, the bighorn female would have to be on average twice the "metabolic body size" of the argali ewe. Since metabolism per unite mass decreases as the three-quarter power of body weight, the bighorn ewe, doubling metabolic size, would have to increase 2.54 times in absolute mass.

At twice the metabolic mass, a bighorn would need twice the foraging area it now requires. That means that it would have to go 1.42 times as far away from the safety of cliffs as it currently does, or spend 50 percent of its foraging time beyond the safe distance from cliffs. In short, the bighorn would have to change from a rock-hopper into an advanced runner, and assume the body shape and functions of an argali in order to regularly bear twins. It would have to do so, because by increasing body mass by 2.54 times, it reduces its power to push its bulk uphill in sprinting. This is due to elementary bio-physics, as muscle power increases only as the square of linear dimensions, but body mass increases as the cube. To be precise, at 2.54 times the body mass it has only 1.84 times the muscle power to accelerate its bulk. That means it would take such a big bighorn a lot more time to reach the safety of cliffs as it would on average have to cover a greater distance at a lower speed. We can safely conclude that a bighorn the size of an Altai argali would be an easier prey to wolves than the current edition. A rock-hopper thus cannot be very large, and if it does grow larger it must become a runner like the giant sheep – or a slow climber like the mountain goat.

However, mountain sheep like other larger mammals, humans included, can vary greatly in body size depending on the quality of forage they are able to secure. The differences between extremes are 3-5 times in body size. That is, the starveling on a range with poor forage and in severe competition with others over the scarce food, will be about one third the size or less of the large individual who swelters in luxury. The latter condition is fairly rare in nature, but does occur when a population begins to colonize vacant habitat and thus feeds on a virgin food resource. Sheep reintroduced successfully to a former range grow larger in body and horn size during re-colonization. This does not last long. Thereafter, body and horn size shrinks on average. The poorer the range, the smaller the maintenance-type sheep on them. Wild climatic gyrations that generate years of abundance and years of scarcity favor a wide spectrum of body sizes, where as uniform environments do not. Note: Asiatic giant sheep, the argalis, are NOT destined genetically to always be huge. In desert regions giant sheep on poor forage may be very small, barely the size of Alaskan Dall's sheep.

Also, as one goes up in latitudes and altitudes into increasing seasonality, so body size increases. A peak in body size is reached at about 60 degrees latitude. Going further north, ungulates again shrink in body size. Since sheep occupy mountains, one must calculate the attitudinal effect, as altitude compensates for latitude. Therefore, bighorn sheep, which occupy high mountain ranges, are expected to reach maximum body size not at 60 degrees north, but at lower latitudes. In fact, bighorns reach maximum body size where expected, namely about 47 to 52 degrees north, in southern Alberta, British Columbia and northern Montana. Further north and south bighorns, as expected, decline in body size, and they do. Since Dall's sheep live way north of 60 degrees north, would you expect Dall's sheep to reach body sizes rivaling large bighorns? Of course not! Dall's sheep live much too far north. Nor do we expect Stone's sheep to grow as large as bighorns because they too live way north of 47 to 52 degrees north. However, we expect Stone's sheep, on average, to exceed Dall's sheep in size. They do. Dall's and Stone's sheep will equal or exceed bighorns living south of 47 to 52 degrees north. Because Siberian snow sheep occupy not only very high-latitudes, but also face a long, cold winters, we expect them to be on average small bodied compared to our sheep, be they thin horns or bighorns. What is magical about 47 to 52 degrees north latitude for our sheep? The best educated guess is that it provides the longest period of availability of high quality food when sheep grow body, hair, horns and produce milk.

Thus there are a number of factors that control body size; though ultimately, they are all tied to quality, quantity and seasonal patterning of the food supply. Bighorns, occupying strip mines rehabilitated to high-quality custom sheep habitat, grow to huge body sizes — north of 47 to 52 degrees north. As expected from earlier work, the body size of the sheep on these productive sites continued to increase for over five generations. That is, an increase in body size in response to superior feeding takes a decade or more to materialize.

There are also intrinsic factors affecting horn size in rams, not only feed and age. Rams that are spirited and aggressive are likely to broom off horns, and may loose significant body weight. That translates subsequently into poor body and horn growth and an early death. On the other hand, rams that participate little in rutting may save a lot of fat that allows them to survive not only the hardships of winter, but also grow larger still the following year. These rams have a good chance of becoming trophy rams because they add more annual segments to their horns, and larger segments as well. This is predicated on ewes choosing security over food and rams choosing food over security in their choices of habitat.

Some argali rams, as we have just discovered, do the opposite. Instead of joining the big rams on fertile meadows and hills far away from secure escape terrain, they stick with the females and young. We have suspected for some time that females crop the forage on their secure, but limited ranges very closely, so much so that young rams must leave their mothers home ranges — or starve. They must seek out insecure but superior feeding sites if they are to grow into normal rams, let alone big ones. The argali rams, however, which choose to stay with the females, grow up to hardly exceed females in size. Their horns are only slightly larger than those of the females, and they lack the long-haired ruff. These rams, old though they may be, look like immatures.

Old and New World sheep have had a very divergent history, as expressed in more than different ways of escaping predators and thus different body forms. Bighorns that came into North America in the last quarter of the major Ice Ages entered a predator Hell. North America's fauna had diverged considerably from that of Eurasia by virtue of evolving a greater diversity and specialization of predators. Despite the specializations, predators had a rough time with the indigenous prey, as it too evolved not only superior escape strategies but also greater pugnacity. The overwhelming evidence for broken and healed bones in old American carnivore species speaks to that. A few Siberian immigrants, however, did manage to establish a foothold in North America about half a million years ago, among them the bighorn, long-horned bison and stag-moose. Bighorns were very limited thereafter, but they appeared in Wyoming about 100, 000 years ago in remarkably modern form. They become common at the height of the last glaciation about 20,000-18,000 years ago, but spread geographically after the glacial maximum. As expected, in the dispersal phase they are huge. They are also more long-legged than current bighorns. Apparently, they were on the way to evolving argali-like body forms. However, the demise of North America's indigenous fauna 12,000-7,000 years ago, and the entry as the Siberian snow dog (gray wolf) with its large, soft paws, put a halt to that evolution. Previously bighorns had to deal with high-speed predators such as the American cheetah and the predacious short-faced bear. A number of species in America grew short-legged in the presence of Siberian wolves, including bison and musk oxen.

North American sheep came out as winners after the great extinctions, which in good part may have been caused by the simultaneous entry of people into North America. Sheep had to face very little competition in the new, species-poor fauna made up of a few old American species that survived and the newly arrived Siberians. Bighorn remains in archeological sites indicate that they maintained numbers in the face of human predation, while elk, moose and bison did not. The latter exploded in numbers after European diseases and conquest exterminated native tribes, and the heavy hand of native people was lifted from North American landscapes. Then "wilderness" exploded for about two and a half centuries, an artifact of European post-Colombian colonization.

The future of wild sheep, be they Old World or New World, is troubling in the face of continued human population growth and economic demands. Old World sheep, given a little protection from excessive killing and some land ungrazed or lightly grazed by domestic livestock, and they virtually explode in numbers due to their high reproductive rate. Biologically, they are easy species to manage. The New World sheep are not. They have to be segregated from domestic livestock, especially domestic sheep. Our livestock carries diseases difficult or even impossible to handle for our sheep. In addition, the ability of our sheep to closely conserve their yearlings mitigates against ready dispersal to vacant habitats. We have to do that for them. This is a lesson learned over a quarter century ago, and having acted consequently on it, we have done a good job reintroducing bighorns. That's a proud success story. So is the rehabilitation of strip mines into sheep habitat. So is the overall, careful cropping of hunted sheep populations. This was the result of public agonizing by interested parties over the direction of sheep conservation in which the Foundation for North American Wild Sheep had such a leading role. Good hands-on management does work.

Valerius Geist, Ph.D.

Valerius Geist, Ph.D., is Professor Emeritus of Environmental Science on the faculty of Environmental Design of the University of Calgary, Calgary, Alberta, where he was first program director for Environmental Science in 1971. He retired from his full-time professorship in 1995 after teaching a wide variety of courses such as environmental science and human biology, wildlife conservatio, large mammal biology, ethology, ecology, evolution, wildlife management and ecology of human affairs. He served as program director for Environmental Science from 1991 to 1994. He is the author of 11 books and the editor two including Buffalo Nation *(1996),* Wild Sheep Country *(1993),* Elk Country *(1991) and* Mule Deer Country *(1990). His first book,* Mountain Sheep: A Study in Behavior and Evolution *(1971) was winner of the 1972 Book of the Year Award by The Wildlife Society. He is currently working on four new books on deer (*The Deer of the World: A Study in Evolution and Behavior*), moose, pronghorn and North American mountain sheep (*Return of Royalty: Wild Sheep of North America*). He is a scientific and editorial consultant for such noted organizations as the British Broadcasting Corporation, National Geographic Society, Boone and Crockett Club and Foundation for North American Wild Sheep. He has been involved with seven documentary films and has contributed 40 articles for 16 different encyclopedias. He is the author of more than 115 popular articles and book chapters, including three chapters for this book. He and his wife Renate are parents to Rosemarie, Karl Alexander and Harold. Valerius holds a B.SC. in zoology from the University of British Columbia, Vancouver, and a Ph.D. in zoology (ethology) from the University of British Columbia. His postdoctoral work was done in Germany at Max-Planck Institut fur Verhaltensphysiologie.*

Photograph by Duncan Gilchrist, inset.

BIGHORN SHEEP

FNAWS GOLD BIGHORN SHEEP AWARD WINNERS

YEAR	HUNTER	SCORE	OUTFITTER
1978	Leland Speakes, Jr.	166	--
1979	John Snyder	187 4/8	Sammy Sands
1980	Armand Johnson	197 1/8	--
1981	Alfred Journey	188 1/8	None
1982	R.B. Swang, Jr.	--	--
1983	Bonnie Ford	193 1/8	None
1984	Hugh MacAulay	--	--
1985	Bud Hartman	--	Randy Babala
1986	Dave Elzwiler	187 1/8	Merritt Pride
1987	Pat Downey	189 5/8	None
1988*	Dave Onerheim	201 1/8	Chuck Christensen
1989	John Gehan	187 2/8	Randy Babala
1989	Chuck Houtz	187 2/8	Larry Smith
1990*	Peter P. Bollinger	200 5/8	Michael Boyce
1991*	Lee Hart	197 1/8	R.L. Neal
1992*	H. James Tonkin	193 6/8	Elvin Hawkins
1993*	Katherine A. Pyra	203 4/8	Percy Pyra
1994	Kevin Williams	200	R.L. Neal
1995*	Richard N. Aznaran	196	Jon Barker
1996*	Gary E. Hansen	204 1/8	Randy Babala
1997	Robert E. Petersen	200 7/8	R.L. Neal
1998	Tim Rushing	190 7/8	None

* Award of Excellence

THE WORLD'S RECORD BIGHORN SHEEP

This bighorn ram was taken in Alberta in 1911 by Fred Weiller. It wasn't officially scored by the Boone and Crockett Club until 1966 when the owner, Clarence Baird, finally sent it in. It scores 208-1/8 points. The lengths of both horns differ by only a margin 1/8 inch with the longest horn measuring 45 inches. Tragically, in 1971, a fire ravaged the Baird's home and this great trophy was reduced to ashes.

Chapter II

Wild Sheep Populations of North America

Rocky Mountain and California Bighorn Sheep

DALE E. TOWEILL, PH.D.

North America was home to seven races of bighorn sheep at the beginning of the 19th century. The most widespread of these races, the Rocky Mountain bighorn (now known to science as *Ovis canadensis canadensis*), may have numbered 1.5 to 2 million animals distributed widely throughout the western United States and Canada.

This animal was found primarily from the Peace River southward in the lofty Rocky Mountain ranges along the Alberta-British Columbia border, through Idaho and western Montana, Wyoming, Colorado and Utah and into the northern margins of Arizona and New Mexico. To the eastward, a separate race, the now-extinct Audubon's bighorn (classified by scientists as *Ovis canadensis auduboni*) occupied the canyons, badlands and low mountains ranges of eastern Wyoming and western North and South Dakota. Westward, the California bighorn (*Ovis canadensis californiana*), occupied river canyons and low mountains of the central valley of British Columbia, eastern Washington and Oregon, southwest Idaho and the Sierra Nevada ranges of northwestern Nevada and California.

As compared with the far more numerous Rocky Mountain bighorns, the number of California bighorns is believed to have exceeded 100,000 animals. There animals narrowly escaped the fate of the Audubon bighorn, now extinct. Just as Audubon bighorns were eliminated from South Dakota in the late 19th century and from North Dakota about 1905, California bighorns were totally eliminated from Oregon by 1915; Washington by 1935; Idaho by 1940; and Nevada by 1946. A very small remnant population of California bighorns persisted in California's Sierra Nevada in the vicinity of Mt. Baxter and Mt. Whitney, and a nucleus population of 1,500 to 2,000 California bighorns survived in central British Columbia.

The remaining four races (*Ovis canadensis nelsoni, O. c. mexicana, O. c. cremnobates, and O. c. weemsi*) were distributed among the isolated desert mountain ranges of the southwestern United States and northern Mexico, and are collectively identified as the Desert bighorn group, discussed later.

The Decline of Bighorn Sheep

Although fossil record is very fragmentary, available data from fossils seem to indicate

A herd of mature Rocky Mountain bighorn rams with 3/4 curls congregates in Gardiner, Montana, area during the rut. Efforts by FNAWS and other wildlife conservation groups have brought back the historic bighorn sheep herds of Montana since the late 1970s.
Photograph by Jack Reneau.

A bighorn ram and ewe graze on grass within sight of the National Bighorn Sheep Interpretive Center in Dubois, Wyoming, and provide an "up close and personal" opportunity for the public to view Rocky Mountain bighorns in their natural habitat without bothering them.
Photograph courtesy of NBSIC.

A herd of Rocky Mountain bighorn sheep pay no attention to visitors to the National Bighorn Sheep Interpretive Center in Dubois, Wyoming. FNAWS and many other wildlife conservation organizations have contributed to the Dubois center so the public can learn more about North America's magnificent animal.
Photograph courtesy of NBSIC.

that the ancestors of all North American sheep evolved in Asia. These ancestral sheep entered North America from eastern Asia by way of a land bridge across what is now the Bering Sea. During the Pleistocene Period (better known as the last great Ice Age) formation of glaciers is believed to have bound up a significant portion of the earth's water in huge sheets of ice spread across the northern hemisphere.

The effect of these vast sheets of ice was to lower the level of the earth's oceans by as much as 490 feet, thus exposing what is currently the floor of the Bering Sea and vast expanses of the continental shelf as dry land, and while the extent of dry land increased and decreased over the period of centuries, many kinds of animals (and most likely humans as well) occupied the vast plains created by rising and falling ocean levels. The animals that occupied the exposed plains and the mountain range that presently forms the Bering Island chain gradually spread eastward into the north. As North America's glaciers reached their greatest extent, bighorn sheep and other species were forced southward (and northward), into forests and plains where the great deserts of northern Mexico and the American Southwest are located today. The distribution of bighorn sheep at the beginning of the19th century is consistent with the pattern that would be expected if the ancestral populations of sheep selected habitat much like that selected by Rocky Mountain bighorns today; habitats that became progressively drier through the centuries following the decline of the ice sheets. The slow change from cold and dry to hot and dry climates occurred so gradually that populations had time to adapt to a broad range of newly-developing habitats.

By any standards, bighorn sheep were among the most successful of the Ice Age mammals. Not only did they survive the changes that occurred in North America following the Ice Ages (while many other species, such as the mammoth and saber-tooth tiger, became extinct), but they adapted to a tremendous range of habitats, from the icy mountain tops surrounded by permanent snowfields to the rocky desert ranges of Baja California and northern Sonora.

Native American tribes avidly pursued bighorn sheep for utilitarian purposes, including meat and utensils, and many revered them for their behavioral qualities as well. Some cultures, such as the "sheepeater" tribes of the western Shoshone, based significant aspects of their culture and livelihood on bighorn sheep. An increasing body of archaeological information indicates that Native American peoples were capable of harvesting wild sheep to the extent that some wild populations of bighorns were depressed by hunting. Images of bighorn sheep painted or hammered into the rocks throughout western North America provide scientists with information concerning the presence and former range of wild sheep.

Rocky Mountain and (especially) California bighorns fared poorly as western North America was settled. Although Audubon's bighorn sheep were first described by Lewis and Clark in 1804, they became extinct almost exactly one century later. Virtually all California bighorn sheep populations were lost within a few decades after that (except for a remant individuals in California and some remote herds in British Columbia), and by 1950 Rocky Mountain bighorn herds survived only in rocky and remote locales in Alberta, British Columbia, Colorado, Wyoming, Montana and Idaho.

The decline of bighorn sheep coincided directly with settlement of the western United States. Bighorn sheep were tasty and often easy to obtain locally, especially since their rocky habitat was easily identified. Many animals, unaccustomed to long-range firearms, made easy targets as they retreated to cliffs and hills within easy range of marksmen intent on providing food for themselves and growing work teams of miners and others attracted by western settlement. The lowland populations — the Audubon's and California bighorns —

were most affected by uncontrolled subsistence and market hunting.

Those herds that survived the hunters faced another, double-pronged threat associated with settlement — the use of bighorn range for uncontrolled livestock grazing. Critical winter as well as summer ranges were grazed heavily, without regulation. Cattle were among the first livestock brought westward with civilization, and although cattle reduced the amount of food available for bighorns, they were not nearly so destructive to bighorn sheep ranges as the domestic sheep that soon followed. And while the settlers fought their own range wars, the unregulated competition for forage resulted in another type of range war. That is, widespread severe overgrazing that decimated critical bighorn sheep habitats. The bighorns that could find forage faced a more subtle and severe threat — parasites and diseases brought by domestic livestock.

Bighorn sheep evolved in Asia, as did the ancestors of domestic sheep. But while domestic sheep were selected over centuries for their resistence to parasites and disease, North American sheep, like the Native Americans who crossed the Bering Strait with them, escaped many common Asian disease organisms and, over time, lost much of their resistance. The effect of re-exposure of bighorn sheep to parasites and diseases of domestic sheep, like the re-exposure of native Americans to influenza and smallpox, was devastating. Entire herds were decimated by outbreaks of scabies and lungworm, soremouth and bletongue and a wide range of other diseases never completely documented.

By the early 20th century, many people believed that bighorn sheep populations throughout the United States and Mexico were doomed to eventual extinction, and that herds in Canada faced grave risks.

The Beginnings of Recovery, 1912-1974

The close of the western frontier — and the beginning of modern big-game management — can perhaps best be marked as 1912, the date Arizona became the 48th member of the United States. Although various state legislatures had passed laws to protect wildlife and establish hunting seasons in the middle to late 1800s, most such laws were rarely enforced. Enforcement was particularly lax in the remote bighorn sheep habitats of the western United States; indeed, many game laws were poorly enforced for at least several subsequent decades. However, with the transition from territorial to state government, the legal framework was in place to establish and enforce a modern system of wildlife management. Just three years prior to Arizona statehood, in 1909, a fellow by the name of Aldo Leopold graduated from the Yale School of Forestry. This forester compiled the tenets of wildlife management developed in the United States by President Theodore Roosevelt and the first Chief of the U.S. Forest Service Gifford Pinchot and in Canada by Prime Minister Sir Wilfred Laurier and his chairman of the Commission on Conservation Sir Clifford Sifton. The basic tenet of modern wildlife management was that wildlife was a renewable crop of the land, as were the trees in the national forests. Wise use of the land could result in annual crops of wildlife, from which a harvest could be taken each year. Leopold published his ideas in 1933 in a textbook, Game Management. Game Management was a classic and a book that outlined the complete basis for wildlife management as practiced today, and it provided a text for the first of the college-trained wildlife managers, who graduated in the late 1930s. The Wildlife Society, the international professional organization of wildlife managers, was formed in 1937 and expanded and refined the underlying tenets of wildlife management.

Those first wildlife managers in the western United States who faced the task of rebuilding wildlife populations, focused first on the more widespread and popular species such as deer and elk. They were aided by the fact that the Taylor Grazing Act, passed in 1934, had put a mechanism in place to begin to restore degraded wildlife habitat. The urgent demands of World War II reduced emphasis on game management, but that emphasis again soared when millions of former servicemen, trained in the use of firearms, returned following the war and created a large and growing demand for outdoor, and especially hunting and fishing, opportunities.

It was sportsmen demands for restoration of big-game herds that forced the state and provincial wildlife management agencies to focus renewed attention on species like bighorn that still existed, although in low numbers. Management of such species rarely "pays its own way" in terms of license revenue, and thus private funds are often necessary to supplement the limited resources of wildlife management agencies. Prodded by sportsmen, most of the western United States had begun bighorn sheep restoration projects by the late 1950s, moving Rocky Mountain bighorns from thriving herds into vacant habitats, and restoring California bighorns to areas from which they had been eliminated years earlier.

A concentrated effort by wildlife biologists, sportsmen and wildlife conservation volunteers in the last 25 years has brought wild sheep populations back from the brink of extinction throughout North America. Here wildlife biologists and volunteers in Montana work to age a young bighorn ram by gently prying its mouth open and examining its teeth. The mask over its eyes keeps the ram from getting overly excited as scientific measurements are taken. This ram is from the Rock Creek herd in western Montana.
Photograph by Fred King, Montana Department of Fish, Wildlife and Parks.

Much of this early work was summarized in a special report, *The Bighorn Sheep in the United States, Its Past, Present and Future,* prepared by a Professor of Zoology, Helmut Buechner (1960). Dr. Buechner examined all of the old records of bighorn sheep distribution in the United States and identified the locations and size of existing remnant populations. The numbers were not encouraging. Viable remnant herds of Rocky Mountain bighorns occurred in only four states (Colorado, Idaho, Wyoming and Montana); two additional states (South Dakota and Utah) had perhaps two dozen animals combined (Table 1). The total population estimate for Rocky Mountain bighorns in the United States was less than 10,000 animals even with the most optimistic estimates. California sheep were virtually gone, with 200 to 400 in California and a few individual animals in Washington. These animals were supplemented by about two dozen animals in Oregon, the result of a transplant from British Columbia in 1954 (Table 2).

Dr. Buechner also documented the problems associated with restoration of herds — limited critical winter ranges (often in poor condition), competition with domestic and wild animals for limited forage, losses due to predation, and widespread infections with lungworm and other diseases. With range conditions, poor and surviving bighorns often chronically stressed, lungworm infections often reached chronic levels in bighorn sheep herds. As a

What a treat to witness such a head-butting session in the wilds of British Columbia! In 1974 when the first group of wild sheep hunters gathered at Mt. Horeb, concern was expressed that bighorn populations were in serious decline. Thanks to financial support from FNAWS and an organized scientific effort to observe and study wild sheep in the last 25 years by Canadian, Mexican and United States' biologists, such fears have all but disappeared.
Photograph by Valerius Geist.

result of chronic infections, many herds were subject to catastrophic die-offs. The message was clear: For bighorn sheep populations to be restored, managers would have to find ways to expand critical winter ranges, improve forage supplies, reduce predation where possible, and reduce the occurrence and impacts of lungworm and other diseases. Bighorn sheep herds in Canada, although typically in much better condition, faced identical challenges.

Armed with new knowledge, wildlife managers went to work to restore bighorn sheep populations. Additional bighorn sheep range was acquired through easement or purchase, and land managers focused on improving range conditions using traditional forest management practices (burning, fertilization, seeding). Livestock grazing levels were adjusted in some areas, and managers identified ways to reduce competition between bighorns and elk on key winter ranges. New drugs to combat lungworm infection were developed, and administered through mineral blocks placed on bighorn ranges. Perhaps most importantly of all, managers examined the now-vacant ranges formerly occupied by bighorn sheep relative to their suitability for transplant with animals from still-healthy herds. By removing some bighorns from herds at or near the carrying capacity of critical winter ranges, managers could reduce the grazing pressure on existing ranges while providing new homes and often better range conditions for the transplanted animals. As a result of these efforts, bighorn sheep began to recover dramatically.

Restoration of California bighorn sheep provides perhaps the best documented example. California bighorns were essentially gone from Western landscapes by 1940, although their habitat remained pretty well intact. California bighorn sheep herds in British Columbia were largely untouched by diseases of domestic livestock, although they were affected by livestock overgrazing. The Junction herd was particularly successful. In part to allow their own ranges to recover, British Columbia allowed some of their excess animals to be used for transplants both elsewhere in British Columbia and in the western United States.

Oregon was the first state to transplant California bighorns from the Junction herd. A total of 18 animals were moved to Hart Mountain National Antelope Refuge in 1954. British Columbia moved more animals to its own vacant or underutilized habitats in 1955 and in 1956 sent another 18 California bighorns to North Dakota. Washington received 18 California bighorns in 1957 and Idaho received 19 in 1963. While Idaho continued to receive California bighorn sheep from British Columbia (nine in 1965, 10 in 1966, and 12 in 1967), the first states that had received animals from British Columbia were already beginning to move the off-

Rams show their dominance towards each other. North America was home to seven races of bighorn sheep at the beginning of the 19th century. The most widespread of these races, the Rocky Mountain bighorn (now known to science as Ovis canadensis canadensis*), may have numbered 1.5 to 2 million animals distributed widely throughout the western United States and Canada. Tables 1 and 2 in this chapter show current population levels.*
Photograph by Valerius Geist.

spring from those first transplants. North Dakota, Washington and Oregon were all actively moving California bighorns to other vacant habitats among themselves (and Nevada) during the period 1959 through 1974.

British Columbia continued to provide animals periodically, as conditions allowed; California received 10 bighorns to reestablish a herd at the Lava Beds National Monument in 1971. The net result of all of this shuffling was to increase the number of California bighorns by at least 50 percent, to approximately 3,000 animals by 1971. A review of California bighorn sheep herds nationwide published in 1970 estimated that British Columbia herds totalled 1,879 animals, with 390 in California, 303 in Washington, 209 in Oregon, 120 in North Dakota, 90 in Idaho and 8 in Nevada. And while harvest continued to be allowed on a limited basis in British Columbia, Oregon first offered six permits for the harvest of California bighorns in 1965 and Washington offered California bighorn sheep hunting permits beginning in 1966.

The Boone and Crockett Club, National Audubon Society and the Wildlife Management Institute sponsored a symposium in Montana in 1974 to get all bighorn sheep management agencies together to assess their accomplishments and to help guide future developments. The results of nearly 15 years of bighorn sheep restoration efforts were impressive: herds of Rocky Mountain bighorn sheep in the United States had increased by 50 percent, from the 8,100 to 9,700 in 1960; and from between 12,060 to 14,130 in 1974. California bighorn population increases were even more dramatic; although herds in California were declining slowly, establishment of new herds boosted the total number of California bighorn sheep in the United States from a mere 200 to 400 to well over 1,100 animals! The lessons learned over this period were equally clear. Bighorns could not be expected to spread and colonize new areas like the more familiar whitetail deer. Bighorn sheep restoration would depend on multiple transplants into vacant habitats. This key message established a management framework for the future.

Bighorn Sheep Management in the Past Quarter Century

This was the setting when a group of bighorn sheep hunters and conservationists met in 1976.

Aware of many of the successes in restoration of wild sheep in North America, they were also aware of the critical problems, first clearly articulated back in 1960 by Helmut Buechner in his book, *The Bighorn Sheep in the United States, Its Past, Present and Future*. He wrote:

A British Columbia bighorn lamb pauses in the deep snow. Recent studies of diseases and parasites, many sponsored by FNAWS, have brought a great deal of knowledge critical to the management of bighorn sheep. *Photograph by Valerius Geist.*

"Accomplishment of the required reductions in competing animals and the research needed for the development of more precise conservation programs seem remote at the present writing. State departments of game are obligated to produce large quantities of harvestable animals and cannot justify the expenditure of sufficient funds for proper management of populations of bighorn sheep. The U.S. Forest Service is well aware of its responsibility in the preservation of bighorn sheep, but democratic procedures and pressures from congressmen and administrators hinder reductions in numbers of livestock. The huge federal ranges administered by the U.S. Fish and Wildlife Service are threatened with curtailment in size and lack of funds to accomplish research and management essential to preservation of the desert bighorn sheep. Altogether, the future of bighorn sheep in the United States is not bright despite the fact that sufficient knowledge is at hand to develop an effective program of conservation. ... Perhaps the greatest need of all is for an informed general public willing to accept the responsibility of seeing to it that good conservation practices and a continued program of research are carried out."

These group of individuals meeting in 1976 grasped the vision of forming a nationwide group of sportsmen and conservationists. Such a group would ensure that plenty of attention was focused on the needs of North America's wild sheep. Further, it could help coordinate opportunities for state and federal wildlife management agencies to help ensure restoration of wild sheep populations, and raise at least a portion of the funds necessary to implement critical research. Armed with this vision, they formed the Foundation for North American Wild Sheep, officially incorporated in 1977, with the motto, "Putting Sheep on the Mountain."

Although the success of the Foundation for North American Wild Sheep is detailed in greater detail elsewhere in this book, the effect of having an organization dedicated to highlighting bighorn sheep management issues, focusing attention on bighorn sheep management concerns and providing funding as necessary to implement bighorn sheep management projects was pivotal in accelerating bighorn sheep recovery efforts. Provincial and state trapping and transplanting programs flourished. Sportsmen, outfitters and guides and conservationists generated funds specifically to benefit wild sheep restoration efforts, and those funds were often made available to state wildlife management agencies. The agencies, in their turn, used the funds to buy capture equipment and helicopter time, ensuring that bighorn sheep could be captured and transplanted more efficiently.

Many transplants were made, and new herds established, using funding provided by the Foundation. For example, state and federal agencies has conducted 43 transplants of California bighorn sheep between 1954 and 1976, a period of 22 years. That number doubled over the next 15 years (1977 to 1990), to 83 transplants, often using funds generated at least in part by FNAWS. As the principal wildlife research biologist for Idaho Fish and Game working with bighorn sheep in the early 1980s, I had an opportunity to participate in numerous releases of bighorn sheep into wild places, most of them devoid of wild sheep for past decades. This was more than just a job for most of us. It was the fulfillment of a vision that had attracted us to the wildlife profession in the first place, a vision of participating in the restoration of a national treasure. Watching healthy bighorns run off into the steep and wild canyons of southwestern Idaho, vacant of wild sheep for decades, seldom failed to bring my heart into my throat even as I reveled in the sheer grandeur of the setting.

In addition to the direct movement of bighorn sheep, FNAWS played a major and direct role in the protection or acquisition of critical bighorn sheep ranges throughout North America. Livestock owners were contacted relative to modifying their operations on public land for the benefit of bighorn sheep herds and, in some cases, private lands were purchased to reduce competition between bighorn sheep and other animals in critical bighorn sheep habitats. Funding was also provided for management of many bighorn sheep ranges. Ranges that had deteriorated due to past abuse or neglect were fertilized or burned to promote the development of high-quality forage and seeding was conducted as necessary.

Bighorn sheep disease and parasite management became (and remains) a high-level priority. I very clearly remember confronting a sudden and unexpected die-off among Rocky Mountain bighorn sheep in Hells Canyon, Idaho, in 1984 and 1985. Funding for surveys and testing of sick animals was critically needed if my agency was to respond to the emergency That emergency funding was provided by the FNAWS. The knowledge gained from this and other emergency investigations on bighorn sheep diseases and parasites was supplemented by targeted studies funded by the Foundation and conducted by some of the most capable wildlife veterinarians in the world. That knowledge proved valuable as new disease outbreaks occurred, such as the most recent die-off in Hells Canyon in 1995. Armed with knowledge from previous studies, what could have been a devastating loss of carefully restored bighorn sheep herds was contained.

These kinds of which were completed simply because there were strong and vocal advocates urging state and federal wildlife and land management agencies to consider bighorn sheep needs and helping to provide funding as necessary, provided dramatic results. Most suitable habitat for Rocky Mountain and California bighorn sheep herds in the United States and Canada is once again occupied by wild sheep, and many herds are still growing. The number of Rocky Mountain bighorns in North America nearly doubled in the 15 years between 1976 and 1991, and the number of California bighorn sheep in North America tripled in that same period (see Tables 1 and 2).

Projections for the Future of Bighorn Sheep

Time has shown that Dr. Helmut Buechner was quite correct when, back in 1960, he identified the elements required for bighorn sheep to recover. The limitations identified in 1960 are still with us. State and federal agencies still must deal with severely limited resources. Land allocation decisions are constrained by concerns about livestock grazing and various kinds of development. Many of the answers to critical biological issues, such as the transmission and impact of parasites and diseases demands additional (and increasingly expensive) research. And so it remains. The future well-being of the nation's bighorn sheep resource still lies in the willingness of conservationists to champion and fund bighorn sheep restoration as a priority.

As a wildlife manager and biologist, I can look back over four decades of wild sheep management. I know many of the people involved in getting us to this point in our history of bighorn sheep management — the people involved in the success and the failures. As I contemplate the future of bighorn sheep in North America, one fact stands out in high relief — success demands commitment. Together, the state wildlife management agencies and federal land management agencies can accomplish great things for wildlife conservation—if and when they operate in concert with strong state and national advocacy groups such as FNAWS.

Some projections for the future of bighorn sheep can be made with a high degree of confidence (which is not to say that they will occur, but rather that the necessary support is in place to help ensure that they do). Rocky Mountain bighorn sheep should flourish in Hells Canyon at the juncture of Oregon, Washington and Idaho within the next few decades. FNAWS has been instrumental in bringing together landowners, federal land managers and the three affected state wildlife management agencies in an effort to rebuild wild herds of Rocky Mountain bighorns in Hells Canyon. This vast area includes a National Recreation Area and Wilderness and huge expanses of public land habitat suitable for bighorn sheep. The habitat is extensive and appears to be in excellent condition. The herd could (and very likely will) become the greatest single herd of bighorn sheep to occur anywhere in the United States, able to rival, in sheer numbers, the great herds of Canada's continental divide. Funding for this monumental effort is critical.

California bighorn sheep populations in the United States will likely double in size one more time before the herds fully occupy all available habitat in the vast expanses of canyon lands and mountain ranges of the northern Great Basin, Sierra Nevada, central Oregon and Washington, and the canyon lands vacated by the now-extinct Audubon's bighorn in North and South Dakota, Nebraska and eastern Wyoming. As their numbers increase, sportsmen and wildlife managers will come to fully recognize the patterns of behavior and adaptation that make them unique and different from Rocky Mountain bighorns.

Studies of diseases and parasites have already brought a great deal of knowledge critical to the management of bighorn sheep. That knowledge base is increasing as new studies are completed, and more importantly, fundamental baseline data is accumulating as blood sera and tissue sample collections are built. These fundamental efforts will enable new techniques and knowledge gained from human research on viruses, prions, and other causes of diseases in humans to be applied to bighorn sheep. It appears likely that tremendous strides will be made in the understanding and prevention of massive disease outbreaks that can (and have in the past) resulted in widespread die-offs among bighorn sheep populations.

All of this can occur if sportsmen and conservationists remain united in their belief that maintaining wildlife, and particularly bighorn sheep, is an important component of natural resource management in North America. It is clear that humans have the capacity to destroy wildlife populations, either directly, or more likely (and more insidiously) by the destruction or benign neglect of critical wildlife habitat. Hunters have formed the core of the wildlife conservation movement in the past and, I believe, they must do so in the future. So I expect hunting to continue as one of the cornerstones of conservation. Hunting is both a noble tradition and a valuable tool of wildlife managers, one which has endured from the dawn of prehistory. Despite those who doubt, I see no other group willing to take on the mantle of responsibility for North American wildlife conservation.

Table 1

Published estimates of the population status of Rocky Mountain bighorn sheep, 1960-1998 (data from Toweill and Geist 1998).

	Population Estimate			
State or Province	**1960**	**1975**	**1985**	**1998**
Alberta	—	9,250	10,000	10,000
Arizona	0	0	0	500
British Columbia	—	1,250	3,000	3,000
Colorado	2,500-3,200	—	6,045	6,995
Idaho	2,400-2,800	—	2,805	1,640
Montana	1,400-1,700	2,250	4,600	4,890
Nebraska	0	0	0	57
Nevada	0	16	70	276
New Mexico	0	300	500	600
Oregon	0	55	250	490
South Dakota	22	150	165	375
Utah	Remnant	0	200	850
Washington	0	20	266	200
Wyoming	1,800-9,700	4,500	6,305	6,725
	8,100-9,700	19,450	34,206	36,598

Table 2

Published estimates of the population status of California bighorn sheep, 1960-1998 (data from Toweill and Geist 1998).

	Population Estimate					
State or Province	***1960***	***1970***	***1975***	***1985***	***1990***	***1998***
British Columbia	1,235	1,879	1,850	2,775	4,685	4,500
California	400	390	195	300	320	170
Idaho	0	90	50	530	1,205	1,460
Nevada	0	8	30	151	484	1,248
North Dakota	0	120	250	—	232	350
Oregon	24	—	210	1,007	1,805	2,890
Utah	0	0	0	0	0	30
Washington	Remnant	303	400	550	586	745
TOTAL	1,659	2,790	2,985	5,313	9,317	11,393

Dale E. Toweill, Ph.D., is a professional wildlife biologist whose career spans more than 20 years and several organizations. He is the state wildlife program coordinator for the Idaho Fish and Game Department. Dale has conducted and published scientific research on California and Rocky Mountain bighorn sheep as well as elk, mule deer, pronghorn, cougar, bobcats, coyotes, ring-tailed cats and river otters. He is co-author with Valerius Geist, Ph.D., the book, Return of Royalty: Wild Sheep of North America, *and co-editor with Jack Ward Thomas, Ph.D., the book,* Elk of North America: Ecology and Management. *Dale has been associated with FNAWS over the past 15 years as a member, scientist, technical advisor, reviewer of Grants-in-Aid proposals and speaker at many national conventions. He is also a professional member of the Boone and Crockett Club and a certified wildlife biologist with The Wildlife Society. Dale is married to Deyanne. They are the parents of a son Jim and a daughter Teale.*

This photo by GarryVan Hee took 1st place in the Live Sheep category during the 1998 convention.

DESERT SHEEP

FNAWS GOLD DESERT SHEEP AWARD WINNERS

YEAR	HUNTER	SCORE	OUTFITTER
1978	Joseph Osterbauer	169 2/8	--
1979	Bill Jeffrey	--	Mexican Govm't
1980	Paul Robey	179 7/8	Mexican Govm't
1981**	Javier Lopez del Bosque	194 3/8	--
1981	Nicholas Falzarano	--	Mexican Govm't
1982	Bruno Scherer	191 5/8	Mexican Govm't
1983	John Harris	188 2/8	Dr. Tom Boggess, III
1984	Jerry Walters	178	--
1985	Curt Carter	--	Mexican Govm't
1986*	Pat Bollman	180 5/8	Mexican Govm't
1987*	Joe Crawley	180 1/8	Mexican Govm't
1988	Everett A. Hodge	185 3/8	Floyd Lee Krank
1989*	Arthur R. Dubs	198 5/8	Floyd Lee Krank
1990	Kenneth A. Brunk	177 1/8	John V. Zenz
1991	Tony Caligiuri	179 2/8	Larry Heathington
1992	Michael J. Pace	178 6/8	George Martin
1993	Steven Besso	190 7/8	Floyd Lee Krank
1994*	Jerry Fletcher	190 2/8	L. Heathington/P. Cimellaro
1995	Ron Norman	178 1/8	Hub Grounds
1996	Lee R. Anderson	188 1/8	Floyd Lee Krank
1997*	Jerry Fletcher	193 6/8	Larry Heathington
1998*	Robert E. Petersen	188 7/8	Larry Heathington

* Award of Excellence ** Special Award

THE WORLD'S RECORD DESERT SHEEP

In 1942, Carl Scrivens acquired the skull and horns of a desert sheep head he found in an old wagon. The vaqueros told him that an Indian had shot the sheep the previous year and left the head behind. In 1951, Samuel Webb measured the ram with a final score of 205-1/8 points. This trophy is on loan to the Boone and Crockett Club's National Collection of Heads and Horns at the Buffalo Bill Historical Center in Cody, Wyoming.

Desert Bighorn Sheep

Raymond Lee

The bighorn sheep of North America descended from the wild sheep of central Asia. From the highlands of Tibet and the deserts of Mongolia, the ancestral sheep crossed the Bering land bridge to North America. Ice age glaciers eventually forced them south as far as Mexico. When the last glaciers receded, many sheep stayed on these southern mountains.

There are two types of mountain sheep in North America. The thin horns, *Ovis dalli*, are found in Alaska and western Canada. Dall's sheep and Stone's sheep are the two races of thin-horned sheep. The bighorns, *Ovis canadensis*, are found throughout western North America. The Rocky Mountain bighorn generally has longer, thicker horns and a stockier body than the desert bighorns. Desert bighorns, so called due more to their habitat than morphological characteristics, are comprised of five subspecies. Baja California has two subspecies, *O. c. weemsi* and *O. c. cremnobates* (which also extend into California). Arizona also has two subspecies, *O. c. nelsoni* and O. c. *mexicana*. The bighorn sheep in Sonora are of the *mexicana* subspecies. California, Colorado, Nevada and Utah have populations of the *nelsoni* subspecies. California also supports a small population of *californiana* in the Sierra Nevadas. New Mexico and Texas have received transplants of both *nelsoni* and *mexicana.*

At their peak, North American bighorn sheep numbers were estimated at nearly two million, ranging throughout most western mountain ranges and extending eastward as far as North Dakota, Nebraska and Texas. Populations were so great that as late as 1885, bighorn sheep were still being commercially harvested. At present, occupying only 4 percent of their historic range, the desert bighorn sheep population has fallen to approximately 25,000. Of these, about 19,000 are found in the United States; with Arizona supporting 6,500, Nevada 5,500, California 3,500, Utah 2,500, Colorado 500, Texas 300 and New Mexico 200. The remainder are found in Baja California (3,500) and Sonora (2,500), Mexico. The decline occurred primarily between 1850 and 1900. Causes for this decline were: persistent and unregulated hunting; competition with livestock for food and water; and, most importantly, livestock-associated parasites and diseases.

The diseases most responsible for impacting bighorn sheep populations were scabies, chronic sinusitis and pneumonia. The causative agent of scabies is mites. The mite that has been widely blamed for causing die-offs in bighorn sheep is *Psoroptes ovis*, which is the species that is common to domestic sheep. Scabies was one of the first diseases reported in bighorn sheep. This is because the effects of the disease are easily observed and a large number of animals were affected. The presence of scabies was so severe that in some areas local residents reported that it was difficult to find bighorn sheep that were fit to eat.

Chronic sinusitis is often the result of secondary infection due to bot fly larvae dying after becoming trapped in the sinuses. Although domestic sheep have been affected by bot fly-related infections, nasal bots are generally considered only minor pests. This is in contrast with bighorn sheep where bot-related mortality is often high. This can be explained by the fact that bighorn sheep have extensive sinuses when compared with domestic sheep, especially within the horn cores, which provide many areas for the disease to become established.

Although the pneumonia complex was among the first diseases described that led to large-scale die-offs of bighorn sheep, it is one of the least understood. One of the difficulties in understanding the role of pneumonia in bighorn sheep populations is determining the factor

Arizona Desert Bighorn Sheep Society and the Minnesota-Wisconsin Chapter of FNAWS work on a joint water development project.
Photograph by Gary Schafer.

A March 1992 water development project in the Kofa Mountains, Arizona, is one activity that has enhanced the growth of desert bighorn sheep populations in the state.
Photograph by Gary Schafer.

that causes the pneumonia. A variety of bacteria have been isolated from pneumonic bighorn sheep. Stress, in combination with bacterial and viral agents, is important in "triggering" the onset of pneumonia in bighorn sheep.

The bighorn sheep's susceptibility to these diseases may be the result of genetic isolation from Asiatic sheep for more than 10,000 years. With the introduction of domestic livestock in the mid to late 1800s, bighorn sheep were subjected to pathogens for which they had no inherent immune defense.

Bighorn sheep are members of the Bovidae family and are in the sub-family Ovidae, or the sheep and goat family. Bighorn sheep differ from goats by having concave foreheads, glands near their eyes and on their hind feet, ears longer than their tails and horns curving laterally. The eye and interdigital glands of the bighorn secrete a waxy, odorous substance used in communication and location. The bighorn's hooves are adapted for their rocky surroundings. The outer edge of the hoof is hard and sharp, the inner is protruding and spongy to provide traction.

Desert bighorn sheep show considerable differentiation between the sexes. Adult males typically weigh between 160 to 200 pounds, while adult females range from 75 to 130 pounds. Ram horns may measure 30 to 40 inches along the outside curl with a basal circumference of 13 to 15 inches. Ewe horns are generally 10 to 13 inches long with a circumference of 5 to 6 inches.

The horn sheaths are of epidermal origin and grow over conical bony cores to form permanent structures. These cores consist of a vascular network that allows the horn to perform thermoregulatory functions. The horn core is honeycombed with chambers, or sinuses, which reduce the weight of the skull. The skull and horns can weigh up to 45 pounds.

Horn growth is rapid for the first four years and then slows considerably. Annual rings form each year during the rut with the fourth year being the deepest ring. Horn growth is stimulated by sexual hormones and annual ring formation is due either to the presence of these hormones or to the physiological stress during the breeding season. Brooming or breakage, occurs during the head-on clashes the rams engage in during the rut. Brooming also occurs when the sheep use their horns as tools to open barrel and saguaro cacti. The size of the horns determines the ram's social dominance position.

Newborn bighorn lambs weigh 8 to 10 pounds. The young are active within minutes after birth and in a couple of days can gallop and follow their mothers nearly anywhere. The young have dark eyes and fuzzy, dark-grey hair. As they mature, their eyes take on the characteristic golden or amber color. After several months, they take on adult coloration; dark brown in their northern range and pale buff in the southern. Color accents are a white muzzle, rump patch, eye rings and edging on the rear legs, with a black tail. Bighorn lambs are weaned at five to six months and adult dentition is completed by age four.

Physiological and behavioral sexual maturity varies. While rams as young as six months may be capable of breeding, they don't, due to the dominance of older rams. In nature, ewes usually aren't bred until 2-1/2 years and rams don't breed until 3-1/2 years. Desert bighorns have a very long breeding period. The *nelsoni's* breeding season extends from early July through October and the *mexicana's* extends from early June through September. Mock battles are fought between the rams until the breeding season occurs. At this time pitched battles take place. Rams collide with such force that the clashes can be heard over a mile away. Desert bighorn sheep don't form harems. The males will travel from ewe group to ewe group attempting to breed with as many ewes as possible. Peak rutting activity occurs in August. The gestation period is 179 days, with *nelsoni* producing young from mid-January to April and *mexicana* from January to March.

Usually one, rarely two lambs are born. Young rams stay with their mothers until two years of age. They then leave the nursery herds and join in the bachelor herds. Except during the breeding season, and sometimes during the spring when early vegetation sprouts, bighorn adults separate according to sex.

Bighorn sheep have a life expectancy of 12 to 14 years, but have been recorded as old as 18. Mortality is initially very high, with approximately 50 percent of the lambs dying within six months. Between ages one and nine, mortality is very low. Beyond age nine, the mortality rate increases more than 10-fold. One cause of this marked increase is tooth decay. Wear begins to affect the teeth at 8 to 10 years, making it difficult to feed properly.

Bighorn sheep ranges are limited by the availability of food and water. While bighorn sheep have been observed to go up to six months without drinking, deriving metabolic water from their food, they will normally water every two to three days. Prodigious drinkers, bighorn sheep have been observed to drink at a rate in excess of 2-1/2 gallons per minute. Bighorn sheep have functional sweat glands, but they secrete only 5 percent as much as humans. Therefore, other methods of heat dissipation are used. Bighorn sheep have developed enlarged water retention organs, can withstand a high degree of dehydration (up to 30 percent of their body weight, which is more than a camel can endure), and their body temperature can reach 107 degrees.

Competition for water is a problem with feral and domestic animals. Cattle will approach bighorn sheep at waters, apparently scaring them away. Bighorns have been unable or unwilling to penetrate salt cedar barriers along rivers and are forced to compete with burros for water only short distances from the rivers.

Each state with desert bighorn sheep has developed partnerships with conservation organizations to help in the construction of wildlife waters. In Arizona, the Game and Fish Department, in cooperation with the Bureau of Land Management and the Arizona Desert Bighorn Sheep Society, and more recently with Desert Wildlife Unlimited, develops up to 10 bighorn sheep waters annually. These water projects vary from simple tinaja modifications to extensive artificial water collection and storage systems. The goal is to develop the most cost-

Bighorn sheep bunch together as they react to the fall of a drop-net.
Photograph by Dave Daughtry.

effective, environmentally sensitive, maintenance- free waters possible. In Nevada, 113 bighorn sheep waters have been developed in this fashion.

In captivity, bighorn sheep will eat 3 to 4 pounds of dry alfalfa a day. In the wild, grasses are important to the bighorn. Bighorn also feed heavily on jojoba. Pincushion, barrel and saguaro cacti provide moisture. Preferred plant species vary with habitat quality, locality and species availability; however, brittle bush, catclaw, desert holly, fairy duster, globe mallow, mormon tea, palo verde, primrose, ironwood and mesquite beans are all utilized as food sources.

Bighorn sheep develop relatively fixed feeding patterns. Young follow their mothers, who follow their mothers; the oldest female with the most descendants leads the herd. Migration and escape routes, salt licks, watering and feeding places are all acquired knowledge. Due to their lack of pioneering instincts, bighorn sheep may be precluded from historic use areas due to disruption of traditional patterns.

Bighorn sheep are adapted to live under stable conditions at the carrying capacity of the habitat. Populations are dependent upon their ability to retain their annual production on the range of their parents and to exploit stable plant communities. This differs from deer, for example, which are adapted to exploit short-lived opportunities for population expansion. Bighorn sheep have relatively low reproductive rates and don't respond well to changes in their habitat; therefore, they can't be harvested at rates appropriate for deer.

Compared to other ungulates, desert bighorn sheep are longer lived, exhibit more density-dependent mortality, and have lower reproductive potentials, with populations stabilizing near carrying capacity. The major factors identified as regulating bighorn sheep populations are density and forage. Densities of bighorn sheep usually vary from less than 1 animal per section in small isolated populations to more than two per section in large healthy herds within preferred habitat. The Black Mountains contain the greatest densities found in Arizona estimated at eight bighorn sheep per section.

In years of good forage conditions and low densities, populations can increase rapidly. This has occurred many times in the western United States when small herds of bighorn sheep have been transplanted to pristine vacant habitats. For example, the Aravaipa Canyon bighorn sheep population is the result of a release of 22 bighorn sheep. Following their release, the herd grew rapidly, and in less than 10 years was estimated at more than 100 individuals. The population doubled every 4 to 5 years.

Mortality rates are higher for rams than ewes. This is because rams occupy significantly larger home ranges than ewes, including habitat without typical escape terrain, leading to increased mortality due to predation, accidents and other factors. In addition, rams are in poor condition following the rut and high mortality during this period has been observed.

Most predation on bighorn sheep is opportunistic and doesn't affect population numbers. Golden eagles, bobcats and coyotes have all been implicated as predators. Mountain lions also prey on bighorn sheep. On desert ranges, however, bighorn sheep and deer are usually too sparse to support a predator of this size. In areas where the deer population is large enough to support a lion population, they can present serious problems to the bighorn sheep population.

Bighorn sheep, particularly the young, have been described as having the best tasting meat of any wild game. It was the preferred game meat for miners, cattlemen, railroad workers and surveyors. To keep from becoming a roast, however, the bighorn was given excellent eyesight, often compared to 8X binoculars, capable of seeing a person 1-1/2 miles away. In desert areas, scent is rapidly dissipated by the heat, hence the evolutionary emphasis on vision; yet, they can smell a human more than 100 yards away. Once danger is detected, the bighorn is capable of speeds in excess of 30 miles per hour, often over terrain most people could not even walk on.

In the Southwest, desert bighorn sheep are found in nearly all of the relatively dry, mountainous areas. They range from northwestern Sonora, particularly in those ranges along the Sea of Cortez north of Kino Bay, and north of La Paz, Baja Sur, to the middle of Nevada and Utah.

Desert bighorn sheep are distributed throughout most of the mountainous areas of western Arizona. Transplanted populations have been re-established in some central and southeastern mountain ranges. The *nelsoni* subspecies occurs primarily in the northwestern portion of Arizona. The *mexicana* subspecies occurs primarily in the southwestern portion of the state. California' *cremnobates* subspecies occurs in the southwestern ranges in San Diego and Imperial counties, and the *nelsoni* subspecies occurs in the Transverse Ranges and the mountains in the Mojave, Sonoran and Great Basin deserts. Desert bighorn sheep occur in the southern and central mountain ranges in both Utah and Nevada. Small populations of desert bighorn sheep also occur in a few ranges in west Texas, a few ranges in southwestern New Mexico and a few ranges in southwestern Colorado.

Merely providing the distribution for desert bighorn sheep tells little about the densities at which they occur. The number of animals observed per unit of survey effort represents a method for comparing bighorn sheep densities. Observation rates vary considerably. In southern Sonora, the mean observation rate has been 40 animals per helicopter hour. Observation rates in both Sierra Viejo and Sierra Kun-Kaak, both in Sonora, exceeded 90 bighorn sheep per survey hour. By comparison, in Arizona, observation rates traditionally average 10 bighorn sheep per hour.

Transplanting bighorn sheep from areas of high density into historic habitat is a management activity used to enhance bighorn sheep populations. Many states trade other big-game species for bighorn sheep. Arizona has been involved in an aggressive bighorn sheep transplant program since 1955. Arizona has transplanted 1,279 bighorn sheep, with 195 being sent to other states. Nevada has been even more active; since 1968, Nevada has transplanted 1,362 animals, with 256 being sent to other states. Starting in 1973, Utah has been involved in the transplant of 528 animals. The Foundation for North American Wild Sheep has been particu-

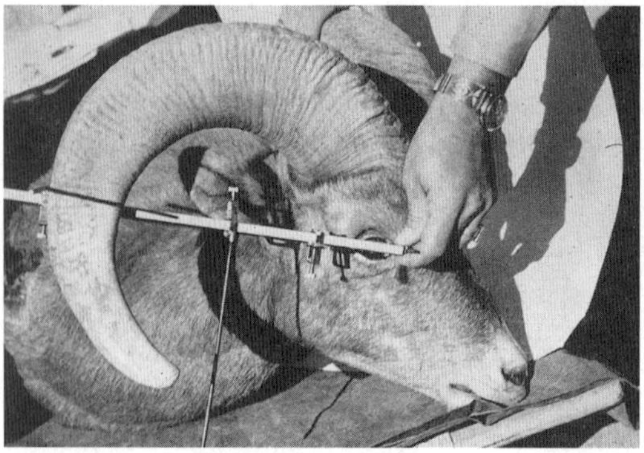

A "Russo ratchet" used in the 1950s and 1960s in Arizona to determine whether a ram was 3/4-curl, and therefore legal. In 1985, Arizona went to an any-ram-rule, precluding the use of this device.
Photograph courtesy of Raymond Lee.

A bighorn sheep release took place July 31, 1983, in the Grand Wash Cliffs, Arizona. Here one of the released animals takes a jump from a truck.
Photograph by Dave Daughtry.

larly effective in supporting transplant efforts. In both Arizona and Nevada, approximately one-third of the hunting permits available are for populations that have been transplanted within the past 20 years.

Bighorn sheep have always been one of the most sought after hunting trophies. From prehistoric man to the modern hunter, the "King of the Mountain" has proven to be an elusive, even mystical, prey. Bighorn sheep, like most North American ungulates, have been hunted since their first interaction with man. Aboriginal hunters avidly pursued bighorn sheep. Bighorn sheep bones have been found in archaeological sites dated more than 8,000 years old. After the invention of the bow and arrow, ungulate harvest, determined from bones found in archaeological sites, increased nearly three-fold. Prehistoric men set up rock walls and drove bighorn sheep past concealed hunters. Several groups even built "dummy" hunters out of stone to help drive bighorn sheep into their traps. Bighorn sheep played an important role in Indian mythology. More bighorn sheep-related petroglyphs are found in the West than any other animal. While early hunters were primarily seeking food, bighorn sheep, especially the desert varieties, have developed into perhaps the most coveted trophy animal of all time.

Due to the marked declines in bighorn sheep populations near the end of the 19th century, most states closed their seasons on bighorn sheep. Arizona was one of the states that limited the take of bighorn sheep. The first game laws enacted by the Territorial government in 1887 provided punishment for the unlawful taking of bighorn sheep from February to October. In 1897, an Amendment of the Penal Code was made to close the bighorn sheep season until such time as judged appropriate by the legislature.

Many western states have once again legalized bighorn sheep hunting. California closed their season in 1873 and reopened it in 1987. Texas closed their season in 1903 and only reopened it in 1988, though the initial season was not available to the general hunting public. Colorado also reopened its hunting season in 1988.

Arizona, Nevada and New Mexico all began hunting bighorn sheep in the early 1950s.

A group of sheep in Aravaipa Canyon, Arizona, was photographed on April 21, 1971, including "Old Grand Dad." This ram, center, died of natural causes and its horns scored more than 191-6/8 unofficial B&C points. The ram is on display as an European mount in the Phoenix headquarters of the Arizona Game and Fish Department.
Photograph by Raymond Lee

A ram in the Black Mountains of Arizona.
Photograph by Dave Daughtry.

Since Arizona's first season in 1953, 2,254 permits have been authorized. A total of 2,200 hunters have harvested 1,922 rams for a hunter success of 59 percent. Hunter success rates have varied considerably over the years. From 1955 to 1965, the hunter success averaged 33 percent. Since 1980, the hunter success has averaged over 91 percent.

In all the United States, fewer than 5,000 desert bighorn rams have been legally harvested from 1952 to 1997. To put this into perspective, desert bighorn sheep hunting opportunities are only found in Arizona, Nevada (with approximately 1,900 total rams having been harvested since modern seasons were opened), Utah (approximately 270), California (85), New Mexico (5), and Texas (5).

For the 1996 season in Arizona, 6,843 sportsmen applied for 108 permits, a 1 in 68 chance of obtaining a permit. Hunters harvested 100 rams. In 1996, hunters in Nevada harvested 102 rams and those in Utah took 29. More limited hunts took place in the other states with a harvest of 14 rams in California, five in Colorado, two in New Mexico, and 1 in Texas. The Mexican Government issued seven bighorn sheep hunting permits for Sonora in 1995. These permits sold for $45,000 and six sheep were harvested. There were 13 permits authorized for the 1996-1997 season. In Baja Sur, four permits were issued in 1996-1997, with three rams being harvested. Permits in Baja Sur were reduced to three in 1997-1998. These tags are available only through auction or raffle. One permit was issued for Baja California for the 1998 season. The Foundation for North American Wild Sheep has been instrumental in supporting bighorn sheep management programs in Mexico and in helping to remove the ban on bighorn sheep hunting in Baja California.

Being considered a trophy animal has led to management principles uncommon for more utilitarian species. While most state game and fish agencies have mandates to provide maximum recreational opportunities for common big-game species, bighorn sheep are usually hunted in a more conservative fashion to maintain trophy quality.

A common concern in many states has been the harvesting of younger age rams. Each state has attempted to solve this problem by different methods. In Nevada, for example, in 1965 a regulation specified that a legal ram must be seven years old or score 144 Boone and

Bolting from the noise and wind of an overhead helicopter, this desert bighorn ram races over desert landscapes in the Las Virgenes area of Baja Sur.
Photograph by Raymond Lee.

Crockett points. Hunters were also required to carry a spotting scope and were given an indoctrination session to familiarize them with the technique of judging the size and age of rams. This indoctrination session soon became mandatory. The new regulation and the indoctrination session appeared to help reduce the young ram harvest, but not to the extent desired.

In Arizona, a similar path was taken. In 1966, the first bighorn sheep hunters' clinic was held. This was an attempt to familiarize hunters with successful hunting methods and give them the ability to judge trophies in the field. The clinic also attempted to instill the desire to harvest older age rams. This clinic soon led to the formation of the Arizona Desert Bighorn Sheep Society; an organization dedicated to the conservation of bighorn sheep.

The bighorn sheep hunters' clinic also provided aspiring hunters with information to help make their hunts more successful. The clinic not only provided a forum for exchange of information, in many cases it enabled a client-guide relationship to be formed. Perhaps the greatest change seen in bighorn sheep hunting has been in human behavior. This, perhaps more than anything else, has led to the marked increases in hunter success rates seen in recent years. In the early 1950s, hunters commonly took to the field, relatively ill-prepared and often by themselves. Today, well-equipped camps are the norm. Frequently, these camps are manned by past year's permit holders for the unit, as well as friends and just plain sheep aficionados, who met at the sheep clinic and want to experience the hunt, if only vicariously. This results in many eyes scanning a mountain range to produce the best possible ram for the current permit holder.

Due to the high demand for permits (in some cases exceeding 400 applicants for each permit), some states place limitations on the number of bighorn sheep that can be harvested by an individual hunter. In Arizona, hunters are allowed to take only 1 desert bighorn sheep and 1 Rocky Mountain bighorn sheep in their lifetime. Other states limit hunters to 1 bighorn sheep every 10 years. Most states have developed operational plans to help manage their bighorn sheep populations. Arizona's plans call for the hunting permit allocation to result in the harvest of either 6 percent of the total number of rams estimated in the population, or 12 percent of the estimated number of older age rams, whichever number is lower. In California, permits are authorized for up to 15% of the estimated number of rams 2 years old or older. Other states have similar harvest strategies. It is doubtful that bighorn sheep populations will ever meet the demand for hunting opportunities; however, sportsmen will continue to have an important role to play in expanding bighorn sheep management programs.

The world's largest desert bighorn sheep, found in Baja California, scored 205-1/8 Boone and Crockett points. Arizona produced a pick-up skull scoring 201-3/8 points, and produced the largest taken by a sport hunter at 197-1/8 points. The various other sport harvest records for each state are as follows: Baja California 192-5/8; Sonora 187-3/8; Nevada 184-6/8; New Mexico 178-7/8; California 173-4/8; and Texas 168-2/8.

The dollars generated by the sale of bighorn sheep permit tags annually could not begin to support even a small portion of Arizona's bighorn sheep management program. In 1996, the 6,843 sportsmen paid the Arizona Game and Fish Department $3 each for an opportunity to enter the computer drawing for bighorn sheep permit tags, a total of $20,529. Of the 106 tags issued, 10 were to non-residents. This produced $7,500 in tag sales ($750 each) from the non-residents and $14,400 from the 96 resident tags ($150 each) for a total $42,429 of revenue to the department. At the current contract price for helicopter surveys, fewer than 80 hours of flight time, not including personnel costs, could be purchased; less than needed to survey the bighorn sheep distribution in only southwestern Arizona.

A relatively recent program that was developed to generate revenue to support wildlife conservation programs is the special permit or governor's tag. Originally started in Wyoming in 1980, it is now used by almost every Western state to varying degrees. Using Arizona as an example, two bighorn sheep permits are issued annually. For the past 12 years, the Game and Fish Department and the Arizona Desert Bighorn Sheep Society have entered into an agreement whereby the Society auctions 1 permit (at the Foundation for North American Wild Sheep convention) and raffles another to raise funds earmarked specifically for bighorn sheep management projects. Since the program started in 1984, $2,285,860 have been raised from the 24 permits ($1,328,000 from the 12 auction tags and $957,860 from the 12 raffle tags). Special tags have been issued for bighorn sheep in virtually every state, Mexico, Canada and various Indian Reservations. Much of the success of the various bighorn sheep management programs is dependent upon the funds derived from these special permits.

The primary limiting factor for most desert bighorn sheep populations is loss of and fragmentation of habitat. This can occur through direct losses to urban expansion into foothill areas, or indirectly through the expansion of human-related recreational activities. Another human-related factor that impacts bighorn sheep is the removal of water from bighorn sheep habitat. The majority of natural springs have been modified to provide water for humans, their domestic animals or for mining operations. By providing protection from humans, the designation of large areas as "wilderness" may prove beneficial to bighorn sheep — as long as these areas do not serve as attractants for large numbers of outdoor enthusiasts and wildlife management activities are allowed to continue.

There are a number of factors limiting the distribution and abundance of desert bighorn sheep in Mexico. Perhaps the most important of these are related to economic difficulties. Use of bighorn sheep to meet basic food needs still occurs in some areas. As the economic value of these animals becomes more widely known and the proceeds more widely distributed, this type of loss will be reduced. Private landowners, who fully realize the value of bighorn sheep, have developed bighorn sheep breeding facilities on their land. Unfortunately, to ensure that the bighorn sheep stay on their holdings, many landowners have constructed fences. This precludes the bighorn sheep from using seasonally important habitat, reduces genetic exchange, and increases susceptibility to catastrophic events, particularly those disease related.

The future for desert bighorn sheep looks relatively bright. Organizations such as the Foundation for North American Wild Sheep represent a large number of people who have been very supportive of programs to enhance bighorn sheep populations and their habitat, and these people have been willing to put both their money and their sweat into seeing that these programs are successful. That combination of money and sweat has put a lot of bighorn sheep back onto the mountains; it will take both to keep them there.

Raymond Lee, known as Ray to his friends, is the big-game management supervisor for the Arizona Game and Fish Department. He has been instrumental in transplanting more than 1,000 bighorn sheep while in that position. He edited with Bill Hook the publication, Borrego: The Fall and Rise of Desert Bighorn Sheep in Arizona, *and received the Honor Plaque from the Desert Bighorn Council for a publication titled,* The Bighorn Sheep in Arizona. *FNAWS awarded Ray the Presidents Award and the Statesman of the Year Award for his continuing work on desert bighorn in Mexico. He also received the Health of the Land Award from the Bureau of Land Management for his work on bighorn sheep and burro interactions and was named the Wildlife Professional of the Year by the Arizona Chapter of The Wildlife Society. He is the author of more than 50 publications and current editor of the professional publication,* Desert Bighorn Council Transactions. *Ray holds a B.S. in biology and a M.S. in zoology from Arizona State University. He is a certified wildlife biologist and certified fisheries biologist. Ray's professional work has taken him to all seven continents. He is a fellow of the New York City-based Explorer's Club.*

This photo by Craig Richardson took 2nd place in the Live Sheep category during the 1998 convention.

STONE'S SHEEP

FNAWS GOLD STONE'S SHEEP AWARD WINNERS

YEAR	HUNTER	SCORE	OUTFITTER
1978	Dan Gildner	--	--
1979	William Boggs	--	Clay Martin
1980	Henry Baddley	--	Stone Mountain Safaris
1981	David C. Coleman	176 4/8	Gary Moore
1982	Romeo LeDuc	181 5/8	--
1983	Dick Sullivan	178 5/8	Dave Weins
1984	Larry Gearhart	174	Garry Vince
1985	Mike Mell	--	Kirby Funnel
1986	Gus Brown	179 1/8	Garry Vince
1987	Johnny Chilton, Jr.	176 3/8	Garry Vince
1988	Gary Bogner	183 4/8	Dave Wiens
1989	Steve Best	176	Glen Drinkall
1990	John D. Chalk III	176 6/8	Dave Wiens
1991	Phillip Gilligan	178	Dave Wiens
1992	Larry C. Fisher	177 2/8	Self Guided
1993	Steven Adams	174	Gary Moore
1994	Kevin H. Olmstead	178 5/8	Prophet Muskwas Ent. Ltd.
1995	Ken Clarke	175 2/8	Dave Wiens
1995	Terry Ridley	175 2/8	None
1996	Kenneth D. Eason	176	Kevin Olmstead
1997	Dennis Lord	183 5/8	Garry Vince
1998	Robert Joseph	179	Upper Stikine River Adven.

THE WORLD'S RECORD STONE'S SHEEP

Taken by L.S. Chadwick in 1936, this ram scores 196-6/8 points and is considered by many the finest trophy ever taken. The Greatest Spread and Tip to Tip Spread are both 31 inches. The length of the longest horn is 51-5/8 inches. This trophy is currently a part of the Boone and Crockett Club's National Collection of Heads and Horns at the Buffalo Bill Historical Center in Cody, Wyoming.

Stone's Sheep

VALERIUS GEIST, PH.D.

Stone's sheep are dark thinhorn sheep (*Ovis [Pachyceros] dalli Nelson*, 1884), occupying the southern portions of that species range. About 13,000 live in northern British Columbia and 4,500 in the southcentral Yukon Territory. They get their name not from rocks or stones, but from the hunter-explorer Andrew J. Stone of Missoula, Montana, who brought back the first specimen from the headwaters of the Stikine River in 1896. In recognition of his contribution, J.A. Allan of the American Museum of Natural History named this new sheep *Ovis stonei* in 1897. Thus the correct spelling of its popular name is not Stone sheep, but Stone's sheep, honoring an explorer, not rocks.

The coat color and patterns may vary greatly between Stone's sheep in the same band, let alone between regions. In the South and West the sheep tend to be very dark. The head is always lighter than the body, the rump patch is white, large and extends well above the root of the tail. In this, Stone's sheep differ from Siberian snow-sheep whose rump patch reaches only the root of the tail. A dark stripe runs dorsally down the rump patch to the tail, and there are broad white stripes running from the rump patch down the rear margins of the hind legs; there are comparable stripes on the front legs. These stripes are wider than in bighorns. Otherwise, the coat pattern and color variability defies a succinct summary. Some sheep have a white belly; some have a dark belly band or virtually no white; some have snow-sheep like faces; some have a light neck; some have white faces etc. Weight, horn size and shape also vary widely. While American sheep are very similar in biology, Stone's sheep differ from bighorns by being less "juvenile" in their behavior, emphasizing ceremony over overt aggression and sex.

Subsequent to Allen, Charles Sheldon published in 1911 about the distribution of Stone's sheep in *The Wilderness of the Upper Yukon*. Carl Rungius, destined to become a legendary wildlife artist, accompanied Sheldon and illustrated the book, making it timeless. Sheldon showed a change in coloration from South to North, with more sheep becoming lighter. His was no statistical statement, but a recognition that there are dark and light sheep in all populations, but that their respective frequencies shifts south to north. In my Cassiar study area about 3 percent of the sheep were light. None were pure white, as even the lightest had a black tail and silvery hair. The last evidence of Stone's sheep integrating with Dall's sheep in the Yukon are black hairs in the tails, back, around they eyes and the bridge of the nose of otherwise white sheep in the western Yukon, central Alaska, and southwestern Northwest Territories. As Stone's sheep become lighter, they tend to isolate pigmentation in a saddle. This variant was formerly declared a separate species, *Ovis fannini*, by William T. Hornaday. Today we consider "Fannins" merely as light Stone's sheep.

Sheldon put Stone's sheep on the map, but while they remained largely unknown to science, they were a trophy that generated legends. In 1936 L.S. Chadwick wounded and Roy Hargrave hunted down the magnificent Stone's ram that has the longest horns not only of all American sheep, but of any sheep in the subgenus *Pachyceros*. Both horns measure more than 50 inches. The Chadwick ram is by far the most famous trophy ever taken in North America. The guides on this hunt along the Prophet River, Frank Golata and Curly Cochrane, became legends on their own. Jack O'Connor wrote vividly of his Stone's sheep hunts with Golata in *Sheep and Sheep Hunting*. My introduction to American wild sheep came in a book detailing a

Range of the Stone's Sheep

Photo by Doug Bell.

Stone's sheep differ from Siberian snow-sheep whose rump patch reaches only the root of the tail. A dark stripe runs dorsally down the rump patch to the tail, and there are broad white stripes running from the rump patch down the rear margins of the hind legs. Here a herd of Stone's sheep crunch through piles of snow in British Columbia.
Photograph by Valerius Geist.

hunting trip by German Count Lothar Graf Hoensbroech who hunted in 1939 on the Prophet River, guided by Curly Cochrane, and by another legend, Reinhold Eben-Ebenau of Slave Lake, Alberta. Upon his return Hoensbroech found the Royal Canadian Mounted Police waiting for him. World War II had broken out and Hoensbroech went into internment in Calgary.

After the war, this great hunter, naturalist and philosopher wrote about his experiences. So did Eben-Ebenau who contributed seminally to an understanding of moose biology. Unfortunately, both books were published only in German. Associated with Stone's sheep is another legend, Skook Davidson. A highly decorated soldier, he led the Canadian Geographic Surveys for decades and guided such celebrities as Prince Abdorreza Pahlavi and Herb Klein. I spent three weeks of the Stone's sheep rut of 1968 living with Skook in his cabin at Terminus Mountain by the Ketchika River. My account in *Mountain Sheep and Men* suggests that Jack O'Connor's description of him as a picturesque old Scot is an understatement.

The first extensive field study of Stone's sheep was my own. Politically it was part of a ploy dreamed up by Tommy Walker, who ran a first-rate operation out of Cold Fish Lake, and by Ian McTaggart-Cowan, then the head of the Zoology Department of the University of British Columbia. Cowan rationalized mountain sheep taxonomy in 1940, a work that remains a standard today. My study was to draw attention to the Spazisi as an area for conservation, which Tommy hoped would be under university control, with himself as manager, controlling access.

With his help we built a cabin in summer of 1961, just off little Gladys Lake, below mighty Mount Will. Here Tommy did not hunt, save for unusual circumstances such as guiding Fred Bear who shot his Stone's ram here with bow and arrow. On a fly-by one February, Tommy had seen a small band of rams. That was all we had to go on. By good luck the cabin was located in the hub of several small winter ranges, inhabited by about 80 Stone's sheep in 1961, and 86 in 1962. I lived nearly two years in that cabin, isolated most of the time as the radio broke down. My closest neighbors were the families of Alec and Charlie Abu who lived about 40 miles away at Hyland Post and looked after Tommy's horses in winter. They were of the Talthan tribe. Their ancestors had netted and clubbed sheep and goats at Gladys Lake. I studied almost two yearly cycles of Stone's sheep before moving on to bighorns and Dall's sheep in subsequent years.

My study was one of the first on the behavior of free-living large mammals. Ultimately, it led to me being so completely accepted by sheep, that ewes attempted bodily to keep me in their bands, while some rams attacked me. Of all that I was innocent in 1961. The daily observations of Stone's sheep and about 35 mountain goats soon generated insights. These were to lay the scientific foundation for much of today's sheep management and point to the future. They were later translated by my student, Beth MacCallum, into physical design of mountain sheep habitat that was shaped from exhausted coal mines in the foothills of Alberta. The insights generated in the field were tested in the design of custom sheep habitat. Sheep loved it. Strange bighorns from Jasper National Park came and refused to leave. The former mine developed into an oasis of bio-diversity, growing not only ever larger bighorns and magnificent rams, but attracting other wildlife, including many species of nesting birds.

Stone's sheep taught us how mountain sheep communicate, and revealed the function of horns. These are more than weapons, sledge-hammer and karate-chop combined; they are also display organs that revealed vitally important information about the biological success of each ram. Horns not only gives clues to ewes, which prefer large-horned over small-horned rams, but also to other rams. These judge the combat potential of a stranger by its horn size.

Stone's sheep are dark thinhorn sheep occupying the southern portions of that species' range. About 13,000 live in northern British Columbia and 4,500 in the southcentral Yukon Territory.
Photograph by Valerius Geist.

Young rams follow the largest-horned ram to discover the locations where he feeds so well as to grow large horns, or lives so securely as to grow so old and long-horned. Mountain sheep, rather than disperse as young adults in search of a place to live, follow a good prospect to very distant habitat patches. Much sheep habitat is too patchy, dispersed and distant, and thus too risky to find by bold exploration. It's cheaper and safer to follow an adult and spy out its places of residence. Adults had up to six seasonal home ranges to which they returned year after year, with mathematical precision. However, young rams chose not any ram, but the one with the largest horns. Such a ram has living proof on his head of its greater successes in finding superior forage and/or greater security from predators. Years later we learned that on superior ranges, where sheep grew exceptionally big, they may also disperse and search for vacant habitat patches individually. Normally, sheep range is too patchy and the job of finding such too costly and risky.

And this made good sense, as sheep evolved closely with Ice Age glaciers. They dispersed and multiplied during glaciations when forests dwindled into remnants, and the land beyond the ice was covered with dusty, fertile, short-bladed grasslands. When the climate reversed and the glaciers withdrew, the timber spread over the land leaving but scattered patches of grassland at high elevations and southern exposures. Now the sheep's tradition of passing on home ranges to youngsters paid off handsomely. In their occupation of the north, sheep had to master glacials lasting 75,000 years and interglacials lasting 25,000 years. Sheep readily dispersed when the land is open. This was demonstrated by remnant Stone's sheep herds in the Yukon after market-hunting depletion in the Gold Rush days, provided they lived on long, open ridges. Populations on choppy, timbered mountains with small patches of habitat usually failed to recolonize lost space. The best way to return sheep to once-occupied ranges is to reintroduce them, rather than wait till nature allows them to disperse - during the next ice age.

Stone's sheep led us to the logic behind the great variation in horn growth within populations. A ram's fate was in his horns. Those that grew rapidly attained big body and horn size sooner, entered earlier into rutting and died young. Those that grew poor horns delayed entry into active breeding or hovered about opportunistically, trying to catch a superior ram off guard and breed his female. Such rams lived to old age. The less rams bred, the longer they lived. I clued in on this after blowing snow fooled me about a Stone's ram size. I stalked two large rams as best as visibility allowed. One raised his head over a snowdrift. It nearly

B.C. rams and ewes.
Photos by the Fort St. John Ministry of Environment.

took my breath away, as I viewed him in the rifle scope. What a fine, fully curled, massive head! He never knew what hit him and I rushed in only to find a 14-year-old dwarf, with horns measuring 32.5x12.25 inches. That matched the highest age reported till then for any American sheep. That tiny old ram would not go out of my mind — nor the fine young rams I had see about Gladys Lake that promised to grow record-size horns but vanished. I was later able to show that rams who died before average age had much better horn growth than those dying after average age.

That sheep differed greatly in size was not new. New was the recognition that they grew just like the high-plane or low-plane domestic sheep studied by animal science. That is, the differences between large and small were exactly of the type that one can induce willy-nilly in domestic livestock by changing the food intake of individuals during their growth to adulthood. The size of sheep in nature thus said nothing about their genetics to grow large or small. Small sheep should give rise to big sheep — food resources permitting. There were real differences in the behavior and reproduction between large- and small-bodied sheep. Large sheep, vigorous and virile, acted like colonizers; small sheep, listless and often barren, merely hung on. They were dispersal and maintenance phenotypes, respectively. Later I showed the same applied to humans, with the phenotypically larger enjoying, among other, better health. Consequently, the biggest and most important book I wrote dealt not with sheep, but with the biology of health. The plasticity of mammals to nutrition was later crucial to debunking bad taxonomy based on comparative measurements of populations that did not distinguish between genetic and environmental effects. It led to investigations, public hearings, international controversies and to the courts. Yet it all began with Stone's sheep.

Embarrassing to me was to discover why young rams are allowed to follow a large-horned male, and thus discover its secret places for feeding and hiding. You see, it makes no sense for an old ram to allow young rams to follow, as the followers, potentially, could eat him out of house and home. A big ram, therefore, should discourage small rams from following. He does not. Why not? I showed that young rams going with an old one act towards him qualitatively and quantitatively like a ewe in heat. Young rams use pre-sexual foreplay on the old ram, which does not object. Moreover, a ram that defeated a rival and gained dominance would not run him off. Rather, it would court and mount the subordinate as if he were a female. Rams insult subordinates with courtship gestures but punish them should they recip-

A large British Columbia ram with a lamb.
Photos by the Fort St. John Ministry of Environment.

rocate. Provided, of course, they could catch the subordinate. These lustily insulted big rams from a safe distance uphill, just beyond where the big rams could reach. In short, courtship between males glued rams bands together, year-round.

It was not the only glue. Rams love to clash any time of the year. Let a fight break out and rams would dash in just to take part, to steal clashes from the two combatants. Large rams would, notoriously, clash into the larger fighter, small rams into the smaller. In real fights, it was the inferior, insecure ram, often the loser who would initiate clash on the dominants. Once a big ram got cross, he would blast the smaller rival with little ceremony.

Not everything we found drew cheers. I noted in the rut that old rams were the last to come and the first to go from the female ranges, the duration of a ram's stay with ewes thus being inversely to his chances of siring lambs. That made sense. It was not in a ram's interest to exhaust himself fighting to breed females and then deprive them of the food to grow his children. Incidentally, this insight is central to explaining mountain goat biology, except that among goats the territorial nannies chase off even the biggest billy. Stone's rams left voluntarily. What righteous indignation met that idea! I did not respond, allowing colleagues time to come to their senses. I am not interested in winning battles, only wars. Later, I discovered the critics did not grasp that the Stone's sheep and goats are imprisoned by deep snow after the rut, crowded onto tiny spaces. Their mid-winter imprisonment ends only when hard sun crusts form on the snow in March, allowing dispersal. I thus took for granted what is not observed on southern sheep ranges.

Watching Stone's sheep daily made them more than objects of scientific curiosity. They became personalities. When old B-ram began to limp after the rut, I watched out for him, snowshoeing to see where he was, observing discretely at a distance, hoping for his recovery. Alas, he deteriorated and a wolf killed him in early March. I put this episode in simple words into my first book, *Mountain Sheep*, hoping that it had a place in a scientific treaty. My editor agreed. Later I received a letter from Jack O'Connor, who praised my writing and mentioned that episode, noting that it had moved him. We met subsequently at a technical sheep meet in Idaho. He graciously invited me to visit, but death took him away. How much I had wanted to tell him that I had studied his books, word for word, after coming to Canada in my teens, and in need of mastering English. I never realized then I had chosen the writings of a master and a professor of journalism. Nor did he learn I had written to him then, obtained the name of Tom Burgess who built my first rifle, a 7x57 mm, inspired by his books. It was an essential tool in the wilderness and I took with it, among others, a 45-inch Stone's ram. I think he might have smiled.

Years later, the Gladys Lake study area, upon application by Ian McTaggart Cowan; Vladimir Krajina, a renowned forest ecologist; and myself became British Columbia's largest Ecological Reserve, now surrounded by the Spazisi Wilderness Area. Here Stone's sheep may live on undisturbed, just as they did in Tommy Walker's day. My cabin still stands, I am told, and the Stone's sheep are doing well.

The biography of Valerius Geist, Ph.D., is located in his first chapter of this book titled, "Wild Sheep: Creatures of the Ice Age."

DALL'S SHEEP

Photo by Alex Rota

FNAWS GOLD DALL'S SHEEP AWARD WINNERS

YEAR	HUNTER	SCORE	OUTFITTER
1978	Harry Scharfenberg	171	--
1979	Neil Courtice	--	Gregg Williams
1980	John Younggren	--	Bill Moynihan
1981	Bill Medley	$172^{3}/8$	Dave Young
1982	James Lingo	--	Bill Moynihan
1983	Floyd Salts, Jr.	$179^{1}/8$	--
1984	Rusty Reed	$175^{2}/8$	Alaska
1985*	Sherwin Scott	$187^{1}/8$	Bill Ellis
1986	Leslie Finger	$173^{2}/8$	Stan Stevens
1987	Gary Martinek	$173^{4}/8$	--
1988	Lionel Heinrich	$174^{4}/8$	Stan Simpson
1989	Beaver Nelson	$173^{2}/8$	--
1990	Carl E. Jacobson	$181^{3}/8$	Scott Milleur
1991	Lyle Foster	$176^{6}/8$	Stan & Deb Simpson
1992	William J. Dunbar	$173^{6}/8$	--
1993	Joe LoMonaco	$174^{3}/8$	--
1994	Jim Bernardin	$172^{5}/8$	Kelly Hougen
1995	William J. Dunbar	$178^{6}/8$	None
1996	Larry M. Bull	$173^{3}/8$	Paul Claus
1997	Steve Rahn	173	Archie Lang
1998	Kelly R. Elmer	$177^{4}/8$	NWT Outfitters

* Award of Excellence

THE WORLD'S RECORD DALL'S SHEEP

Harry L. Swank, Jr.'s seven-year quest for a World's Record Dall's sheep ended in September of 1961 when he took this ram scoring 189-6/8 points. It received the highest award given by the Boone and Crockett Club, the Sagamore Hill Award, given to trophies worthy of great distinction in memory of Theodore Roosevelt, Theodore Roosevelt, Jr. and Kermit Roosevelt.

Dall's Sheep

Wayne E. Heimer, Ph.D.

Since the United States purchased Alaska from Russia 132 years ago, Dall's sheep have been ignored, exploited and managed by varying approaches. Present-day management practices range from total protection in state-created viewing areas and congressionally established national parks, through biologically sound harvest management on state and selected federal lands, to highly permissive management for subsistence uses on other federal lands. How this spectrum of management practices evolved is a fascinating, sometimes rough-and-tumble, human adventure in biology and politics.

The Foundation for North American Wild Sheep (FNAWS) happened on the Alaska scene just as dramatic changes in sheep management began to take place. Though only 20 years old, FNAWS has played significant roles in both the biological and public policy areas in Alaska. Typically FNAWS strives to "put sheep on the mountain" by funding "hands-on" biological management and research, but FNAWS activities in Alaska have varied slightly from that pattern. In Alaska, major FNAWS funding has been primarily focused on assembling and presenting already-existing data to influence public policy to maintain and enhance sheep hunting opportunity. Perhaps paradoxically, the role of FNAWS funding in Alaska has not been one of "putting sheep on the mountain," for in Alaska sheep still exist in pristine habitats. Rather, FNAWS has played the seemingly contradictory role of "keeping more hunters on the mountain." The National FNAWS organization has met the primary task of preserving the opportunity to harvest mature rams on federal public lands in Alaska.

Public hunting opportunity has been the historic linchpin of North American wildlife conservation. Where the opportunity for public hunting has been maintained, wildlife populations thrive through active management programs. World experience shows that where public hunting is not allowed, public interest in wildlife suffers and wildlife conservation becomes an exclusive, expensive "government function" instead of an individual human activity. Broadly stated, preserving hunting opportunity is, at present, the best hope for "putting" or "keeping" sheep on the mountains.

In contrast to the role played by the National FNAWS organization (maintaining hunting opportunity), the active membership of the Alaska FNAWS chapter has been a major influence in Alaska's Dall's sheep management program. Without its funding support and aggressive advocacy of modern biologically based ram harvest practices, Dall's sheep management would probably be on a different, less biologically secure, course in Alaska today.

Early History

Except for those populations heavily exploited by aboriginal hunters, Dall's sheep were basically untouched when Alaska was purchased from Russia in 1867. Early accounts recall that Eskimos harvested significant numbers of sheep from readily accessible populations in the Northern Brooks Range for their own use and for barter to whalers and explorers. During this period, aboriginal use of Dall's sheep was part of the nomadic seasonal cycle or as a back-up source of meat when other sources failed. The harvest activities of these aboriginal hunters may have had significant negative impacts on locally used Dall's sheep populations. At least one modern anthropologist, John M. Campbell of the Maxwell Museum of Anthropology in Albuquerque, New Mexico, suggested (in a 1974 article) that aboriginal overharvest of

National Geographic *writer Tom Melham and Anchorage, Alaska, high school teacher Dave Larsen, a FNAWS volunteer, help sheep biologist Wayne Heimer paint neckbands for Dall's ewes in sheep camp.*
Photograph by John Abrams.

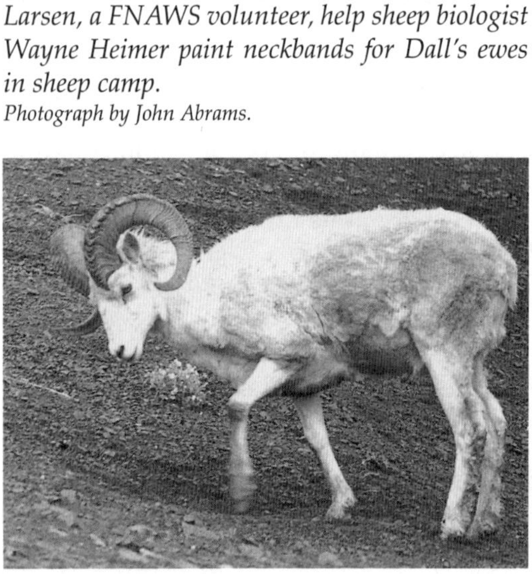

A young Dall's ram.
Photograph by Robert Cassell.

Who me? An ear-tagged Dall's ram stares at the cameraman, surrounded by Dall's ewes along the Robertson River, Alaska, study area mineral licks.
Photograph by S. Watson Keller.

Are "ewe" in love? Dall's sheep tromp through the snow.
Photograph by Valerius Geist.

Dall's sheep was responsible for the scarcity of Dall's sheep documented in the Brooks Range during the late 19th and early 20th centuries.

As the Territory of Alaska developed, explorers and naturalists became interested in Dall's sheep. It was the explorers/naturalists who gave Alaska's wild white sheep their rather intriguing name, Dall's sheep. One of these early explorers, a fellow named Nelson, had an interest in mammals, and named many of them throughout Alaska. Nelson offered the first Latin scientific name for these white sheep in 1884. It was simply *Ovis* (the Latin name for the sheep genus) plus *montana* (for mountain) as the species name. Nelson also proposed the subspecies name, *dalli*. Tradition says that Nelson offered the *dalli* subspecies name in honor of another early Alaskan explorer, William H. Dall. Dall was primarily a mariner (he did describe Dall's porpoise) and river traveler. There's no record of him ever having been in Dall's sheep country. In 1897, another namer of animals, J. A. Allen, changed the species name from *montana* to *dalli*, and it has remained so ever since.

William H. Dall's name continues to be associated with Alaska's white sheep because the accepted rules for the scientific naming of new species call for it. These rules say the species name (in this case *dalli*, the Latinized form of W. H. Dall's last name) should be "de-Latinized" and its possessive form (Dall's) used as the scientifically accepted "common name." Under these rules, the scientifically correct common name would be "Dall's sheep."

This technical naming rule causes some confusion where Alaska's wild white sheep are concerned. No wild white sheep were ever possessed by W. H. Dall, and it's unlikely he had anything to do with any wild sheep. His name remains linked to wild white sheep because of Nelson's admiration of Dall as an explorer/naturalist. In addition to aboriginals and explorers/naturalists, Dall's sheep were exploited by the trappers, miners, market hunters and homesteaders who followed. Some sheep populations were virtually decimated because of harvesting by humans during the early 20th century. Alaskans and others sometimes refer to the sheep as "Dall" but the correct name is Dall's sheep, in honor of W.H. Dall.

Dall's Sheep Management During Territorial Days

Within a decade after some populations had been decimated, the first closed seasons and restrictive bag limits were established (in 1926) as the emerging science of wildlife management began to influence the Alaska scene. Under this scheme, the open harvest season was scheduled during early fall, primarily because of traditions attending wildlife management in other places. The fall season also made sense to early managers because Dall's sheep were not breeding at that time, and were at their fattest. The bag limit was restricted to two rams. Ram-only hunting appeared to be biologically sustainable because a single mountain sheep ram will breed many ewes. The regulation was also consistent with the "males only" harvest seasons characteristic of early conservation efforts.

During this period, knowledge of Dall's sheep was primarily local. No systematic population monitoring or regulatory program existed, and managers applied local knowledge of Dall's sheep populations over what today would be unthinkably huge areas. For example, an Alaska Territory-wide closure of Dall's sheep hunting in 1942 was based on the assumption that the harsh weather that decimated Dall's sheep populations on the Kenai Peninsula that year occurred throughout all Alaska. Ram harvest was again allowed in 1943. The one-year closure in 1942 was probably an ineffective management action. Modern studies of population dynamics and weather patterns suggest it was unlikely Dall's sheep had suffered throughout the entire Territory. Additionally, we now know that Dall's ewes don't usually reproduce until age three or four years, and twinning has never been seen. Hence the one-year closure

A mature Dall's ram gives us the eyeball.
Photograph by Valerius Geist.

Young Dall's ram in a herd.
Photograph by Robert Cassell.

was insufficient time for population recovery.

Nevertheless, when the harvest season was re-opened in 1943, the general bag limit was more conservative, one ram, except in the Brooks Range where the limit remained at two rams. Although ewe harvests were prohibited, the more liberal bag limit in the Brooks Range provided greater harvest opportunities for the Eskimos and Indians with a history of Dall's sheep use for food. These seasons persisted until 1951 when the bag limit was revised to an even more conservative harvest regulation.

In 1951, only rams with horns that had grown through 3/4 of a "curl" (the circle described by the horn helix) were legally harvestable. This regulation went a step beyond "males only" hunting, and appears to have been designed either to assure harvest of rams with larger horns or protect young rams from harvest. The 3/4-curl rule was initially a federal regulation that came to the Territory of Alaska from Washington, D.C. via Wyoming, where it was first applied to bighorn sheep in 1930. Under the 3/4-curl rule, the bag limit in the Brooks Range remained two legal (3/4-curl) rams in the Brooks Range until 1970.

By 1970, sheep hunting in the Brooks Range by non-local Alaska residents was rapidly increasing, as was use by guided non-resident Alaskans. This increase was at least partly due to the Dall's ram season opening two weeks earlier in the Brooks Range than other hunting seasons throughout the rest of Alaska. Naturally, this drew commercial sheep guides to the area. These increased harvest pressures generated the perceived need to limit overall ram harvests for conservation purposes, and this need apparently outweighed providing additional harvest opportunities for Brooks Range Indians and Eskimos in the minds of wildlife managers of that period. Additionally, it was common knowledge that the Indians and Eskimos of the Brooks Range had never limited their harvests to rams as specified in regulations 20 years earlier. Managers of the day chose to regulate those who would comply and benignly ignore illegal local uses.

Going back to 1930, the legal size/age of rams for harvest has been defined by the portion of the "curl" attained by each ram's horns. This still works because horns are never shed, and continue to grow throughout a sheep's life. The outside surface of unbroken ram horns describes complete circles, or full-curls (when viewed down their helical axes) at maturity. The horn size/age limit commonly applied throughout North America was arbitrarily set at 3/4-

curl in an effort to provide both biological safety and maximum ram harvests.

While the 3/4-curl regulation may have been compatible with bighorn sheep biology, it was poorly suited to Dall's (or thin-horn) sheep biology. Bighorn and thin-horn sheep are behaviorally different. One significant difference is in the frequency and extent of ram horn brooming (the breaking of horn tips usually in dominance fights). Virtually all bighorn rams broom both horns back to 3/4-curl by maturity. Brooming in Dall's rams is common, but Dall's rams seldom broom both horns as severely as bighorns. Consequently, many 3/4-curl bighorns are mature, but virtually all 3/4-curl Dall's rams are simply young. This meant that the 3/4-curl rule, when applied to Dall's sheep, allowed harvest of juvenile rams. As sheep hunting increased in Alaska, application of the 3/4-curl rule to thin-horned Dall's sheep caused subtle, long-term negative effects to the heavily hunted populations where mature rams were completely removed. It took the newly formed State of Alaska decades to discover and rectify this situation.

Evolution of the State's Biologically-Based Harvest Regulations

The history of Dall's ram harvest regulations in Alaska is historically linked to the re-emergence of Rocky Mountain bighorn sheep in the American West. Prior to settlement of the Western United States by Europeans, bighorn sheep were abundant in suitable habitats (although unsubstantiated estimates of more than a million bighorns appear grossly inflated). Settlement of the West greatly reduced their numbers, and by 1900 only small relict populations remained in mountainous areas that had limited appeal to stockmen and miners.

The introduction of domestic livestock diseases to bighorn sheep was the primary factor in their demise, but the decline was also aggravated by competition with domestic livestock for grazing lands and overhunting for the frontier meat market. When wildlife managers began to practice conservation and restoration of bighorn sheep, the first approach was total protection throughout the Western states.

This total protection (along with control of predator populations to benefit the livestock industries) eventually brought many bighorn populations back from the brink of extinction. As sheep populations returned to viable levels, managers sought a balance between protection and use. This meant providing some harvest opportunity while still allowing herds to grow. Wildlife managers understood that herd growth would produce benefits. These benefits included the conservation-funding revenues produced by license and tag sales after passage of the Federal Aid to Wildlife Restoration Act (a.k.a. the Pittman-Robertson Act of 1937). Benefits also included economic growth attending development and maintenance of the guiding and outfitting industries, and the high public interest in conservation which comes with public hunting.

A review of mountain sheep hunting regulations across North America shows the most common attempt to balance herd growth and hunter use was harvest limited to surplus males (rams). Rams, which could be removed by hunters without noticeably compromising lamb production, were defined as a harvestable surplus. The conclusion reached by most sheep managers was that harvesting rams at the youngest acceptable age (before natural mortality removed any more of them than necessary from the shootable population) would give the greatest sustainable harvest.

This data-free assumption, in 1930, came 14 years before the first study of mountain sheep survival. This landmark Dall's sheep research by Adolph Murie in 1944 (later verified in bighorn sheep) revealed a consistently low ram mortality between one and seven years of age in unhunted populations. Even though the research data indicated low ram mortality after

Dall's in the snow.
Photograph by Valerius Geist.

recruitment at yearling age for Dall's and bighorn sheep, the 3/4-curl regulation persisted as the dominant rule governing harvest of Dall's and bighorn rams. This turned out to be serious. Here's why:

When managers formulate regulations, it is understood that the resource must be conserved even if the allowed harvest takes place at the maximum level. It was in this respect that the 3/4-curl regulation originally developed for bighorn sheep in Wyoming failed when it was applied to thin-horn sheep in Alaska.

When (and where) sheep harvests by human hunters are biologically insignificant, it doesn't matter what the regulations are, or even if any regulations exist. Hence, before the human population of Alaska and the interest in Dall's sheep hunting increased to the point that harvests could negatively affect local Dall's sheep populations, regulations were largely irrelevant. However, when hunting pressure increased to the point where virtually all legal rams were removed from hunted Dall's sheep populations, the 3/4-curl regulation was insufficient to protect the population from negative impacts.

The 3/4-curl regulation provided some benefit in that it protected ewes from harvest, but the overharvest of old rams resulted in social disruption that was harmful to Dall's sheep (and hunter) welfare. Lamb production declined, and mortality among young rams greatly increased.

Mountain sheep in general, and Dall's sheep in particular are intensely social herd animals. In mountain sheep society, each individual has a clearly defined social position keyed to horn size. The behavior, reproductive success and nutritional condition of each sheep are determined by this social rank. This powerful influence on Dall's sheep population dynamics was not recognized when the 3/4-curl regulation was implemented.

The two important aspects of Dall's sheep ecology, which were disrupted by near-total harvest of rams down to the 3/4-curl limit (at the average age of 4.5 years), were lamb production and immature ram survival. When virtually all dominant rams, those fully mature rams aged seven to eight years and older, were removed, lamb production declined significantly. Some lambs were still produced, but the frequency of breeding by 18-months old ewes increased (even though it resulted in stunted lambs and ewes). Also, prolonged lactation (ending in April instead of October) associated with low ram abundance and/or the absence of mature rams resulted in rare consecutive-year lambing success by ewes. Low lamb production and low young ram survival might have been explained by poor range conditions if ewes had not bred at unusually young ages (18 months) and fed lambs on milk for an unusually long time. Both early breeding and extended lactation periods are usually considered signs of high-quality range.

Survival of young rams was also seriously compromised because immature rams became active breeders and began to pay the energetic and mortality costs associated with dominance before they were behaviorally and physically mature. This decreased immature ram survival. The cumulative effect of population disruption resulting from mature ram overharvesting was lowered sustainable ram harvests.

To see if ram harvests could be increased, an experimental full-curl regulation was established in a heavily hunted portion of the Alaska Range in 1984. We wanted to see if limiting harvest to only mature rams (which was expected to increase lamb production and immature ram survival), would actually lead to increased harvests of mature rams. Alaska Department of Fish and Game funding for this effort was supplemented by the Alaska Chapter of FNAWS. Additionally, FNAWS members contributed as volunteers to the field research involved. After the change to full-curl ram harvest, lamb production doubled; and immature ram survival apparently increased dramatically. Eventually harvest rates for full-curl rams exceeded those sustained under the 3/4-curl rule.

Following our discovery that the negative effects of ram overharvest could be reversed by harvesting only mature rams, and that hunters could actually harvest more rams as a result, I proposed an end to the 50-year span of 3/4-curl ram harvests in Alaska for the Alaska Board of Game meeting of 1988 through Alaska Department of Fish and Game channels. This proposal was the first to factor the subtleties of animal behavior into harvest management of big game in Alaska, and was squelched by the Department of Fish and Game before the public got a chance to respond to it. Sheep hunters, primarily FNAWS members, were furious and subsequently introduced the proposal to change to full curl harvests independent of the Department of Fish and Game. Because of Alaska Department of Fish and Game resistance, the FNAWS proposal became highly controversial. It was defeated.

The next year, long-time guide and FNAWS member, Bill Waughaman, introduced the proposal again. As a result of compelling public testimony coordinated by FNAWS and Tom Kirstein of the Alaska Professional Hunters Association (also a FNAWS member), the Alaska Board of Game passed the present regulation limiting ram harvests to full-curl or eight-year-old rams, or rams with horns broomed on both sides.

This change assured the presence of mature rams in even heavily hunted Dall's sheep populations. As stated above, the result of this change in age/size of rams at harvest was increased harvests of larger rams. It also simplified sheep management. Hunter success was not affected. With few local exceptions, thin-horn sheep throughout North America are now managed for the harvest of full curl rams. Legal definitions differ between states and provinces, with Canadian definitions allowing the harvest of younger rams than the Alaska full curl regulation.

So, we see it took 40 years of Alaska statehood to rectify the biologically incorrect 3/4-curl regulation established by a well-meaning, but biologically naive Federal Territorial Government. The State of Alaska now attempts to manage Dall's sheep in accordance with a published, biologically based working management hypothesis I published in the Northern

Wild Sheep and Goat Council Symposium proceedings in 1988, and vigorously supported by the FNAWS-Alaska membership. Unfortunately for sheep hunters, the 3/4-curl regulation was not the end of federal intervention limiting sheep hunting opportunity in Alaska.

Federal Intervention and Participation by FNAWS

Development of the vast oil reserves in arctic Alaska during the late 1970s caused the most extensive federal intervention in state wildlife management in U. S. history. Involvement of FNAWS funding in Alaska grew directly out of this federal intervention into sheep management.

This story began with historic Native land claims in Alaska. After the oil discovery at Prudhoe Bay, it became clear that a pipeline required to get the oil to market could not be built until these Native claims were settled. Not everyone thought building the oil pipeline was a good idea. Alaska Natives joined the environmental preservation community in opposing the pipeline. Compromises between Alaska Native/environmental preservation interests and oil-development interests to allow building the trans-Alaska oil pipeline resulted in a Congressional mandate to place huge tracts of federal public land in federal conservation systems. In these compromises, the Alaska Natives and the environmental community agreed to stop fighting pipeline construction in exchange for a billion dollars and 44 million acres of land (what the Alaska Natives got) and transfer of at least 80 million acres into new wildlife refuges, national forests, wild and scenic rivers and new national parks (what the preservation interests received).

Initially, lands selected for transfer to the National Park Service contained a large fraction (up to half in some proposals) of the Dall's sheep habitat in Alaska. Everyone understood that if these lands were to become national parks, there would be no opportunity to harvest sheep on them in the future. Consequently, the State of Alaska resisted these land transfers to the National Park Service, at least partly to maintain sheep hunting.

An intense, protracted and bitter Congressional debate followed. Compromises ending this debate ultimately resulted in a 25 percent loss of sheep hunting in Alaska and federal recognition of a subsistence harvest priority for rural residents of Alaska. However, the struggle was epic, and FNAWS played an important role.

It was during the Congressional debate in the mid to late 1970s that FNAWS became active in Alaska. Ken Fanning, a concerned sheep guide and FNAWS donor, initially contacted me with a request for donation of surplus Dall's ram horns. Ken and others sectioned these horns with a band saw, attached an engraved brass plate to each one, and donated 40 of them for auction at the 1978 FNAWS convention (as belt buckles). The total raised was about $10,000. With this seed money FNAWS supplemented the State's effort to maintain as much sheep hunting in Alaska as possible.

The State's effort was successful in preserving general hunter access to sheep on about 75 percent of the state's sheep habitat. The remaining 25 percent was classified as national park land where only rural subsistence hunters were allowed to harvest sheep. Others permanently lost Dall's sheep hunting opportunity on these lands. Subsistence seasons on federal lands are typically lengthy (lasting from seven to nine months), and bag limits are liberal (up to three sheep, including pregnant ewes and lambs). Subsistence harvest reporting is voluntary. As anecdotal evidence accumulates, it's beginning to look like this liberal approach to management conflicts with Dall's sheep population conservation.

The most notable success in maintaining sheep hunting opportunity was in the North Wrangell Mountains where 12,000 Dall's sheep were kept available for hunting on federal land. Over the 15 years since ram harvest opportunity was preserved on these sheep habitats, ram harvests from this area have averaged approximately 250-300 rams per year, a cumulative harvest of more than 4,000 rams. The statewide harvest has averaged about 1,100 rams per year during this time.

This first effort to preserve hunting opportunity set the pattern for FNAWS funding activity in Alaska. In a departure from normal FNAWS activities, funding was not provided to "put more sheep on the mountain," but rather to "put more hunters on the mountain" through preserving Dall's ram harvest opportunity. To this end, FNAWS funding focused primarily on maintaining hunter access to Dall's sheep. Early FNAWS contributions were made in support of an umbrella group of outdoor organizations engaged in gathering and presenting data to preserve sheep hunting opportunity in the face of federal land classification changes.

Other major FNAWS investments in Alaska have focused more specifically on "keeping nonresident hunters on the mountain" through support of the guiding industry. An accounting of money FNAWS has spent in Alaska shows the organization has granted almost $400,000 to benefit sheep management in Alaska. Of this amount approximately 10 percent ($32,100) has been spent directly on sheep management or research. The remainder has gone toward maintaining hunter access to sheep. As stated in the introduction, history shows that an interested, participating hunting public is vital to effective and efficient wildlife conservation.

In addition to this funding from the National FNAWS organization, the Alaska Chapter and other State Chapters where there are sheep hunters but no sheep (notably Minnesota, Wisconsin and Iowa), have contributed to biological research and management projects within Alaska. The most expensive research project ever supported by FNAWS in Alaska was a study of the economic value of sheep hunting in Alaska. I justified this project by arguing that Alaska would be more likely to preserve Dall's sheep hunting and set relevant fines for regulatory violations if the dollar value of Dall's ram hunting were known. As I designed this project in 1982, it had a price tag of $22,000. The National FNAWS organization contributed $13,500, leaving a significant amount to be raised elsewhere. To do the job, we needed a consulting economist, which we could not afford at that level of funding. To solve this problem, the biologist doing the work, Sarah Watson (Keller) enrolled in a two-year graduate resource economics program at the University of Alaska. This delayed completion of the project, and required that Sarah pay her own graduate tuition, but it did allow us to complete the project on a reduced budget. Other project expenses were covered by FNAWS-Alaska in what produced an unexpected benefit to the project.

To raise money, the Alaska FNAWS Chapter raffled a $5,000 cash prize. Proceeds above the value of the prize were slated for the economic valuation study. As a FNAWS member, Sarah sold tickets to everyone who would buy them, including her family in California. At the drawing, her uncle, "U.J.," was the winner, and generously donated the $5,000 prize back to his niece's project. Thanks to the efforts of a great many FNAWS members, the project costs were eventually covered, and the project completed.

The project proved to be a ground-breaking effort that foreshadowed other economic assessments of hunting in Alaska. Updated estimates of the dollar value of Dall's sheep hunting in Alaska are approximately $12,000,000 per year. Approximately 80 percent of this money comes from through the guiding industry. Non-resident sheep hunters are required to have a guide to hunt sheep in Alaska.

Additionally, fines as high as $5,000 per ram for illegally taken Dall's rams have been levied in Alaska as a result of this FNAWS-funded economic valuation study. Although this fine is less than half the defined "market value" of a Dall's ram, it is 10 times higher than the traditional $500 that had been imposed prior to Watson-Keller's work. Again, the project did not put more sheep directly on the mountain, but showed the economic value of keeping both sheep and hunters on the mountain. FNAWS-Alaska has also expended funds and efforts in public education about Dall's sheep natural history and management, funded surveys in the Kenai and Chugach mountains, and volunteered time for field work with the Alaska Department of Fish and Game.

Present Conditions and Future Prospects

During the 1990s, Dall's sheep populations throughout much of Alaska have declined because of difficult weather. Additionally, predator pressure, primarily exerted by wolves on adult sheep and coyote predation on lambs, are additive factors. They are certain to slow population recovery.

At the present time, hunting prospects are still good, and during the 1990s some outstanding rams have been harvested from unlikely places as hunters have tried, ever harder, to find "the big one." In 1990, two new "top 10" Boone and Crockett rams were harvested from the Brooks Range, which had formerly been assumed incapable of producing "high book" rams. The Chugach Mountains adjacent to Anchorage continue to produce outstanding (and some "book") rams in one of Alaska's two trophy management areas. Both trophy areas require lucky hunters to draw a permit to participate in these unusually good hunting experiences. Additionally, Alaska's oldest trophy management area, the Tok Management Area (established 1974), continues to produce outstanding trophies and high-quality hunting experiences.

Fall of 1995 was unusually good in the Tok area with a harvest of 59 rams reported. These rams averaged 37 inches at 9.2 years of age with 32 percent of the harvest greater than or equal to 39 inches and 19 percent of the harvest 40 inches or greater. The largest ram harvested in this area in 1995 was over 46 inches and scored 176 Boone and Crockett points. He was the largest ram harvested from the Alaska Range during the last 25 years.

In other areas harvests have declined significantly during the 1990s, and the short-term outlook is bleak because of weather-related lamb production failures and unabated predation by wolves and coyotes. The time may have come when FNAWS can benefit sheep hunting in Alaska by actually putting more sheep on the mountains as well as trying to maintain hunter numbers. The most likely way to do this will have to focus on decreasing environmental resistance to population recovery. This means predator reductions, the only management-alterable factor that has not already been manipulated to favor Dall's sheep.

Both state and national FNAWS organizations have shown willingness to support sheep recovery through predator reductions. Each organization put significant (enabling) funding into economic incentives for trappers to harvest wolves in what was called the Fortymile Caribou Calf Protection Program in the Tanana/Yukon Uplands. This area contains relatively few sheep, although they were likely to benefit from wolf population reduction. FNAWS invested in this program to see if trapper incentives could actually reduce wolf populations. The thinking was that if trappers really could lower the wolf population to benefit the caribou calves (and the few sheep present), a similar program might be of great benefit in better sheep habitats. The economic incentive program was highly successful in lowering wolf numbers. Unfortunately, opposing a ballot initiative sponsored by animal rights activists to prohibit snaring (the way most wolves are harvested by trappers) required the attention of the Alaska Trapper's Association, which, administered the successful wolf trapping program this year. Application of economic incentives to reduce wolf and coyote numbers in habitats with great sheep production potential now appears to be at least two years away.

It may be difficult, but FNAWS has an outstanding record of putting sheep on the mountains for hunters and others to enjoy. I remain optimistic for the longer-term future.

One tangible reason or optimism is Alaska's participation in the FNAWS governor's permit auction at the national organization level. Alaska is the latest jurisdiction to offer a governor's permit for auction. Through the hard work of dedicated members of FNAWS-Alaska, Alaska Department of Fish and Game Biologists and administrators, and the Alaska Legislature it became possible for the governor to donate a governor's permit to FNAWS for auction at the end of the 1996 Alaska Legislature.

An ear-tagged Dall's lamb is held by Wayne Heimer at the Robertson River study area's mineral lick on Sheep Creek in Alaska.
Photograph by C. Harms.

Alaska's governors have had governor's permits available since statehood, but their issuance became socially unacceptable after abuse of permits granted during the late 1960s. Reinstating the governor's permit mechanism and maximizing the value of the permit for auction required legislative action. After a 32-year moratorium and more than a decade of work by visionary biologists and FNAWS members, the first governor's permit for Dall's sheep in any area where sheep can be legally hunted was auctioned at the National FNAWS convention in February of 1996. The winning bid was $200,000. Of this amount, approximately $20,000 went to National FNAWS to "put more sheep on the mountains," another $20,000 to FNAWS-Alaska to improve sheep hunting in Alaska, and the remaining $160,000 into the Alaska Fish and Game Fund to be spent maintaining and enhancing sheep hunting opportunities in Alaska. Improvements in sheep hunting and management are expected to come through future projects to be jointly agreed upon by FNAWS-Alaska and the Alaska Department of Fish and Game.

The 1997 permit brought $50,000 which will be apportioned by a similar formula. Even though the permit brought less money in 1997, a functional mechanism now exists for assuring increased money to be spent on sheep management in Alaska. If this money can be wisely spent, Alaska's sheep and Dall's sheep hunters in general should benefit.

Wayne E. Heimer, Ph.D.

Wayne E. Heimer, Ph.D., was born on a sheep-free farm in northeastern Colorado, and graduated from what is now the University of Northern Colorado at Greeley with undergraduate majors in chemistry and biology. From there he went to Washington State University for a master's degree in what is now called molecular biology. Wayne arrived in Alaska to pursue a doctoral program through the University of Alaska's Institute of Arctic Biology at the University of Alaska in 1967. At the end of that program, Wayne entered public service as Dall's sheep biologist for the Alaska Department of Fish and Game, and served Alaskans and Dall's sheep for the next 25 years before being retired from active state employment. Wayne may be unique among wildlife biologists in having studied life at all levels from the molecular to that of human politics. This breadth of perspective shows in Wayne's prolific production as a writer of both scientific and non-technical articles about sheep research and management. Nobody seems to know exactly how many papers Wayne has written, but the total approaches 100. Wayne was on the job in Alaska before FNAWS was formed; but knowing a good thing, has been a staunch FNAWS member, cooperator and supporter. He received the State Statesman Award for his contributions to Alaskan sheep management in 1994, and reviews Grant-in-Aid requests for FNAWS. Wayne says he is working on several books, at least some of which will be about Dall's sheep.

Chapter III

Putting Sheep on the Mountain: The Role of the Hunter

Daniel A. Pedrotti

"What is the greater tribute to a `Grand Old Ram' that has survived years of evading predators, droughts, harsh winters and dangerous rock slides, has led his herd, created his progeny many times over and has now completed his job? To fall to disease, starve to death, be taken by wolves or freeze on a mountainside -- his remains to be devoured by vultures, his bones eaten by rodents and his dust blown away in the wind never to be seen again or to be taken by a hunter, provide nourishment to his family and be immortalized in a museum or in a trophy room where he will be admired forever." (Daniel A. Pedrotti, 1994)

Where would wild sheep be today without the passion, dedication and monetary support of hunters?

"A good news story, and a great one, is North America's continental system of wildlife conservation. It is the only sizable system of sustainable natural resources development. It is a significant creator of wealth and employment and a great contributor to our quality of life. It is the greatest environmental success story of this century and an outstanding cultural achievement of North American Society. While widely admired and imitated beyond North America, it is poorly understood and ill appreciated in the U. S. and Canada itself, including wildlife biologists and managers." (V. Geist, Professor Emeritus of Environmental Science, The University of Calgary, Alberta Canada, 1998)

The above quote from one of the foremost writers of scientific information on wild sheep in North America and an immigrant from Europe who sums up the view held by learned men from outside of our own continent. Why is it that the media and public here in our country do not understand and appreciate this success story that could not have become a reality without the tremendous contribution of energy, resources and volunteerism on the part of hunter-conservationists like you and me? The media chooses to ignore this phenomena under the guise of political correctness but there is no excuse for us.

"The greatest environmental success story of the 20th century is the return of wildlife — and wild sheep — in North America. The wildlife we currently enjoy and take for granted is not merely a gift of nature but is wildlife restored and returned due to the effort of three generations of North Americans." (Dale Toweill, *Return of Royalty*, 1998)

This quote coming from one of our leading wild sheep biologists is something for which we can all be proud. Spurred on by our great conservation leader, Theodore Roosevelt, who formed the Boone and Crockett Club in the late 1800s, the proliferation of organizations formed by hunter conservationists, dedicated to the mission of restoring and conserving wildlife and their habitat, reads like the A B Cs: A-Audubon Society (1897), B-Boone & Crockett Club (1887), C-Camp Fire Club (1897), D-Ducks Unlimited (1937), Foundation for North American Wild Sheep (1977), Rocky Mountain Elk Foundation (1984) and the list goes on. You fill in the blanks. There is a highly motivated hunter conservation group for most every letter in the alphabet and this is the remarkable phenomenon we want to highlight.

Combined, these organizations represent millions of dedicated volunteers who are willing to supply the energy, resources and time that has been needed to bring wildlife in North America from the brink of extinction to its present state of abundance. Who are these people and what moves them to this great cause?

You have already read a great deal about them in the preceding chapters. Mostly, they are hunters, but not all. Their economic status ranges all the way from those who only afford a hunting license to our great philanthropic donors who relish in their ability to literally give millions for this cause. They are volunteers, who come from all walks of life, who dedicate much of their discretionary time to serving on the boards of these societies, writing for their publications, going to meetings to spread the word, doing hands-on projects. They are biologists, wildlife managers, state and federal employees, and ranchers. Many are company employees and there are executives as well.

Despite the valiant efforts of Roosevelt, Audubon, Pinchot, Hornaday and the other pioneers, the decline of wildlife numbers continued into the early 20th century. This situation called for extraordinary measures and sportsmen used their political influence to create a powerful source of funding to help wildlife. They arranged for a bill in Congress that earmarked a tax on sporting guns for the purchase of hunting preserves. Fortunately this bill, modeled after Old World hunting ideas, failed, but the idea lived on and later became the Pittman-Robertson Act that was signed into law in 1937.

Ironically, this bill, that restricted the use of hunting license fees to purposes of administration of state game and fish agencies, fulfilling the mandate proclaimed by Theodore Roosevelt 50 years earlier "that management of wildlife was to be based on the best science and scholarship available" was signed by one of his descendants, Franklin D. Roosevelt. It precluded the use of these funds for enforcement activities and required wildlife restoration using habitat acquisition, development and most importantly, research.

Thus hunters not only provided resources through hunting license fees but also taxed themselves in order to create the essence of this success -- the science of wildlife biology. Wildlife biology in the 1930s was a new science in need of basic research and education. Pittman-Robertson and other laws created and funded professional schools and research programs. Newly trained scientists soon began providing key information on different species that advanced management and began the recovery of American wildlife.

You would think that such an efficient and successful approach to the science of wildlife biology would flourish as long as man exists and wishes to conserve our habitat and environment, but this institution, unique in the world, is in danger. As we read this essay, the greatest assemblage of wildlife scientists the world has ever known faces an uncertain future. Budget cuts in all the wildlife agencies throughout the continent are forcing hard choices and it is ever more tempting to reduce research and thus jobs for these highly qualified and motivated scientists.

So, in addition to population explosion, habitat loss to ranchettes, pressure from the radical animal rights groups and the large core of our population that is apathetic to our cause, we now have a threat to the scientific community that has helped achieve this era of abundance.

What are we to do to assure that this legacy, developed in the 19th century, handed over and greatly enhanced by us in the 20th, will be passed on to the generations of the 21st century? There are many possible scenarios and I am sure each of you have thought about this dilemma and have solutions of your own.

It would be impossible to ascertain the value provided by hunter-conservationists to the recovery of big game and especially wild sheep in North America. It is enormous, to say the least, and has also contributed greatly to non-game species as well. There is a saying "those that have the resources call the shots." We have provided the resources and have earned the right to call the shots so let's get on with it.

Here is what I think. Today with millions of dedicated sportsmen supporting many diverse aspects of conservation, our best hope and our next challenge is to unite these diverse and dedicated efforts. Yesterday's obstacles were mainly ignorance and disregard for wildlife. Today's obstacles, in addition to those enumerated above, are diffuse goals and even dissension among those who care deeply for wildlife. Many efforts have been made to bring this immense force together but some failed and others have seen only partially successful.

Imagine the power of such a dynamic force, created by the unification of the millions of members and resources of our "alphabet" of conservation organizations. This effort would have access to huge sources of energy and the political impact could be astounding. We would then be fully equipped to overcome all obstacles. This force would assure us that generations of the future could enjoy a world characterized by the abundance of wildlife that we have worked so hard to conserve.

This unification of effort must happen if we are to preserve and bequeath our "Wildlife Cultural Heritage"* to succeeding hunters and wildlife enthusiasts.

Who can provide the leadership as we roll into the next millennium? Can the Boone & Crockett Club, where it all started over 100 years ago, be the catalyst? Perhaps we ought to ask them.

* *This phrase was coined by Ken Davis, a director of the Rocky Mountain Elk Foundation.*

This by David Beardall photo was 3rd place in the Live Sheep category during the 1998 convention.

SECTION III

The Stories of the Hunters

This by Erik Ruf was on the Spring 1997 cover of Wild Sheep.

Chapter I

Hunting Stories by FNAWS Members

One Day on a Sheep Hunt

Bob Karel
From *Wild Sheep*, Fall 1990

Bob Karel with his hard-won prize.

The alarm clock didn't awaken me because I had a restless night and was fully conscious waiting for 3:30 a.m. to arrive.

Martin Lamoureaux, our guide and outfitter, reached over to still the noise of the clock. My friend and hunting partner, Casey Gerber, groaned and sat up in his sleeping bag. We were excited and anxious. Two days ago we saw a pair of legal rams at the very top of the mountain we planned to climb that morning. The three of us hunted Stone's sheep in the Sifton Ranges of the northern Canadian Rockies, 300 miles north of Prince George, British Columbia.

My confidence was high because the rams did not show themselves on our side of the mountain yesterday. We expected them to be on the side opposite the morning's climb. We reached the top before they did and looked down at them sometime later today.

In the confusion of getting a quick bowl of oatmeal, packing our gear and preparing for the day's hunt, I broke the frames on my hunting glasses. Foolishly, I hadn't packed a second pair.

A real plus for any sheep hunter is the spectacular scenery and a view from the Finlay River.
Photograph by Bob Karel.

Our jeans were wet to the knee at 4:35 a.m. as we walked in the gray light of early dawn through dew-soaked grasses. The water line moved up to mid-thigh as we side-hilled on a steep and slippery stretch that took much more out of us than we expected. It was difficult to keep the rifle balanced on my backpack. I held onto small trees and kept from falling at the same time. Our concentration and our focus was interrupted at the sight of a very large porcupine waddling below us on its way to doing whatever porcupines do at the start of a day. Casey and I now began to realize the immensity of this mountain. We reached the base of it and were already glad for our physical conditioning, but both of us wondered if we prepared well enough.

The ice cold mountain stream water provided a welcome refreshment. It was 5:20 a.m. This portion of the climb featured snow, which in itself is a treat for two sheep hunters from Florida. Half crawling, half walking up the snow wash, we hear "whistlers," the rock marmots that inhabit these cliffs and can ruin a stalk with their high-pitched calls. Walking behind Martin, I marveled again at his stamina and his ability to climb with a fused right ankle, the result of a bad fall years ago. I recalled the moose hunt he outfitted for me in 1986, and how he has what all guides wish they had — the ability to instill in his hunters a quiet confidence they will take home their trophy if they persevere.

This was the toughest part of the climb. We had been at it for more than two hours. It was now 7:15 a.m., and both Casey and I are surprised at the steepness and the danger. We clung with all four limbs to a rock ledge that was an absolute sheer drop-off. My thoughts returned to the intense preparation in advance of this hunt; running 3-1/2 miles a day, five days a week for the last 2-1/2 years; and biking 30 minutes a day, four days a week, for the last three months. All the preparation was necessary and helpful. However, I didn't prepare for the real fear that griped me. Looking at Casey, he was predictably direct.

"Bob, we bit off more than we can chew. We could die up here! I want to see my kids again."

I concentrated on not looking down, and on the hopeful thought the rams would be there. We made it to the top, and were careful not to skyline ourselves as we belly-crawled to good glassing positions. Even though we didn't see any rams, we felt a deep sense of accomplishment at finishing the climb: 8:30 a.m.

That, combined with the spectacular view, the enormity of the mountain range we were glassing, and a feeling of distinct insignificance, made this moment truly special and memorable.

As we searched in vain for rams, we spotted 35 mountain goats on various surrounding peaks. These animals were not as majestic as sheep, only because they lacked a ram's magnificent headgear. Nonetheless, they represented genuine wildness and a mountain challenge that beckoned any hunter. I knew I would return to hunt goats.

Watching the mountain goats at long range was magic enough, but now we saw a single, big billy walking through a narrow pass 300 yards below us. This was the first close-up look we had at the big-game animal since we arrived four days ago. With our binoculars, we saw specific details — eyes, ears, horns and dirty yellow coat — that lacked fullness because it was still summer. Seeing him gave me the hunter's "edginess."

Martin spotted a single ram at 9:30 a.m., a distance of 600 to 700 yards. He was alone on an intimidating table-top ledge that, on one side, appeared to drop into nothingness. Our side of the ledge featured a treacherous rock slide beginning just below him. Since we both wanted to take a sheep, we searched diligently and patiently for his companion, albeit unsuccessfully.

My attention returned to the ram in sight. He was clearly legal. Beyond that, he was an exceptional animal, with thick bases and flaring horns. There was an overused metaphor about how a "majestic ram oversees his kingdom from a loft perch." The sight of him took my breath away.

The wind picked up, and it was either that or the presence of game that gave me a severe case of the shivers. I put on all the clothes I packed up the mountain, including a wool cap, gloves, a down vest and jacket. As a primitive desire to take this beast starts to assert itself, I realized how open it is between him and us. Nothing but air separates us. In our favor was the fact that we were above him. He looked the opposite way. Indecision about the open stalk won out and we held our position.

No second ram appeared and we decided to begin an admittedly risky stalk at 10 a.m. A rock looked to be about 250 yards from where he lay became our destination. The sun shone on him now and his horns appeared golden. Sometimes he laid his head down and looked to be sleeping. At other times his head was erect and alert, but he always looked down, away from us. We began our approach with the wind in our face. Martin quietly cautioned us not to dislodge any rocks.

Long ago, Casey and I agreed I would take the first shot. We made this stalk together hoping another ram became visible. This hasn't happened yet. The time was 10:35 a.m. Martin instructed me to slide a bullet into the chamber. Closing the bolt gently, I realized for the first time this goal could be achieved. A surge of adrenaline coursed through my veins.

We covered just more than 100 yards, mostly by crawling slowly across uneven, rocky terrain. We heeded Martin's advice about dislodging rocks, but the truth was we made some noise. It was virtually impossible not to. The gods of the hunt were with us, as was the wind. The ram continued looking away from us. No "whistlers" have announced our presence to foul things up either. My involvement in this hunt was absolutely intense. My head was in the game to the degree I saw, felt and heard with precise clarity.

When you routinely look at animals through spotting scopes or binoculars at extreme ranges — animals you cannot see with the naked eye — you tend to lose perspective. Looking at this ram at 300 yards, without the aid of optics made him seem closer than he was.

The Sifton Range held a treasure of sheep for the three of us who hunted Stone's sheep, 300 miles north of Prince George, British Columbia.
Photograph by Bob Karel.

Consequently, I have what was probably a misplaced sense of "I can bag him from here."

Martin said we should get closer. It was now 11:20 a.m. He was encouraged by the fact the next 50 yards walked with relative ease as the mountain flares provided our first cover. The ram hasn't turned his head. Martin and I proceeded without Casey, having abandoned thoughts of a second ram.

We arrived at the shooting place at 11:40 a.m., and I asked for the chance to catch my breath. With a history of buck fever, not the debilitating kind but the kind that required a little self-talk before shooting at a big-game animal. I needed to gain control of myself. As I laid my rifle in a mossy gap between rocks, the ram finally turned his head. We later felt he saw or heard my coat. Maybe he just decided to look behind him. The object of my fantasies stood and looked directly at me at 250 yards. Though I was not sure why, I remained deadly calm. My sight picture and cross hairs were perfectly still. I felt the shot was entirely possible.

At 11:41 a.m. Martin whispered to wait until the ram turned broadside. By now my "deadly calm" feeling of one minute ago started to erode. I was getting nervous. Time for more self-talk, I thought.

My .270 was zeroed in at exactly this distance. The cross hairs were immobile on his chest, just below his neck. I stopped breathing. The sound of the shot surprised me. I worked the bolt unconsciously. The ram showed signs of being hit, but still stood.

From Martin I heard, "Shoot again."

I did, and the second 140-grain hand load hit him close to where the first one had. The ram was still on his feet, and we don't want him to fall the wrong way off the ledge.

In the excitement, and without my glasses, I found I crowded the scope and it cut my forehead twice. Wiping blood from my eyes, I fired a third shot that was a clean miss. The ram lay down briefly, struggled to his feet and turned broadside. The fourth bullet put him down hard.

Looking to Casey behind me, I raised my arms like a referee signaling a touchdown. Casey raised a clenched fist in response.

We retraced our stalk and had lunch where we earlier shed our backpacks. Martin applied two butterfly bandages to my forehead. The scar serves as a reminder to carry a second pair of glasses on future hunts. I barely tasted the canned meat sandwiches and candy bars in my anxiety to return to the ledge where the ram lay and touch him.

By 1:25 p.m. I was pulling myself over a jagged stone just below the ram. I was finally able to stroke his horns. Lifting them with both hands, I appreciated the satisfying weight and length, and the appealing shape and symmetry. Stroking his neck and back, I felt the muscles beneath his skin. I realized the kind of prize I possessed. How much I respected this animal. My thoughts and feelings raced in every direction from sorrow at having ended his life to the magnificence of this environment, to the rightness and the evolution of our predatory instincts.

In the book, The Green Hills of Africa, Ernest Hemingway takes a rhinoceros with an outstanding running shot at 300 yards. In truly accurate terms, Hemingway describes how he always was able to recall the shot and the moment, how it would give him pleasure when he thought about it and how he was completely certain his pleasure would not diminish over the years.

That is how I feel now. This thing I have done, this moment, this achievement, will be with me always. The ram will make me happy when I think of him, and of the hunt, until I rest in the bone yard.

I almost completed the mundane chore of caping, skinning and quartering the ram by 3:30 p.m. As we began to load the quarters into our backpacks, we heard rocks fall and something climbed through the pass below us.

Immediately we thought, "Casey's ram!"

It began to rain lightly. We were tensed trying to find the source of the sound. Moments passed and a lone billy goat, probably the one we saw earlier, showed himself at less than 100 yards from where we sat. He was a beautiful trophy, but we were not hunting him. We were glad for his being here, though, and enjoyed watching him for several minutes.

The rain made the descent of the mountain slicker and more treacherous than the ascent had been, and by 5:10 p.m. we took another of our frequent and welcome rests. Casey and Martin took every opportunity to glass for another ram, but we had not spotted any other sheep this day.

At 6:20 p.m. we continued down the mountain. I was tired and my quadriceps ached, but worse than that, the fear returned. The rocks and the mountain itself were genuinely slick and dangerous. Casey and I shared glances without words that said volumes about our state of mind. As we clung to the mountain, sometimes desperately, I knew where Casey's thoughts were because mine were there as well.

By 7:45 p.m. we were in the snow wash again, and the whistlers started up at about the same location. Water was just ahead, and we were thirsty beyond description. I never realized the muscles going down a mountain were taxed in about the same manner as they were going up.

After an hour's rest, and refreshed from the mountain stream, we began side-hilling in the opposite direction from the morning. Casey and I agreed side-hilling was the most physically demanding part of the entire day's efforts. I was amazed Martin could do this at all. I was not really sure what a fused ankle was, except Martin's right foot seemed to turn in unnaturally. He was constantly compensated for that as he walked. I was thankful for his friendship, his determination and his skills.

Arriving in camp at 10:30 p.m. I slid the pack off my back as I savored the excitement of the hunt. From our camp, we viewed the Finlay River Valley, as well as the mountains beyond it. Once again, I realized how insignificant these mountains made me feel and how humbled I was to be here. The fear was gone now and the humility was comforting and pleasant. I needed to come back again, as often as my income and my circumstances allowed.

Sheep backstraps fried with mushrooms, cheese potatoes and a drink of Tang made up our almost midnight meal. This was an exclamation point to the day's activities and to the fulfillment of my dream. The backstraps were delicious, and I knew if they tasted like sawdust I would still enjoy them because of what they meant to Casey, Martin and me.

Comfortable in my sleeping bag, my head resting on a rolled up coat that served as a pillow quite nicely in these circumstances. Before drifting off to sleep, I recalled an article I recently read in an outdoor magazine. In it, the author wrote of a Stone's sheep hunt being "an experience that can change your life."

At the time, I thought he had overstated it. I was wrong.

The Extra Effort Ram

DON MACE
From *Wild Sheep*, Spring 1991

As we viewed this rugged hanging basin high in the sheep mountains of Wyoming, I heard the clattering of rocks.

"This is the one we have been looking for. Take him!" said Greg.

As my eyes searched wildly for the ram, my mind flashed back over the many things involved in getting me to this point.

Greg Peterson of Pinedale, Wyoming, was guide for Don Mace when Don took this outstanding ram that had a green score of 175 B&C points. This was a ram of a lifetime.
Photograph courtesy of Greg Peterson.

It all started at the 1990 FNAWS convention in Reno. I had already taken a Stone's and desert bighorn, and felt the most difficult leg in my quest for the Grand Slam of sheep would be the Rocky Mountain bighorn. I did considerable research on bighorn areas and outfitters, and heard many good things about Wyoming and Greg Petersen Outfitting.

Greg was young, but had hunted sheep for 12 years. By serving only a limited number of hunters and guiding all of his sheep hunters personally, he had built an impressive record with a high percentage of records book-type heads. I met Greg at the convention, and after convincing him of my good physical condition and backpacking abilities, he suggested hunting in Wyoming Area 23 that contains Wyoming's highest peaks, and is generally considered to be the most difficult terrain in North America.

As you guessed, I did draw a permit and found myself in this awesome and inspiring area on September 1, 1990. The terrain seemed to be made up of solid granite cliffs and glaciers, with little in between. We covered much territory, ranging in elevation from 9,500 to 13,000 feet. After four days and many miles we saw 40 rams, but not the granddaddy of which I dreamed. Greg felt the sheep had altered their normal pattern. He suggested we continue hunting later in the season, taking advantage of probable snowfall and the resultant sheep movement.

One evening about three weeks later, I received a phone call from Greg saying he located some good rams. This is what I had been waiting for. At 3 a.m. the following day I drove from Salt Lake City to the Green River Lakes near Pinedale, Wyoming. There I met Greg who already had the horses saddled.

Joining us on the trip was Greg's father, Fred, who volunteered to go along as camp cook and wrangler. Fred completed his Grand Slam in 1980, and over the years had taken three bighorns in this area, including an unbelievable 41-inch ram. I silently wondered if Fred

Don Mace took this ram in Wyoming Area 23 that contains the state's highest peaks and is considered by most as difficult terrain.
Photograph courtesy of Greg Peterson.

thought being a wrangler and camp cook was anti-climactic to earning his Grand Slam and past sheep hunting experiences.

It was a pleasant ride on the 12-mile trail that followed the Green River to its headwaters. While riding, Fred explained his "love affair" with sheep and the sheep mountain was never ending, and actually increased with time. He definitely was willing to accept any position in order to be included on a sheep hunting trip. As I listened to the father-son conversations, I comprehended their great love and respect for this mountain and the sheep that reign supreme. This common interest seemed to intensify their already strong father-son bond.

My confidence in my guides and my enthusiasm for the hunt were both increasing with each mile traveled on my well-mannered horse. Fred, who was leading the pack horses, reined up and said, "End of the road." I was ready.

Greg wanted to get partway up the mountain that evening. We quickly assembled our backpacks and headed up the hill. Fred remained below, tending to the horses and setting up a base camp. We reached Greg's spike camp in two hours of good hiking just as the sun was going down. As we set up our backpack tent and prepared dinner, I relaxed and reflected. Less than 24 hours ago, Greg had phoned me in Salt Lake City. Now, here I was, high on a rugged and majestic mountain, anticipating my hunt of a lifetime.

Long before daylight the next morning, I heard Greg moving about and boiling water for our breakfast of oatmeal and coffee. During breakfast, Greg was quiet and obviously contemplating something. Midway through our second cup of coffee, the silence was broken. Greg explained the sheep he had previously located were a fairly easy sidehill stalk from our spike camp around to the next small drainage. He explained the late-season storms had moved the majority of the rams to this intermediate elevation, but often times a few of the really big, old rams stayed at the high elevation and ran single. He said most of his past hunters were physically unable to accomplish this type of a feat, but he felt I could. This was a definite boost to my ego, but looking straight up at the huge mountain above us, and then thinking of the respectable rams close at hand, I wondered.

I thought, "Here is a guide who doesn't even have the permit, and he is willing to go that extra mile!"

I said, "Let's go."

The snow was 8 to 12 inches deep in the draws and northern exposures, and was partially bared on the southern exposures. The terrain was extremely steep as we went up, up, up. I was thankful I had trained for this hunt. After 3-1/2 torturous hours of climbing, we broke out of the last scrub timber and over the lip of this almost vertical mountain. There before us was the most beautiful sight I had ever seen. Directly in front of us was a small lake surrounded by gently sloping banks. Above and in back of the lake was a spectacular hanging basin, consisting mostly of boulders and slides. We inched our way to a rock that could serve as a shooting rest. Then we heard the rolling rocks.

The flash back of all my sheep-hunting rigors seemed like an eternity, but probably lasted only a couple of seconds. I came back to reality as Greg was sliding his down jacket between my rifle and the rock on which it rested. Greg had one hand firmly on my shoulder and the other pointing toward the location of the ram. I picked up the movement of the big ram as he glided effortlessly up through the shale slide.

"Get on him and wait until he stops. Range is about 375 yards," Greg said.

The ballistics chart of my .270 automatically clicked into my mind, and as if a prophecy, the ram stopped and turned broadside to view his intruders. The shot rang out and echoed through this magnificent basin on top of the world.

The ram measured a full 16 inches around the base and 37 inches around the curl. He scored 175 points, green. As we stood admiring this trophy of a lifetime, I felt a great exhilaration, mixed with a very deep and humble feeling. Only those who have been there can totally understand.

After many pictures, Greg caped the ram for a life-size mount and boned out the meat. Our packs weighed a good 100 pounds each. As we made the treacherous six-hour descent, I could tell Greg was anxious to share our experience with his father.

We arrived happy and exhausted at the base camp as the sun eased over the horizon. We were greeted by Fred and the horses as we set our backpacks against a fallen tree. We silently admired the massive ram head protruding from the backpack. Fred put his arm around Greg and gave him a proud and approving smile.

He didn't have to say anything.

Changing of the Guard

Gene White
From *Wild Sheep*, Winter 1992-93

Lee Trimble took the author on his first sheep hunt in 1975 in the Yukon. This 11-year-old Dall's ram was downed on the Yukon-Northwest Territories boundary line.
Photograph courtesy of Gene White.

"This is my favorite part of the hunt," whispered my guide and outfitter Ray Woodward, as we lay glassing four bedded mature Dall's rams 400 yards away.

They were stately posed atop a high Northwest Territory vista surveying the beautiful MacKenzie Mountains. The waiting part wasn't my favorite. Anticipating that all-important shot was making me shaky; butterflies were getting out of control. For some reason my subconscious brought up every sure shot I'd blown in the past. I was starting to shiver. I didn't know if it was the cold, wet moss we were laying on, or the excitement, but it was more than I could handle. The longer we waited and watched those rams, the more I shook. At that long range I wouldn't shoot unless I had a good rest and a steady hold of the cross hairs that I wouldn't have if I lay there much longer.

"Ray, I think I can make it from here."

Ray's hesitation didn't hinder my optimism.

"Remember three years ago I took one at this range under similar circumstances?"

Ray was obviously considering my request and the three hard days we'd spent pursuing these sheep. He crunched down a bit lower behind the clump of grass, hiding our presence from the sheep, looking over my way, and giving me the okay.

"The big one is lying right on the top at the very edge of the cliff, so anchor him good. It's a steep drop on the other side."

Ray had owned, outfitted and guided the Arctic Red River area since 1977. This hunt was to be his last, but not because his plane had fallen from the sky as many other noted sheep outfitters have met their end (Gary Powell, Frank Stewart and Lynn Castle, to name a few). There was no wilderness accident similar to that which took the life of Lloyd Zeman when his horse fell off the mountain with him on it. Ray has retired as an active sheep outfitter. That's right — retired! A fairly uncommon feat in this addictive field. I felt very privileged to be with this legend on this his last outfitted hunt.

In past years with Ray, I'd taken a 40-3/4-inch and a 39-1/2-inch ram, but more than the inches on the horn or the majesty of the country, it was the way Ray and his crew made the hunter feel at home. Arctic Red River had always been a family operation. Three of his sons and stepson acted as guides. Jackie, his wife, and Jerry, his sister-in-law, kept the campfires

The author's last hunt when he took a Dall's sheep with the help of Ray Woodward and Kelly Hougen.
Photograph courtesy of Gene White.

warm and delicious with food on the table at base camp. The good news was Ray has sold to Kelly Hougen, another energetic, personable sheep man who will keep the area up to the same high standards Ray had maintained.

I took my ram, but not in the classic manner as Jack O'Connor would have done it. Jack would have laid the ram over after a well-placed shot and pierced the top portion of the ram's heart. My first shot, aimed at that same portion of the heart, missed the ram. The sound of the shot ricocheted off the rock outcropping beyond the rams confused them as to my location, and off they came directly at us at full speed. When the biggest ram came by at 50 yards, I shot him again — this time in a vital spot. Ray didn't say anything about the shot, but he did congratulate me on a nice ram. The mixed feelings of a successful hunt, the death of a beautiful animal, the knowledge this hunt may be my last sojourn on this hollowed ground, all flooded my mind at the same time. If you could bottle this feeling to recapture the moment again, illegal drugs would be overshadowed and outdated.

Back at camp, after the meat packing is behind you, you can relax and give your feet, legs and back a much-needed rest. Ray's area is backpack only. If you take a ram here, you've earned it. Even the hard plywood beds at base camp felt like Hilton's best. Base camp was on a high mountain lake named Goober. Where the name came from I didn't know, but this 10-by-14-foot wooden shack was a welcomed vacation retreat especially compared to the backpack tents of a 7-by-4 foot used in the mountains. Right after Ray and I got into base camp, the skies closed in from rain, fog, etc. As everyone knows, when the weather closes in, you do everything possible to take your mind off the boring wait. But there's usually not much to do on these high mountain sheep hunts except wait.

The first Northwest Territories hunt in the early 1980s when I was about 30 pounds lighter. The weather was great after three days of being socked in. Today, Gene White is a guide for giant tarpon fish in Flagler Beach, Florida. He said he'd be glad to trade a tarpon fishing trip for a sheep hunt any day. Call him at (941) 964-2232!
Photograph courtesy of Gene White.

So, in between meals of sheep meat cooked every way imaginable, bannock and the cabin stock of goodies, Ray in his easy going manner laid out, at my request, the beginnings of Arctic Red River.

In 1969 when the Northwest Territories first opened to hunters, the area was owned by Johnasen. He was a local man who apparently was more a naturalist and trapper than a hunter. For years the area went unhunted even after it was transferred to Fred Carmichael, a pilot and outdoorsman, who took limited numbers of sheep and what he did take were small. He only ran it a few years. Then in 1977, Ray took over the area and transformed it into one of the greatest trophy sheep areas in the Northwest Territory, but it wasn't without sacrifice. Ray's first wife left him with three small boys, but it didn't slow him down at all. Ray is a man of great faith in God and himself. Within a few years he met and married his present wife, Jackie, She also had three small sons, and together they started over.

At 6 feet, 6 inches, and 280 pounds, Ray is a mountain of a man. Jackie is the size of a petite Kathy Rigby, well under 5 feet and under 100 pounds, but a better match you'd never find. They both share the love of the outdoors and of each other.

The few people of this remote mountain area don't have television or radio as we know it. They have what the old miners used — single side band. It's no longer used by miners, but it is a lifeline and, in many instances, it's the only contact with the outside world to many in these backwoods areas. In times of bad weather, this "entertainment" was better than Regis and Kathy Lee Gifford. We listened to a woman who was frantically calling for a helicopter to help retrieve a man who had either fallen or was shot off a mountain ledge, apparently directly across the lake from her. When the helicopter finally did arrive, nobody was ever found or any evidence to verify her story. Only an empty liquor bottle was found at her cabin, but while the search continued there was a lot of anxious moments.

In recounting some of Ray's favorite memoirs of past hunts, he didn't center his conversation on the 42-2/8-inch ram taken in his area some years ago, or the famous outdoor writers who had hunted with him. Instead he recounted two incidents. He told of a 74-year-old hunter named Bob White (no relation to me). He met Bob at a Louisiana outdoor show. Bob had never hunted big game before but he, like everyone else, wanted a 40-inch sheep. He asked if Ray thought he could help him get one. Ray didn't make any promises, but told him he had a decent chance. So the next year Clark, Ray's stepson, who also guided me on my first trip, took a pack dog to help carry grub and guided him on a 40-mile pack trip. They encountered snow on the third day, but Bob was successful in taking a 40-1/4-inch ram just barely making the 40-inch mark. But after all was said and done, Bob was more excited about all the great experiences he had than the inches on the horn of his beautiful ram.

Eric provided a pleasant memory for me. Eric wasn't a hunter seeking a great hunt. He was a young fellow who called on Ray one winter and volunteered his services free of charge as a packer. He wanted to learn the ways of the outdoors. Eric did. He worked out quite well, despite a rookie's mistake of putting wet boots too near the fire and burning them; a mistake that could be very inconvenient depending on where you are from base camp. He later advanced and became a competent guide for Ray. These were just a few of Ray's rich array of pleasant and not-so-pleasant memories.

Ray loved his 14 years in this northernmost sheep hunting area in the Northwest Territories. Kelly Hougen, Arctic Red River's new owner, is a real energetic, intelligent, personal fellow in his mid-30s, from White Horse and has an extensive hunting background. What impressed me about Kelly while I was there was his love for the outdoors and his great desire to give his hunters a great hunt and a trophy ram.

Ray hasn't left the hunting scene by any means but has simply moved his operations to two completely different fields — goose hunting on the Texas Gulf Coast, for greater, lesser, speckle bellies and snow geese, and in the off-season, fishing, bird watching and wildlife viewing in the Barren Grounds of the Northwest Territories. In Ray's new area, happy fishermen have taken lake trout up to 32 pounds and arctic grayling up to 4-1/2 pounds. Kelly didn't do too bad during his first year along the Arctic Red River. His hunters took six rams over 40 inches including two that went over 170 B&C points to make the all-time records book.

His summary of the season sums it up very well: "We're real happy with our first year in the area. More importantly, all of the hunters went home happy."

At the ripe old age of 47, I hope I can make at least one more trip back to the MacKenzie Mountains. Next time I hope I can take my 14-year-old son, Jack, whom I hope someday will share my love of the outdoors!

Reflections from Within

Frank R. Kusher
From *Wild Sheep*, Spring 1994

Frank R. Kusher holds his desert bighorn sheep that was taken on November 17, 1992, on the Ireteba Peaks in the Eldorado, Nevada. Frank began his efforts to collect a Grand Slam of North American wild sheep when he collected his Dall's sheep ram in 1979. His desert sheep was the trophy that completed his quest.
Photograph courtesy by Frank R. Kusher.

This short story isn't going to begin with the standard exclamation of surprise as I was notified I had received the coveted Nevada desert sheep permit. No indeed. I have to go back in time much further in order for everyone to appreciate and share with me in my moments of glory.

So if you want to read a tale about a work-day type of fellow who collected his Grand Slam, sit back and I'll share with you some of the highlights, and bend your ear just a little.

Born and raised in Pennsylvania, I've hunted since I was a child. As far back as I can vividly remember I followed my dad with his beagles to the woods to hunt rabbits. He never forced me to go with him. Something inside me drew me to the woods. I can vividly remember listening to the beagles barking and sometimes refusing to give up even hours after the last light of day was gone. Standing in the dark on a cold November night quickly made this little 5 year old hungry and chilled to the bone. You know something I kept going back. The wonderful outdoors, the thrill of the chase and being with my dad were the greatest feelings.

As the years went by I continued to enjoy hunting in Pennsylvania, but I kept saying to myself I wanted to go on a hunt out West. One evening while watching a nature program on television with a lady friend who later became my wife, I talked about my desire to hunt mountain lion. At that time I was working in the coal mines.

My future wife said to me, "You have a decent job. Why don't you go?"

Before I knew it I was booked for a lion hunt in Colorado. The year was 1977, and I was fortunate to collect a beautiful tom that was 9 feet, 4 inches long.

After that trip I was not excited, hooked or whatever catchy word you want to use. I was ADDICTED. At that point in time I came to one stark conclusion: There was no way I would have the luxury to hunt every species on the planet.

Out came my old hunting magazines and books. One of my favorite books by Jack O'Connor kept coming to the top of the pile. Dreams of becoming O'Connor the Second echoed in my mind. Hell, I was already halfway there because I owned a Pre-64 model 70 in .270 caliber. I continued my research and questioned myself. Do I want to hunt one of the large bears? Impressive, yes, but not for me. Perhaps a hunt for a heavy-antlered elk or

Frank's bighorn ram was taken by him on October 30, 1986 in the Absoroka Mountains, Wyoming.

A Stone's sheep ram taken in the Turnagain River area of British Columbia by Frank R. Kusher on August 7, 1984.

---

moose? Equally impressive, but again, not in my program at that time. Finally, the light came on. A ram, the most majestic animal on the face of the earth! Well, sheep hunting was what I wanted to do next.

Rather than ramble, I'll endeavor to be brief. In 1979 I spent 13 days hunting Dall's sheep in Alaska's Nutzotin Mountains and was lucky enough to harvest a nice ram on the next-to-the-last day of the season.

In 1982, a trip to British Columbia was unsuccessful. This was devastating to me but I got over it. Instead of sulking, I dedicated that hunt to all the people who say booked hunts are all guaranteed. It's truly amazing what a guy will do to prove a point. Then, like a sailor on shore leave who had been over-served, I talked the outfitter into letting me go back. After a lively episode with a drunken guide, I was the one who finally sobered up.

In two years I tried again. Call it the proverbial sheep fever. The same drive that had me out in the woods with my dad in those early years told me I had to hunt another ram. I hunted with the late Coyne Collison and took a very respectable, heavily broomed ram. I passed up a ram on either side of the broomed ram that was about 39 to 40 inches each, in exchange for the old-timer. That proves size doesn't always become the measure of a true trophy.

Two rams were on the wall, and that was all the rams I ever expected to get my sights on. A letter from Wyoming in 1986 notified me I drew a permit. I thought that was the hard part. However, this hunt turned into a marathon, to say the least. My hunt was booked with Rick Felts of Grizzly Ranch. He and I really developed a friendship. Due to lack of luck, it was necessary to hunt for 30 days at the risk of losing my job with the coal mines. I just couldn't give up the chase. I believe Rick was ready to change my last name to Felts. After all that time, another broomed ram became one of my most prized possessions. I only regret I didn't keep a diary of all the miles both walking and horseback riding, combined with the many unusual sights I witnessed. One spectacular sight deeply etched in my mind were two eagles locking talons and somersaulting through the air. It only lasted a moment but the memory will always be with me.

The following February I married my lovely and understanding wife, Barbara. We honeymooned in Wapiti, Wyoming, with Rick Felts at Grizzly Ranch. I finally got to see parts of Wyoming I hadn't seen in my earlier 29 days of hunting. Another humorous event happened after we went to Cody with Rick and stopped for a few cold ones.

As we were driving back to the ranch, Rick looked up at the many peaks in the Absaroka Range and said to my wife, "See all the mountains up there, Barb," as he pointed. "Frank and I have been on them all at least twice!"

Now, I had three fine examples of the king of the mountain. The last one seemed impossible in my quest of the Grand Slam. My wife kept telling me to keep applying for the desert ram. I likened getting a desert sheep permit to getting hit with a meteorite.

In June 1992, the meteorite missed but the permit struck. I truly felt like a celebrity. Outfitters were contracting me by phone and mail. After many conversations with various guides, I finally chose Ron Biggs of Wildlife West of Fallon, Nevada. My permit was for Area 265. It didn't take a rocket scientist to figure out a ram would not be waiting as soon as I arrived in that desolate area.

However, I had the permit, a reputable guide and the biggest desire anyone could pack into a 188-pound frame. On November 17, 1992, and 12 years of applications and hunts, my wildest dreams were fulfilled. With the help of Ron and the great eyesight of Mike Ryle, our sub-guide, the 9-1/2-year-old broomed ram was mine. I'm not proud to say I was genuinely overcome with tears of joy. Even though one horn was damaged due to an infection from the bott fly, I wouldn't trade him for any ram living or dead in the state of Nevada.

Ron and Mike set up a well-equipped and comfortable camp among some huge and picturesque boulders. Ron sent me a bundle of pictures to show previous successes. However, I could tell in an instant his sheep-savvy was a-plenty. As anyone who has hunted sheep knows, sheep camp isn't always the most luxurious. Ron had everything well-equipped and well organized as if I was on a vacation and not a hunt. Waking up rested and having nutritious, good food, certainly helps to maintain stamina during the physical demands of the hunt. I tip my hat to Ron. A more sincere and honest person does not exist.

A bonus to the hunts was very special to me. In our hunt area there were petroglyphs I was able to view and photograph. Some of these pictures in rock were of desert sheep. Knowing that I, in 1992, had the privilege of hunting the same animals the noble native peoples had hunted centuries ago gave an extra special meaning to my hunt.

My impossible dream came true. My trophies are at my home to be shared with family and friends. One thing, however, has overshadowed this personal accomplishment. The birth of my beautiful daughter, Francine, on March 18, 1993, has now turned another chapter in my life.

Will I ever hunt sheep again? I certainly hope so. Possibly, just possibly, Francine might tote my .270 up a mountain some day, just like her ol' dad.

Gentlemen Move Over, the Ewes Did It

Duncan Gilchrist

From *Wild Sheep*, Summer 1994

Once again, 1993 was a great year for Rocky Mountain bighorn ram hunters in the state of Montana. By my tally, 13 rams were shot in Montana with scores exceeding 190 Boone and Crockett points. This figure is 10 points higher than is needed for inclusion in the all-time records book. I call these super rams. Four of these 13 persons were women hunters. I thought that was interesting, but upon investigation found the feat totally fascinating.

My special interest in women sheep hunters began at the 1992 FNAWS convention where I was working at a booth. I had heard rumors of an outsized Alberta ram scoring 203-4/8 points being shot by a woman. One day a woman with a tag saying she was from Calgary was standing in front of the booth. I seized the opportunity to ask if she knew who this person might be. The answer?

"Yes, I am that sheep hunter," she said.

I soon learned Ann Pyra and her husband, Percy, were total bighorn fanatics with an enviable track record. Percy has a 191-1/8-point ram listed in the Boone and Crockett Club all-time records book, *Records of North American Big Game, 10th Edition.* Ann also has another ram that will almost reach the 190-point mark. Since that day we have shared video footage, and they do indeed have an intense desire to be around wild sheep.

Mavis Lorenz scored big with her 200-1/8-point Boone and Crockett Club bighorn ram taken when she was 65 years old. The ram ranks the 16th largest bighorn in the world, 6th in Montana.
Photograph by Duncan Gilchrist.

One evening last year in late September I had a call from a woman, Mavis Lorenz, who announced she had a Montana ram tag, was 65, and was looking for information to help her judge the horn size of a wild ram. I explained to her I was working on a ram-judging video, but the narration was still incomplete. I offered to sell her a copy for a modest price.

A few days later, after watching the video, she called again with some more questions on how to judge sheep horns. I offered to go hunting with her for a couple of days. Mavis declined.

"I prefer to hunt alone. Men usually don't like to hunt the way I do. They aren't patient enough to glass for long periods," Mavis said.

It was only a few days later when Mavis called again excitedly announcing she had scored in a big way. Her ram had 44-inch horn lengths and a B&C green score of around 200 points. I was told that when Mavis saw this ram what she had learned from the video flashed through

Sheri Dorathy took her 10-1/2-year-old bighorn ram with the aid of her .270 Winchester in the Paradise area of Montana (Unit 124). The ram has a green score of 194 B&C unofficial points.
Photograph by Duncan Gilchrist.

her head. Instantly, it was obvious a 190-class ram stood before her. The official score of her ram is 200-1/8 B&C points and it ranks the 16th largest bighorn in the world, 6th in Montana. The basic measurements are horn lengths of 44 on the right and 44-1/8 on the left and bases of 16-1/8 inches on the right and 16-2/8 inches on the left. Both second quarters are 14-1/8 inches.

In all, this solo woman hunter put in 11 days in the field and passed up lesser rams waiting for a true trophy. Mavis backpacked her camp high into the mountains, leaving it there until she scored. In addition, she managed to skin her ram for a life-size mount. After meeting Mavis, I found I would be proud to hunt with her anytime if allowed to come along.

Not all the super successful women hunters in 1993 were older. Sheri Dorathy is 21, and she drew her coveted Montana tag in the Paradise area of Montana (Unit 124). Sheri and her family moved to Montana some nine years ago, and she has been hunting ever since. Her father, Van, told me Sheri often hunts by herself and has many solo deer and elk kills to her credit.

Unit 124 is not an easy area to hunt especially early in the season before the rut, as most rams live in dense timber. Most persons hunting in this unit glass from Highway 200, spot their quarry, cross the Clark Fork River in a boat and begin the stalk. Sheri and her father used a different technique by glassing the timbered valleys and draws from the top. They saw many rams and each was discussed as to if it would reach Sheri's goal of 190 Boone and Crockett Club points. They were never quite sure, so in the end each was passed by.

On the fourth day they were studying a group of bighorns when some new rams were spotted running. This time it only took a single glance to tell that here was the ram of her dreams. Soon she had any hunter's trophy of a lifetime on the ground with the aid of her .270 Winchester. A good lesson can be learned here. A top trophy looks like a top trophy when compared with others of the same species.

Sheri's 10-1/2-year-old bighorn ram green scores 194 B&C points, but is not entered in the Boone and Crockett Club as of August 1998. The basic measurements are: length, 41 inches; base, 15-4/8 inches; first quarter, 15 inches; second quarter, 14 inches; and third quarter, 11 inches. This is a very massive ram.

Shelley Goodman is surrounded by her family as she displays her outstanding bighorn ram that was officially scored at 191-6/8 B&C points. Husband Kirk is the left and her daughter is to the right. Brother Monte Ishler is behind Shelley.
Photograph by Duncan Gilchrist.

Let's look at a highly successful Dall's ram hunter from Alaska. In 1975, a girl by the name of Sue Arthur accompanied two friends and me on a Dall's sheep hunt to the Brooks Range. At the time Sue was working as secretary to the director of game in Juneau, Alaska. This was her first Super Cub trip, her first time to the Brooks Range and her first sheep hunt. Accompanied by Rick Furniss, she shot a fine ram. Since that time she has brought her Dall's ram tally up to 14. This past year she killed a ram taping more than 39 inches.

In a recent communication with Sue, I noted at 57, I was slowing down in the hills. In a few days she responded by saying at 51, she had not slowed down a bit, but she could not pack as big a load as she used to. In the next sentence she tells of packing 100 pounds from spike camp to the base camp. That's slowing down?

A while back I was visiting a taxidermy shop in Missoula, Montana, when I saw a huge set of horns lying in the corner. I was told they belonged to a woman, Shelley Goodman, and the horns were officially scored at 191-6/8 B&C points. It didn't take long to find this woman who is filled with glowing thoughts about hunting bighorn sheep. She told me in no uncertain terms this had been the greatest hunt of her life.

Shelley comes from a family of six boys and two girls, all of whom were brought up to hunt. Shelley is the mother of three and she and her husband, Kirk, take to the field each fall with their children. The youngest is going to start hunter safety training soon.

Shelley harvested her ram close to home in Montana's Petty Creek area (Unit 203). She applied in Petty Creek since it is close to Missoula. In addition she knew huge rams lived there. About five years ago, she saw a top live ram many call "Old Moses."

I asked Shelley what she used for horn judging criteria. Much of her advice came from taxidermist Dale Manning at the Big Game Connection in Missoula and from the books, *Montana: Land of Giant Rams* and *Montana: Land of Giant Rams, Vol. II.* They looked for horn drop below the jaw line and big inner holes. A 60x spotting scope was relied upon to evaluate each ram.

In all, Shelley hunted some 21 days with either her younger brother, Monte Ishler, or her husband, Kirk. This sheep brought she and her brother closer as friends than they had ever been. During 21 days afield, they roamed over the unit by hiking and horseback riding. In all, more than 100 rams were inspected.

One day a huge ram was spotted that Shelley was going to shoot until the ram turned its head. About a third of one horn had been broomed. She considered shooting this ram, but in the end, the opportunity was passed.

Shelley uses a Winchester .30-06, a 4x scope and 180-grain Sierra boat-tail handloads.

Successful people generally have some sage thoughts they pass on the others and Shelley is no exception. With her obvious good sense of humor she would like to pass on these words to other women hunters.

"Be patient. Choosing a ram is like choosing a husband. You'll know him when you see him."

Another Missoula woman who killed a ram is Tami Kirkland. I have known Tami and her late husband, Shane, for several years. We have some mutual friends in New Zealand. Tami is far from new to the world of hunting as she has taken chamois and feral goats in New Zealand, with lots of deer, antelope and elk in Montana. She was lucky once before in the Montana drawing and killed a moose.

Tami has been applying every year since 1978 for a bighorn sheep tag. Depending upon the quality of hunting, she has changed her preferred unit several times. This year she drew for and received a tag in Petty Creek (Unit 203). Just like Shelley, she chose Petty Creek as it is close to home and the unit has a reputation for producing large rams.

As with many other western Montana sheep hunters, Tami called John Ottman and Dale Manning soon after she had her tag. These men are the owners of Big Game Connection. John has taken a 190-plus ram and is considered to be very knowledgeable about bighorn sheep and sheep judging. These men maintain a supply of photographs and videos to help permit holders. Tami also relied heavily upon the book, Montana: Land of Giant Rams.

On the first day afield, Tami hunted by herself and saw nothing. After that, she hunted with a succession of companions. Lots of people wanted to sample the joys of sheep hunting by heading into the hills to help Tami search for a ram. Weekends were spent looking for rams and, in all, about 10 days were consumed in hunting.

On the day of her kill, Tami was hunting with John Ottman and another past successful super ram hunter, Carl Schmidt. The season was nearing its end, so her standards were lowered a little. In all, on that last day they saw about 40 rams, including a very large trophy. It evaded them, but in the end she shot a quality ram unofficially scoring 178 B&C points and a long-horn length of 38-2/8 inches along with a 15-3/8-inch bases.

Now that her tag is filled, Tami Kirkland is not shy in saying, "This was the best hunt of my life."

Tami's husband, Shane, encouraged her to hunt by herself so her words of advice to other woman sheep hunters are not surprising:

"Have confidence. Do it. Don't be afraid to go alone."

Tami is raising her two girls to be hunters.

While gathering material to write this piece, I learned there are many woman who take the sport of hunting as seriously as any man who ever carried a rifle. Those I talked with have grown up around hunting as youngsters. I truly believe there would be many more women who love to hunt if dads would take the time to expose their daughters to the sport. Just maybe this is the best way to keep our sport of hunting alive and reverse the trend of declining numbers of hunters.

Dads, introduce your children to shooting and hunting! It may be easier to go afield with your buddy, but by taking your son or daughter, you may be making a buddy out of them.

A New U.S. Number One Ram
The Jim Weatherly Bighorn

DUNCAN GILCHRIST

From *Wild Sheep*, Fall 1994

There was not a single 200 Boone and Crockett point bighorn ram killed in North American from 1955 to 1987, when David L. Onerheim killed a 201-1/8-point ram in southern British Columbia in 1987. That trophy was just the start.

In 1990, Lester A. Kish took a Montana ram near Anaconda, Montana, which scored 200-7/8 B&C points. The next season a 200-2/8-point ram was taken by Eugene R. Knight in Montana's Missouri River Breaks. In 1992 the 200 mark was broken again in Montana by Richard B. Wiant, with a ram scoring 202-3/8 points. During the same season an Alberta woman, Katherine A. Pyra, took a giant officially scoring 203-4/8 points.

The "Good Old Days" must be today. In 1993, the United States record was broken again in Montana with James "Jim" R. Weatherly's trophy officially scoring 204-7/8 B&C points. Its herd mate was taken three days earlier by Mavis M. Lorenz of Missoula. That ram officially scored 200-1/8 B&C points. In addition, a third top ram was taken in Montana by Kevin E. Williams in 1993 that officially scored 199-6/8 points.

Jim Weatherly holds the Boone and Crockett Club trophy horns his new United States Number One bighorn ram that scores 204-7/8 B&C points. Jim this outstanding ram in 1993 in the Upper Rock Creek area of Montana near Missoula where he lives.
Photograph by Duncan Gilchrist.

Whenever a trophy is taken of the quality of the Weatherly ram, usually there are some lessons that can be learned by the rest of us. Most top trophies are taken by good hunters. There was no exception to this with the new Number Four bighorn ram in the world, replacing the famous James Simpson head. (The Weatherly ram is officially ranked in 1998 by the Boone and Crockett Club as Number Four in the world and Number One in Montana.)

Jim Weatherly has been hunting all his life, having been born and brought up on a farm in a small central North Dakota town some 48 years ago. While attending engineering school at North Dakota State University he spent his summers working in Montana, so it was natural that he wanted to move to the "Big Sky" state. Jim and his wife, Susan, have been living in western Montana since 1971, and here they have raised their two children. Today Jim owns his own civil engineering firm, WGM Group, and is a workaholic by his own admission. His wife confirms this.

Jim had been applying for a Montana ram permit for 21 years, and even had five points when the preference system was phased out. For many years he applied in the Thompson Falls area, but as the population quality declined, he began trying for an Upper Rock Creek area permit.

Even though Jim has been a hunter all his life, he did not consider himself a trophy hunter — until he drew his Montana ram tag. A goal was set to take a 190-point ram. He soon went to work learning how to judge and determine a 190-point ram. When asked what was used as judging criteria, the following concepts were noted:

(1) The bottom of the horns must drop below the bottom of the jaw line.
(2) The size of the inner hole must be large — grapefruit size or larger.
(3) Must be overall mass out to the third quarter.
(4) Must be able to evaluate all three sides.

Not being a gun fanatic, Jim uses factory ammunition in his .270 Remington M700 that he bought used in 1963.

So Jim now had a tag. What next? His long-time friend Mike Moore offered to show him the hunting area. Mike has a successful track record in hunting elk, sheep and deer in the Rock Creek area. During his past hunts he had covered the sheep haunts, including some spots that few others knew about. Jim planned to go all-out for a top ram, and he was willing to hunt the entire season, if needed. Five days of scouting were spent before the season opened, along with a full seven days after the season started.

Rather than carry gear into the back country for a spike camp, Jim elected to start from the valley bottom each day, hiking to ram country before light. He soon learned the secret of finding mature rams in western Montana is to hunt in the trees, in fringes behind the open parks. Each time he saw sheep in the open they were either ewes or immature rams; that is, until the fateful day.

Here is the story of harvesting the new Number One United States record bighorn ram in Jim Weatherly's own words.

"As I stepped from the timber at the bottom of a large open park, I could just make out two sheep about a half-mile above me. A look through my binoculars confirmed they were rams; however, it was too dark to judge them. It was overcast and a cold wind was blowing across the open meadow. Since big rams are always in the timber, I was not carrying my spotting scope in order to reduce the weight of my backpack. As I climbed the ridge, three trees were noted on a ridge from which I could crawl out and look at the sheep without being seen.

"As I approached the first tree, I could see that one ram was immature. The other ram would not lift his head from the grass for me to judge him. The little ram was the only one constantly checking for danger. When I crawled out to the second tree, I got several glimpses of the second ram as he lifted his head and grazed upwards, obviously to arrive at tree line shortly after daybreak. This ram looked to be a potential shooter. I crawled back over the ridge, keeping the lone tree between me and the sheep. Once again I climbed the far side of the ridge until I was level with the ram that was about 175 yards away.

"Between the sweat from my climb and the cold wind, my hands and wet shirt began to freeze. For the first time I realized here was a large, shootable ram if his opposite horn was equal to the near horn. He still would not look up or around. The ram merely raised his head to walk two steps and graze again. Finally, he turned his head to look back at the smaller ram. The ram's right horn was equal to or better than the left, and the horn's mass was greater than I had dared imagine.

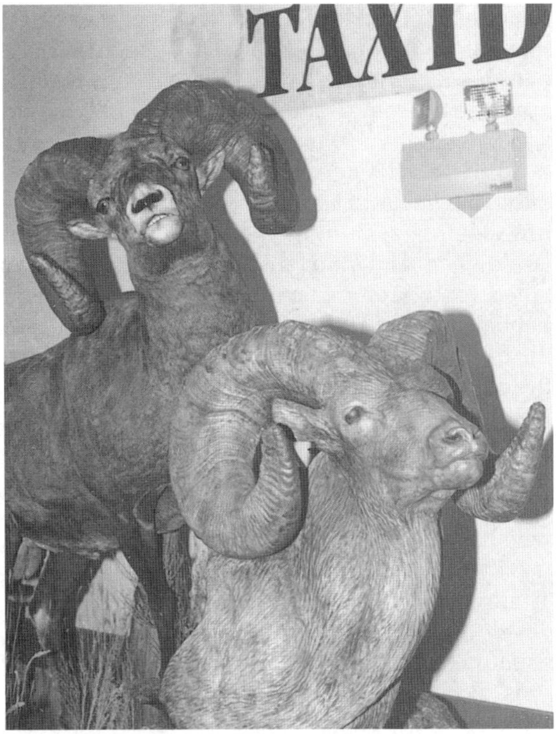

Jim Weatherly's mounted Rocky Mountain bighorn ram is shown to the left and a bronze of his trophy is in front of it. His ram made hunting history with its massive measurements and is ranked Number Four in the world and Number One in Montana.
Photograph by Duncan Gilchrist.

"He was definitely over 190 points, and out in the open where I would get at least two more shots if I only wounded him. He was right above the trail, which would make packing the head and meat easier. At the instant I decided to take him, I looked through the scope while the rifle was resting in the fork of a tree. I began to shake uncontrollably. I put my rifle down, took off my day and fanny packs and sat down. How nervous I was. I made a decision to put on my jacket because of the cold, then decided it would be cumbersome and to leave it off.

"I raised back up as the ram was digging his right horn into the ground. My rifle was up again only to find that two small twigs had to be broken away in order to better rest the rifle. I took aim at the heart area, told myself here goes and squeezed off a round. The ram hunched, turned straight away from me and gingerly started walking away. All I could see was a rump and the back of a massive set of horns, neither of which I wanted to hit on a second shot. The size of the horns was beginning to sink in. The ram walked 20 feet. He was hit hard and if he tried to run uphill he would afford a broadside shot. Suddenly he spun 180 degrees and faced me so fast I couldn't get a side view. Another 180-degree turn and he fell."

During 12 days afield about 70 rams were seen. All but three rams were in groups. This trophy of many lifetimes was the only large ram seen in the open. The remainder were found in trees, generally within 200 yards of open grassland. Jim shared a bit of information he learned that should help any sheep hunter working timber. When hunting from above, rams are uneasy about escaping uphill and become confused, giving the hunter added time to observe.

Jim saw several rams that he thought would break the 190 mark, but only on one other occasion was a ram seen that he would have taken. If a man is to take a top trophy, then good trophies must be rejected. The day of the other opportunity was very warm for early fall. It was a difficult day to hunt as it was noisy afoot as well. As Jim was slowly working through the trees a mule deer doe was jumped. His binoculars were quickly up and following the running deer. Suddenly four rams were in view. Slowly the Remington .270 was laid on the ground so a better view could be had. One of the stationary rams moved and instantly, without question, Jim knew there was a giant ram in front of him. A self-imposed rule had been mandated. All three sides had to be seen before a shot would be attempted. Jim became more nervous as time passed. He attempted to find a clear shooting lane while trying to fully appraise the ram. Suddenly, the herd bolted and instead of four, there were a dozen rams. It was too late.

What advice does Jim have to pass on? As soon as his tag had been drawn, the first order of business was to learn the country. A field trip was made with an experienced local hunter. Jim soon knew of hiding holes that few others realize are there. He talked to all major local landowners and knew just where he could go or not go on private land. He purchased detailed topographic maps and was able to determine the location of the bulk of the major habitats. As with other knowledgeable western Montana hunters, he says trophy rams live in the timber and not in the open. When hunting the timber, try to keep from walking through the main bedding grounds. Travel at a distance and look in. If you spook a group of rams from their bedding grounds, they may not return for the rest of the season.

When I asked Jim what his goals for hunting were going to be in the years to come, he produced a stack of big-game regulations and applications from every state and province where sheep are found. At the least, he is going Dall's sheep hunting. He is one more man who has been smitten with "sheep fever."

By now you are probably wondering why Montana is producing so many large rams. There is no definite answer, but I have my own ideas. It is a well-known fact that animals will develop to the best of their genetic ability in an expanding population. The state of Montana usually keeps sheep populations in check with the issuance of a large number of ewe tags, plus taking additional sheep from the population through the capture of transplant stock. In these ways the effects of an expanding population are prolonged. There is always plenty of food. All of Montana's large rams come from introduced herds that have been blessed with a succession of mild winters, favoring the best of horn and body growth.

Jim's ram now has been officially scored at 204-7/8 Boone and Crockett Club points. It is the largest Rocky Mountain bighorn ever taken in the United States and the fourth-largest bighorn ever listed in the Boone and Crockett Club records books, including *Records of North American Sheep, Rocky Mountain Goats and Pronghorn,* published by the B&C in 1996.

Buffalo Peaks Ram: A Bighorn Adventure

Anthony J. Conte

From *Wild Sheep*, Fall 1994

"Rams," Glen whispered. "I see four rams."

I zeroed in on the area he had been glassing and after searching the nooks and dark crannies, I finally spotted four white rump patches. They were two miles across the valley and about 1,500 feet above our elevation. The place was southcentral Colorado, Unit S-12 to be exact. This sheep unit is a panorama of beautiful western land, including the Buffalo Peaks, a natural arch, and the Arkansas River, which is a fine trout resource.

After 14 years of rejected applications, a Colorado bighorn permit sets the stage for an unforgettable hunt resulting in a 3/4 slam that began almost a quarter of a century earlier.
Photograph courtesy of Anthony J. Conte.

The date was October 6, 1993, and at a half-hour after sundown the bighorn season and my hunt would come to an end. For the past 12 days, Glen Roberts, my guide out of Buena Vista, Colorado, and I had been glassing, walking, riding and scaling the Colorado Rockies in search of the elusive bighorn ram. Each of these days began the same with breakfast before daylight, a climb to a suitable viewing area and a systematic glassing of the opposite slope in search of sheep. Rams were spotted on most of the days and several stalks were attempted, but something always seemed to go awry. Either the rams would simply move on us or they would remain totally obscured in the timber. Everything we tried appeared to be doomed to fail through no fault of our own. The sheep were just not being cooperative and despite all our efforts I still had my nonresident tag to fill.

The high elevation (10,000 to 14,000 feet above sea level), rough terrain and lack of success had taken its toll on my body and mind. The four rams spotted earlier in the morning were just the medicine needed to renew my spirits and rejuvenate my body. We were due for a change of luck and perhaps it was about to happen. We continued to glass, alternating between binoculars and the spotting scope. Soon the sun poked over the eastern summits, spreading light into shadowed areas that were black voids moments earlier. Glassing or patient thorough glassing is difficult and extremely important.

The utmost problem facing the hunter is locating and identifying the sheep. Many hours are usually spent on a sheep hunt with a good set of binoculars and a 20-40 or 60x spotting scope before suitable rams are found. Glen possessed not only top quality optics but also the patience, fine eyesight and concentration needed to carefully study an area and find sheep body parts. It was uncanny how he could search each square foot of a given area and detect a small portion of horn, a dark neck or the white rump of a ram bedded in the rocks or screened by brush and timber. Although I was using an excellent set of 9x40 Bushnell binoculars, I had to rest my eyes frequently. My eyes would not tolerate prolonged periods of glassing.

After painstakingly studying the rams' headgear, Glen decided the four rams were of legal size. In Colorado, a legal ram is a male bighorn sheep with a horn or horns that have one or both tips grown through one-half or 180 degrees of a circle. A line that bisects the eye and the base of the ear will intersect the horn at half curl. We decided to monitor the ram's movements until approximately 9 a.m., hoping they would bed. Two finally decided on a rocky outcropping that afforded them a view of the timber below. The others disappeared into the timber to the left of a small aspen patch. Glen explained the sheep in this area would very often spend most of the day in the trees, possibly to avoid the hot afternoon sun. On some days, he continued, the bedded sheep would get up between 12 noon and 1 p.m. and move to get a drink. They would then bed again but often in a different place, sometimes on the opposite slope.

Glen made it a practice to glass at this time of day to catch them on the move. Today we hoped to get to the four rams before they decided — if and when they decided — to relocate for a midday drink. Our approach was relatively simple. The plan was to climb up through the timber on the east slope, using the trees for some concealment from the rams' searching eyes. The wind was in our favor, blowing from the sheep towards us. All we had to do was stay in the timber until we reached the exposed rock ledge, keeping to the right of the aspen grove and out of sight line of the sheep. It seemed like a pretty basic plan. Two hours later, after a steep and treacherous climb, we pulled ourselves to an elevation just below the rams last-known position. It was a grueling climb, marked by lots of puffing, but I had made it again.

Glen motioned to me to hold tight as he moved farther ahead to improve his view of the rocky area. After relocating three or four times to get a different glassing angle, he discovered the rams were no longer there. Glen was quite sure they had not spotted us. They must have just ambled to higher country. Since the season was coming to an end, we had no choice but to follow the rams. Even though there was a good chance we would spook them out of the timber, we planned to follow them to the top. I was still huffing and puffing badly, so Glen decided to "take five" and give me a breather before we began this risky stalk. I huddled in some boulders, taking advantage of the time to get my wind back and to let my pounding heart settle down. I faced the reality this could be my last day of sheep hunting ever, and I could be going home without a ram.

My mind drifted back to early May when Glen phoned and offered his congratulations. I had finally drawn a Unit S-12 bighorn ram permit. Over the past 14 years I had applied 23 times (14 times in Wyoming without success and nine times in Colorado) before drawing the hard-to- come-by bighorn permit. Although during the past 12 days I had experienced the complete spectrum of sheep hunting, success had managed to escape my grasp. Glen and I climbed until our bodies cried out. We left the peaks in complete darkness and suffered through several aborted stalks and rode for countless hours, often by moonlight. These long hours in the saddle left me sore and weary. My mountain horse, Berry, and Glen's mule, Strawberry, were remarkable animals that took us through heavy timber stands, up and down steep grades and across streams. I found it hard to imagine a horse or a mule could negotiate in such difficult country. A guide once told me a good mountain horse was God's gift to the alpine hunter. I totally agree.

Anthony J. Conte painted, "Buffalo Peaks Bighorns," and donated it to FNAWS for the 1995 Phoenix convention auction. Anthony painted it to reflect the beauty of the Colorado Rocky Mountains and the majestic sheep that rule this kingdom to pay tribute to his hunt. Anthony has donated two paintings to FNAWS auctions.
Photograph courtesy of Anthony J. Conte.

As I reviewed the past 12 days I realized how much effort was put forward on each and every day of a sheep hunt. I had been hot, cold, tired, body sore, sweaty, thirsty, hungry and emotionally drained, sometimes all in the same day.

One day in particular was most discouraging. After a long ride and an especially demanding climb and stalk, we finally got on a ram just before dark. I attempted to shoot while I was still short-winded and ended up pulling the shot just enough to miss the ram to the left. This wasted — and presently only — opportunity weighed heavily on my mind. I had waited so long and worked so hard for this moment. A puff of cool breeze brought me back to the present and my eyes found Glen.

"Take a few minutes," he urged. "We're in no hurry."

I nodded and my thoughts again drifted back this time to 1969 and my first sheep hunt. It had taken place in the Wrangell Mountains in southeastern Alaska. My guide and outfitter was Don DeHart, master guide Number Five in Alaska. Don was a pleasure to be with in the wilderness and became a lifelong friend. On the fifth day of the hunt, I took a fine Dall's ram. In 1986, I was fortunate to hunt sheep with Ed Dolhan in northern British Columbia. On the 14th day of the hunt I managed to score on a magnificent Cassiar Mountain Stone's ram. Ed had an expression I have never forgotten.

"Come as a stranger, leave as a friend," he said. I did just that.

I was brought back to the present by what sounded like a rifle shot echoing from the slope below to the high country above. The loud crack prompted Glen to spin around and glass below.

"Well, I'll be," he whispered. "Those rams are well below us, right back where we came from. They're just having fun being rams and butting horns."

I had never heard the sound of two rams smacking their horns together and I shall never forget the sharp clack that interrupted the mountain stillness. This classic bighorn duel is a ritualistic spectacle that begins as the breeding season approaches and quite often it establishes the order of dominance among the rams of a particular band. Glen quickly laid out a plan to move toward the rams, using the timber to hide us from the sheep's outstanding eyes.

"We'll move slowly and quietly," he advised, "using the crack of their horns to lead us to them. Slide one into the chamber and be ready as we get closer; one mistake and this stalk could blow up in our faces."

Over the next half-hour we moved only a few hundred yards, mapping our route by the periodic skull cracking of colliding bighorns. My heart was beating loudly and my throat became dry with excitement as I anticipated the possibility of a second chance. We slowed to a snail's pace as we approached the spot where the last horn slamming took place. Out of the corner of my eye, about 80 yards to the left, two rams were engaged in eye contact.

They reared up, cocked their heads and propelled themselves together with a skull-cracking smash. What a rare treat it was to witness this event. I was indeed privileged. Almost immediately one ram turned towards me. He must have caught some movement on my part, and in a second they were both streaking downhill. I dropped to one knee and raised my Browning BAR .300 Winchester magnum to my left shoulder. A third ram tried to follow the first two. I swung the rifle to an opening in the timber and cheeked the stock as my eye picked up the escaping ram in my Leupold 3-9x Variscope. As the ram's head moved through the crosshairs, I squeezed the trigger, and he collapsed in his tracks. Glen let out with a "Yes!" and we grabbed each other in a glorious moment of victory I shall never forget. I had taken a bighorn ram — the trophy of a lifetime — on the last day of the hunt, which was the perfect ending to a most exciting hunt. I now had my third ram and a three-quarter slam of North American sheep, over a period of time that began 24 years earlier.

After many photos and lots of happy talk, we skinned the ram for a lifesized mount. The 180-grain Remington Core-Lokt bullet entered the ram's neck and severed his spinal column, putting him down instantly. Hours later, as we packed the ram's head, hide and boned meat down to the valley, I counted my blessings and realized what a lucky man I was at this moment. Sheep hunting is very demanding, both mentally and physically. I had, at 63, met this challenge and triumphed. After years of rejected applications, the odds turned in my favor, and I had drawn and filled my tag. My luck had definitely changed.

First, I would like to express my sincere thanks and gratitude to the Colorado Division of Wildlife for making nonresident bighorn permits available and giving me the opportunity to harvest a Colorado bighorn in the beautiful Buffalo Peaks area.

Second, I feel most fortunate to have had Glen Roberts guide me on this most memorable hunt. His commitment to sheep and sheep hunting is exhibited by his participation in the lung worm medication programs, sheep transplants, census taking, fertilization of winter sheep range and feeding programs in Colorado. His aim is to put more sheep on the mountain than he takes off, a goal he definitely meets. His expertise contributed directly to my ultimate success and his companionship was most appreciated on this difficult hunt. In the view of his involvement with sheep and his endeavors on behalf of Colorado bighorns, I have given Glen Roberts the title, "Grand Ram Master of Colorado."

I feel a great sense of accomplishment in harvesting three North American alpine prizes. Maybe there is still time to beat the odds, one more time, and draw a desert ram tag. Maybe, just maybe, a Grand Slam?

Anthony J. Conte is a pharmacist, pharmaceutical chemist and clinical monitor by profession but a sheep hunter by passion. He attended Long Island University where he graduated with a B.S. in pharmacy. He later received a M.S. in pharmaceutical chemistry from Columbia University. He is a current consultant to Shaman Pharmaceutical Company, Inc. of San Francisco, California. In his spare time he enjoys oil painting, hunting, fly fishing, writing, skiing, figure skating and listening to classical music. He and his wife of 40 years are parents to two adult children, Charles and Lisa. Anthony lives on the north shore of Long Island, New York, where he has resided for more than 40 years.

California's First-2nd Alternate to Hunt Desert Bighorn Sheep

JOHN "JOHNNY" L. CAREY
From *Wild Sheep*, Fall 1996

What are the odds of drawing a California desert bighorn sheep tag? Multiply that by being a first alternate to get a tag, only if someone turns it down. Now factor in being the second alternate and having two hunters in front of you.

"Desert bighorn sheep tag, hmmmmm. No thanks," you'd say.

That's exactly the scenario that led to my sheep hunt in the fall of 1995.

The state of California's one-day orientation had passed prior to my notification I would receive a tag. So, I met with Andy Pauli, a California wildlife biologist in my hunting zone, The Old Dads, and received my personal orientation on the tailgate of his truck. Andy was most helpful and enthusiastic about the up-and-coming ram season.

Johnny Carey with his desert bighorn sheep.

The first words Andy spoke were, "Do you have any idea how lucky you are?"

With only weeks left prior to the opener, I scouted one day with my brother, Tommy, and located 15 ewes and one monster ram. The impression this animal left on me is one of admiration and respect. Unfortunately, I could not locate this monster after the season started.

Opening weekend found me in camp with my hunting buddies who were my brother Tommy, Don and his son, Dean, from Palm Desert, and Ed Hilliard from Big Bear. We spent our time glassing the Old Dads and Kelso Mountain Ranges. We spotted and stalked many sheep that turned out to be ewes or small rams. I spent one day hiking over the top of the Old Dads and down the opposite side. This experience gave me total respect for these magnificent creatures and their habitat.

My second trip out to my zone had me chasing a nice ram for three hours. This guy must still be laughing at me because he gave me the slip just at dark, miles and miles from my truck. The beauty of the desert at night made the hike back a lot more bearable. This was not going to be the slam dunk I anticipated weeks before.

On my third trip out, with my wife Diane, things started to pick up. We came upon Steve Talmadge, another tag holder, who had two rams in his spotting scope about one an a half miles out. Steve tried to decide if they were big enough to go after. I quickly set up my 30x and told him, "If you don't want to stalk, I do."

Needless to say, he went for a closer look. Diane and I held the two rams in our scope for the 90-minute stalk. Steve took the larger ram of the two with a perfect double lung shot at

Field photograph of Johnny Carey and his outstanding ram.

300 yards. We helped him retrieve his trophy from the mountain back to his truck. This was the first bighorn I'd ever seen harvested and it was a real treat.

The next morning Diane dropped me off on the west face of the petroglyph mountain range. Steve had spotted four rams high up on the range at sunset the night before and was convinced one of the rams would make the records book. See what the sheep bug can do to you? Steve already had his ram and was still glassing till dark — just wanting to see them.

Diane drove around the range to the east as I zig-zagged my way through the range. Dropping down the backside into a deep drainage, I jumped the four rams Steve had spotted. Unfortunately, they had already busted me and were on a full run uphill at 150 yards. Hoping they would stop before going over the top, I tried to locate the big boy in my rifle scope. To no surprise they didn't cooperate with my plan and vanished over the mountain. Twenty-five minutes later I was going over the ridge where I last saw them. Crawling on my belly, I glassed the four rams at 350 yards. They were standing on full alert in the tightest group I'd ever seen. All the rams were legal to harvest, but one absolutely dwarfed the other three. Bodies too close, I chose not to shoot for fear of hitting a smaller ram. In a flash they were on a dead run, and I watched them as they left, for not a new canyon, but a new mountain range. Definitely not a slam-dunk.

It would be three weeks before I could get back out to hunt. When I got back, I had to give it all I had for the relatively long season was winding down. I was actually entertaining the thought of hunting from an F-16. I called an old friend in Bishop, California, Steve Yeager.

"Steve, this is Johnny. I've got a desert bighorn sheep tag I'm dying to fill and want to know if you want to help."

Steve was elated to hear of my tag. "Do you have any idea how lucky you are? I'd give my left whatever."

Needless to say, Steve was in. He is a member of the Society for the Conservation of Bighorn Sheep and FNAWS, and encouraged me to join both. I did. Steve is a guzzler captain in the White Mountains and knows more about sheep and their habits then anyone I know.

"After we get your desert ram, you must go for the Grand Slam," Steve hinted on the phone.

"Oh, great!" I thought to myself, now he wants to go to Denny's (restaurant). My true sheep education truly began on January 14, 1996, when Steve met me on the Kel-Baker Road in the heart of my zone. Steve and I hunted and glassed that Sunday until dark, seeing only three ewes for our effort. A two-hour hike back to the truck after dark made me realize Steve would go to all lengths with me to find a ram. Steve's enthusiasm (always saying, "We'll find them, don't worry) paid off the next morning. At 10 a.m. Steve scoped four rams trotting uphill due east of us at two miles. In less than a minute he spotted two more rams in the same drainage much lower down and a half-mile closer. We both set up our spotting scopes for a good look. One ram was super heavy with broomed horns. The other had long horns that flared out on the ends.

"These were the rams you want," Steve offered with a huge grin.

"Oh, buddy," I shot back. "Let's go get one."

"Hold up," snapped Steve. "Let's watch them and see what they do. We've got all day."

Forty-five minutes later the rams bedded down and the stalk began.

It was almost two hours since we last saw the rams. Finally we were coming up to the top of the last finger that separates the rams from us.

"Use the cactus," Steve said, "as a shield to peak over the top."

We looked out at the huge drainage from above. No rams were in sight.

"They should be right here," Steve said as he glassed.

I looked across the canyon with my rifle score and, BINGO, the heavy-horned ram was walking down a small sandy bottom at 1,200 yards. He vanished behind a 20-foot berm, heading to our left. I moved my rifle up the sandy bottom and the long horn was in his bed to our right at 1,500 yards. An hour later the long horn was up and feeding uphill to our right and away.

"Okay, they're definitely separated. Which one do you want to go after?" whispered Steve.

"We haven't seen the heavy horn for an hour, but he should still be down there," I said. "But, the long horn is still in view and is a dandy, so let's go after him."

We slid back off the finger and used it for cover to make it 600 yards uphill and to our right. Over the top, again using a cactus and there he was at 750 yards, tearing apart a barrel cactus. As he fed down towards the sandy bottom and into the shadow.

Steve whispered, "Let's go over the top where the finger saddles."

"Has he lost his mind," I thought. We hadn't put more than our heads over the top for four hours and now we were going to walk over.

"Steve, are you sure about this?" I asked.

"Yeah," he replied.

Steve had us put our arms around each other's shoulders, totally bent over at the waist and taking small steps, side by side right over the saddle.

"Think like a sheep and don't let go," Steve repeated. When we cleared the saddle. We hurried to the bottom and up the other side of the last ridge. Would he still be there or had our sheep impersonation drove him to stateline? Cresting the top, we spotted the ram. He was still feeding, slowly downhill towards the bottom.

"He doesn't know were here," said Steve, "and remember you've got all the time in the world."

I laid my .270 on my backpack and settled in while Steve set up the 30x on a tripod.

"He's out over 400 yards," Steve whispered. "Do you want to get closer?"

"No, I can make the shot," I answered back. "He's in the scope. I'll tell you exactly where your bullet hits," Steve assured me. I prayed for the Lord to bless this effort. I placed the scope two feet above his neck and squeezed off the shot.

"You got him!" yelled Steve as he jumped to his feet. The bullet hit the ram in the pump for a perfect harvest.

Steve and I met with Andy the next day, who told me my ram was 9 years old. His horns were about three feet long and flared out beautifully. I don't look at this as a hunt of a lifetime, but rather as one that will prompt me to see a lifetime of hunts. Special thanks goes to all the people who helped me on this hunt including Andy Pauli, Donna Davis and Steven Hill. Steven stopped by my camp opening weekend to share knowledge and encouragement. To the original tag holder and first alternate who passed it along. Please call. Dinner is on me! To Tommy, Ed and Don who make every hunting trip a great adventure. To Diane, my bride of 18 years, who has out hunted and fished me for the past four years and to you, Steve. Without your help I'd be California's only second alternate with a tag instead of a bighorn sheep.

Hunting the Old Dad Kelso Zone

STEVE TALMADGE

From *Wild Sheep*, Winter 1996-97

There were five hunters selected to hunt the Old Dad Kelso Zone in the 1995-1996 season, and I was one of them.

After getting over the initial shock of being drawn for a California Nelson desert bighorn sheep tag, I began receiving calls from professional guides, all wanting to assist me in my hunt. After careful consideration, I decided to pass on all of the guides and hunt the bighorn on my own.

The first hunt I planned was a four-day trip in mid December. My son, Bryan, was with me. I was really hoping to see some bighorns, as this may be the only time Bryan would be able to go on a hunt like this. We began looking through the Old Dad area and a silhouette of a large bighorn ram was visible. He was about a mile away and my son and I were both able to get a good look at him with the spotting scope before he moved out of sight. Wow, an unforgettable trip already!

Steve Talmadge of Castro Valley, California, with his "Old Dad Kelso Zone" ram.
Photograph courtesy of Steve Talmadge.

We set up camp in Jackass Canyon and hunted the next two days without success. Bryan spotted an ewe on the north side of the mountain. We captured two bobcats on videotape, but could not find another ram.

At the end of the third day we saw Andy, the fish and game biologist, sitting outside his truck, looking through his spotting scope. He was watching two legal rams. We set up our spotting scope and I also tried to see them through my 10x Redfield binoculars. I learned a very valuable lesson. Although we were able to find them in the spotting scope, my binoculars were not powerful enough to spot them. Since I had been using the binoculars the past two days there may have been many more rams I missed!

As I watched the rams, Andy mentioned one looked like a nice ram. I thought briefly about going after him but chose not to. Andy had found him, not I. I preferred to wait and find my ram myself. The next morning we broke camp and hunted our way out, using only the spotting scope to no avail. No rams that day.

The day after Christmas I left home at 3:30 a.m. and arrived at 17 Mile Point at first light.

I planned to hunt my way into Jackass Canyon and set up camp. I brought a lounge chair, blacked out the outsides and left lens of a pair of shooting glasses and settled-in comfort to watch for rams with my spotting scope. At about 1:30 p.m., before reaching camp, I spotted two rams in the lower foothills east of Old Dad Mountain. They were about 2.5 miles away and both appeared to be legal in size, one slightly larger than the other. I watched for about 15 minutes and planned my stalk. I knew this might be a once-in-a-lifetime opportunity, was not particularly concerned if it would be of trophy size or not and did not want to blow the chance. About this time John Carey, another lucky hunter, drove up. He approached and looked through my scopes as I told him my hunting plan. John stayed with my scope to watch my hunt.

I drove down the road and out of sight of the rams. They were now about four miles away. I hiked 3.5 miles to a ridge top. With the sun at my back and a slight breeze on my face, I began my stalk. After 1.5 hours, much of this in pain on my hands and knees on desert rock, I was within 400 yards of the rams located just across the draw. The rams had not seen or sensed me yet. I was as close as I could get without alarming them. I chose to shoot.

I engaged the bipod on my Ruger .25-06 and cranked my scope all the way to 12x. I sighted in on the largest of the two rams. Lying prone, using a bipod, I was shaking too much to shoot. The thrill! The excitement! The pressure was almost overwhelming. I rolled to my back to collect myself. After about five minutes I rolled to line up my sights again. The ram had moved and was facing away from me. It seemed to take a lifetime for the ram to move broadside again, but once he did I was calm enough and my crosshairs were now steady.

I slowly squeezed the trigger. The shot echoed across the canyon. The bullet hit my target, and the ram went down five feet from where he was hit. The other ram ran to the top of the ridge, stopped and looked in my direction, then ran over the edge.

I walked to my trophy. His horns were larger than I expected. I sat, filled out my tag and realized I had just joined the elite club of successful desert ram hunters on my first ram hunt. I felt a sense of pride and accomplishment. I had done this on my own without anyone to guide me.

As I ran to retrieve my camera before nightfall, I met John Carey, heading toward me. He offered to help me with the photographs and in packing out the ram. It was dark when I finally set up camp.

The next morning I called Andy, the biologist, to take measurements. The ram was going on his eighth year and had a broomed-off, two year's horn growth measuring 150-5/8 points. Andy told me this was the 74th ram taken in California since the hunts began.

Before the hunt I wondered what it would be like to hunt rams. It was more enjoyable and exciting than I thought. I am now hooked and looking forward to accomplishing a Grand Slam!

Passion

Art Froese
From *Wild Sheep*, Spring 1997

Erik and his mountaintop ram after 10 days of weather, travel and hunting.
Photograph by Art Froese.

In the deep shadows of darkness, the four horses labored their way from the bowels of the rocky creek bottom towards the brim of the high mountain meadows above. The mosses, grass and dwarf birch on the valley sides gave them solid footing to carry the riders and their weighted packs upwards to the sparse trail that threaded its way to Half-Moon Camp.

This camp is located deep in the heart of the Mackenzie Mountains on the Redstone River system in Canada's Northwest Territories. The riders were caught deep in their own thoughts of the events of the last seven hours and the underlying resolve to return to the small comfort of their tent camp almost two hours away.

As the first rider, Guide Brent Sinclair, topped out on the meadows edge, a pale blue glow lit him and his mount from the moon beyond. The next two riders, Guide Dustin Ekhart and my son, Erik, were met by the same escape from darkness in the thin light over the crest of the valley. The weight of the Dall's ram, horns, cape and backstraps in my pack felt somewhat lighter as I joined the trio in the wide, rolling spaces below the dark outline of the peaks around us. The dark spruce tree forms below and the distant reflection of the moon in a small alpine lake showed the general way home. As Brent and his mount sought a safe route around bogs and saturated alpine ground, I returned to my thoughts of what had brought us here, the adventure of the last nine days and the events of the last few hours.

Two years ago I booked a Dall's sheep and caribou hunt for some Belgian friends and me with Dave and Carol Dutchik of Redstone Trophy Hunts. Work and a sudden health problem kept me from joining my friends. Sometime later, Dave and I worked out the arrangements for a 1996 hunt with my son Erik. In his early youth, Erik had always joined me for several days of elk or deer hunting at a small cabin in the foothills of the Rockies west of Calgary with rancher Roy Copithorne. He loved the stories, the cooking and the cold November nights.

The early morning start on his horse, Tubby, which followed us in the darkness, are strong and clear memories for him. Football, basketball and city adventures had kept him away for several years, but at age 17 and now in his senior year of high school, with plans for attending the university, I felt compelled to offer him one mountain wilderness trip for the great North American wild sheep. I know college and his personal growth will not allow us this opportunity again for a number of years. Both of us had worked on our shooting skills and conditioning prior to leaving for the Northwest Territories on August 18, 1996.

Arriving at the Dal Lake base camp we were met by Dave and Carol Dutchik, guides and cooks and snow. We set our minds to the joy and demands of the two weeks ahead. The hunting groups split direction the next morning. Erik and I joined hunter Joe Brodresser from Florida and guides Brent and Dustin and Joe's native guide, Arnold Payo, on a long day horse pack deeper into the mountains and away from our final destination at Hook Lake.

The mountains held a spattering of snow as we set up our small tent camp in the early evening. The next morning snow and low clouds stayed with us all day and kept us from hunting the creek valleys and mountain basins around us. An early morning rise the following day gave us a quick start as the hunter and guide combinations split up and wound their way out of sight in the trees for the climb to the high basins beyond. Brent and I found rams, but none in the class I wanted. I already had two Dall's rams taken on previous hunts. The hunters and guides straggled into camp between 10 and 11:30 p.m. with a few less pounds and tales of wonderful country, but no rams worth hunting. Brent and Arnold made plans for the next day and hopes were high as our eyelids closed.

Mother Nature seemed to be testing us as the morning rain still came down. Brent became concerned about our ability to cross the Ravens Throat River if the rain continued for several more days.

"It could rise between 4 and 6 feet, and then we would have to wait four or five days before we could cross. We are better off using this bad weather as a travel day, crossing the Ravens Throat and then hunting the valleys and basins in that area," said Brent.

So we left, satisfied that Brent and Arnold were right, but with an unspoken regret for the valleys we would not hunt since Dave had chosen a particularly rugged area for us that held great expectations. The decision to move camp became even more correct as we crossed the river later that day with a 10- to 12-mile per hour current raging under our horses' bellies. Several hours later we arrived at Pond Camp to join Dave and Carol, their son Jessie and assistant, Dallas, and their two hunters. Both of their hunters had taken sheep the day before. The success showed in their faces and lifted our spirits. The best part was Dave and Carol's greetings, the heated tents and the ability to dry out some of our clothes and boots. There is an uncanny ability for some outfitters to make you feel welcome and confident in tough conditions. You know you have chosen your outfitter well when they can create such an atmosphere. Dave and Carol are such outfitters.

The continuing rain the next day kept us in camp once again with Dave and the guides carefully assessing strategy and plans in relation to the weather. It was decided Dave and his group would forge ahead to the prime caribou hunting area around Hook Lake while our group would hunt separately in the valleys. The youngest legs, Erik and Dustin, would take the toughest country that held high hopes for big rams. The other two hunters would split into different drainages and continue their hunt. It was evident Erik's spirits were good and he was at peace with his surroundings as he and I fished and caught Arctic grayling in the overflow pond that evening.

The next day, Brent found 16 rams with what appeared to be seven in a bunch on a distant mountain located seven or eight miles away. That spot by Brent showed why some consider him to be one of the finest guides in the business.

"Those white dots are all the same size," he explained. "They must be rams."

Brent guides for Dall's sheep in the Northwest Territories, Stone's sheep in northern British Columbia, spring grizzly in the Yukon Territory and red deer in Argentina — all in the same year. On top of this, he outfits for archery elk and then mountain lion in Alberta in the fall and winter. At 52 years of age, I thought I'd seen and lived most outdoor skills, but Brent would show me more later.

Erik and Dustin saw rams with one in particular sounding like a real keeper. Their eyes and bodies belied a very hard day walking creek bottoms and climbing high basins, but their resolve stayed with them. Their future references to that one ram showed the recurring review of a hunting decision all hunters have experienced at some time in their hunting careers. Arnold and Joe had found rams but not the size Joe wanted. The guides resolved we would move farther up the valley so Brent and I could hunt the seven rams and the others could look over some more country.

By 6 p.m. the next day we had our new camp set up. Shortly after, the four of us left on horses for the mountain we thought held seven rams. It was possible there were more than one good sheep in the bunch, in which case Erik and I would both have decisions to make. Throughout all this travel the mountains with their grassed shoulders, boulder creeks and high talus slopes and cliffs had transpired to become home. By now none of us hunters felt any doubts about long trips on game trails that threaded through forest, bog and mountain passes. We left only footprints. Our hunting gear became minimal as it does for all who spend day after day in the outdoors.

After crossing a deep creek, we climbed out of the creek bottoms to the high meadow covered with mosses of all colors, dwarf birch, some willows and blueberries, interspersed by clumps of small spruce and the diggings of the resident grizzlies. From a contour in the meadow, we found the seven rams feeding in sparse grass below the upper stone shoulders of the mountain peak. Even from a mile away, the leader looked heavy and long in the horn.

"He's got to be 38 inches or better and he carries his weight well," offered Brent.

There was another ram in the bunch that was heavy and wide but probably less than 35 inches. On the ride across the meadow to the mouth of the creek at the base of the mountain, my mind ran into overdrive. It was the ninth day of the hunt. I had so hoped Erik would get a chance at a good sheep.

While Brent and Dustin tied the horses to some scrub spruce in the creek bank, I turned to Erik and said, "You and Brent can make the final stalk for the ram if things turn out."

His perplexed look and lack of an answer had me confused as we climbed quickly up the mountain to the shoulder across from the rams. As Brent set up the spotting scope, I finally came to my senses and spoke to Erik again.

"I'd like to see you go with Brent, but I'll go if it will make you feel right."

His eyes lit up and he replied, "You go."

What a great and wondrous surprise to discover one's son had figured out the most fundamental ethical issue for the hunter. As a father, I could give him the opportunity to hunt, but I could not give him an animal to kill. A hunter must hunt the hunt on his own. He must be part of finding the game, assessing the strategy, and when things work out, to carry out the stalk and kill.

I thought, "Kid, you've shown me something."

Now, the hunt switched focus and gears. We watched as the rams bedded high on the downwind side of the rock shoulder across the drainage from us. It was impossible to approach them in this location as they had the wind and the view. We'd have to wait to see if they moved or leave them to another day. Just as we started to think about camp, the lead ram started down off the ridge into the creek bottom behind the shoulder of the mountain we were on. Brent and I immediately sprang into action and climbed higher on the shoulder and hugged contours around and down towards where we thought the rams were moving. We ran down a talus slope to a lonely 3-foot spruce and a tangle of rock. After a few minutes Brent slowly raised himself and then fell to the ground, the rams were right there 30 yards away.

A sudden swirl of the wind and Brent blurted, "They're gone."

Instinct started me running up the mountain to look for the rams going high as they most often do. Every fraction of a second counted. Then Brent's call and a sideways look showed the rams crossing the bottom of the drainage on their way up the rock shoulder across from us. I grabbed Brent's pack and rammed it onto a rock and asked for directions on the big ram.

"Lead ram," replied Brent.

"Left or right group?" I implored.

"Left," came the reply.

At the sound of the rifle shot, the hunt was over. The ram came to rest in the lower talus slopes in the waning daylight. Brent and I gathered our things and he shook my hand with a "Great..." look.

I had let him in on my discussion with Erik and he'd said, "Great kid you got there."

I savored the many strong emotions that ran through me from the poignant sadness of the kill, to the intensity of the hunt, to my great love for the country in which I stood and the other thoughts that fathers have. I did not notice the moon rising in the pale sky as we took pictures after Erik and Dustin arrived. There was much to do and darkness threatened.

Brent gave the best demonstration of caping sheep I've ever seen. He used his thumb in front of his fist to separate the tissue between the skin and the fat, all the way to the neck. Under the glare of flashlights, we took front and hind quarters, the backstraps, ribs, neck and tenderloins. I had the right to carry the head and cape and backstraps while the others split the rest of the load.

The long and quiet walk on the creek bottom to the horses allowed us to collect our thoughts. Now as we rode in the moonlit meadow, the northern lights began to play in the expansive sky to our left. At the sight, Erik remarked in wonderment at the world around us. When we finally entered our tents, 3 a.m. had come and gone.

The next day, Erik and Dustin hunted toward the large group of mountains to the west while Arnold and Joe rode to another basin in the north where we'd seen four other rams. My job was to keep the fire burning, the coffee hot and the guide fed while he caped and fleshed, turned lips and ears and later salted the cape. Brent and I swapped stories, jokes and thoughts on a sunny day with the calm dark blue lake framed by mountains before us. The colors in the high landscape had begun to shift to deep oranges, reds and purples and the sky held high fluffy cumulus clouds. Late that day the other hunters came back to tell us that the rams Arnold and Joe pursued had gone and the ones Erik and Dustin found were too small.

So, once again decision time arrived in camp. Dustin and I would head out in the morning to Horse Meadow camp to hunt caribou while Brent and Erik would hunt away to the southeast and Joe and Arnold would go east after rams. When these separate groups left in the morning I could see the resolve in their eyes and posture. If the weather held, they had a great chance for rams.

The ride to Horse Meadow held adventure for both of us as neither one had been here before. For the most part the trail held long views of great high meadows flanked by groups of mountains on the sides. We were entering the great fall gathering area of the mountain caribou — an area also loved by the wolf and grizzly. The note in the cook tent told us Dave and his hunters had taken caribou and were now settled in at the base camp at Hook Lake.

A rare and special moment in the hunting times of father and son. We thought Erik's caribou, "Not bad for a last day bull."
Photograph courtesy of Art Froese.

The night rain and early morning wet snow and low clouds kept my thoughts with Erik and Brent and what the weather did to their plans. We hunted the high sloping plains to the east and found more than 100 caribou but none larger than the one I had at home.

Eventually, the weather got the best of us and we headed back to the comforts of a heated tent and a planned feast of more sheep backstrap. At the break in the river valley, we could see action in camp and the binoculars showed Erik's red coat and Brent busy outside the tent. I held my breath and my words as we entered the campsite to see Brent fleshing a sheep cape. Brent's calm story of a decision of a good sheep, framed by pending bad weather, gave the context. Erik's excited description of a mountaintop stalk with a steep slide down the back of an old cirque stirred me as a hunter and a dad. Steep downhill shooting created some problems but his obvious satisfaction with the ram pleased me to no end. The ram was old, wide, good-looking and scored just over 155 Boone and Crockett Club points. Most importantly, Erik found the age in the horns to meet his criteria for an old sheep.

"I want an old one," he said. "It doesn't matter what he scores."

The deep joy of our small pocket of civilization became even higher when Arnold and Joe arrived in the rain with a tight 1-1/4 curl ram they had taken the day before. This company of adventurers were now solidly bonded with the common knowledge of hard-won success after prolonged effort. Joe at 58 and Erik at 17 had developed a "Jake and the kid" affection for each other that showed in their talk, their jokes and their laughter. It also showed in the joy of each other's success.

We had three days of caribou hunting left that held the ritual of finding a high ridge or knob from which to survey the countryside for caribou and other game. The four of us spent a day watching these animals from a high knoll while Arnold and Joe worked north into secret haunts. The grizzly spotted by Dustin provided several hours of attention and comments as it nonchalantly ate berries, roots and generally moseyed around the country. One big boar had me wishing the authorities would reopen the season on these animals. No one found big bulls that day, so we decided to split up the next day to improve the odds of finding a big bull.

This strategy worked. Joe and Arnold came back that night with a mountain caribou

The author and his ram looking back to the mountain that held the adventure of the events of the previous day and night.
Photograph courtesy of Art Froese.

scoring in the 390s that would put him in the Boone and Crockett Club records book. We shared in Joe's satisfaction. Dustin and I had seen more than 300 caribou and stalked one for several hours I later turned down. The horses, sinking in bog holes deep enough to fill the saddle bags with water, challenged us even more. Late that night the plans for the last day of hunting were made. Erik, Brent and I were to hunt again the high plains to the east while Arnold, Dustin and Joe would clean up camp, organize and go to the base camp at Hook Lake.

We left early in a light drizzle and climbed into the large basin and plains that now held a mosaic of sweeping fall colors. Binoculars and spotting scopes kept finding caribou until Brent proclaimed in a strong whisper the one in my scope was a shooter. It took more than an hour to get around the bogs and lakes before we had this group of bulls in our glasses again from 500 yards away. After analyzing their movement it came time for Erik and Brent to be off on a downhill crawl through the dwarf birch and hummocks towards the group of eight bulls and harem of cows and calves.

I watched the caribou continue feeding and could only guess where Erik and Brent were by the two small bulls to their right, that now stared into the hillside at what I believed to be their location. The roar of a rifle and the solid "ka-thunk" told me the bull was down. The other three shots turned out to be as good but unnecessary as I made my way towards them. Brent's enthusiasm made me think it must be a good bull and he was, as he scored over 400 points to put him well up in the Boone and Crockett Club's records book.

After showing respect for the animal, we let the solid satisfaction of "not a bad bull for the last day" grab hold of us. Following pictures, Brent again showed me some new ways of handling the cape and carcass. We brought the packhorse over, loaded him and were headed out on the long ride to base camp at Hook Lake and the slow end of our hunt. The feeling of satisfaction and great love for this country and the hunt kept me occupied in my daydreams on this homeward journey.

The reunion with Dave and Carol and the other hunters brought stories of high adventures, great effort and difficulties mixed with the joy of success and one sad tale of disappointment. All these stories held us in suspense. They will undoubtedly be told again in kitchens, bars and over campfires. It is a tribute to Dave and his guides that the last guide-hunter combination arrived in camp at midnight on the last day with a fine bull caribou. Even on departure day, on my way to an early morning breakfast, I found Erik and Joe locked in the wilderness, fishing the mirrored lake that reflected the richly colored tapestry of the lower mountain slopes and the peaks above in a peaceful scene of people who belonged and loved this great adventure we call sheep hunting.

Erik earlier had allowed that, "I'm glad I came this year rather than when I was younger as I appreciate it more." Later he stated, "I don't really want to leave."

That evening in Norman Wells at the McKenzie Hotel the stories and camaraderie continued as Erik played some Eric Clapton and Otis Redding tunes on his guitar for us older types.

My great surprise was the warm welcome, hope and long hugs and kisses from my wife, Donna, for both of us when we arrived back in Calgary with grizzled and weather-beaten faces and wide smiles.

I did not expect a father and son sheep hunt to draw the entire family into its web. Old-timers know the greatest gift you can give someone is your time. Parents know values and ethics are of utmost importance. Athletes and hunters thrive on the passion of the adventure. For Erik and I, the 15 days with Redstone Trophy Hunts gave us more that we ever expected. He has a passion for wildness and the outdoors that is powerful and deep.

Whether he will continue to share the passion I have for hunting, and particularly sheep hunting, only time will tell. My daughters, Hannah and Kirsten, and my sports therapist have me largely convinced I might get a few more miles out of this tattered body. Maybe they're right. I once again feel the deep silent call of the wilds in the country of high mountain sheep.

Escalante Bighorn Adventure

Garfield Thomas

From *Wild Sheep*, Fall 1997

What a team! From the moment of that July 1996 midnight phone call from Pete Cimellaro, "You drew the Utah Escalante Unit desert bighorn tag!" to carrying that ram out of the canyon area, I was privileged to plan, prepare and hunt with an undaunted, persevering and the great-humored team of Bill Mamalus, Byron Wiley and Pete Cimellaro.

With early inquires to Don Peay of the Utah FNAWS Chapter and many others, we soon learned access to the best ram country of the Escalante Unit was difficult, if not impossible. Steep canyon walls would require two days of backpacking into a canyon, up a river and then up the other side with a few places to ascend the sandstone walls. We were told it is the toughest desert bighorn hunt in the Southwest.

After years of applications to states offering a desert bighorn sheep permit drawing, Garfield Thomas finally was selected and took this outstanding ram in the Escalante Unit of Utah.

Photograph courtesy of Garfield Thomas.

Thanks to Bill Mamalus' efforts a couple of years earlier, he found a way out of the Escalante river bottom to the high plateau to the east. Last year, while scouting this high plateau, Bill saw a couple of rams that would score first place in the Utah records book for Ovis candensis nelsoni. We hoped to be able to find one.

On a warm October Friday we left the truck and started the trek for eight or nine miles, as the raven flies, but more likely double that with the ups, downs, twists and turns of canyons and the river. The scramble took us through some of the most fascinating geography I'd ever seen. The rock formations, hues, sheer walls and stark beauty were awesome. The Escalante river bottom was lush with willow, tamarisk, cottonwood and various other vegetation. Mule deer and beaver tracks decorated all the banks.

Partway up the east canyon wall we spike camped by a rain-filled pothole. The next day we made it to our base camp location, some eight miles and 2,000 feet higher than the river bottom. The high plateau of rolling sandstone knobs bordered by steep canyon drops offered a multitude of cuts and depressions where the sheep could travel. Here and there were spots of sand for tracking, but not much.

That first afternoon we glassed a couple ewes and lambs within the plateau area we could access. Much more landscape was within glassing range, but steep canyons separated us. For a few seconds we saw a nice, "class four" ram that disappeared into the smooth dome-like rocks without a track or clue as to where he went.

Steep canyon walls made the retrieval of fresh meat especially challenging on this hunt. Here the men on the hunt repel a backpack filled with meat.
Photograph courtesy of Garfield Thomas.

The hunters four include, left to right: Pete Cimellaro, Garfield Thomas, Bill Mamalus and Bryon Wiley. Garfield Thomas had to wait until he past 65 years of age to draw a desert bighorn sheep permit which goes to prove you need to keep in tip top shape because you never know when you'll go sheep hunting! The wait was worth it.
Photograph courtesy of Garfield Thomas.

As nice as that ram looked, Bill said he had seen bigger horns last year. My hopes were flying high but after two more days of covering all the area south of camp, fresh sheep signs were nonexistent. I was starting to become concerned.

Bill would say, "Gar, be patient. When it happens, it happens."

The local ewes provided some entertainment. From a distance, one saw us glassing from a high point and decided to investigate. Stopping about 150 yards away she was becoming uneasy. Pete encouraged her approach with a grunt. Closer — grunt, closer — grunt. From only a few feet away she decided we were not to her liking and trotted off in a big circle.

That fourth evening, having filled up on some rather acceptable freeze-dry, we made plans to search the area north of camp. If we found good sheep signs we would then consider moving the camp. So far the weather had been about perfect. The next morning the starlit sky gave us adequate light for a quick breakfast and an early start.

Not far from camp the two local ewes showed and Byron and Bill held back to see if they might attract some ram company. Pete and I moved farther north to set up. A high ridge offered a shadowed ledge, just under the skyline, with the sun just rising over the rocks behind us — a perfect spot.

After glassing for a few minutes and lowering my 10x28s, there, only 100 yards out, appeared a ram, head-on. About the same time I whispered, "There's a ram." The sheep showed in Pete's 10x50s. The ram was looking straight toward us and into the sun, a beautiful class four for a Utah "Nelsoni."

As I untied my rifle from my pack frame and slid a .280 into the chamber, Pete whispered, "Wait until he turns."

Advice from Garfield Thomas: Glassing is 90 percent of a sheep hunt.
Photograph courtesy of Garfield Thomas.

An award-winning photograph of three friends sharing a moment beneath Utah natural arches in the Escalante Unit. Left to right you see Pete Cimellaro, Bryon Wiley and Bill Mamalus. The photograph received third place in the 1998 Reno convention photo contest.
Photograph courtesy of Garfield Thomas.

The ram didn't move but my scope developed the shivers.

Pete said, "I think you pole-axed him; keep your scope on the spot. I'll circle for a look." The seconds seemed like hours before Pete let out a whoop. My whoop brought out two more from Byron and Bill who were coming up on the double. As Bill had cajoled yesterday, "Patience; when it happens, it happens."

After years of applications to states offering a desert bighorn drawing, I finally had my ram, and in some of the most majestic landscape in the world.

It sure was great to have an exceptionally skilled taxidermist as part of the team. Bill Mamalus of Custom Taxidermy of Salt Lake City removed the cape without a shred of flesh left on the hide. Our Escalante ram was in beautiful condition. His coat was full and thick, horns slightly broomed, but solid and heavy (green score of more than 163 B&C points) and a thick layer of fat on his rump.

With great exuberance, we had the horns, cape and meat back to base camp in short order. Repacking everything into heavy packs, the team moved out over the slick rock and down the vermillion canyon walls to our camp of four nights earlier. Fresh loin and oyster, fried in the rump fat, made a tasty addition to the freeze-dried dinner. In the morning we faced the tough pack trip down more sculpted canyon walls and cliffs to the river and then up the great sand dune.

Wednesday's trek through this country of fantastic forms and spectacular color was exciting, but the loads were down right work. Ascending the 600-foot elevation of the sand dune was especially tough. Each step slid back six inches. Once on top of the canyon wall, we could see the truck in the distance. Only two more miles of slick rock and blowing sand. Cheeseburgers and Coke, anyone? What an experience! What a hunt! What a team! And the best part is that this ram was taken just as I celebrated my 69th birthday.

This proves that you have to keep in "sheep shape" waiting to be drawn for a sheep tag.

Chapter II

Hunting Stories of Auction Permit Holders

The "Special Sheep Permit" Auction Has Been a Huge Success for Wild Sheep

DANIEL A. PEDROTTI

Governor Edward Herschler of Wyoming, in 1979, created a "Special Sheep Permit" to hunt a ram in any of the designated sheep harvest areas in Wyoming during the 1980 season. This permit was given to the Foundation for North American Wild Sheep to be used for raising much-needed funds for the well-being of wild sheep in North America.

Dan and Carolyn Pedrotti at a 1997 FNAWS convention reception.
Photograph courtesy of FNAWS archives.

The Foundation chose the auction format as a means of maximizing the proceeds from this gift and indeed the permit was sold for $23,000. The entire amount was retained by the Foundation to be applied to wild sheep Grant-in-Aid projects and thus began a program that has now spread to 13 states, two Canadian provinces and Mexico and culminated this past year in Reno, Nevada, with gross proceeds of $1,968,000.

Ninety percent of the funds raised at auction go directly back to the states issuing the permits to be used specifically for sheep projects, except in the case of Wyoming where the Foundation retains discretionary control of all the funds. Sheep projects include habitat protection, water enhancement, transplants to reintroduce sheep to their now-depleted ancestral ranges, animal disease studies and many other needed works that not only benefit wild sheep but also all the other species inhabiting the wild lands of our continent.

To date the Special Sheep Permit Program, funded through the FNAWS auction procedure, has generated $12,115,150.00 for wildlife enhancement to be enjoyed by outdoors-oriented sportsmen and sportswomen as well as all other outdoor enthusiasts; funding that would otherwise be unavailable.

It is truly unfortunate this program has become controversial and especially since some of the criticism is coming from our own diminishing hunting community. Despite the ever-present threat on our hunting heritage by the "antis," we now find ourselves defending our right to hunt from hunters themselves.

Rocky Mountain bighorn majesty in British Columbia. The special sheep permits help put sheep on the mountains of North America.
Photograph by Valerius Geist, Ph.D.

Listed are some of the arguments being used to undermine the great work that has been accomplished through this program:

(1) "Special permits" reduce my chances of being drawn for a sheep hunt.

(2) Most states grant "special permit" holders privileges beyond that which other permits allow.

(3) Such high dollar hunts make hunting an elitist sport and create a bad image for hunting.

(4) The successful bidders do not always hunt with "Fair Chase" standards.

(5) The State Fish and Game Departments do not need these funds and often do not use the funds for sheep projects.

(6) These funds could be better utilized for other more worthwhile projects.

Perhaps, as a first step to answering these criticisms, it would be appropriate to ask "Just who are the people who pursue the special permits?"

A review of the bidders who have participated in this process since its inception reveals that no less than 70 different individuals have successfully purchased Special Sheep Permits. Undoubtedly more than 10 times that many actually participated in the bid process. Successful bids have ranged from the $17,000 paid for the North Dakota permit in 1986 to the new record of $405,000 for the Alberta permit in 1998.

It is impossible to fully understand what motivates each of the permit buyers but it is known from personal experience that all have one goal in common and that is the improvement and conservation of the wild sheep of North America. They are also dedicated to a number of the same principles and goals possessed by the rest of the hunter-conservationists; i.e., preservation of the right to hunt, protection of wildlife habitat, conservation of the other wild animals of North America and a belief that the funds generated will indeed be used in the proper stewardship of our wildlife resources.

To illustrate the nature of these high-profile hunters let me cite a few examples:

(1) Doc Thurston, purchaser of a number of the earlier sheep permits, on more than one occasion after finding his trophy, said to his guide: "Thank you for showing me this nice ram. I think I will leave him for some other lucky hunter."

(2) Bill Leech, purchaser of the first FNAWS Utah sheep permit, arranged for the capture of his ram, then flew it to Texas where it was released to become an integral part of the desert bighorn restoration project in Texas.

(3) Several of the permits were not filled simply because a ram of suitable trophy quality was not found. Often these holders of unfilled tags returned in succeeding years to purchase a subsequent permit.

(4) A recent permit holder was not allowed access to his preferred hunting area but still came back the next year to be a major under-bidder for the same state permit.

(5) The successful bidders for the Wyoming permit in San Antonio in 1994 and again in Reno in 1998 donated these permits back to the Foundation where they were sold again for near-record prices.

Now to address the criticisms cited above:

(1) Special permits reduce my chances of being drawn for a sheep permit.

The increases in total sheep-hunting permits in the participating states refute this conclusion. The auction permit program has actually increased the chance of drawing sheep permits in most states that participate. Take Montana, for example, where some 70 rams have been harvested that scored 190 B&C points or higher. Of these, about 60 have been taken since 1980 and most came from introduced populations. These introductions were made possible largely through FNAWS funding and more than $1,924,200 returned to the state, since 1986, as a result of Montana sheep permits auctioned by FNAWS. Without this program, many areas would not have huntable populations and therefore no permits would be available.

One permit per state participating in the program amounts to less than 1 percent of the sheep tags issued each year. In most states there are already 10 or more applicants for every permit available to residents and in the hundreds for nonresidents. The "special permit" is normally over and above the quota established for the draw. Is it realistic to assume that one unlimited permit seriously affects your odds on drawing a permit?

(2) Most states grant "special permit" holders hunting privileges beyond that which other permits allow.

The extra privileges sometimes criticized include: unlimited area to hunt; permit is for entire season; permit holder allowed time to hunt beyond the regular season; special permits may be acquired at more frequent intervals than those allowed under the states drawing rules; and the question of the right to transfer the permit to someone else.

The purpose of the special permit is to maximize the funding available to wild sheep projects; therefore, is it not logical to make this hunting opportunity as attractive as possible? Would the Arizona or Montana permits have earned in excess of $300,000 each for wild sheep projects if the hunt in those states were restricted to one area and only 10 days allowed for the hunt?

Granted, the season should not be extended or a special season arranged for the high-dollar permit holder and that transfer of this right only be permitted under extreme circumstances. The sale of a special permit for gain must be strictly forbidden, but if a person wishes to bid and contribute large sums to wild sheep annually, he or she should be encouraged to do so.

(3) Such high-dollar hunts make hunting an elitist sport and create a bad image for hunting.

Big-game hunting, and particularly, sheep hunting, has never been and never will be available to every hunter unless he or she happens to be privileged to reside in one of the Western states, Canada or Alaska. One of the reasons for the high value of the "special permit" is that non residents face extremely long odds or are virtually excluded from the drawing process even though most of the sheep habitat is located on government lands owned by all hunters.

Trophy hunting for big game will continue to be under attack as long as we hunters remain on the defensive and try to alibi for our hunting heritage. Justifying the right of a person in an economic bracket that allows him to spend $300,000 or more to pursue his quest may be easier than trying to explain the motives of one who deprives his family of basic needs each fall in order to pursue his.

Granted that the sheep permit auction creates a high-profile situation. Who questions the hunters who pay $50,000 to $100,000 and more to hunt the exotic places on this earth such as Africa and Mongolia as well? Either hunting is a basic human right or it is not and it must not be limited for political or economic reasons.

If the hunters are sensitive to outside criticism because someone may have the big dollars to purchase a bid permit then should they not be sensitive to the fact that some people can afford a better lifestyle? Just where is this reasoning heading? Has not anything been learned from the worldwide collapse of socialism?

(4) The successful bidders do not always hunt with "Fair Chase" standards.

All hunters are paying the price for the bad publicity generated by the so-called "slob hunter." There may be cases where permit buyers used less than "Fair Chase" methods in pursuit of their trophy; however, this has not to my knowledge been documented. Should the goal of the buyer be to obtain the new "Number One" or even a ram that stands high in "The Records Book." This may very often be the case; he simply cannot be guilty of taking it by methods that do not qualify as "Fair Chase" as defined by the Boone and Crockett Club if the hunter wishes to be listed in "The Book."

Should this individual or any other hunter be denied the right to use the guide of his choice in pursuing sheep or any other "Big Game"? For many who did not grow up in the North American wilderness, having a proper guide is a life or death situation. Must they put their lives in the hands of the elements or that of an unknown person in order to qualify as true sportsmen? Do you not call your family doctor when you are seriously ill; or choose your own mechanic when your car won't run?

Not all permit bidders are seeking records book rams. For example, the 1994 Texas permit was sold for $70,000 in San Antonio, yet at the time it was a well-known fact there were no Texas rams that would score high in "The Book." These hunters must be motivated by an intense desire to give something back to the species that has meant so much to them.

There is another solution to this problem. In Texas the permit hunt is conducted by Texas Parks and Wildlife personnel. This method could be adopted by other states as well.

(5) The state fish and game departments do not need these funds and often do not use the funds for sheep projects.

The Foundation for North American Wild Sheep enters into a contract with each state that grants a permit. This agreement provides that the funds generated must be used specifically for sheep projects. Each year a report is submitted by the participating states to the FNAWS office in Cody, Wyoming, outlining the use of permit funds.

Do the states need these funds? Yes, in fact several states did not even have a sheep program until it was funded through this system. As a result of the 1994 sheep permit auction, the state of Washington now has an additional $90,000 to fund a wild sheep program.

If the goal of wildlife management is to conserve all wildlife species then we have a win-

win situation, because funding the sheep programs through the "special permit" system creates a windfall of funding for game departments to use for other wildlife projects.

(6) These funds could be better utilized for other more worthwhile projects.

Those who bid on the auction permits are spending their money to enhance wild sheep. There are other people in the world that might want to fund AIDS research, enhance the elk of North America or perhaps endow a hospital or university. These select "Wild Sheep People" have chosen to spend "their money" on enhancing an animal they love. The freedom of choice is theirs and not another man's.

With the ongoing population growth, the desire for open spaces, and the loss of habitat confronting all our wild species there can be no more worthwhile use of conservation and education funding than these programs designed to ensure that wild life will be available to be enjoyed by generations to come.

Had not Theodore Roosevelt, George Bird Grinnell, Gifford Pinchot and many other far-sighted individuals set aside their energies and resources in the late 1800s we would likely not have to worry about sheep or any other wild animals.

Isn't it far more desirable that these generous donations be utilized to propagate and conserve wildlife for future generations than to have the funds dedicated to one of our museums merely to preserve the past?

In my opinion, to attack the special permit program that has to date generated more than $6,000,000, inclusive of other species permits, for wildlife improvement, particularly the wild sheep of North America, is to attack my very basic right to pursue the hunting of wildlife in a sportsmanlike manner — one of the most cherished rights handed down to me from generations past.

We all know that hunters provide the only source of funding for wildlife. Aren't these hunter's dollars that we are talking about? Where else can we go to get funding of this magnitude?

Yes, the benefits to wildlife far out weigh the negatives cited as criticisms above. We must continue the "special permit" program and expand it to include all states and the provinces of Canada where sheep populations exist.

As we now pay tribute to the permit buyers, I encourage you to read and enjoy the excitement of the sheep hunts generated by this program as told by the purchasers themselves. It would be impossible to include all of the hunts so consider this a fair sample. Each participant has done a wonderful service to our goal of "Putting More Sheep on the Mountain." We salute them and thank them for being so generous.

Daniel A. Pedrotti served as president of the Foundation for North American Wild Sheep from 1996 to 1997. He served on the FNAWS Board of Directors six years and is current president of the Boone and Crockett Club beginning in 1997. He has been an active member of the Boone and Crockett Club since 1989. Dan is a geologist and president of an independent oil and gas exploration company, Suemaur Exploration, Inc., in Corpus Christi, Texas. Dan is a founder and past president of the Texas Bighorn Society and a director of the International Sheep Hunters Association. He and his wife Carolyn have been married since 1955 and are the parents of four adult children, Daniel, Patti, Diane and Mike, and grandparents to 10. Daniel and Mike, as well as four grandchildren, are FNAWS life members.

California Desert Gold

Lee R. Anderson
From *Wild Sheep*, Summer 1994

I must admit that when I purchased the California auction permit February 1993, I had only two thoughts. First, I wanted a chance to shoot a desert sheep and, secondly, I saw it as a chance to get a respectable-sized sheep. I really did not see a state record as a probability nor was it a burning obsession with me to set a record. Actually, the sheep I have successfully taken to date have been rather average. This is by way of background for my hunt story and what turned out to be "California Desert Gold."

Immediately after winning the 1993 California tag for desert (Nelsoni) sheep, I engaged Floyd Krank of Aravaipa Outfitters to scout and guide for me. Floyd has had more experience in desert sheep hunting and more in California than any other outfitter, so I went for the "learning curve" guy, and I must tell you now I made the right choice. Before I even started the hunt I knew I was going to really like Floyd and his crew, wife Mary and assistant guides Rick Jones and Chris Harlow. Both Rick and Chris had made trips to California over the summer to scout what would be the best area, (you are able to hunt three different areas) but unfortunately they did not see anything to brag about. They were concerned about where the desert rams might have headed. In the Kingston-Clark area which borders the Nevada-California state line, the sheep travel to the Nevada side for winter grazing and the concern was that they may already be over the ridge and into Nevada. Most of this discussion took place on the phone in the days before the start of the hunt. My excitement level was too high to be very concerned at this point, and besides I was used to working hard to find sheep and was prepared for some big time climbing and scouting.

Floyd and Mary met me at the airport in Las Vegas and we proceeded down the interstate, 75 miles to camp. I couldn't get over this point, since I had recently traveled around the world to the Pamirs Mountains of Tadjikistan for Marco Polo sheep where the trip itself is the hardest part of the hunt. Camp was a new travel trailer with all the amenities. Mary and Floyd insisted I take the bedroom. I haven't ever had it that good on a sheep hunt, but I must admit I haven't paid my dues on the hardship side, so I wasn't complaining. Andy Pauli showed up from California Department of Fish and Game and gave us our orientation. Andy was accommodating and helpful. He explained to us how California has built up its sheep population to the point where permit hunting is allowed on a limited basis. He was obviously very proud of the work his department has done. He promised to swing back through in a few days and left the names and numbers of four people in his department in case we needed help. Believe me, I felt good about buying the permit because Andy made me feel so good about how the money was spent on sheep habitat and studies.

Our first day was into the Kingston Mountains, the largest group of mountains and the most rugged of the areas allowing hunting. We covered a lot of territory that first day and saw several smaller rams and ewes. The rams have plenty of water in the Kingston, and all the sheep I observed were healthy and fat with good looking hides. However, we failed to see a good ram, and I couldn't help but think back about Floyd's comments on not seeing any good rams that summer. The highlight of the day was seeing my first white desert sheep. In fact it was the first white sheep anyone had seen in the Kingston Mountains. The first day was completed by walking out in the dark with flashlights, a quick meal by "Chef Mary" who in addition to her cooking is just about the nicest person you could ever hope to meet. A few fun moments around the campfire with Floyd and Rick, and off to bed as that 4:30 a.m. alarm comes early.

Day two and we are off to the Clark Mountains. Today a little easier climb and Floyd and Rick know the area pretty well. They hunted it quite a bit the previous year. In fact, they informed me it is the home of "Slippery Willy," a nice ram they failed to harvest the proceeding year because he kept giving them the slip. Our day started at dawn with a long climb up a wash to the three quarter point of a large mountain covered with pinon trees and cedars. Floyd suggested we spread out because the visibility is limited between the trees. We could see better if we spread out and cover more area. We proceeded to do this and within a half hour Rick spotted two ewes in the valley below, and got a quick glance of a large ram who had walked out of sight.

We set up scopes and could just catch the tops of a ram's horns behind a rock. We needed to change position quickly, but without spooking the sheep. We quickly circled the area where Rick saw the big ram. We came up over a ridge separated by a draw between the sheep and us, and set up our scopes. We thought we hit pay dirt. Directly across from us were three large rams, several small rams and a dozen or so ewes. The ewes were quickly eating barrel cactus and three rams were bedded down. We were seriously considering one of these rams, and a lot of debate went on between Floyd and Rick about age and size. One thing we knew for sure, one of them was at least 175, but he was only 6 years old. Ultimately it was decided that since it was only day two, and because the ram might not be quite 175 (175 is the state record for California), and only 6 years old, we left them be. If we needed to later in the hunt, we can maybe find them again. We continued to glass them and marveled in the 100 power video/scope image that we looked at. I again reminded Floyd and Rick that I was not bent on a record, but would not want to shoot at anything under 170. Another 30 minutes passed and Floyd whispered he'd spotted a fourth big ram 100 yards up the mountain, and 40 yards to the right. We didn't believe what we saw. Laying there all this time had been the "Monarch" of the herd, the self-appointed leader, who by virtue of his age and size chose not to join the group. Who would not and did not join in the male shows of strength typical of younger rams? We quickly set our scopes on this new ram, who we estimated to be 9-1/2 years old, with very little brooming and certainly 175 plus.

Floyd was concerned he could only view the ram from behind and claimed he would not score a sheep unless he saw it from all sides. Rick was just as convinced that we were looking at a new record Nelsoni desert sheep. A rather interesting debate took place between Floyd with years of experience in scoring sheep, and Rick who was also experienced but in his mind saw a records book sheep. Floyd again reminded me it was day two and perhaps we should look some more.

Rick adamantly said "you have to shoot that ram." Both men were right for all the right reasons, and I had to make the final decision — the moment of truth. All of us agreed it was big. Not all agree it was record. I again reminded Floyd and Rick that my obsession was not the record, but a good representative sheep. Having said this one more time, I announced I would try to shoot this ram. At that very moment discussion ended about whether or not to shoot and discussion started on where and how.

We looked over the situation and decided we couldn't get closer. We were about 350 yards from the ram and there were some good flat rocks to crawl out on, and not much wind to allow for. I routinely practiced at 300 yards and felt good about the shot. I placed the crosshairs of my .300 Winchester magnum, made by Les Webernick at Rifles, Inc., on the shoulder hump and figured on a 10-inch drop at that range. The shot rang out through the valley and the reported thud of a hit came back to us. A great moment of respect happened at this point. Many thoughts went through my mind. My mind was on the "Monarch" who gave up his life so other sheep could exist in his place, through future permit hunts. We took our time getting over to the ram. We discussed our strategy again about whether we should have shot or not. With great anticipation, we climbed the other side.

We decided not to score the ram before we returned to camp, but decided we would measure the horns for length. Floyd held the tape on the base and Rick measured the length. We were at 36 inches with lots more to go. Rick held out four fingers, and Floyd, who couldn't see the tape said, "34?" Rick said, "No, 40." Both Floyd and Rick said they had never seen a 40 inch Nelsoni before. The other horn measured 38-1/2 inches. The trip down the mountain was a happy event that only times like this can create. A lot of good natured kidding took place, and each of us was the brunt of some of it. I was amazed at how close I felt to these people after only two days, and how much I was going to miss not hunting the full 16 days with them.

At camp we set up the camp table for measuring. We still weren't positive we had a new California record until Rick added up the measurements. Everyone threw a dollar in the pot, and Mary won. Our "Desert Gold Ram" scored 179 points. We called both Vern Bleich and Andy Pauli from the California Department of Fish and Game, official measurers for FNAWS. They measured the ram at 177-3/8 unofficial Boone and Crockett Club points. At that time, it was the largest desert bighorn sheep to be taken in California by a hunter, according to CDFG records. Both men were happy with our success and felt that a new record would inspire more interest in the California's program next year. I guaranteed them it would.

In closing, I must admit I feel my record will be beat. I am just thankful I had a chance of a lifetime to see and take a great sheep. I even know where that potential new state record is, and he is only 6 years old now. Wait until he grows up!

Idaho Lottery Sheep Tag 1995
An Epic Search

Jon Barker

From *Wild Sheep*, Spring 1997

As the last shot went off, the ram disappeared at a run into the heavy timber without any sign of being hit. Our hearts sank. The ram had half a mile of heavy timber below him, and it was late in the afternoon. This was a ram we had known for several years. We had been watching him, constantly, for the last several weeks. We had already put two stalks on him that day and we were a long way from camp. I decided to send Joe Schram, my client, back to camp with a couple other people. Then Shane, one of my guides, and I hiked down to look for the ram.

At the time I had no idea what tough decisions I would soon have to make or what course of events would unfold over the next month. When we reached the timber, Shane picked up a few spots of blood where the ram had entered the timber. We worked a couple of patterns down through the timber to the bottom of the drainage. It was getting towards dusk when Shane and I started back up. It was almost dark and we were 2,000 feet above the bottom when we looked back down and saw a very large ram in a small opening, that we had walked through earlier. We were shocked but there he was. His horns seemed identifiable as the same ram we were chasing. With high hopes we returned to camp in the dark.

Early the next morning we worked our way down the ridges and through adjacent timber patches until we reached the area where we had seen him the night before. In this area we found four beds, each with a fingernail-size spot of blood in them. We split up with three guides going through different sections of timber and creek bottoms for 15 hours that day and never came up with any sign. The only thing we scared up were a few black bears.

Next came one of the hardest things I've ever done. Joe naturally wanted to keep hunting and I had to go to him and say, "Joe, I know you have hired me for this hunt and I can't find any sign of your sheep, but I want to ask you not to hunt anymore while I spend a few days looking for your ram." Like the true sportsman he is, Joe agreed to drive 300 miles home and wait to return at a later date.

Two guides and myself spent the next several days in the immediate area where the ram disappeared. We back-tracked to the rams' beds several times and walked up and down through all the adjacent timber. One guide remained up high to observe the other guides as they walked through the trees watching for movement or any sign of birds. We came up empty. Now it was time to get serious.

I would go out for a few days at a time, first looking at the distant end of the unit to see if the ram had grouped up with other animals. Then I would start crawling up and down the creeks, near where we had injured the ram with my machete, figuring he had crawled down into the bottom somewhere and escaped up. I would be convinced I had the answer to where he was, go out, and crawl through all the blackberry brambles for days until I was too tired to go on. With no energy to continue, I would go home exhausted, dejected and wondering where he could be. Within a day or two I could not stand it. The pressure built within me and I could not hold myself back. I would come up with another theory about where the ram went and head out in search of him for several days.

This went on for a long time. I continually thought of the ram and what commitment I had to him. Was he injured and suffering? If he wasn't seriously hit, where did he go? Did I have the right to take Joe after another ram in a small population of sheep without knowing if this ram was dead or not? Why could I not find him? I had crawled repeatedly through every creek bottom. We had watched closely, but did not see any birds in the area to indicate a carcass. I had never expected to look for a wounded animal as long as I did, but something would not let me stop.

Gradually I began to see all the rams that were originally with this ram. They reappeared without him and I knew then that something was very wrong. Luckily, I had guides with great conviction to convince me to keep looking in the first place and to help me in the search. Not everyone would have the time or the resources to continue with such a search. I did.

One the 28th day of the search while I was walking along once again near where we found the rams' beds, I saw an oddly even curvature on the ground under a pine tree. As I went over to investigate I could see that it was a horn sheath laying on the ground. As I looked further I found quite a bit of sheep hair in a pile with lots of bear scat. Next I found three lower leg bones and a chunk of backbone about 16 inches long. These parts and a second horn sheath I found over a 20-yard radius, all half buried under pine needles and dirt.

About 30 yards downhill I located the ram's skull. It was the cleanest skull I have ever seen with virtually to hair or meat on it. It did not even have an odor. All the bones were completely stripped clean. Even the ligaments along the backbone were all gone. All that remained were impeccably clean white bones.

There was only one apparent answer. The ram had been devoured by bears. I mean completely devoured. Most of its bones were missing. Perhaps if the bears caught the ram while it was still alive it would help explain the efficiency with which they cleaned his carcass. More amazing is that this must have happened that first night that Shane and I saw him. We had been there the next morning walking within 10 to 20 yards, or closer, of this site on an open, thinly treed hillside. I believe we did not see him because he was not there, not his carcass anyway but only a few half-buried bones. How could those bears have torn his horns off the cores of the skull? Ask a taxidermist. This is a very difficult task.

As I packed the skull and horns out of the canyon bottom, I began to reflect on the events during the last month and what had led me to this point. I began to realize more than ever what this ram or any wild animal means to me; what the essence of a hunt is; and how that animal's spirit is represented in our lives. I thought a long time about what responsibilities and commitments we all have to the game animals that we hunt and our responsibilities to pursue those animals in an honorable and ethical fashion.

It is well worth a moment of reflection to ask ourselves how we would handle any given situation that does not turn out as we planned while hunting. We must ask ourselves if we should take a shot from where we are or just wait and make a closer stalk, striving for a greater chance at a respectable, clean and quick harvest, while increasing our challenge and excitement of the chase as well.

How prepared are each of us to search for a wounded animal? I often asked myself during the search what I would have done had the ram been a whitetail doe in a heavily populated area instead. As they should, these answers will vary for all of us. We are all different human beings.

I hope my experience may help everyone stop and think a little more often about their decisions in the field. It is guaranteed that as these times progress, the ethics of hunting along with our desires and challenges will be questioned in stronger and stronger ways. We will be defined as a group by our actions and our thoughts on the more difficult questions we face. For our own sake, and certainly for the sake of the animals we pursue, we should prepare ourselves beforehand with the answers that are within us.

A Pair to Stand On
Two auction tags yield back-to-back monarchs for Bob Petersen.

COL. CRAIG T. BODDINGTON, USMCR

This is not just a tale of two rams. It is also the tale of a dedicated hunter, an equally fanatical outfitter, and how they worked together — using the Sheep Foundation's auction tag program to achieve a fitting capstone to a distinguished sheep hunting career.

I should say right here I don't figure in this story, except as a scribe, but I've known the key players for a long time, and I've watched the auctioning of prime sheep tags for many years. I'll step out of the story in a moment, but first let's set the stage.

The auction tags: Like most of us, even the least desirable of the auction tags is beyond my means. So, like most of us, I've watched the gavel sound with some varying mixture of envy, horror, shock, amusement and sadness. The amount of the first three of these ingredients depends a lot on who buys the tag and how much it brings. But regardless of who or how much, there's always a trace of amusement that someone would actually deem a hunting license worth so much. And there's always sadness, for like all hunters of limited means, I hate to see the continued cost escalation of quality hunts.

Bob Petersen's huge Montana ram, one of the best bighorns of all time, had horns more than 42 inches long and bases of 17-1/4 inches. He carried the mass all the way. He looks pleased, and who wouldn't be?
Photograph by Larry Heathington.

There's no escaping the fact auction tags aren't for everyone. It would be easy to decry them as a blatant sale of public-owned wildlife. This is far too simplistic. The auction tags, customarily valid in a choice of open areas throughout a generous season, offer exceptional opportunity, but no guarantees. It is this exceptional opportunity that is the lure; the auction tags offer an unusual chance to search for the very biggest rams. With such potential it's hard to settle for second best; many of the auction tags go unfilled. But whether a ram is taken or not, there's also no escaping the fact that the auction tags put a lot of sheep in the mountains.

The hunters who purchase the tags are sheep hunters, of course, but they're also benefactors of the sheep they hunt, and no single program has been able to provide the wherewithal for FNAWS to put more sheep on the mountain. All of the auction tags are wonderful, but over the years the most prized have usually been Montana's Rocky Mountain bighorn tag and Arizona's desert sheep tag, obviously because both states have ideal habitats that produce some of the finest rams of each variety.

The outfitter: Larry Heathington of Sheep, Ltd., of Kingman, Arizona, is a trophy hunter's

Some of Montana's sheep country is deceptively open — high sagebrush hills that the rams drift into during the rut.
Photograph by Larry Heathington.

trophy hunter. He sets high standards, and his soft-spoken manner belies bulldog tenacity. He's a sheep hunter, yes, but he has also been extremely successful in finding exceptional pronghorns, elk and deer for his clients. Regardless of the game, he seeks for the very best. In this regard he is no different from many fine outfitters. But unlike most, Heathington hasn't confined his search to a limited area or concession. He thinks big, and he wants to go where the big ones are — wherever they are and can be legally hunted.

In this regard he was probably shaped by his native Arizona. For reasons not altogether known, Arizona produces some of America's finest big-game trophies. Not only desert sheep, but also big elk, pronghorns, mule deer, Coues whitetails and even black bear. However, wildlife resources are fragile in this arid state and it has a large population of hunters. It's a given sheep permits are few and hard to draw, but virtually all of Arizona's permits are tough draws. In Heathington's incessant scouting he occasionally located superb animals but often didn't have hunters fortunate enough to draw. Naturally, he gravitated to the auction tags and, just as naturally, he expanded his search far beyond Arizona.

The hunter: Robert E. "Pete" Petersen is a true American legend. He grew up in the California desert, shooting jackrabbits and building stock cars. These two passions, guns and cars, eventually became the bulwarks of one of America's largest publishing empires — but the road wasn't easy. In the late 1940s Petersen started *Hot Rod* magazine in his garage, passing them out at stock car races. The gamble paid off, and as the years passed there were more magazines. Many more magazines, among them *GUNS & AMMO* (1958) and *Petersen's HUNTING* (1973). By then Petersen could hunt the kind of game any boy dreams of when shooting jackrabbits in the desert — elk, elephant, bear, lion and, yes, mountain sheep. And, ultimately, he could bid on permits that offered opportunity to hunt the very finest of the mountain sheep.

When I hear that one of these prized auction tags goes to a hunter like Bob Petersen, of course there's some envy. But there's yet another emotion as well. It just plain tickles the hell out of me when such a good hunter obtains such an opportunity. I've worked for Bob Petersen for 20 years, and I think the thing I admire most about him was — whether he's hunting elk, elephant, bear, lion or sheep — he's still a boy shooting jackrabbits in the desert. Petersen just plain loves it, and loves it equally, whether he's chasing quail through the chaparral in California, stalking buffalo in Zambia or climbing a mountain after a ram. At the time Petersen hunted these two rams, 1996 and 1997, he was 70 and 71 years old, respectively, and still not only able to climb the mountains to reach his ram, but able to make any shot required for a successful conclusion. Bob Petersen never asks anyone to make it easy on him, and he remains one of the finest field marksmen I have known.

Petersen appreciates a quality trophy, and this led him to Larry Heathington and, ultimately, the most prized of the auction tags, but Pete doesn't hunt for recognition or for the book. He hunts because he loves to, and this makes him the kind of hunter that, I believe, most hunters would like to see successful in taking world-class rams.

Over the course of many hunts together, Larry Heathington and Bob Petersen became close friends. At Heathington's suggestion, a few years ago Petersen purchased the Oregon bighorn tag at auction. He and Heathington took a superb ram, finally ending his quest for the North American Grand Slam. This goal of some decades, marked by several unsuccessful hunts for Rocky Mountain bighorns, had now been reached. Scoring about 192, the Oregon ram was truly spectacular. But after several seasons, and with the mountains still beckoning, Petersen asked Heathington if there was any chance of finding a still larger bighorn. Heathington believed it was possible in just one place.

In Montana's bighorn country, the rams are rarely far from timber. Once they're in the trees stalking is almost impossible, so you must bide your time and catch them in the open.
Photograph by Larry Heathington.

At the 1996 FNAWS convention Petersen purchased the Montana auction permit and donated $220,000 to the Foundation in the process. Heathington knew the Montana sheep country well, but he believes in doing his homework. He started scouting in April and May, concentrating on the Perma Plains and Rock Creek areas. As he expected, in spring-green meadows he found four or five huge rams. The best of these was a huge ram in the Perma Plains that he believed would score above 202 inches. This was the ram he wanted for Petersen, but Fate intervened. In the late spring this grand old ram was hit by a train, and only one horn was found.

This tragedy turned Heathington from the Perma Plains to "Plan B" — Rock Creek. Rock Creek is a much more difficult area, comprised of about a million acres of timber and very few openings, but it has some of the finest bighorns in the world. Working with local outfitters Larry and Barbara Clark of the T-Heart Ranch, during the spring and summer Heathington identified three different rams that, plus or minus, should score close to the magical 200-inch mark.

On a quest such as this it's essential to know that the quarry exists, but in Rock Creek that doesn't do much good. The great rams are fairly visible through the spring and early summer, but in the heat of August they vanish into the big timber, and they're still there when the season opens on September 1. Heathington brought Petersen in for the last 15 days in October, when (hopefully!) cooler weather would bring the big boys out into the open again. On October 15, when Petersen started his hunt, no big rams had been seen since early August.

The weather was still warm, and although they hunted hard for a week they saw only young rams and ewes. Then, almost on cue, the first big snowstorm hit on about October 22. Now things would get interesting.

Indeed they did. The day after the storm the hunt nearly ended in disaster. Riding along

Outfitter Larry Heathington's tent camp in Arizona's desert sheep country. Heathington prefers to set up a tent camp for a base of operations, relying on the good old "whitewall" tents that have been the outfitter's standard for generations.
Photograph by Larry Heathington.

a steep sidehill covered with deep snow and cantaloupe-sized boulders, Petersen's horse stumbled and went to its knees. The saddle, lacking a back cinch, was catapulted over the horse's head, and Petersen along with it.

He landed among the rocks flat on his back — a back that has troubled him for years. Heathington, fearing the worst, ran to him. As Petersen described the exchange, Heathington stood over him and asked, "Are you dead yet?" Unprintable reply. He wasn't dead nor, miraculously, was he seriously hurt. But he was shook up and badly bruised, and the hunt continued at a much slower pace.

But it did continue, and with the snow the big rams started to appear. The next day the sheep were in the meadows in scattered bands, and they glassed a wonderful ram with evenly matched 45-inch horns. Despite the great length, however, he had little mass and Heathington judged him at about 190. They passed. That same day they found a near-perfect ram they judged to score about 197 or 198. This one looked good, especially to a hunter who hurt in every part of his body.

They made a long uphill stalk and had the ram unaware, in easy range at about 220 yards. Petersen laid his pet sheep rifle, a Kenny Jarrett .300 Winchester Magnum, over Heathington's pack. This rifle, veteran of numerous hunts on several continents, refused to fire. The mechanical reason was that debris — perhaps accumulated in the fall, perhaps during the stalk — had lodged in the trigger assembly. The real reason, I suspect, is that the rifle knew this wasn't Pete's ram. They found the same ram later, confirmed him at 197 to 198, and this time Petersen, recovering from his fall and feeling better, turned him down.

Rock Creek is an extremely difficult area — steep and rugged and cloaked in thick timber with few breaks. As Heathington puts it, "Rock Creek is not a place where you get lucky. It's hard, but there is no place in the world where you can see rams like this." Now, with the snow, they were seeing big rams. And then they found the right ram.

They were glassing not very far from Larry Clark's headquarters. From a ridge they spotted a big ram across the canyon on the lower end of the inevitable timber. By now they had passed some very fine rams, but the big ones looked big, and this ram looked very big.

They closed to 200 yards, and while Petersen lay his rifle across a pack Heathington studied the ram. He was on the edge of a ewe herd with four other rams. He had big bases, good length and carried the mass well. He might score 200, but he was clearly bigger than the Oregon ram. Heathington told Petersen to take him.

As Heathington described it, "I no sooner whispered than the rifle went off, not once but twice—'whack, whack,' just like that. Nobody shoots as well or as fast as Petersen. The ram folded, and that was that."

Not quite. Typical of Rock Creek rams, this wonderful monarch had attained his spectacular growth early and was just 8-1/2 years old. His bases were 17-2/8, the horn length 42-4/8, with the last quarter measurement fully 10-1/8 inches. His final score was a whopping 200-7/8 unofficial B&C points, and the finest trophy Bob Petersen had ever taken in a long lifetime of serious hunting.

Unlike most modern sheep hunters, the desert bighorn was not Petersen's stumbling block along the road to his Grand Slam. He took a nice ram in Mexico many years ago, and he'd taken yet another nice ram with the Nevada auction tag. But now that he had a huge Rocky Mountain sheep, one that he couldn't possibly better, he wanted a really fine desert sheep to match. Heathington knew of only one option. At the FNAWS convention in 1997, fellow sheep hunter George Harms pushed Robert E. Petersen all the way to $300,000, but Petersen walked away with the Arizona auction tag.

Arizona's desert sheep country is Heathington's home territory and he knows it well. But at this particular time he knew it exceptionally well; he'd guided the 1996 auction tag winner to a spectacular 193-6/8-inch ram. That ram was gone, but he already knew the whereabouts of two other great rams. The best of these was in Unit 22, out of Apache Junction. The other was in Aravaipa Canyon.

Petersen started his hunt in October, working the Lakes Chain primarily by glassing from a boat. Although the "Lakes Ram" had been seen twice during the summer, now that the season was open they couldn't find him. Hedging his bets, Heathington and Petersen kept looking while their friend, Bernie Lawrence, went to Aravaipa to look for a big ram for them.

On about October 20 Lawrence phoned that he had found the big Aravaipa ram. With nothing big showing in the Lakes, they packed up and drove to Safford. Bernie Lawrence had indeed found a big ram, and was able to locate it again for Heathington and Petersen. But after close inspection Heathington concluded it was not the same big ram he'd seen. The configuration was identical, but it lacked the distinctively massive bases.

But now they were in the Aravaipa, so they continued hunting this rugged canyon country. On the third day, glassing from one ridge to another, they found a band of 17 sheep up on the canyon rim. There were nine rams, and despite the distance one of them looked right.

There was no easy approach. They couldn't go around; the only option was to drop down and come up from below — and this route was barred by a 200-foot cliff of nearly sheer sandstone. As Heathington related, "It looked bad enough to me, but Pete was 71. Still, it was the only approach I could figure."

Bernie Lawrence, Larry Heathington and Bob Petersen climbed up from the bottom, then started the assault on that forbidding cliff. Progress was an inch at a time, scrabbling for holds with fingers, elbows, knees and toes. "I crabbed my way up," said Petersen. But he made it up, and now they were slightly above the sheep with the wind in their favor.

The canyon rim here is scattered cedar and mesquite with some of Arizona's ever-present cacti. They worked along the rim slowly, glassing from cedar to cedar. Then one ewe moved from behind a cedar down in a little hollow about 150 yards below them. They were trapped behind a mesquite, Petersen literally hanging in the tree and finding a makeshift rest, Heathington hunkered down and looking for the rams.

There they were, to the left behind some cedars, closer still—just 120 yards. Now the sheep were milling, mixing and chasing right in front of them. Now hidden, now visible, there were eight ewes and nine rams. Three of the rams were over 180 inches. A fine ram that Heathington estimated at 182 stepped clear of the milling sheep and looked up the ridge at them.

"Should I shoot?"

"No."

"He's big!"

"Hang on!"

That ram faded into the mass of sheep and cedar, and then another good ram, this one about 176, stepped out. "Hang on!"

Close range, shifting sheep, multiple big rams and lots of tension. Mistakes can be made—or the whole bunch can simply move off. Or the gods can smile. All of the sheep moved off to the right in a body. All except one. The big ram stood alone, casually nibbling on a bush.

"See that lone ram? Shoot him." A single shot instantly rang out from the Jarrett .300, and the huge Aravaipa ram staggered and rolled into some bushes below.

This great sheep had 38-inch horns, well-broomed. His distinguishing feature was his heavy bases and the way he carried his mass all the way; those astonishing bases were 16-3/8 inches, and the last quarter measurement was a whopping 11-1/2 inches. The ram scores 189-4/8 unofficial B&C points.

It's unclear which ram Bob Petersen prizes the most, that magnificent Montana bighorn or the stunning Arizona desert ram. It's also unclear which was the most frightening experience — the fall from the horse or the sandstone cliff. But Petersen knows he cannot possibly beat either ram, and he's now all done with bighorns. In fact, one might suspect that Petersen is done with sheep hunting altogether, content to rest on his laurels. Me, I'm not so sure. He and Heathington are now partners in a great Stone's sheep territory in northern British Columbia — and if I know Pete, he has no intention of sitting in camp while there are great Stone's sheep up in the high basins!

Craig T. Boddington of Paso Robles, California, is senior field editor for Petersen's HUNTING and GUNS & AMMO magazines. In May 1996, Craig was promoted to Colonel in the U.S. Marine Corps Reserve and has served in the U.S. Marine Corps as an infantry officer since 1974. Craig is the author of numerous adventure hunting books including many through Safari Press of Huntington Beach, California. He joined as a professional member of the Boone and Crockett Club in 1985 and is a regular contributor to their national magazine, FAIR CHASE: The Official Publication of the Boone and Crockett Club.

California Adventure and a Wyoming Hunt with J. Curtis Newman

JOHNNY DRIFT
From *Wild Sheep*, Winter 1994-95

During the 1994 FNAWS convention in San Antonio J. Curtis Newman of Newport Beach, California, was the successful bidder of two outstanding auction items: Research and Management Adventure donated by the California Department of Fish and Game and the Wyoming Bighorn Permit, donated by Wyoming Governor Mike Sullivan. Good friend Johnny Drift accompanied J. Curtis Newman on both of these exciting adventures. On behalf of J. Curtis Newman, Johnny submits the following.

Our first adventure was a week assisting the California Department of Fish and Game in a desert sheep survey and capture effort. This adventure started south of Interstate 10 near the Arizona border. The area was bordered on the east by the Colorado River. Using a Jet Ranger helicopter with four observers including the pilot the mountains were flown using the new advanced technology of a GPS system.

All the information was recorded on a detailed map and later stored in a computer for future use. Every desert sheep, deer or burro was recorded exactly where they were observed from the helicopter, quite a change from surveys done just a few short years ago. This information was needed to determine the number of sheep tags to be available next year for this particular mountain range. It was also useful to see the condition of the sheep range, count how many burros were using it and see if there were places that could be further developed with man-made waterholes. We all agreed that where there was water, there was wildlife, be it sheep, deer or burros. We flew vast areas that were barren of desert sheep because there was no available water suggesting that all the money FNAWS has allocated towards water projects should be producing desert wildlife. While camping in the desert we met a very interesting gentleman and fellow FNAWS member Leon Lesicka of Brawley, California. Leon has devoted much time and effort towards building waterholes. He was the founding president of Desert Wildlife Unlimited Inc. This non-profit group has devoted many weekends using Leon's excavating equipment and knowledge of the California deserts to build waterholes that benefit sheep and deer in the surrounding areas. They install a water catchment system in a suitable dry wash run pipe to a cement holding tank that is designed to allow the entry of sheep and deer as the water level recedes. Any rain coming down that dry wash is directed to the tank to help keep it full. Leon and his volunteer crew have completed a project in 10 to 12 hours. Most are done on weekends and when completed they are very hard to see from the helicopter. The site is cleaned up, by this I mean all tire tracks are raked out and there is no sign of any disturbance. He deals effectively with most of the "red tape" that is associated with many projects submitted by the Department of Fish and Game.

After flying our initial area to the satisfaction of Vern Bleich and his associates at the California Department of Fish & Game we moved camp to another mountain range located about 75 miles west and just east of the Salton Sea. Priorities for this new location was to net gun 14 desert sheep, install radio collars, ear tags, collect blood samples and document other vital information. J. Curtis Newman really enjoyed this part of our experience especially since it was very near his birthplace and where as a teenager he got to know the surrounding mountains, waterholes and natural springs. He even knew of a couple that were not documented. The Department of Fish & Game used his knowledge of the past to compare with what is here now.

The California Department of Fish and Game will once again be offering this opportunity during the 1995 Phoenix convention. Anyone looking for a real adventure, flying in a chopper with the doors removed, about 150 feet above the desert floor, net-gunning wild desert sheep and assisting in all the data gathering would do well to bid on this auction item. It is a real fun filled week living in the California desert, enjoying the companionship of great people. Its really informative and makes one aware that water is the critical part of desert sheep numbers. If we could only get some sensible predator management for the mountain lions of California, the future of desert sheep would really move in a positive direction.

Completing this adventure we drove from California to Wyoming, to begin our bighorn sheep hunt. We arrived a few days early to allow us both to get used to the higher elevations. We spent quite some time looking over wildlife on the North and South Fork drainages of the Shoshone River. Ewes were moving to winter range along with countless mule deer. This offered many opportunities for some great wildlife photos. B. Joe Coy of Yellow Creek Outfitters provided his outfitting and guiding services for this hunt. We left the trailhead at the 5,000-foot elevation and within four and one half hours of steady climbing we arrived at base camp at 9,600 feet, one and one half feet of snow and a strong wind gusting over 40 m.p.h. This camp was complete with a big cook tent tables chairs propane stove with oven wood heaters and everything else to make it as comfortable as possible. The sleeping tents had aluminum cots and wood heaters. All the tents had canvas floors clean and first class! No one lacked for anything including meals. The first day out B. Joe spotted two outstanding rams but the snow storm and high winds was the deciding factor in waiting it out until the next morning. It seems like the old boy upstairs was looking after J. Curtis Newman as the following morning the weather was beautiful no wind. As soon as B. Joe started to glass he spotted the rams about a half mile from the previous day. After a three-hour stalk, Curtis got his chance to fill his tag. He harvested a heavy broomed 11-1/2-year-old ram. The ram will be mounted life-size to compliment the great California desert ram he already has. We never scored this ram because no one really cared how many points it totalled especially J.C. Newman! Some people are losing sight of what hunting really is. After returning to base camp we decided to take advantage of the good weather and get back down to lower elevation. The next day after three hours of poor footing on a steep snow covered trail we got back to dry sage brush and no wind. Perhaps my recollection of this hunt makes it to appear to be a piece of cake. Just climb up and harvest a great ram and come back down. I failed to mention however that J. Curtis Newman is 83 plus years old. He is an inspiration to anyone that knows him.

In closing we both would like to thank B. Joe Coy for a great trip. He did everything possible to make the hunt the enjoyable success it was. His camp, horses, equipment and knowledge of the area left nothing to be desired. Thanks, Curt, for sharing these adventures with me.

Hawaii 1986: Filmmaker Dubs Pays Record $79,000 For Montana Permit

From *Wild Sheep*, May 1986

Arthur Dubs of Medford, Oregon, spent $79,000 on the Montana permit, shattering the previous record of $67,500 handed over for a Nevada desert sheep permit in 1984. Crazy, you say? Arthur Dubs has more than one reason for his actions.

A little "Indian scouting" in Montana some time ago was the shaping of Dubs' plan. On this particular trip, he spotted a tremendous ram that may become the star of a new film Dubs is planning. Dubs' movie will show the harvesting of not only a records book bighorn, but, hopefully, a new World's Record bighorn.

It is not only the film Dubs wants. In the past, he found himself wrapped tightly in the grips of the business world producing feature motion pictures.

"I want to start unwinding," he said, "I want to get back to nature."

Another reason Dubs pursued the coveted permit is the benefit it provides sheep. The amount Montana currently spends on wild sheep habitat has almost doubled due to Dubs' contribution according to Jim Ford, regional supervisor for Montana Department of Fish, Wildlife and Parks and FNAWS board member.

"The institution of the auction permit and the money this and future permits will generate for sheep management in the state will be a real plus for our program," said Glenn Erickson, special projects chief for the Department's Wildlife Division in Helena.

Erickson explained the money will play an important role in helping the Department meet its currently planned bighorn sheep management objectives. Among these objectives are increasing research efforts to further the Department's knowledge of population trends and better determine allowable rates of harvest, protecting and preserving key bighorn habitat by conservation easements, leases, land exchanges or purchases, and transplanting bighorns into suitable, unoccupied habitat.

"Intensifying land uses are decreasing the quality and quantity of bighorn sheep habitat in Montana," Erickson continued, "so we see a real need to learn more about the sheep we do have and to protect the habitat they require. Dubs' contribution to the program will be a definite asset in working toward these goals."

Art Dubs has the Grand Slam. Now, he wants to make it the highest scoring Grand Slam ever taken by one hunter. Currently, his bighorn is the lowest scoring ram of his Slam. That will soon change when Dubs shoulders his .378 Weatherby magnum, given to him by Roy Weatherby who wanted Dubs to film an African safari and travels to Montana's interior. Art's bighorn hunting season runs from September 15 to November 30, 1986.

Aravaipa Desert Bighorn

Arthur R. Dubs

Wild Sheep, Winter 1989/1990

It was an eerie, almost mystical feeling that engulfed me as I made my way along the jagged rim of Aravaipa Canyon. I couldn't help but think that long ago an Indian hunter must have crept along these same paths as he stalked the ancestor of my prey — "Old Chiphorn" — a magnificent desert bighorn that had eluded sportsmen for years.

Aravaipa, which is Apache for "little running water," is an eight-mile gorge that slashes through some of the most breathtaking and rugged country in southwest Arizona. Its spectacular grandeur often leads those who venture into its depths to compare it to the Grand Canyon.

As I have done in the past, I wanted to capture the history of Aravaipa Canyon's wilderness, and the excitement and spectacle of the hunt on film. There's no way I could accurately depict the lump I got in my throat when I first saw "Old Chiphorn" through my spotting scope. From a mile away I could tell he was a fabulous ram. He looked extremely top-heavy carrying his massive horns. Never did I dream I was looking at the largest desert bighorn ever to be taken by a sportsman.

Since producing my first full length feature film, "American Wilderness," in 1968 I was enthralled with both hunting and filmmaking. To be able to accomplish both, to me, is to have the best of both worlds.

Because of my movie-making ventures and other business interests, though, I hadn't been able to devote any time to hunting the last 20 years. I stepped back and examined where my life was going. I was successful and I had a lot to be thankful for, but I wasn't having much fun anymore. Then three years ago, I heard about "Old Scarface," a magnificent Rocky Mountain bighorn of great proportion. He lived in the Wallowa Mountains of northeastern Oregon, 300 miles from my home, where he had managed to escape hunters and now was a ripe old age of more than 14 years.

When first hearing of this legendary ram and his elusive ways, it again triggered the challenge and inspiration to pick up the chase. While tracking him for nearly two months, I was able to capture the old fella on six hours of movie film before his natural death at nearly 15 years of age.

After my filming episode of Scarface, I was hooked again, and at the auction of the Foundation For North American Wild Sheep, I purchased a permit to hunt desert sheep in Aravaipa Canyon, which was made available thanks to the Arizona Game and Fish Department. Later, when outfitter Floyd Krank of Globe, Arizona, called me about having seen Old Chiphorn, I couldn't wait to pack my camera along with my .300 Weatherby rifle and head to Arizona. I was determined to not only attempt to bring in Chiphorn, but also to share the experience and drama of the adventure on film with my fellow sportsmen.

Floyd had first spotted the ram three years before and named him Chiphorn because of a large piece of horn missing on the back side of his left horn. The wily old ram managed to escape all of Floyd's efforts to track him down whenever he had a hunting client. Much like Old Scarface, this ram seemed to know when hunting season started and when it was time to lay low. Floyd finally caught another glimpse of him on a scouting trip the previous summer during the height of the rutting season. The ram was lean and haggard from running off young upstarts. At the time, Floyd had serious doubts whether the old fella would be able to make it through another Arizona winter.

I arrived in Aravaipa on December 1, 1988, and met with Floyd a day before the season opened. He knew Aravaipa Canyon extremely well. He began surveying sheep in Arizona about the time of the historic Old Granddad transplant in Aravaipa. Floyd had already set up a comfortable camp and we both had a congenital dose of sheep fever. We sat late in the evening exchanging ideas on how we might be able to outsmart Old Chiphorn. We had 21 days until the end of the season.

The time of 5 a.m. came early. Before sunrise, Floyd and I found ourselves dodging cactus and slipping on rotten volcanic rock as we made our way through the jagged rims overlooking Aravaipa Canyon. Soon golden rays of sunlight started to spread a brighter path for us to follow along the canyon rim. It was a spectacular sight, one which was rewarding and I thought had already paid all the bills.

As we cautiously peeked over the high rim, we were treated to a spectacular view of the canyon and we could hear Aravaipa Creek bubbling along 2,000 feet below in the bottom of the canyon. A group of a dozen ewes and lambs were contentedly grazing on a rocky outcropping only a few yards from us. We skirted around them, leaving them undisturbed as we moved along under the rocky rim glassing every draw and crevice. The sheep were scattered and we saw several small rams, but no Chiphorn. We arrived back in our cozy camp two hours after sunset with a few new scratches and an accumulation of desert cacti.

After three days of dedicated hunting, from daylight till dark, we ended up with far less tread on our boot soles and no leads on Old Chiphorn. Floyd and I were both beginning to wonder if the old ram was still alive. We had passed up and photographed several trophy rams that probably would score well up in the record book but we were determined to hold out for Old Chiphorn. We knew he would outclass them all. We decided to move camp.

It was a wide awakening when at sunrise on the fifth day of our hunt from the top of the rim overlooking Aravaipa Creek in the main canyon, we spotted a heavy-horned ram with a group of other sheep through our binoculars. He was feeding on a barrel cactus, over a mile away, on a talus slope at the base of a high rock cliff. He was in one of those special places I call the "Ram Hole."

We rushed to set up our spotting scopes and with closer observation what we saw took the wind right out of us. There was no question of identification. Even from the great distance, we could see the ram's massive broomed horns. We both knew that it was Old Chiphorn. I remember Floyd making the comment: "Art, you're going to love 'em."

The ram had not only survived the summer's rutting season, but also he had bulked up considerably and was robust and healthy. He was banging heads with two of his younger cohorts showing his prowess in view of a half dozen ewes and one small lamb. He looked out of proportion holding up his huge load of yellowish horn growth and I was amazed how such a small bodied sheep could carry such a big set of horns with such agility. It was later that I came to understand why.

After quickly taking a series of film sequences, we wasted no time in moving around the rim to get in closer position of the old ram. An hour later, after weaving around the head of several rocky scabrous side canyons, we found ourselves 600 yards away looking down on Old Chiphorn 1,000 feet below us. It was then that the moment of truth began to plague us. How would we be able to retrieve our trophy if I bagged him?

It's pretty well known that I have been wanting to upgrade one of my Grand Slam's of

North American sheep I accumulated more than 20 years ago. I knew if I could nail and retrieve Old Chiphorn, this addition along with my best Dall's, Stone's and Rocky Mountain bighorn would far exceed my goal of bagging the largest Grand Slam of North American wild sheep.

After backtracking around the last side canyon, Floyd and I found ourselves again overlooking Old Chiphorn now not more than 250 yards below us. We were lucky. He was still with the other rams, ewes and the young lamb. Much closer now, we could see through our binoculars he was continuing to dominate the herd, flirting with the ewe with the small lamb.

I quickly set my tripod and camera in position and ran off a few more feet of movie film. Not chancing the ram outsmarting us, I chambered a 180 grain bullet into my rifle barrel. Then I slowly crawled out to the edge of the high rim as far as possible overlooking Chiphorn and his gang below us. As my rifle came to rest on a crop of rock, the cross hairs of my scope found the great ram. I held my breath and took my best shot. It was a good one.

Floyd yelled, "You got 'em!"

As the echo of the rifle continued to riffle throughout the rocky canyon walls, the rest of the herd scattered into the cactus and rimrock overlooking Aravaipa Creek in the canyon below.

After congratulations, the real challenge began, which was the climb down the high rim to the talus slope below to retrieve my trophy. After back-tracking through several side canyons and losing our footing on rotten rock, we had to change our routing. Dropping down the backside of yet another canyon, we discovered a faint trail that exposed some fresh sheep tracks. I have taken several Rocky Mountain bighorn, and these tracks were as large as any Floyd or I had ever seen. We knew they had been made by Old Chiphorn.

We continued to follow the tracks down the canyon through a narrow passage hidden by tremendous boulders and a jungle of undergrowth. We knew if Chiphorn and his gang could make their way through the canyon onto the talus rim below, we had a chance to do so as well.

In our struggle to get through the rocky passage, sometimes crawling, we finally broke out into some of the most beautiful terrain we'd encountered in Aravaipa. As we made our way deeper into the canyon, it was refreshing to begin seeing autumn foliage and I was surprised to hear so many desert birds singing melody with the loons. Soon we knew why, as we came across several crystal clear water holes. The canyon was an oasis — a paradise for birds and many snakes as well.

Breaking out of the canyon, we came across a large colossal Indian cave. Floyd commented that "between the sheep and the Indians, the cave could tell us quite a story." It now had been almost three hours since we last saw Chiphorn from on top of the rim. We knew he had to be somewhere, perhaps just around the next rim.

We knew we had taken a huge ram, but we weren't prepared for just how big. The experience was like comparing it to winning a Gold Medal in the Olympics and taking an Academy Award, all rolled up in one. We were looking at a fabulous desert bighorn.

I broke out my measuring tape and made a quick measurement of more than 42 inches on the length of both horns. More impressive was the massiveness of his horns. Both horns were broomed off as big as coffee cups, with more than 16-inch bases. The old ram had four teeth left and all were loose but one. He appeared to be at least 12 years old. After a closer observation, Floyd and I began to realize how such a small bodied sheep could support such a huge set of horns. He was built like a bulldog, with tremendous muscular neck and shoulders. Muscle tone tippled throughout his beautiful, dark-furred body. We came to the conclusion that his huge front hooves were essential in supporting the weight of his horns that he had to carry over the years.

I knew immediately I wanted Old Chiphorn as a full mount. Floyd and I took a complete set of field measurements before dressing him out and then again later, with no hide on the carcass. I wanted the end result to be the truest representation possible of the ram as he appeared in the wild. This proved invaluable to taxidermist Michael Boyce, who later mounted the ram, naming him "Steppin' Proud!"

A few days later, the horns of Old Chiphorn (in the Green Score), measured 198-5/8 Boone and Crockett Club points. After a drying period, nearly four months later, they were scored at 197-1/8 by a Boone and Crockett Club official measurer. They were also measured by Safari Club International and certified at 198 points, a new SCI World's Record. I plan to enter the horns for final certification at the Boone and Crockett Club's next 21st Awards presentation. (The ram officially scored 197-1/8 B&C points and is ranked Number Four in the world by the Boone and Crockett Club in their new, *Records of North American Sheep, Rocky Mountain Goats and Pronghorn*. Nearly five inches larger than Javier Lopez's ram of 192-5/8 points, taken in 1979 in Baja, California, Old Chiphorn appears to have enough extra inches of horn to continue "Steppin Proud" and claim the exclusive title of being one of the largest desert bighorn taken by a sportsman.)

In the meantime, Chiphorn has made a few national appearances and Floyd and I have been quite busy. Regrouping, we met again in Aravaipa Canyon to photograph and interweave the colorful history and background of the Apache Indians' role in hunting the Aravaipa sheep for survival. We also retraced and filmed the true story with Duard Sanford, who at the age of 13 homesteaded in the canyon with his father in 1922. He was instrumental in helping the Arizona Game and Fish Department establish desert bighorns back into the canyon with the transplant of the Old Granddad ram, as a lamb, in 1960. The famous ram (the largest in Arizona at the time), went on to dominate a herd of more than 20 sheep, eventually siring many champions, which it is believed also included Old Chiphorn.

After several months of editing and scoring, I completed a feature-length video entitled, "Aravaipa Desert Bighorn." Thanks to the efforts of the Arizona Game and Fish Department personnel and others who previously filmed the transplant, including Old Granddad, I have been able to document this memorable account, including some highlights of the 1989 FNAWS convention into the video. I have dedicated the video film to the Foundation for North American Wild Sheep, for its enduring and successful efforts in sheep restoration in North America and for making possible my permit to hunt this magnificent animal.

My company, Pacific International Enterprises, has commissioned Dennis Jones, one of America's foremost wildlife sculptors, to create a limited edition bronze of my ram entitled, "The Dubs Ram — Old Chiphorn." The bronze and video cassettes, with a preview, will be available at Jones' FNAWS booth numbers 108 and 151 at the convention. A bronze will be donated for auction at the convention as well.

** This was not a story about an auction permit hunt, but Arthur Dubs is a long-time auction supporter so he is recognized here.*

Old Scarface Remains Forever Wild

George R. Elsom, Jr.
From *Wild Sheep*, Fall 1987

The hot sun was scorching the rugged Wallowa Mountains as Arthur Dubs, Don Turner and his 9 year old son Sy crested yet another high alpine ridge. It was one of those magnificent spring days that magnifies all of nature's beauty. Suddenly Dubs, a trophy hunter and owner of an independent motion picture company, was taken back by the sight before his eyes. On the next ridge was the largest bighorn ram he had ever seen. Quickly dropping to a prone position, Dubs sighted in. He squeezed the trigger of his movie camera and began to memorialize the legendary ram known as "Old Scarface."

"Scarface," also referred to as "Spot," the nation's largest recorded Rocky Mountain bighorn ram in existence, scoring 202-7 / 8 Boone and Crockett points, is also the most renowned. He has recently become a national legend with his universal popularity stirring the nation's sportsmen into "sheep fever."

With one eye missing and a scar on his nose, Old Scarface had the face only a mother could love. His horns showed the obvious wear of years of battles with his rivals.

Since the age of six, Old Scarface has horn measurements record books are made of. As a ram of such incredible magnitude seldom goes unnoticed, Old Scarface was no exception and was often the subject of zestful conversation between big-game hunters and sheep enthusiasts alike. His legend was made of the many stories of futile attempts by hunters to hang his great head on their wall.

"He was sought after every year we had a season up there," said Oregon state police lieutenant Gerald Del Fatti. "Some hunters were going to hold out for Scarface, but they could never find him while hunting. Before and after the season people could locate him, but no one could find him during the hunting season." He was also the focus of illegal plots to poach him. The elusive ram always denied the hunter his trophy.

Almost two years ago, Old Scarface became the subject of conversation between a fellow neighbor and a big-game hunter with Arthur Dubs. He was told of the legendary bighorn ram living in the high country of the Wallowa Mountains. A photograph revealed what the spoken word could not accurately describe, the horns of gigantic proportion.

Dubs was instantly intrigued by Old Scarface. With movie equipment he traveled 300 miles from his home to southern Oregon to northeastern Oregon where he met Don Turner and his son, who lived in the foothills of the beautiful Wallowa Mountains. Dubs and Turner had both heard hunters' stories of the ram's ability to elude his pursuers, as "Old Scarface" was often the target of the hunter's bullet and the poacher's greed. Nevertheless, Dub's goal to film the old monarch on motion picture film was an adventure he looked forward to. Many had photographed Scarface over the years, but never before had the great ram been filmed by the motion picture camera.

Dubs, often referred to by his competitors as the dean of wildlife film producers, often accomplishes what others only dreamed of. Once again this held true. After he and Turner had climbed thousands of feet into the high Wallowas carrying their camera gear, they spotted "Old Scarface." He was standing in all his glory on a far ridge. The sight was breathtaking, one which sent a captivating feeling through Dubs. It was a feeling he had felt 20 years before, and he recognized the symptoms. It was sheep fever. He knew once it set in there was no stopping it for he had experienced the fever previously on numerous sheep hunts.

As a boy Dubs developed a love for nature. By age 16 he already had a collection of trophies. As he grew older, his delight in the challenge and adventure of filming his hunts became a hobby and common practice. This allowed him to more fully enjoy the wilderness he so deeply enjoyed.

It was in 1966, while still in the construction business, that Dubs embarked on the ultimate of big-game hunts, the Grand Slam, the tagging of four species of North American wild sheep. Such a hunt is coveted by sportsmen as one of the most difficult of big-game hunting. But Dubs felt such a challenge alone was not enough. He believed filming the entire adventure with a movie camera would be even a greater endeavor. Even of more difficulty was his goal to film and bag four record-class sheep in less than one year. With a camera in one hand and a gun in the other, Dubs set out for his unparalleled adventure.

His expedition led him 50,000 miles from the barren deserts of Baja to the vast glaciers of Alaska. He became the first to bag a record-class Grand Slam of North American wild sheep in a period of less than eight months. In addition, he was successful in filming his entire adventure as it actually happened.

Over the next two years Dubs cut and molded the 40,000 feet of film he had shot into "American Wilderness." The film was an instant success, playing in theaters across the nation and grossing close to $12 million ($48 million by today's standards). "American Wilderness," a successful hunting and nature film, quickly pushed Dubs into the motion picture business. Dubs then incorporated Pacific International Enterprises, his wholly-owned production and distribution company that soon became one of the most successful independent motion picture companies in existence. The company became distinguished internationally for its family-oriented films. Some of his more popular titles he produced and distributed include, "Adventures of the Wilderness Family," a three film series, "Across the Great Divide," "Vanishing Wilderness" and "Windwalker."

Eventually the success of his film company and the entanglements of the corporate world caught up with Dubs, smothering his hunting fever. His sheep fever slowly became subdued, but not forgotten. Over 20 years had gone by and now "Old Scarface" provided him with another opportunity to return to nature.

The first sight of "Scarface" on the far ridge rekindled the sparks of sheep fever in Dubs, but now in a different way, for Dubs had left his gun behind. Dubs and the Turners began tracking and photographing "Old Scarface." Little by little they were able to gain the ram's confidence. Dubs was able to place his camera within a short distance from the old ram while Turner shot captivating stills.

As he continued his filming, Dubs noticed something peculiar and funny about the ram. It was his unnatural chewing habit. Curiosity recently provoked Dubs to question Oregon Department of Fish and Wildlife biologist Vic Coggins about the matter. After the old ram had died a natural death, Coggins examined the ram's head and stated that "Old Scarface" had an abnormally long upper tooth which caused him to eat and chew his cud in a rotating fashion. By using both sides of his jaw when chewing "Old Scarface's" lower teeth had not worn as normal. This unusual characteristic was probably the reason for the ram's near perfect teeth and longevity.

For over two months Dubs and the Turners continued tracking and filming the old ram until Dubs was stopped by a serious back injury.

"There grew a mystique about this old ram," said Dubs regarding Old Scarface's uncanny knack to sense his friends as well as his enemies. "I have photographed many bighorn sheep in my life, but I have never seen a ram so elusive and smart. When he really wanted to avoid us, it would sometimes take days to relocate him."

As with each fall, hunting season soon came to the Wallowas. The hunters again prepared to attempt to find the elusive ram. Several of the six hunters who were fortunate enough to draw tags to hunt, even participated in the special hunting seminar designed to better equip them with the knowledge to find Old Scarface. But as usual the evasive old ram was never seen by any of the guides or hunters.

With winter came a chill as cold as death. The deadly virus, Para influenza, Type 3, took a heavy toll on the Lostine sheep herd. Old Scarface moved on to greener pastures in the sky. Whether from the virus or just old age, Old Scarface had died a natural death at almost 15 years old.

Six months later, while backpacking in the high country in late October, Washington state resident, Don O'Callahan, miraculously found the ram's remains in a high alpine basin in the Wallowas. A rare find, O'Callahan turned the huge ram's head over to the Oregon Department of Fish and Wildlife. Upon first sight of the horns, department biologist Vic Coggins quickly recognized them as the horns belonging to Old Scarface.

There is now several thousand feet of dramatic motion picture film and slides of Old Scarface in Dub's studio that were the last to be taken before the ram's death. Dubs has plans to finish his production and release a feature-length theatrical motion picture of the old ram entitled, "Forever Wild." He is looking forward to welcoming this film to his library of family oriented motion pictures.

"I am happy to have captured the memory of the legendary ram on film, so it can be preserved for future generations to enjoy," Dubs stated.

Always keeping an ace in the hole, Dubs went on to say, "`Forever Wild' will be an exciting surprise for everyone. The film will appeal to sportsmen, wildlife enthusiasts and the general public alike. `Forever Wild's' story is an action adventure drama that will interest audiences of all countries. We will probably be making the initial press release announcing the film very soon." "Forever Wild" is a memorial of a great ram, the largest bighorn in the country.

** This was not a story about a auction permit hunt, but Arthur Dubs is a long-time auction supporter so he recognized here.*

Anatomy of an Auction Hunt

Jerry Fletcher

As I slid forward into a prone position on the black rock outcropping I felt the weight of long expectation set firmly down on my shoulders. Out in front of me I knew the ram of several lifetimes was in the boulder pile 150 yards away. Now I had to make the shot.

This entire process began on Saturday, February 27, 1993, at the Foundation for North American Wild Sheep convention in Reno, Nevada. When Stanley Boots of New York, and I took the auction to new heights on the Arizona desert bighorn sheep permit there was a hush in the banquet hall like someone sucked the air out of the room when the permit reached $300,000, and then went on.

When the dust settled, I had purchased an opportunity to hunt desert bighorn sheep in Arizona for $303,000 the highest sportsmen-related auction item ever sold. Stanley Boots finished second only in the fact that he didn't get the permit; his commitment to wild sheep and the Foundation over many years is to be commended and applauded.

Since that auction I have probably been asked 100 times why I paid so much to hunt one sheep. Believe me, I didn't walk into the auction that afternoon planning to spend that amount of money. But I wanted this permit, not another desert permit in another state. Arizona is home and it has been extremely good to me for a lot of years. Plus I was fully aware that all of the money spent on the permit would go back into the sheep program in Arizona for water development, research and transplants to put "more sheep on the mountain" for my family and others to enjoy in the years to come. Finally, I felt that with an Arizona permit I could spend more time on the preseason scouting trips, looking at wildlife and beautiful country, which is what I enjoy doing besides spending time with my business and family.

Immediately following the auction I told Larry Heathington of Sand Tank Outfitters to start putting a team together to guide and outfit this hunt. I already knew from earlier conversations with him that he and Pete Cimellaro would be concentrating their efforts in Unit 22 north of Phoenix and 40-A south and east of Gila Bend where the rams had the potential to build large bases, which is what it takes to produce big sheep.

Over the spring and summer Larry and Pete spent in excess of 30 days in the field identifying rams. Some days Larry would go alone into the blast furnace of Arizona's hottest desert, with temperatures commonly reaching 115 to 120 degrees in 40-A outside Gila Bend, the nation's hot spot as often as not. Pete concentrated his efforts on the lake chain of Saguaro, Canyon and Apache, which make up Unit 22, either by boat or on foot as the situation dictated. When I could break away from work for a day or two my friend Jack Smith and I would tag along so I could see the rams they were classifying.

The scouting trips started off slowly as Arizona was having a wet spring, creating feed everywhere, which allowed the sheep to scatter into isolated pockets. We saw ewes, lambs and small rams, but the big boys were eluding us at every turn. It quickly heated up both literally and figuratively. As the temperature rose above 100 degrees we began to see more and more large rams as the rut got started, and they became dependent on available water sources. For a guy who had seen some desert sheep in his life, but not many, looking at rams that would make Boone and Crockett — only to be told he's not in the class we are looking for — was mind boggling.

During these scouting trips Larry Heathington told me for the first of many times to come, "If you want a great ram, you have to be able and willing to pass good ones." That statement became easier to take as we saw rams of all shapes and sizes in both 22 and 40-A. Some of the larger rams we nicknamed Darkhorn, Flare, The Papago, Rocky Mountain, Lambtips, 182 and others, simply as a means of reidentifying a ram when it was seen on another trip through characteristics and distinguishing marks. When I was unable to go on a scouting trip and Larry or Pete would call to tell me how it went. I knew exactly which ram they were discussing when they mentioned these nicknames.

The other great advantage to these scouting trips was in getting to know my guides and outfitters personally. In every guide-client relationship there is an adjustment period as you feel each other out and make the transition from strangers to friends. As these trips progressed we began to laugh, joke and play while putting on some hard miles looking for the "Best of the Best."

Summer turned into fall and we narrowed our choices down to four or five rams out of more than 100 we had looked at, all of them well above the 180 point mark. Based on the vast storehouse of information we had available to us we decided to open the hunt in Unit 22 since as two or three of the largest rams we had seen were along the lake chain, knowing full well we still had some exceptional rams in 40-A should we need them for any reason.

I arrived in a very comfortable tent camp early on the afternoon of November 29, 1993, with my friends Jack Smith and Joe Jordan. I was met by Larry and his normal crew of Richard Mitchell, Todd Ralls and Michelle Bemis, who would be doing all the cooking for the hunt. Pete Cimellaro was due to arrive later in the day with Royce Wood of Nevada, who would be along on the hunt assisting Pete.

I would be hunting daily with Larry and Richard, while Pete would help by looking for additional rams in other areas. He had another client to guide to a ram of his own, a young gentleman, Tod DeMaree of Phoenix. Tod had been extremely lucky to draw one of the two general permits for this coveted area, and would be hunting out of the camp until he was successful and had to return to college.

That first afternoon we went looking, trying to see two large rams Larry and Richard had already located. In less than three hours, I was able to see a large number of sheep, including both of these tremendous rams we estimated would break 180 B&C points by a very wide margin. How many hunters ever have the opportunity to see one desert ram this large, much less two, in the same afternoon? I had visions of a very successful and short hunt dancing in my head that night as I tossed and turned, trying to get some sleep. We still had one more day to scout prior to opening day, most of which would be spent keeping track of both rams; hopefully, they would stay put.

Opening day: December 1 opened to a typical, cool, clear desert morning, which found all of us moving up the lake by boat to the "Darkhorn" ram, one of the two or three best rams in the unit. As the sun broke the horizon we quickly located the ram and his two smaller companions on a high bench, placidly feeding on palo verde and brittle brush. As the sun climbed, so did the ram. It finally topped out into a myriad of hanging valleys and plateaus that make up the bulk of the unit. The visions of a short hunt quickly disappeared along with the ram.

For the next week everyone looked for darkhorn or a ram we call Sluff Horn, from the lake or walking back into the back country to good vantage points where you could see lots of country. On December 5, 1993, well after dark, Pete, Woody and Tod DeMaree brought in Tod's ram that is one of, if not the best-looking desert sheep everyone in camp had ever seen. We had seen him on a number of scouting trips and earlier that morning nicknamed him

"Rocky Mountain" based on the configuration of his horns. As I looked at this magnificent ram, I knew that had he been mine I would be totally satisfied, no regrets. But he belonged to Tod and I still had to hunt for a ram of my own.

Late one evening, Joe Bill Pickrell arrived back in camp to announce he had found the darkhorn ram.

Early the next morning, Pete, Matt Dominey and Joe Bill all relocated the ram in an isolated canyon quite a distance from the lake. By the time we were all able to complete an approach to this canyon, the ram had again vanished into the surrounding countryside. We could find other sheep in these canyons, including a huge ram with his right horn broken back to only 10 inches, but no Darkhorn. All of the rams we located were moving quite a bit. They were in the final stages of the rut, just walking and looking for receptive ewes. We would see a ram one day and then see him seven miles away the next, still walking, still looking.

Based on the mobility of the rams and with lots of time left to hunt, I made a conscious decision to move to 40-A with Larry, Richard and Jack for a few days to see if we couldn't locate some of the larger rams Larry had seen earlier, while Pete, Joe, Bill and Jerry Walters would continue to look in Unit 22 for the two or three rams we were interested in.

If you like wild, uninhabited country, 40-A is right up your alley. It has vast stretches of Sonoran desert habitat with saguaro cactus, palo verde, brittle brush and mesquite flats surrounded by rugged mountain ranges that push up out of the desert floor. There is a lot of country to look at with a few scattered sheep throughout it, but no really large concentrations.

Every day we were in the unit, we saw sheep; not many, but sheep. As often as not, Larry or Richard would locate an old ram off by itself that earned a closer look. Each time Larry and Richard would find it lacking for one reason or another — based on one, length on the other, and one horn broomed back too far on the last one. But believe me, it was probably the hardest thing I've ever had to do was to pass on two of these rams that were absolutely spectacular to look at, within rifle distance and were sheep any sportsman would love to possess.

But I kept hearing Larry say, "If you want a great ram you have to be willing and able to pass good ones."

Jack Smith kept saying, "You guys are crazy." Our search continued.

On the night of December 12, 1993, we received a telephone call from Pete at Larry's house, which is only an hour and a half from the unit, indicating that he and Jerry Walters had located the Darkhorn ram in Unit 22 long with another spectacular ram that we had never seen before. We quickly made arrangements to meet him there before daylight the next morning.

December 18 found Larry, Pete, Jerry, Otto Walters, Jack and I perched on a high ridge overlooking a huge basin in Unit 22. We could all see one ewe and four rams northeast of us, including a great ram with chocolate colored horns, but no dark horn ram. It again disappeared into the landscape. Larry and Pete could only see one solution to this problem, which was to walk into the country where the ram had been last seen and try to locate it as we went. If we were unable to find him we would still be in position to stalk the other ram from a better angle.

As Larry, Pete, Jack and I made the mile to mile-and-a-half walk into some sandstone ridges where Darkhorn had been, we could periodically check the other ram and his companions. They seemed quite content to stay where they were in the morning sun. For no reason we could ever determine, out of the blue a helicopter flew into this basin and lit on a ridge directly above the four rams in violation of wilderness guidelines that were in effect. We all cringed as the rams got up out of their day beds, milled around and began to move off. The helicopter stayed put for a few minutes then picked up and flew out of the basin, never to be seen again. Although jittery and spooked, the rams moved a short distance and calmed down, to everyone's relief.

The going got extremely slow as Larry and Pete closely checked out each little pocket before crossing any ridges, trying to locate Darkhorn. We would closely examine every inch of these small ravines prior to moving across them to peer into the next one. Four ridges later, Pete finally located a single ram facing away from us in his day bed, 700 yards away, lying behind some brush that obscured our view. Was this Darkhorn?

A few minutes later as he turned his head back and forth, we finally had our answer. We had found this ram that had eluded us on more than one occasion. However, moving into a shootable position was not going to be easy as we had to cross some open country to get there. We set up both spotting scopes; one on Darkhorn and one on the chocolate horned ram to try and compare them. Which was bigger? To this day, I don't know the answer to that question. They were both tremendous rams to be admired by anyone. We made the decision to try and harvest Darkhorn, if we could get it done. We had looked at him on a lot of occasions and were fairly certain he would meet or exceed our expectations.

In the middle of the day, Darkhorn got up and moved higher on the slope where he had been bedded on and moved into a boulder pile that obscured his vision in our direction. We quickly gathered our gear and crossed a major drainage between us and the ram, which put us totally out of his sight for the duration of the stalk. There was only one more decision to make, based upon where the ram was located. We could either stalk into 25 yards of him by coming out on a ledge above him, or I could shoot from a small ridge 150 yards away. I chose the longer option as I felt it would be less risky and would give me a chance at a better rest.

We crawled onto a ledge through a cholla patch that we were all wishing wasn't there. By the time I got into a solid prone position with the rifle resting on a day pack as solid as I could possibly make it, the ram still wasn't in sight behind the pile of boulders. As I waited for him to appear, the stress amplified until I felt like I had a 400-pound gorilla setting squarely on my shoulders. We had all come a long way to get here, 10 months after the auction.

As if by magic, the ram appeared on the largest boulder in the middle of the slide facing me and quartering slightly to the left. I put the crosshairs directly on his right shoulder, and when it felt exactly right, I gently squeezed the trigger. By the time the recoil had subsided, the ram was gone. As if in slow motion the ram moved out from behind the boulder pile — barely moving — toward a 100-foot cliff directly in front of him. I shot again and the ram calmly laid down right on the edge of this shear drop-off. Everyone held their breath, hoping against all hope that he wouldn't fall off into space. I should have known better. After all, this was the disappearing ram. Several seconds later, he rolled gently to his right and off the edge into thin air. As he free-fell into space, all of our spirits plummeted with him. We could all see broken and splintered horns after all the hours and labor that had been put into this one ram.

Miraculously, when everyone got to the ram, he had some chips and dings along the length of the horns, plus a badly splintered left horn tip, but they were still intact and repairable.

Later that same evening, Jerry Walters, an official FNAWS measurer, green-scored the ram at 191-4/8 gross; 190-2/8 net. This same score was confirmed by the Arizona Game and

Fish Department at their mandatory check station on Monday morning, making it the second largest ram ever checked out in Arizona at that time!

Was it worth it? Every dollar went into sheep management, to produce other transplant herds like the herd in Unit 22. I got to pursue my dreams of taking a great ram over 10 months in my home state. You bet it was! Would I do it again? Probably.

I would like to thank FNAWS, the Arizona Desert Bighorn Sheep Society and the Arizona Game and Fish Department for "putting sheep on the mountain." Thanks goes to everyone who helped on my hunt for one day or for the duration of the hunt, as they all were there solely for the purpose of helping make my dream come true, doing whatever it took, from chores to looking until their eyes hurt, all for the betterment of the auction program. I would especially like to thank Larry and Pete for their commitment to this hunt to the exclusion of other priorities. Without their time and effort, the fulfillment of my dream would not have happened.

The Boone and Crockett Club official score stands at 188-2/8 points for the entry score as of September 1998 and because the ram is so large two additional official measurers must confirm the final score before an official B&C ranking is given.

Montana 1994

Jerry Fletcher

I purchased the 1994 Montana bighorn sheep auction license at the San Antonio convention. On November 27, 1994, the last official unit in Montana closed, which in effect ended my season. No ram was harvested during the 40 plus days we spent in the field in pursuit of the "Best of the Best."

Please do not assume that because I didn't harvest a ram that the hunt was unsuccessful or unfulfilling. On any given day I could have taken a ram in excess of 190 points and on at least seven occasions, rams that scored between 195 and 200 points. However, we chose to hunt two rams solely that would score well in excess of 200 points, excluding all others — win, lose, or draw. We saw these particular rams on almost a daily basis. Unfortunately, they spent the majority of their time out of the legal hunting zone and we morally and legally wouldn't cross that barrier, no matter how tempting it might have been. It comes down to the simple fact that after you have seen these rams, "No other ram will do!"

Throughout the hunt we laughed, joked, played and had a great time while in pursuit of these two magnificent rams. We got to watch some massive rams rut, a lion charge a band of rams, large rams feed unconcerned 50 yards of us and other great rams in their natural environment. All of these experiences are what hunting is all about. It's not just the harvest of an animal for your trophy room. Everyone benefitted from this hunt in some form or another. The people of Montana have Arizona dollars to "put more sheep on the mountain," motels, gas stations, convenience and sporting goods stores, etc., all gained needed revenue. While the people who assisted on the hunt gained through experience and / or monetarily. I had the time of my life, storing memories that will last long after the fact of not harvesting a sheep is forgotten.

In closing, I would like to thank everyone who helped on the hunt in some fashion, from outfitters to cooks. Lee and Sharon Bridges, Larry, Barbara and Neal Clark, Richard Mitchell, Jack Atcheson, Jr., Wayne Taysom, Virginia Frehner, H. Howard Larson and Julia.

Special thanks to my friend Jack Smith who went on the hunt as moral support and to handle needed errands to make the hunt successful. Plus, I want to thank Larry Heathington who coordinated everyone's efforts and spent 60 plus days in Montana looking for the "Best of the Best," through all of the hunt days, rain, sleet or snow.

Last, but not least, I thank my family who without their support and love, none of this would have been possible.

Alberta 1995

Gary Hansen
From *Wild Sheep*, Winter 1995-96

Those of us who have been bitten by that pervasive sheep bug have, at one time or another, day-dreamed the perfect sheep hunt. To many of us, the dream contains similar elements such as Rocky Mountains, great sheep terrain, large numbers of quality rams and, of course, the eventual harvesting of a mature and record book Rocky Mountain bighorn.

Most of us who have a special love and admiration for the Rocky Mountain bighorn have the imagination to dream about it but we rarely, if ever, get the opportunity to live the entire dream. During the 18th annual FNAWS convention held last March, I was fortunate enough to purchase the 1995 special Alberta sheep permit. It was during this convention and auction my dream and quest for the classic Rocky Mountain bighorn began, except on this go-around, it would not be a daydream!

Going in to the auction, I felt the special Alberta permit held the highest potential to provide its holder the opportunity to harvest a 200-plus-inch ram. The record books also supported my opinion the biggest rams have lived and still live in Alberta. The fact this was the first time ever Alberta was offering this permit, and that it was a November or rut hunt, made this an opportunity of a lifetime. Two other factors influencing me was my feeling the permit would go much higher in 1996 if a 200-inch ram was harvested in 1995, and also the possibility a special permit may not even be offered up in 1996 by Alberta. In any event, my final bid offer held, and I was well on my way to fulfilling a dream.

Shortly after the convention, I began talking to friends in Alberta who helped me map out a game plan with various goals in mind. Three trips were planned out and eventually made to the Alberta slopes. The first in April during the green up, the second during early August, and the third and final trip in mid-November for the hunt itself. In total, 26 days were spent in the field. My long-time friend and taxidermist, John Lewton, offered to accompany me throughout this experience and spent the 26 days helping me film and observe these exceptional rams.

In April, we were able to observe large bands of rams throughout their winter range. In August we found exceptional rams in the high sunny alpine valleys well above timberline. Our scouting trips took us as far north as Wildlife Management Unit 438 (WMU 438), and as far south as WMU 414. In both extremes, we identified rams we felt would score near or above the 200-inch mark. Some of those rams were later located and identified in November. During the 26 days afield, we observed approximately 300 rams: 20 to 25 of these were book class, while 10 to 12 were in the 190-point plus class, and two or three in that rare 200-point plus class.

For days at a time, during these scouting trips and the hunt itself, we would live with these exceptional rams, filming and observing their every move and characteristic. Hundreds of photos were shot and hours of video taken of rams with massive bases and long sweeping curls flaring well above the bridge of their nose. Massive rams with enormous necks and huge bodies — bodies so big it made visual scoring difficult. Often during this experience, I would stop and think how lucky I was to be living out this experience. Words cannot express my appreciation and admiration for these animals that have provided me memories to last the rest of my life.

The Professional Outfitters Association of Alberta (POAA) was the driving force behind the issuance of this special permit. Lorne Hindbo spearheaded the effort for the POAA and with the sponsorship of the Rocky Mountain Elk Foundation (RMEF)-Canada, influenced Ty Lund, minister of Environmental Protection and Natural Resources, authorized the issuance of this special permit. RMEF-Canada, with significant help from FNAWS, maximized the value of this special permit. This joint effort truly "put sheep back on the mountain." Some of the key members of POAA were also very helpful to me in helping me accomplish my goals on this hunt. I cannot say enough regarding their unselfish willingness to do whatever possible to assist me throughout the process of preseason scouting and the hunt itself. I would like to take this opportunity to express my gratitude to Lorne Hindbo, Bobby Turner, Gordon Utri, George Ostashek and others in the POAA who helped me in this quest. Special thanks goes out to a man who knows his sheep, Randy Babala, my "S-Outfitter" and guide on this venture, and also to my good friend, John Lewton, who was unwavering and by my side throughout.

The entire effort resulted in the harvest of a true monarch. The stalk took a little over three hours and ended on a steep wind-blown slope well above timberline. It was a classic end to a classic hunt and left one indelible mark in my mind: Alberta is home to the monster bighorn. The shear size and beauty of these rams, in my opinion, cannot be equalled anywhere in North America.

Thank you, Alberta, for the memories of a dream come true.

1992 Montana Bighorn Auction Permit Hunt

TIM MAGNESS
For *Wild Sheep*, Winter 1992-93

The 1992 Montana governor's bighorn sheep auction permit was purchased by a Valley City, Ohio, farmer by the name of Glenn Smith. He paid $88,000 to the Foundation for North American Wild Sheep for the privilege of harvesting one Montana bighorn ram from any area legal to hunt.

Sheep hunter and outfitter, Jack Atcheson, Jr., of Butte, Montana, and myself, Tim Magness from Anaconda, Montana, along with several other sheep fanatics, offered to help the auction permit holder find the best of the best. Between all of us, we spent days on end, in the mountains, adding up to weeks of effort in helping to find the best ram possible.

On April 21, 1992, Jack Atcheson, Jr., was on a scouting trip for Glenn when he found a giant of a ram, dead. This ram is known as "The King." The largest bighorn ram known to live south of the Canadian border. His net B&C score is 203-5/8.

Glenn is an enthusiastic hunter, who spent many days in the field looking at rams. We dedicated two, week-long hunting periods to preseason scouting. We scouted the upper Rock Creek area, the Anaconda area and the Highlands area, 30 miles south of Butte. In the Rock Creek area, on the 4th of July weekend, we found three contenders. We named them "Chip," "Dent" and "The Stud." All three were world class rams. On August 1, 1992, we went into the Highlands and found a great herd of 63 rams. We were able to watch them for several hours from a distance of 300 to 350 yards. In all, about 20 would make the B&C records book. Four would be in the 190 plus point class.

After running out of names for all the big rams we were finding, we called the three best, Number One, Number Two and Number Three. Through the spotting scope, we felt Number One was the best ram Jack or I had ever seen. We thought this ram might score close to 200 points and could have the potential to be the new state record for a hunter-killed ram. He also was only two or three miles from where Jack found "The King." Number Two was an old friend we had seen several times before. We estimated him to score in the mid 190 range. Number Three looked like an outsized Dall's ram with his divergent curl. Jack thought his horn tips dropped so low he would plow the ground when feeding. We later referred to him as "The Snow Plow Ram." Taxidermist Bob Maier said we were watching a ton of moving ram horns and that we actually were!

Glenn made a decision. "Number One will be my ram."

On August 23 my nephew, Andy Stinson, moved into the high country to keep track of Number One until the season opened on September 1. It snowed four inches while we were erecting camp. Number One was not seen during this period. Once the snow came, the rams became hard to locate. They vanished into the timber. Glenn and I decided to move into camp three days before the season opened. Slippery rocks had injured Andy's heel. We were becoming discouraged.

I thought, "It might be a long season looking for a single seldom-seen ram."

Each day we saw more and more rams as the weather improved. They finally returned to their normal haunts. The day before the season, Butte taxidermist Bob Maier came into help us scout. We still could not find Number One. At last light, the night before the opening, we spotted 13 rams, three of which looked like top heads.

We departed for the field long before first light, on opening morning. Our 30 days of waiting, since first seeing the three big rams, were over. Thoughts ran through our minds, "Is Number One as big as we think?" Soon we spotted two small rams 100 yards below us, feeding. Suddenly they were gone. It was time to hike again so as to peer into a hidden pocket I knew about. A half-hour later we were staring at three top bighorns, although Number One was not with them. There was one that would score 190 plus. They were bedded 125 yards from a rocky knob upwind from us. We were close but no cigars yet.

The morning of September 2 found us looking for Number One in the trees. We saw rams but not what we were looking for. By the afternoon, we had moved to a new basin, next to where we had originally located Number One in August. We spent the remaining hours of the day in the basin. As darkness fell, a new 190-point ram we had never seen before stepped out. Our hoped were up, but darkness swallowed him.

On day three we decided to look at a new area. By the process of elimination we hoped to locate the home of Number One. It was a long side-hill walk to reach the pocket I call the "Hell Hole." Thirty minutes after legal shooting light, six rams were located in the very bottom of the pocket. What luck. Both Number One and Two were there. Our stalk was planned to take us through a small finger of pine and juniper trees. Our confidence soared. What could go wrong? Both of us still wonder what did.

At 300 steep yards, the decision was made to stop. Number One was bedded facing away on the opposite side hill. Glenn found a good rest and held in the middle of the ram's back. It had been a long road to this point and we knew that some sort of victory was close but not for certain. After the shot, we learned our only victory was dust over Number One's head. All six rams left in haste, one more desperation shot was fired. Nothing was hit. Suddenly they were gone and all was very quiet. Two empty brass cases, two empty hearts and a long walk out of the "Hell Hole" was our reward.

After that it was hard to rise above sheer depression. We decided not to return to camp. We would attempt to figure out what the rams had done. Where had Number One and his fellow rams gone? Glenn's rifle had been sighted in to be two inches high at 100 yards. We checked it at an old cabin site. The rifle was on. Glenn was relieved to know the riffle was okay. He knew he was responsible for the miss. While at the cabin, I nearly stepped on a rattlesnake but all my luck had not left. It did not bite me. Thoughts of where the rams had gone raced through my mind. There were only two choices.

"Maybe they traveled a long way to a new area, or they might be in the basin next to the Hell Hole."

A hike to a ridge gave us the answer. We found the six rams silhouetted against the skyline, including our sought-for trophy.We ran from the ridge to a rock point below the rams. A new stalk was on. An hour later we were looking upward 175 yards at the bedded sheep. Only their heads, necks and horns could be seen. Not enough for a shot. Our only hope was that they would rise and start feeding again before darkness ended our day. At 7:15 p.m. a shot was heard in the far distance. Show time was about to begin at last. Number One and his companions sprang from their beds and stood looking about — like sentinels staring into space. I thought maybe their minds were filled with memories from the morning. Number One was standing facing us with his brisket and shoulders in full view. Glen moved below me positioning himself at a rock that was to be his shooting rest. For the second time of the day, Glenn had Number One in his scope. Would terror strike again? His confident voice whispered.

"I feel good with this shot, Tim."

An instant later all that I was able to see in the spotting scope was four hooves in the air. I could only hear the report of Glenn's rifle. All was quiet. Number One was ours — the result of a beautiful shot. A happy permit holder and his guide silently shared a quiet moment of respect for this fine ram. A ram the local sheep fanatics held in awe and respect — a feeling that is hard to express physically or mentally to those who do not understand.

During our now completed project, we estimated a minimum of 500 man hours were spent in the field, as well as, at home trying to decide on an area, ram, technique, people right for the job and the way to travel when leaving camp each morning.

Glenn Smith, in my opinion, took this ram as a ram should be taken. He kept all his options open, including the choice of the best areas and best people interested in his success. Thanks, Glenn Smith, for letting me be involved with the 1992 Montana bighorn sheep permit and being with "Wild Sheep on the Mountain."

The Foundation for North American wild Sheep returned $79,200, from the auction of this permit, to the Montana Department of Fish, Wildlife and Parks for their sheep management programs. The Foundation will again be auctioning a Montana bighorn sheep permit during the 1993 Convention.

A Records Book Ram on 900 Acres

Tom Pawlacyk
From *Wild Sheep*, Winter 1994-95

In 1987, the Washington Legislature adopted a rule authorizing the Department of Fish and Wildlife to auction permits to generate revenue. One or two permits for each of several big-game species could be auctioned each year. The revenue would be used to purchase or enhance habitat, transplant animals or conduct research on the auctioned species. Public response to this issue was overwhelming, and in October 1993 the Washington Wildlife Commission unanimously approved the concept of auctioning a bighorn sheep permit for the 1994 hunting season. The Foundation for North American Wild Sheep and its 1994 convention fundraiser was the arena selected to auction this first-ever auction permit. Bidding for this permit was tremendous with FNAWS Life Member Tom Pawlacyk the successful bidder. Through the generosity of Tom as high bidder, the Foundation returned $90,000 to Washington for its sheep management programs. Tom tells his hunting story to us.

My hunt for a Rocky Mountain bighorn sheep started at the FNAWS auction in San Antonio, Texas. There I purchased the first-ever state of Washington auction tag, which entitled me to hunt in three management units in the Blue Mountains located in the southeast corner of the state.

After the convention I contacted Jack Atcheson, Jr., to guide and outfit my hunt. Jack agreed and contacted Fred Hendrickson, a Washington resident, who scouted for us and made contact with local landowners to hopefully provide us with access to the sheep. The Northwest went through a terrible dry spell, lasting for four months. This dry period changed the sheep habits and located most of them on private lands near the Grande Ronde River. Fred Hendrickson, along with Dean Miller, a local resident, who operates a sportsman landowners group, contacted up to a dozen landowners along the Grande Ronde River to hopefully write an agreement that would include trespass fees. A tentative agreement was reached and we all thought we had a deal.

Jack Atcheson, Jr. and Fred Hendrickson arrived in the hunting area one week before the season to finalize the agreement between the landowners. Prior to their arrival, and unknown to them, a mega landowner on the Oregon side of the border contacted the Washington landowners and helped destroy our potential agreement that would have given us access to the sheep.

After our agreement fell apart, we started to negotiate with individual landowners. In order to scout their land for a big ram we had to pay $20,000 trespass fee even though there might not be any rams on their land. Fortunately there were a couple of high points off of public roads where we were able to glass private lands for rams. We were lucky enough to see a couple of quality rams on private land adjacent to and south of the Grande Ronde River. We contacted the landowner a day before the season started and negotiated a $5,000 trespass fee. Mr. Kinney was a reasonable person. He did not agree with the high trespass fee that Charlie Singer, another landowner, was charging Oregon auction tag holders. At last! We had approximately 2,000 acres to hunt, and hopefully, we would be able to find at least a record book ram. Due to the small piece of ground we had permission to hunt on, we had very little hope to harvest a 190 plus ram.

On opening day, September 1, 1994, Jack Jr., Fred and I started our hunt on the Kinney property. We spotted three rams first thing that morning but the rams were travelling east and eventually left the Kinney property. His east boundary was easily recognized and we continued our climb along it to the south. Jack, Jr. spotted seven rams moving fast from east to west at about 600 yards from us entering the Kinney property. We hurried to a high point and were able to watch the rams moving very fast. It was obvious that someone had spooked these rams. Their tongues were hanging from their mouths, and they had no intentions of stopping. Two of the rams in this group were very large, Jack Jr. was sure this was the last time we were going to see them.

Later that afternoon we spotted three rams bedded down and stalked to within 400 yards of them. One of the three was estimated to be a 180-class ram, with 15-inch bases and 40-inch horn length. Under the circumstances he looked good to me, but Jack thought with any luck we could do better.

His last comment of the day was, "When you see a records book ram you will know it's a records book ram."

At his point we had hunted about 900 acres of Kinneys' land and my chances of harvesting a truly large ram did not look good! The second morning of the hunt took us to a knoll on the north end of the property, where we could glass the 900 acres we had hunted the day before. We were hoping the seven rams we had spotted had holed up or would return. About 9 a.m., Jack Jr., spotted two rams bedded at the bottom of a draw and decided they deserved a better look through the spotting scope.

After about 10 minutes he turned to me with a grin on his face and said, "Do you want to know what a record ram really looks like?"

I couldn't get over to his scope fast enough, and I wasn't disappointed when I did. This ram was three fourths of a mile away and looked very large in his scope. We decided to move up above the ram to allow for a closer look and measure his horns. At 11 a.m. we were 450 yards above the ram. This allowed Jack a real good look. He estimated the ram's B&C score to be 188 points. Under the circumstances this was more than acceptable to me. At noon, the rams bedded and it was time to make our move on them. We moved very slowly to the south, climbing higher to a point where we could come down on the rams. They had bedded in some brush and were out of our sight. As we slid down on top of them a few inches at a time, the wind was in our favor, and by 3 p.m., we were in a position on an open grassy slope 250 yards above the rams. At 3:30 p.m. the rams left their beds and walked into the open. I put the big ram on the ground, ending 6-1/2 hours of nervous anticipation. My ram scored 188, exactly what Jack estimated it to be. His official score after the 60-day drying period was 184-5/8 B&C points and is ranked fifth largest in the state of Washington, according to the 1996 book by the Boone and Crockett Club, Records of North American Sheep, Rocky Mountain Goats and Pronghorn. Mike Boyce will lifesize mount the ram and hopefully have it on display during the 1995 FNAWS convention in Phoenix.

Mike Boyce called me upon receiving the horns and cape at his studio and said, "The Boone and Crockett score will never reflect just how beautiful your ram is!"

Jack Atcheson, Jr. and I believe there are some very large beautiful sheep to be harvested in southeast Washington. However, there are some very serious access problems that need to be resolved. If you are interested in purchasing the Washington tag, feel free to call Jack or myself and we will share some ideas with you.

Idaho Sheep Hunt in September 1989

Dan Pocapalia

In as much as this was my 16th sheep hunt during the past ten years, I felt somewhat confident, even though some of my previous hunts had been unproductive and some disastrous. With everyone's help and consultation, I selected what I considered to be the best sheep guide and outfitter in Idaho, Stanley Potts of Hailey, Idaho, to serve my special needs, and, in turn, he selected the area that offered the best opportunities for bagging a ram in the 180 plus scoring range.

Stan picked me up on schedule at the Hailey airport three days before the opening of the sheep hunting season on September 2. Since my special permit allowed me to hunt any areas that were legally open to sheep hunting, I left it up to Stan to select the area that offered the best opportunity for locating a ram that would score 180 plus. By any stretch of the imagination, this was a big order, and Stan felt confident Area 537 would be the right place to hunt, where only three permits were issued this year, and, as I understand it, the area had only been open for hunting the last four years, thus allowing the rams to mature to 10 to 12 years old, for larger trophies with 16-inch plus bases that are needed for a high score, and we both agreed on this. The weather was extremely hot, dry and windy on my arrival.

I had never met Stan Potts prior to his picking me up at the airport, and frankly, I wondered every now and then just how my blind date with a new outfitter was going to turn out. There's always a degree of speculation. During the two-hour drive to our camp site we talked about many subjects. I was really getting the feeling I made a good guide selection with a first-class operation. It has been my experience over the years the first sight of your camp can have a lasting impression and, to say the least, sometimes a big surprise, and not a good one.

Stan was carefully maneuvering his 4-wheel drive up a narrow logging road when he finally announced, "We are there!" What I saw was a small meadow with foot high grass, four horses grazing in the meadow, with a small igloo tent set on the far side of the meadow. I finally located the main, or cook's tent (the most important one), or what appeared to be the main tent. It was a mess. Stan stared in disbelief and embarrassment. I looked at him and he looked at me and neither one of us said a word. He walked to the tent and opened the tent flag. Everything was scattered all over the floor. He assigned me the igloo tent and after 45 minutes of getting my things organized, I joined Stan, who had made a complete survey of the tent situation. He concluded the grazing horses had gotten tangled in the guide ropes that held the cook's tent up, and once the guide ropes were broken loose from the anchoring stakes the tent collapsed. This made me feel much better. After 45 minutes he had the tent in tip top shape and orderly.

It turned out to be a great camp. The camp was dry and all water was brought in from a spring some miles away. The camp site was situated at an elevation of 8,800 feet, based on my altimeter, just a few feet below timberline.

The next two days prior to the opening of the hunting season were spent scouting and glassing different areas. We spent a great deal of time glassing mountains one to five miles away. We located many rams, but none were in the Boone and Crockett category. I also got the impression Stan was toughening up the "dude" to see what I could handle at an elevation of 10,000 feet. I don't mind admitting there just wasn't very much oxygen at that elevation, but after a couple of days I managed quite well. I had passed my annual physical examination.

On opening day we located four legal rams, none of which met my requirements. On September 3 we were working a high, barren ridge high above timberline. The ridge was covered with flat, sharp-edged shale and frost. Stan looked down into the bottom of the canyon and located three rams feeding. We instantly lay down on the frozen shale so our outlines would not be silhouetted on the skyline and immediately tried to determine if one of the two legal rams would make the book.

For eight hours we were pinned down in the one spot with the rams 1,500 yard from us. They would feed for a short period of time; then lie down. There was no way to back track out of sight.

One of the rams was a possible Boone and Crockett "keeper," but we were too far to determine if the marginal ram would make the book. We both expressed our doubts and estimated he would score 175 to 178 maximum; not good enough, but felt we would like to take a closer look at him. Finally, in late afternoon, we managed to back away from the rams' line of vision without spooking them.

The game plan for the next day was to attempt to locate the three rams, make a sneak close enough so we could make an accurate estimate on how he would score. We had pre-established I did not want to bag a ram unless he made the book, since I have several in the 175 and 178 category. At 6 a.m. the following morning we were worming our way up the trail we named "Shoe Goo," because after several days up and down this steep shale, our mountain boots were coming apart and were in need of repair with the stitched seams coming undone, as well as the soles. I just happened to have a tube of "Shoe Goo" that saved the day. We both managed to repair our boots with the stuff.

Once on top of "Shoe Goo Ridge," Stan immediately located the three rams we saw the previous day. Stan suggested that we start our stalk immediately. The rams were several miles down the mountainside feeding on a small, rolling plateau. We figured it would take us several hours to get to the approximate area where Stan guessed they would be for a final inspection at a close range.

We were on our final leg of our stalk, and had just crossed this large, flat meadow and worked our way up a low, flat bench with gently sloping plateaus. As we topped the bench, we paused for a breather and, once again, agreed the ram must score 180 plus or it was a no go. Stan was in agreement and started to brief me.

"Now, Dan, remember in Idaho if this is the ram we are looking for, you must be prepared to shoot at 400 yards or more," he told me.

We concluded that I must aim 6 inches above where I expected impact to be. Stan asked, "Dan, can you handle it?"

I replied, "No problem with my 5-pound, .284 smokestick I can handle it."

I was all psyched to aim 6 inches high. This was going to be the longest shot of my life.

Stan said, "Now get your smokestick ready. We will cross the shallow gully and once we crest the ridge on the far side, we should be able to see the rams. Let's go!" I was all psyched up and hoping he will make the book.

Stan said, "Don't shoot unless I tell you."

We took ten steps and Stan tapped me on the shoulder and pointed down the shallow gully and there the three rams were, looking straight at us, very nervously, 40 yards away. I went down on one knee for a better gun rest, but I couldn't see the rams because a small ridge was impairing my line of sight, so I stood up in full view of the rams. At this point I am positive the ram would not score over 178 but it was a very respectable ram for any sheep hunter. Stan never uttered a word of confirmation the ram would score 180 or more, and, at that point, I just instinctively squeezed the trigger at the super close, easy shot. Even though it was made off-hand, standing up, believe it or not I just flat missed that ram. I did not want the ram and I was very disappointed I missed the easy shot, but I was happy since he would not have made "The Book." My conclusion is he was just too close. I must have just aimed 6 inches high and missed. It was a storybook stalk.

Several days later Stan spotted a ram late one evening I was convinced would have made the Boone and Crockett Club's records book. However, it was too late in the evening to make a stalk and we decided to try for him the next day. We spent the rest of my hunting days looking for him, and never saw him again. Keep in mind, it's the smart one that lives to grow those trophy horns. It's the dumb ones that are hanging on someone's wall. During my 10-day hunt we located 25 legal rams.

This is sheep hunting at its best! At this point in time I have no definite plans to return to Idaho for another hunt, but am giving the possibility some consideration if a good ram is located.

The Special Kofa Sheep Hunt

DAN POCAPALIA

This was a once in a lifetime opportunity for bagging the unusual desert bighorn sheep trophy. Standards and criteria for this trophy were set very high at the inception of the hunt.

On October 2, 1984, I was picked up at the Yuma airport by Arizona Game and Fish Officer Robert Remington, who issued the special sheep tag, then drove me to the base camp at the foot of the Kofa Range where I met my guide, Danny Nasca, and his assistant, Tom Saad. It was pointed out this special hunt was going to be very unusual in many ways. The trophy sheep I was seeking had to be in the top 5 percent class because it would be difficult to explain a small, run-of-the-mill trophy, especially when you consider the number of dollars donated to FNAWS for the special Kofa Range sheep permit issued by the Arizona Game and Fish Commission. This was a special opportunity for doing the unusual and everyone was very excited and certainly everyone expected the unusual. Every now and then, when I stared up at those rugged Kofa Mountains I had to climb, I began to ask myself, "Why?"

There never was any question in my mind, with the extremely high standards set forth, that the risk of going home empty-handed was very real and was perfectly acceptable. However, Nasca, my guide, kept insisting and reassuring me there would be no problem meeting my trophy sheep objective, age and horn size, as there was a large number of big rams in the Kofa.

We started the hunt on October 2, 1984. In the beginning, the field trips were short and easy, with lunch at the main camp, and as each day went by, the "dude" (that's me) became a little tougher, and so did the field trip. The grueling grind of climbing the Kofa peaks day in and day out can be quite an experience, especially with no weather breaks to offer an excuse for a rest and Nasca's constant go-go attitude. What a guide!

The first 20 days of the hunt were spent taking inventory of the western Kofa Range in an effort to locate a suitable ram. Richard Remington and Lowell Wittaker from the Arizona Game and Fish Commission were very helpful in every respect. These two gentlemen are certainly to be commended for their dedication toward their work and are a credit to the Arizona Commission.

After the first 12 days in the field it was necessary for me to leave due to business commitments, and I returned on October 19 to continue the hunt. Once again, Danny Nasca and I were up at daybreak and returning at dark. As each day passed unsuccessfully, the odds of bagging this unusual trophy was becoming more doubtful. One must keep in mind we saw an average of five legal rams each day and on one particular day we saw 13 legal rams, so there was no shortage of rams. But, the ones we were looking for were extremely difficult to locate in those rugged Kofa Mountains.

On Friday, October 26, while descending off the difficult spiral peaks in Tunnel Mine Canyon, at approximately 5 p.m., a long ram appeared on a ridge approximately one mile away. The ram was small, but within minutes two other rams appeared on the ridge horizon and one was "Old Balloon Horns!" He is the one we have been trying to find during the past 20 grueling days. We saw him once on the first hunt. It was too late in the day to make a stalk, so all plans were set for the following day, October 27.

Incidentally, during the first 20 days we did locate two other respectable sheep. One was labeled as "limpy" because he limped when he walked, the other ram had an extremely large piece of the backside of his left horn missing and his horns looked like elephant tusks. We decided on "balloon horns" as my trophy of a lifetime. His horns were huge at the base and extended extremely heavy to the tip. He was a magnificent ram.

The next day, Saturday, October 27, early in the morning Nasca and a friend of his by the name of Robert Grey from Phoenix, Arizona, and I were climbing a low pass where we could canvass and observe the Tunnel Mine Mountain and Basin in an effort to locate "Old Balloon Horns." About 7:30 a.m. Nasca had located "Old Balloon Horns" in the spiral peaks. We kept an eye on him until he bedded down in the top of the spiral peaks. The game plan was on, we were carefully making our stalk.

Nasca commented, "Dan, he's in the box, we've finally got him!"

The three of us were carefully making our way up Tunnel Mine Mountain and were three quarters of the way up, in the final stages of the stalk of a lifetime for a very, very unusual trophy ram. At approximately 10:35 a.m. we were startled by the sound of flap, flap, flap. It turned out to be the U.S. Fish and Wildlife and Arizona Commission helicopter, who were starting their annual desert bighorn sheep survey count in the Tunnel Mine Basin. The chopper spent approximately two hours taking sheep inventory and, needless to say, this was the end of the hunt. The Kofa Range consists of 660,000 acres and the survey just happened to start over the stalk of a lifetime—why??

We spent the next three days trying to locate "Old Balloon Horns," but to no avail,

End of story!

Alberta 1996 with Bernie Fiedeldey
The Good Ole' Days

Jim Priday
From *Wild Sheep*, Spring 1997

There was a lot of excitement at the 19th FNAWS convention in Reno in 1996. Several extraordinary rams had been taken in 1995, and the first-ever Alberta minister's permit had produced one of the largest bighorn taken ever. Bernie Fiedeldey, an avid mountain hunter, was no stranger to this convention and was the lucky high bidder for the 1996 Alberta permit — the permit that will assist in Alberta's renewable natural resource, the Rocky Mountain bighorn.

Bernie is also no stranger to the alpine sheep meadows with 11 wild sheep to his credit. He simply said, "Just to have the opportunity to be there is something special!"

He is referring to Alberta with spectacular scenery and one of the very few places that could offer a man a shot at a 190-point plus inch ram or even every hunter's dream of taking a Boone and Crockett World's Record ram, a benchmark that most hunters can only dream about.

Bernie said to me, "when talking about his hunt, "The reality is that rams that go 190 points plus inches do not grow on trees."

Part of what he was excited about was the possibility of having a true Alberta Rocky Mountain bighorn with its own special look, its thick neck, its balanced weight and its heavily-broomed horns. I could tell Bernie was really looking forward to this hunt.

After talking to several of the most qualified members of the Professional Outfitters Association of Alberta (POAA), Bernie selected Glen Willsie of Wild Horse Outfitters and Skip Selk of T-Y Guiding and Outfitting to help accomplish his personal goals of his hunt. This consortium would prove to be two very hard-working men who never gave up and put a lot of pressure on themselves to find a tremendous sheep.

Bernie said it was never, "Let's go take a sheep." It was always, "Let's go find that big sheep!" Little did they know they were headed for the worst weather since the winter of 1984 and would experience brutal January weather in November.

Skip Selk made the first scouting trip in July and found several nice rams in the winter range. His video showed two of them were exceptional rams. Skip and Glen Willsie then spent the week before Bernie arrived scouting the area. Bernie arrived in Calgary on November 8, 1996. They had decided Cadomin (Unit 438) was the best place to start and would maximize the opportunity this special permit offers.

Upon arrival in Calgary, a snowstorm immediately dumped 12 inches of snow and prevented Bernie from getting to Cadomin that day because the roads were closed. They did make it to Cadomin the next day but it was too late to hunt, so on Sunday they traveled to the Cardinal River Mine. There they found a lot of sheep and since you can't hunt on Sunday, they took several photographs and shot a lot of video footage. It was exciting to see so many sheep concentrated in one area as they fed in the grass through the snow. Monday began with another six inches of snow and after glassing for sheep off of the mine for three more days, they had another dump of snow that made a total of 24 inches of snow on the ground. No sheep were moving off the mine and things were getting a little disappointing after six days of glassing. The hope was the rut would start, and the sheep would move from the mines to the rutting slopes — one of the true advantages of this permit. Although there were a lot of young rams chasing the ewes all over the mountain, the older rams knew it wasn't time and sat tight in the mines and parks. They actually watched one ram start up into a basin and after walking about a quarter mile in belly-deep snow, the ram turned around and walked back to the mine in the same tracks. He was pushing snow with his belly. Another interesting observation was watching wolves on a fresh kill. Bernie saw a couple of rams walk within 20 yards of the wolves but circled them to give them to give them some space and respect.

They decided to change areas and moved farther south to the Burnt Timber and Panther drainage's as well as Sheep Creek. They would also check out the Ya Ha Tinda Ranch that is a big wintering ground for elk and also a tremendous number of sheep congregate there. It is now about day 10 of the hunt.

The weather has taken a terrible turn for the worst as temperatures plunged to minus 30 degrees Celsius. They hunted these drainages with snowmobiles for four or five more days, which was very uncomfortable because of the brutal cold temperatures. Thee weren't great numbers of sheep coming out of the park nor were there any special sheep showing up yet. They found about 50 sheep on the ranch, but a couple of the larger rams would only score in the 180-inch range is size. This hunt was about a 190-point B&C ram or better, so they dumped the snowmobiles, gathered their horses and went back to Cadomin. The temperatures were still bitter cold. The snow was still 24 inches and drifting, so they decided it might be best to take tents and set up a spike camp at timberline, which would allow them to occasionally return from the top of the mountain to warm up. That didn't work out because once they had fought their way to the top of the mountain in two feet of snow, they didn't feel like returning to spike camp with the climbing so slippery and difficult.

They hunted this country west of the mine for a few days, but not many sheep were moving, so they decided to try the White Horse Creek area. Ship went home one afternoon and returned about 2:30 a.m. with snowmobiles to go farther into White Horse Creek the next day. It is now Wednesday of the last week of the hunt. Glen and Bernie went to the area west of the mine again to look for sheep while Skip was checking out White Horse Creek. Ship had traveled about 15 miles into White Horse Creek and returned with the news he had found a really nice ram. They all returned on Thursday but couldn't find the sheep. They had apparently moved to the back side of that mountain and eluded them. On Friday they returned to White Horse Creek and after another empty search, unanimously decided that afternoon to return to the Cardinal River Mine. Finally, there were a few sheep coming off of the mine Friday evening and about two miles from the mine area they found a few rams to bed down.

Saturday, November 30, they decided to go back to the mine, rode the horses through the mine and timber as far as they could take the horses, and spotted a band of 14 rams with two of those rams worthy of a closer look. They left the horses at timberline and climbed up the mountain in the deep snow by pulling themselves with buck brush and limbs. It was so steep and the snow so deep they couldn't climb without pulling on the branches. Reaching a tree that was the last bit of cover between the hunters and the sheep, they were able to determine one of the rams was exceptional and was what they were looking for.

The largest ram walked out to the edge of a high rock ledge and was looking out over the basin as sheep often do. He was huge and sophisticated looking — a ram that carries his massiveness out farther along the curve than most rams do. It was about 250 to 275 yards away. Bernie laid his backpack on the uphill side of the three, laid down in the prone position and found the ram in his scope. The guides, Skip and Glen, wanted to get closer, but there was no way to do that without being seen by the band of rams, so the decision was made to go ahead and take the magnificent ram from that tree. It was about 11:15 a.m. on the last day of the hunt. The guides were bracing Bernie with their knees so he wouldn't slide in the snow while setting up from the shot. Bernie made one well placed shot and as the huge ram plunged over backward, the men shouted and screamed with joy. This was the culmination of more than 20 days of hunting in temperatures that would have discouraged most men or women from even being on the mountain. They had finally taken a 190-point plus ram on the last day. Bernie noted that in spite of the bad weather, neither guide had ever mentioned taken a lesser ram. They worked tirelessly and never showed signs of being discouraged.

Last year Gary Hansen said he saw two, maybe three, 200-plus inch rams, and he collected the largest one. Later he reported the other one (Number 51) had been found dead after locking horns with another large ram and falling off a cliff. Perhaps if the weather and deep snow had not kept the sheep from moving earlier, another 200-inch ram could have been found this year. But, for Bernie Fiedeldey, a man who is always striving to make his "good better and his better best," this classic, huge, 190-class Rocky Mountain bighorn is a dream come true. It beats his first 184-inch bighorn and completes his second Grand Slam of wild sheep.

His legacy of sheep hunting accomplishments for his second Grand Slam includes this 1996 Alberta Rocky Mountain bighorn taken with Glen Willsie and Skip Selk, a tremendous 42-inch Stone's ram taken in August 1996 with Myles Bradford in British Columbia; a beautiful one-of-a-kind white desert ram taken in California on the last hunting day in January 1996 with Aravaipa Outfitters — a 1995 auction tag used just a few weeks after knee surgery — and a heavily-broomed old Dall's ram taken in the Mckenzie Mountains of Northwest Territories in August 1994 with Gana River Outfitters.

Bernie exemplifies today's generation of tough and dedicated sheep hunters. The last significant part of this story would be to recognize how touch and dedicated some of these men are who purchase these auction tags. They are men like Bernie who would rather go home empty handed, then to taken a lesser ram. They will spend as many days as is necessary battling the elements in the mountains to find that special ram, often the entire number of hunting days in a given season. This story is a tribute to Bernie's Alberta ram and to his enormous commitment to the conservation of North American wild sheep.

As Bernie and I sat and talked about his three 1996 sheep hunts at the end of a wonderful evening, he looked at me, smiled and quietly said, "You know, Jim, these are the good ole' days of sheep hunting."

Just look at the size of the rams taken in the last few days, and I think you'll agree with Bernie that these really are "the good ole' days." Happy Hunting!

I was November 30, 1996, and Bernie Fiedeldey had taken his magnificent Alberta monarch in spite of deep drifting snow, high winds , cold temperatures and frostbite. The ram has the classic large body, long, dark, burly hair and magnificent horns that are heavy and broomed. They measure an estimated 42 inches and 7-1/2 inches at the tips. The massive horns carry their weight with 14-1/4 inches at the quarter and 14-1/4 inches at the half. The ram green scored at a little more than 191 Boone and Crockett Club points and if he had 16-inch bases instead of 15-inch bases, would have easily pushed the 200-inch mark the way he proportionately carried his weight.

A "Charlie" Ram

Sherwin Scott

"Old Charlie" was what I called him. He was irascible and a bit churlish. Sometimes he was downright grumpy. He was decidedly lazy. He was my guide when I was hunting Stone's sheep in British Columbia in 1969, which was my first time to hunt Stone's, I might add.

Charlie and I hitched up together when Bob Kjos, the outfitter, flew me into his base camp deep in the Cassiar Mountains of northern British Columbia. Bob's home was near Toad River Lodge; kind of a highway weigh station, not all that far from the Yukon border. At any rate, after my arrival at his lodge by rented taxi, we flew in his Super Cub to main camp that was located near the headwaters of the Racing River.

The river itself upon arrival looked reasonably placid being only about 20 yards across and about as deep as a horse's ankles. Its color was somewhat more fascinating because it was not the clear water as you might expect from a high mountain stream. Instead, it was more of a milky color. When I inquired as to the reason why, I was told it was because the river was glacier fed, and when glaciers melted, they produced a milky-colored runoff. Regardless, the water was COLD. So cold, in fact, that if you tried walking in it your feet would go numb before you managed to go even a few feet.

Bob had a lovely camp comprised of a cook tent and sleeping tents for the hunters, guides and wranglers that was near the banks of the river. The river itself coursed its way down a fairly wide valley. We were told moose were frequent visitors to the willow patch that was close and visible from the cook tent. One of my camp mates was a doctor from Texas named Jim who, if memory serves me correctly, completed his Grand Slam on this hunt. What transpired after supper the day of our arrival was that we found a small group of Stone's rams on a massive ridge across from camp. The ridge itself was a series of small valleys sloping from top to bottom with each little valley or draw created by rain and time's endless erosion.

The slopes were a most verdant green, nearly free of bushes, and covered in a short green grass quite obviously preferred by sheep. Jim found these rams and all of us took a look through the spotting scope. For me, it was the first time I had seen a Stone's ram alive, and I must say I was rather surprised at the color variation exhibited by and between each of the rams. They varied from a light and whitish gray to being very dark gray, almost black.

The ram Jim was interested in was very dark and quite long of horn. He was so excited he found a log to sit on and watch these sheep until it was too dark to see through his spotting scope, a 15x to 60x Bausch & Lomb, I think.

Prior to daybreak, Jim and Bob crossed the river on horseback and proceeded across the flood plain to the cliff that led up to the ridges and valley above them. From our vantage point in camp, it looked like the cliff was about 250 feet high. Jim and Bob had their work cut out for them.

We could see them making very slow progress up those steep slopes. Hand-over-foot was their mode of movement. Off to the left of their ascent was a particularly beautiful, tumbling waterfall that must have fallen for at least 80 to 100 feet. Apparently, there was a stream created by glacial melt that lay between the ridges Jim was attempting to reach.

The sheep they were after were quite low on the ridge above them, and, what's more, the sheep were on the other side of one of these rounded mini ridges that was going to keep the hunters hidden from them. It seemed it was going to be an easy stalk and from camp we watched the whole affair unfold as if you were sitting in a movie house with the script in hand.

Once they made it to the top of the cliff, they wormed their way towards the sheep by crossing over the little ridges. I could see them stop periodically as they reacquired their bearings on the location of the rams that chose to remain in their hideaways, or so they thought. At last, Jim could be seen lying down and putting his rifle over his pack for stability. All the sheep ran, save for one. After a while, off in the distance a muffled shot could be heard. Jim had his ram, and what a ram it turned out to be. When measured back in camp, those horns were right at 41 inches in length — a fantastic sheep!

Well, "Old Charlie" and I left the next day to do some fly-camping. We had as our entourage the two horses we rode plus four packhorses to carry our belongings, food, tent and camp utensils. Coursing our way downriver, we crossed it several times as what little trail there was wound back and forth in and among the trees, brush and river plain. Mostly it was a pretty ride. We were able to keep cool since we were in the river bottom among the trees and occasional willows.

My horse knew he had a tinhorn on his back and made no pretense of acknowledging this fact by constantly lagging behind. I would take a stick to his hindquarters or try to dig my heels into his flanks to cause him to "giddy up," but instead of having the desired effect he would bow his back and lay his ears down as if to say, "Do not bug me or you will get your comeuppance." I did at his behest later in the hunt.

Frankly, I think that horse and Charlie were somehow related, given their penchant for having similar personality traits. Most guys who have spent time hunting sheep can probably relate to this since they have had at least one experience not unlike mine. We found a tributary stream entering the racing river. This was where Charlie said we were to head up since there were sheep to be found. A pretty little stream it was, maybe four steps across and clear as a bell.

We rode upstream for about four hours when Charlie said, "We may as well camp here. The sheep will be up top the mountains."

We set up our two-man tent, hobbled the horses and made camp hospitable. Later that afternoon I saw some goats toward the head of our creek and suggested we go take a closer look. Charlie was at best indifferent to this proposal. His opinion was the goats would still be in the same area next morning. It was getting late and supper needed to be made. With grudging disappointment, I acquiesced. (Years of experience later would teach me to be more assertive.)

The next morning, we looked for the goats, but they had indeed moved out of the country. Up to this point in time, I had never shot a goat, nor for that matter had I hunted one previously.

The weather gods were not especially helpful for the next four days. It rained incessantly. The clouds hung so low as to make glassing for sheep next to impossible, and Charlie had no affinity for getting wet. For me, I was so wet behind the ears I did not know quite what to do. Hence, we spent most of our time in camp just listening to the patter of raindrops on the roof of the tent. This was fine if one's proclivity was to be somnambulant. I was raring to go and Charlie was not much of a talker. He more or less liked to express himself by permeating our tent with his passing of ill-fated wind.

Four days of less-than-perfect hunting, and I was ready for a change of scenery. We saw no sheep, only goats, on the day of our arrival in spike camp. One day we did have two wolverines pass by us on a mountainside. When they saw us they both snarled simultaneously and high-tailed it out of the country. I had no idea those nasty little fellows could run so fast. They were gone and out of sight in the blink of an eye.

As we departed spike camp and headed back towards main camp, I noticed our little tributary stream rose quite a bit because of the rains we incurred. Everything was going quite well. Charlie was his laconic self, heading up the pack string. I was content to be the last in tow. When we arrived at the confluence of the tributary stream and the erstwhile low, but quick-flowing Racing River, we got the shock of our lives.

The river was no longer ankle deep on the horses. It was roily, turbid, muddy, whitish-gray in color, ripping along at maybe 25 m.p.h. It was, to say the least, a frightening looking torrent of water that was not the least bit tamed.

I asked Charlie with trepidation plainly evident in my voice what he thought we should do under the circumstances. He turned sideways on his horse, looked over his shoulder at me and exclaimed, "We'll just have to make our way back to camp the way we come." This meant recrossing the river on horseback several times if we were to follow the trail. Besides, there were places where the canyons were so narrow there were no other options available for us.

Charlie, being first, entered the river. His horse suffered from an immediate case of high timidity, and correctly so. As Charlie cajoled and spurred him on, the horse looked as if he walked on proverbial thin ice or muskeg because there was so little sure footing.

The river nipped at the horse's front feet. We went farther into the river. It rose above his knees. A little farther, and the water was almost at his belly, tearing at his legs. The horse was in a terrible fright. Charlie held not only to the horse's reins but also to the rope of the front packhorse. You could see the wild look of fear in that horse's eyes. He knew if he lost his footing in the rampaging frozen waters, there would be no survival. The current was much too strong for him to regain their feet.

Charlie made it over, but not before the water reached almost halfway up his horse. It was up to me to cross. I cannot really tell you how I felt. All I know is that I knew, as did the horses, that a slip or fall would be curtains! Even if I could swim across, the cold water would have incapacitated me in seconds. We made it across but being on the back of that horse as he gingerly felt his way across the slick, rocky river bottom as the current increasingly threatened to flip him over was akin to being in a kitchen chair suspended from beneath a motorized hang glider. We ended up crossing that rain-swollen river eight more times on our way to main camp. All I know is, it wasn't my time to go!

After the harrowing ride, we spent the rest of the day in camp and proceeded to hunt the ridges and upper bowls where Jim had taken his ram. We hunted much higher up because it turned warm, and we expected the sheep to try and avoid the biting bugs by being on the wind-swept slopes.

As usual, our luck was holding. We could find no rams nor even a hint of them. Nonetheless, the Man Upstairs was looking out for me again. Charlie and I were crossing or traversing our way just under the lip of a very steep mountain and across a ridge covered by grass and underlying bb-sized gravel or rocks when I lost my footing. Down I went on my stomach, feet below me and toes dug into the mountain, hands and fingers gripping and tearing at anything to break my rate of fall and all the while I was inexorably sliding faster and faster down the mountain to be dashed into a jumble of boulders dozens of yards below.

I went down the mountain and its slope increased, which sped my impending fall. There was one tiny little bush on the side of the ridge, and as I was going by I reached out, nearly uprooted it, but it stopped me. Frozen in fear and trembling uncontrollably, I couldn't move for several minutes. Charlie was way above me and in no way could lend assistance. I'd have to make it back up by myself if I could. Inching my way spread-eagle and on all fours over the next half-hour, I reached Charlie. At no time was I ever able to get on hands and knees, let alone walk, since there was the continual risk of my sliding back down again.

Charlie sagaciously decided we should hunt above camp where another hunter, a banker from California, and his guide had found a big ram. I was told the hunter was sufficiently corpulent and out of shape. He couldn't make the climb to where the sheep were living. The morning was delightfully clear with warm sunshine on our backs as we mounted our horses for the ride upstream. The river subsided some, but like its name Racing River, it was still racing. After mounting my horse, I immediately noticed a certain intransigence in his behavior. He appeared to have awakened on the wrong side of the bed. He lowered his ears and I knew it was a precursor to a possible unpleasant ride. My ride lasted another two seconds before I found myself communing with the Canada jays in the trees. The damned horse threw me for a country mile.

Confidence was shattered with appropriate bumps and bruises, but no broken bones. I was determined to get back on the horse. Cowboys say it's important to always get back on a horse if you fall and or get bucked off, lest you be so afraid you never get on a horse again.

Charlie asked, "Scotty, are you okay?"

Still I could tell there was a look of sardonic mirth about him that overrode his meretricious concern.

"Yup," I said. "I'm okay." I looked at Charlie. I was determined to ride. I said, "Charlie, you can now ride my horse."

Charlie went back to the corral and secured another horse for himself, so I guess he did not want to embarrass himself. I still think he set me up!

We left camp and followed the river for maybe two or three miles and proceeded to do some glassing on the ridges and slopes across the river from us. The lower part was covered with timber whereas the upper slopes were in brush (shin tangle) and grass. In no time we found the solitary ram heretofore found but not taken by the other hunter. He was gray of color and his horns appeared to be well past a full curl. That was plenty good enough for me. We could not cross the river with the horses. The sheep would see us, so we decided to wade across the river.

Charlie proceeded to take off his boots, socks, pants and the bottoms to his long johns. This was a sight with his scrawny white chicken legs I'm sure had not seen the sun in years emanating from his shirt. He had this grungy top to his long johns that barely hung over his bottom and a mottled brown and bedraggled hat sitting on his weather-beaten head. He timorously stepped into the frigid, milky river, a staff in one hand and bundled clothes in the other, looking for all the world like a character right out of "Uncle Remus."

Now it was my turn to laugh, which I did with uncontrolled restraint. Charlie did not take kindly to my humor at his expense judging from the mutters seen crossing his lips along with the scowl planted squarely on his face. I crossed the river after him, but left my clothes and boots on, and glad I was, because after the first eight or 10 steps my feet had lost any feeling, so cold was the water. In bare feet, I doubt if I could have made it without falling.

The ram was about a half-mile above, and the trees and brush would block us from him initially. When we broke out into the brush and open we were able to work our way towards him by staying in a draw until we were almost opposite his location. After that it was somewhat a piece of cake because we were able to close the distance to about 75 yards without being detected.

We watched him graze for a period of time. I took aim with my 7mm Remington magnum loaded with 160-grain Noslers and closed out what had been a hard, fascinating hunt for my first Stone's sheep with a guide not soon forgotten...Charlie.

Incidently, the ram ended up scoring in the low 160s. His horns were just shy of 38 inches in length and his age was estimated at between 10 and 11 years of age by the British Columbia hunting authorities.

** This was not a story about a permit hunt, but Sherwin Scott is recognized in this section for his support of the permit program.*

Photograph by John W. Herbst

Section IV

The Next Twenty-Five Years
for
Wild Sheep
or
Perhaps One Hundred

Judy Whitcombe's photo, shown above, was on the Summer 1998, cover of Wild Sheep.

Chapter I

The Next Decade
Their Future is in the Hands of FNAWS

Dean R. Picl

I first became acquainted with the term Ovis pyrexia, or sheep fever, in Lloyd Bare's book, *High Country Hunting*. Upon that initial encounter, I considered it to be the novel turn of phrase that might be expected from one who had grafted a devout love of sheep hunting onto a degree in pharmacology. Several years later, after I had agreed to write this chapter, I again read Lloyd's book and that phase seemed to attach itself to my mind and wouldn't go away. It nagged me for days and I couldn't figure out why.

What was it about, this term Ovis pyrexia, that had caused it to take on such an unusual significance? Why was it bugging me? Lloyd had called it the "dread disease" with which sheep hunters are afflicted. I began to sense there might be something about that description that was bothering me. But what?

Lloyd certainly didn't mean that he had stumbled across some strange virus or bacteria that required the immediate and undivided attention of medical science. While I haven't asked him, I am sure he was talking about something akin to a primordial urge that pulled those so diagnosed from hearth and home to a place, not dramatically but succinctly, called "sheep country." And, suddenly, there was the answer! It was in that "pulling urge." And, not just a "pull" to a place, but a "pull" to be "part of the place" and to share some kind of almost inexplicable "oneness" with a majestic wild creature whose strength was mixed, unfortunately, with a vulnerability that we may have had a hand in creating. The reduced habitat and, as yet, uncontrolled diseases that we have foisted upon these wild creatures seem to have created a resolve to seek solutions for the problems we created. But, Ovis pyrexia is not the product of guilt. It stems from a less easily quenched feeling more akin to a mixture of respect, admiration and devotion. It feeds, not only upon those who trod the high ridges and glass across deep valleys, but inflicts, as well, those who toil on the sidelines on behalf of the genus Ovis.

Nowhere are more of these inflicted and contagious souls visible than among the ranks of the Foundation for North American Wild Sheep (FNAWS). Its members, directors, officers, office staff and volunteers are the identifiable carriers of this "dread disease" and the source of the serum that will guarantee the future of the wild sheep of this continent. The "pull" is the "push" that will permit us to do those things, which are essential to viable, healthy and expanding sheep herds in North America.

A smiling June Picl enjoys the Ladies' Luncheon and Auction in 1992.
Photograph courtesy of FNAWS archives.

The Foundation of the future, which will undertake the tasks before us, is not exactly the Foundation of the past. It is assuming a new face, employing new skills and is imbued with a new resolve.

What does the Foundation look like that will take us into the new millennium in the year 2001? It is slimmer and more flexible. Overhead is being reduced by the elimination of expenditures that results did not justify. Its "voice," Wild Sheep magazine, is receiving increased attention. A general "face lift" will make it more appealing to members and advertisers alike. Headquarters in Cody is being made visitor-friendly with an emphasis on putting our best foot forward. Cooperation with the Buffalo Bill Historical Center and its Museums will bring more people to our door, and, hopefully, into our ranks. New ways are being sought by the Board of Directors to better administer the funds that are put at their disposal.

Exciting innovations have been made in the manner in which the Foundation will attempt to acquire the funds necessary to discharge its stewardship over the wild sheep of North America. Professional assistance, long overdue, has been obtained to lead us in this task. While still needing and grateful for the donors of goods and services whose contributions, upon auction, have funded our Grant-in-Aid, income from investments is viewed as a necessity of the future. Direct contributions of money to establish a substantial trust fund are being sought as a method of guaranteeing the work of the Foundation during years to come. Under consideration is a Board of Trustees, composed of some of our country's business leaders, to make important financial decisions based upon their combined experience and business acumen. Such an endeavor is seen as an enhancement of the Foundation's strength. The addition of this assistance from Trustees should increase the funds available to achieve our goals and permit the Board of Directors to concentrate upon the running of the Foundation.

The office staff has been found to be a wellspring of previously unused talents. A new atmosphere has invaded the Cody office in 1998 in which the women are encouraged to use their imaginations and creativity to better perform the tasks over which they have jurisdiction. The "Information Highway" now runs through our office, so essential and current data will be immediately available to the Directors so their decisions will be based upon what "is" and not upon what "was" or "might be."

Will there be changes in the way our officers and directors lead? It must come to pass if all of the other changes are to have any meaning. Fundamental amendments to our bylaws, presently under consideration, will increase the influence of the executive element in the Foundation's administrative process. Leaders will be given not only the opportunity but also the mandate to do what the name implies. It is critical to our future that those who "take the

Dean Picl, left, discusses FNAWS business with Jim Glass and Pete Cimellaro at the 1992 FNAWS convention in San Diego at the Town and Country Hotel. The theme for the convention was, "The Partnership Continues."
Photograph courtesy of FNAWS archives.

Staffing the FNAWS information booth at the 1990 National Rifle Association annual meeting were, left to right, Ron McMillen, FNAWS President John Harris and Dean Picl.
Photograph courtesy of FNAWS archives.

wheel" know how "to steer" and that those who entrust them with that responsibility are aware of the importance and consequences of that decision. It is for the members of this Foundation, each year, as they cast their ballots for Directors, to send to the board room, only those who are equipped to make those kind of decisions and to fulfill those kinds of rolls.

Every decade in every century is important to wild sheep — those of the past as well as those of the future. Those past years only have significance and importance if they have taught us something we are able to use in making this Foundation more responsive to the needs of today and tomorrow. The future of those that we have undertaken to husband requires we keep one ear tuned to the past, with the other ear and both eyes searching ahead for ways to do it better.

Ovis pyrexia must so infect us all that wild sheep will abound and strains of the "dread disease" will be virulent in the children of children of generations yet unborn.

Dean R. Picl is chief legal counsel for the Foundation for North American Wild Sheep, a position he has held since 1985. He is an attorney from Covina, California, who offers his knowledge of FNAWS and its direction at all Board of Director meetings and other FNAWS gatherings. He is often seen as a Master of Ceremonies at FNAWS national conventions, and volunteers his time at national conventions of other organizations distributing information about FNAWS and its projects to all who will listen. Dean was admitted to practice in June 1957 after sandwiching two years in the U.S. Army between undergraduate and law degrees from the University of Southern California. His predominate area of practice has been as a criminal defense lawyer, which has included the defense of numerous cases involving alleged violations of state and federal hunting laws. He and his wife, June, are the parents of Kimberly who is the mother of four and lives in Utah; and Greg who is married and the father of none who owns many toys in Claremont, California. His representation of the Foundation is now in its 15th year, the highlight of which he considers to be the friends he has garnered across this country and Canada.

Chapter II

Giving Sheep a Voice: The Politics of Wild Sheep

PHYLLIS CARLSON AND PAULA KARRES

FNAWS began with a group of hunters who wanted to give something back to the resource they so dearly loved. At that time, they sought to prevent the extinction of wild sheep. Little did they know a storm was brewing that would threaten the policies of sound conservation management. The Animal Rights movement was gathering speed, and began to focus on those who choose to hunt.

About the mid-1970s voices were heard opposing trapping. New laws were enacted requiring certain types of traps, banning some trapping locations and in several cases, restricting the number of trappers in particular areas. Next came the gun debate that has escalated to front pages, "Dear Abby" and even the comics. Firearms used for hunting are in the midst of this debate, with archery and fish hooks not far behind. Animal rights groups even have blamed local wildlife agencies' programs geared to involve young people in hunting as a cause for the rising violence among our children.

The Need to be Heard

"Putting Sheep on the Mountain" has always been FNAWS' mission and it became apparent this could not be accomplished through our Grant-in-Aid program alone. Our voice needed to be heard in Washington, D.C. The law is clear of how much a 501-(c)(3) organization can spend toward lobbying efforts. In general, FNAWS can expend 20 percent of its first $500,000 of exempt purpose expenditures, 15 percent of the next $500,000 of exempt expenditures, 10 percent of the third $500,000 of exempt expenditures, and 5 percent of additional amounts after that. An "exempt expenditure" is defined as the amount of money FNAWS spends on projects relating to our charitable purpose to "put sheep on the mountain." These limitations created a need for wildlife "watchdog organizations and lobbyists dedicated to protecting hunters" rights and sound wildlife management. By pooling our resources with these organizations we have become a powerful force.

The Lamb and the Wolf

Wildlife management politics first became a major issue in 1987. A concerned member informed FNAWS President Dr. Jerry Waite that the Department of the Interior was considering the reintroduction of the gray wolf into Yellowstone National Park. Dr. Waite took a stand for FNAWS and asked the Board to draft a resolution stating the official position of the Foun-

FNAWS Executive Director Karen Werbelow, far left, met with Vice President Al Gore and a variety of conservation organization leaders to discuss wildlife management issues in March 1997. This historic meeting opened communication lines with the major conservation organizations and the Executive Branch of the White House.
Photograph by Michael V. Terrell, Executive Office of President, Council on Environmental Quality.

Pete Cimellaro, an Arizonan and past national president of FNAWS, led the charge against Proposition 200 that would have banned hunting, fishing and trapping in Arizona if passed by the voters in 1992. Pete was treasurer and chief spokesman for Arizonans for Wildlife Conservation that defeated the proposition by raising money to buy advertising on radio, television and newspapers to clarify their message of support for trapping, hunting and fishing to the public.
Photograph courtesy of FNAWS archives.

dation. A campaign to assemble information followed with the help of Jack Atcheson. Even the Ministry of Environment in British Columbia was contacted for its experience in dealing with wolf populations.

The evidence was overwhelming. Reintroduction of the predator would imperil the fragile wild sheep herds in and around the Park. We needed action. Successive presidents of the Foundation supported the 1987 resolution by testifying against the reintroduction on Capitol Hill. The House Subcommittee on National Parks and Public Lands heard from Arnie Johns in 1989, and from Howard Pollock in 1990. FNAWS' member Arlene Hanson, spearheaded a group called "The No-Wolf Option Committee," which fervently opposed the reintroduction. From 1991 to 1994, FNAWS approved grant-in-aid monies to help fund the No-Wolf Option Committee in its quest to collect data, alerting the public of the ramifications of wolf reintroduction.

The official *Federal Register* notice published in August 1994 substantiated our fear that the protection of wild sheep had been overlooked. Only qualified biologists could confirm depredation of wild sheep herds and funding for any such personnel to monitor the herds was nonexistent. Comments were submitted restating our position. When the Final Environmental Impact Statement was issued, FNAWS chose to support option Number 1 — no reintroduction. Despite the Foundation's concerns and Arlene's efforts, and under protest of animal rights groups in Canada that actually wanted to keep the wolves, the U.S. Fish and Wildlife Service transplanted 40 gray wolves from Canada to Yellowstone National Park.

The media, along with the National Park Service, have elevated the predator to a spiritual symbol. Tourists are encouraged to view the wolf in its new habitat, and purchase the merchandise that supports groups such as The Defenders of Wildlife. They never reveal the destruction of wildlife left in the wake by the wolves. Litigation is ongoing.

Smith & Wesson's Public Relations Manager Chris Dolnack, left, presented a $2,500 check to Wildlife Legislative Fund of America's Vice President Rick Story, center, to help defeat Arizona's Proposition 200. Pete Cimellaro, right, represented Arizonans for Wildlife Conservation at the check presentation.
Photographs courtesy of FNAWS archives.

To commemorate the Bureau of Land Management's (BLM) 50th anniversary in 1996, FNAWS joined BLM in the Golden Anniversary Partnership Agreement and identified 12 specific areas on public lands containing outstanding sheep habitat. BLM's Acting Director Mike Dombeck, left, congratulates FNAWS Executive Director Karen Werbelow after Karen signed the agreement and attended BLM's birthday celebration on July 18, 1996, in the historic Department of the Interior office building in Washington, D.C.

In late September 1994, FNAWS First Vice President Leland Speakes, Jr., right, FNAWS Treasurer Phyllis Carlson, center, FNAWS Past President Howard W. Pollock, rear row, and FNAWS Executive Director Karen Werbelow, left, stood on the U.S. Capitol's steps after attending the 5th Annual Congressional Sportsmen's Foundation Dinner and Auction at the Hyatt Regency in Washington, D.C. FNAWS hosted the Caucus breakfast with Wildlife Forever. Secretary of the Interior Bruce Babbitt was keynote speaker at the breakfast.

The 6th Annual Congressional Sportsmen's Foundation banquet and auction in September 1995 attracted several notable FNAWS members and staff. Left to right on the U.S. Capitol steps in Washington, D.C., include FNAWS President Pete Cimellaro, FNAWS Executive Director Karen Werbelow and FNAWS Treasurer Phyllis Carlson. The Congressional Sportsmen's Foundation was established to promote and preserve our historical and traditional hunting and fishing rights in America.

Wildlife Management by Popular Opinion

The Wildlife Legislative Fund of America, and their subsidiary, The Wildlife Conservation Fund of America, approached the Foundation in 1984. They offered to represent the interests of FNAWS, protecting wild sheep management programs from adverse interference and our right to hunt. They have a full-time staff in Washington, D.C., that actively searches for wildlife issues, sometimes hidden in proposed legislation. The WLFA State Services office does the same for local issues that affect wildlife management. In turn for our membership, they keep the Foundation informed, acting as campaign coordinators and managers to defeat anti-hunting propositions.

In 1991, Arizona Proposition 200 reared its ugly head putting sportsmen to the test. Though touted as an anti-trapping proposition, hidden language to manage wildlife by "humane and non-lethal methods" also endangered the future of all hunting and fishing in Arizona. FNAWS' member Pete Cimellaro was instrumental in connecting the WLFA and FNAWS with Arizonans for Wildlife Conservation. With the help of the WLFA, FNAWS' members were guided through a grassroots effort, raising $25,000 for the Arizonans for Wildlife Conservation, to defeat this precedent-setting proposition.

Since then, ballot issue propositions became the order of the day for animal rights groups. The Humane Society of the United States garnered its forces, and in 1995 collected enough signatures to place eight anti-hunting propositions on ballots across the country. Organizers portrayed the bear and mountain lion as cute and cuddly mystical creatures whose populations should not be managed by regulated hunting. Disguised as an attempt to manage wildlife in a "humane manner" (birth control shots), the underlying issue was to end all hunting and fishing. Idaho attempted to end the use of hounds and baiting practices to hunt bear. California citizens contested the wisdom of their fish and game agency to manage the California mountain lion. Oregon implored its residents to end the hunting of bear and cougar with hounds. Word came of an 11th hour proposition in the state of Washington to do the same.

Grassroots organizations that opposed the Humane Society tactics solicited funding from the FNAWS who, in turn, appealed to the WLFA. Even with joint funding and the invaluable strategy from the WLFA, most propositions passed. The campaign led by Idaho's Sportsmen Heritage Defense Fund in Idaho borrowed a technique used by Arizonans for Conservation. They mailed the FNAWS membership information about their plight along with a plea for contributions, ensuring sufficient funds to win their campaigns.

A Good Offense . . .

The result of the Humane Society's frenzy to put wildlife issues into the hands of the public, rather than educated biologists, gave birth to a new organization, the Ballot Issues Coalition (BIC). During the 1997 North American Wildlife and Natural Resources Conference in Washington, D.C., Safari Club International invited FNAWS, along with representatives from 13 other hunting, trapping and wildlife management associations, to join forces in an attempt to counter the anti's new tactics. By pooling money, resources and professional expertise, they would aid local grassroots organizations in the fight to preserve sportsmen's rights. Because of the valuable lessons learned in the 1996 elections, the BIC encourages local conservation and hunting organizations to create their own ballot initiatives, guaranteeing their rights.

In Utah, members of the Foundation have taken this concept one step further. Don Peay, cofounder of the Utah Chapter of FNAWS, is currently the executive director for the Sports-

In spring 1995, FNAWS Treasurer Phyllis Carlson, left, presented a $10,000 check to Wildlife Forever's Executive Director Doug Grann. FNAWS and Wildlife Forever joined forces in 1995 to support the Oregon Department of Fish and Wildlife's annual California bighorn sheep capture and transplant program.

Who are those smiling men behind those signs? Idaho Chapter members who demonstrated their support for the six Idaho Department of Fish and Game commissioners in Boise, Idaho, in spring 1995 include, left to right, Herb Myer, Bill Louderback and Dr. Robert "Bob" DiGrazia.
Photographs courtesy of FNAWS archives.

Celebrating the defeat of Proposition 2 in Idaho are, left to right, Brenda and Wayne Schwabrow, Don Clower and Burke Mantel. On November 5, 1996, the hunting community of the United States had a tremendous victory in Idaho with the defeat of this proposition. Proposition 2 was an anti-hunting initiative filed by the Idaho Coalition of Bears (ICUB) that was largely financed by the Humane Society of the United States whose goal is to eliminate all sport hunting in the United States.

Sportsmen demonstrated in favor of the Idaho Department of Fish and Game commissioners in spring 1995 and sent a strong message to Idaho Governor Phil Bolt that they didn't want the commissioners replaced.

men for Fish and Wildlife. His leadership and access to Utah's legislators helped to produce a ballot initiative to end all ballot initiatives. An amendment to change the Utah State Constitution will be voted on in November 1998, guaranteeing the right to hunt. FNAWS is supporting this amendment through the Ballot Issues Coalition.

On the same day FNAWS and the U.S. Forest Service signed the master MOU, another agreement with Region II of the Forest Service was signed in Washington, D.C., with FNAWS President Arnie Johns, left, and USDA-FS Region II representative, Dale L. Wills, right.

FNAWS' Voice Is Heard

The Congressional Sportsmen Caucus was founded in 1989 by members of Congress, to promote unity between state and federal legislators for the advancement of outdoor sports. Past President Howard Pollock arranged for FNAWS to become active participants. FNAWS has sent representatives to every annual Congressional Sportsmen Caucus banquet since its inception. The September banquet is a time when WLFA coalition members and other national associates gather to meet and discuss the challenges facing the sportmen's community in the coming year. It is also a chance for FNAWS representatives to meet the senators and members of Congress with legislation pending in the wild sheep states.

Between the WLFA and Congressional Sportsmen Foundations' briefings, The Wildlife Management Institute arranges forums with U.S. Department of Agriculture Forest Service (USDA-FS) and BLM officials. The schedule is tight. All free time is spent meeting United States senators and representatives about current issues facing FNAWS; i.e., the Bomb Range in Idaho, National Fish and Wildlife Foundation Grants and Hells Canyon. Meeting face to face with legislators and their aides has greatly increased the recognition of FNAWS.

Past President Phyllis Carlson has represented FNAWS at the Congressional Sportsmen's Caucus from 1994 to 1997. Her recollections capture the attitude of wildlife conservationists.

"Hunting, fishing and outdoors people, myself included, try to avoid crowded places. I have always felt the nation's Capitol had that `stuffed shirt atmosphere' and was apprehensive of what it meant to be `politically correct.' However, I was impressed with the many aspects of Washington, D.C.

"While walking the ominous halls of the Senate Dirksen Building and the Senate Russell Building, I was overwhelmed with the feeling that here was little old me, carrying a message of great importance on behalf of our membership. Our majestic wild sheep are a valued resource in the United States and we as hunters feel the responsibility of protecting this resource.

"The U.S. delegates we would meet were well prepared. Their aides tracked down every bit of information about FNAWS they could find. Some had even found our web page on the Internet and studied it carefully. Most of the legislators we had appointments with were from the Western sheep states, both wild and domestic, and were aware of the many concerns we had.

"FNAWS Past President Howard Pollock is a former Alaskan Congressman and well-respected by his peers. Leland Speakes, Karen Werbelow and I had the honor of having him as a tour guide when we visited the Capitol. Senators, congressmen and federal officials stopped us along the way and greeted Howard with great respect. I was in awe as Howard took the time to introduce our group. He continued our "Red Carpet Tour" by pointing out famous paintings and sculptures, and gave us time to read historic speeches engraved on brass plaques. Finally, he took us to the balcony of the Senate where we watched and listened to the last arguments of a bill `on the floor' and ultimately voted on. (All of us, including Howard were dismayed at the procedure, and he commented that things had changed since he was in office.)

"At first, I felt fortunate to have this guided tour of the impressive buildings and monuments. Yet, walking up the steps of our nation's Capitol with the stars and stripes waving overhead, my feelings changed. I was one who avoided heavy populations, and was leery to visit that foreboding place that controls the many aspects of our lives. I began to sense a kind of spirituality grow deep inside me. I felt an immense lump rise in my throat while driving through Arlington National Cemetery and later touching the name of someone I knew inscribed on the Vietnam Memorial. They built those marble and granite buildings to stand strong throughout time and represent the freedom so hard fought. It is this freedom that allows me to represent wild sheep and ask our nation's leaders for their help.

"Our appointments included visiting with the senators in Idaho, Washington, Oregon, Wyoming and Alaska. We explained and urged them to support the wild sheep projects in their states. The compatibility between domestic sheep and wild sheep was a major issue in 1997. FNAWS had requested funding from the National Fish and Wildlife Foundation for disease research as part of the Hells Canyon Initiative. This required approval from Washington, Idaho and Oregon delegations. Our meetings must have been successful, for we did receive a $75,000 grant to study Pasteurella in Hells Canyon.

"Seeing Alaska Congressman Don Young's office was especially fun. He is an Alaskan through and through, with Dall's sheep, caribou and a huge brown bear rug in his office. Many different Alaskan native artifacts also decorated his office. His pride is there for everyone to see.

Alaska Senator Frank Murkowski was very informed about FNAWS to the credit of his hunting partner, Urban Rahoi. Senator Murkowski and his entire family spend at least one week every year in Urban's 'culvert camp'. . . it's bear proof! FNAWS was honored to have Senator Murkowski speak at our 1997 convention.

"Each year I have had an opportunity to speak with Minnesota's Senator Rod Grams and his aide Kiel Weaver. They get a kick out of the fact `the sheep lady' who is so dedicated, lives in a state with no wild sheep. They are fascinated that the Foundation's first office was in Minnesota, and that we do our investment banking in their state.

"Montana's Senator Conrad Burns is a fun and feisty fellow. He is an auctioneer and leads the auction at the Congressional Sportsmen's Caucus banquet. When you want to talk sheep, hunting or animal rights, this guy is right there supporting you all the way.

"During the CSC trips, many other meetings are arranged to make good use of time. I have participated as a FNAWS representative in meetings with the WLFA, the Wildlife Management Institute and side meetings with other partners within WMI. We have discussed each other's problems, accomplishments, directions and even conventions.

"Perhaps most interesting was meeting with the directors of the various federal agencies. Jack Ward Thomas, director of the USDA-FS at the time, has an incredible view! His office is closest to the Washington Monument and very impressive; though I admit, I would rather see a mountain vista.

"I truly enjoyed the past director of the U.S. Fish and Wildlife Service, John Turner. He not only is from a Wyoming family that ranches and outfits, he loves our FNAWS pins. Something as immaterial as a pin made a difference and made a friend. I wish life was that easy everywhere.

"Leland Speakes takes great pride that the Senate Majority Leader Trent Lott is from Mississippi. Both Leland and Senator Lott were in the U.S. Air Force, and can look at the Mountain Home Air Force Base situation from both sides. He keeps the senator informed of the loss of sheep and the alarming drop in lamb birth rates due to overflights by the U.S. Air Force.

"Leland's political side is stronger than I originally thought. Mississippi Governor Kirk Fordice, consented to be a speaker at the 1998 Saturday night banquet, that is, after Leland convinced him he should become a life member of FNAWS," concluded Phyllis.

Another organization based in Washington, D.C. that has been extremely beneficial to FNAWS is the Wildlife Management Institute (WMI). By becoming a partner in this organization, we became a team with the Izaak Walton League of America, Rocky Mountain Elk Foundation, the Boone and Crockett Club, the Ruffled Grouse Society and the Wild Turkey Federation. WMI, a nonprofit scientific and educational organization, committed to the conservation and professional management of North American wildlife, fits our organization like a glove. Not only do they watch legislation on the hill, the policies of USDA-FS, Bureau of Land Management and the Fish and Wildlife agencies are closely monitored.

Paul Hansen, director of the Izaak Walton League of America, arranged a meeting between the Partners of the Wildlife Management Institute and Vice President Al Gore in March 1997. Executive Director Karen Werbelow attended this historic meeting that opened communication lines with the major conservation organizations and the Executive Branch of the White House. The meeting concluded with the theme that wise use of natural resources is a supported concept that recognizes the needs of the American people, and ultimately contributes to a stronger economy.

The Wildlife Management Institute alerted FNAWS about potential severe budget cuts for the Forest Service. Executive Director Werbelow submitted comments to the House Appropriations Committee, testifying to the adverse effects that wild sheep would suffer. This led to an invitation to participate in the Bureau of Land Management's budget hearings. When Senator Craig (R-Idaho) introduced the Public Land Management Improvement Act, he asked her to state the Foundation's position personally in front of the Subcommittee on Forests and Public Land Management and the Committee on Energy and Natural Resources.

FNAWS vs. the U.S. Air Force

Wild sheep completely disappeared from the Owyhee Canyon in 1939. Through the efforts of the Idaho Fish and Game, the California bighorn was reintroduced into the canyon in 1963. The herd flourished, providing transplantable wild sheep to nine states until 1993, when the populations diminished. Overflights by the Air Force, sometimes lower than 1,000 feet, caused the sheep to move from prime habitat to less suitable habitat. Past President Dr. Robert DiGrazia challenged the Air Force and led FNAWS into a nine-year battle that may never be resolved.

Topping the list of potential military base closures in 1989 was Mountain Home Air Force Base in Idaho. The announcement to change its mission to a composite wing gave the base new life. An "Enhanced Training Range" was added where Air Force personnel could train

North Dakota Governor Edward T. Schafer in spring of 1993 signed a bill into law that provides for a nonresident bighorn sheep license to be made available through the lottery drawing. Those present for the bill signing included: (seated left to right), Representative Al Carlson, Governor Schafer, Representative Dan Austin, and (standing, left to right), Senator Donna Nalewaja, Game and Fish Director Lloyd Jones, Game and Fish Wildlife Division Chief Ron Stromstad and FNAWS Member Dave Currier.

President George Bush, center, spoke to FNAWS members at Red Butte Gardens on July 18, 1992. FNAWS Member Don Peay of Salt Lake City, Utah, was asked to participate in the discussions. Pictured during this meeting, left to right, are Johnny Morris, Mr. Bass, Utah Lt. Governor Val Overson, President Bush, Don Peay and Jay Wangsgard of Ducks Unlimited. President Bush stood behind all sportsmen in support of fishing, hunting, camping and recreational use of public lands in America.

On November 16, 1988, FNAWS and the U.S. Forest Service entered into a partnership agreement by signing a Master Memorandum of Understanding (MOU) in Washington, D.C. Pictured signing the MOU is FNAWS President Arnie Johns, seated left, and USDA-FS Deputy Chief James Overbay, seated right. Those standing are representatives of various conservation organizations.

fighter pilots to fly at supersonic levels and practice the use of live ordinances (bombs).

The original proposal, made by the Idaho governor, placed the range directly on wild sheep territory. Talks between the Idaho Fish and Game and the Air Force followed, and they moved the range to a location apart from the base. Dr. DiGrazia argued that since this was public land, due process was required. Both the existing base and the proposed range needed to be combined for the necessary studies and Environmental Impact Statements. Major issues included lambing seasons and the right to hunt. Air Force overflights in April, May and June caused stress in ewes during the lambing season. Overflights were also detrimental to the many recreationists using the area for hunting, hiking or rafting.

Lawsuits were filed and the courts ruled in favor of Dr. DiGrazia and the Owyhee Canyonland Coalition. A high-ranking general in the Air Force, General McCarthy, traveled to Idaho and announced there would be no further plans for expansion of the Mountain Home Air Force Base. A deep sigh of relief was felt throughout Idaho.

In 1994, the Air Force announced a new training range proposal. The public commented on scoping statements, Draft Environmental Impact Statements and Final Environmental Impact Statements. Renowned wildlife biologist Valerius Geist, Ph.D., submitted his comments opposing to the bomb range, along with many others concerned with the health and well-being of once-thriving Oywhee herd. Overall, 90 percent of the comment received denounced the proposed training range. Still, the Air Force record of decision was to utilize the range.

Despite negotiations and agreements not to fly during lambing season, Fish and Game personnel within the Air Force contend those overflights during lambing season are necessary to decide if they affect the wild sheep. Public outcry against the Air Force was overwhelming, so overwhelming that Senator Kempthorne (R-Idaho) included the range as an amendment to the 1999 Defense appropriations' bill. The area would be turned over to the Air Force, lock, stock and barrel, depending on how the Senate votes. At the time of this writing, a campaign to contact senators and lobby Capitol Hill is in progress.

Politics through Grant-in-Aid

Outfitting and wildlife conservation go hand in hand. Through the generous donations from outfitters and outfitter organizations, FNAWS is able to raise considerable revenue for our Grant-in-Aid program. FNAWS reciprocates by supporting the outfitting industry.

The Alaska Professional Hunters Association has been a leader in the preservation of hunter's rights since 1983. They have successfully campaigned to reopen prime hunting areas, defend outfitting as a viable industry and influence federal legislators to create an organized state guide system. Almost every year during the last 15 years, the APHA has been funded through the FNAWS grant-in-aid program, continuing their fight for hunting and outfitter rights.

British Columbia, with its diverse and abundant wildlife populations, and thousands of square miles of wilderness has, in recent years, become an appealing target for anti-hunting activists. The actions of the antis have sometimes reached near radical proportions. In the early 1990s, activists mailed outfitters letters containing razor blades, supposedly laced with rat poison, as a scare tactic and deterrent to hunt. However, these actions have lost them more public support than any gained.

The Guide Outfitters Association of British Columbia has asked for FNAWS' assistance since 1984 to counter these actions. The GOABC's funding requests include a number of initiatives to counter the misinformation of the anti's. Everything from public educational material and a province-wide scholarship program for Grade 12 students have been undertaken to inform the public and elected officials of the integral role that hunting plays in wildlife management.

The ballot referendum, equal the Ballot Initiatives in the United States, is also an avenue being used in Canada. Groups devoted to protecting the grizzly bear constantly petition to eliminate bear hunting, where bear populations are plentiful. By manipulating census information from the Ministry of Environment, animal rights groups use the media to convince the public that the bear is in danger. The GOABC, along with the BC Cattlemen's Association, BC Trappers Association and other like-minded organizations, have established a group called The Wildlife Management Alliance. Together they have created a potent force to counter the media blitz and were successful in stopping the referendum from reaching the ballot.

Native land claims have threatened to close large areas to nonresident hunting in the Yukon and Northwest Territories. The Yukon Outfitters Association and the Association of MacKenzie Mountain Outfitters have received grant-in-aid funding to help them with their legal expenses, as they fight for their livelihood. Slowly, most of the land claims have been settled, and they have retained the hunting rights for the nonresident.

The Politics of Education

All who belong to FNAWS believe that regulated hunting is a legitimate tool of conservation.

However, the public's view of hunting has been influenced by anti-hunting smear campaigns, characterizing hunters as blood-thirsty poachers. The public's lack of education prompted the FNAWS Board of Directors to sponsor school teachers to attend the Safari Club International's American Wilderness Leadership School. Teachers from the United States and Canada are selected based on their experience and class curriculums. Those without any conservation education and from the inner cities are preferred, with the hope they will develop an open mind toward the theory of wise use. For a week they are exposed to current environmental education trends, issues and concerns of multiple use. They meet with federal and state wildlife experts and are introduced to conservation programs that emphasize wildlife geared toward children. Hands-on outdoor skills are introduced including fire arms safety and survival techniques. The results have been phenomenal! Thank you letters to FNAWS have included such phrases as "I never knew how important conservation is" and "I now know how legitimate hunting plays an crucial role in wildlife management." All the teachers who have attended comment about the unforgettable friendships that develop during their outdoors experience, much like the camaraderie between hunters.

In 1991, FNAWS began the "Youth Essay Contest" to further educate our children and provide an opportunity to hunt for those who otherwise might not be able to. Every year a topic is chosen with an emphasis on wildlife management or ethical hunting. Teenagers between 16 and 18 years old would have the chance to hunt for caribou in the Northwest Territory courtesy of Kelly and Heather Hougen of Arctic Red River Outfitters. A strong advocate of teaching today's youth the value of hunting, fishing and trapping, Kelly also initiated a "Outdoor Education Camp" in conjunction with the Yukon Department of Renewable Resources. Both these programs are still successful today, providing quality experiences to those who might not otherwise have the chance to hunt or fish.

Epilogue

The Governors Hunting Heritage Symposium brings representatives of hunting organizations together with state government officials to reaffirm hunting as a tool for conservation. In the midst of several speakers commending the many conservation accomplishments of all the associations, NBC correspondent Roger O'Neill stood at the podium and delivered a resounding wake-up call. He described the anti's as a well-organized group, who were "college graduates" compared with us still in kindergarten. This analogy enraged many of the attendees until he further explained. The anti's are experts at manipulating the media, and playing on the Public's sympathy. This control filters to our elected officials who base their platforms on public opinion. Their strategy was to target the smallest groups of the hunting community first, trappers and archers. Then they attack wildlife management policies, eliminating hunting as a part of conservation. Meanwhile, the hunting minority sits idly by as our rights are slowly stripped away. To protect our ideas of conservation, wildlife management and our hunting heritage, we must form a partnership with other conservation and hunting organizations. Only by joining together and combating the anti's with full force will our rights remain intact. While our politics may not always agree with those who govern our country, we must always take the opportunity to meet with them and voice our concerns. The Foundation for North American Wild Sheep not only "puts sheep on the mountain," we must also provide the "voice for sheep on the mountain."

Phyllis Carlson, the first woman president of FNAWS, was raised on a farm where conservation practices were something you naturally did. Eating what you raise, either plant or animal, was a way of life. Before marrying her high school sweetheart, Doug Carlson, many of their dates were spent target shooting with a borrowed .22 or .410. Their first gun purchase was a .22 Remington nylon 66 with a Sears Roebuck scope. Since then, she has become active in trapshooting and has participated on a trapshooting team for more than 20 years. In the early 1960s they started their own home building and remodeling business. Phyllis and Doug have hunted together for over three decades, and have hunted extensively in Alaska, many Western States and Canada. Her pride and joy is a Belgium made .270 Browning bolt action. When they have a chance, they fish for walleye and pan fish between hunts. During the "Game Fair" in Minnesota, she was honored as "Outstanding Big Game Hunter of 1985." It was on a 1973 Dall sheep hunt in the Alaskan Wrangell Mountains that they met Harvey Kadlec, who would six years later be president of the Minnesota FNAWS Chapter. Harvey invited Doug and Phyllis to a meeting, and they asked Phyllis to be their secretary. She was an active member and officer of the Minnesota Chapter (which evolved into Minnesota-Wisconsin Chapter) for 11 years. She than ran for a seat on the FNAWS National Board and was elected for three consecutive years. Phyllis culminated her term as a board member as its president in her ninth year. After raising three children, she now enjoys her seven grandchildren who constantly visit. During the spring, she is planting flowers in her garden, between working with Doug in the construction business, which she dearly loves. Phyllis is a member of the NRA, SCI, North American Hunting Club and a life member of FNAWS.

Paula Karres is a native of Los Angeles, California, and was raised believing hunting had no use in today's society. After moving to Wyoming, she was reeducated, and is living proof that a little education goes a long way. Once a year, she and her horse make a pilgrimage to British Columbia and trains with former Captain of the Canadian Three Day Equestrian Team, Robin Hahn. During the summer she competes at Horse Trials throughout the Rocky Mountain Region.

Paula has been with FNAWS for five years, working as a secretary to the executive director and as chapter affiliate coordinator. By collaborating with FNAWS Past President Dr. Robert DiGrazia, FNAWS member Arlene Hanson and GOABC Executive Director Dale Drown, she was able to compile the history of FNAWS involvement in politics. She lives in Cody with her husband, David of 18 years, two children, Kelly and Amanda, six dogs and five horses.

Congressional Sportsmen's Caucus and Foundation

The Congressional Sportsmen's Caucus (CSC) was founded in 1989 to preserve and promote the traditional rights of America's 80 million hunters and anglers. Prompted by the increasing attacks on hunting, fishing and the shooting sports, this action was a watershed event for America's sportsmen and women. Today, this influential group has become the largest ad hoc caucus, presently numbering nearly half of both houses of the U.S. Congress.

The Congressional Sportsmen's Foundation (CSF) was chartered in 1989 as a 501-(c)(3) nonprofit organization. Its mission is to ensure that current and future generations will have the right and opportunity to hunt and fish by having a strong voice in decisions that affect the outdoors.

Working to protect the integrity of the hunting and fishing conservation trust funds, and preserving hunting and fishing opportunities on public lands, the CSF educates and informs members of Congress, federal officials and the public on timely issues. Often referred to as the "sportsmen's link to Congress," the CSF has a strong and productive relationship with the Congressional Sportsmen's Caucus, which serves as a legislative strike force on key issues affecting hunters and anglers.

Since 1989, the CSC and CSF have successfully worked to thwart government proposals to eliminate or reduce access to public lands and to halt attempts to raid various sportsmen's trust funds. Working together they have strongly supported wildlife habitat programs such as the Conservation Reserve Program, the North American Wetlands Conservation Act and the National Wildlife Refuge Act.

As of May 26, 1998, Jay McAninch was selected as the executive director of the CSF. He is a certified wildlife biologist and received a B.S. in science in fisheries and wildlife biology from Iowa State University. He earned a M.S. in zoology from Ohio State University. Jay has been recognized nationally for his work in developing and implementing urban deer management programs and for landmark studies of bowhunting. He has coordinated and managed the development of a coalition of wildlife and sportsmen's organizations aimed at winning ballot initiative campaigns. Jay has also help guide the planning efforts of major industry groups and sportsmen's organizations to examine their present status and strengths and initiate activities to benefit sportsmen and natural resources.

FNAWS is one of many organizations that supports the work of the CSF. To learn more about CSF or to receive a current list of legislators who are active in CSF, write to them at 1730 K Street, NW, Suite 1300, Washington, D.C. 20006. They can be reached by calling (202) 785-9153.

Foundation for North American Wild Sheep Board of Directors Meeting June 12 and 13, 1987

REINTRODUCTION OF THE WOLF INTO THE YELLOWSTONE ECOSYSTEM

WHEREAS, Yellowstone Park and surrounding area is public lands and belongs to all of the people of the United States and not to a select few, special interest groups; and

WHEREAS, the Foundation for North American Wild Sheep is dedicated to the enhancement and preservation of wild sheep in North America; and

WHEREAS, there are 13 distinct native sheep populations in and adjacent to Yellowstone Park; and

WHEREAS, some of these native populations are declining and are in danger of extinction; and

WHEREAS, native sheep populations should be preserved, protected, and enhanced; and

WHEREAS, wolves are known to prey extensively on wild sheep; and

WHEREAS, wolf depredation is known to be detrimental and a limiting factor to wild sheep; and

WHEREAS, there are no wolf populations in or adjacent to Yellowstone Park; and

WHEREAS, the species of wolf that was native to the Yellowstone area is extinct;

NOW, THEREFORE, BE IT RESOLVED that the Foundation for North American Wild Sheep is opposed to the reintroduction of any wolf species into Yellowstone Park or adjacent areas.

This resolution was adopted by the Foundation for North American Wild Sheep at the Board of Directors meeting June 13 and 14, 1987, in Reno, Nevada. A copy of this resolution was sent to the following organizations:

Yellowstone National Park, National Park Service, U.S. Department of the Interior, Montana Department of Fish, Wildlife and Parks, Wyoming Game and Fish Department, Idaho Fish and Game Department, Gallatin National Forest, Shoshone National Forest, Teton National Forest, Montana Stockgrowers and Montana Woolgrowers, Idaho Stockgrowers and Idaho Woolgrowers, Montana Outfitters and Guides Association, Idaho Outfitters and Guides Association and the Wyoming Outfitters and Guides Association.

Chapter III

For Two Decades FNAWS Has Fought the Anti-Hunters — and Won!

RICK STORY

Among species conservation groups, there has been no greater defender of the outdoors tradition over the past two decades than the Foundation for North American Wild Sheep. The staff, board and membership of FNAWS has been in the thick of every key fight for sportsmen's rights since the early 1980s.

As an officer of the nation's principal organization devoted to defending and advancing outdoor sports and wildlife management, I have been proud of FNAWS' involvement on behalf of its members' right to hunt over the years. Whenever anti-hunting influences attempt to infringe upon the rights of hunters, we have come to count on FNAWS' active involvement in campaigns to fight back. Two FNAWS past president's, Craig Johnson and Pete Cimellaro, serve on the board of The Wildlife Conservation Fund and Wildlife Legislative Fund of America and we have been fortunate to have enjoyed the ongoing counsel of the FNAWS professional staff and board members, including great friends like Phyllis Carlson, Dr. Bob DiGrazia, Arnie Johns, Dan Pedrotti, Howard Pollock, Karen Werbelow, Garry Vince and Lloyd Zeman.

The WLFA and WCFA were formed to fight the anti-hunting movement in 1978. Since almost the very beginning, FNAWS has stood as a lynch pin member of our national coalition of national, state and locally based organizations united for the defense of outdoor sports. Today, FNAWS is joined by some 1,400 organizations, in as many communities coast-to-coast, which join in campaigns to fight the anti-hunting movement.

The first involvement by FNAWS in campaigns to advance outdoor sports occurred in 1982 during the presidency of Duane Smelser. With this action, the organization became part of a national coalition of organizations to win back hunting rights on 12 million acres of Alaska lands closed to sportsmen with passage of the Alaska Lands Act of 1980. Under the direction of Executive Director Lloyd Zeman, FNAWS marshalled its members for the campaign, which won editorial support of half of the major market daily newspapers in America.

Although the bill introduced to win back hunting rights was never voted upon, the campaign was successful in halting further action by Congress to lock up even more Alaska lands to hunting. Sportsmen had made their mark in the nation's capitol and FNAWS was beginning its long history of support of sportsmen's rights.

In 1984, the Humane Society of the United States, the nation's largest and most danger-

Anti-hunter protestors in Virginia attracted television cameras and other media attention in 1990 at a wildlife refuge near Washington, D.C. The Wildlife Conservation Fund of America works with a variety of organizations and individuals to inform them about the value of hunting and wildlife management practices. The Foundation for North American Wild Sheep has a long tradition of supporting the educational efforts of WCFA.
Photograph by Jack Reneau.

ous anti-hunting organization, sued to stop hunting on the National Wildlife Refuge System. The unsuccessful suit was followed by a series of congressional bills designed to boot sportsmen off the nearly 100 million-acre system — lands which had been largely purchased with hunters' dollars.

The level and intensity of the anti-hunting movement's attack was beginning to escalate and FNAWS' defense effort was stepped up to meet the challenge. In that year, FNAWS made the first of several contributions to enable the Wildlife Conservation Fund to develop and field a major program of education designed to muster the combined strength of sportsmen's organizations throughout the nation.

The program was called "Protect What's Right." Today, nearly 1,000 sportsmen's organizations in communities throughout America are involved. These organizations are provided a series of public education materials to condition local public support in favor of hunting. The clubs also receive an array of self-help materials which teach them how to rally their memberships in support of hunters' rights at the federal, state and local levels.

These clubs form the nucleus of the nation's principal lobbying force in support of sportsmen's rights. Through these organizations, brought together because of FNAWS' vision and foresight, an army of some 1.5 million hunters, anglers and trappers stands ready to defend and advance sportsmen's rights throughout America.

By the late 1980s, the anti-hunting movement was becoming a regular fixture on network news reports and on the editorial pages of the nation's major, daily newspapers. The impression that was left with the public and with policy makers was that animal rights was becoming a mainstream movement. Perhaps this is the reason that various government agencies began to downplay the importance of hunting from a conservation standpoint and, in some cases, became down right hostile to sportsmen.

An example may be the case of four international sheep hunters, all FNAWS members, who in 1988 were detained by law enforcement officials of the U.S. Fish and Wildlife Service upon entry into customs from their hunting trip to China. They were accused of attempting to import an endangered subspecies of Argali sheep. The hunters' trophies were seized and held for almost two years. A federal court in Texas finally ordered the trophies returned to the hunters and charges were dropped.

A small minority of anti-hunting protestors have motivated a core group of pro-hunting groups to activate themselves. The first involvement by FNAWS in campaigns to advance outdoor sports occurred in 1982 during the presidency of Duane Smelser. With this action, the organization became part of a national coalition of organizations to win back hunting rights on 12 million acres of Alaska lands closed to sportsmen with passage of the Alaska Lands Act of 1980.
Photograph by Jack Reneau.

This case brought to a head a recent series of instances in which hunters had been treated poorly by law enforcement agents. FNAWS President Arnie Johns encouraged FNAWS to join WCFA's coalition of hunter groups working to reform USFWS law enforcement policy, bringing consistency and clarity to law enforcement. A federal commission was convened which, in fact, identified a series of deficiencies in law enforcement practices by the USFWS. Most of these had earlier been identified in a years-long study by the WCFA and its coalition, led by FNAWS and other hunter groups. Reforms to establish a mission statement by the Fish and Wildlife Service, to establish a hunters' assistance office at the Fish and Wildlife Service and a series of other reforms were adopted in 1990.

In 1992, FNAWS was one of the leaders in a nationwide effort to save hunting, fishing and trapping, the fate of which was slated to be decided by Arizona voters in November of that year. Deceptive ballot language, ostensibly written to ban trapping, called for a ban on all outdoor sports. It proscribed managing wildlife in Arizona by "lethal means." Then-FNAWS board member Pete Cimellaro managed the day-to-day operation of the pro-hunting campaign to defeat the measure.

FNAWS itself contributed and raised over $100,000 — some 10 percent of the entire campaign budget. It contributed directly to the campaign from its coffers and called upon members to contribute to the Arizona effort. FNAWS did a fantastic job of getting out the word to members and other sportsmen and even helped inform and motivate other national conservation organizations to enter the fight. Other major national organizations that contributed to the fight included The Archery Manufacturers and Merchants Organization, Safari Club International, The Wildlife Legislative Fund of America and the Rocky Mountain Elk Foundation.

The key to this dramatic victory was the size and availability of the campaign war chest. Some $1 million was raised from sportsmen throughout America and the majority of the available dollars were on hand by July 1992, in plenty of time to secure television advertising time for the November election. Campaign Treasurer Pete Cimellaro, who ran the campaign office and worked with his staff of volunteers to amass an impressive war chest, and dedicated to making the voters of Arizona understand the dangers of the proposition.

Despite pre-election surveys that forecast a major loss for sportsmen, the well-funded, pro-hunting campaign resulted in an overwhelming sportsmen's victory, by a 2-1 voter margin.

FNAWS has since been involved in a number of other voter issues to defend hunting. These include the successful 1996 issue in Idaho in which anti-hunters threatened to outlaw bear hunting with hounds and bait. Again, thanks in no small part to FNAWS' support, sportsmen logged a big victory. Congress, and Washington, D.C. in general, can be a hazardous place for sportsmen. FNAWS has joined the WCFA and WLFA for years in hosting a table at the annual Congressional Sportsmen's Foundation banquet, in part as a means to gain access to influential members of Congress who decide hunting's fate on a daily basis.

In 1993, and again in 1995, FNAWS also became directly involved in direct and grass roots lobbying efforts to defend and advance sportsmen's rights in the nation's capitol.

The federal Endangered Species Act came up for re-authorization, as it does every three years, in 1993. Anti-hunting influences, with great pomp and circumstance, were making big noise about amending the act to make it more restrictive. The principal targets of their campaigns were sportsmen and wildlife managers. They had even gone so far as to propose that candidate species be provided full protection until such time as their true status could be ascertained. Because endangered species status frequently results in the declaration of "critical habitat" — areas on which virtually all human activity (including hunting!) are stopped — the proposal imperilled hunting.

FNAWS once again joined the WLFA and WCFA coalition to fight back. Because we have long subscribed to the notion that the best defense often is a strong offense, a series of pro-hunting, "Common Sense Amendments" to the endangered species act were proposed. They reaffirmed hunting's role as a wildlife management tool and cleared the way for traditional wildlife management to occur in the face of the restrictive endangered species act.

The tactic worked. The no-hunting lobbies spent all of their time fighting the pro-hunting amendments and had little time to work for their own amendments. To date, the common sense amendments have been featured in every one of the important endangered species act re-authorization bills to be considered in Congress.

Hunting came under fire again in 1995, as the Clinton administration entered into a lawsuit settlement with environmental organizations which threatened hunting. The settlement, on an unrelated case, forced the US Fish and Wildlife Service to agree to perform what amounts to environmental impact assessments for each of its hundreds of National Wildlife Refuges every year. A red flag went up at the WCFA offices in that the procedure provided a ripe opportunity for anti-hunting organizations to sue if they determined that the assessments were inadequate.

Of even greater concern, the Fish and Wildlife Service in subsequent weeks began making even bolder moves to stop hunting on refuges. In response to a budgetary directive by the USFWS director, refuge managers in California, Mississippi and other states proposed closures of hunting on refuges under their jurisdiction.

The WCFA had informed FNAWS of the threat. FNAWS responded with funding to establish the WCFA's Sportsmen's Legal Defense Fund, which threatened to sue. Congress intervened in 1995 and ordered no hunting closures on the refuges, but the coalition established to defend hunting on the refuges and other public lands wanted a greater assurance.

Legislation to elevate hunting's status on the refuge system was introduced. A storm of protest went up among environmental organizations. FNAWS President Pete Cimellaro led a group of other conservation group officers in signing a letter to members of Congress and the pro-hunting bill passed the House of Representatives by a veto-proof margin. In the next session of Congress, a similar bill to protect hunting on the refuges cleared the full Congress with only one dissenting vote among hundreds cast by representatives and senators.

The support and active involvement of FNAWS and its members have been crucial to hunting's defense all over America. Since 1982, the organization has provided ongoing support to the work of the WCFA, which has proudly listed FNAWS among its most important supporters.

FNAWS has proven that the terms hunter and conservationist are far from mutually exclusive. The work it has done in defense of outdoor sports is rivaled only by its record of support of sheep and other wildlife.

For more information about the WCFA and WLFA, contact Rick Story, vice president, 801 Kingsmill Parkway, Columbus, OH 43229. Phone (614) 888-4868. FNAWS is a long-time supporter of the Wildlife Conservation Fund of America (WCFA). The WCFA protects the heritage of hunters and wildlife management by providing legal services and by performing research and public education work in defense of the right to hunt. The WCFA's companion organization for legislative work is The Wildlife Legislative Fund of America. It operates in the 50 states and on Capitol Hill.

Rick Story is vice president for The Wildlife Conservation Fund of America (WCFA) and its companion organization, The Wildlife Legislative Fund of America (WLFA). Rick is a 17-year veteran of WLFA's battles to defend and promote hunting, fishing and trapping throughout the United States. He is one of a team of professionals from WLFA who include lawyers, lobbyists and public relations experts. Rick holds a B.A. from The Ohio State University in journalism and public relations. He specializes in the public relations field of "issues management" and is an avid outdoorsman. He is a former newspaperman, outdoor columnist and marketing director. He is a veteran of the U.S. Navy and lives with his wife Becky and two daughters, Emily and Erin, in Pickerington, Ohio.

Chapter IV

Desert Bighorn Sheep in Mexico

RAYMOND LEE

In December 1992, I had the incredible good fortune to fly into Mexico with a helicopter to survey the various mountain ranges in Sonora for the presence and abundance of bighorn sheep. The funding for this survey was provided by the Arizona Desert Bighorn Sheep Society. The commander at the airport of entry asked what we were doing.

I told him, "Looking for bighorn sheep."

He said, "You might as well go home. They have all been poached."

I told him that since we had come this far we'd go ahead and take a look around. His expression was a classic, "Loco gringo," or "Crazy White Guy." Five days later, after observing nearly 600 bighorn sheep in virtually every range we flew, we departed Mexico from the same airport.

The airport commander stated, "You didn't see any did you?"

"On the contrary," I said, "we saw hundreds." His expression was again classic. This time the expression was one of disbelief.

Based upon these positive results, the next year, with funding from FNAWS and from Mexico (FNAWS received a check for $6,000 from Mexico. They told me until then they had never received any money back from Mexico.) I returned to Sonora. This time we were flying primarily in areas we had not surveyed during the first survey, and some that we had to establish population trends, as well as distributions. After again classifying hundreds of bighorn sheep in dozens of mountain ranges, these flights eventually took us to Tiburon Island in the Sea of Cortez. A handful of bighorn sheep had been transplanted from the mainland to Tiburon Island in 1975. Since that time little had been done with the population. Mexican biologists had conducted ground surveys the year before and estimated a population of 80 animals. We classified more than 80 animals during the first hour of the survey. Eighty more during the second hour, and another 80 during the third hour. Out of gas, but not out of island, we returned to the mainland. I thought the Mexican biologists along on the flight would be as ecstatic about the results as I was. Instead they were quite upset.

"What will we tell our boss," they asked.

"That we found the mother lode of bighorn sheep?" I offered.

"No," they said. "You see, our boss is an animal preservationist, and she didn't want us to find any animals on the island so she could promote the story that the bighorn sheep had all been killed by poachers."

In 1995, the Arizona Game and Fish Department (AGFD) and the Ecological Center of

Desert landscapes and a desert bighorn ram in the Sierra Viejo area of Sonora, Mexico.
Photograph by Raymond Lee.

A desert bighorn ram among the rocks in the Las Virgenes area of Baja Sur.
Photograph by Raymond Lee.

Sonora (CES) shared a National Fish and Wildlife Foundation grant of $80,000 that was part of a restitution payment made by a United States hunter caught poaching bighorn sheep in Mexico. The U.S. Fish and Wildlife Service (USFWS), with a little guidance, decided it would be a good use of this money to conduct a bighorn sheep study in Mexico. As a result of this funding, the AGFD and the CES continued helicopter surveys in Sonora and captured and radio-collared a number of bighorn sheep for movement and mortality studies as well as conducting a habitat utilization and evaluation study.

In 1995, FNAWS funded a survey in northern Baja California. In this case we had a recent survey conducted by the Bighorn Institute with which to compare our results. While our survey was not as costly as the one conducted by the Bighorn Institute, we found sheep at a faster rate. Our results indicated the bighorn sheep population in Baja California was at least stable, if not increasing. The results of this survey were used to dissuade the USFWS from listing the peninsularis sub-species of bighorn sheep in Mexico as endangered, as they did in California.

In 1996, we flew the El Vizcaino Biosphere Reserve in Baja Sur. The results of this survey indicated the presence of a healthy population of nearly 300 bighorn sheep. Four hunting permits were issued. These permits were sold at a special auction conducted in Phoenix, Arizona. These permits produced $262,000. Four members of FNAWS purchased these permits and eventually harvested three rams in the reserve. The uniqueness of hunting in the Reserve, I feel, is not well appreciated in the United States. The El Vizcaino Biosphere Reserve is a United Nations designated protected area. One of the few similarly protected areas in the United States is Glacier Bay National Park in Alaska where absolutely no hunting is allowed! After helicopter surveys were conducted in 1997, three permits were issued for 1998. Two were auctioned at the FNAWS national convention and another raffled, with a total return of $224,000.

In 1997, we flew other areas in Baja Sur, including areas near Loreto and further south near El Mechudo. Again, good numbers of bighorn sheep were classified. This will lead to the issuance of additional bighorn sheep hunting opportunities in Mexico. These surveys were funded by the National Commission for the Understanding and Use of Biodiversity (CONABIO), a Mexican conservation organization. We have additional surveys scheduled in

Skull and bones from a ram, found near Pico Johnson, Sonora, Mexico.
Photograph by Raymond Lee.

Sonora and Baja Sur in 1998. These surveys will be funded by a variety of sources. Almost all as the result of revenues received from permits auctioned or raffled at FNAWS.

I well remember my first FNAWS Board meeting on February 26, 1993. I was asked to attend a discussion of Mexico. The representative from Mexico was asked to explain for what the money he was asking for would be used.

He stated, "Are we not friends? Friends do not ask each other what they use their money for. You should simply trust me as a friend and give it to me." And they did! To say that I was surprised would be an understatement.

Now we come to the more current relations between FNAWS and Mexico. First, and extremely importantly, one must realize that Mexico is not simply Mexico. It is comprised of individual states, with numerous political/governmental subdivisions, just like the United States. To say that the bighorn sheep management programs in Sonora, Baja California and Baja Sur, are the same is like saying the bighorn sheep management programs and the people involved are the same in Arizona, California and Texas. There could not be more difference. Yet we tend to paint them with the same brush.

In Arizona, for example, I am considered by many to be the prime manager of bighorn sheep. However, within the Arizona Game and Fish Department, we have a number of regional and district biologists who have responsibilities for the management of bighorn sheep within their units. Some of these units include lands administered by the U.S. Forest Service, the National Park Service, the Bureau of Land Management, the U.S. Department of Defense or the U.S. Fish and Wildlife Service. Each of these agencies have biologists on staff who feel they have a legislative mandate to manage the wildlife resources on their lands. Arizona also includes five Indian tribes who have bighorn sheep populations that at sometime reside on their reservation lands. Each of the United States with bighorn sheep have similar overlapping administrations. So do the various states of Mexico.

It sometimes seems that members of FNAWS have memories like elephants. When a FNAWS president met with Mexican biologists from Baja Sur in Mexicali, the president brought forth information that was more than 20 years old. One of the biologists supervising the new

A lone desert bighorn ram from Tiburon Island, Sonora, Mexico, surveys his territory from a rock outcropping.
Photograph by Raymond Lee.

bighorn sheep program in Baja Sur asked me later, "I was 6 years old when FNAWS was making those decisions, why are they punishing me now?"

As a sometimes intermediary between various Mexican groups and the FNAWS Board I have been in the a unique position. I have observed the deep distrust by individual FNAWS Board members for almost all things Mexican. On the other hand, I have observed the dissatisfaction of FNAWS by some of the biologists from Mexico. On more than one occasion, the FNAWS Board did not fulfill its agreements to the Mexican programs.

In general, these dissatisfactions are created more by misunderstandings and a lack of communication. For example, the Reserve in Baja Sur wanted the return of their money from the auction of the tags to pay the salaries of the wardens they had hired to patrol the sheep areas. FNAWS was reluctant to release the funds, asking for a detailed accounting of all program expenditures before they would send the money to Mexico. FNAWS sought additional information on various management programs and on the permits associated with the hunting programs. Sometimes this has been painfully slow in coming. Are these major issues? Not really. Could they easily be blown out of proportion such that they do become major issues? They have. Communication is a two-way street. Both sides must be willing to fully and openly articulate their concerns.

While some choose to think of the Baja peninsula as half empty of sheep, I prefer to think of it as half full. In 1998, FNAWS received approximately $90,000 from the auction and raffle of bighorn sheep tags in Mexico. This is unprecedented! All of the tags were purchased by FNAWS members. All of the money raised is being spent on bighorn sheep conservation programs or social development projects with FNAWS representatives having a say in the expenditure of these funds. We are holding the new Mexican programs to a higher standard than we are the various state programs, and they are trying to meet these standards. Is the program a success? Judging by the 1998 results, overwhelmingly. Did it take longer to reach this point than we would have liked? Certainly, but we did.

None of this could have been accomplished without the efforts of brothers Drs. Manuel and Pablo Chee. Their Mexican Foundation for the Conservation of Wild Sheep (MFCWS) provided a forum for FNAWS Board members to discuss bighorn sheep management issues directly with the appropriate leaders in Mexico. They were directly responsible for three presidents of FNAWS to meet with Rodrigo Medellin, and subsequently Felipe Ramirez, who

Jumping from rock to rock is a desert bighorn ram photographed in the El Mechudo area of Baja Sur.
Photograph by Raymond Lee.

holds the position of General Director of Wildlife in Mexico - the position responsible for all wildlife management programs in all of Mexico. These are the types of interactions necessary to build awareness of each others' issues and trust on both sides.

In 1998, a permit for Baja California was auctioned at the Safari Club International for $141,000 and purchased by a life member of FNAWS. Remember, no less than the president of Mexico issued a decree banning bighorn sheep hunting in Baja California in 1990. In less than a decade we were able to overcome a presidential decree. Do you think that could have been accomplished in the United States?

Will we continue to have difficulties dealing with Mexico? Certainly. The differences in language and culture will always be with us. Can the issues be resolved? Of course, with communication, and with respect for each others' positions. Mexican managers must realize that FNAWS is not just a cash cow to be milked for money to be spent unaccounted for below the border. FNAWS must realize that it is truly a North American foundation, and that a significant number of bighorn sheep reside south of the border; and that with a little support, many more sheep can be put on mountain ranges in Mexico to be enjoyed by hunters from any country.

Raymond Lee's biography is found in the chapter he wrote for this book on desert bighorn sheep titled, "Desert Bighorn Sheep: North American Survivors."

Chapter V

Wildlife Conservation: What Shall We Defend?

Valerius Geist, Ph.D.

North America's backbone of nature conservation is a continental system of wildlife conservation. It came into being some 80 years ago, the child of two befriended heads of states: President Theodore Roosevelt of the United States and Prime Minister Sir Wilfrid Laurier of Canada. They were ably assisted by both country's foremost citizen who pitched in to save wildlife and to give it a legal, lasting basis for existence.

Here lie the roots of the Boone and Crockett Club and the American Bison Society. We owe a great debt to the likes of Gifford Pinchot, Roosevelt's able lieutenant, or Clifford Sifton, multimillionaire, civil servant, developer of the Canadian West, who chaired the long-lasting Commissions on Conservation (1911-1919) that held public hearings on natural resource conservation.

Where would wildlife have been without a fire brand like William T. Hornaday, or the thoughtful, able C. Gordon Hewitt and his generating of the justly famous migratory bird treaty? It was way ahead of anything comparable elsewhere. It still is! Ours is a populist system in which the ownership of the wildlife resource de jure, but above all de facto, is being exercised by the public. Until recently, this public had successfully opposed political forces to make wildlife private property. Our system of wildlife conservation is a great cultural achievement of North American society, being the only large-scale system of natural resource use that has not merely sustained the resource but has increased it steadily. Wildlife conservation has greatly contributed to North America's quality of life, while generating rich economic rewards in the creation of both wealth and employment.

At an annual one-time continental spending of about $70 billion, and some 50,000 jobs per billion of throughput, wildlife conservation has been important in continental welfare, and an excellent, informative example of how to protect and enhance a natural resource while maximizing economic benefits. While the resource is owned by the public, the creation of wealth is done by the private sector producing goods and services for the users of the resource. How significant the "Public Good" is in daily life is well illustrated by our system of wildlife conservation and management. What three generations of North Americans achieved here can only be judged by comparison to other systems of wildlife conservation and how poorly they are doing.

A central characteristic of the North American system of wildlife conservation is its use of blue collar and rural support for active nature conservation and environmental protection. Of course, one did not use such fancy names in the past, but that's what it boils down to. By

Richard R. Reneau, 16, works with Rocky Mountain bighorn sheep in Rock Creek, Montana, and learns first-hand about wildlife management practices that were put into place by America's first wildlife biologist Aldo Leopold and others after him. Involving youth in the wildlife management process insures the future of our wildlife heritage.
Photograph by Jack Reneau.

Volunteers from the Foundation for North American Wild Sheep and other conservation organizations have dedicated themselves to the expansion of wild sheep populations in North America and the enhancement of their habitat. These lambs symbolize that success.
Photograph by Valerius Geist, Ph.D.

protecting wildlife and its means of existence, a wide-range of species and great bio-diversity was achieved. The broad support by rank and file working people was and still is the source of its strength. Without broad and effective blue-color support for wildlife as a "Public Good," there cannot be public wildlife conservation or the high economic returns associated therewith.

Wildlife without blue-color support is destined to slip into private control, for private exploitation, but also for very private mismanagement and misuse. It not only makes us all very much poorer economically or takes away from our quality of life, but it also leads to very bad conservation practice. Those are the unhappy lessons history teaches again and again.

Our system of wildlife conservation captures the passions entailed in hunting in a highly constructive fashion. It allocates by law the opportunities to hunt in an egalitarian fashion. That implies an armed public. Those that choose to participate spend money to look after wildlife, lobby politically on behalf of wildlife, by spending time afield they continually monitor the land and respond with indignation to large-scale environmental damage. They lobby and pay for land to be managed for wildlife and form the basic — the core — of North American nature conservation. They pay their dues annually, and they bring a plethora of concerns to the attention of like-minded people and politicians. They are not single-cause organizations, and they are not into sensational environmental coverage. Yet no endangered species act is meaningful without this basic conservation work, that depends so much on people to people communication, by those that hunt and fish. And they do it for truly modest return in animals taken, as the success statistics indicates.

The strength of our system of wildlife conservation resides in a handful of fundamental policies that have been translated into laws within all North American jurisdictions. These are public ownership of wildlife, with private ownership allowed only at the discretion of the sovereign. It makes the sovereign responsible through elected representatives to consider the fate of this public resource. Secondly, we outlawed markets in dead wildlife. This may be the single most important policy protecting wildlife. Yes, it has been eroded as have others, but that's what we had, and it did plenty of good. Thirdly, we allocated wildlife by law, not by the market place, or social position or land ownership. We insisted wildlife be only killed for cause — for its meat, furs or when it endangered our lives or destroyed our property. That put an end to thrill killing and affirmed a certain reverence for life. We recognized that migratory wildlife needed cooperative management between states and we established sound working treaties to that effect. We enshrined science as a source of knowledge for the welfare of wildlife and honored a great president by calling this the Roosevelt Doctrine.

Within a decade of its inception it led to a new uniquely North American profession, the wildlife biologist, the first notable one being Aldo Leopold. This cluster of policies generated public support and concern. Conservation organizations mushroomed to safeguard the fate of wildlife. Where but in North America is the membership in wildlife conservation organizations greater than in all political parties combined? There arose the only system of sustainable natural resource use, the greatest environmental success story of this century. None of those that struggled to create it saw its success. They worked hard and clever on our behalf, the generations unborn, believing that wildlife was worth saving.

And in that they were right, and for that we owe them lasting thanks.

A biography of Valerius Geist, Ph.D., is found at the end of his first chapter in this book titled, "Wild Sheep: Creatures of the Ice Age."

Epilogue

Presidents Speak!

When the first effort to complete this book began more than five years ago by Dr. Roy A. Schultz and Duncan Gilchrist, letters were written to all past presidents asking them to tell their stories about how they became involved with FNAWS and its many projects.

This chapter contains letters from past presidents who did not attend the 1998 national convention, including a letter from Lloyd Zeman written in 1977, as well as the taped interviews Susan C. Reneau did with past presidents at the 1998 national convention or on the telephone after the convention through late July 1998. These are exact transcriptions of what past presidents said when interviewed in Reno in 1998 or on the telephone. Their comments are profound and thought-provoking.

You will note a similarity of purpose and direction from all past presidents. A thread of continuity prevails. This impressed me about the interviews because none of them knew what other past presidents said. Many commented to Susan during the interview they thought their opinions didn't agree with anyone else but, if fact, most past presidents expressed similar concerns.

The interviews and letters are published in this chapter in order of a person's term of office as president with their years of service under each name.

Listen to their voices and their advice for future generations of sheep hunters and sheep enthusiasts.

Lloyd Zeman
President, 1977-1980

A Message from Your First President
Summer 1977

Dedication, determination and a sincere interest in North American wild sheep. That's the kind of sportsmen who are the backbone of our Club. Most of us have had the privilege of collecting one or more trophy rams — matching wits with the greatest trophy in North America and probably the world — in the remote areas of the wild and steep high country where these magnificent animals make their home.

Spotter, Carlos Sanchez, takes a bid for Lloyd Zeman during the 1988 auctions.

How We Got Started

Our Club began with a very informal gathering of 14 (Jim Froelich in his report to Bob Housholder said it was 12 sheep hunters) sheep hunters at Mt. Horeb, Wisconsin, during a stormy weekend in November of 1974.

"Died-in-the-wool" sheep hunters came from Iowa, Michigan, Wisconsin and Minnesota. The following April we met again at the Playboy Club at Lake Geneva, Wisconsin. Our numbers increased to 46. Our gatherings were mostly social and an exchange of sheep hunting stories and pictures.

Then in April of 1976, we met in Des Moines, Iowa, and the word began spreading — more than 100 persons attended. And the action began. We organized, electing a seven-man team to head the informal Club and set a purpose.

Rather than just being a social group, we decided to start doing something for the animals we all care so much about. Those of who have had the privilege of collecting trophy rams decided we wanted to "give something back." So we planned the Detroit National Convention, which was held last March. We had 265 people at the Saturday night banquet and some 300 persons attended at least part of the convention.

The Detroit convention was our first effort at raising funds for worthy sheep management and conservation projects. In a small way at least, we plan to do for North American wild sheep what Ducks Unlimited does for ducks. Our first try at fundraising brought in over $12,000. In early January 1978 we plan to deliver a 3/4-ton 4-wheel drive pickup truck, loaded with all the necessary equipment, and donate it to the Mexican Game Department. It will go to Mexicali where it will be used by game officials to "work" waterholes for sheep, patrol for poachers and perform various other duties such as checking on the sheep in the northern Baja.

Where We Are Headed

Recently, we incorporated as a nonprofit organization under the name of the Foundation of North American Wild Sheep. We have eight directors and will elect more at our convention in Memphis, Tennessee, February 2 to 4, 1978.

Anyone who wants to join our organization can by paying $25. Charter memberships (a one-time fee and then $25 annually) are also available for the first 100 persons wanting to do so. We have only one class of members, but Charter members will get a special pin that has "Charter Member" on it. Please send in your donation as soon as possible and I'll send you a membership card with your number on it.

Although you don't have to become a member to attend our convention at the Hyatt Regency in Memphis February 3 to 4, you do have to pay a registration fee of $35 per person or $50 per couple. Once you see what we're doing, we're sure you'll want to join.

Any money we take in over expenses will be donated to carefully selected projects. We have a host of good sheep management projects that we'd like to help finance in Canada, the United States and Mexico. But we need your help in raising the money. This will be our way of "giving something back" and perhaps giving our children and grandchildren the opportunity of enjoying the privilege of seeing and hunting magnificent trophy rams in their breathtaking, rugged and beautiful home in the high country of the wilds of North America.

We ask for your cooperation and support!

Thank you.

Dr. Roy Schultz
President, 1980-1981

Tell us about the start of FNAWS?

The Foundation for North American Wild Sheep was founded by dedicated sheep hunters who wished to give something back. Most had been to the "top of the mountain" and wanted "the experience" they had to be preserved for their children and their children's children. I see that same dedication in members of FNAWS today.

Jan and Dr. Roy Schultz during the 1991 convention.

The enthusiasm generated in the original gathering of 13 in Mt. Horeb, Wisconsin, poured over into the second meeting of 100 dedicated sheep hunters in Des Moines in 1975. At the end of the meeting, Lloyd Zeman took the podium and asked the gathering, "What should we do to continue the momentum?"

Those present chose Lloyd Zeman, myself, John Babler, Duane Smelser, Frank Bays and Dale Leonard to set up an organizational framework to get it done! For various reasons, Frank Bays and Dale Leonard, eventually drifted away from the organization, but not until they had left a considerable impact on its early years.

Lloyd, Duane and I first set up phone calls, then conference calls with Bob Housholder's Grand Slam Club. Bob had founded the club and he ran it nearly single-handedly. It is understandable that he wanted to continue the control. I now feel he did not realize the neccesity of establishing a 501-(c)(3) Foundation to raise sufficient monies necessary to work with biologists, government agencies, and sheep hunters to preserve and expand historic wild sheep ranges.

After Bob turned us down in being a Midwest chapter of the Grand Slam Club, we pursued a joint venture of being a sheep chapter of Safari Club International.

This proposal, too, was turned down by SCI's Board of Directors. Even though the clubs had common goals, there was competition between Lloyd Zeman and Mac (C.J. Mc Elroy). In the early years, there was considerable competition in raising of funds, especially donated hunts from outfitters. I tried to get the organizations together again when I was president. Representatives of both SCI and Game Coin were invited and spoke at the FNAWS Convention in Phoenix. Bob Housholder of the Grand Slam Club was also in attendance and was presented an award. Except for the Grand Slam Club, the other organizations have not continued this relationship.

These two negative events turned into a positive. It led us to organizing the Foundation for North American Wild Sheep in June of 1977 as a 501-(c)(3) Iowa Corporation with Lloyd Zeman, John Babler, Dr. Roy Schultz, Frank Bays and Dale Leonard as the incorporators. From this beginning, the Foundation for North American Wild Sheep was history.

Are you still active in FNAWS?

Yes. I have attended every convention since the inception, except for the one in Philadelphia. I purposely missed that one to make a point. I think I was successful. I'm actively working on the history. I was on the Board of Directors for a total of 12 years and never missed a board meeting. My present activity revolves around the Past President's Council. It is in our bylaws that past

Presidents are required to meet in a forum at the convention and give constructive input as to the long-range direction of the Foundation. At first the Board took this seriously and welcomed our inputs. In recent years, the meeting has been relegated to a "Cocktail Party" and no serious consideration is given to ideas of the past presidents.

In the early days, all directors and officers paid their own way to the meetings and the convention. I think we'd have a stronger organization and more committed people if this were still true. Many past presidents have this same opinion.

Is there the same purpose as before or do you think the tone is changing?

I think the purpose is still the same, which is putting sheep on the mountain by raising monies necessary to support research, habitat acquisition, transplants, disease control and state agencies to continue research and control harvest of the excess population. I think the purposes are as visionary as they were, but the way we raise money has shifted dramatically. Now the Foundation is a major marketer of permits from different states to support each state's game programs.

This helps sheep programs in those states and is good as long as the monies continue to stream in. It's my opinion that states should not depend solely on this money, because if suddenly the take decreased, their programs would be in serious trouble. Past presidents and board members could be used as a tremendous resource. They have seen what works and what doesn't work. The Board needs to listen to the legacy and the people who come before them so they'll not repeat the same mistakes. Philadelphia is a good example.

Although the relationship with the Grand Slam Club is currently good, the Board turned down an offer to incorporate the club into FNAWS for one dollar. They did not even consult with any past presidents or former board members. With the rapid turnover of members of the board, continuity is even more important.

Do you see a future for FNAWS as a growth organization monetarily or in membership wise?

I think FNAWS, unlike Ducks Unlimited, Safari Club International or the National Wild Turky Federation, which have literally hundreds of thousands of members, will be stabilized at 6,000 to 10,000 members. It has stayed fairly stable at 4,000 to 7,000 members over the last 10 years. I don't see much growth because of the species-specific nature of the organization. Per member, we raise more than any other conservation organization. The passion for wild sheep keeps our heart in it.

What advice can you give to future FNAWS members? If you could speak to a member in 2020 what would you tell them about FNAWS, and what they should do or not do?

Listen to the membership! Listen to the legacy of past presidents, officers and directors.

Have vision and follow your heart. Be proud to be a hunter and promote hunting as a tool for proper conservation. Make lifetime friendships. My good friend Keith Johnson served on the board for a short period of time. I truly respect and admire his attitude toward hunting and conservation. He was a great asset to the Foundation and brought common sense to the Board in the short time he served. He continues to be a strong supporter of FNAWS. I'm proud to call him my friend.

Should FNAWS become more politically active?

Our original direction was to put legislative monies with other organizations such as Wildlife Legislative Fund of American (WLFA) that has full-time lobbies in Washington. I met Danny Galbreath on a sheep mountain in Baja and convinced him to join FNAWS. He was a WLFA officer and convinced the Board and, rightfully so in my opinion, to let experienced people do our lobbying for us. It has been effective. With only 5,000 to 6,000 members involved in FNAWS, it is hard to convince legislators we have political clout. I think we are better off working directly in areas for the sheep and their habitat and joining with others who have the numbers for political clout. It's good to set the money aside, but let someone else who is a professional do the job. Historically, WFLA is professional and good at what they do.

How do you get young people involved with putting sheep on the mountain?

We need to educate and interest young people in hunting and hunting as a tool for conservation. If they hunt big game, they will eventually become sheep hunters. If they are dedicated sheep hunters, they will join and support FNAWS. After the first sheep hunt, you are either a dedicated dyed-in-the-wool sheep hunter or you won't be caught dead on another sheep mountain.

Pat Latham with Dr. Roy Schultz in San Antonio.

Jerry Christian
President, 1981-1983

How did you come up with the name "Foundation for North American Wild Sheep?"

First of all, as a sheep hunter you want to be completely removed from domestic sheep. We actually had people who believed that the sheep club was for domestic sheep — woolies — and that really infuriates hunters. We sat and sat and sat. Everything we came up with had to have "wild sheep" in the name. To meet the criteria of the government we also had to have the word "foundation" in it, so we came up with what we called "FNAWS" or Foundation for North American Wild Sheep. It was horrible, but we couldn't think of anything else. A few years ago when you'd say you were from FNAWS, they'd look at you like, "What is that?" What is a FNAWS?"

Jerry and Sandy Christian, left, with Don Mangels at a Saturday Evening cocktail Party.

The first real official meeting as FNAWS was the minute we finished in Dearborn, Michigan, with the Midwest Chapter of the Grand Slam Club. After that we made our checks out to FNAWS. When we went to Nashville, that's when we were officially FNAWS and we could say that. Before that, Bob Housholder of the Grand Slam Club had a real strong grip on that. He was the founder of it, and so we couldn't dare use that name. Lowell Baier, a lawyer, was a very instrumental man on the technical part of this.

Lowell told us we couldn't use this name because Bob could come in and say, "Hay, I want these monies because it's my club you're operating under."

We said, "Well, okay Bob, we'll operate with you but we want this money to go to sheep."

Bob said, "I'll give it to them. You can't."

That's when we said no, we will. That's why we created FNAWS.

Was there a particular event in wildlife management that made you say now is the time to get FNAWS started?

It was a string of events that convinced me that something needed to be done. My good friend and hunting partner, Roger Hill, met socially with dedicated hunters and conservationists like Lloyd Zeman, Doc Schultz, Duane Smelser and Jerry Waite. They told of going on bighorn hunts several times before even seeing legal rams, let alone harvesting a mature ram. They shared stories about risks of hunting in Mexico and told of the great pressure on the Stone's sheep of British Columbia from wolves and overgrown vegetation.

We would talk to the Wyoming and Montana people. We could see that at the rate they were disappearing, we needed better management and better disease control. We decided we better do something now.

Our first recipient of the Outstanding Achievement Award went to Lanny Wilson who was a big game biologist in Idaho. He had the same feelings we did. These wild sheep needed protection from domestic sheep, and we needed to put money into research. As it turned out later, we learned you could put certain medications into sheep to get rid of a disease that will kill wild sheep. We saw back then that a lot of historic ranges of wild sheep had disappeared, like the Badlands in South Dakota. We put wild sheep back into the Badlands and in the Buttes of Western Nebraska.

The last bighorn from the Badlands were no longer in existence, so we took the California bighorn, which is the closest genetic relation to the Badlands bighorn. This was a subspecies we put back into the Badlands. We worked closely with the wildlife biologists in the different states. We knew we had to start some place, so we went to the biologists. We knew the states wouldn't allocate funds to help the wild sheep, so we knew the hunters would have to carry the whole load. We could see our monies and multiply them with federal money.

I was in the cattle business and spent time in Wyoming and Montana. I was applying to Wyoming Fish and Game for a bighorn permit (of which I've never been drawn) and learned of the domestic sheep and the lung worm disease killing off bighorns. The economics of sheep ranching was losing battle, but the domestic sheep were grazing in the high mountain meadows of the Rocky Mountains. The game commissions were doing a good job, but the grazing leases were as old as the West itself and politically influenced, leaving their hands tied and budgets very tight.

We knew something had to be done and done quickly. With funds from our own pockets, we contacted outfitters from British Columbia like Garry Vince, Gary Moore, Gary Powell and Bobby Kjos who Doc Schultz and other people had hunted with. They were good friends. They were very helpful when we told them of our dreams of a "Sheep Club" and even though they were somewhat dubious, they jumped in head first and gave us mailing lists, charter memberships, personal time and, most importantly, they gave us credibility and moral support!

Personally, we made many trips to the western United States, Canada and the Yukon, joining them at their outfitter and game department get-togethers, telling them of FNAWS and what we were going to do. We would start to run into people in our search for hunters who had heard of us as a wild sheep organization that was really dedicated and every bit of the money they made went back to the sheep and not to a director. These hunters would say they'd like to be a part of this organization. D.J. "Doc" Thurston of North Carolina bought some of the first hunts and spend tremendous amounts of money. Dan Galbreath (Daniel M. Galbreath) would leave bids with us for $30,000, $40,000 or $50,000 for paintings because he knew the money was going for sheep. It was hard to turn an artist's product into real cash money. Artists are a special breed of cats. The government looks at them in a funny way. They don't get the deductions like the hunting permits get. We were working on the same theory as Ducks Unlimited where we would get a double deduction. If you buy this hunting permit you can deduct if from your taxes. We're a non-profit organization so there is actually no tax on those dollars. We thought this was the route to go.

Another good member who sought us out was the R.J. and Hildur Harris family. He purchased our first Grand Slam hunt and numerous hunts since then.

It started growing slowly at first. We had our bumps. We, our board, went out and collected a lot of dollars. You notice that I don't say "I." In those days we didn't care if you were president or secretary or the youngest board member. We all seemed to work together. We never agreed, but we always worked together. We had a common goal and we didn't think we were always right.

Is there a theme that developed as you were progressing with the organization? Did you have an underlying theme of where you wanted to go? Did you have a targeted population of sheep to focus upon?

We were in desperate need of a theme and logo all along. We spend most of our energies on making FNAWS a recognizable trademark. The old saying, "A picture is worth a thousand words," prompted us to select the world-famous Chadwick ram painting by Gary Swanson. As you know, our motto, "Putting Sheep on the Mountain," came from Dr. Jerry Waite and his Board of Directors. We knew exactly where we wanted to go. We wanted our children and grandchildren to have the opportunity to pursue the magnificent bighorn rams of North America — plain and simple!

We didn't target any one species but since the Dall's sheep were very stable, we concentrated on the Stone's sheep, knowing that other than a small population in the Yukon, the majority were located in northern British Columbia. These outfitters had their act together and we figured if they supported us, the rest would fall in line. What really convinced me we were on the right road was a statement from Gary Moore.

He said in front of the country's top biologists, "Get the exploding wolf population under control and burn off some overgrown mountainsides and in a few years I'll produce a population of harvestable rams that will compare to the 1960s." And, as we now know, that is exactly what happened!

Garry Vince added, "You know what needs to be done. We know. Now we have to convince those that make the decisions from an office in the government and the misinformed animal rights groups."

This is a very touchy subject when you're dealing with the balance of nature. Here we were getting right in between the wolf and the wild sheep. This was in the 1980s. FNAWS members were instrumental in a change of policy on predator control. We couldn't change it. They had to do it. We had no political clout in Canada. Canada was its own country.

When did the staff get started?

When I was president, our staff consisted mainly of Jane Babler as secretary working out of donated space at the printing shop she and John owned in Minneapolis. These were two very dedicated people! We tried to pay her with what money we had. Also, board members did a lot of their own mailings, etc. Then after we started making a few dollars, we got our own office. After numerous land donations in the Cody, Wyoming, area, we chose to rent a building in Cody right down from the Buffalo Bill Historical Center. This was in about 1982. I was in Wyoming at our office numerous times in the early to mid-1980s. My wife, Sandy, and I helped in the office along with Barb Reed, B. Joe and Lori Coy and Don Schmalz. Lloyd Zeman, Basil Bradbury and Director Arnie Johns were paid staff during our years of establishing a home office.

I think after 1985 FNAWS hired a gentleman from the east coast, Ted Bryant, who was quite popular with Ducks Unlimited. He lasted a year. He was a hunter, but not a sheep hunter. We found out there is a difference. There are elk hunters. There are antelope hunters.

There are whitetail hunters — now there's a dedicated group. This was when we learned we needed a sheep hunter.

We were scared to have someone in there like C.J. McElroy of Safari Club International because of their tremendous ability to earn funds but to keep part of it for themselves. We didn't want any of that, so that's probably why there is a lady in the office. That may not be the exact reason, but it boiled down to the fact that we wanted someone to be able to take care of the office. We wanted the money going to the sheep. We were adamant about having somebody in there who would work for the common goal and not for themselves.

We always found that when we got a dedicated sheep hunter, nine chances out of ten, he was a good man to have on the board. Tom Saad took us from shoebox accounting to accounting for everything. He's a very good man. He was the one to get us up to date with our accounting. To give you an example, I was president-elect in Phoenix. The convention was over. Lloyd had a few medical and personal problems at the time, so everyone packed up and left Phoenix the day the convention was over. All the proceeds from the Phoenix convention, a convention that was quite successful, was in a shoebox. We're talking thousands and thousands and thousands of dollars, so I asked for the key to the safe so my wife and I could count the money. My wife is a perfectionist and an ex-banker, not a president but a clerk, so my wife and I sat there and counted this money. We put everything in the right stacks, and that was how we were running it.

We're talking a multi-thousands of dollars in cash! The money was very easy to steal. We needed proper accounting. Computers weren't really being used then. We were running it like a bunch of farmers and sheep hunters, which we were. We were so dedicated and we trusted everybody. A guy might say he'd like to make a donation and he'd give us $1,000 in cash. He'd receive a receipt on the back of a napkin. The member would take that and send it into the Sheep Foundation. Nobody in the world knew about this donation except the man and the member. We depended upon each other's honesty, and Tom Saad really got us on the right track. He lost friends by doing this. He split the board on this. The board said they could do this and not do that and we don't need to that. We needed the accounting. Tom Saad was the one who aimed us in the right direction.

When was the first permit given to FNAWS to auction?

We always got a hunting permit from the governors of Wyoming and they have never asked for money back. The first permit was from Wyoming in 1980 when Lloyd Zeman was president. We even blotched the first permit. The governor from Wyoming, Ed Herschler said all we had to do was come to the office and get it signed off. We sold it to a precious medals man from Texas (Dr. Vogt), but we couldn't get our representative in Wyoming to go down to the governor's office. All he had to do was sign that paper so he could go hunting. We sold it for quite a bit of money ($23,000), but he sat in camp. He never signed for the permit.

The next year, Lloyd Zeman, Doc Schultz and I contacted Governor Ed Herschler. I knew the governor from the cattle business. He said we could have hunting permit.

When we asked how much he wanted back, he said, "You keep it all. If we need it real bad, we'll come to you."

Every penny we've collected for that Wyoming permit has always come back to the Sheep Foundation, 100 percent, no strings attached. Now days, all of the permits have strings attached, except Wyoming. We can spend that money anyway we want.

My favorite story on permits involves Dan Pocapalia, Doc Thurston, Sam Pancotto and other hunters from around the world who came to the FNAWS conventions because they liked our group. I was running the auction with Harvey Kadlec. I had the Wyoming permit and one of the outfitters said he thought he could get another one if we could sell it. I told him I knew I could sell it. I was pretty brassy then. I think it sold for $50,000 or $51,000 (the permit sold in 1983 for $50,000 to Dan and to Doc Thurston for $49,500). I looked down at Doc Thurston when he bought it and asked him how many permits he would like.

"I'll take two," he said.

He was a real Southern gentleman. Doc Thurston bought one and so did Dan Pocapalia. Isn't that interesting how a man will sit there and ask us to auction the permit off again, maybe for $30,000 or $35,000? This just goes to show you that these people like Dan Pocapalia and Doc Thurston weren't interested in saving any money or being cheap. They wanted to help the sheep. I think Dan was given an honorary life membership, but he never really joined. You could see these men were dedicated. They would tell us they wanted to give us some money.

How do you find those kinds of bidders?

There was a taxidermist in Cheyenne (Wyoming) who knew the governor. The taxidermists were named Roy Ostermiller and Joe Nevins who owned Frontier Taxidermy.

Roy said, "Jerry, if you ever get out here, I'll let you use my shop and I'll bring in some guys. There's a guy from California who has hunted a sheep but he wants to get a big one, and he has the facilities to do it."

So we started having what we called the "Governor's Party" in Wyoming. This was never funded by anything from the Sheep Foundation. We funded this on our own. For instance, Tom Hinders went to Winchester and asked them for a fully engraved rifle to give to Governor Ed. We gave the rifle to his wife, Casey, because if you give it to him it stays with the state of Wyoming. This was in the late 1970s or early 1980s. This was really the first time, until we received the Arizona permit, that we had anything besides Wyoming to work on. Wyoming really put us on the map. You really have to boil it down to the generosity of the governor of Wyoming. We were also doing a good job for them.

Where do you see FNAWS going in the next millennium?

I would be foolish to think that FNAWS will grow and get big. It's not. Dedicated sheep hunters are very few, and the sheep population is growing. We're never going to be a large membership but we will be a dedicated membership made up of dedicated hunters. The funding is probably the secret to this whole thing. We have guys like Pete Cimellaro who is a very intelligent man who turned around the referendum in Arizona from a minus to a plus and saved a certain type of hunting. That's what our membership is all about.

Our numbers will always remain stable. If we ever reach 7,500 members, that would be the max. The average guy working 8 to 5 cannot afford to hunt sheep. They can if they keep putting in for a permit but your chances of drawing are one in a million. We still need to keep people like that and not shut them out. We need the young man who chops wood for a living to draw a Wyoming or Arizona permit. We'll always leave a permit for him. We won't let it become the way of Europe where the king was the only man to hunt. We don't want money to be king. We need the $40 and $50 memberships in FNAWS, but to fund projects like Hells

Canyon we have to go to another source. Just think about it. That's a lot of money to ask from our membership every year, and now their budgeting for $10 million in Hells Canyon.

If I tell that a guy earning $12, $15 or $20 per hour, he wouldn't even want to join if he knew about a $10 million project for wildlife. We're going to have to go to the big hitters. They're not going to give the money to you if you are squandering the money. The other aspect of this is convincing people with large sums of money that they should invest in the projects of FNAWS. It would be interesting to see how many of the Fortune 500 companies have owners who are even hunters. I'll bet there would be very few.

A good man is Bob Petersen, former owner of *Petersen HUNTING* magazine. He's putting money in and buying permits from us. He's been here all along. He's been watching us like a hawk. People who are successful don't throw their money away. They got rich for one reason. They watch it. When they study and see that we're doing the right thing, and they do some calling around, they know we have a good reputation. We need to keep it that way.

Our biggest problem as we look into the future we have to find the high ground that will keep a board of directors that is dedicated to sheep They have to have time. They can't be short on time. They'll donate time that's just as important as your business or your job. We still need some people of a financial background who can help us make more. Can we invest this money? We need to go to somebody besides our members to get this money. Maybe we can take our money and have somebody who is very financially sound help us direct this money.

A big donor will be mentioned in the "LA Times" or the "Wall Street Journal" or the "New York Times." If the donor is president of a Blue Chip company with 99.9 percent of the stockholders anti-hunting or non-hunters, that donation could hurt the business. If they donate something it must be crystal clear that it is going to the right place and that it is not going under the table. We need a board of directors that is beyond reproach and our staff in Cody has to be educated. We have to keep advancing.

In closing, I remember what Red Duke told me.

"Jerry, when you start mentioning names you inevitably forget someone."

In my 25 years with FNAWS, two of them as president, I'm sure that number is well over 100. So, I apologize to those I've forgotten to mention and to thank you all for letting me serve you and the wild sheep. But most of all I want to thank my wife, Sandy, who with me has never missed a convention, has maintained things on the farm and has kept our four children on the straight and narrow path. With her support, help, encouragement and patience, it has been a big part of our past and with the growing interest in hunting from our youngest son, Jason, we look forward to the pursuit of bighorns with both rifle and camera in the future.

1990, Jerry Christian, right, with Dean Picl, left and Dan Pocapalia, center.

Duane Smelser
President, 1983-1984
Life Member Number One

Letter dated January 17, 1994, to Dr. Roy Schultz

I will first summarize our beginning. We were members of the Grand Slam Club, which was a personal function of Mr. Robert Housholder from Arizona. Mr. Housholder produced a newsletter for his members. Mr. James Froelich wrote a note in Bob's newsletter sometime early in 1974 asking the members if they would like to meet in Mt. Horeb, Wisconsin, to socialize and swap sheep stories.

The dates picked for this outing turned out to be the worst possible. It was not just a stormy weekend. It was a blizzard the morning my wife Doris (Billie) and I were leaving home. I plowed my driveway but by the time I got Billie to the car, my driveway was drifted over again, so I re-plowed it (approximately 12 inches of light, soft snow), jumped in the car and started our trip from Detroit, Michigan, to Mt. Horeb.

There were times on the expressway that it was a complete white out. I had to guess where the road was, but we got to Mt. Horeb. Coming home was a different story as we came through Chicago. The snowbank plowed to the side of the road was higher than our car and disabled cars were everywhere. Before we got to Detroit, the police closed the expressway and we all were put up for the night by kind, concerned people.

The blizzard kept us confined to our hotel (November of 1974), which was all right, as we had just come to swap sheep stories anyway! There were 14 sheep hunters at this meeting (Jim Froelich reported 12 in the Grand Slam Club newsletter), including Mr. Lloyd Zeman who I had sheep hunted with in Mexico.

Our group agreed to meet again the next April in Lake Geneva, Wisconsin, where 46 were present. Our group met again in 1976 in Des Moines, Iowa, and was chaired by Mr. Lloyd Zeman with 100 people in attendance. Here we decided to call ourselves "The Midwest Chapter of the Grand Slam Club." As our group had continued to grow, and there was no reason it would not do the same in the future, we needed a home. Lloyd asked for the podium.

"What should we do? What is our future?"

So the floor nominated seven people to explore our future and report back at our next meeting, which would be in Detroit (Dearborn, Michigan). This meeting was chaired by Mr. Frank Bays and co-chaired by Mr. Al Perry and myself, Duane Smelser. Nominations at this time to explore our future were Lloyd Zeman, Duane Smelser, Jim Froelich, John Babler, Dr. Roy Schultz, Dale Leonard and Frank Bays.

We approached Mr. Robert Housholder of the Grand Slam Club, asking if we could get a 501-(c)(3) tax status so we could raise funds to support a conservation effort for the Grand Slam Club since we were all members. His answer was, "No."

We then asked C.J. McElroy of Safari Club International (SCI). He sent two men to Chicago to meet with us — us being Lloyd Zeman, Roy Schultz and Duane Smelser (John Babler was also there). They being Keith Bates from Chicago and editor of the SCI magazine and Jack Leads from Minnesota who was the SCI Conservation Club president. We discussed the possibility of our sheep group becoming a sheep chapter of SCI. They took this suggestion back to their board of directors, who turned us down.

The Detroit meeting was chaired by Mr. Frank Bays, who did his homework getting the Director General de Flora y Fauna Silsestres of Mexico City and Senior Mario Luis Casio to attend. He also brought along four desert sheep permits for our auction, which brought sheep hunters in from 1,000 miles away. Frank also went to New York to pick up the Chadwick ram, the largest ram taken on the North American continent, which improved our ram display tremendously.

The Detroit fundraiser was a success with many new members plus money to fund conservation or put sheep back-on-the-mountain.

We then wanted to incorporate our club, naming it, "The Foundation for North American Wild Sheep." Incorporators were to the best of my memories: Mr. Lloyd Zeman, Duane Smelser, Roy Schultz and Frank Bays (also John Babler and Dale Leonard starting on June 17, 1977). Mr. John Babler's office in Minneapolis, Minnesota, (actually St. Paul, Minnesota), became our home and our office. John was generous enough to offer it and Lloyd was just across the border in Iowa. He could, as president, drive to work at the office when necessary.

My term as president of the Foundation came back to back with my presidency of Safari Club International. I was very active with Mr. Ken Fanning, Keith Johnson and Ron Sumervill [sic] from Alaska on Alaska land legislation, which closed more prime sheep-hunting acreage than has ever been closed in all of North America. Sheep live in mountains and mountains are pretty, and where there are mountains, the legislation made national parks. This drastically curtailed the sheep hunting in Alaska, but being active allowed us to help draw the boundaries that were used. We reduced the acres from 46 million acres down to 24 million. It did include such notable big sheep areas as the Bernard Logan Glacier in the Wrangler [sic], St. Elias, Lake Clark, Denali mountain ranges, the Gates of the Arctic, and the Brooks Mountain Range that was the largest concentration of Dall's sheep in the United States. Along with Mr. Jim Glass of the Wildlife Legislative Fund of America, I, Duane Smelser, testified as an authority on wild sheep habitat before the U.S. House of Representatives and Senate committees.

Accomplishments, notable events and highlights of my presidency:

* October 5, 1983, Gary Powell was killed in an airplane accident. Gary and his wife Olive were strong supporters of FNAWS for many years.

* We played a large role in opening 12 million acres of parkland in Alaska to Park Preserve so hunting could be allowed.

* FNAWS had a role in getting a sheep management bill in California.

* Horn-marking pins in Montana.

* FNAWS put the wheels in motion for the building of our educational center and office 18 miles west of Cody, Wyoming. Great plans that never happened, as changes developed.

The Town and Country Hotel in San Diego was the site of our 1984 convention. Our keynote speaker Saturday evening was an old friend, Secretary of Interior James Watts. As president of Safari Club International I had interviewed Secretary James Watts. I also attended a special breakfast for a few friends in Washington with Mr. Watts. The discussion was of Alaska National Interest Lands Conservation Act. Millions of acres were closed off to sport hunting. Moreover, the act locked up 55 percent of the big sheep population in Alaska.

Those prime sheep habitats will have to do without the benefits of flora and fauna studies to protect them from disease. Mr. Watts was in favor of park preserves over parks as the first allowed sport hunting and flora and fauna studies. We also discussed the 1981 John F. Lacey Act revision (named for early 20th century United States senator, friend of Theodore Roosevelt and wildlife conservationist), especially the provision that could trap an American sport hunter returning home who may inadvertently have taken under the importation violations to federal felonies ($20,000 fines and/or a five-year imprisonment sentence) regardless of the violation's technical nature.

Now, back to our San Diego convention. Working with our friends in Alaska, we put together 25 donated sheep hunts for our record-breaking convention. My wife Doris (Billie) Smelser hosted the first Ladies' Fashion Show and Auction. Her models included Joni Saad, Past President Tom Saad's wife; Jill Fanning, wife of Ken Fanning of Alaska; also male model Skipper Bettis, an outfitter, and others. The FNAWS San Diego convention set records held at the Town and Country Hotel. It grossed in excess of $1.2 million. The auctions generated more than $882,000 and the convention netted roughly $745,500 — all figures were new records!

Clark Engle, another friend from Anchorage, Alaska, who was a real supporter of the Foundation, was killed in another plane accident.

Tom Hinders
President, 1984-1985

With the help of Dr. Roy Schultz, Susan C. Reneau was able to meet Tom at his home in Ames, Iowa, in July 1998 for an evening of conversation, salmon steaks and grilled pork chops with his 16-year-old son Matt and Doc Schultz. The following interview about his involvement in FNAWS in the early years is recorded below.

Tom Hinders at work, 1998.

Tell us about the shirts and other items you sold to raise money for FNAWS in its early years? I understand you were instrumental in starting the merchandizing for FNAWS.

We did shirts, caps, ladies' purses, aprons, jackets, towels. We created embroidered decorations on them from Gary Swanson paintings. That included the Chadwick ram, a bighorn, a mallard duck and a lion. Roy Schultz invested $10,000 in this company and I invested $10,000. The name of the company was "Yavonka." Erv Netcott also invested $10,000. We weren't making any money and we had a corporation. The reason we started doing these things was because FNAWS didn't have any money to pay for any of this. I guarantee we never made any money of this. All we ended up with were a few towels. I still have some purses, shirts, towels and aprons somewhere. Roy Schultz has a denim vest.

How did you pick the material? Where did you go?

Jerry Waite told us he could sell all this, so we met some buyers. Jerry said we could use 30 dozen of this, and 20 dozen of that, and we bought all of that. We bought it all. That's what some of the $30,000 went for. We sold everything with a percentage over our cost for everything that was sold. Don't hold me to a number but I think it was 10 percent over our cost. Jerry Waite would say he wanted three dozen towels or 50 dozen to me. Roy Schultz only put the money in and was a silent partner. Roy recognized the value of the Gary Swanson paintings and wanted to help FNAWS raise money.

I was dealing with a lady from Tucson, Arizona, named Yavonne Pope. I'd fly to Tuscon. We produced these very special items to be sold only at FNAWS conventions to raise money for projects. I never made a dime. (Dr. Schultz said they ended up donating the whole corporation to FNAWS who continued the venture. At least we were able to write it off on our taxes.) There was a catalog showing these items. Now they are collector items.

I met Yavonne Pope at one of the Safari Club International shows where she was displaying all the fancy jackets she had made by hand. This is how it all came about. She was an artist. I walked past her booth, saw her stuff, and said, "This is nice." This was about 1983 or

1984. I said to her, "I'm going to take you to dinner." And she said, "Great." So I took her to dinner and told her, "First, I'm going to make you some money because you are starving to death. You can't sell enough of these things and you're too talented!" I asked her what we could do to produce her jackets in mass. Then she started talking to me about a new embroidery machine just in from Japan called "Tigima." I told her to find out more about this machine. She told me she would have to reproduce the Gary Swanson art in different colored threads and number every thread color. It took weeks and weeks and weeks to do this. I paid her to reproduce Gary Swanson's art in thread to make the embroidery that went on the jackets, hats, aprons, etc. The only thing Gary was paid was $1 per item as a royalty for the use of his artwork (Editor's Note: Subsequently Gary donated the copyright for the Chadwick ram painting to the Foundation for its continued use as our logo.)

(Roy Schultz speaks: Tom Hinders gained more for us on donations than anybody else ever had before. Tom was able to get donations from outfitters and twist their arms long before he became president.)

How did you get involved with FNAWS? What was your motivation?

I booked a sheep hunt with Earl Boose in Des Moines and that's when I got started in FNAWS. There was Bill Butler and about two or three other outfitters were there. I was going to hunt sheep, that was all. I booked a hunt with Earl Boose. My back was bad. I had back surgery. I ended up in Canada with back surgery. I called up Earl and told him I couldn't go because of the back surgery. Then at the New Orleans convention in 1982 it was very controversial.

The Board was then expanded to 17 people in 1982 and I was elected as a member of the Board. That's what happened. The shirts and things started after New Orleans.

Now I'm in power and I have a job to do. Right? I serve on the Board. I became president after Duane Smelser. (Roy speaks: "Smelser was a little controversial and I tried to protect him because he was president of Safari Club International at the same time. The Foundation was very much against Safari Club at that time because they had turned us down. It was one of those interim deals. I said Duane does a great job, but the Board said if you missed two meetings, which Duane had, you're out. Duane had a really good excuse and I said they should not put him out because he was a great leader and he'll help us out.")

What else did you do for FNAWS?

I followed Duane. I served as president-elect and I did everything for him. I was THE MAN. I got the hunts; I traveled; I met; I talked; I cheered; I raised funds; I was convention chairman. Doc Schultz was there then, too. I went North and raised 32 fully-donated sheep hunts. That's when FNAWS got all the money. Never happened before and will never happen again. This was when Duane was president in 1983. I also chaired all funding committees. I read every proposal for funding, everything. I dealt with every biologist. I read them all.

Tom Saad
President, 1985-1986

Tell us how you first got involved with FNAWS?

My interest grew from my involvement with the Arizona Desert Bighorn Sheep Society. I was president of that group in 1967. I became aware of FNAWS when it was brand new around 1980 when it was becoming a little more well known. My life membership number is 369 so I was pretty early in the membership. My interest as I became aware of FNAWS was to try to bring them together with the Arizona Sheep Society to do some kind of joint partnering for desert bighorn sheep. I think this was one of the first alliances that was created between the two groups and it became a model for FNAWS in the future for linking with all the other groups.

1989, Past President Tom Saad, right, confers with Howard Pollock.

Tell us about the "Betty Lou" water project?

It is in the Plomosa Mountains, incredibly good sheep country. In fact it was where my wife took her ram. Part of our joint get- together to get know each other was this water hole concept. That's what our local sheep group was known for. That is, the water hole projects. We wanted to introduce the FNAWS people to a water hole project and show them firsthand how it was done and what was done with the money, once it was raised. That was 1981 when FNAWS came to Phoenix for their early convention.

An interesting little aside. As I was driving into this water hole project on that weekend I was coming up a dirt road and it was pretty rough. I saw a Mercedes car parked off to the side of the road and there's a fellow there. I pulled up and it was Dan Pedrotti. He had driven his car as far as he could go, so I picked him up and hauled him in the rest of the way.

The Betty Lou project was the first water hole project we ever did with FNAWS. There were about 50 people, split between the two groups. The local group probably had 30 people there; all the officers I think from FNAWS were there including Lloyd Zeman. This was an excellent joint project and a great campfire discussion about the possibilities of what we could do together in the future to help wild sheep, especially desert bighorns. In my judgement the desert bighorn are the most threatened of all the species. If we lose the opportunity to hunt any wild sheep it will be the desert bighorn. They are in such a fragile environment. The habitat is shrinking beyond belief.

What was your progression of activity in FNAWS to get you to the level of Board member and president?

I went to the Houston convention in 1980 along with another fellow from the Arizona Sheep Society, Dean Bowdoin. As a result of the Houston convention and having met all the officers there was an interest to come to Phoenix for a convention, so I volunteered then to head up the convention effort in Phoenix in 1981. I really jumped in with all four feet, did I ever, and not really knowing what I was doing or what I was in for, but I did. I became the convention chairman. I got the local group to handle all the logistical things that needed to be done. It was quite exciting, really. I think it was the highest gross they had ever experienced to that point in time. I think it set the stage to show the possibility of these major fund-raising convention kinds of format. I think our gross was more than $750,000.

I was elected to the National FNAWS Board of Directors in Phoenix at that point in time after serving as the convention chairman in 1981. I served from 1981 to around April of 1986. I was elected president in 1985 and served as president just before Dr. Red Duke.

What activities were going on and what were you proud of during your time on the FNAWS Board of Directors?

Soon after I joined the Board there was another little turmoil on the Board that I won't go into in any depth so I was elected treasurer. I was elected at a meeting in Minneapolis-St. Paul around 1981 after the New Orleans convention when I was elected treasurer. That was in our early history and what you found was that our finances were very elementary in terms of how the money was being handled and where the investments were going. It was very simplistic. There were no controls of any kind. Because of my background and appreciation for working for government all my life, accountability and control was Number One. That's when I really started to make a real contribution to FNAWS.

I had a professional concern about tightening up the financial reins on FNAWS and I have a pretty strong work history concerning administrative details and how to run a business or any kind of an agenda-producing operation. I'm good at getting business items worked through an agenda. I've attended hundreds and hundreds of professional level board meetings, which meant I knew how those types of meetings worked.

What did you instigate as treasurer of FNAWS?

The first thing I did was to maximize investment earnings because we had a lot of cash coming through. Previous to 1981, the money was put into a bank in a regular savings account for the most part or Cds (certificates of deposit) or in a cigar box. They had a lot of cigar boxes, that's for sure. We had kind of a little investment committee group and then we aligned ourselves with an investment firm. There was a bank in Cody, Wyoming, we worked through for quite some time but I don't remember its name. It was actually like a brokerage function through the bank for a while and then we went to an outside investment firm. They invested the money in low risk kinds of things like government securities, as I recall, but certainly better than passbooks or CD rates at regular banks. These were low risk investments that had some kind of a warranty or guarantee to them. They were always pretty secure and no speculation — FDIC insured.

Were you ever worried about what was going on with the savings and loans at that time? What else did you do as treasurer?

No, we really didn't. It just wasn't a factor. In fact we stayed away from them because they were getting into problems. The other thing that I did that helped FNAWS throughout the years was to establish what I call the "fund accounting approach or system." As part of that system, we had dedicated revenues that were assigned to key functions that were in place. The office operation had to be self-supporting with its own sources of revenue.

How did you get the office operations self-supporting?

What we did was assign the normal membership dues as dedicated to support the operation of the office. Then there were a few smaller raffles that were also used to support the office, too. This was to support the salary of Lloyd Zeman and a secretary and Basil Bradbury. Brad was rather pricey. He even started out pricey. His salary was even good for now, let alone 1985. One of my goals in the evolution of all this was to upgrade the overall quality of the magazine so we were willing to pay what it took to get Basil Bradbury in there, knowing that he could do it. Brad was brought on board in early 1984 to get an annual done, too. We wanted to bring someone on board who could make the magazine go from a one-color newsletter up to a real magazine.

My philosophy, and I think the philosophy of everyone else on the board, was that a good quality publication was necessary to cement the membership together and give them something that was tangible for their involvement and their contributions. We saw that as a key marketing and recruiting tool. It actually worked very, very well in the early stages.

Were you still president in early 1984 or was that Tom Hinders?

That was still Tom Hinders in early 1984 and I came in after Tom Hinders and tried to carry it to a level that we all envisioned it to be. I think we did in fact reach that level before I stepped down. In fact I was looking at those old issues. One of the ones when I was president was a very nice, slick production that we all hoped to get through.

Do you have any comments about the two properties in Cody that at one time were slated to become the headquarters?

I supported the purchase of a pre-existing structure. We got into a very extensive evaluation of building our own headquarters on the donated property. One of our Board members was a wealthy developer who had his own architect giving us different design concepts without charge. But, what happened was we got up to about $1 million for what this new building would cost to construct and so the Cody property was actually a house at a steal kind of sale price (around $125,000 to $150,000, in that range), which was down from the $1 million that we had talked about in terms of building our own headquarters outside of town. It was a beautiful site with the mountains in the background but my feeling was, and I was successful getting people off the idea of building a new headquarters, to get into a nice building at a price we could afford and keep it in Cody. I'd like to think I was somewhat instrumental in making all that happen.

When did you put a sales contract down on the building that is currently the headquarters?

I think it was in 1985 during my presidency. It was all such a big issue and we got derailed there trying to consider Boise and trying to build our own building. As I recall, when I evaluated the whole thing the best opportunity was to stay right there where we had our roots and where we could get into a building that we could afford. That all happened in 1985.

You also had some staff changes going on in 1985. Can you tell us about all that?

It was during my presidency that Lloyd Zeman exited for a variety of reasons. Actually Tom Hinders and Arnie Johns were sent to Cody to run the operation and to stabilize things. I remember all that pretty clearly during my term.

They did an incredible job because it was a highly explosive situation. Lloyd was a very strong personality. There was some concern that people might even be shot, quite frankly! Hinders and Arnie were very nervous. My God, they had to go to Lloyd's house and repossess Foundation property. It was a very tenuous and a scary situation. We were all sad about the things that occurred, but you kind of had to be there to understand the progression and why we felt we had to take the action that we did. It is somewhat akin to removing Bill Gates from Microsoft. Lloyd Zeman was so much a part of FNAWS. He was FNAWS, no question about that. I know we all have our regrets about past things and they stay with us a long time.

This was a very tumultuous time for FNAWS when you were president. What are your reflections now that you look back on it from a distance of time?

This was a pivotal time in the history of FNAWS. It represented a change of direction there from what we had been to what we had evolved to and to try and get it to the next step. That is, more of a business and fund-raising organization rather than just a "good old boys" hunt club with a few dominant personalities running the show. This was a sense of maturity. It was a growing pains thing. Lloyd's personality was so wound into the fabric of the organization that the perception was that he was controlling everything and all the decisions. People thought he was the puppeteer of the Board and the Board followed orders from him. After a while the Board resented that kind of feeling from the membership. A lot of that feeling was well-deserved because Lloyd kind of just took control and did a lot of things. I don't mind having this in unless you think it is too antagonistic. (No, it's not.) This was an honest reflection about how we felt. I think what I just said explains what we did and why we did it. There may be people who don't agree, but for those of us who were involved I think what I just said encapsulates very well why we did make the change.

Why did you make the change?

It was more a matter of moving away from a dominant personality running the business to more of a corporate board philosophy where the elected board members were free to make these decisions without always running counter to our paid executive director.

Were you involved with Ted Bryant who was selected as the executive director after Lloyd Zeman?

Yes, I think I was president when we hired him. He was a nice man but I think he was a little overwhelmed. It is important to note that this Foundation Board has always been comprised of strong-minded leaders. Mixing that many together and then trying to work for that many different kinds of individuals is difficult. The thing we finally insisted on that he report to the president and get individual board members away from direct contact so that he wouldn't be pulled in 10 different directions.

What were your impressions of FNAWS after you started fading out of the presidency and turning the leadership baton over to Red Duke?

The things I look back on in terms of my accomplishments are some of the things we have discussed. The membership was in excess of about 6,000 when I transitioned out and a lot of that was due to some of the things that I tried to accomplish, such as elevating the magazine as a marketing recruiting tool and beginning some innovative membership recruitment ideas and recruitment drives. As I transitioned out, I thought most of the terrible things happened during my regime and it was all behind us.

When Red came, he had kind of a nice clean slate and a new attitude. We could all move forward in a very strong, positive way. I had a lot of regrets and concerns about the problems we had but felt good that at least we got through them and we survived and now we were on the right road.

What advice do you give to the members of the new millennium?

There is a great amount of competition in the conservation field in terms of fundraising. Back in our early days, we were one of the few that existed. Now we have so many different organizations raising funds to support their own causes. There is a lot of crossover. I think together we do a lot of good, in that regard, but sheep hunting is such a specialized endeavor. We share a certain bond, but there aren't a lot of us. My concern is that we won't be able to sustain our ability to raise meaningful dollars into the future to do the kinds of things we need to do.

It certainly takes incredible dedication and commitment from the Board members who are running the organization. In terms of the Board members' involvement, they must have oversight over the revenues they are bringing in and to make sure we maximize the return back to the resource (wild sheep and its habitat) and not draw off monies to support the magazine and the administrative overhead. Those are my biggest critical complaints.

My advice is to always be conscious of the overhead kinds of money that is going out that is not supporting conservation, education and research in a direct way. They need to do more volunteer kinds of things. In our early history there was a lot of volunteer effort in terms of running the office operations. I guess what I remember is that Board members made an incredible commitment to go out into those outfitters' camps and these various meetings of outfitter groups. They made an extreme effort to get fully donated hunts. Now they just sit back and expect these things to come to them. Our percentage of payback is very high.

Another thing we've gotten ourselves into are these special state and provincial auction permits where we get very little revenue directly back to support the entire FNAWS operation. Looking back, that was a giant, terrible mistake, quite frankly. We should have structured the special permits better. What bothers me is that some of these states get this money and not all of it is spent on sheep management. It gets a little defused and harder to track in some of these individual states. The bottomline is that we need x number of dollars to operate FNAWS and should be capturing this revenue for our overhead. Legislatures expect 100 percent return back, and I understand that, but FNAWS has overhead. State programs would not have the money from the auction unless FNAWS existed. States need to be reminded of that.

Any other advice?

Certainly the encroachment of humans on sheep habitat, especially encroachment on desert sheep populations, will continue to be of concern. It is a significant problem and I'm not sure how we can hold that off. I think civilization is on a course that is somewhat irreversible. There needs to be some innovative kinds of ideas here in terms of helping these wild sheep transition to smaller habitats. To do that, we need to control predators and disease a lot better than we have done in the past. I think that is one of the big challenges. How are we going to manage these sheep and continue their survival with smaller habitats? In one way we have got to deal with the predator situation and also with the disease situation that decimates the sheep populations like Hells Canyon. Once the disease takes off, there is nothing you can do.

Dr. James "Red" Duke
President, 1986-1987

What motivated you to become active in the Foundation for North American Wild Sheep in the first place?

My first involvement with a wildlife conservation organization was with the Safari Club International chapter in Houston in early 1973. I went to the first 10 meetings of Safari Club in Las Vegas. In 1976 I killed a really big Stone's sheep and somehow heard about this bunch (FNAWS). I submitted it and they gave me an award. I beat out Jerry Waite. We first met each other in Minnesota or in Memphis. I really got involved with FNAWS in Memphis.

Master of Ceremonies, Dr. Red Duke, during the 1990 Saturday Banquet.

When did you start on the FNAWS Board?

I started on the FNAWS Board around 1981 or 1980. Probably the other event was in 1979. In 1981 I organized the Texas Bighorn Society at the FNAWS meeting in Phoenix. About 12 guys got together and that's where the Texas Bighorn Society came from. During that time period there was all types of craziness going on within the Foundation Board. Those early guys, the past presidents, were feeling left out and wanted to have more control. Probably the most important thing I ever did for the Foundation was to make a motion that no past president would ever be on the Board again. That saved me from getting back on the Board again, and it also demanded that new blood would keep being infused into the organization.

I became president in 1986. I became president-elect the year before. The only reason I agreed to be president was because for several years Dr. Roy Schultz and I had reviewed all these applications for money. He and I are maybe pseudo-scientists, but not sheep biologists. It became evident that over and over again this organization had been funding the same thing, which was stupid. I could see the idea that we would develop a center for sheep studies. My thought was that we would use that center as a place to get the best brains together to ask the important questions and then take what money we had and fund appropriate studies with RFPs (request for proposals). We'd send out these RFPs and fund it or get funding from other places. We sent RFPs to 23 universities and got back a bunch of proposals. I got some pretty hotshot academics in Houston to help me pick out three people who had nothing to do with the whole process who, in turn, picked out three biologists to tell me the best three projects that existed. They picked three projects, and I thought I could finish this in two years during my president-elect and president years. That was my purpose.

My strongest contribution to FNAWS was pulling it together financially and getting it more organized. Pat and I really hit that ball hard and we also cut out of frills and got down to business.

One other thing I got a kick out of during my presidency. In 1987 we got the 1903 law changed in Texas that said if we get enough sheep and the biologists find a good ram, the governor can offer a permit. We got the permit in Nashville and made $61,000 and the guy never went hunting. The next year we got the permit again, and we got $42,000 for it, and the guy never went hunting. We kept trading antelope for sheep, so that's why we have 500 free-ranging sheep in Texas. You acquire through these years of working a very intangible, almost inexplicable, situation and feeling of identity with these people. Like Gallo wine and 100 year old wine, old friends are better. I've known some of these old fools - Schultz, Christian, Babler, McMurray - those are rich old friends.

What is the future direction of FNAWS? What do you see as positive and what do you see as negative?

The people are the positives - their commitment and apparent willingness to sacrifice. That's all an organization is. The only criticism I've got is that we're still not doing what I was trying to start at the Center for Sheep Studies as is every organization. That is, we're not seeing the funding processes achieve all for which they are capable. We still need hard, cold prospective controlled research. It's all a deliberate step-like thing like a mosaic of stone. You keep putting pieces in.

I have a scientific background and I live with it all the time. At this time that organization is still lacking. There aren't very many people that are involved in this who have that kind of interest or background, not that I'm anything special. I'm just another soul.

I'm not at all criticizing the way they run the bank books. The only thing you've got to figure out a better method of analyzing what you're going to fund. The way to do that: Number one - start simple, KISS (keep it simple, stupid). You can't eat a whole hog in one bite. Create something such as a scientific gathering of some of the best people you can get and ask them what are the big questions. Get those sheets out and keep writing and keep hanging and put them on the damn wall and keep working at it until you get it down to some kind of a two or three point thing to do, to the point of even saying these are three things that we'd really need to know. Then get those scientists, pay them a little bit for their time, and get them to write up a hypothesis and send it out to all the universities and see what you get.

If you were talking to someone in the next millennium about FNAWS, what's the message you want to share?

You must take care of the habitat. The sheep will take care of themselves. The most important thing we can do to devise techniques to prevent the multiple division of large ranches in the foothills of their native wild sheep ranges. We don't need damn houses out there - ranchettes.

To get young people involved with FNAWS, take them out in the field. Coordinate, depending upon the nature of your projects, with kids and get them involved. Then you've got to worry about insurance which is why it gets expensive. When we were doing the National Forest Foundation thing, which we're still doing, the first two years we were taking these CCC camp people from the inner city and putting them in the national forests. They were running five grand a piece. You had to pay them minimum wage. We wouldn't have to pay minimum wage, but you've got to cover it.

Kids love dirt. Get them involved capturing sheep, building water holes, building fence posts out there doing something that's meaningful, so they can see something they're doing. That's not hard to figure out. It's a matter of having the time and the support to get the money to do it. But, that's a doable thing. We've got a finite amount of land. We're the stuards of it. We're been given that opportunity. Our ancestors have screwed it up quite a bit. I would hope that we would, in general, assume that each generation will make it a little better than the way it was found. I think we have in some ways, but there's been erosion going on in some other places, like the urban sprawl.

What is the place of FNAWS? Do you think the place of FNAWS is with habitat and wildlife improvement or in the political arena?

You can't be a lobbyist because of those five or six little numbers - the 501-(c)3. Organizations like that can't lobby.

Individuals can do anything they want to. I always think the best message is the positive message. I think FNAWS could improve its message by not making such an issue of all these photographs of animals taken rather than photographs and articles about what has been accomplished. Good PR is taking credit for your good work. We don't do that very well. But, I think we're waking up. We don't have a big picture. We're always looking internally. I think if you're talking about a political point of view of being involved in the process, the only way you can do it is by individuals or by having a hell of a good reputation. And, when they want to argue with you, you show them your good works. I'd take two thirds of the trophy pictures away. I never agreed with that in the magazine. They should emphasize the projects. Emphasize the achievements, not the big rams. It's sales, plain old sales.

It would be wise for the FNAWS Board to develop a standing committee in which they have a five- or three-year plan that's constantly updated, a strategic plan to promote the good work and achievement of FNAWS. Secondarily, that will bring you more members. But you've got to keep the strategic plan updated. There's no business plan, marketing plan, scientific plan. In any of these plans you can handle it the same way. You have can get all these good thoughts written down and you can come to some conclusions. That would work for all of them.

Dr. Jerry Waite
President, 1987-1988

Jerry, you were one of the founders of FNAWS and probably was one of those guys who was in Wisconsin in 1974 and was one of the first. Is that true? Were you one of the first?

Dr. Jerry Waite, right, with Helmet Gruel in 1989.

That's true. I was one of the first. I missed the Mt. Horeb meeting but we had one in Lake Geneva, Wisconsin. That was the first I attended. Ten of us got together and showed our slides on the wall. This was an informal get-together. We decided at that meeting to do something for wild sheep like Ducks Unlimited does for ducks.

Then somebody said, "We don't have any money for stamps."

So, 10 of us each threw $100 on the bed and we started this organization with $1,000. That was in 1974.

We knew each other because we had all hunted sheep and Doc Schultz introduced me to Lloyd Zeman. I got to know Jim Froelich and a lot of the founding fathers. I went around to them and started the Sheep Foundation. Most of the founders are here at this convention.

Some of the original founders were Lloyd Zeman, Duane Smelser, Fred McMurray, Jerry Christian, Jim Froelich, Dr. Roy Schultz, Dr. Jerry Waite and others. We were really committed to bringing good habitat back to wild sheep. We just wanted to return something to the mountains where the wild sheep were harvested. We wanted to do something so our sons and daughters could hunt them in the future. Some of our original founders went to a conference at the University of Montana to discuss sheep and sheep habitat.

The first meeting was at Mt. Horeb, Wisconsin. Jim Froelich called that one. There were 14 people there. The second meeting was at Lake Geneva, Wisconsin, at the Playboy Club. A bunch of us still don't know if we went to the Playboy Club or we went because of the wild sheep. Maybe a little bit of both. This was blizzardy and icy so that was in the winter time. I want to say January or February of 1974. Then we went to Des Moines, Iowa, for the third meeting, and Detroit, Michigan, in 1977.

We called ourselves the Midwest Chapter of the Grand Slam Club. They didn't want anything to do with us, so we had to come up with another name. Lloyd Zeman and the powers that be named us Foundation for North American Wild Sheep right after the Detroit meeting in 1977.

What were some of the overriding principles that you as founders wanted to establish from the beginning?

We wanted to maintain the habitat, transport animals to historic ranges and study diseased animals. Doc Schultz and I are both veterinarians so we were interested in the biologists who specialized in wild sheep. At the first convention in Detroit, we grossed $70,000. We thought we were on the top of the world! We had an auction with three desert sheep permits. Now one of them is worth three or five times that but money had a different value then. We knew we were on to bigger and better things. We refined things and we made mistakes along the way but as a good Scotsman like I am, you don't make the same mistake twice. We recruited the early members by word of mouth. If you had a friend, we'd say bring them along. Sign them up. That's how everything go started.

I was on the Board for 14 years before I became president. Due to elections and the changes in bylaws I qualified for the Board for 14 years. I didn't want to be president because I was having too much fun. The Board of Directors elected the president, so they were the ones to call the shots.

What are the successes of FNAWS?

Of course, our successes are that we've transplanted so many sheep that in some states we're running out of habitat. People where they live stress the sheep and historic ranges are now covered with houses. This is especially true in California where the desert bighorn are. Humans think they're smart, but all we do is louse up Mother Nature. I'm extremely proud of the transplants, the disease studies and the habitat. We've co-funded habitat purchases and turned them over to the states. FNAWS does not own any of the land. They turn it over to the states. Every project we fund has to have a written progress report and written end report recording the success or failure of a project. They're not all successful. I was the one who coined the motto, "Putting Sheep on the Mountain." That was my theme when I was president. I figured I would be president someday. Everybody on the Board had to okay a theme for a new president. I would never bring that one up because I was saving it. I just thought it summed it up, "Putting Sheep on the Mountain."

What is the future of FNAWS and where do you see it going? What advice can you give?

As far as membership goes, we're never going to be 100,000 members. There will never be that many sheep hunters. This is more a limited group, as far as memberwise, but we maintain an excellent membership. We raise a lot of dollars to put sheep on the mountain. Over the years we've found that 5,500 to 6,000 members is about where we stand.

The management has to change with the changing times. If you stand still, you're dead in the water. Things will always change. Look at the history of what's been done. It's all written down. Talk to the past presidents and the founding fathers who are still around. They can tell you about the mistakes we've made. We don't go down those dead-end roads anymore.

I think they should plan conventions where the city is very receptive to us. The name of the game is to raise dollars to put the sheep on the mountain. The Long Range Convention Planning Committee has a recommendation the convention site should be researched very well. That's the one thing I would recommend. Reno is an excellent place. We have always made tremendous amounts of money here.

Keep up the good work and support the wild sheep. Thank the members and donors of the Sheep Foundation. They're the backbone of the organization. The members, the donors and the buyers are the three groups that we have. No one group needs more attention. They all deserve credit. Let's do anything to put sheep on the mountain.

Let the other clubs work around us. We're staying in Reno to make money every year. It takes all three groups of those people to make it work.

Arnie Johns
President, 1988-1989

Letter in 1994 to Dr. Roy Schultz

President Arnie Johns received this FNAWS belt buckle as a gift in 1989.

I became involved with FNAWS in 1977. A meeting was organized in Des Moines, Iowa, by the late Lloyd Zeman for the purpose of starting a sheep hunter's club or organization.

Eighty enthusiastic hunters, outfitters and guides from many Canadian provinces as well as Alaska and the lower 48 states attended this meeting. It was decided to form a club, naming it the Midwest Chapter of the Grand Slam Club.

At that time Bob Housholder of Phoenix, Arizona, was executive director of the Grand Slam Club. When he was approached to give us permission to use the name and affiliation with his club, he denied us the use of his club's name and affiliation. A short time passed and a new organization was born, and chartered in the state of Iowa called The Foundation for North American Wild Sheep (FNAWS) — a nonprofit.

I immediately started to attend board meetings and was involved from a member's standpoint. I was appointed to serve on committees well before I became a Board of Director member. I was elected to the Board in 1983 and served as treasurer, vice president, president-elect and, finally, was elected president in 1988.

During my term as president I traveled to Washington, D.C. to participate in a document signing ceremony with the U.S. Forest Service and the Bureau of Land Management (BLM) signing a Memorandum of Understanding (MOU). On another occasion I testified at a U.S. House of Representative's sub-committee hearing on the reintroduction of wolves in the Yellowstone ecosystem.

As a FNAWS Board member and officer, I served on every committee formed and was involved with the annual convention planning for eight years as convention chairman and program chairman. Today some 17 years later, I remain active as Award of Excellence and Special Awards chairman.

One of my achievements while I was president that I feel good about was the completion of the twice life-sized metal sculpture of the Chadwick ram by Cody, Wyoming, artist Bill Feeley, and erecting the sculpture at our international headquarters in Cody, and finally, the dedication of the sculpture garden and headquarter's building (dedication ceremonies took place June 17, 1989).

Looking back at the years of growing pains the Foundation had in its earlier years, I feel good about the fact that so many hardworking people made a difference to the enhancement of sheep populations in North America.

Harvey Kadlec
President, 1989-1990

How did you get involved with FNAWS? Tell us about your presidency.

We did a lot of changing of our accounting system. Our ways of raffles became more secured. We went to cash registered. We left a paper trail for everything that was done.

I started out by going to the first convention in Memphis. I was asked to help out with the auction, and then it ended up I did the auction for the next 15 years. We went to Minneapolis, and John Babler was in charge of the convention there. He asked me if I would do the auction, so I did it there. It just kept on going. I kept on doing it. I'm the one that put everything together and cataloged it and gave it over to the auctioneer. I was in charge of the auction from 1977 to 1991. My son, Bradley, did it a couple of years and I helped him. He did the auction when I was president.

President Harvey Kadlec and his wife Sue trip the light fantastic after the 1990 banquet.

Tell us about your experiences with auctions. Have there been any that stood out in your mind?

The one that really stands out is Vivi Crandall's painting, "Downhill Racer." That was the auction I remember the most. She wanted certain stipulations on how to sell it, and I said we didn't make any deals. Whatever it sells for, that's what it sells for. If you don't like the price, then you buy it back yourself. She wondered what it would sell for and I guessed $5,000. She wanted at least $10,000 but it actually sold for $17,000.

When these permits went for a lot of money, Rich Roan used to jump on tables and everything else to get people to bid. Some people didn't like that because they thought it was unprofessional, but it gave color to the auction, and he brought in a lot of money. I just like the way he does the auction. He puts color into it. I've been to auctions, and they stand up there like a recording. Rich over the years knew the people. He knew who the bidders were. He could get money out of people when they were done, but they weren't done.

He'd say, "Say, come on. I'll throw in an extra $10 or $100 if you'll just bid once more."

Usually, he didn't have to because that guy would bid once more, but the guy who wanted it would bid again. Rich would just keep it going.

What were the accomplishments you had in your presidency?

I think my biggest accomplishment was the changing of the accountant and the accountability of FNAWS. I'm not going to say the negative, but there was a lot of it. We got a CPA on Board. We had more people running it. We had backups so you didn't have questions about who was doing what. When you're going through this amount of money, and only one person is handling this, how much was disappearing? We knew a lot of it disappeared, but we couldn't really prove it.

In the next year, there was so much more money, and we had about the same volume. We used cash registers instead of a box. There was a paper there that said how much was taken in and what it was taken in for. You have the trail of all the items you had. You knew how much you had on the raffle. It's all rung up on a tape. It's all there. You don't have to pick it apart.

When you are thinking of FNAWS, why is it such an important part of your life?

It's important for what it does for the conservation of wild sheep and putting more of them on the mountain.

Where do you see FNAWS going for the new millennium?

I think it is headed in the right direction. There are glitches in it, but there are glitches in everything that goes on. You're always going to have to repair things and try to fix them to make them better but as long as it keeps on progressing and they do correct the things as they come up instead of letting things go, then I think it will go okay.

Is there advice you can give to the new generation of FNAWS members?

I know one thing is that as the Board changes over, there is a lot of time reinventing the wheel. It has been done in years past. You don't have the continuity. New people always want to do their own thing, but there isn't enough research about what has been done and what works. They try not to do it over or they come up with some idea and they say, "We're going to do this." If there was enough research on an idea they'd find out that way back when it was tried and it didn't work. That's not to say it really won't work, but if it has been tried for a year or two, it's not going to work.

Do you think FNAWS will continue to grow or stay the same?

I don't think it is going to grow that much because there aren't that many sheep hunters. It will grow slowly as the population grows. There are more people all the time, but I don't think it will grow in leaps and bounds because it is centered on one species unlike Safari International Club where they hunt all over the world. It's more expensive to sheep hunt then hunt elk. There will be more members of Rocky Mountain Elk Foundation because there are so many elk hunters, just like deer hunters. Most places you can just go. FNAWS is more of a restrictive club.

Do you still come to all the meetings? Are there any conventions that stick out in your head?

I come to all the national conventions. I've only missed one convention since 1977. I remember the ones that were a complete flop like Hawaii and Philadelphia. Hawaii was a great place to go for the people but it was horrible for a the convention because the members didn't want to be inside the hotel. With Philadelphia, nobody wanted to go there. People don't want to go to the East unless they have to. People go there for business, but normal people don't want to go there. You hear people say we've been to Reno before, but it's so much easier to work a convention being connected with the Reno Hilton for so many years. When you go back to the same place, you get most of the same people there. They don't change all the time. The facilities are there. You don't have to fight unions. It is so much easier.

Were you one of the founders of the Minnesota-Wisconsin Chapter of FNAWS? Tell us about that.

We founded the chapter around 1981. This last spring at the banquet we had to turn people away because they could only serve 430 and they had too many people who wanted to come. It started in Minnesota. There are so many hunters there, and the chapter has taken off as part of the original Foundation. The original Foundation moved away from Minnesota, and the chapter started up and kept getting bigger. The Eastern Chapter of FNAWS may be bigger because they have a much bigger area but the Minnesota-Wisconsin Chapter is one or two as far as the chapters go.

Somebody asked me if we should get bigger facilities for the chapter banquet because they're selling out. I said no. If you can create a demand, you'll get more people. They'll buy them ahead of time. If you get it too big, they'll change their mind before it comes. If they pay their money ahead of time, they're going to go. I think since the chapter started they have donated more than $600,000 to fund FNAWS projects.

President Harvey Kadlec welcoming everyone to the 1990 convention.

John Harris
President, 1990-1991

What comes to mind when you think of your presidency?

President John Harris with daughter Ashley and wife Barbara in 1991.

The one really big issue that changed or occurred while I was president was that the desert sheep hunts in Mexico fell apart right before convention.

In December I talked with the executive director (Karen Werbelow) on the telephone while I was on vacation for Christmas. She indicated that the Mexicans had pulled the permits and were not going to have them at the convention. We had been out selling them and talking about them in sheep camps right before convention. This was a major fund-raising event for us. We tried to look at ideas to replace that money.

I spent most of the Christmas holiday on the phone with Bob DiGrazia, Garry Vince and Karen Werbelow discussing that issue. We came up with the idea of doing a sheep camp. It's tough to get a trip for that coming year. We wanted something that was significant enough to replace the desert sheep hunt, so I bought a Stone's sheep hunt for the Sheep Foundation from Garry Vince. We put that on the sheep camp.

We got lots of criticism when we did it because they thought it would never make the money with the desert sheep camps we had. We made more money that year than we had the previous year with the desert sheep. The person who won the Stone's sheep hunt when they raffled it off was a Canadian outfitter. He donated the hunt back and we auctioned it off at the Saturday auction and got another $18,000. It was actually a double fundraiser. We were able to put together something out of virtually nothing and out of disaster. I'll never forget that Christmas because we were really trying to figure out what we were going to do to replace that hunt.

What are some other highlights of your presidency in 1990?

One of the things that was unusual was that I was elected president during my first term, during my third year in office. There had been a small amount of turmoil. We were restructuring the accounting system. I was the treasurer the second year in office when Harvey Kadlac was president. Harvey and I worked very hard to restructure the way we accounted for the funds that were made. We put in an accounting system and the use of cash registers. Harvey actually did that under his presidency. He and I worked together on that, and we continued that process under my presidency. We wanted to be able to show where the money was coming from and show that we were responsible for it. It is virtually the system we have in place today. You buy raffle tickets and you'll see we have cash registers and an accounting system. We changed accountants at the same time, and the Board was split on that.

By having a strong accounting system, what did you notice changing in FNAWS?

One of the things was that we had always heard rumors from people who would question what happened to the money in the past. We'd have members who questioned what occurred. We also tightened up the drawing system for raffles. That's why you see the drawings occurring in front of people on Saturday night. The drawings had previously occurred behind closed doors. The problem with that is that it gives people the chance to speculate that something is wrong or something is rigged. We wanted to be open with it so when people spent their dollars they knew they had a chance to win something. They'd see it.

We were trying to get a system in place where we stayed in Reno two or three years and then if we wanted to move, we'd move, but only move out for a year and then come back to Reno. We always wanted to use Reno as our base. We always made money here because of the gambling and because of the other things the hotel could offer. We get cheaper rates on all the things that we do. Nevada is not a union state, so we can do a lot of work ourselves. We are a volunteer organization, and we do a lot of hands on including unloading a truck and things like that. While unions are not bad, if you are a union state you can't do those things. The costs of putting on a convention are sky high for us when we are working in that kind of environment.

What do you think FNAWS has to offer today? Has it changed?

I think FNAWS is progressing. One of the things I do see that I think is encouraging is we see change of faces on the Board. With that you get new ideas. I will always work for FNAWS because I believe in what they are doing, but I think the people running it ought to change on a regular basis. New ideas ought to come in. I think it's being operated for us as more of a business now than it was when it started years ago. Not that it was bad years ago, but I think with the environment today, we need to operate as a business. We're in a tough market. We're competing against other markets now.

In the past, Safari Club and FNAWS were the only two out there. The Elk Foundation came on Board, and they grew rapidly. In fact, this weekend they're having a major convention right now so there is competition for the same dollars that are out there for conservation organizations.

One of the things I also think was very important under my presidency was that we formed an alliance with Safari Club. Warren Parker* was the president of Safari Club at the time. We would do a lot of things together. We would go to outfitter meetings rather than both organizations sponsoring a hospitality room. The two organizations would go together and sponsor a hospitality room with both of our members who were in attendance working at the hospitality room. We cut costs and we did a lot of cooperative ventures. Some of the funding we were doing was cooperative. We started working a lot more with the political climate in Washington, D.C. Warren and I were both pushing our organizations to do that. Those activities are continuing and they are very important.

Do you see a difference in the FNAWS of yesteryear and now? Has there been a significant change in the tone and purpose of FNAWS or is it about the same?

The purpose is exactly the same. We raise money. To coin the phrase of Jerry Waite, we "put sheep on the mountain." That's what we're still about. It's still done with very limited paid staff. It's a volunteer organization, and that has not changed. We still have people who are willing to work and to get involved. They are a very active membership.

What do you think the next millennium will hold for FNAWS? Where do you think FNAWS is going?

What we need to accomplish in this organization is to work together. We have had some problems in the past and in the past few years there has been conflict on the Board, to some degree. I think different ideas are important, but you need to be able to come together and work together. That makes the organization stronger.

The thing that is important for the future is education of young people. We need to put an all-out effort there. We also need to be looking at the political arena because in many areas that's where we are being beaten now. We need to educate the general public as well. We have these ballot box initiatives that change the focus of what we do.

What is the biggest threat to FNAWS in the future?

The organized group of anti-hunters that are looking for preservation, rather than conservation, offer the biggest threat. If we don't do everything we can, which includes educating our youth, then we're going to get beaten. We need to build our organization to keep it strong from beginning to end. It should be a lifetime commitment. We can't rest at all, ever. One thing that distinguishes us from everybody else, including the other conservation groups, is our membership itself. Our membership has the willingness to pull together and work as a group. We have outfitters to millionaires. They all do the same work. We need to always focus on our membership.

President John Harris hangs out with Mark Twain during "Welcome Back to Reno" night in 1991.

Howard W. Pollock
President, 1991-1992

How did you get involved with FNAWS?

Let me back up a little. I was on the board, elected first in 1985. I served two, three-year terms. My last year was the year I was president, which was in 1991. I was originally invited by a former president to attend the Board meeting just to observe. When I did, I found several things that bothered me. There seemed to be no parliamentary procedure at all, and there seemed to be a need for some order in the meetings.

I learned parliamentary procedure and have used it all my life. I've tried to encourage other people to learn it. I have written a number of books on the subject, one for FNAWS. FNAWS gives this booklet to all new Board members. I was very fortunate having this kind of background, talent and technique in parliamentary law, for it resulted in my being elected nine times as president of national or international organizations. The National Rifle Association of American is one, of course, but some of them were United Nations organizations; some were Alaskan organizations; a whole bunch.

Past-President Arnie Johns, left, receives the 1992 Presidential Award from President Howard Pollock.

When I first visited the FNAWS Board, I realized there was a need for acquainting the members of the Board with the fundamentals of running a meeting, how to make a motion, how to second it, how to make a substitute, how to bring a vote to a conclusion, what limits there were on the number of times people can talk, how long they can talk. That's what is in my booklet now.

I published this booklet for FNAWS, but it was written by Howard W. Pollock, parliamentarian/president. I am still thought of as the parliamentarian for FNAWS. When I walk into a room, people will look at me and ask if that motion was okay or the president will look over at me and give me a positive nod as they wait for an answer from me. One of the things I learned is to never stop anything on a technicality if everything is moving forward. That's not the purpose of parliamentary procedure.

The purpose of parliamentary procedure is to help things move along, not to say the book says this. The book follows Robert's Rules of Order, newly revised, so it's always updated. It's very difficult in that thick book to find an answer to anything. You have to go to three or four places. That's what I was trying to get away from. In the little booklet that I wrote, it talks about a certain subject and gives a definite answer you need for that subject and goes to another subject. It's easy to read.

Obviously, parliamentary procedure is very important to you for bringing order to FNAWS and shortening the meetings. But, in terms of your accomplishments during your presidency, what were they?

One year I thought I was in line to be nominated for president, but we had a deadlock. The current president wanted to serve a second term, so our current president, Phyllis Carson, whom I dearly love, tried to break the deadlock. She nominated a third person who was elected. I waited another year in order to be nominated by the Nominating Committee and to be elected as president. Unfortunately, on the Board of 12 there were three people who seemed to be dedicated to making sure things would not work smoothly. I won't name them, but they were really a pain.

I've been in many organizations, and still am very active, but I think the organization closest to my heart is the Sheep Foundation. I just like the informal atmosphere that everybody has in the conventions and meetings. It's very nice. But, as president I had problems because every time I would propose something at a meeting, I knew this little clique would oppose it. If I said black, they'd say white. If I'd say green, they'd say purple. I went through the whole year that way. It was difficult. The majority controlled, so I got things through, but it was always done the hard way.

One of these three eventually became president and came up to me privately several years later to apologize, saying he now understood what difficulty I had trying to get a consensus and letting everybody say what they think. He said he now realized that you can disagree without being disagreeable. I still find that a little problem on the board, but it's getting less and less and less. Now, when I go to a Board meeting I sit and watch the people and observe who's contributing something, who's sitting there quietly thinking, and those who start out saying that something won't work. Eventually they have to be persuaded before a vote. A lot of it has to do with personal attitudes, which is okay. I would not like a Board where the president made all the decisions and people just said yes with a stamp of approval. It's not supposed to be that way.

Why were you so passionate about FNAWS?

I think because I'm a sheep hunter. I understand. When people go through the trials and tribulations, the work of packing and climbing and getting one brief moment to get a good shoot and harvest a trophy sheep. You develop a camaraderie that's pretty hard to find in other organizations. You find it in other organizations, but it is a unique trait in FNAWS.

I'm so proud of our current president, Phyllis Carlson, because to me she's the First Lady of the Foundation for North American Wild Sheep. She's not only the first lady of FNAWS, but she's the first lady president we've ever had. That's significant. I happen to think women are every bit as good as men — sometimes far better, and sometimes not. But, I treat them all as equals. Men tend to be the rugged outdoorsmen, and they think they have to carry the women along. I don't think that's true, certainly not with Phyllis. Phyllis is unique, and I just love her. She and her husband are just super. They are such great people. I'm here now to just help. I'll work at the booths or do whatever I can, wherever I can, to help. When I meet in our annual closed meeting with past presidents, I'm going to talk about cooperation. I'm going to mention something they probably know. You can be in the opposition. You can oppose something. You can give all your most persuasive reasons, but if the vote goes the other way, then as a matter of loyalty you should say, "I support the vote and the decision of the majority."

That's not always easy. If someone asks you, they'll say, "Well, I'm so surprised. You were so much against this or that. Why are you supporting it now?"

And I will answer, "Because the majority made a decision after we heard all the arguments. Only God is perfect, so I have to assume that they made the best decision."

Since you have been a leader in FNAWS, what has been the area of discussion that keeps coming up that may have caused some dissention?

Sometimes it's that and sometimes it's other things. I have a personal belief that the Board of Directors should not be the Nominating Committee. I think there should be non-board members who are members of FNAWS who are on the Nominating Committee. The Board has never seen fit to agree with me on that idea. The Board sits in the capacity of a Nominating Committee, nominating their own members, which I don't think is the best way to do it. They've made the decision. I've brought this up several times, and they think differently. They like the power and control of it. That's one of my concerns, and it still is.

The other is the annual meeting of members. I don't know whether we don't put the right emphasis on it or we don't schedule it at the right time, but the members walk out and couldn't care less about what passes or fails at the annual meeting of members. We keep trying. We've tried putting it immediately after the big breakfast, and that helps a little. A few more people stay. We're flexible and I think that's good, but I don't know what to do about it. On the other hand, if there were problems, the membership would be there in flocks. They're not, so they must be contented with the job the Board is doing.

Is there a particular area of emphasis that continues as a theme through FNAWS?

Hells Canyon has certainly been a big item for the last several years. That didn't happen to be the project during my presidency. We have always been dedicated to the concept of putting more sheep on the mountains. That means a low administrative overhead, a very successful convention to raise money. Everybody knows that the money is very well spent by FNAWS to put sheep on the mountain. People attend the FNAWS conventions because they love sheep and sheep hunting and the outdoors. FNAWS will never be a large organization, memberwise. There are just not enough sheep hunters. But, dollarwise, we're Number One.

What do you see as the future of FNAWS?

I don't see any phenomenal growth in FNAWS simply because there are a limited number of sheep hunters, spouses and supporters. I don't see a phenomenal growth in income because we've really taken an enormous amount of money proportionately to each member. It's most important that we operate efficiently. We have a marvelous staff, all women by the way, in Cody, Wyoming, who do an exemplary job that keep the costs down. Everyone of them is a hard-working person. If someone came there and didn't work hard, they wouldn't last long. They'd be worked to death, to begin with. These ladies do it because they love this organization. They know what we're trying to do to help the wild sheep. They learn to really to respect and admire the sheep hunters as a group.

Do you see any concerns developing about the recruitment of younger members?

That's a constant concern for any organization. All of us are concerned that the new generations are the future, so we've got to have programs for the youth. We've got to have father-son, father-daughter, mother-son, mother-daughter kind of programs and relationships. We need to have safety training for our youth. We need to do a lot of work with photographs and videos to keep the kids excited. One of the big obstacles that all sportsmen have is the formal school systems. A lot of them are the teachers. Because of lack of experience and sheer ignorance, they are anti-hunters. Somehow we need to convert teachers. We need to really emphasize teacher programs.

Both SCI and NRA have summer programs where different chapters and national headquarters put money into summer camps and give scholarships to teachers. In SCI we take a teacher out of the Washington, D.C. area, which is probably the worst area to do this, and we find that a teacher really becomes knowledgeable and interested in the outdoors and the children who are his or her students. Our SCI Board will send them to camp for two weeks. They're taught everything. They're taught firearms safety. They're given rudimentary courses in biology, in the proper food for animals, the water systems. I'd love to see that happen in FNAWS.

We're doing some of that. We're not doing as much as we could. We have limited resources. A lot of the money we raise is with a compact with the different states. We'll sell off a ram hunt and make a lot of money from it, but we give most of that right back to the states. If we could work out programs with the state, not just to fund their fish and game organization, but to create programs to educate teachers and children, this would be great. We're then looking toward the future.

We need to always look toward the future, and the future lies in our youth. There are a lot of organizations dedicated to youth. We need to become friends with them. We can sit down at a table with somebody who absolutely disagrees with us, but if we talk rationally and explain our situation and tell them that we believe in utilizing the game species, then we will make headway.

I am a big-game hunter. I am proud of it. I don't shoot aimlessly. If I shoot, I shoot what I think is a real trophy. I will never, ever waste the meat. That's what I mean by utilization of the species. If we want to hunt, that's fine, but make sure we eat the meat or give it to somebody who will eat it or give it to some cause. You have to teach people that if they are going to hunt, they have to utilize the entire animal. That means utilizing the trophy, the skins, the meat, everything.

Nationally and internationally, the movement is going the other way. There are more tree huggers and prairie fairies and bunny cuddlers because that's the way they're educated. They don't know. We need a lot more public relations work to educate the public. That means educating the teachers as well as the students.

Dr. Robert "Bob" DiGrazia
President, 1992-1993

What was your motivation for getting involved with FNAWS and when did you first get involved?

President Dr. Bob DiGrazia, left, receives a bronze sculpture from from President-elect Craig Johnson. Sculptor Rick Taylor is in the center.

I got involved in 1980. I was the second generation of FNAWS. I loved to hunt. I had never taken a sheep when I joined FNAWS. I was in an outfitter's camp in 1980 and picked up their magazine and I tore the little label out of the back of the magazine and sent it in. I was a $25 member.

We first got organized in the state of Idaho in 1985. Tom Hinders was president of the Sheep Foundation. He came to Boise and contacted a few of us who were members in Idaho. We organized the Idaho chapter in 1985. I was still just a $25 member. We had our first fundraiser in Boise. I was president of the Idaho chapter about two years later.

The Sheep Foundation wanted the chapter members to carry before the Idaho legislature the legislation to authorize one of the auction permits from Idaho. Idaho was one of the few states that didn't have an auction permit at the time. The battle was to get the state of Idaho to give a permit. That's where I got my notoriety before the FNAWS membership.

We have a legislature in Idaho that is agriculturally dominated. The relationship between the hunting community and the agricultural community is not a friendly environment. There is so much public land in the state of Idaho that ranchers and farmers utilize. When they utilize public land for ranching and farming, the competition is wildlife. We were trying to generate funds for wildlife, and bighorn sheep specifically. The ranchers and farmers felt threatened. They felt threatened because the money from the sale of the permit would generate an astronomical amount of money to perpetuate wild sheep habitat and wild sheep.

Wild sheep carry a price tag for ranchers. We have a disease problem, especially to the wool growing industry. Wool growers are tremendously threatened by wild sheep. The management agencies, the BLM and the Forest Service, don't like the wild sheep and the domestic sheep coming together because they don't want to take the liability for the die-off. Our legislature is controlled by the agriculture industry. The battle was pretty immense.

When we first went into the legislature, we were crucified in 1987. We got nowhere, so we backed off. We wrote the legislation. We asked for two permits; one for auction and one for lottery with 90 percent going to bighorn sheep management and 10% retained by FNAWS. What we came out of the legislature was one permit for auction. It had a minimum bid of $30,000 with a 5 percent commission to FNAWS and the rest of the money went to the Idaho

Fish and Game Department. The first tag was sold in 1988.

When we came into this venue with a bighorn sheep permit with a minimum bid on it, it was embarrassing. We were very up-front about it. Jerry Waite was the president. We were very concerned about it and whether the FNAWS public was going to accept or not accept it. The membership accepted it for two years. On the third year it was brought up for bid at FNAWS; they refused to bid on it. The FNAWS membership refused to be told there was a minimum bid on a sheep permit. They refused to be told by the state of Idaho that they are going to have to pay that. It didn't sell and it was embarrassing to the state.

We decided to go back to the legislature and lay it before the legislature. We did this twice. You can't imagine the efforts I had to put into it, not once, but twice. We had to go back to the legislature and reintroduce the legislation and explain to them that it was embarrassing for the state of Idaho. We needed to take the $30,000 minimum bid off. What we also did was we added the lottery ticket and put pressure on the agriculturally dominated legislature by dedicating lottery funds to wildlife research, specifically to solve the problems between domestic sheep and wild sheep. It put it in a position where the agricultural interests couldn't vote against it. That was in 1991. When I wrote that legislation, people were looking at me oddly. They just thought I was playing political games, which I was, but I felt in doing that there were some solutions to be solved here and to be dealt with.

Where this came to fruition was in 1996 when we had the die-off in Hells Canyon, we had the research facility in place to do something that had never happened before. That's when we pulled the 76 sheep off the mountain and brought them to the hospital and pulled that blood out of them. The sheep all died, but we got absolutely unbelievable amounts of data off of those sick sheep. They died from Pasteurella multocida. When they did the research they proved that it wasn't a domestic sheep problem. These wild sheep didn't die because of domestic sheep. All of a sudden these ranchers who didn't like us started thinking that maybe we did a good job for us, which we did!

We had the facility in place. We had personnel in place. We had the ability and the money to pull the sheep off the mountain, and study them. All of a sudden we had an astronomical amount of data and they found out that this die-off was NOT related to domestic sheep. It worked out. It was a win-win situation. All of a sudden instead of being an odd bodkin out there, people were coming up to me and saying I pretty much had this figured out. The problem was Pasteurella multocida rather then Pasteurella haemolytica. It was a different bacteria, but it wasn't related to the domestic sheep industry.

The scientists still don't know what happened. Now with the money that's coming out of the auction permit and the money coming out of the lottery permit, Idaho has the leading wildlife research facility in the nation. That's because of the Idaho Chapter of FNAWS. The ranchers and farmers are still fighting us, but we're working pro-actively with them. We still have an agriculturally dominated legislature. It's a battle every year.

After I got all this legislation passed, my name was going out before the FNAWS membership quite a bit. I don't even know who nominated me to the national Board, but I was nominated in the summer of 1987 and was elected and started my term in 1988. I was still a $25 a year member. I was like a second generation leader. I had a strong grassroots background and a strong legislative and political process background. That was my attribute. When I came down here, I started to learn how FNAWS worked. It took me about three years before I really felt comfortable. I got re-elected for the second term, and in the middle of my second term I was elected president in 1992 from May of 1992 to May of 1993.

During your presidency I assume you continued your fight in Idaho?

Oh, yes. When I came into FNAWS, I addressed the Board with a whole agenda of what I wanted to do. One of the things I wanted to do was continue the dialogue with the domestic sheep industry. My whole point was that instead of fighting each other, we needed to find some things and solve some problems. We did sign a memorandum of understanding with the wool growing industry, the BLM and FNAWS. This was just BLM lands, not Forest Service lands. The compromise that we came to was that we would respect each other's rights to have wild sheep and domestic sheep, and that we would entrust the BLM to manage the two species to keep them a part. Their goal was to keep them a part by 50 kilometers or natural barriers as they exist. If there is a cliff there they wouldn't be allowed to get past that. It was a discussion that still continues to work pretty well today. This is an agreement where a rancher who has a grievance against wild sheep encroaching onto grazing land they can go before the BLM and say shoot the wild sheep or visa versa. If a rancher leaves one of his sheep on public lands and it wanders off and gets into bighorn habitat, then the BLM can either capture it and give it back to him, or shoot it and get rid of it. And, the rancher doesn't get reimbursed for that. There's no fingerpointing, either. If there is a die-off and we did the best we could to keep the sheep a part, then there wouldn't point any fingers.

That's what happened in Hells Canyon. We were patient enough to let the facts come out. You have to realize that when that die-off happened, it could have been a war. We waited until the data came out to have the scientific facts of what happened. The conflict calmed down. Hells Canyon is a new ballgame for us now. FNAWS is dedicated right now to generate about $10 million in the Hells Canyon initiative to put wild sheep back in that canyon. There's more habitat there then any place else in the continental United States. No one bats an eye of numbers of 5,000 to 10,000 sheep in that canyon. We just put 40 in there and that's a start. We just hired a fundraiser to raise $10 million in 10 years.

The land in Hells Canyon has been set aside. What they need now are sheep. The hard part is finding Rocky Mountain bighorns. They just went to Canada in December 1997 and brought 40 bighorns down and released them. That was the first big move. They're hoping to do something like that on a yearly basis. It still is a matter of generating the money. The second part of that is managing the disease process. We're talking about pure Pasteurella and finding more answers to that through blood testing, monitoring, collaring the sheep.

Hells Canyon is on the border between Oregon, Washington and Idaho. That's the neatness of all this. Where else in wildlife management have we been able to get three state agencies and two federal agencies to sit down at the same table with a private agency, FNAWS, and come to a solution to work together. The two federal agencies are the BLM and the U.S. Forest Service. The Board of FNAWS and personnel in the agencies are managing this project. There is a core group of about 10 people.

I do remain active as a past president but I try to keep my mouth shut.

What were several proud moments as president of FNAWS?

There are two things. One was the agreement with the BLM I already talked about. The other thing was the FNAWS Board. We worked very well together. I was very proud of the Board I had to work with. We had a convention that year that the gross revenue tripled in one year in Reno. To put this in perspective, the gross on our convention prior to 1993 was bumping along around $1 million a year. We came to Reno and I felt the Foundation was about to explode financially.

I sat the Board down and told them we were in a pickle. The pickle was we had three major conventions this same hotel. We had Safari Club here. We had Rocky Mountain Elk Foundation here, and we had FNAWS here. We were last. I know guys, we've got to have our "t"s crossed and our "I"s dotted. We've got to be organized because we're going to get one chance. If people see we're not organized, and we're not doing something properly to deal with the membership, we're going to flounder here. If it gets out to the public that we're arguing with each other and not working well with each other, then they'll go to SCI or RMEF. We basically have the same membership. SCI is a major competitor. We're last so let's make them come to our convention. Let's trump everybody. Let's bring in a big name entertainer. I told them this is what we should do. We should try to work all summer to get a big name entertainer and announce it at the last minute so no one else could react to it. We'll announce it at the end of October that we'll have this big entertainer.

Karen Werbelow and I worked extremely well together. She and I talked to a couple of management agencies. We hooked up with a man from Denver and said we wanted a western singer. We wanted this person to come in on Wednesday night so the FNAWS membership would say we're sticking our necks out to compete, and they're not afraid to compete. It got down to the last minute in October. We signed Patti Loveless for $36,000 for one night. This had never ever been done in the hunting community before. My name was all over this thing, and my fingerprints were all over it. I was criticized very severely. I had membership calling me and asking me why I was spending FNAWS money to bring her in. They said she had nothing to do with sheep.

I said, "Look guys, we're in a pickle here. We have to make people come to FNAWS to spend their money. We're going to set the mood. We're going run a good convention, and where the chips fall that's where they fall."

Patti Loveless pulled every dollar we could possibly pull in here early. People did sign up to come for the whole convention. She set a mood for the convention and they really enjoyed her. People were very pleased with it, and the pocketbooks just opened up. The convention revenue jumped from $1 million a year to $2.8 million a year. That's a lot of growth. Then, all of a sudden, people were calling and telling me that was a good idea. Tom Brown who was on the Board of Directors with me from Colorado criticized me but he is a very gracious man. He called me after the convention to say he needed to apologize to me.

He said, "You knew what you were doing. When you can make a business decision like that and get that kind of cash flow in return, you did a good job."

We tripled the revenue FNAWS had to deal with in one year.

What's the future of FNAWS?

The future of FNAWS is all political. The battle right now is in Washington, D.C. Everybody is bickering for money for wildlife and management. We need to have a major presence in Washington, D.C. I'm a firm supporter of FNAWS hiring somebody or sending somebody to D.C. to sit there and take our issues before the management agencies back there. It's all money. It is he who speaks the loudest that gets the money. FNAWS needs to get in that game. Funding is important but you have to have the money to fund lobbyists. You can't have a lobbyist unless you have the money to pay for it. I think Karen Werbelow should go back there. I think they need to take the executive director's position and split it. We still need an executive director for the management and administration of the Sheep Foundation, but we need to have a political consultation in D.C. on a full-time basis. Karen has a lot of credentials back there. She's earned her dues. She had an appointment with Vice President Al Gore. She was asked to sit on the wildlife budgets a few weeks ago. They already know her.

Is there anything in the positive or the negative about FNAWS that you can share with us?

The positive is the volunteer level in FNAWS. That's a strength that no other organization in the nation can deal with. I can get on the phone and ask anyone to do something for me for nothing. Or, they can call me, and I'll do the same thing.

The negative for FNAWS, there are some strong opinions and very strong concerns. We have territorial issues that divide people at certain times. FNAWS is no different than the federal government. We only have a certain amount of money. It would be nice to fund everything in Nevada that the people of Nevada would like to fund. It would be nice to fund everything in British Columbia that we should fund, but the money is just not there. When you fund in B.C., you tick off people in Nevada, or you fund in Nevada you tick off people in Idaho. Sometimes it's hard to heal those wounds, but we go on.

The membership is precious. There are some great people in FNAWS. The bond that binds us all are that we hunt sheep. No matter if you're rich or poor, or if you're in extremely good shape or if you're in poor shape, you still have to walk to the top of the mountain. Everybody has that bond. We all experience the same thing. We all come together for sheep.

FNAWS Accountant Jim Reilly, left, and Dr. Bob square off for a velcro foot race during the 1995 convention.

Craig Johnson
President, 1993-1994

What did you accomplish as president of FNAWS in 1993?

We had one of the biggest conventions ever in San Antonio. This was one of the most fun convention ever. The Texas hospitality and the convention were great. The only negative was that the convention cost a little more. We had a decent year financially, but wasn't as good as it could have been.

We had the exhibitor / donor party out at the Lone Star Brewery. We had a collection of trophies out there that was pretty marvelous. The Wednesday night party was around the river walk with a Tex-Mex theme. On one side was the Mexican-type foods and the other side was the Texan-type food. We had a border in between and you had to get between the "border guards" to pay our dues. I can't exactly remember what we did but every time you went back and forth across the border and didn't do it right, you would be "thrown" into jail. It was a really fun party.

Janet and Craig Johnson at the 1994 convention.

We gave money to Wildlife Legislative Fund of America (WLFA). In the 1993 election was Proposition 200 in Arizona, which was an anti-trapping, anti-hunting initiative. This was defeated by a 2 to 1 vote margin. FNAWS with its chapters was the largest contributor for defeating that initiative with well over $100,000 in contributions. The money was used for television and printed literature. The money went for the whole campaign. Pete Cimellaro was one of the three who was responsible for the day-to-day operations of the campaign. This was before he was president of FNAWS.

When did you start with FNAWS?

The first time I went was in Des Moines in 1976, so I go back a long way with FNAWS. There were approximately 100 at that meeting. I think it was held at a Ramada Inn. It was changed at the last minute. About a week before we were to meet, the original hotel burned down, so they had to move it. With men, women and children, there were about 100 total.

I was motivated to join because I was a sheep hunter. I started sheep hunting in 1973 with California bighorns. I never saw anything that year -not a ram, a ewe or anything. In 1974 I took my first Dall's sheep in Alaska.

This was before there was a FNAWS. This was just a get-together. The first convention was in 1978. I went to Detroit in 1977, and it was still an informal organization. It was just to get together with hunters and outfitters were starting to come and talk to us about future hunts. The early years were all projects — transplant or habitat improvement — after the Foundation formally started. I wasn't really into politics in the early 1980s. In 1987 was when I first ran for the board. I had been to all the conventions and had been fairly active. Businesswise, I was in pretty good shape and I was able to devote more time to it. I was elected for a three-year term. I wasn't reelected. I ran, but I was not elected. That was in 1990. Then we had a resignation from the Board in 1990, so I came on later in 1991.

When I first came on the Board we were in the process of redoing all the bylaws. We'd go to meeting after meeting after meeting until 8 at night. We had a set of bylaws that included a president-elect but the president-elect had to be confirmed the year the president was elected. We call it a president-elect, but technically a president-elect becomes president automatically. Our bylaws said it had to be confirmed. Unfortunately, Howard Pollock was president-elect one year, but the Board did not make him president. That's the only time that's happened. The Board decided to amend the bylaws to read that the president-elect won't become president and that anyone on the Board could become president. We made it a first vice president. This was a mistake. We should have a president and a president-elect.

In 1994 the first vice president thought he was in line to become president. There was a major conflict at the end of the meeting and the Board elected Garry Vince president. The thing we learned from this is that a president-elect is the best system. That's something that should be changed in the future. Right now its first vice president and all people on the Board have the same chance of being elected president. I think the president-elect system would eliminate conflict plus since we only have presidents for one year it would be a better grooming process for the president if you knew you had one year. That makes a lot of sense.

State chapters were becoming bigger and better in 1993. They started really raising a lot of money. The chapters raise the money, and they want credit. If national raises the money, then they should get credit. I agree with the state chapters that they should get credit for the money they raise.

I'm chairman of the convention steering committee right now. This is an advisory committee established in July 1997 for a two-year term so I will still be the chairman of this committee until 1999 unless I resign. This is a good committee that needs to be expanded with more non-board people. It doesn't do a lot of good to just have Board people on that committee doing the same thing they did on the board. An advisory committee would be better to have more people who are not on the Board who would look at things differently.

At this time it can't be done because that's not the way it is set up. Right now there's only two non-board members out of six. There are two staff on it, two past presidents and rest are current Board members. The steering committee came up with some proposals this year, which were passed by the board. One was a basic philosophy which was very important. The philosophy is that all net income from all auctions and raffles should be available to fund sheep projects. Some of that money has been used to cover other expenses. One of the major problems is the cost for the convention. The amount we are charging does not cover the expenses. We need to reduce the costs for putting on a convention or we need to increase the registration fees. It has been passed that we increase the registration costs for the 1999 convention. Until all the bills come, we don't know if everything was covered. The Board of Directors and the management of FNAWS have a responsibility to the donors to see that their money goes to sheep projects. I am a fiscal conservative.

In terms of projects to be funded and the expansion of FNAWS, do you see in the future more state chapters going the way of other state chapters where they are wanted to give proper credit for what they do?

Until just recently, almost all of the projects the state chapters funded were projects that came directly to the state chapters. As they're getting more money and National FNAWS didn't have money to fund their Grant-in-Aid projects, the state chapters tried to pick some up. The chapters felt that if they are using their own money, they should get credit for it. They didn't just want to send their money to national and not get credit.

I think its tough to expand national. Unfortunately, there are only so many sheep hunters, and we are a single-species organization. There's room for some expansion, but it's tough to have tremendous amounts of expansion.

If you were to speak to the membership of FNAWS, what would you want to say?

I think FNAWS needs to look at upgrading the magazine. It isn't the same class as some magazines. It's a little more technical, and I think a lot of people don't read it. The magazine needs a face-lift, a redesign. FNAWS needs to look at what their registration costs are. I think Hells Canyon is something we need to continue but at the appropriate pace. What does that mean? Let's not do transplants into areas where there were potential die-offs. Let's wait until the research is done. Let's make sure it is done slowly. It may be slower than we prefer because if we have domestic sheep in an area, you may have to wait five years or so. If you rush that you may risk die-off.

We need to emphasize the recruitment of youth. It's tough to do. Most of our programs we try to get sons and daughters involved with our projects. How? If I had the answer, I'd try to do it.

Garry Vince
President, 1994-1995

What are the memories of your presidency that stand out in your mind?

1989, Dr. Roy Schultz looks on as Garry Vince, center talks to Dan Pocapalia about Stone's sheep in British Columbia.

I was appointed, not elected. I was very proud that I was the first FNAWS president from Canada. When I first came on the Board, I was asked by one of the founders, Dr. Roy Schultz, if I thought I was very much out of place to be a director of the Foundation for North American Wild Sheep, let alone becoming a president. For a mountain boy from the North, it was very much a privilege. Being an outfitter and a kid in Stone's sheep country, I thought I had a lot to offer from a different point of view on sheep and sheep management. It's not just the transplanting of the sheep or the water holes. It's getting rid of the wolves and the predators. It's the burning. We were allowed to do this in the early years that made an upswing in the sheep populations in northeast British Columbia. I was a big part of collecting funds from different places.

Did you also work with the Forest Service and government organizations to allow controlled burning?

Controlled burning was done with a lot of funding from FNAWS. There was reticence on the part of federal workers to allow controlled burns. This is what we had to deal with in Canada. We finally got our wildlife biologist, John Elliott, to see that controlled burning was good management practice. He saw that wolf control was increased from controlled burning. He was the wildlife biologist for northeastern British Columbia. He did his thesis on the grunt of the African lion in Tanzania, if you can believe this. He knew how predators worked over there, but once he saw how Mother Nature worked in B.C. with the tough winters, high numbers of predators and the limited range, he saw that there needed to be a change. He introduced wolf control and the burning plans. He started the policy of shooting the wolves. The early days of British Columbia wildlife control they used what was called 10-80. They had poisoning programs throughout the program, but political pressure came on, so they had to do other things. This issue was what I brought to the FNAWS Board. This was one of the accomplishments that I was much a part of.

When did you get involved in FNAWS? Why did you get involved?

I can't remember. I was one of the top charter people (number 91). There was a good mixture of professional men and outfitters. It was a good organization to belong to because most of the money they made they gave back to wildlife and wildlife programs. Plus, it was mostly a volunteer organization, which I liked a lot. The other organizations I was acquainted with didn't put out nearly the money for wildlife as FNAWS put out. If you raise the money in FNAWS, the money goes directly back to wildlife.

Was there a particular direction that you saw in FNAWS that you became involved in? What's the glue that keeps FNAWS together?

We tried to get more members. There are a limited amount of hunters who are sheep hunters. I don't think we'll ever go above it. We tried every way that we could to introduce new members. You will find that each president tried to increase the membership, but it is very hard.

Sheep permits that come from different states and provinces are a big plus for the Sheep Foundation. FNAWS takes a very small percentage of the money that comes to the states and provinces.

FNAWS is a very tight-knit group. They all come here to this convention. Everyone digs in to help. This is for fun and everyone has a good time. Everyone comes to raise money for sheep.

In the nine years you have been involved with FNAWS, what kind of evolution of purpose have you seen?

We gained all the way along with dollars raised. We also gained respect throughout the various government agencies and state fish and wildlife departments. We have funded a children's education program in British Columbia. Funding of all these projects is what I am most proud of like the transplanting and water hole projects. I was involved with a Utah Chapter transplant three years ago. I see FNAWS continuing the way it is, but over the years there have been disagreements about where conventions should be held. You always have to look at your bottomline. Nevada is where FNAWS makes money. We've lost in other places even though people always want to move it seems like. The main thing the Sheep Foundation is geared to do is to raise money to put sheep back on the mountain. When we say putting money in to put sheep on the mountain, every time you help sheep you help other animals.

Do you see the necessity of FNAWS getting into lobbying and political activity?

No, I don't, because we do not have the bucks to fund lobbying and political activity. Safari Club International is probably the best lobbyist of anyone. They have the money. My clients and members of FNAWS will agree that if we just stay focused on raising money for wildlife and wildlife management, we'd be a lot better off. Let Safari Club do the lobbying for the hunter.

Would you want Safari to be the voice for FNAWS in terms of politics?

Safari lobbies for the individual hunter, worldwide. This is what FNAWS has tried to stay out of. Stay away from the politics.

Please reflect upon the strengths and weaknesses of FNAWS.

The strength is that we look at every state or province that has sheep and try to do something for them. We want to continue to try to work with all branches of the government. We have such a good name that anybody who needs anything will come to us. I'm just proud that I'm tied up with this group and all the things this group has done over the years.

The weakness of this organization is that there are a few people who have their own agendas. Many years ago Dan Pedrotti came up with an idea that each section of the United States and Canada would nominate a director from each section so that the leadership is spread around the whole continent. Everything that has ever gone wrong in FNAWS is people disagreeing on that Board.

I've always thought there was a conflict of interest to have a guide and outfitter on the Board. There should not be any outfitters on the Board, and I'm an outfitter. Every director who runs should be very carefully screened. He or she should come to the Board with a plan for what he wants to do. If he runs long enough, he's going to get on the Board. They should be on the Board with no agenda and nothing to be gained by being on the Board.

Who is the ideal FNAWS Board member?

The doctors, the lawyers, the contractors. They have to be good businessmen and they have to have a true heart for wildlife. Men or women on the Board is fine. Years ago, FNAWS had an outfitter advisory Board. They had an outfitter who sat in on the Board of Director meetings. This was a good idea. That way you don't need an outfitter on the Board. If you look at these past Boards, you'll see three or four outfitters.

Where to you think FNAWS is headed?

If we keep moving around the convention, we're going to lose. We have to sit down with the Rocky Mountain Elk Foundation and Safari Club International to get these convention dates all set up. We need to get their lawyers involved with ours. They have to have agreement. We're not only hurting one another but we're hurting the wildlife. We have goodwill ambassadors along with legal council. There are very good gentlemen on all sides, and we need to hash this out. We need to make a long-range plan because some of these groups have reservations three and four years in advance. We are hurting all organizations by overlapping.

My advice is to take directors from all different locations of North America. We definitely need to have Canada better represented. We shouldn't have an outfitter on the Board, but we need to have an outfitter sitting at all the meetings and listening. When I first came on the Board, Jerry Waite was the president. He did that. We gained a ton of information. We had an extremely good working relationship. Most of our work was done after the meetings were over sitting up in our room.

Pete Cimellaro
President, 1995-1996

What were your reasons for joining FNAWS in the first place and tell us about your presidency?

Cathy and Pete Cimellaro in 1995.

I got involved with FNAWS 20 years ago through the Arizona Desert Bighorn Sheep Society, a FNAWS affiliate. I began to come to conventions and meet people. My own sheep hunting evolved and I went throughout North America hunting sheep. It just grew on me. It became the disease so prevalent among sheep hunters - Ovis unitus, I call it. In 1989 I ran for the Board and was elected. That started my next six years of service.

I guess I could say it was my calling during that time because I viewed it as extremely rewarding work. I loved the people. They were all dedicated. It didn't matter if they had 10 cents in their pocket or 10 million, it didn't matter. They were consumed by this love of the outdoors and this love of these magnificent animals. That was my direction. In all my years working with Arizona, I worked with all agencies and individuals and did a tremendous amount of project work. I've been on more than 130 work projects — transplants, water hole development. I was elected president during my second term on the Board.

As president of FNAWS, what were some of your goals to achieve when you were president? And, did you achieve them?

There were two things that really stood out that were consuming me for many years. One was my work in Mexico. There had been many problems over the years with sheep conservation programs in Mexico. Indeed, they were really non-existent. FNAWS embarked on a project, which they let me coordinate as the Mexican liaison over the last six years. We tried to establish a new program in Mexico, a credible program that worked for the sheep and that worked for the people and was something we could clearly demonstrate was not just another Mexican giveaway. The program in Mexico involved complete management. It's been very slow and painstaking. Last year we sold four permits for Baja Sur in the Vescajano Bysphere Reserve. That money has been transformed into an ongoing management program. We are in the process, as we speak, of opening Northern Baja to hunting. Hunting there has been closed for 10 years. The complete management program is based upon the money raised from the sale of hunting permits. Saturday night at the convention in 1998 two Tiburon Island sheep tags were auctioned. This is the first time these have ever been auctioned.

The Hells Canyon project began during my presidency. This project is still evolving. It is so large, involving so many states and people, that it's in that process of evolution. I can't say now where it is going or how it's going to get there, but certainly that is a project that FNAWS can be very proud of and certainly will be successful, long term. It has brought a lot of people together. This cooperation is very important because what we're doing is managing sheep, not on a state-by-state basis, but we're managing sheep using the big picture in a large block of habitat. Its crossing state lines; its crossing jurisdictional lines from private land to BLM land to Forest Service land to state land. We're looking at the big picture and managing for the big picture. I think, long term, that's where we are going to have to be with these types of projects. I don't think we're going to be successful if you continue to look at very small areas of management. These big cooperative projects are going to be a great benefit. Bottomline $10 million for 10 years is the number we targeted early on because it certainly would open a lot of doors. We're going to be doing several things.

One, FNAWS has brought on board a gentleman who is a professional fundraiser. Murray Blackwelder is his name. Murray is as we speak meeting with large FNAWS contributors. He is also evaluating the way we do business. I can tell you we don't do business as we should, in my mind. We can do better. That is one of the things that I think Murray's relationship with us will point out. There are things now that we currently do that cost us rather than help us. I think that is going to come out in Murray's evaluations. I think that over the next year or two I'm going to be very optimistic and say that we WILL be doing things better. We need to. We'll be looking at the process of raising money and looking at the way we might be hurting that cause, administratively. Bringing a level of professionalism that is sorely needed. We've always relied on volunteers; we've prided ourselves on that. I'm not advocating a change in that, but what I am saying is that for our own office and for the Board of Directors we need to kick it up a notch. Our accountability has got to be beyond reproach. It has to be a situation that we will have to demonstrate that we are doing the right thing with the millions of dollars we are handling. It's important that we really do take care of business.

What advice you would give to future members of FNAWS and current members with your wealth of experience in various aspects of sheep restoration?

Cooperation really stands out. It's easy to have an agenda and to believe in what you're doing and to close off your mind to other ideas. I'm guilty of it. We all have been. It's been a problem for FNAWS. We need to look at things openmindedly and not who brought it to the table or what it is but what does it do. I think FNAWS really needs to adapt because we've had that problem in the past. There needs to be better cooperation between the Board of Directors, the officers and the FNAWS staff. We need to listen more. We have a wealth of great people in our membership and a lot of qualified people who can aid us. We really need to take things and try to chew on them and decide if that might be the right way to do things.

We need a little bit more flexibility. We need to listen to the people who support us - the chapters and affiliates. They are our strength and they are FNAWS away from the annual convention. We don't do a good job with our chapters and affiliates. We never have. It falls on the FNAWS Board. We've done such a lousy job over the years. That's why we don't currently have the kind of chapter system we would like to have and the relationship with the chapters that we should have. It all goes back to the FNAWS Board taking on the responsibility of developing a working relationship that works for the chapters and national. They have to take it upon themselves. I challenge them to do that. I recently wrote a letter to the Board, and that was one of the main issues I talked about.

What do you envision for the future of FNAWS?

I'm concerned. I do believe we have not done as good as we could and in order to do better we need to become more professional in all aspects of the organization. That is my biggest concern right now. I think it is imperative that we begin to be more professional at the administrative level and at the Board level. We need to make decisions about the right information. We need to do things for the right reasons. It's difficult for us because emotion has such a great effect on what we do. We have to leave a little of that out of it and look at it from the business perspective.

We tend to take things to heart too much because we're passionate about this sort of thing. That passion is a wonderful asset and at the same time it is somewhat of a curse because we sometimes think from the heart. I have optimism that things will turn around, and professionalism is the key. We need the right people to do the right jobs is paramount. Professionalism is defined as training - having the skills, the knowledge to administer an organization the size of FNAWS. We are a multi-million dollar corporation. We raise four to five million dollars in just two nights of auction at convention.

We all know to be successful, long-term planning is important. It is incumbent that you do long-term planning. That is part of the professionalism I am speaking about. You develop a program. You find out where you want to be in five, 10, 15, 20, 100 years. Then you put the programs in place that get you there. That's what we need to be looking at. Right now we're a little too "hand to mouth." We live from convention to convention. We don't have sight of that goal out there and where we need to be and where we need to be with sheep in the future.

Current Board members don't listen to former past presidents and former Board members, and they think they are inventing something that has already been invented so there's a lot of wheel spinning. The "has beens" are looked upon as doing a great job but now their time has passed and we're the current Board and we need to take care of business. You don't learn enough on the Board. You have a lot of turnover. By the time you learn something, you're off the Board.

I really don't believe the current Board will make the changes necessary. I think its going to have to be an involvement with professional managers and long-term goals outside the FNAWS Board.

The commitment from all of the past presidents and all FNAWS members is there. We want to do this. We're not turning our backs on the organization we love. We are there to do it but we just need to be utilized and not forgotten.

Daniel A. Pedrotti
President, 1996-1997

Why did you become involved with FNAWS? What was your motivation for joining FNAWS?

1997, President Dan Pedrotti with his wife Carolyn. Carolyn got her name placed on this plaque honoring the presidents' spouses.

I went on my first sheep hunt in 1977. Mike Simpson, a taxidermist, encouraged me to go to the FNAWS convention in Minnesota, which I think was the third convention. Subsequent to that I joined FNAWS, and went to all the conventions except Hawaii. In Houston, Red Duke was president. He was putting together a team to put on the Houston convention. I made a couple of trips to Houston. Red asked a team of people to put on the convention in Nashville, Tennessee. I, along with Tommy Caruthers, Lowell Baier and a small group of people were on the team. Lowell Baier was program chairman. Tommy and I worked under him on the logistics of the program. From there, I went on to become program chairman and was involved in every convention after that until I became president. I came on the Board for three years and ran for reelection. I was unseated. I was asked to come back to run the following year and was reelected. From there I became president.

Initially I joined FNAWS to get acquainted with the outfitters and talk to the hunters and find out where the good sheep hunting was located and who to book with. Conservation has something that's been in my life since a child. When I was asked to get involved with FNAWS, I felt a great sense of love for this organization and what they stood for which is putting sheep on the mountain. My involvement went beyond the hunting aspects and the conservation aspects took over.

What were your accomplishments?

Unfortunately, my main goal, to restructure the board election process failed.

Soon after being elected to the FNAWS Board in 1990, I became aware that an imbalance existed as to membership representation on the Board. It wasn't until my third year that I realized how severe it was. Upon being nominated for a second term and realizing that a campaign against my re-election had been successful, I did a study of the geographic distribution of our membership and found that the new board would consist of three members from the Minnesota-Wisconsin Chapter, one member from Mississippi and seven from the Western states and Canada. I also noted that the two states with the largest regular membership, California and Texas were not represented. Further analysis revealed that the four Western states and Canada with Board members, represented less than 20 percent of the membership. The remaining 80 percent were dependent upon the Board members from the Minnesota-Wisconsin Chapter and Leland Speakes, Jr.

In an effort to cure this inequity, broaden the views of the Board and to enable qualified candidates from outside the Western states to be elected to the board, I devised a plan whereby North America would be divided into geographic districts based on membership numbers. The plan was very simple. It provided that only two members from each district could serve on the Board at the same time. This was modeled after our bylaw that prevents more than two members from the same state and/or three outfitters from serving concurrently. The plan was rejected, and I was asked by President Bob DiGrazia not to bring it up at the members meeting during convention.

This condition persisted throughout my second term on the Board, despite my continued lobbing for a change in the election procedure. In 1996, I pleaded with the membership to nominate qualified candidates from California and states east of the Mississippi. I was successful in recruiting some great nominees. However, our election system, based on popularity and poor voter response prevented these exceptional members from providing the much needed leadership as we move into the 21st century. Since that time the situation has become more skewed to the West with only two Board members from east of Colorado. Now nine members are from six states and one provinces of Canada that comprise less that 20 percent of the membership and it could get worse.

This is my most serious frustration with FNAWS. We have some members with great dedication, resources and qualifications who could make tremendous contributions to wild sheep but they are precluded from serving due to our antiquated election system. They won't even run because nobody needs to have their good intentions slammed in their face. FNAWS will never attain its deserved position of greatness in the conservation world unless and until our most respected and deserving members are afforded the opportunity to serve on this Board.

What are the strengths and weaknesses of FNAWS as we finish 25 years?

Our strength — the sheep. We have a species that creates more attachment than any other in the world. Sheep of the world are loved so much by so many people. We have a built-in constituency. It's limited because of the cost of sheep hunting. We will never be an organization as large as the Elk Foundation because it's focused on sheep. Multitudes of people are just not available. Our strength is in this unique species and the emotional attachment that the hunters have for it.

One of the weaknesses is the limited number of people who participate in sheep hunting. We'll never have 100,000 member organization because there just aren't that many people. Our election process precludes us from attracting top people on to the board, people of stature, who can open the doors to corporate America and who would streamline our business approaches to things and who could effectively speak for us in the halls of Congress. The election is a pure popularity contest and its impossible for dedicated people who are not well-known throughout the FNAWS membership to be elected.

We need all geographic areas of our organization represented. For example, for the past 10 years, 80 percent of the Board represents 20 percent of the membership, mainly western. California, which has the most active membership in FNAWS, has not had a Board member. Pennsylvania, one of the strongest eastern chapters in the country, does not have representation on the board. Minnesota and Wisconsin have continually had representation. Texas has off and on had representation over the years, but there's nobody there now and there's nobody in Texas now who is well-known enough to think about running on the Board at this time.

Where do you see FNAWS headed and do you think it is headed in the right direction?

FNAWS has been cyclic in its direction. We've had ups and downs. We came off a serious down period during the San Antonio convention. We had a faction on the Board that absolutely had the wrong agendas and banned together to thwart every positive effort we tried to make. Those people were finally eliminated from the Board by normal attrition. We had a very strong focus and a targeted direction. After I left the Board in April of 1997 there seems to be a slippage and a factionalism recreated on the board. Frankly, the new Board scares me to death.

Dedication will always be there. I see a resurgence of interest among the former presidents and Board members. This wake-up call so to speak will be one of the most positive things to happen and we will regain our momentum to figure out a way to build a Board that has respect it deserves in the conservation world.

What advice can you give to future FNAWS members?

We pride ourselves on putting more percentages of the dollars we received on the mountain than the other conservation organizations. This is because of a strong volunteer Board and a minimal staff at the FNAWS office. It may be time to shed that image and become more professional about hiring the executive director or staff who are more qualified and so there are more of them. The staff now work very hard, but they need help. Maybe we need a larger staff and an executive director of stature.

Do you see FNAWS getting more and more into the political arena, or do you think your strength is raising money to go into conservation projects?

The move toward more political activism is part of our problem today. More attention is spent in Washington, D.C. than on putting sheep on the mountain. I think that is the problem we have today. We support WFLA and the Wildlife Management Institute. These people are the front runners. These are the people in the trenches who should be doing our lobbying for us and for us to use our energies and resources for that is a mistake. FNAWS will continue to raise hundreds of thousands of dollars for conservation, barring a complete collapse of the organization. We're seemingly on a downsize curve at this point and if its not corrected FNAWS will become a third-rate conservation group and funds will dry up.

I think my priority is not hiring a fundraiser. My priority is to create a prestigious board, obtain a professionalism from everyone involved, get back on our feet and start competing on a level playing field. I would like to see Boone and Crockett Club regain the high ground and become the umbrella organization that all these huge resources would look up to. We need to pull together to achieve this awesome question of what's going to happen in the next 100 years with wildlife, with habitat, with our right to hunt, with human population, with our right to hunt and all aspects of conservation. I see a need for leadership that will help everyone pull in the same direction.

The greatest threat to wildlife and wildlife habitat as we know it is the urbanization of rangeland of North America. I have a dilemma as a firm believer in free enterprise. The Nature Conservancy and the federal government take over lands that protect the habitat, yet its frightening to admit those very people who control these lands prevent us from using them for recreational purposes.

How can your skill and knowledge be better put to use in FNAWS and are they being utilized properly?

The past presidents and past Board members as a group are the most under-utilized pool of knowledge and experience that we have. Safeguards were put in to prevent control by former members who had personal agendas. But, I see a pool of recent and former past presidents and directors who would be willing to step up and provide this added energy and knowledge that's just not being utilized. There's a tremendous desire by a lot of the former Board members of the Sheep Foundation to regain the high ground that it deserves and that they are willing to step forward as soon as possible. They have allegiance to the animals and to the cause. They believe in putting sheep on the mountain. We need to understand our roots and our resources.

President Dan Pedrotti, second from the right, and Clayton Williams, right, give a thumbs-up to the mariaachi band during the 1997 convention.

Phyllis Carlson
President, 1997-1998

What have been your accomplishments as president of FNAWS?

Doug and Phyllis carlson at the 1992 FNAWS convention.

We started the year with a convention in Philadelphia that was with all good intentions but not successful. Along with Grant-in-Aid (GIA) overfunding in the last three or four years, and especially in the previous year as we funded an extreme amount to Mexico, went into the year with poor convention turnout and no dollars to fund. It was a real low point. Plus, we had a situation in Hells Canyon where we were trying to raise some funds and had not raised any. Our executive director, Karen Werbelow, raised $58,000 for Hells Canyon, but our fund-raising efforts did not.

This year we've restructured the Hells Canyon Initiative Committee. We got that back down to a manageable number of individuals. We cut it back down from 35 to 40 people to less than a dozen. It's pretty hard to make a decision when you have 35 or 40 individuals. So we got that under control and back down to basics to where these 12 people can have their own committees. We had a successful sheep transplant from British Columbia to Hells Canyon — 39 sheep into Hells Canyon into the three states of Idaho, Washington and Oregon. That was successful.

We're still dealing with Mexico with some permits for the 1998 auction. We do have three tags in Baja Sur. We'll auction those between the regular Friday and Saturday auctions. By tomorrow noon, we'll know if we have two Tiburon Island tags. Tiburon Island animals are regulated by the fish and wildlife director of all of Mexico. But the Seri Indians own the island, so there was a confrontation but we have all the players under one roof at this convention and we are pounding out a contract as we speak. (In reality, two hunting permits were auctioned at the Saturday banquet in Reno in 1998 and $395,000 was raised from the sale of the two Tiburon Island permits.)

What was it that attracted you to FNAWS in the first place and what's your history with FNAWS? When did you first get involved?

I guess I got involved because of the guy I married in the first place. He always said it was in the marriage vows that he would go hunting with his brothers; I would not make Spanish rice; and, we would move a lot. We accomplished all of those things only I went hunting with him instead of his brothers. Once you go antelope hunting, then you need to go elk hunting and pretty soon you need to go sheep hunting. We went Dall's sheep hunting in

1972 in the Wrangell Mountains, the mountain next to where the Swank ram came from. (Harry L. Swank, Jr. took the World's Record Dall's sheep in 1961. It scored 189-6/8 points.)

When we came out of that camp, we ran into Harvey Kadlac going into hunt in the same area. We didn't know Harvey from Adams' ox. He said he was from Minnesota, and we were from Prior Lake, Minnesota. The following year, or close to it, we got a card from Harvey that he was starting a Minnesota chapter of the Foundation. We had already heard of the Foundation and had been to their Minnesota convention in about 1978 or 1979. The meeting was out toward Prior Lake. They had a map that told us how to get there and I called up Harvey and told him we would be there but the number of the road he said to turn off on was not right. With that we talked for an hour and a half. This is March or April of 1979.

When we were at the meeting, Harvey asked me to take some notes. At this point in my life, unless I had been formally introduced to somebody, I wouldn't talk to them. I was scared of my own shadow. I was about as shy as the sun on a cloudy day. I've always thanked the Foundation for helping me come out of my shell. As you can tell, I don't have that problem anymore! So that's what started it. By the end of the day, when I was taking notes, which was about three sentences because 16 people were watching a movie, there's not a lot to say. At the end of that meeting they asked if I would be their secretary. I was their acting secretary for nine of the 11 years I was with that Board. That was the Minnesota Chapter of FNAWS which turned into the Minnesota-Wisconsin Chapter today. Then I was elected to the national Board of FNAWS in 1988 or 1989. I am now concluding my 9th year, which is three terms of three years each. I count it as 20 years with the Foundation.

Hunting and shooting sports got us starting. It started out as shooting sports because my husband and I were so young that the only thing we could afford was a box of .22 shells and go out and shoot tin cans. That's what we did for dates.

When you look at the 20 years you've spent with FNAWS, have you seen a more complicated development of purpose or is it still a singular purpose?

The purpose is to keep the sheep on the mountain in the first place, to make them healthy, have enough habitat for them, and that way to increase the herds. We want them at a sizeable number since we are hunters and want to hunt them. We want to make sure there is always a viable herd of the different species. There are more complications to what it used to be. Before, you could spend all your dollars on habitat, biological improvement, transplant, water hole projects, things for the sheep. Now you have to defend hunting. You have to defend the habitat. You have to defend yourself because sometimes animals have more rights than humans do.

We've wanted the money to go towards direct sheep projects, but now we have to spend them in Washington, D.C. We have to spend them to combat initiatives to wipe out hunting. Its not just our species that we have to defend. It's all of them which is like the story of the Lutheran German minister during World War II who wouldn't defend the Jews because he was Lutheran. Then the Nazis went after the Catholics but the minister didn't defend them because he was Lutheran. And then the Nazis went after the Lutherans but by then there wasn't anybody to defend him. You have to defend all the wildlife. As a single species organization, we struggle with that. We struggle with bear initiatives. We struggle with the trapping and the hound hunting debates. But, if we don't help these other groups, who will help us when attention is turned to wild sheep?

What do you see as the future of FNAWS?

I have a renewed confidence that the Foundation for North American Wild Sheep will continue and continue in such a way as being a leader of putting sheep on the mountain and putting your money where your mouth is. We're going to do this. We've always stood up for what we believe and stuck our neck out there. We have hired someone to raise funds, Murray Blackwelder. In just one day there are dollars in the bank that we didn't have before. There were people out there that were waiting to be asked. When you are elected as a director, you bring many talents from these 11 people but asking somebody for money to put into the bank for an endowment or as a grant is not always an easy thing for us to do. There's money out there for a great cause so we've hired a professional person to help us achieve that goal.

I've seen outfitters and artists being solicited from ourselves, our chapters, SCI, their chapters, the local rod and gun clubs and other conservation groups. They all want something from those people, and these people want to give, but you don't want them to go broke. They have to provide for their families and their other interests. I see that dwindling for all of us if we're not careful. There are wealthy people out there who want to support us and have been waiting to be asked in the proper way, to know where their money is going to go, how it's going to grow and who's going to take care of it. Now they are finding out through the language they understand what type of financial means we have and what type of banking investment people we use. I see in the future using that type of money to do our funding to cover our administrative costs and GIA projects; not as much from the outfitters and artists themselves.

We're going to progress. I believe the Foundation, along with Safari Club International and the Boone and Crockett Club, have had their opportunities to fail, but for some reason they didn't. Maybe the people in charge weren't the most creative or clever in this particular area but somehow they persevered. I think we've had our struggles. We're still going forward. I have all the confidence in the world of our Board and our future Board. There is history here. There are people who can give the Board history. That's where our strength comes from. We have things in place for the future boards.

We're getting a handle on a lot of things like in Washington, D.C. The anti-hunting factions have worn these people out but they're coming back. Lobbying isn't a dangerous direction to go but what we've done in the past we have funded groups that were lobbyist for us. We'll always have to participate in the political arena. We need a strong presence in Washington, D.C. In our bylaws policy, a 501-(c)(3) nonprofit organization can only spend a certain percentage on political activity. We'll spend up to the maximum. When we have done that in the past, we have turned to our chapters if something really needs to be pushed in Washington. Karen Werbelow has been invited to the White House twice. Maybe they've been anti-hunters themselves, but the fact is we were there, and we were invited.

How do we reach out to the next generation of hunters? How do we get younger people involved to carry on the tradition?

There's a program called "Becoming an Outdoors Woman." I have participated in two sessions of that in Minnesota as an instructor. The enthusiasm I see there is interesting. The first time I was an instructor I couldn't get over the age of women. How many older women there were, even into their 70s. This was in May. The following September a lot of younger women came to the class. I tried to figure out why they came. Many of them were engaged to guys who liked to hunt or fish or camp. I was really encouraged that they knew nothing about it and had not been involved with it through their parents. They're going to have children. Now they have both parents involved. Those kids are going to be there.

Even the grandmothers should get involved. I'm always trying to tell my grandkids about hunting and fishing and the outdoors and the appreciation for all animals. That's where I think woman play the big role in teaching their children an appreciation of the outdoors. Fishing seems to be an easy thing to start kids on. We also started our kids on dog training with throwing the dummy into the water. In shooting sports, we had them shoot a bb gun into a cardboard box so they could collect their bb's and do it all over again. From there, we started them hunting squirrels and chipmunks and we moved up to antelope and deer. You need to get them involved and a connection to the activity or they won't continue. The future is with the next generation.

Is there anything I haven't asked that you think I should ask that reflects upon your presidency?

This is still a man's world in the hunting arena. However, it started out with tolerance with women in camp. There is a real acceptance now. I still see that when it comes right down to climbing the mountain or running the Sheep Foundation, there's some reluctance.

I think this past year has been very successful now that it's about over. I'm getting so many hugs; so many pats on the back and handshakes from some women but the men, because its a man's world, I've really noticed how many men have come up and told me what a good job I've done, and many have called me "trailblazer."

Concluding remarks?

Past presidents grumble about being sent out to sea and not coming back to the docks. I see it a little differently than the past presidents, so I'm going to practice what I preach. I'm going to be supportive. I'm going to let the next Board handle things. I'm going to try to find my own niche and take it from there.

Leland Speakes, Jr.
President, 1998-1999

How did you find out about FNAWS?

1994, Leland Speakes, second from left, wife Monnie, left, daughter Lucy Capocaccia and her husband John.

One of the very first directors was Bill Walters. He was only 45 minutes from Memphis. We're located in what is known as the Mississippi Delta. The ground here is as flat as the top of your desk and the richest agricultural soil in the world. We have a group called the Delta Sheep Hunters that was made up of maybe 12 to 20 sheep hunters. We'd have a little meeting every year. We'd meet at somebody's hunting camp and show slides of pictures and talk about sheep hunting. Through that I got to know Bill Walters. Bill Walters was convention chairman in Memphis. This was all going on about the same time as the Memphis convention. Bill invited me to be the hospitality chairman in Memphis. My love of sheep and sheep hunting and being around a great bunch of folks led me to become more active in FNAWS. I've been active from the very beginning.

What has been the driving force of FNAWS? Has there been a change in its purpose?

The accomplishments that FNAWS has made in protecting and increasing sheep populations has been a real revelation. It just makes me feel good to be a part of that. The restorations, the transplants, the enthusiasm of the people for the sheep and plus, sheep hunters are different. They're different because they always want to be on top of the mountain. For any lesser monies, you never get to the top of the mountain.

How would you describe your passion for wild sheep?

I would describe it with a little story. Whenever I abdicated to go hunting for anything else in the western United States - elk, mule deer, anything else - I find myself thinking, "I ought to be sheep hunting." It takes precedence over anything else.

What role have your played in FNAWS?

In the early years, actual hands-on projects, no. I haven't been a mover and a shaker for actual hands-on projects. I've been involved in raising money for these various sundry projects. At one time I was very active in the Mexico effort, way before Pete Cimellaro, in the late 1970s or 1980s. FNAWS was very active then, too. We gave them trucks, radios, binoculars. The situation down in Mexico has not been monitored. It has not been checked. The problem has not been solved.

If I had a person who was as easy pickings as we are, I'd quit everything else and pick that person all the time. We just believe the sun will come up every morning and it will go down, and we think the Mexicans are just like that. They have a hand out continually. If it was not for the fact that they have such a wonderful resource, I would be glad to write them off. They have such a large population of desert sheep. It is the most wonderful resource you can imagine.

At one time the game department managed it properly along with the guides and outfitters who did a marvelous job. We need to see if we can have surveys done to determine if there is a huntable herd of sheep. We have spent an absolute fortune down there trying to satisfy those people.

What is the significance of FNAWS in the last 20 years?

I think we have been instrumental in opening sheep hunting and increasing the healthy populations of wild sheep in the United States and Canada. It's only in recent years that it dawned on us that there were diseases that were wiping out these sheep almost as fast as we could put them in. This is the pasteurella. Early on, I never heard of pasteurella. FNAWS has been instrumental in getting antibiotics to the animals in recent years, absolutely. We're not through with this problem. We have a long way to go. This is certainly something the scientists at the Caldwell Center in Boise, Idaho, are studying. Dr. Dave Hunter is the director and Karen Rudolph is our scientist.

When all these sheep got sick in Hells Canyon in 1995 we rescued 70 of them and brought them back to Caldwell. All of them died, but this was the first time we had tissue samples of recently dead animals. We have a tremendous volume of information. We are running DNA tests on it. She is also looking for a gene that would make offspring to be immune to the disease. She's looking for a cure. We have two centers. One is called the Cian Center. It was the original University of Idaho veterinary clinic. They built a new one and they gave this facility to the Idaho Fish and Game Department.

In the last three years, we've put $250,000 into Mexico. My favorite expression is, "We finally have the Mexican deal worked out . . . once again."

In terms of your presidency, do you have particular goals?

Hells Canyon is utmost in my mind. Getting some viable money program going for Mexico. A disease study is right at the top of my list. Maybe fundraising will be my flagship. We have struggled over these many years with hand-to-mouth operation. We give away all the money that we have and then we're broke. We generate more money and give it away. That's the way a nonprofit organization is supposed to operate. You're not supposed to have any money.

We've hired Murray Blackwelder, a fundraiser from Iowa. I am very impressed. He will direct the fundraising. We've had nonprofessional fundraisers every other year. We pride ourselves in being run by volunteers, which is fine to a certain point, but I look around at our efforts to raise big bucks for sheep, and it never works. I am convinced that Murray is going to guide us. He took it upon himself at this convention to meet with the major donors to FNAWS. The people are so anxious. That's what's so gratifying about this organization. The people on the bottom rung financially all the way up to top rung are all willing to give what they can. He'll give us a report. I'm not going to speak until I see the report.

If you were to give advice to people for the future of FNAWS, what would you say?

I think it is absolutely and totally essential that we develop an endowment trust fund. We need to have this organization financially able to run itself so that our total income from the convention goes to grant-in-aid. We need our running of the office, running of the organization, running of the membership, magazine, all of that taken care of by itself with memberships and through income from endowments. Murray has it broken down as to how endowments would be used.

We are stymied in membership growth. I have been the membership chairman the last two or three years. The first year we were at 4,500. We finally got up to 6,000 and it has stayed there. We might recruit 400 or 500 new members, but we lose 400 or 500 members. We need 10,000 members but I have not been able to do that. I'm going to appoint somebody besides me to be membership chairman and maybe get some new blood in there. We've had membership campaigns and given away prizes. Our life memberships drive have been very successful.

Everything hinges on money and memberships. Everything hinges on those. If you have those all these other problems will be solved. We also need to be politically involved. I've been going to the Congressional Sportsmen's Caucus every year. And, as a result of that I don't like being in Washington and I don't particularly like the bureaucracy up there but I've gotten to know those people and I've gotten very comfortable up there. If you want something done, it's nice to be able to pick up the phone and say, "Hay, Joe, this is so and so" and then you can tell him what's on your mind.

Should FNAWS fund a full-time lobbyist in Washington, D.C.?

I don't think we need a full-time lobbyist, no. There are a dozen different groups and it seems like we've joined too many. Maybe we need to centralize that, but still be involved. You have to stay politically involved.

Leland Speakes poses with Philadelphia native, Ben Franklin, during the 1997 convention.

Interviews with Early FNAWS Members

Three long-time members of FNAWS, Keith Johnson, Fred McMurray and Bill Walters who never became presidents of FNAWS, were interviewed by Susan C. Reneau at the 1998 convention or in the summer of 1998. Their contributions to FNAWS were mentioned by past presidents and other FNAWS members as people who should be interviewed. Susan's questions to them and their responses are reproduced.

We begin with Keith Johnson who was one of the first guides and outfitters to step up to the plate to be counted as a supporter of an informal group of avid wild sheep hunters that became FNAWS in 1977. He helped the early members to locate names and addresses of men and women who might want to join FNAWS or attend early FNAWS meetings.

Throughout the writing of the 25-year history of FNAWS, the editors of the book mention the volunteer work of Fred McMurray. Fred was at the first informal meeting in Mt. Horeb, Wisconsin, in 1974 and continued to dedicate his passion for wild sheep to this growing organization we now call FNAWS. Even though he never was elected president, Fred was cited by numerous current members and founders as someone who has a profound perspective about the mission and purpose of FNAWS. His words serve as guidance for future generations of wild sheep fanatics.

Bill Walters stepped forward at the first FNAWS convention to set a tone of success for the fledgling organization. Many past presidents and current presidents recognized Bill for his continual contributions to FNAWS, so his interview is included in this section about "Movers and Shakers."

Hundreds of other outstanding FNAWS members and supporters have been interviewed and included in this book through its dozens of chapters, but these three interviews seemed to represent the overall tone of what it means to "put wild sheep on the mountain."

Keith Johnson
Charter Member 98

Please tell us how you got involved with FNAWS.

I met Duane Smelser at another show and he advised me that I should be in this little clique of people. He tried to explain it to me. I didn't really understand what it was except that it was a group of sheep hunters who got together. I joined and gave him the money for the charter membership at the Safari Club show. He was collecting the money for this group.

The first convention-type meeting I attended, I think, was at a Holiday Inn in Minneapolis during a bad blizzard. There were 530 people as I remember attending the festivities including seven guides from Alaska. I was one of the guides.

That was the show when Ed Stedman, Jr. had just killed the Number Two desert sheep in Mexico (now ranked Number 15, 187-6/8 B&C points, taken in 1976) and came busting in the door to that convention and brought that sheep horn in there. Now it's probably ranked a little lower but everyone was very interested in the desert sheep hunt coming off the Baja.

Tell me some crazy stories about the early years of FNAWS.

Two small chickens were auctioned off during the ladies' auction. They were chickens that were donated to be auctioned off to be hung as decoration in a kitchen. They weren't live chickens. These were long-necked chickens. Skipper Bettus ended up getting stuck with them. The chickens were his property coming out of the show.

As he walked by he turned to me and said, "Come here. Come here. Take these. Take these. Give me $100."

And I said, "What the heck is the $100 for?"

And he said, "Just give me $100. Just give me $100!"

Joe Bishop and I were together. The two of us kept walking down the aisle, and Joe kept asking what this was for. I was fumbling around, and Joe Bishop said something.

Skipper said, "Joe, give me $100 right now."

So Joe pulled a $100 out and handed it to him and asked, "What's that for?"

Skipper said, "Here, take these chickens and hold onto them." As Joe held onto them, Skipper started to walk away, and he said, "Congratulations, you just bought two chickens."

Joe Bishop chased him all the way down the hall, and today they are still hanging in Joe Bishop's kitchen. And, he's still talking about his $100 chickens.

I remember the Texans started passing a hat around and they started auctioning the hat. I can't remember what it sold for but it was an outrageous fee. It became a competition between people from Minnesota and Texas. They were trying to buy this man's hat. Of course, Rich Roan (the auctioneer), fell right into the line of it, and he started goading people into buying this hat. He'd get more people to pay outrageous prices for a hat. They got the wildest number. They finally bid that thing up so high that the Texans finally won their bid on it but I do not remember how much they paid for it. I think it was only auctioned the next year after but it wasn't as big a splash as that first year.

What does FNAWS mean to you?

Probably the most important thing to me has been the people I have associated with and enjoyed being with who are sheep hunters. They seem to be the finest people in the world. They are the honest, hard working, straight-forward, up-front people whom I have enjoyed all my life and these people are very unique to be associated with. I really come to these conventions for the comradery with friends like that. I've been in FNAWS since its beginning, but I really enjoy everybody who is in FNAWS. I've been an outfitter in Alaska for many, many years. I hunted sheep for an awful lot of years and then I sold out my sheep operation and hunted bear, moose and caribou on the Alaskan Peninsula. I sold that out in 1984 and keep hunting bear, moose and caribou. I've been an exhibitor since the beginning in Minneapolis when only seven guides from Alaska attended.

Are there any words of advice you can give to future members of FNAWS who aren't even members now?

I don't see very many young people in here and I'd like to see the younger generation brought into this organization because they'll enjoy it. We need to promote the younger generation coming here. My own children don't come here anymore because there's none of their peers here.

I think people who are devoted sheep hunters and devoted people to this organization should bring their own kids. Go out on a limb. Pay their way. Bring them in. Bring them something to come here for. My kids range in age from 23 to 36 years, and they really don't have an interest because there's really not anyone here who is their own age. My boys have hunted with me and have participated. Jerry has been one of the guides and so has Robert hunted with me but their interest has fallen by the wayside. They are young people just getting started in business so it's hard for them to come.

In the beginning, as I understand the rules, we were organized as a traveling convention. There's been some animosity for coming to the same place all the time and traveling all the time. I realize that in Reno we make more money because they give us a good deal on booth space and rooms. It's easier to have a convention in Las Vegas or Reno, but our traveling should include Los Angeles, Las Vegas, Houston, San Antonio, Dallas. When I was on the Board there was a committee elected to go investigate these different places.

I would like to see us go to Alberta, but we investigated that and went up there and the people from Canada welcomed us with open arms. We got a wonderful situation. We found a place to go in Calgary. Come to find out the place where we got stopped from bringing gold, guns and bronzes was across the border. It was not a problem going into Canada but the problem was the U.S. Customs to come back. They are the ones that stopped us. I'd like to see that program to see if they'd allow a convention to happen. They squealed like a mashed cat when we tried to do it back then. We lost out on the Alberta adventure. We wanted to honor the Alberta people and the British Columbian people and the Northwest Territory people. We were trying to hold it there to bring more of those people into FNAWS, but it didn't work. I think Dallas and Denver would be good places to go but I'm not on the Board anymore. If we travel I'd like to see those two cities investigated.

Fred McMurray
Founding Member of FNAWS
Charter Member 9

Tell us how FNAWS began.

Jim Froelich really got things going. That was in Wisconsin. He called Duane Smelser and myself in 1973. This was before we tried to associate with the Grand Slam Club. We were at Mt. Horeb first. We were at the Playboy Club after that and then we were in Minneapolis-St. Paul after that. We were first in Mt. Horeb, Wisconsin, on a snowy, blizzardy weekend. We met at a small hotel, I think a Holiday Inn. It must have been in February. On the way home they took us off the expressways and put us in schools because they had all the expressways closed. We were in southern Michigan when we got called off the highway going back home. I'm from Florida now, but I was originally from Michigan.

Fred McMurray at the 1997 Men's Lunch.

What was the motivating factor for coming together? Did you know each other?

We didn't know each other. Jim Froelich had called Duane Smelser and Duane had mentioned the meeting to me. Duane was a friend of mine and was a member of Safari International Club at that time. Jim said he was just calling a bunch of sheep hunters together just to have a meeting and talk sheep. Everybody was asked to bring their pictures with them and if they had movies, bring those, too. That's exactly what happened. Everybody brought their pictures of sheep and movies and sat there all weekend and talked sheep and sheep hunting. It originally got started just to share pictures and stories. We ended up enjoying ourselves so much that weekend that we decided to get together next year. We talked about where to get together and we decided the Playboy Club in Wisconsin in 1974. We all decided we'd try to bring another sheep hunter with us.

I think Detroit was when we really started to become an organization in 1977. That was when we got our organization moving. This was when we realized we were attracting hunters. We had met Mike Valencia who had gotten us a couple of desert sheep hunts. One was auctioned off and one was raffled off. So many people heard about this. We had a crowd of about 200 people at that time who came to that convention. We had it at the Hyatt in Michigan outside Detroit.

What was discussed at the Detroit convention? Did you talk about a mission or a purpose?

We realized then that we were an organization. I was on the Board. That's when we approached Bob Housholder if we could get affiliated with him and the Grand Slam Club. He said yes at first but then he changed his mind when he didn't think we were strong enough to do him any benefit. He didn't think it would have worked out together.

We decided to call ourselves Foundation for North American Wild Sheep in Memphis in 1978. I think we went to Omaha before we went to Detroit in about 1975. Our initial interest was to conserve sheep. We thought we were a success from the very beginning because it just seemed to blossom. Every year it doubled and doubled and tripled and tripled. I was always a pusher on the Board to move our organization around. Some of them wanted to keep it in one spot, like Las Vegas, and I always said we should pick up more hunters. I figured the only way we would pick them up was to move to different areas. I think this paid off, and that's how we spread out so fast. Reno is good. The Canadians can come down without going long distances. That's why we always liked Reno.

From your perspective as someone who has been with this organization from the beginning, are you happy with the way things are going?

I think their intentions are still the way we intended. They are pure. We have some political problems, but that's growing pains. Every club has that. I think it is run well now. It's just a little different now. It's not the close group you knew. Everyone knew each other. Now you just come in here and it has gotten so big. It's unbelievable. I'm so proud to be a part of this. It gives me chills every time I think about this. It started as a small friendly group and now ... I have some great memories of the early years.

We were on a carousel and I had fallen asleep. My head was turned back and every time we came by this table of girls, they started pitching popcorn at me. I don't remember which carousel, but I think we were in Omaha. I was sleeping, so I don't know.

I think one time we took brown wooden menus at a restaurant and we bet who would have enough nerve to take one out. We made bets on this and the money would go towards the Sheep Foundation. One guy did it and the other guys lost, so the money went into our fund. And, I think that menu went around at our auction for two or three years.

The other thing that was pretty exciting was Duane Smelser's hunting hat. I took his hunting hat and we auctioned it off.

He said, "You can't auction off my good luck hat."

He was raising cain about that hat, so it brought that much more money. The auction price got up to $40 or $50 and then he said his hat was worth more than that. So the price went up higher. That hat went up for auction for four or five years in a row. I think Duane finally bought it back or someone gave it back to him. It just kept getting passed around to create more fun.

Are the original founders still active in FNAWS?

Two or three have passed away. One is Lloyd Zeman. Jim Froelich. He's gone.

Do you think FNAWS is headed in the right direction?

Yes.

Can you give future FNAWS members about what they can do? Any advice?

Our main goal is to put more sheep on the mountain. That's it. We need more sheep on the mountain for our kids and for the generations after that. I think we have accomplished that and we're still doing that. You can't believe the funds we raise here. We're now up to over 5,000 members. In the beginning we never thought there were 200 sheep hunters in the world. I think in the Grand Slam Club at that time when I applied for it, I was number 56. Mainly we concentrate on sheep and anything to help sheep. We started out small, and now look at us. We commit so much funds, and that's it.

Bill Walters
Charter Member 5
Convention Chairman, 1978
Memphis, TN

What was the success?

We agreed to do that at the previous year's meeting of the Midwest Chapter of the Grand Slam Club in Dearborn, Michigan. That was held in early 1977. We formed the Foundation for North American Wild Sheep shortly thereafter and I agreed to be the convention chairman for the first convention. There were other meetings, but that was the first time the group was called FNAWS. We incorporated it and got it approved as a tax-exempt organization with the Internal Revenue Service code 501-(c)(3). We did all of that.

Why did you get involved with FNAWS?

I'm an avid sheep hunter. I just love the animal. I was impressed with what Ducks Unlimited did for ducks of North America, and I just wanted to do the same thing for the North American wild sheep. As a matter of fact during some of our initial meetings we had extensive discussions about as to whether or not the objective of the foundation would be sheep of the world rather than sheep of North America. After several meetings and discussions, we decided to restrict it to North American wild sheep because it was just too broad of a directive to go for the whole world and too many governments involved. It would have spread us out too much.

Lloyd Zeman was the spearhead of the thing. He started out as the volunteer executive director. Then, as we got some money, we hired him on a salary. During that first year, and even into the second year, we had meetings of the Board and Lloyd told us if they didn't get the dues money in we'd have to come out of our own pockets and keep the thing going until we could get to the next convention. But we never did. Some of those initial people were Dr. Roy Schultz from Iowa, Dr. Jerry Waite from Iowa, Jerry Christian from Iowa, and John and Jane Babler from the Minneapolis area. Jane wasn't a director but she was of great assistance to John. They ran FNAWS out of their office for a while. Lloyd had his office in Minneapolis for a while. Frank Bays was another director. He hasn't attended in years. Michael Valencia is another one of the originals. He still comes to conventions. Last time I saw him was two years ago. Another couple of early guys were Frances Stubbs of Atlanta and Fred McMurray. I saw Fred 30 minutes ago. I think he was in that Mt. Horeb meeting. Duane Smelser was another one of the originals. I haven't seen him in years.

What was it you wanted to accomplish in FNAWS?

I wanted to do something, particularly in range management. During the early years, Canada had the Olympics and they totally devastated a winter range for bighorn. Building the facilities for the winter Olympics, I wanted to be involved with that. The Canadian government seemed to want to propagate wolves. Wolves can have such a devastating effect on wild sheep. Mainly, habitat protection was what I was interested in. I felt like the sheep could take care of themselves if the habitat was taken care of. FNAWS could help by raising money and selling hunting permits. We put sheep on the mountain. We took that money and fought to save the habitat and to help the outfitters to perpetuate habitat in their areas. Americans have made a lot of mistakes in the American West. Canada is 30 to 100 years behind us. I was hoping the Canadians wouldn't make the same mistakes as Americans.

What did you learn in the early days of FNAWS that didn't work that you vowed you wouldn't do again?

In the early days we had a very difficult time doing anything with the Alberta government. A lot of the states, particularly Wyoming, accepted us with open arms and gave us sheep permits. Even Mexico gave us permits. Matter of fact, the very first convention in Memphis we had five desert sheep permits to sell from Mexico. We had five desert sheep permits to sell when such permits were really hard to get. Alberta was extremely tough to deal with. It has only been in later years they have given us a sheep permit. I think they felt like their sheep herds were in such good shape and the range was in such good shape they didn't need us. They may have been gun-shy from preceding organizations, not us. I never understood why they didn't accept us. Some of the problems have been overcome, but only in the last five or six years. Mexico is way more difficult now than it was back in the early days. I don't know exactly what the problems are, but I think it has something to do with anti-hunting organizations down there.

Where do you see FNAWS going in the next few years and the direction for the new millennium?

Despite the activity from the anti-hunting organizations, I think FNAWS will continue to grow and to continue to meet its objectives and continue to be a very visible force in wildlife conservation. I think FNAWS has a very good future. I continue to see young hunters coming in. We went for a period of time when we didn't see young people during the late 1980s, but now, as I continue to come here, guys my age in their early 50's and these young guys continue to come. And now there are lots of women. We always had a lot of women around, but they are more prevalent now than before. That's good. That shows more interest on their part.

Is there any advice you can share with the future generation of FNAWS members? What are things that do work and don't work that you'd want to tell us?

If they basically just get involved. If you believe in what FNAWS is doing, then get involved with it and volunteer to help. Do whatever somebody asks you to do that can help the organization. Try very, very hard not to use it as a self-serving job for your own personal joy or your own personal satisfaction. That's hard to do sometimes, but sometimes in an organization like this there are people who try to use it for their own benefit, and I don't think that's good. One thing I think is significant about FNAWS as opposed to other volunteer conservation organizations from the very beginning, FNAWS directors never took any salary and were never paid for the work they did. The only thing they ever got were their travel expenses paid. I think that's good because it shows the intent of the directors to work for the wild sheep and the goal of the organization. I was on the Board of Directors for a long time. I got involved at the 1977 Midwest Chapter of the Grand Slam Club. The Board of Directors are the heart of FNAWS. Sheep hunters are conservation minded.

Give us some recollections of what the 1978 convention was like compared to the 1998 convention.

Of course, the crowds in 1978 were much reduced. We did have some exhibits and people from all over North America. We had a contingent from Mexico. We had several outfitters come from the Yukon, British Columbia and Alberta. I do remember that our total auction take was about $40,000 which included five very highly coveted Mexican desert sheep permits that at that time on the open market would have gone for $5,000 a piece. We had a very good auctioneer from our area named Rodney Garrison who was equally as good as these guys still used today. We had an outstanding desert sheep ram that we displayed that we called the Ed Stedman ram, which was at that time one of the largest sheep to be killed in Mexico. Ed kindly let us use that and he shipped it in for us. (The Stedman ram is ranked 15th in the world by the Boone and Crockett Club with an official score of 187-6/8 points). We also had for the banquet on Saturday night a very outstanding Dall's sheep from Alaska. I remember being very upset about that because I had made the arrangements through an outfitter in Alaska to send it to us 10 days before the convention, and on the day before we started it still wasn't there. I was responsible and I was trying to figure out what to do about this guy's ram. Delta Airlines said they knew nothing about it and they couldn't do anything about it. I was really upset. I kept calling air freight at the Memphis airport and they kept telling me they didn't have the sheep or anything for the Sheep Foundation. They said they didn't know anything about it. The banquet was about to start at 8 p.m. At 5 p.m. they called my room and said they had the sheep. They wanted to know what I wanted to do with it, so I told him I would pick it up. The sheep had been there the day after the outfitter sent it but it was stuck back in a corner. I was relieved. I had to guarantee that hotel for the Saturday night banquet 200 plates. We got down to a few hours before the dinner after selling advanced sales. We had 160 tickets sold so we were 30 or 40 short of our guarantee. We had to pay her for the 200 plates. We didn't know where to start. There was nobody to help. We just did the whole thing.

Appendix

FNAWS is more than a conservation organization, it's people working together.

FNAWS CHARTER MEMBERS

(1) Lloyd Zeman
(2) John Babler
(3) Duane Smelser
(4) Roy Schultz
(5) Bill Walters
(6) Frank Bays
(7) Mike Valenica
(8) Robert Pagel
(9) Fred McMurray
(10) Jerry Waite
(11) Lloyd Bare
(12) Roger Hill
(13) James Froelich
(14) Lowell Baire
(15) Don Lesier
(16) Curtiss Judd
(17) J.D. Maiden
(18) Dr. Paul Burbank
(19) C.F. Munk
(20) William Falls
(21) Roman H. Hupalowski
(22) Bud Callahan
(23) L. Irvin Barnhart
(24) J. Daniel Thill
(25) Robert Zachrich
(26) Hobart Pardue
(27) Watson T. Yoshimoto
(28) Alfred Winchester
(29) Sam Sanders, Jr.
(30) James Gray
(31) Dr. Robert Bullard
(32) Robert Moses
(33) Marlin Grasser
(34) Paul Weimer
(35) Marshall Rush
(36) Edward Busche
(37) Richard Wehling
(38) Warren Parker
(39) Gary E. Hines
(40) Ted Walters
(41) Wayne Weber
(42) Charles Erickson
(43) John Schwabland Jr.
(44) W.E. Moynihan
(45) Buron Griffin
(46) John Mills
(47) Jerry Malavos
(48) Jerry Eckward
(49) Sam Winston
(50) Jack Thorndike
(51) Don Harstad
(52) Jim Babala
(53) Chancy Ernst
(54) Carl Bailey
(55) J. Dorsey Smith, MD
(56) Larry Kolb
(57) Harvey Kadlec
(58) S.G. Stephan
(59) Elvin Hawkins
(60) John Epstein
(61) Win Condict
(62) Russ Kloosterboer
(63) William Townsend III
(64) Mike Mills
(65) Dennis Nastrom
(66) Earl Boose
(67) Ed Stedman, Jr.
(68) Mahlon T. White
(69) Don Haney
(70) Glen E. Park
(71) Johnny Moore
(72) Burt Jaeger
(73) Jerry Smith
(74) James Schmalz
(75) Casey Jones
(76) Donald Prince
(77) James Hicks
(78) Daniel Galbreath
(79) Frank Pachmayr
(80) Abner Clements
(81) Bert Dickey III
(82) Sam P. Ingram
(83) Gary Koopman
(84) Lyle Warner
(85) James Endlich
(86) Harold Schultz
(87) Lanny Wilson
(88) Gaylord T. Soli
(89) Richard Van Vorst
(90) Thomas J. Radoumis
(91) Garry Vince
(92) Clayton Martin
(93) Charles Ruperto
(94) Patrick Bollman
(95) Paul Asper
(96) Dr. Edward M. Kowaleski
(97) Marvin Gaston
(98) Keith Johnson
(99) Bill McKenzie
(100) Jerry Christian

Award Winners of the Foundation for North American Wild Sheep Convention Location and Year Presented

OUTSTANDING ACHIEVEMENT AWARD WINNERS

Mario Luis Cossio	Detroit	1977
Lanny O. Wilson	Memphis	1978
Dr. Valerius Geist	Minneapolis	1979
Wyoming Game and Fish Department	Houston	1980
John Batten	Phoenix	1981
Carl Mahon	New Orleans	1982
Arizona Desert Bighorn Society	Reno	1983
Marvin Wood	San Diego	1984
John Elliot	Reno	1985
Dick Weaver	Hawaii	1986
Robert McQuivey	Nashville	1987
Dr. Tom Thorne	Reno	1988
Dr. Larry Irwin	Reno	1989
Dr. Ian McTaggart-Cowen	Reno	1990
Raymond Lee	Reno	1991
Glen Roberts	San Diego	1992
Reed Steinmetz	Reno	1993
Arlene Hanson	San Antonio	1994
Dr. Robert DiGrazia	Phoenix	1995
Donald Peay	Reno	1996
Don and Leona Clower and Pete Cimellaro	Philadelphia	1997

FEDERAL OUTSTANDING STATESMAN AWARD

Dr. Valerius Geist	Minneapolis	1979
Ron Sommerville	Nashville	1987
Thomas McMillan	Reno	1988
Stephen Mealey	Reno	1989
President George Bush	Reno	1990
Dick Shultz	Reno	1991
Duane L. Shroufe	Reno	1993
James Baca	San Antonio	1994
Bob Richmond	Phoenix	1995
Tim Schommer	Reno	1996
Jack Ward Thomas	Philadelphia	1997
Melanie Woolever	Reno	1998

STATES STATESMAN AWARD

Senator Robert Presley	Reno	1988
Senator Ken Fanning	Reno	1989
Governor Mike Sullivan	Reno	1990
Dale Wills	Reno	1991
Robert Tully	San Diego	1992
Bob Stevens and Herb Meyr	Reno	1993
Wayne Heimer and Jack Kilpatrick	San Antonio	1994
Craig Stevenson	Phoenix	1995
Vic Coggins and Lloyd Oldenburg	Reno	1996
Kevin Hurley and Felipe Ramirez	Philadelphia	1997
Lyn Nielsen	Reno	1998

OUTSTANDING CONTRIBUTOR AWARD

Helmut and Gerda Greul and Rich Roan	San Diego	1984
Doc Thurston	Hawaii	1986
Arthur R. Dubs	Reno	1989
Robert Senter	Reno	1990
Stanley Boots	Reno	1991
Greg Stires	San Diego	1992
Stanley Boots	Reno	1993
Jerry Fletcher	San Antonio	1994
Jerry Fletcher	Phoenix	1995
Lee R. Anderson	Reno	1996

The Outstanding Contributor Award was renamed the Outstanding Conservationist Award beginning with the 1997 Philadelphia convention.

OUTSTANDING CONSERVATIONIST AWARD

Jerry Fletcher	Philadelphia	1997
Lee R. Anderson	Reno	1998

PRESIDENTIAL AWARD

M.F. Wood	New Orleans	1982
Raymond Allemand and James Allemand	Reno	1983
Howard Pollock	Reno	1985
John Turner	Reno	1989
Charles Jagou	Reno	1990
Dean Picl	Reno	1991
Arnie Johns	San Diego	1992
Ron McMillen	Reno	1993
Pete Cimellaro	San Antonio	1994

PRESIDENTIAL AWARD continued

The Foundation Administrative Staff: Executive Director, Karen Werbelow Publication/Advertising, Sandra Nicola Exhibit Hall Coordinator, Terry Ziehl Membership/Merchandise Coordinator, Kim Nieters Grant-In-Aid Coordinator, Becky Layne Convention/Public Information, Paula Karres	Phoenix	1995
Alberto Tapia, Dr. Pablo Chee Rodriquez, Dr. Manuel J. Chee Rodriguez and Ray Lee	Reno	1996
Tommy L. Caruthers	Philadelphia	1997

SPECIAL ACHIEVEMENT AND RECOGNITION AWARD

Jim Froelich	Houston	1980
D.J. Thurston	Reno	1983
Jim Froelich	Reno	1985
Dick Weaver	Hawaii	1986
Teresa Heinz	Phoenix	1995
Dr. Jerry Waite	Reno	1998

PINNACLE AWARD

Duane Smelser, Lloyd Zeman, Jerry Christian and Dr. Roy Schultz	San Diego	1984
Tom Saad	Reno	1985

GARY POWELL MEMORIAL AWARD

Doc Thurston	Reno	1985

CHADWICK RAM AWARD

Duane Smelser	Reno	1985

GORDON EASTMAN'S GRASS ROOTS AWARD

Ed Pribyl	Reno	1988
Tom Brown	Reno	1989
Dean Bowdin	Reno	1990
Bill Leech	Reno	1991
Richard Mitchell	San Diego	1992
Ron Ball	Reno	1993
Pat and Gene Holloway and Harold Eshelman	San Antonio	1994

GORDON EASTMAN'S GRASS ROOTS AWARD continued

John and Jane Babler	Phoenix	1995
Leon and LaVelle Lesicka	Reno	1996
Bill Hook	Philadelphia	1997
Oscar Carlson	Reno	1998

"THE FRANK GOLATA AWARD" OUTSTANDING OUTFITTER OF THE YEAR AWARD

Myles Bradford, Dalziel Hunting, Ltd. and Bradford & Company Guide Services, Ltd.	Reno	1996
Urban Rahoi, Ptarmigan Lake Lodge	Philadelphia	1997
Garry Vince, Muskwa Safaris, Ltd.,	Reno	1998

YOUTH ESSAY CONTEST WINNERS

Charles Darrell Hoovestol	Reno	1991
No Contest Held in 1992		
Cody Voermans	Reno	1993
Marissa Johnson	San Antonio	1994
Angie Wanamaker	Phoenix	1995
Marcus Hite	Reno	1996
Justin Miller	Philadelphia	1997
Ryan Swanson	Reno	1998

Due to commitments Ryan Swanson was unable to take the winning hunt. The hunt was taken by 1st runner up, Garrett McKitrick.

1989 Desert Bighorn Archery Gold Award presented to Jim Ryan, hunter, and Floyd Lee Crank, Outfitter.

President John Harris and Dan Pedrotti present the 1991 Grass Roots Award to Bill Leech.

PATRONS
PURCHASERS OF SPECIAL PERMITS AT THE NATIONAL CONVENTIONS OF THE FOUNDATION FOR NORTH AMERICAN WILD SHEEP

Nevada Permit

JIM BROCKIE,1984 - Nevada Permit $25,000
JOHN TURNER,1985 - Nevada Permit $42,000
DAN POCAPALIA,1986 - Nevada Permit $42,500
MIKE BOYCE,1987 - Nevada Permit $57,000
DAN POCAPALIA,1988 - Nevada Permit $35,000
FRED MUSSER, JR.,1989 - Nevada Permit $29,000
ELDEN KRISTIANSEN,1991 - Nevada Permit $45,000
CARL G. HOEHNER,1992 - Nevada Permit $65,000
JOHN PAUL,1993 - Nevada Permit $75,000
THOMAS LUTSEY,1994 - Nevada Permit $79,000

Nevada Antelope

LEN GULDMAN,1995 - Nevada Antelope $8,250

Montana Permit

ART DUBS,1986 - Montana Permit $79,000
ART DUBS,1987 - Montana Permit $109,000
ART DUBS,1988 - Montana Permit $93,000
GUY MILLER,1989 - Montana Permit $74,000
GREG STIRES,1990 - Montana Permit $61,000
GREG STIRES,1991 - Montana Permit $80,000
GLEN SMITH,1992 - Montana Permit $88,000
SHERWIN SCOTT,1993 - Montana Permit $205,000
JERRY FLETCHER,1994 - Montana Permit $310,000
JERRY FLETCHER,1995 - Montana Permit $281,000
ROBERT PETERSEN,1996 - Montana Permit $220,000
LEE R. ANDERSON,1997 - Montana Permit $238,000
ANONYMOUS,1998 - Montana Permit $300,000

Montana Moose

JIM RYAN,1989 - Montana Moose $13,000
JOHN BRANDIS,1992 - Montana Moose $9,000
MEL KINSEY,1993 - Montana Moose $13,000
JACK WILKINSON,1994 - Montana Moose $9,000
JOHN R. WILSON,1995 - Montana Moose $14,000

Arizona Permit

DAN POCAPALIA,1984 - Arizona Permit $64,000
DENNIS HANKERSON,1985 - Arizona Permit $42,000
DENNIS HANKERSON,1986 - Arizona Permit $27,000
JACK HAYS,1987 - Arizona Permit $39,000
ART DUBS,1988 - Arizona Permit $47,500
JIM RYAN,1989 - Arizona Permit $53,000
STANLEY BOOTS,1990 - Arizona Permit $48,000
BOB CURTIS,1991 - Arizona Permit $52,000
DOUGLAS M. JOHNSON,1992 - Arizona Permit $67,500
JERRY FLETCHER,1993 - Arizona Permit $303,000
STANLEY BOOTS,1994 - Arizona Permit $245,000
LEE ANDERSON,1995 - Arizona Permit $233,000
JERRY FLETCHER,1996 - Arizona Permit $285,000
ROBERT E. PETERSEN,1997 - Arizona Permit $295,000
BERNARD FIEDELDEY,1998 - Arizona Permit $140,000

Wyoming Permit

DR. VOGT,1980 - Wyoming Permit $23,000
JOHN TURNER,1981 - Wyoming Permit $29,000
DOC THURSTON,1982 - Wyoming Permit $43,000
DOC THURSTON,1983 - Wyoming Permit $49,500
DAN POCAPALIA,1983 - Wyoming Permit $50,000
DOC THURSTON,1984 - Wyoming Permit $52,000
DOC THURSTON,1985 - Wyoming Permit $40,200
DAN POCAPALIA,1985 - Wyoming Permit $40,200
TOM MCMILLAN,1986 - Wyoming Permit $25,000
BERNARD QUINTILIANI,1987 - Wyoming Permit $35,000
OLIVER A. JONES JR.,1988 - Wyoming Permit $13,000
FRANCIS BOUCHARD,1989 - Wyoming Permit $24,750
EUGENE W. BILL,1990 - Wyoming Permit $30,000
DAN POCAPALIA,1991 - Wyoming Permit $29,000
LARRY BUCHER,1992 - Wyoming Permit $38,000
GARTH FREHNER,1993 - Wyoming Permit $55,000

Wyoming Permit continued

J.CURTIS NEWMAN,1994 - Wyoming Permit $45,000
J.CURTIS NEWMAN,1995 - Wyoming Permit $38,000
RON SIMEK,1996 - Wyoming Permit $27,500
RON SIMEK,1997 - Wyoming Permit $35,000
DAN POCAPALIA,1998 - Wyoming Permit (Redonated) $35,000
BOB KELEZ,1998 - Wyoming Permit $30,000

Wyoming Moose

BOB BOBBITT, 1997 - Wyoming Moose $12,500
JOHN S. RIGGS, JR.,1998 - Wyoming Moose $7,500

Utah Permit

BILL LEECH,1983 - Utah Permit $32,000
DOC THURSTON,1984 - Utah Permit $33,000
PETE PAPAC,1985 - Utah Permit $27,000
LEE ANDERSON,1993 - Utah Desert $45,000
CARL HOEHNER,1993 - Utah Bighorn $55,000
JAY WESLEY JONES,1994 - Utah Bighorn $62,500
ARTHUR CRATE,1994 - Utah Desert $51,000
DONALD G. COX,1995 - Utah Bighorn $72,000
RICHARD GARZOLI,1995 - Utah Desert $41,000
CARL HOEHNER,1996 - Utah Bighorn $40,000
JERRIE EATON,1996 - Utah Desert $62,000
DONALD SCHNEIDER,1997 - Utah Bighorn $39,000
SCOTT LINDERS,1997 - Utah Desert $47,000
CHARLIE ANDERSON,1998 - Utah Desert $44,000
GARY RIGOTTI,1998 - Utah Bighorn $36,000

Utah Moose

MICHAEL D. JUSTICE,1995 - Utah Moose $10,750

Utah Antelope

TOM VACCARO,1995 - Utah Antelope $3,750

North Dakota Permit

JIM ROCKSTAD,1986 - North Dakota Permit $17,000
JACK WANAMAKER,1987 - North Dakota Permit $21,000
CA. AGRICULTURE,1988 - North Dakota Permit $17,000
ED STEDMAN,1989 - North Dakota Permit $25,000
STANLEY M. BOOTS,1990 - North Dakota Permit $34,000
BOB FLETCHER,1991 - North Dakota Permit $37,000
SERGE M. d'ELIA,1992 - North Dakota Permit $30,000
DAVID IRWIN,1993 - North Dakota Permit $28,000
HOSSEIN GOLABCHI,1994 - North Dakota Permit $47,500
KATHRYN IRWIN,1995 - North Dakota Permit $31,000
DAN POCAPALIA,1996 - North Dakota Permit (Redonated) $35,000
RON MILLER,1996 - North Dakota Permit $32,000
VIRGINIA FREHNER,1997 - North Dakota Permit $25,000
CARL HOEHNER,1998 - North Dakota Permit $30,000

Texas Permit

DOC THURSTON,1987 - Texas Permit $61,000
BERNIE FIEDELDEY,1994 - Texas Permit $70,000
ROBERT JOHNSON,1996 - Texas Permit $77,000
DAN BOONE,1997 - Texas Permit $51,000

Mexican Permit

DAN PEDROTTI,1984 - Mexican Permit $18,050
AL BARONE,1984 - Mexican Permit $18,050
JOE LO MONACO,1984 - Mexican Permit $18,050
LEE HICKS,1986 - Mexican Permit $25,000
ROBERT OLANDER,1987 - Mexican Permit $23,000
DANIEL SMITH,1987 - Mexican Permit $25,000
FRANCES BOUCHARD,1988 - Mexican Permit $27,500
NORMAN MEINTS,1988 - Mexican Permit $28,000
GREG WAMBOLD,1989 - Mexican Permit $28,000
CLIFFORD SENTER,1989 - Mexican Permit $21,000
ARCHIE NESBITT,1990 - Mexican Permit $27,000
DAVID DRETTMAN,1990 - Mexican Permit $29,000
ANONYMOUS,1998 - Mexican/Tiberone $195,000
JERRY FLETCHER,1998 - Mexican/Tiberone $200,000
RON SCHAUER,1998 - Mexican/Baja $58,000
JERRY TYRELL,1998 - Mexican/Baja $51,000

Yukon Permit

CRAIG SKAGGS,1988 - Yukon Permit $28,000

Idaho Permit

BOB SENTER,1988 - Idaho Permit $68,000
DAN POCAPALIA,1989 - Idaho Permit $36,000
JERRY FLETCHER,1990 - Idaho Permit $30,000
ED PARISEAU,1992 - Idaho Permit $51,000
EARL LUNCEFORD,JR.,1993 - Idaho Permit $31,000
GLENN THURMAN,1994 - Idaho Permit $50,000
ANONYMOUS,1995 - Idaho Permit $70,000
ANONYMOUS,1996 - Idaho Permit $101,000
LARRY BULL, JR.,1997 - Idaho Permit $33,000
BILL TILLY,1998 - Idaho Permit $55,000

California Permit

JIM RYAN,1988 - California Permit $59,000
ROBERT SENTER,1989 - California Permit $40,000
HOSSEIN GOLABCHI,1990 - California Permit $37,000
J. CURTIS NEWMAN,1991 - California Permit $42,000
STANLEY M. BOOTS,1992 - California Permit $61,000
LEE ANDERSON,1993 - California Permit $100,000
ROBERT E. PETERSEN,1994 - California Permit $110,000
BERNIE FIEDELDEY,1995 - California Permit $110,000
ANONYMOUS,1996 - California Permit $101,000
LEE ANDERSON,JR.,1997 - California Permit $84,000

Grand Slam Hunt

R.G. HARRIS,1986 - Grand Slam Hunt $47,500

Alberta Permit

ANONYMOUS,1995 - Alberta Permit $225,000
BERNIE FIEDELDY,1996 - Alberta Permit $200,000
THOMAS PAWLACYK,1997 - Alberta Permit $200,000
SHERWIN SCOTT,1998 - Alberta Permit $405,000

Hualapai Permit

PETE PAPAC,1986 - Hualapai Permit $22,000

Colorado Permit

ROBERT SENTER,1989 - Colorado Permit $43,000
JOHN DIEDRICH,1990 - Colorado Permit $31,000
STANLEY BOOTS,1991 - Colorado Permit $40,000
MICHAEL BOYCE,1992 - Colorado Permit $45,000
FRED MC LAREN,1993 - Colorado Permit $36,000
ANONYMOUS,1994 - Colorado Permit $67,000
CARL G. HOEHNER,1995 - Colorado Permit $67,000
HUGH JACKS,1996 - Colorado Permit $55,000
ROBERT JOHNSON,1997 - Colorado Permit $32,500
ANONYMOUS,1998 - Colorado Permit, ... $56,000

Colorado Goat

SIMON ARAGI,1989 - Colorado Goat $7,000
TOM VACCARO,1990 - Colorado Goat $15,500
GARTH FREHNER,1991 - Colorado Goat $8,000
VIRGINIA FREHNER,1992 - Colorado Goat $7,000
TOM VACCARO,1993 - Colorado Goat $8,500
VIRGIL LAIR,1994 - Colorado Goat $9,500
TOM VACCARO,1995 - Colorado Goat $10,000
BUTCH KUFLAK,1996 - Colorado Goat $7,500
BOB MAYS,1997 - Colorado Goat $7,800
STEVE WEINBERG,1998 - Colorado Goat $5,850

Colorado Moose

VIRGIL LAIR,1996 - Colorado Moose $19,000
VIRGIL LAIR,1997 - Colorado Moose $9,750
STEVE WEINBERG,1998 - Colorado Moose $6,300

Maine Moose

PETE PAPAC,1996 - Maine Moose $8,000
GENE HOLLOWAY,1997 - Maine Moose $7,200

New Mexico Permit

DAN POCAPALIA,1990 - New Mexico Permit $70,000
PAMELA ATWOOD,1991 - New Mexico Permit $46,000
STANLEY M. BOOTS,1992 - New Mexico Permit $67,000
HOSSEIN GOLABCHI,1993 - New Mexico Permit $55,000
LEE R. ANDERSON,1994 - New Mexico Permit $58,000
TOM PAWLACYK,1995 - New Mexico Permit $123,000
GUY MILLER,1996 - New Mexico Permit $100,000
DON SNYDER,1997 - New Mexico Permit $78,000
RON SCHAUER,1998 - New Mexico Permit $77,500

Oregon Permit

SHERWIN SCOTT,1992 - Oregon Permit $57,000
ROBERT PETERSON,1993 - Oregon Permit $102,000
LEE R. ANDERSON,1994 - Oregon Permit $110,000
STEVEN C.WALKER,1995 - Oregon Permit $73,000
BEN HOLLINGSWORTH,1996 - Oregon Permit $50,000
FRANCIS BOUCHARD,1997 - Oregon Permit $30,000
ANONYMOUS,1998 - Oregon Permit $59,000

Washington Permit

TOM PAWLACYK,1994 - Washington Permit $100,000
LEE ANDERSON,1995 - Washington Permit $67,500
STANLEY BOOTS,1996 - Washington Permit $55,000
CARL G. HOEHNER,1997 - Washington Permit $35,000
DONALD SCHNEIDER,1998 - Washington Permit $27,500

Washington Elk

JOE BALMELLI,1996 - Washington Elk $9,500
VIRGIL LAIR,1997 - Washington Elk $8,500

Navajo Nations

DONALD SNYDER,1995 - Navajo Nations $36,000
PAUL SCHNEIDER,1996 - Navajo Nations $33,000

Alaska Permit

JERRY FLETCHER,1997 - Alaska Permit $200,000
HOSSEIN GOLABCHI,1998 - Alaska Permit $50,000

Alaska Caribou

WM CLOYD,1998 - Alaska Caribou $2,200

Just a few of the many permit purchasers. We gratefully thank all purchasers and bidders for their participation.

Participants in the Workshop on the Management Biology of North American Wild Sheep The University Center University of Montana June 18 to 20, 1974

(1) Don Aldrich
The Wildlife Federation
Missoula, Montana

(2) John H. Batten
Boone and Crockett Club
Racine, Wisconsin

(3) James A. Blaisdell
National Park Service
Klamath Falls, Oregon

(4) Walter L. Bodie
Idaho Fish and Game Department
Salmon, Idaho

(5) Gerald W. Brown
Student
Thompson Falls, Montana

(6) Roger Bumstead
U.S. Forest Service
Missoula, Montana

(7) Karry Constan
Montana Department of Fish, Wildlife and Parks
Livingston, Montana

(8) Mario Luis Cossio
Director General of Wildlife
Mexico

(9) Zarate Elena Cossio
Mexico

(10) William I. Crump
Wyoming Game and Fish Department
Lander, Wyoming

(11) Dennis A. Demarchi
Secretariat
Environment and Land Use
Victoria, British Columbia

(12) Ray A. Demarchi
Fish and Wildlife Branch
Cranbrook, British Columbia

(13) Dave Dunaway
U.S. Forest Service
Concord, California

(14) Glenn L. Erickson
Montana Department of Fish, Wildlife and Parks
White Sulphur Springs, Montana

(15) Jim Ford
Regional Coordinator
Montana Department of Fish, Wildlife and Parks
Missoula, Montana

(16) Charles Funk
Klamath Falls, Oregon

(17) Robert Garrott
Missoula, Montana

(18) Valerius Geist
University of Calgary
Calgary, Alberta

(19) Tommy L. Hailey
Texas Parks and Wildlife Department
Breckenridge, Texas

(20) John W. Hanes, Jr.
Boone and Crockett Club
Great Falls, Virginia

(21) John T. Harris
University of Montana
Missoula, Montana

(22) Jack Helvie
U.S. Fish and Wildlife Service
Las Vegas, Nevada

(23) Bill Hickey
Idaho Fish and Game Department
Salmon, Idaho

(24) Brian Horejsi
University of Calgary
Calgary, Alberta

(25) Maurice Hornocker
University of Idaho
Moscow, Idaho

(26) Rodney T. John
Utah Division of Wildlife Resources
Draper, Utah

(27) Rolf L. Johnson
Washington Department of Game
Tenino, Washington

(28) Judith M. Johnston
Rocky Mountain Forum
Stevensville, Montana

(29) Fred L. Jones
U.S. Bureau of Outdoor Recreation
Falls Church, Virginia

(30) Reuel G. Janson
Montana Department of Fish, Wildlife and Parks
Missoula, Montana

(31) Marvin R. Kaschke
Montana Department of Fish, Wildlife and Parks
Moiese, Montana

(32) Warren E. Kelly
U.S. Forest Service
Placerville, California

(33) Bob Klaver
Cooperative Wildlife Research Unit
Missoula, Montana

(34) C. Eugene Knoder
National Audubon Society
Denver, Colorado

(35) L. Jack Lyon
U.S. Forest Service
Missoula, Montana

(36) James V. McKenzie
State Game and Fish Department
Mandan, North Dakota

(37) David M. Marshall
U.S. Fish and Wildlife Service
Washington, D.C.

(38) W.V. Masson
Oregon Wildlife Conservation
Hines, Oregon

(39) Robert L. Means
Bureau of Land Management
Arlington, Virginia

(40) Gale Monson
Stevensville, Montana

(41) James K. Morgan
Stevensville, Montana

(42) Cliff Morrow
National Rifle Association of America
Washington, D.C.

(43) Peter Nash
Ball State University
Muncie, Indiana

(44) William H. Nesbitt
National Rifle Association
Washington, D.C.

(45) Lyman Nichols
Alaska Department of Fish and Game
Cooper Landing, Alaska

(46) Errol Nielson
Idaho Fish and Game Department
Boise, Idaho

(47) Bart W. O'Gara
University of Montana
Missoula, Montana

(48) W.L. Pengelly
University of Montana
Missoula, Montana

(49) Floyd W. Pond
Society for Range Management
Missoula, Montana

(50) Daniel A. Poole
Wildlife Management Institute
Washington, D.C.

(51) F.C. Pullman
Boone and Crockett Club
Chicago, Illinois

(52) John P. Russo
Arizona Game and Fish Department
Mesa, Arizona

(53) Leo Rutledge
Stone's Sheep Workshop
Hudson's Hope, British Columbia

(54) Allen Schallenberger
Choteau, Montana

(55) Gene G. Schoonveld
Colorado Division of Wildlife
Fort Collins, Colorado

(56) D.M. Shackleton
University of Calgary
Calgary, Alberta

(57) J.A. Sime
National Parks of Canada
Calgary, Alberta

(58) Jim Simpson
Outfitting
Jasper, Alberta

(59) Bruce L. Smith
Missoula, Montana

(60) Walter Snyder
New Mexico Department of Game and Fish
Santa Fe, New Mexico

(61) J. Juan Spillett
Cooperative Wildlife Research Unit
Utah State University
Logan, Utah

(62) Glenn St. Charles
Pope and Young Club
Seattle, Washington

(63) John G. Stelfox
Canadian Wildlife Service and Northwest Territories
Edmonton, Alberta

(64) Pan T. Stiver
Las Vegas, Nevada

(65) Robert G. Streeter
U.S. Fish and Wildlife Service
Fairfax, Virginia

(66) Robert L. Taylor
University of Nevada
Reno, Nevada

(67) Tom Thorne
Wyoming Game and Fish Department
Sybille Game Research Unit
Wheatland, Wyoming

(68) James B. Trefethen
Wildlife Management Institute
Washington, D.C.

(69) George K. Tsukamoto
Nevada Department of Fish and Game
Reno, Nevada

(70) Richard A. Weaver
Arizona Game and Fish Department
Yuma, Arizona

(71) Diane Wilcoxson
Forest Service (FSL)
University of Montana
Missoula, Montana

(72) E. Earl Willard
University of Montana
Missoula, Montana

(73) Lanny O. Wilson
Bureau of Land Management
Boise, Idaho

(74) W. Wishart
Alberta Government
Edmonton, Alberta

(75) Al Woolf
New Florence, Pennsylvania

(76) Philip L. Wright
Boone and Crockett Club
Zoology Department
University of Montana
Missoula, Montana

(77) Jim Yoakum
Bureau of Land Management
Reno, Nevada

Life Member Listing

November 1998

1. Duane Smelser
2. Fred McMurray
3. Harvey Kadlec
4. Randy Babala
5. John Babler
6. William Townsend, III
7. Win Condict
8. Clint Bentley
9. Walter White
10. Dave Wade
11. M.F. Wood
12. Gary Swanson
13. Watson Yoshimoto
14. William Boggs, Jr.
15. Roy Schultz
16. Garry Vince
17. Thornton Snider
18. Hobart Pardue
19. Don Kirn
20. Daniel Smith
21. Allen Keller
22. Edwin Cox, Jr.
24. Lee Hicks
25. Larry Kelly
26. Howard Gilmore
27. Dan McCarty
28. Bill Leech
29. George Vogt
30. Oscar Carlson
31. Jacques Henri Jean
32. William Phifer
33. Johnny Moore
34. Mahlon Butch White
35. John Poehling
36. Harold McEuen
37. Red Sorensen
38. Donald Cox
39. George Auer
40. Irvin Barnhart
41. David Swank
42. Rod Hardie
43. John Cadario, Jr.
45. Gary Jorgensen
46. Joe Carter
47. Nicholas Falzarano
49. H. W. Culp
50. Hank Culp
51. Don Harstad
52. Richard Ullman
53. Hale Kreycik
54. Don Corley
55. Don Beattie
56. Dwight Farr
57. Herman Bennett
58. Barclay Townsent
59. George Gamble
61. Earl Lunceford
62. John Turner
63. Howard Pollock
64. Paul Broun
65. Wayne Phelps
66. Jerry Suk, Jr.
67. Leland Speakes, Jr.
68. Robert Logan
69. James Wetzel, Jr.
70. Sharon Kubick
71. Bill Dudley
72. Gregory Stires
73. Jim Rockstad
74. Arnie Johns
75. Nicolas Franco
76. Daniel Pedrotti
77. Buron Griffin
78. Walter Ford, Jr.
79. Ozzie Davis
80. Robert Smitherman
81. Chancy Ernst
82. Ronald Simek
83. Carlos Sanchez
84. Robert Kubick
85. Andy Hagberg
86. Delbert Strickling
87. Richard Mandeville
88. George Hixon
89. Robert Keen
90. Urban Rahoi
91. Torrence Hunt
92. Bob Leslie
93. Joel Pat Latham
94. Max Williams
95. James Stuart
96. Jimmy Barho, Sr.
98. Cy Ford
99. Larry Poland
100. James Skidmore
101. Bill Moynihan
102. Bob Wood
103. George Adolph
104. Jack Beal
105. Dick Weaver
106. Dennis Hankerson
107. George Cooper
108. Jack Cook
109. Robert Anderson
110. Walter McDonald
111. Duane Nelson
112. John Lindberg
113. Belle Desrosiers
114. Lad Shunneson
115. Don Wenger
116. Joe Bishop
117. Thomas Dirks
118. Duke Leduc
119. Raymond Allemand
120. Buck Allemand
121. Leon Naccarato
122. Toby Johnson
123. Dale Marks
125. Richard Edwards
127. Doug Van Zandt
128. Keith Johnson
129. James Goforth, M.D.
130. Glenn Napierskie

131. David Sebold
132. Raymond Bartram
133. Dean Buntrock
134. Wayne Rivard
135. William Waugaman
136. Al Farace
137. Billy Dunbar
138. William G. Reed
139. Ken Fanning
140. Alden Glidden
141. Roger Schladetzky
142. Richard Rohrer
143. Robert Hawke
144. Frank Lahnston
145. Mac Lewis
146. Thomas Bunch
147. Tom Vinson
148. Dan Duncan
149. Gordon Rockefeller
150. Richard Smith
151. Helmut Greul
152. Doug Wattenbarger
153. Donald Harrell
154. C.J. Fuller, M.D.
155. Don Carper
156. Kennon Cantley
157. Alan Sackman
158. Wayne Pocius
159. Olive Powell
160. Lavon Wenger
162. Chet Brown
163. Kim Adamson
164. Jack Dickie
165. Roger Johnson
166. Bill Gillespie
167. James Willoughby
169. William Padilla
170. Daniel Crook, M.D.
171. Travis Roberts
172. Larry Cameron
173. Adrian Erickson
174. HIH Prince Abdorreza
175. D.L. Cloudt
176. Marvin Smith, M.D.
177. Dan Pocapalia
178. Peter Toggenburger
179. Forrest Burnett
180. Bubba Glosson
181. John Sauls
182. Jeff Reynolds
183. Jerry Christian
184. John Macivor
185. Carey Mock
186. James Grace
187. Bruce Keller
188. Jim Hill
189. Jon R. "Dick" Gray
190. George Houser
191. Tom McMillan
192. W. E. Hartman
193. Joseph Lomonaco
194. A Clinton Lindburg
195. Ed Weatherby
196. Lance Lupton
197. Lloyd Miller
198. Daniel Newman
199. Walt Sikes
200. Jim Bush
201. Floyd Foster
202. Skipper Bettis
203. Mike Telesmanic
204. John D. "Jack" Frost
205. Ned Schroeder
206. Dave Wolfe
207. Phyllis Wolfe
208. Ken Trudell
209. Neal Stevens
210. Glen Thurman
211. Derril Lamb
212. Charles S. Pearcy, II
213. John Diedrich
214. Paul Webster
215. Randy Webb
216. William Lenheim
217. Lloyd McMahon
218. R. Douglas Yajko, M.D.
219. Al Cito
220. Thomas Hinders
221. Ross Peck
222. Jimmy Barho, Jr.
223. Steve Bruggeman
224. Tom Pliura, M.D.
225. Paul Dyer
226. R.A. Shoberg
227. James Warf
228. R.G. Harris
229. Lee Wallace
230. Rodney Gussman
233. Nolan Wright
235. Perry Finger
236. Leslie Finger
237. Bruce Hutcheon
238. Larry Butler
239. Daivd Collis
240. Donald Perry
241. Darrel Nelson
242. Robert P. Ellingson, III
243. Dan Steed
244. Michael Hamrick
245. Jim Conklin
246. Martin Rigoletti
248. James Young
249. Arthur Jee, D.M.D. P.A.
250. Curtis Townsend
251. G.J. Salomons
252. Marvin Epling
253. Boyd Altaffer, Jr.
254. James Hughes
255. Gerald Begnaud, Jr.
256. Bob Senter
257. Terry Enfield
258. Burk Mantel
259. Carolyn Mandeville
260. Gary Ingersoll
261. Gerald Curtis
262. Larry G. Plog
263. Eddie Godfrey
264. Fred Musser, Jr.
265. Alan Stone
266. John Harris
267. Stephen A. Mihal
268. Edward Lemons
269. John Elfervig, M.D.
270. Sandra Vince
271. Monroe Beecher

272. Leon Francisco
273. John Russell, Jr.
274. Ted Dalzell
275. Arthur R. Dubs
276. Pete Papac
277. Larry Rivers
278. James Ramsay
279. Rick Taylor
280. Archibald Nesbitt
281. William End
282. Branden Powell
283. John Mussell
284. Bill Stevenson
285. Troy Graziadei
286. Fredrick Holliday
287. Jeff Englebert
288. Rick Furniss
289. J. Bongiovi, M.D.
290. Joe Hardy
291. George Snyder
292. Don Leiser
293. Larry Chavez
294. Ronald Pomeroy
295. Bonnie Plog
296. John Gregg
297. Robert Gregg
298. Len Leduc
299. Ed McArthur
300. John L. Hendrix
301. Kern Chew
302. Bob Diedrich
304. Richard Gunlogson
305. Frank Geduhn
306. Ralph McClintock
307. Jeff Sorg
308. David Daughtery, Jr.
309. Tom Hoffman
310. Ron Hurlbert
311. Sonny Burnett
312. Merle Knuth
313. Wayne Clark
314. Curtiss Judd
315. Bruce Creyke
316. Brad Kadlec
317. Rick Drinnen
318. Roger Card
319. Terry Marcum
320. Ronald Jenkins
321. John Poston
322. Richard Larson
323. Brenton Scott
324. Dick McClung
325. David Boone
326. Ray Herrmann
327. W.G. Hawes
328. John Higbie, Jr.
329. James Brooks, Jr.
330. Judith Lupton
331. Edward Panus
332. Joe Clayton
333. Richard Pierce
334. Steve Tripp
335. Rick Guinn
336. Michael Gorno
337. Tim Krause
338. Alan Ljunggren
339. Matt Caldwell
340. Jurgen Knuth
341. Dale Muller
342. Richard Cotta
343. Daniel Alegre
344. Gary Brown
345. Carl Larson
346. Fred Fortier
347. Jim August
348. Terry Russell Hunkins
349. Dr. C. Michael Irvin
350. Raymond D. Bond
351. Andy Gordon
352. Greg Williams
353. Thomas Tietz
354. Carl Brent
355. David Neighbor
356. Harold Grinde
357. Keele Johnson
358. Thomas Griffin
359. William Butler
360. Richard Mitchell
361. Ron Barker
362. Terry Evans
363. James Campbell, II
364. Gary Benmark, Ph.D.
365. Victor Hugo O'Farrill
366. Cam Drinnan
367. John Ollenburg
368. David Gentry
369. Tom Saad
370. Bob R. Hess
372. Arnold Alward
373. Pat Connell
374. James Duke, Jr.
375. Wayne Henderson
376. Walter Eder
377. Lloyd B. Swiedom
378. Marcia Voisine
379. Jack Hume
380. Franklin T. Crooker
381. Jim Matarozzo
382. Jim Pollard
383. J.L. Hayes
384. James DeForge
385. David Sandstrom
386. Alfred Barone
387. Harold A. F. Starke
388. Wayne Webber
389. Charles Purnell
390. Si Pellow
391. Tony Simek
392. Dick Callahan
393. Guinn Crousen
394. Donald P. Robertson
395. Mike Papac
396. Jerry Waite
397. Michael Borel
398. Terry Ruder
399. Robert Hartman
400. G.H. Brandenburger
401. Dr. Juan Varela
402. James Walden
403. Simon Aragi
404. Ron Kutella
405. Wells A. Moon, Jr.
406. Butch Hautanen
407. Keith Graham
408. Christa Ryan

409. Carl Jacobson
410. Reed Steinmetz
411. William Hanna
412. Donald Prince
413. Stephen M. Powell
414. Frederick L. Wood
415. Dr. Ken Rueter
416. David Colclough
417. William Powell
418. Martin Benchoff
419. Hank Konrad
420. Pat O'Neill
422. Albert D. Seeno, Jr.
423. Ralph Wilkerson
424. J. Dorsey Smith, M.D.
425. James E. Auer
426. Joseph Lora
427. Nicholas Coussoulis
428. John D. Chalk, III
429. Gail Winnie
430. Peter E. Seda
431. Joe Federl
432. Raiford Hancock
433. Don Giottonini, Jr.
434. John M. Gates
435. Wayne A. Dixon
436. Alex G. Ramoz
437. Richard Buck Pope
438. Earl Boose
439. Douglas Kerr
440. Ernest Righetti
441. Tom Lang
442. Rodney E. Fye
443. Peter Bollinger
444. Marco Pastrana
445. James Erickson
446. Thomas J. Tracy
447. Ronald S. Gabriel, M.D.
448. Edward C. Joseph, D.D.S.
449. Jimmie R. Ryan
450. Dennis Gaines
451. Ron Willenborg
452. Gerard Rudolf Damm
453. David Witmer
454. Paul S. Inzanti
455. Robert DiGrazia, D.D.S.
456. Pete Cimellaro
457. Richard M. McDrew
458. Norbert Bremer
459. Wesley Starn
460. Ron Hinman
461. Ron Riley
462. Larry Hunts
463. Maria Novak
464. Robert M. Lee
465. Javier Matar
466. Dale Prochnow, Sr.
467. Topper Frank
468. Timothy E. McEuen
469. Nelson Puett
470. W. Kapp Perry, Jr.
471. Dr. Leroy Lance
472. Johnny Bliznak, M.D.
473. David Myrup
474. Gary Grant
475. Santiago Villarreal
476. John Wilson
477. Royce D. "Woody" Wood
478. Tim Morrison
479. Don McDonald
480. Tom Pawlacyk
481. Gary Kruger
482. Robert I. Anderson
483. Alton Ryan
484. Rocky Crate
485. Craig Johnson
486. Hanns J. Kohler
487. Dean Collins
488. Larry Bucher
489. Albert Anton
491. Hossein Golabchi
492. John S. Rohrer
494. Max Morgan
495. Michael R. Justice
496. Rafael Tames
497. David B. LeFevre
500. Robert S. Rogers
501. James L. "Bucky" Wolford
502. David E. Hernandez
503. Joseph Shelmet
504. Joe Wells
505. Don Wall
506. Ted Mellotte
507. John Paul
508. Joseph Kalt
509. Virginia Frehner
510. Gary Bogner
511. King Cavalier
512. Phyllis Carlson
513. Jerry Fletcher
514. Alan Hamberlin
515. Gerald E. Meyer
516. Drake Shogun
517. Del R. Brady
518. James E. Egger
519. Donald R. Sheffield
520. Jim Tonkin
521. Jane Hunts
522. Norman Etchison
523. Jim Hollister
524. Wesley Hansen
525. Bob Howard
526. Wesley Streed
527. Dominick Calobrisi, M.D.
528. Jimmy Low
529. Jack Brittingham
530. Kevin Olmstead
531. Kurt Gerken
532. Paul Prudler
533. William Stroecker
534. William Treichel
535. Robert Hodson
536. John Tiscornia
537. Jim Voeller
538. Guldbrand J. Kramer
539. Lester G. "Jack" Graf
540. Daniel Currier
541. Dave Currier
542. Jeffrey A. Batchman
543. Jim Travis
544. Je Leta

545. Sherwin Scott
546. Kevin T. Pawlacyk
547. Marvin D. Hill, Sr.
548. Pitt Sanders
550. Kenneth Cherry
551. Ross Messerly
552. L.D. "Jo" Josephson
553. J.C. Spies
554. John C. Crawford
555. Debra L. Card
556. Christopher Ring
557. David Gonzalez
558. James D. Verbrugge
559. James Frickert
560. Michael Wallace
561. Leo W. Mack, Jr. M.D.
562. Craig Wilcox
563. Steven M. Wilcox
564. Curtis Newman
565. Ed Stedman, Jr.
566. Joe Coleman
567. Rudy J. Stecich
568. John C. Graham, Jr. M.D.
569. Archie Lang
570. Doug Bell
571. Bedo Hobbs
572. Bernard Sippin
573. Brian Drettmann
574. John Azevedo
575. L. Dale Gaugler
576. Ralph W. Gitz
577. John C. McConnell, M.D.
578. William Tott
579. Dennis Dunn
580. John Woodmark
581. Michel Kuri Slim
582. Mark R. Herfort
583. Rod Krebs
584. Martin Cotanche
585. Al Gibson
586. Robert J. Johnson
587. Ry Tipton, M.D.
588. Butch Kuflak
589. Gary Hansen
591. Steven Futch
592. Esteban Larragain
593. Nicholas Berg
594. Frank Ruffino
595. Robert L. Fletcher
596. Jerrie Eaton
597. Darwin Baucum
598. Don Hanks
600. Stanley M. Boots
601. Gerald Sutherland
602. Mel Kinsey
603. Sam Pattillo
604. Adrian T. Sada
605. Mark E. Campbell
606. Sam T. Scaling, M.D.
607. James Fero
608. Robert Lynch
609. Edward D. Yates
610. James R. Moriarty
611. Jack Chambers
612. Ralph Adams
613. Andrew A. Samuels, Jr.
614. Ronald Rando
615. Tommy Best
616. David E. Combs
617. Larry Schweitzer
618. Jim Mull
619. Phil W. Long
620. Marlin Grasser
621. Tom Bolack
622. Dr. Harry B. Cook
623. Manuel Chee
624. Rado Drum
625. Ronald Krump
626. Michael J. Miller
627. Paul A. Gardner
628. Curt Babler
630. Jeff Barnett
631. Scott Linders
632. Greg Bond
633. Jay E. Link
634. Daryl E. Haid
635. Edward Spies
636. Larry V. Battarbee
637. Dan W. Bonnell
638. Johnny Rodrigues
639. Andreas Bauer
640. Virgil Lair
641. Jerald Mason
642. Larry Hansard
643. Alex Dousias
644. Betsy Whitson
645. Kirk Jones
646. Ed Gardner
647. Milton Schultz, Jr.
648. Andy Anderson
649. George Harms
650. Duane A. Walker
651. Louis F. Spadaccino
652. Neal H. Skinner
653. J. Craig Holworthy
654. Donald W. Dean
655. Roger Beer
656. Bennett McMillian
657. Roger C. Hooper
658. Martin L. Johnson, M.D.
659. Rod Schuh
660. Robert Jones
661. Ronald C. Miller
662. Tom Liming
663. R. Keith Hite
664. Nico Acquarone
665. Jim Nichols
666. D.J. Kuchta
667. Bill Slattery
668. David F. McCarthy
669. Kenneth L. Barr
670. Elizabeth Johnson
671. Thomas E. Crandall
672. John Kautzman
673. Howard L. Christensen
674. Hugo Kilian
675. George Law
676. Bob Manger
677. Robb Hitchcock
678. Jesse Smith
679. A.D. Hancock
680. Rick Garzoli
681. Dr. Gerald Warnock
682. A.C. Frith
683. Garfield Thomas, Jr.

684. David Hanna
685. George Fry
686. Gregg R. Ulveland
687. Dow Dollar
688. Damian Harrington
689. Stephen Fullmer
690. Kip Slaugh
691. Val Goldthwaite
692. Carlton Goldthwaite
693. Steven Goldthwaite
694. George Williams
695. Don Turcke
696. Charles Drechsel
697. John B. Brelsford
698. Ronald Arndorfer
699. Donald Snyder
700. Ray Barger
701. Joe Diedrich, Jr.
702. Mace T. Coleman
703. Jack W. Ahart
704. Jerome Knap
705. Casey Wilson Bishop
706. Scott M. Ackleson
707. R.E. Sweeney
709. Helmuth Pfennig
710. Paul W. Asper
711. Benny Kirkham
712. Clair C. Wilcox
713. Tex Small
714. Tim Chestnut
715. Louie Swift
716. T.J. Shaw
717. Ron Andrews
718. Harold Haraghey, Jr.
719. Jack Wilkinson
720. Daniel A. Pedrotti, Jr.
721. Chester Mjolsness
722. Ray Robertson
723. Dr. Richard L. Day
724. Wesley R. Blakeman
725. Edwin L. Johnson
726. General Joe Engle
727. Scott Torgerson
728. Jerry Tyrrell
729. Robert Model
730. Frank Bower
731. Donald C. Smith
732. Chet Piazzo
733. Cathy Cimellaro
734. Xavier Aznarez
735. Theodore Schutte, Jr.
736. David B. Hamilton
737. Ronald P. Grunwald
738. B.B. Hollingsworth, Jr.
739. J.L. Gomer Erickson
740. Stan Cosart
741. Sanford Gelzer
742. Earl E. Morgenroth
743. Brian T. Pelczar, M.D.
744. Ray Sindlinger
745. Louis Rupp
746. Paul G. Dressel, Jr.
747. Barry Smith
748. Dan Fox
749. Wilson Reynolds
750. Bernie Fiedeldey
751. Barry Gallup
752. Vern Sluyter
753. C.W. Wallace
754. Scott T. Farr
755. Robert E. Coons
756. Curtis S. Chastain
757. Michael G. Farnham
758. James J. Liautaud
759. Arlene P. Hanson
760. Robert Hanson
761. Bill Pritchard
762. Dale A. Kiriaze
763. Shane Evans
764. Glenn Smith
765. David J. Starkey M.D.
766. Dennis Campbell
767. Steve MacKenzie Grieve
768. Keith Brents
769. James McIntyre
770. Donna Bessette
771. Betty Sowden
772. Drew Browne
773. Eugene Bell
774. Everette Madson
775. Paul Kokes
776. Curt Thompson
777. Dennis Pahlisch
778. Dennis M. Lord
779. Brian Sanders
780. Fred Hatcher
781. Paul A. Young
782. Dr. John R. Thodos
783. Alan Schafer, D.M.D.
784. Marshall Collins
785. Bob J. Newton
786. Robert E. Renner
787. Jake Shinners
788. George Knoop, Jr.
789. John Holland
792. Steve Hornady
794. Kevin Hurley
795. Robert Ritter
796. Tommy L. Caruthers
797. Tim Nolan
798. Jeff Perrella
800. Noel Poux
801. Greg Sawchuck
802. John C. McEwen
803. Rocky Keen
804. Mark Brown
805. Don Miller
806. Haus Guido Riegel
807. Major F. Miller
808. Daniel Kutscher
809. Noel Moore
810. Sam B. Williford
812. Thomas Rutherford
813. James Zimmerly, M.D.
814. Joan Harris Foust
815. William Cloyd
816. C.B. Holloway
817. Bert Kizer
819. Donald P. Chase
820. Dick Jacobs
821. Francis Bouchard
822. Michael S. Pedrotti
823. Norman Stewart
824. Tom Gannon
825. John Lowery

826. Jamie Schumacher
827. Mark Zastrow
828. Trev Grimshaw
829. William W. Dodgson, V
830. Robert Harms
831. Kevin Harms
832. Greg Dixon
833. David McNeill
834. M.S. MacCollum
835. Gene Majors
836. Robby Robinson
837. Dean A. Nelson
838. Douglas Carlson
839. C. Keith West
840. Bill Shields
841. Jake Meadows
842. Charles Eddy
843. Max G. Bauer, Jr.
844. Jim Schomberg
845. Brian Nicely
846. Curt Werner
847. Richard P. LewAllen
848. Donald T. Marshall
849. Mike Weeks
850. Brown Delozier
851. J Carmen De Jesus De La Vega
852. Dean Palmer
853. Robert M Hall
854. Len Wurman
855. Gary Vander Pol
856. Donald E. Wenner
857. William A. Mcafee
858. E. Jerome Carlson
859. Bernard R. Harvey
860. Allan Reishus
861. Charles Robertson
862. Kirby Korth
863. Daniel A. Pedrotti, III
864. Michael J. Pedrotti
865. Scott (Pedrotti) Yoachum
866. T.J. (Pedrotti) Conlee
867. Robert Carlson
868. Kenneth Neff
869. Rich Roan
870. Dyrk Eddie
871. Bill McKenzie
872. Ted Gator Meyer
873. David W. McMurray
874. Al Holland
875. Howard Knutson
876. John Pearson
877. Tino Manossa
878. Graham Weiss
879. Jack Schwabland
880. Ken Wright
881. Lou Misterly
882. Dave Pagel
883. Fred Koken
884. Terry Sieveke
885. Hanspeter Giger
886. James D. Roseman
887. Sr. Eliseo Araujo
888. John M. McAteer
889. Rene J. Dube, Jr.
890. Bob Mays
891. Tommy Ford
892. Doug Schippers
893. O. Everett Rom
894. Dr. Scott DeMars
895. Danny McCallum
896. Kevin C. Rinke
897. Robert Petersen
898. Thomas R. Van Every
899. Roger Ebnet
900. Robert F. Merchant, Jr.
901. Trent Wall
902. Harold W. Armstrong
903. Paul Harvey
904. William A. Keebler
905. Virginia Hanna
906. Mike Davis
907. Larry M. Bull
908. Noah Horn
909. Tom Plooy
910. Harold Billings
912. David B. Colclough, III
913. Dustin Colclough
914. Doug Hogan
916. David Bull
919. Robert Keenan
920. Daniel Amatuzzo
921. Don K. Callahan
922. Jim Edwards
923. Robert R. Cooper
924. Bruce J. Lutz
925. Timothy F. Brammer
926. Brian A. Palmieri
927. Raymond R. Mong
928. Douglas Eck
929. Herb Larson
930. Marian P. LeSage
931. Dean Picl
932. David Potter
934. William Kiefer
935. William H. Tilley
936. Vivi Crandall
937. Fred Boyer
938. Guy Coheleach
939. Douglas Van Howd
940. Jesus Gutierrez-Aja
941. Donny MacDougall, D.V.M.
942. John E. Zagst
943. Grant Powell
944. Richard Wiseley
945. Larry Nelson
946. Billy Vaden
947. Warren Halloway
948. David M. Wells
949. Gerald R. Amys
950. Bruno Scherrer
951. Jerry R. Miller
952. Steven H. Stans
953. Bob Kellogg
954. Scott Anderson
955. Ferlin Koma
956. Robert Hillson
1432. Joseph Lora, III
1951. John Bax, M.D.
2247. Frederick Parker Gregg, M.D.
2867. David L. Mode
7298. S. Ty Looney

The history of the first 25 years of the Foundation for North American Wild Sheep has been a great tribute to hunter-conservationists, who from the "grassroots" level have made a difference in the wild sheep populations of North America. May history record that those who follow continue on the same dedicated path of *Putting Sheep on the Mountain.*

Dr. Roy A. Schultz
November, 1998

Special thanks are extended to the following individuals:

- Hayden Lambson of Pocatello, Idaho, for his original sketch of wild sheep used for the end sheet in all editions of this book.
- Rick Taylor for his original medallion of the Chadwick ram for the super deluxe limited edition book.
- Gary Robert Swanson for his original oil painting for the cover of the regular edition book and the print for the limited edition books.
- Dave Popelka of Ames, Iowa, for design and layout of the book.
- Elizabeth Reneau for her original sketch of a bighorn lamb and Dallen Lambson of Pocatello, Idaho, for his original sketch of wild sheep.

Special thanks are extended to the following organizations:

- Blue Mountain Publishing, Inc. for coordinating the book project.
- Montana Silversmiths of Columbus, Montana, for casting gold medallions for the deluxe limited edition book.
- Publishers Press of Salt Lake City for printing and binding the regular edition book.
- Roswell Bookbinding for production of three limited editions.
- Sigler Printing and Publishing, Inc. of Ames, Iowa, for typesetting and photograph scans.